# PLANT ANATOMY

# PLANT ANATOMY

Pijush Roy MSc, PhD
Ex-Reader in Botany, Vidyasagar College for Women, Kolkata
Ex-Guest Lecturer on Plant Anatomy, Calcutta University

New Central Book Agency (P) Ltd

*Co-Published by*

Bijay Publications

E-mail : bijaypublications@gmail.com

*Exclusively Distributed by*

Bijay Pustak Bhandar (P) Ltd

90/6A M. G. Road, Bhawani Dutta Lane, Kolkata-700007

✆ : (033) 4605-7856, 8777898576, 8274916104

E-mail : bijaypustakbhandar@gmail.com

**NCBA**

**REGD OFFICE**

8/1 Chintamoni Das Lane, Kolkata 700 009, India
email: ncbakolkata@gmail.com

**OVERSEAS**

NCBA (UK) Ltd, 149 Park Avenue North, Northampton, NN3 2HY, UK
email: ncbauk@yahoo.com

**CORPORATE OFFICE**

212 Shahpur Jat, New Delhi 110 049
email: delhincba@gmail.com

**BIJAY PUBLICATIONS**
90/6A M. G. Road, Bhawani Dutta Lane, Kol-7
E-mail : bijaypublications@gmail.com

**BIJAY PUSTAK BHANDAR (P) LTD**
90/6A M. G. Road, Bhawani Dutta Lane, Kol-7
E-mail : bijaypustakbhandar@gmail.com

**PLANT ANATOMY** • Pijush Roy



First Published: 2006
Thoroughly Revised and Enlarged Second Edition: 2010
Reprinted: 2012, 2016, 2018, 2025

PUBLISHER • PRINTER • TYPESETTER
New Central Book Agency (P) Ltd
8/1 Chintamoni Das Lane, Kolkata 700 009

CO-PUBLISHED BY
Bijay Publications
90/6A M G Road, Bhawani Dutta Lane, Kolkata 700007

DIAGRAM
Author

PROJECT TEAM
Swapan Sarkar and Prabhat Jas

ISBN: 978 81 7381 513 3

Price ₹ 765.00

Dedicated

to

*Babu* and *Baba*

# Contents

# Preface to the Second Edition

It is a pleasure for me to present the second edition of *Plant Anatomy*. In this edition seven new chapters are introduced and six chapters rewritten to make the tome more suitable for students. Many figures are redrawn. I preferred to draw most of the diagrammatic figures in simple line-drawing with topographical labelling as it appeared to me an effective contrast against shaded drawings and in some cases essential for clarity.

I express my sincere gratitude to Professor Sunirmal Chanda, Emeritus Scientist, Centre for Study of Man and Environment, Kolkata, and Professor Nandadulal Paria of Calcutta University for enthusing me. I also express my sincere gratitude to Dr Satadip Sarkar, Reader, Presidency College, Kolkata, Dr Nandani Chakrabarti, Reader, City College, Kolkata and Dr Milan Kumar Maity, Ex-Reader, Raja Peary Mohon College, Uttarpara, for manifold help.

I am thankful to Professor Ambarish Mukherjee of Burdwan University and Dr Aparna Pal, Reader, Victoria College, Kolkata, for their encouragement. I am also obliged to Dr Sudha Gupta of Kalyani University for her cooperation.

I am especially grateful to Mr. Amitabha Sen, BCom, MBA, LlB, Director of New Central Book Agency (P) Ltd, for inviting me to prepare the second edition. For any reproduction the permission of the Director is final, but the author deserves acknowledgement.

Cordial thanks are also due to the helpful all who are associated with this project, especially Mr Prabhat Jas and Mr Swapan Sarkar of New Central Book Agency (P) Ltd.

My wife Dr Chhanda Roy, Reader in Bengali, Calcutta University, was fully supportive as usual. I am indebted to her.

1st January 2010
Kolkata

**Pijush Roy**

# Preface to the First Edition

*Plant Anatomy* is written based on the syllabi of Indian Universities and Colleges. Although a large number of excellent books on Plant Anatomy are available, they may not be quite appropriate for the requirements of teachers and students of our country. In most Universities inadequate classes are allotted to accommodate this discipline. The dominant objective of *Plant Anatomy* is to represent in a reasonably balanced and comprehensive way the topics that are needed by the students. This book is primarily written for students and has been made purposely brief. For advanced students there are extensive citations of references—both old and recent, to illustrate the historical background of the subject.

An effort has been made to represent the subject in a simple and easy-to-read style. It is not an original work. It is a compilation of available information on the topics prescribed to the students. For this purpose free consultation of different textbooks and publications have been made. The names of eminent authors of the textbooks and publications along with the names of publishers are mentioned in the references. Indebtedness to all authors along with the publishers for information was gratefully expressed. It is the sole responsibility of the author for errors that might have crept in.

My sincere thanks are due to the Director of New Central Book Agency (P) Ltd, Mr Amitabha Sen, BCom, LlB, MBA, for not only undertaking the publication of *Plant Anatomy* but also permitting to reproduce many illustrations from College Botany, Vol. I, written by H. C. Ganguly, K. S. Das, and C. Dutta.

I am also thankful to Mr Soumen Paul of New Central Book Agency (P) Ltd for various helps throughout the progress of the book. Thanks are also due to Mr Sukumar Patra of New Central Book Agency (P) Ltd for his help in the correction of proof.

Cordial thanks are also due to mywell-wishers—Prof. Balen Nundy, Burdwan University, Dr Satadip Sarkar, Reader in Botany, Presidency College, Kolkata; Prof. N. D. Paria, Head of the Department of Botany, Calcutta University; Dr Milan Kumar Maity, Ex-Reader in Botany, Raja Peary Mohon College, Uttarpara; Prof. Kashinath Bhattacharya, Visva-Bharati University; Dr Swati Gupta Bhattacharya, Reader and In-Charge, Division of Palynology & Environmental Biology, Bose Institute, Kolkata; Prof. Sudhendu Mondal, Director, National Library, Kolkata; Dr Nandini Chakrabarti, Reader in Botany, City College, Kolkata; Ms Raina Roy of Port Melbourne; Ms Neelam Roy, Colesmyer, Melbourne; Mr Prithwiraj Roy, Accenture, Melbourne; and Dr Chhanda Roy, Reader in Bengali, Calcutta University, for their various helps and encouragement.

**Pijush Roy**

# CHAPTER 1

# Cell

A cell is the fundamental unit of all living organisms in terms of structure and function.

History of discovery: Robert Hooke in 1665 first discovered plant cell. The term 'cell' was coined to describe the small walled units that were observed in the sections of bottle cork under simple microscope. Between 1831-'33 Robert Brown

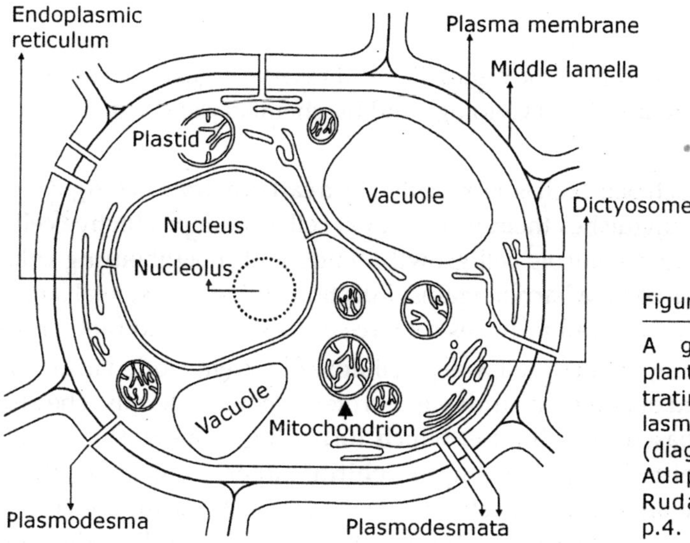


Figure 1.1

A generalized plant cell illustrating protoplasmic contents (diagrammatic). Adapted after Rudall, 2007, p.4.

described the nucleus as small spherical body, which is normally present in every plant cell. In 1864 Mohl gave the name protoplasm to designate the thin mucilaginous layer present inside the cell wall. Shimper and Meyer in 1880 discovered plastids. Thus by more than hundred years ago the largest structural elements of a plant cell like cell wall, nucleus, protoplasm, plastid etc. were described. In later years especially after 1950s with the introduction of electron microscope like SEM, TEM, improved techniques like enzyme cytochemistry, autoradiography, x-ray defraction study, biochemistry etc. the structural and functional elements of a plant cell have been revealed.

**Structure:** Two main regions can be recognized in a plant cell, i.e. the cell wall and protoplasm. Protoplasm consists of nucleus and cytoplasm, and the latter contains a variety of organelles, vacuole, salts and various organic molecules. The various components of a plant cell is outlined in the following table:

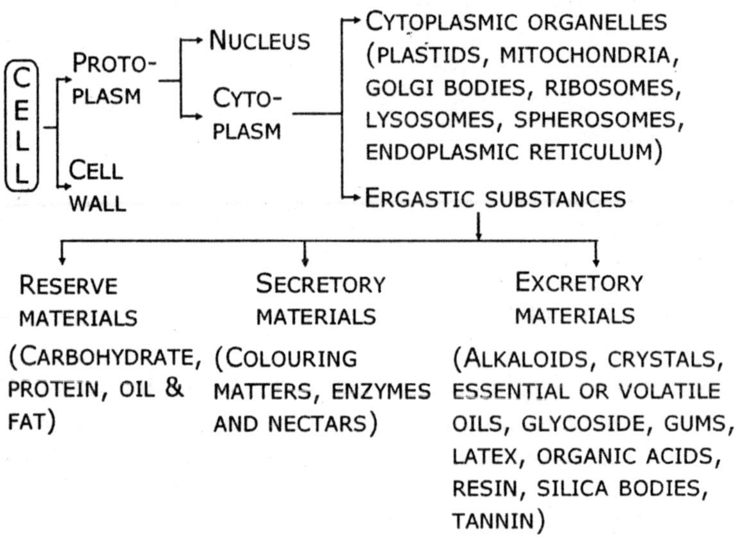


The details of cytoplasmic organelles and reserve ergastic substances are fully described in the textbooks of Cytology and Biochemistry and will not be discussed in details here.

**Cell wall:** Each plant cell is surrounded by a carbohydrate rich rigid wall termed cell wall that distinguishes them from animal cell. (The reproductive cells have no walls). It limits and controls cell growth, binds with neighbouring cells to form tissue, forms a protective barrier against infection and lends skeletal support to the whole plant. (The cell wall provides a source of food, fibre and fuel for man).

The main component of cell wall is cellulose fibrils embedded in an amorphous matrix composed of polysaccharides and proteinaceous materials. The architecture of cell wall is based on cellulose, which forms the framework of wall and the other components like pectin, lignin, chitin, suberin etc. inlay the cellulose framework. The amicroscopic cellulose molecules (8Å width) are combined into an elementary microfibril (100Å width) that forms a bundle called microfibril (250Å thick and contains 2000 cellulose molecules). Electron microscopic study revealed that the units of microfibril aggregated into coarser fibril—termed macrofibril.

**Protoplasm:** The substance present within cell wall including plasma membrane is generally called protoplasm. Normally this term is used to describe the portions left of a plant cell after the removal of cell wall. It is colloidal in nature and consists of nucleus and cytoplasm.

**Nucleus:** Robert Brown in 1831 first described plant cell nucleus from the stamen filaments of *Tradescantia*. It is universally present in all cells except mature sieve tubes and mammalian erythrocytes where nuclei are present at the early stages of development. It is the largest cellular organelle of about 10 μm in diameter although considerable variations exist in size between cells of different species.

In meristematic cells nucleus is roughly spherical in shape and forms 75% of cell volume. The shape may be flattened in differentiated cells where a central large vacuole is present and the nucleus lies against the cell wall. Cylindrical nucleus is seen in elongated cells.

The nucleus consists of a number of components: (a) nuclear envelope or membrane, (b) nuclear sap or karyolymph, (c) nucleolus and (d) nuclear reticulum.

(a) **Nuclear envelope or membrane:** The nuclear contents and nuclear sap remain enclosed by nuclear envelope. This membrane separates the nuclear sap with its contents from cytoplasm. Electron microscopic study reveals that this envelope consists of two unit membranes separated by a gap. This gap is known as perinuclear space. There are many pores on the membrane and the pores may occupy 8% area of the membrane (Fig. 1.2). The inner and outer pore margins show distinct eight globular sub-units called annular granules. Granule may be present at centre also, which suggests that it regulates the free communication between nucleus and cytoplasm.

(b) **Nuclear sap or karyolymph:** At interphase the nucleus contains granular and homogeneous fluid that is karyolymph. The fluid contains chromatin materials (Euchromatin and Heterochromatin), nucleoli and protein materials some of which are in the form of nuclear enzymes (nuclear protoplasm).

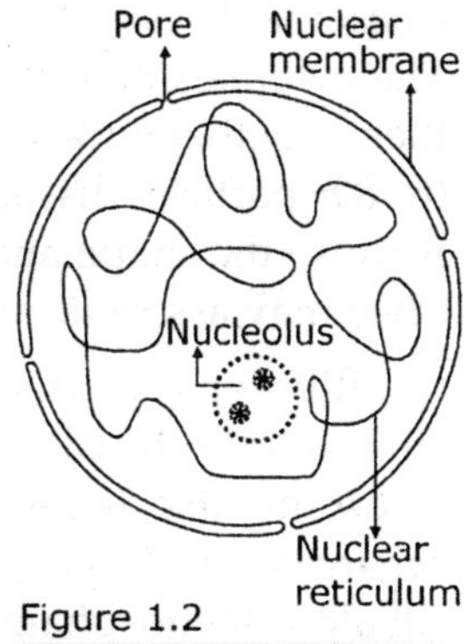


Figure 1.2 General morphology of an interphase nucleus (diagrammatic)

(c) **Nucleolus:** It is roughly spherical in shape and is present in every non-dividing cell. It is not bounded by a membrane and can be clearly seen even in light microscope. In dividing cells nucleolus disappears and its reappearance is due to the activity of a special region of certain chromosome known as nucleolar organizer. Though more than one nucleolar organizer may be present the developing nucleoli fuse to form a single nucleolus at maturity. Electron microscopic study shows distinct regions within the nucleolus: (i) a peripheral granular region that is thought to be the precursors of cytoplasmic ribosome, (ii) fibres composed of ribonucleoprotein and (iii) chromatin.

(d) **Nuclear reticulum:** The undividing nucleus shows a reticulate structure made up of chromosomes and are bounded by nuclear membrane and remains embedded in karyolymph. During the process of nuclear division or mitosis this chromatin reticulum is organized to chromosomes, which are the bearer of the hereditary characters and 'the biological unit of heredity control'—the genes are located in them. Each plant species possesses constant number of chromosomes. DNA, RNA and various proteins compose a chromosome. Electron microscopic study revealed the globular repeating units in a chromatin—termed nucleosome.

**Function of nucleus:** As the nucleus is the major site of DNA, it is responsible for the replication, expression and preservation of genetic information. It also regulates cell division, growth and differentiation. The hereditary characters are transmitted through nucleus.

**Cytoplasm:** It is the portion of protoplasm that is inside the plasma membrane and outside the nucleus. It is a heterogeneous mass containing granules of different sizes and shapes. The mass of cytoplasm contains many organelles and aqueous substance in which many inorganic or organic particles remain dissolved or suspended. The aqueous portion of cytoplasm is termed as hyaloplasm where the organelles are embedded. In the differentiated cells cytoplasm possesses one or more vacuoles that contain cell sap. The cell sap is composed of different salts, sugars and in some cases the specific pigment anthocyanin. The cytoplasmic organelles include ribosome, mitochondria, plastids, microbodies, Golgibodies, lysosomes and endoplasmic reticulum.

**Ribosome:** Robinson and Brown in 1953 first noted ribosome in bean roots. They occur both free in the cytoplasm and attached to endoplasmic reticulum. They are also present in the chloroplastids. In many cells a thin thread joins several ribosome. These aggregates of ribosome are known as polyribosome or polysome. The cytoplasmic ribosomes are composed of two subunits and $Mg^{++}$ ions bind them together with a sedimentation constant of about 80S. When $Mg^{++}$ is removed the 80S particles dissociates into 60S and 40S subunit. The 'S' values are not additive because the shape and size of the particle may change the rate of sedimentation. They may associate again when $Mg^{++}$ is restored.

Ribosome consists of approximately equal parts of protein and RNA. The ribosome is the site of protein synthesis. The mitochondrial and chloroplast ribosome also incorporate amino acids into protein. Sometimes ribosome of organelles and cytoplasm function in a coordinated way. The enzyme protein Ribulose biphosphate carboxylase is synthesized within chloroplast and consists of large and small subunits. The small subunit is synthesized at the cytoplasmic ribosome and then it moves into chloroplast to form Ribulose biphosphate carboxylase.

**Mitochondria :** Altman in 1894 first observed these organelles and called them 'bioplasts'. Benda in 1897 called them mitochondria (Greek: mitos = thread, chondrion = granule). Their occurrence in plant cells (ex. *Nymphaea*) was first reported by Meves in 1904.

They are just visible as small rods or spheres under light microscope. They can be observed easily in a phase contrast microscope under dark field illumination when stained in Janus green B. In later years they are isolated by differential centrifugation, ultracentrifugation etc. and are widely studied with electron microscope (Fig. 1.3).

The shape of mitochondria is not constant and this variation occurs during certain functional stages. They may be filamentous, granular, or they may assume the shape of a club, a tennis racket or vesicular. The shape is more or less constant in similar type of cells that perform same function.

Mitochondria are present in all cells that respire aerobically. Their number varies from cell types. The number may be as high as 5000,000 in the protozoon *Chaos chaos*. In a particular cell type the number is more or less constant and the plant cells contain lesser number of mitochondria than animal cells. In general, mitochondria are distributed uniformly throughout the cytoplasm. They may however accumulate preferentially at the proximal, basal, peripheral side of a cell or around the nucleus. During mitosis mitochondria aggregate near the spindle and at the end of division they are distributed in the daughter cells in approximately equal number.

The size of mitochondria is variable. The width is about 0.5 μm and it is relatively constant in most cells. However the length may be as high as 7 μm. Electron microscopic study reveals that mitochondrion consists of two membranes – an outer and an inner membrane. The outer membrane is smooth, about 6 nm thick and it is the limiting membrane that surrounds each mitochondrion. The inner membrane may form plate like folds called mitochondrial crests (cristae) or tubular projections termed microvilli that penetrate the matrix of mitochondrion. The inner membrane is also about 6nm thick.

The arrangement of cristae varies. They may be parallel or perpendicular to the long axis of mitochondrion. They may be in the form of vesicles or branched to form a network. Sometimes they are arranged concentrically within the matrix. The underlying purpose of the enfolding is to provide an increased surface area within the mitochondrion for enzymatic activity. This inner membrane divides the mitochondrion into two chambers – (i) inner chamber or inner membrane space that is bounded by the inner membrane and (ii) outer chamber or intermembrane space that is represented by the intermediate space between the two membranes and core of the mitochondrial crests.

The outer chamber is filled with watery fluid and is 140 Å to 180 Å in width. The inner chamber is filled with dense proteinaceous matrix. The matrix contains granules of phospholipids, ribosomes, which are slightly smaller than those in the cytoplasm and DNA (mitochondrial DNA). The microvilli and cristae are, in general, incomplete septa and so there is continuity of the matrix.

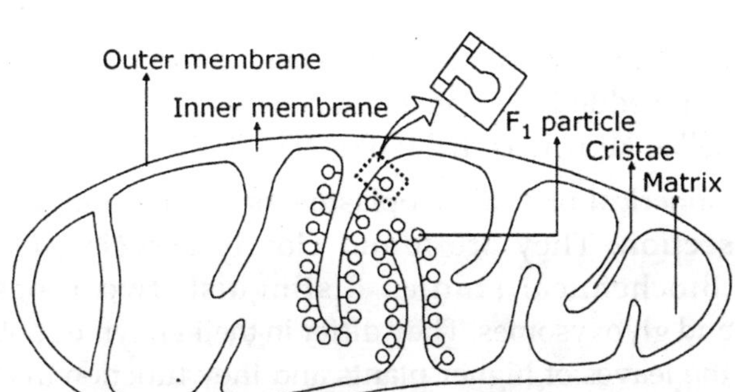


Figure 1.3

Diagram representing a mitochondrion cut longitudinally. The inset shows a single $F_1$ particle with head and stalk.

On the inner side of inner membrane, i.e. the side facing the matrix, there are attached 'elementary' or '$F_1$' particles or oxysomes of 8.5 nm in diameter. These particles are shortly stalked and remain attached to the inner membrane at regular intervals of 10nm. There may be $10^4$-$10^5$ number of elementary particles per mitochondrion.

Mitochondria are regarded as the 'power house of a cell' and they are concerned with energy conversion. Several enzymes of respiratory cycle and all the enzymes of Krebs cycle are found in mitochondria. Electron transport system, oxidative phosphorylation and ATP synthesis occur here. Mitochondria may move and carry ATP wherever required. The movement is less in animals than plants. Lipid synthesis and elongation of fatty acids occur in mitochondria.

Mitochondria are semiautonomous organelles and they can synthesize protein due to the presence of DNA, ribosome and RNA in the matrix.

**Plastid:** Plastids are small, variously shaped subcellular organelles present in cytoplasm of plant cells. Their number varies in different plants. They develop from pro-plastids. They are classified on the basis of colour and function, e.g. chromoplast (non-green and non-photosynthetic pigment plastid), leucoplast (colourless plastid), amyloplast (starch accumulating plastid), proteinoplast (protein storing plastid), elaioplast (fat accumulating plastid) and the very prominent chloroplast (green, contains chlorophyll and other photosynthetic pigments). Chromoplast contains carotenoid and the colour may be red, orange or yellow. They are generally produced from amyloplasts or chloroplasts. Amyloplast synthesizes starch and stores one or more starch grains. The protein containing plastids proteoplast (or proteinoplast) are devoid of grana and contain few thylakoids. Elaioplasts, in some species, are produced from chloroplastid.

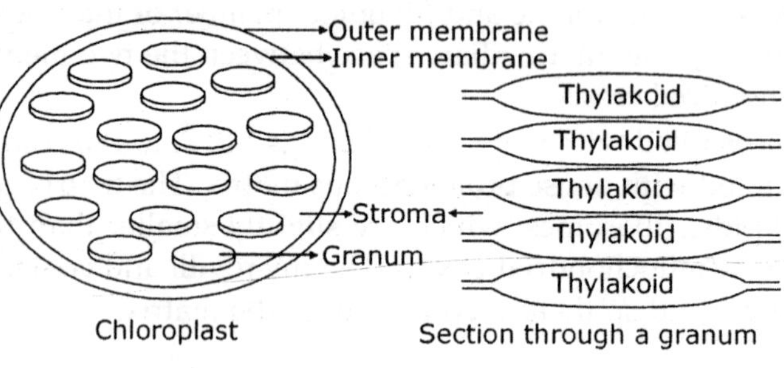


Figure 1.4 Diagrammatic representation of chloroplast and section through a granum.

Chloroplastid consists of a double membrane enclosing homogeneous stroma and grana. Granum remains embedded in stroma and consists of flattened, disc-shaped vesicles forming lamellae known as thylakoids (Fig. 1.4).

**Microbodies:** They are spherical or oval cytoplasmic organelles but usually appear circular in cross section. They are found closely associated with endoplasmic reticulum. Biochemical studies distinguish two types of microbodies – peroxisomes and glyoxysomes. They differ in their enzyme content. Peroxisomes are present in the leaves of higher plants and they function in close co-operation with chloroplastids. They are the sites of photorespiration.

Glyoxysomes are abundant in germinating fatty seeds and they contain enzymes for the breakdown of fatty acid.

**Golgi body or dictyosome:** Golgi body or dictyosome (as it is frequently called in plant cells) is composed of stacks of smooth membranes (Fig. 1.5) and disc shaped sacks called cisternae. Camillo Golgi first described these in 1898. They are usually found scattered throughout the cytoplasm and their number varies from one to several hundred per cell. In sections cisternae appear like flattened sacs and their number may be as high as twenty in each Golgi body. Some cisternae are dilated towards the periphery and appear as vesicles. The outer edges of cisternae consist of network of tubules. The major function of Golgi complex is secretion. Other functions are acrosome formation during sperm maturation, release of zymogen granules, storage of protein or lipid in vesicles and vacuoles, formation of plasma membrane, polysaccharide formation for the new cell wall and primary lysosome formation.

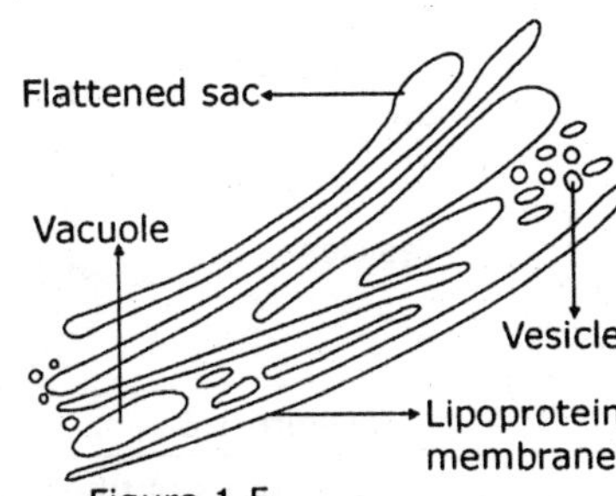

Figure 1.5

Diagram representing the morphology of Golgi complex. After Cohn, 1964, p. 78.

**Lysosome:** They are sub-microscopic and single membrane bounded particle. They lack any internal structure and contain hydrolytic enzymes. They can be observed under electron microscope only. They are the main sites of hydrolytic enzymes and so can hydrolyze protein and carbohydrate. They originate from Golgi complex.

**Spherosomes:** They are also called oleosomes. These are small spherical organelles of 1 μm in diameter. They are surrounded by a half-unit membrane rather than a complete unit membrane. They consist of lipid droplets. They also contain a variety of enzymes namely phosphatases, lipases, RNAase and DNAase, endopeptidases etc. They originate as oil containing vesicles from the endoplasmic reticulum. They are one of the sites of fat and oil synthesis. They store fats. The number and frequency of occurrence of spherosomes vary in different cells. Large number of spherosomes is reported from the guard cells of *Campanula persicifolia, Paphiopedilum* etc.

**Endoplasmic reticulum:** The cytoplasm of a cell permeated by complex elaborate system of membranes termed endoplasmic reticulum (ER). Electron microscopic study discovered the existence of ER in 1945 by Porter *et al.* and the term ER was used by Porter and Kallman in 1952. Now a days ER can be observed in a living cell by the use of fluorescent dyes.

ER consists of double membranes that enclose spaces. They may be in the form of tubules, vesicles, vacuoles and flattened sacs (cisternae) that at several points are intercommunicating. When the tubules were first observed, the micrographs showed that the tubules did not reach the periphery of the cell and hence the term 'endoplasmic' was coined. The diameter of the tubule may be

50 μm to 100 μm. The vesicles are more or less round in shape and range in diameter from 25 μm to 500 μm. The cisternae are long flattened units, 40 μm to 50 μm thick. They are arranged in parallel stacks (Fig. 1.6).

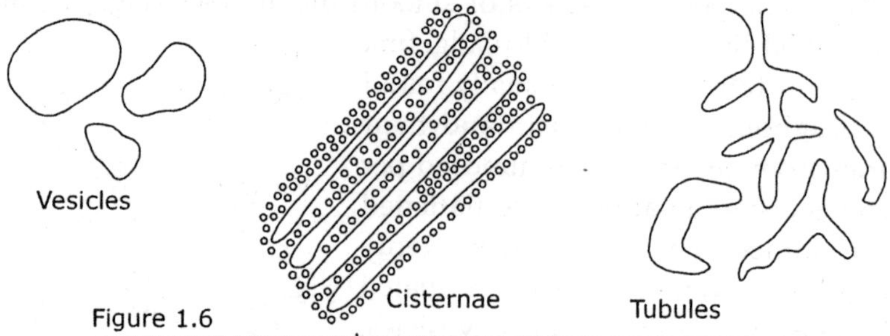


Figure 1.6

Diagrams of the morphology of endoplasmic reticulum. Adapted after Cohn, 1964, p. 61.

Endoplasmic reticulum exists in two forms – smooth and rough; the two forms may be interconnected with each other. The rough form is due to the presence of numerous small particles of ribonucleoprotein on the surface of ER. The functions of ER are suggested as: (1) the membrane provides increased surface area for metabolic activities; (2) the spaces formed by ER are the site of collection of metabolic products; (3) the rough ER are the site of protein synthesis; (4) the smooth ER are present in those cells that synthesize cholesterol, glycerides etc.; (5) ER forms an effective transport system for protein etc.; (6) it has role in the formation of cell wall and (7) it transmits impulse.

**Ergastic substances:** The nonliving cell inclusions produced as a result of metabolic activity of cell are known as ergastic substances. These substances are found in cytoplasm, vacuoles and cell walls. Some of these have great importance in plant life while others are absolutely by-products. These substances are grouped as (1) Reserve material, (2) Secretory material and (3) Excretory material.

**1. Reserve material:** The autotropic plants by photosynthesis synthesize carbohydrate that along with other metabolite forms proteins, fats and oils. These substances may be used up to make new protoplasm and to provide energy for cellular activities or may be stored as reserve food materials for future use. The reserve materials are carbohydrate, protein, oil and fats.

(A) **Carbohydrate:** Carbohydrates (= carbon hydrates) are composed of carbon, hydrogen and oxygen where the ratio between the latter two is two : one. They are often referred to as saccharides (Latin, *Saccharum* = sugar) because the simpler members of the carbohydrate taste sweet. The elements of carbohydrate are obtained from atmospheric carbon dioxide and water during photosynthesis. Carbohydrates are classified into monosaccharides, oligosaccharides and polysaccharides on the basis of number of simple sugar molecule produced on hydrolysis. The product of hydrolysis may be of same or different sugars.

i. **Monosaccharide** (simple sugar): These are composed of single unit of carbohydrate. They cannot be broken into simpler carbohydrates on hydrolysis. Example: glucose (present in grape sugar), fructose (found in honey) etc. They are also referred to as hexose as six carbon atoms are present in these sugars.

ii. **Oligosaccharide:** They contain two to ten units of monosaccharide in a molecule. The oligosaccharides containing two units of monosaccharide are called disaccharide (ex. sucrose); those containing three units are called trisaccharide (ex. raffinose). Sucrose of cane sugar yields glucose and fructose, one molecule each on hydrolysis. It is the table sugar and is extracted from sugarcane. Another disaccharide—maltose yields two molecules of glucose on hydrolysis. The trisaccharide raffinose yields glucose, fructose and galactose on hydrolysis.

iii. **Polysaccharide:** Each molecule of polysaccharide is made up of more than ten units of monosaccharide. These include the storage polysaccharide and the structural polysaccharide inulin, starch, cellulose, pectin etc.

(a) **Inulin:** This polysaccharide yields only fructose on hydrolysis. It is found as a storage material in some members of Asteraceae (tuber of *Dahlia, Helianthus tuberosus*), Campanulaceae and monocotyledons. Inulin is spherical or star shaped crystal. Incompletely formed crystals can be seen as fan-shaped structure. It appears only in cell sap solution of storage organs. They may be precipitated when thin pieces of *Dahlia* are treated with glycerine or alcohol for about a week. A thin transverse section of the treated tubers will reveal the spherical or star shaped or fan shaped crystals (Fig. 1.7).

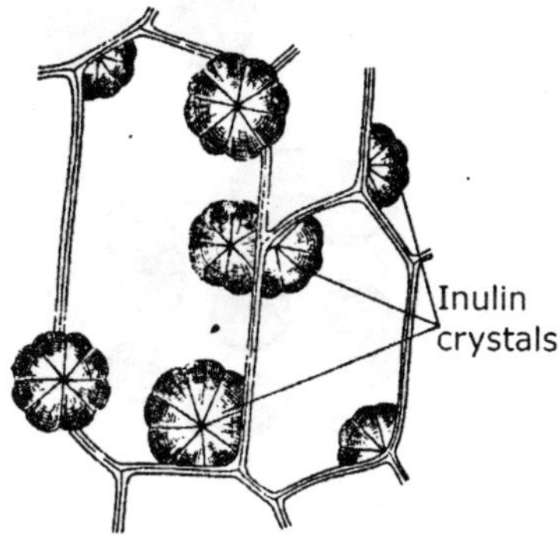


Figure 1.7

Cells from root-tuber of *Dahlia* with inulin crystals.

(b) **Starch:** The glucose units are the building blocks of this insoluble polysaccharide. They are found in all parenchymatous tissues mainly in storage organs like rhizomes, endosperm, corms, tubers, cotyledon of seeds etc. They are also present in latex tubes of stems and roots, endodermis, seeds of cereal, grains of rice, wheat, maize etc. The assimilatory starch, a temporary product is synthesized by the chloroplasts. This temporary starch is subsequently broken down to sugar that is transported to the storage organs where storage starch is synthesized by amyloplast.

Starch appears in the form of grains that stain bluish black with potassium iodide solution. Each starch grain consists of a centre, called hilum that is the centre of origin of the grain. Carbohydrates are deposited in layers around the hilum. This gives starch grain a striated appearance. The deposition of layers depends upon endogenous rhythms. In cereals, single layer is deposited in a day, i.e. the number of striations denotes the number of days' growth. The hilum may be centrally situated as in concentric grains where carbohydrates are laid down concentrically around the hilum (e.g. *Triticum durum*) or the hilum may be eccentric as in case of eccentric grains where the layers of carbohydrates are deposited on one side of hilum (ex. *Solanum tuberosum*) Fig. 1.8.

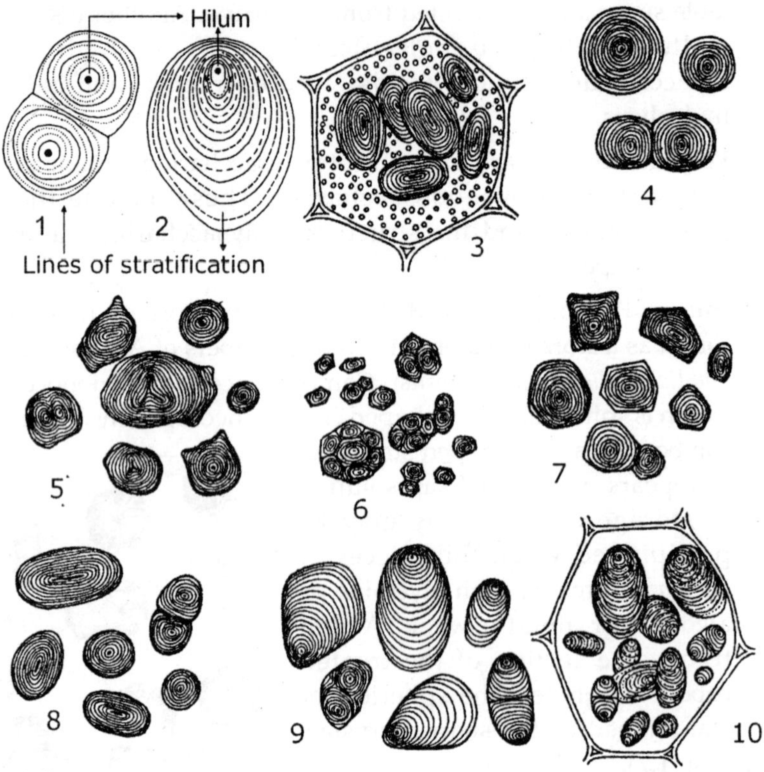


Figure 1.8

Starch grains. 1. Compound starch grain. 2. Simple starch grain. 3. From cotyledon of *Pisum* seed. 4. From flesh of *Musa*. 5. From tuberous root of *Ipomoea batatas*. 6. From endosperm of *Oryza* grain. 7. From endosperm of *Zea* grain. 8. From cotyledon of *Cicer* seed. 9. From tuber of *Solanum tuberosum*. 10. Same in situ.

Starch grains may be classified as simple, semi-compound or half-compound and compound. In simple type each grain remains singly with solitary hilum whereas in compound type two or more grains are aggregated together (ex. *Oryza sativa*, *Ipomoea batatas*, *Avena sativa*) with their separate hila. The half compound grains (e.g. potato) consist of two hila with their lines of stratification aggregated together around which several common layers of carbohydrate are present.

Starch grains vary in shapes, sizes, position of hilum, solitary or in aggregates and in general appearance. Their extensive morphological variations make it possible to identify starch containing plant species.

(c) **Cellulose:** This insoluble structural polysaccharide is the major constituent of cell wall of higher plants. The building blocks of cellulose are glucose units that are held together by β (1-4) linkages. In cellulose molecule approximately 15,000 molecules of glucose residues are present.

(d) **Pectin:** This structural polysaccharide is chiefly present in the middle lamella and primary walls of dicots. Pectins are rich in galacturonic acid, rhamnose, arabinose and galactose.

(B) **Protein :** Proteins are composed of amino acids that are linked by peptide bonds. They are one of the vital ingredients of living protoplasmic bodies. They sometimes occur as structural proteins of cell wall and storage proteins. Storage protein may be in the form of definitive bodies called aleurone grains or they may be amorphous or crystalline. Amorphous protein is found in the aleurone layer (the outermost layer of endosperm) of *Triticum*. Crystalline protein possesses both crystalline and colloidal properties. So these crystalline proteins are referred to as crystalloids. Proteins, in the form of cuboidal crystalloids are observed in the peripheral parenchyma cells of potato tuber and in the fruit of *Capsicum*.

The reserve proteins of potato are mainly composed of lysine and it is removed when the potatoes are peeled. Aleurone grains (Fig. 1.9) are smaller in contrast to starch grains. They stain brown with potassium iodide solution. They are found in the cells of seeds, embryos, endosperm etc. They may occur in specific layers as is noted in the aleurone layers of caryopsis of cereals. Aleurone grains are bounded by a proteinaceous membrane. The membrane may enclose amorphous protein, globoids, and crystalloids (e.g. *Myristica fragrans*) or crystals of calcium oxalate (Umbelliferae). The aleurone grains of *Ricinus* usually contain one globoid and one crystalloid, bounded by a single membrane.

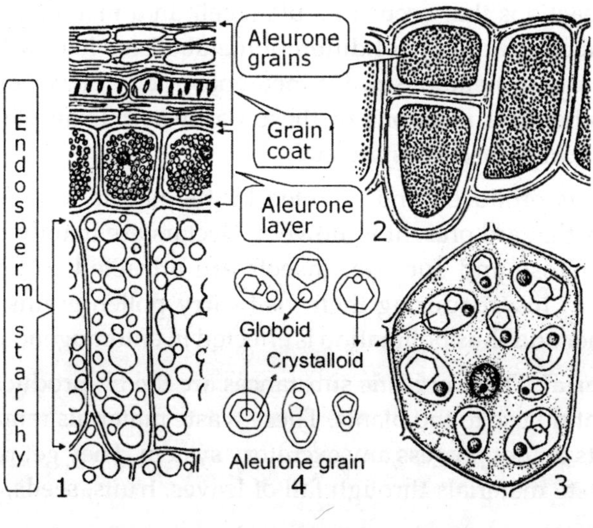


Figure 1.9

Aleurone grains.
1. In *Triticum*.
2. In *Zea*.
3. In *Ricinus*.
4. Diagrams of a few aleurone grains.

(C) **Oil and fat:** They are the glycerides of fatty acid. The oils are liquid and the fats are solid at normal temperature. In contrast to protein and carbohydrate they provide more calories and so they are valuable reserve food materials. They are abundant in the cells of endosperm or perisperm of seed, fruits, spores, embryos, and meristematic cells and in differentiated tissues of vegetative body. They occur also in the cytoplasm of cell sap as dispersed or aggregated large masses. They are produced either by elaioplast or spherosomes. The essential or volatile oils may occur in all tissues (e.g. Conifers), in petals of roses, in the fruit skin of orange, in the special layer (i.e. second layer below the epidermis) of *Elettaria cardamomum* seed. Specialized secretory tissues produce these essential oils. Wax, cutin and suberin are other fatty substances, which usually form a protective covering of the epidermal cell walls. Oils and fats stain a reddish colour when treated with Sudan III or IV.

**2. Secretory material:** These substances, though not concerned with nutrition are valuable ergastic substances. They are produced during plant metabolism and perform specific roles in plant life. These include (i) colouring matter, (ii) enzyme and (iii) nectar.

i. **Colouring matter:** Leaves, petals and fruits of a plant exhibit various colours. This is due to the pigment present in the cell. The leaves and aerial organs possess chlorophyll that imparts green colours to them. This pigment is responsible for the phenomenon—photosynthesis. The other pigment carotenoids, i.e. carotene and xanthophyll impart orange to yellow colour of many flowers and fruits. The red, blue, violet and pink colour is due to anthocyanin—the water-soluble pigment. Due to the presence of carotenoids and anthocyanins, flowers and fruits become brightly coloured. As a result they attract agents to bring about pollination and fruit dispersal, which are the vital phenomenon in plant life.

ii. **Enzyme:** Enzymes are regarded as 'biological catalyst' that catalyzes all the chemical reactions that occur in a plant cell. In a metabolic cell there are many metabolic pathways of different metabolisms like carbohydrate metabolism, protein metabolism, fat metabolism etc. Enzymes catalyze every step of chemical reactions of these pathways. Example: zymase, lipase, protease, invertase, diastase etc.

iii. **Nectar:** Nectar is produced and secreted by the nectar secreting glands called nectaries that are present in flowers. Nectars are mainly composed of glucose, fructose and sucrose. Insects are attracted to flowers for obtaining nectar. Insects during their visit carry pollen grains from one flower to another and thus pollination is effected in entomogenous flowers.

**3. Excretory material:** These ergastic substances are the by-products of plant metabolisms and are of no use to the plants. These waste materials remain stored in the dead cells. Plants do not possess any excretory system. They get rid of some of the excretory or waste materials through fall of leaves, fruits, seeds, barks etc.

Some of the excretory materials are very much valuable to man. A few important excretory materials are mentioned below:

(a) **Alkaloids:** These are nitrogenous waste substances and are found in roots, leaves, barks, seeds and other parts. Example: quinine occurs in the bark of *Cinchona*, morphine is present in the fruit of opium poppy (*Papaver somniferum*), atropine in the leaves of *Atropa belladonna* etc.

(b) **Crystals:** These are the depositions of waste products within a cell. They are composed of either calcium carbonate or calcium oxalate, which is more common. The crystal-containing cell may be similar to neighbouring cell that lack crystals or it may be easily distinguished from the other cells by their shape and size. This special crystal-containing cell is termed as idioblast. Crystals originate within vacuoles. Crystals may remain single or as aggregate in the cell. Sometimes a large crystal is present filling up the entire cell cavity. Usually they lie loose in the cells, but some are found to be suspended from the cell wall.

i. **Crystals of calcium oxalate:** They appear in definite forms that are as follows (Fig. 1.10).

*Prism*: It is rectangular or pyramidal in shape and occurs single or as twin prisms. Example: leaves of *Citrus, Vicia sativa, Begonia* etc.

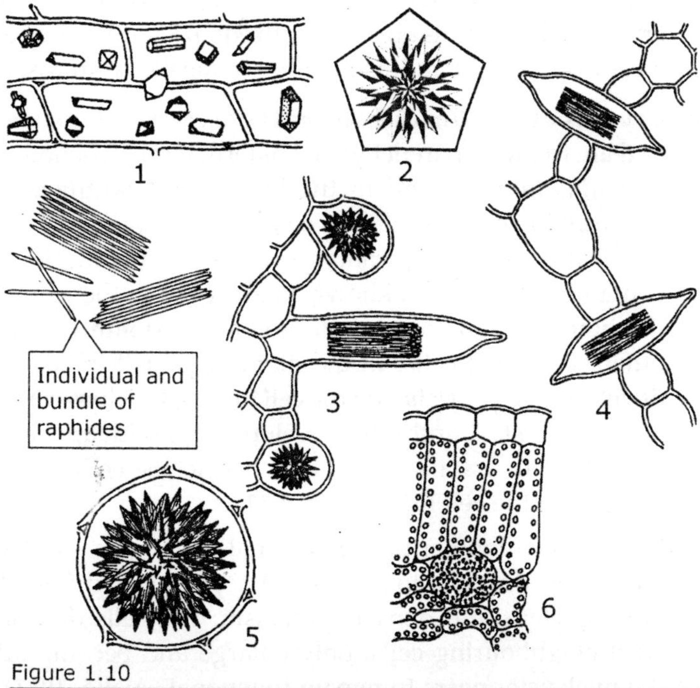


Figure 1.10

Calcium oxalate crystals. 1. Solitary crystals in the tunic of *Allium*. 2. Sphaeraphides in *Nerium* leaf. 3. Raphides and sphaeraphides in the leaf of *Pistia*. 4. Raphides in the petiole of *Eichhornia*. 5. Sphaeraphides in the petiole of *Carica*. 6. Crystal sand in the leaf of *Atropa belladonna*.

*Druse*: It is more or less spheroidal in shape. Many prism or pyramidal crystals aggregate to form a druse. In these crystal-aggregates some projecting points are observed all over the surface. Druse is also known as conglomerate crystal or sphere-crystal or sphaeraphide. They are found in the leaves of *Datura*, *Nerium*, petiole of *Carica* etc.

*Raphide*: It is thin elongated needle shaped or acicular crystal that is tapered at both ends to a tip point. It usually occurs as aggregate bundles in idioblast, which is much larger than the neighbouring cells. Raphides are found in the petals of *Impatiens*, leaves of *Colocasia* etc. The raphide containing cells may contain mucilage. A matrix of polysaccharide may surround raphides as is found in some members of Araceae. Electron microscopic study reveals the presence of excrescences in the form of minute bristles in the raphides of *Colocasia*, whose edible species when consumed causes irritation of mouth and throat. It is assumed that the irritation is due to scratches caused by bristles and pointed ends of raphides (Sakai and Hanson, 1974).

*Pseudoraphides* or *styloids*: These are elongated prismatic crystals that are tapered into blades at both ends. A cell may contain solitary or a few styloids. They are found in some members of Liliaceae, Agavaceae etc.

*Rosette*: In Umbelliferae, the aleurone grains in seeds contain rosette crystals. The crystals are large with equal lengths. They deposit around a centre along the radius in all directions to from rosette.

ii. **Crystals of calcium carbonate:** These crystals are known as cystolith and are found in parenchyma, epidermal cells, trichomes or hairs (e.g. *Humulus lupulus*) and in the leaves of Acanthaceae, Urticaceae, Cucurbitaceae etc. A cell may contain solitary or more than one crystal. The leaf of *Momordica* contains double cystolith. The epidermal cell containing cystolith is known as lithocyst. In the leaf of *Ficus elastica*, the lithocyst that contains the solitary cystolith is much larger than other epidermal cells. In the lithocyst a stalk made up of cellulose develops and projects into the cell lumen. Calcium carbonate deposits around the stalk and fully developed cystolith appears as grape-like clusters with a stalk that is attached to the cell wall from where it hangs into the cell lumen (Fig. 1.11).

During development all epidermal cells have similar appearance. Later, cells destined to become lithocyst can be differentiated by their large nucleus and denser cytoplasm. The cells do not divide along with neighbouring cells, only enlarge and become lithocyst where the nucleus appears to remain functional (Ajello, 1941).

*Crystal sand*: They are very small prismatic crystals and occur frequently in masses in the cell. They are found in Solanaceae and in *Atropa belladonna*; each crystal may be spheroidal or wedge shaped.

Dormer (1962) pointed out the taxonomic significance of crystals. In an extensive study with 112 species of the genus *Centaurea*, Dormer was able to segregate the species into two groups on the basis of crystals.

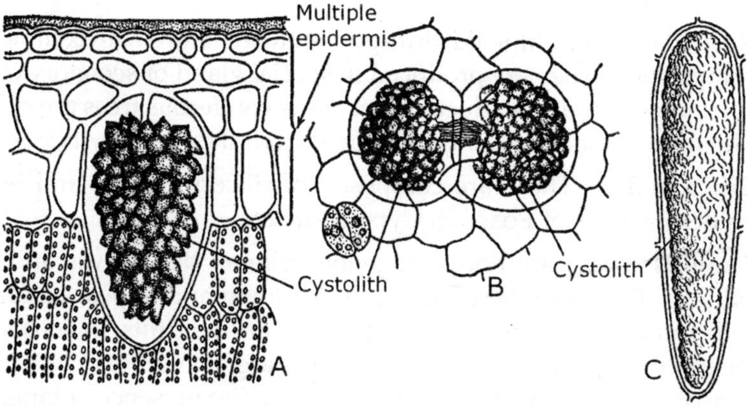


Figure 1.11

Calcium carbonate crystals. A. In the leaf of *Ficus elastica*. B. Double cystolith in *Momordica*. C. Elongated cystolith in *Ruellia*.

(c) **Essential oils or volatile oil:** The odoriferous organs of a plant contain essential oils. They are produced in special cells or glands. They occur in flowers, leaves, and fruits etc. Mention may be made of oil of peppermint from *Mentha piperita*, clove oil obtained form *Eugenia caryophyllata*, eucalyptus oil found in *Eucalyptus globulus* etc. Though volatile oils are excretory products, yet they have the biological significance to attract vectors to bring about pollination and dispersal of seed and fruits.

(d) **Glycosides:** They are composed of glucose and some other substances, e.g. sinigrin—sulphur-containing glycoside present in *Brassica nigra*, digitogenin present in *Digitalis purpurea* etc.

(e) **Gum:** Gums are exudes in the form of thick juice from the bark of *Acacia arabica, A. modesta* etc. The cellulose of cell wall by decomposition forms gums. Usually gums are exuded naturally, but sometimes they are exuded in response to injuries to heal up the wounds. The gum arabic of commerce is obtained from *Acacia senegal*. It is used in medicine, confectionery and sizing and finishing materials in textile industry (Sing and Jain, 1996).

(f) **Latex:** Latex is a fluid secreted by special group of cells termed laticifers. They are present in the families Asclepiadaceae, Apocynaceae, Moraceae, and Euphorbiaceae etc.

Latex may be colourless or coloured as milky white or yellow to orange. It consists of carbohydrate, alkaloids, organic acids, oils, resins, rubber, starch etc. Papain, the proteolytic enzyme is present in the latex of *Carica papaya*. The alkaloid opium is rich in the latex of *Papaver somniferum*; *Hevea brasiliensis*, a member of Euphorbiaceae, yields about 2240 kg rubber per hectare annually.

(g) **Organic acid:** Different kinds of organic acids are found in plants and some of them remain concentrated in leaves and fruits. These are produced during different types of metabolisms. Some common acids are tartaric acid—found in tamarind, malic acid present in apple, citric acid in oranges etc.

(h) **Resin:** These may be solid, brittle or liquid in nature with complex chemical composition. They are produced in special gland or secretory cells present surrounding the resin ducts. Most of the coniferous trees are rich in resins. The resin Canada balsam is obtained from the conifer *Abies balsamea.*

(i) **Silica bodies:** These are composed of silicon dioxide and occur in the epidermal cells of Poaceae, Cyperaceae etc.

(j) **Tannins:** These are non-nitrogenous phenol derivatives and present in cell sap. They may be yellow, red or brown and appear as granules in a cell. Tannins are found in seed coats, in unripe fruits, in leaves and in the tissues infected by pathogens. Tannins are also present in periderm and heartwood; as a result they become hard. The presence of tannin prevents pathogenic attack. Tannins are used in medicine, in leather industry, in dye industry etc.

## Plasma membrane

In plant cells the membrane that lies between cell wall and cytoplasm, serves as boundary of the latter, is designated as plasma membrane or cell membrane or plasmalemma. It is ultrathin, elastic and porous membrane that controls the entrance and exit of molecules and ions.

Plasma membrane is so thin that it cannot be readily resolved under light microscope. If at all visible it appears as a very thin line just inside the cell wall.

**Chemical composition:** The membrane is mainly composed of lipids, proteins and carbohydrates. Water makes about 29% of total weight. Robertson (1959) proposed that plasma membrane is three-layered structure where proteins form the outer and inner layers of membrane that encloses lipids to form a unit membrane.

• *Lipid:* The lipids identified in the plasma membrane consist of cholesterol, phospholipids and galactolipids. The phospholipids include phosphatidylcholine, phosphatidylethanolamine and sphingomyelin. The phospholipids are found to be associated with the outer protein shell in the plasma membrane. Glycerol and fatty acid constitute lipid molecules. In the membrane the lipid molecules consist of two parts—a head and two tails. The head is composed of glycerol and is hydrophilic where as the tails are composed of fatty acids that are hydrophobic. The head and tail are usually designated as polar and nonpolar end respectively. The proteins, in the membrane, are present in two layers and the lipids occur in between them. The lipid molecules are oriented in two (bimolecular) layers with their hydrophilic polar ends directed towards protein and the hydrophobic nonpolar ends face each other. Hydrogen bonds, ionic linkages or electrostatic forces bind the protein and lipid components.

• *Protein:* In the membrane it is present as enzyme protein, carrier protein and structural protein. The enzyme proteins have catalytic activity. The carrier proteins help to transport materials in and out of the cell across membrane. The structural proteins play an important role to form the structure of membrane.

The membrane proteins are classified as integral and peripheral protein, and the criteria for distinguishing them are as follows.

| Peripheral protein | Integral protein |
|---|---|
| 1. Also termed as extrinsic protein. Present on the membrane surface only. | 1. Also termed as intrinsic protein. It is also present on the membrane surface but a portion of it lies within lipid layer. |
| 2. It is free of lipids. | 2. It is associated with lipids. |
| 3. It is present in small amount in comparison to integral protein. | 3. It represents more than 70% of the membrane protein. |
| 4. In it the hydrophilic polar amino acids predominate on the surface (ex. tyrosine, tryptophan etc.). | 4. In it the hydrophobic nonpolar amino acids predominate on the surface (ex. methionine, proline etc.). |
| 5. It is soluble in aqueous solution and so can be separated by mild treatment. | 5. It is insoluble in aqueous solution and so drastic procedures are required for separation. |

• *Carbohydrates:* They occur in the form of glycolipids and glycoproteins. Both of these forms are confined exclusively to the external membrane surface. Bell (1962) is of opinion that polysaccharides confer some stability to the lipoprotein complex in the membrane.

**Origin:** Bell suggested that the phragmosomes, those appear during middle lamella formation and are composed of polysaccharides, help in the formation of new plasma membrane on reaction with lipoprotein.

**Morphology:** The plasma membrane is not readily visible under light microscope. Electron microscope shows that the plasma membrane is 6 nm to 10 nm thick. It consists of three layers—two outer dense layer, each about 2 nm thick and a middle less dense area of 3.5 nm across, for a total thickness of 7.5 nm. On the basis of these findings Robertson (1959) proposed the unit membrane structure to the lipoprotein organization of plasma membrane. According to this concept the outer dense layers represent the protein and the enclosed less dense are the lipids. Very small pores, none of which are of greater diameter than 5 nm, interrupt the continuity of the membrane.

There are different models to explain the structure of plasma membrane. The unit membrane model of Robertson consisting of protein-lipid-protein layer is previously mentioned. The fluid mosaic model of the membrane as proposed by Singer and Nicolson (1972) is now generally accepted. This model assumes that the membrane is a semifluid structure where lipids and integral proteins are

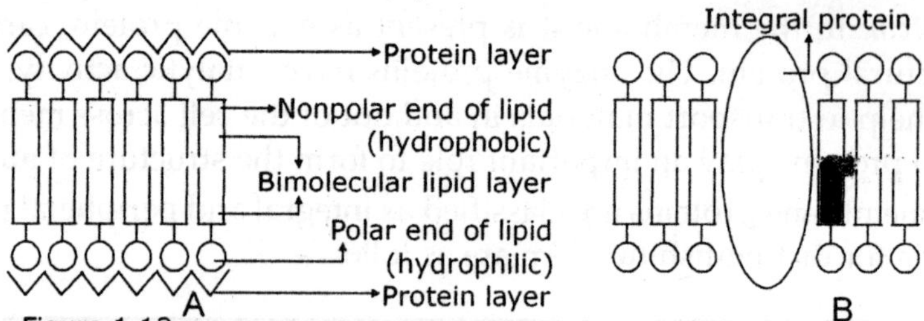


Figure 1.12

Schematic representation of molecular organization of the plasma membrane according to (A) Robertson, 1959 (the unit membrane model) and (B) Singer and Nicolson, 1972 (the fluid mosaic model).

present. They are arranged in a mosaic manner, i.e. the integral proteins are embedded to a greater or lesser extent in the continuous bilayers of phospholipids. The integral proteins are amphipathic molecules, i.e. within the same molecule hydrophilic and hydrophobic groups occur. The hydrophilic groups protrude from the surface while the hydrophobic groups remain embedded in the lipids (Fig. 1.12).

**Functions of plasma membrane:** (i) It is selectively permeable membrane; (ii) its principal role is to regulate the flow of materials in and out of the cell; (iii) carrier proteins in the membrane are involved in the transport of certain materials across the plasma membrane; (iv) in some cases it protects the inner cytoplasmic inclusions; (v) in some cases it perceives the chemical stimulus.

Difference between cell wall and cell membrane of plants.

| Cell wall | Cell membrane |
|---|---|
| 1. It is the boundary of a cell. | 1. It is the boundary of cytoplasm. |
| 2. It is completely permeable. | 2. It is selectively permeable. |
| 3. It is composed of cellulose. | 3. It is composed of protein, lipids, carbohydrates etc. |
| 4. It is rigid and comparatively thick. | 4. It is elastic and comparatively thin. |
| 5. The walls are sculptured. | 5. It is not sculptured. |

CHAPTER **2**

# Diversities of Cell Types in Higher Plants

In higher plants, the great majority of cells conform the general pattern but there exist considerable variations. Microscopy and biochemistry – these two disciplines along with the reactions of cells to different histological stains has helped to differentiate cell types. The variation is usually related to the function of the particular cell type. The cell types are distinguished by their size and shape that is attributable to the cell wall, composition and structure of cell wall, position of cells and cell contents.

**Size:** The size of mature cells varies considerably within quite broad limits. Majority of cells fall within the range of 20 μm-300 μm either in length or diameter. The diameter of parenchyma cells usually ranges from 10 μm to 100 μm. The length of fibres usually 1 mm to 8 mm and they may be 55 cm long as in *Boehmeria*. In certain monocotyledons and Urticaceae the length of the fibre may be 250 mm-550 mm. The latex cells seem to be the largest and these branched coenocytic cells grow almost indefinitely. Usually, the dividing cells of root meristem are the smallest and a considerable increase in all parameters is observed in mature cells.

**Shape:** Each cell has a typical, more or less rigid shape that is attributable to the cell wall. Usually the function and stresses operating on cells during development determine the shape. Mature cells may be oval, cylindrical, ellipsoidal, flattened, lobed, star shaped, prism or fibre like and of various other shapes.

At the onset each cell is spherical as is revealed by isolated cells grown in culture. In meristematic tissues they are polyhedral due to contact with other cells and lack of intercellular spaces. Measurements on isodimetric cells of root apices and stem pith in a number of different plants have revealed that fourteen-sided polyhedral cells are the most common. Although cells with 12, 13, 15, 16 or more faces are found, the average number of faces per cell is close to fourteen. Matzke (1956) found that cells with pentagonal walls are most prominent although there exist tetragonal and hexagonal faces.

**Cell types:** Plant tissues contain a variety of cell types that is revealed in longitudinal and transverse sections, maceration or by other techniques. There are about forty cell types in higher plants (Lyndon, 1990, p. 136), a very brief account of the major types is discussed below.

- *Antipodal cells*: These cells are present in the normal embryo sac, generally three in number and lie at the opposite end of micropyle. They have a nucleus and cytoplasm; in several plants no cell wall is formed and the antipodal nuclei remain free. It is haploid.
- *Bulliform cell*: It is an enlarged group of cells present in the upper epidermis of grass leaves. The cell wall is thick and the cell is hyaline, vacuolated and lacks chlorophyll that distinguishes it from other epidermal cells. It is suggested that this cell expands or contracts to cause opening and closer of leaves.
- *Cambium* (vascular): This is a layer, strip or cylinder of meristematic cells that normally occur in the dicotyledonous plants. They are present in open vascular bundles as fascicular cambium and between the vascular bundles termed interfascicular cambium. They divide to form secondary vascular tissues, i.e. secondary xylem and phloem towards interior and exterior respectively. This lateral meristem causes secondary thickening in most dicots.
- *Collenchyma*: This mechanical cell is present at the hypodermis of many herbs and petioles of leaves. The cells may be round or oval in cross section and their walls are thickened by pectin, hemicellulose and cellulose, which deposit mainly at the corners or at the tangential walls. The cells are living and can elongate.
- *Companion cell*: This cell is found only in phloem of angiosperms, where it remains associated with sieve tube with which it has common origin. It is a nucleated cell with dense cytoplasm and apparently plays some role in the functioning of sieve tube.
- *Cork*: Also known as phellem and is derived from the cambium called phellogen on the outside of roots and stems of dicotyledonous plants. A cork consists of suberized and dead cells and is protective in function. It replaces the epidermis and is impervious to gas and water vapour.
- *Cork cambium*: It is also called phellogen, and composed of a layer of meristematic cells, present on the peripheral side of roots and stems of angiosperms. It cuts off phelloderm on the inside and phellem or cork on the outside. The three layers—phellem, phellogen and phelloderm are collectively termed as periderm.
- *Egg cell*: The functional female reproductive unit is egg cell. It and the other two cells termed synergids compose the egg apparatus in the embryo sacs of angiosperm. The cell remains towards the micropyle and contains prominent nucleus, cytoplasm and vacuole. It is haploid.

- *Endodermis*: It consists of a layer of cells that delimits the cortex from the stele. It is considered as the innermost layer of cortex and in root it surrounds the radial vascular bundles. The endodermal cells possess the characteristic casparian strips on their radial walls.
- *Endosperm cell*: It is nutritive and food storage tissue of seeds. It is developed in the embryo sac of angiosperm as a result of fusion between two polar nuclei and one male nucleus. It is usually triploid in angiosperm and provides nutrition to the developing embryo.
- *Epidermis*: It is outermost layer of any plant parts. The outer wall of epidermal cell may be thickened by the formation of cuticle. Epidermis may bear hairs, stomata etc. Usually it is uniseriate; multiple epidermis is noted in orchid root and *Ficus* leaf. It is the protective layer of primary tissues.
- *Fibre* (Sclerenchyma): It is elongated, narrow cell, tapering at both ends with lignified cell wall, mostly without living protoplast. Usually they lie parallel to the long axis of the organ in which they occur and add mechanical strength.
- *Gland cell*: This cell is thin walled, densely protoplasmic, and has secretory function. Sometimes gland cells surround a cavity where the secretions are collected. They usually secrete a specific substance.
- *Hair*: It is uni- or multicellular epidermal outgrowth. Hairs are present on all parts of a plant including stamens (ex. *Tradescantia*) and seeds (ex. *Gossypium*). They may be absorptive (e.g. root hairs), protective or secretory (e.g. glandular hair).
- *Idioblast*: These cells markedly vary in size, wall thickness, cell content and function from their neighbours, e.g. lithocyst, raphides, guard cells etc.
- *Laticifers*: They are simple or branched, broad, elongated and aseptate cell where latex is secreted.
- *Meristematic apical cells*: These cells occur at root tips, shoot tips and leaf tips. These are undifferentiated and actively dividing cells, which give rise to permanent tissues. These cells are generally isodiametric, contain large prominent nuclei and less intercellular spaces.
- *Mesophyll*: These are parenchyma cells containing large number of chloroplasts, occur between the upper- and lower epidermis of leaves. In isobilateral leaves the mesophyll cells are all alike in contrast to dorsiventral leaves where the cells towards the upper epidermis are cylindrical, arranged at right angles to epidermis termed palisade cells; the mesophyll cells towards the lower epidermis are spongy mesophyll cells, which are loosely packed with large air spaces and contains fewer chloroplasts.
- *Palisade*: These are elongated parenchyma cells containing large number of chloroplasts, set at right angles to and toward upper epidermis of dorsiventral leaves. They are concerned with photosynthesis.

- *Parenchyma*: It is thin walled, vacuolated cell with living protoplast; it is of various shapes and occurs in all parts of plant organs, i.e. pith and cortex of roots and stems, in leaves as mesophyll, in the flesh of fruits, in the endosperm of seeds, in the conducting tissue as xylem and phloem parenchyma etc. It forms the basic and ground tissue of plants where other tissues remain embedded.
- *Pericycle*: It usually consists of parenchyma cells, which form a cylinder surrounding the vascular tissues in root. It lies immediately inside the endodermis and is usually single layered.
- *Phloem sieve tube*: It is one of the phloem elements of angiosperms. The cell is living but contains no nucleus at maturity. Sieve tubes are set in longitudinal files and there is continuity between them through sieve pores, present on their vertical end walls. Translocation of food materials occurs through it.
- *Phelloderm*: These cells are more or less like the neighbouring cortical cells and present around the cortex. These cells are formed from the cork cambium – phellogen on the inner side. It is the inner part of a periderm. Phelloderm may be considered as secondary cortex and sometimes contains starch and chloroplast.
- *Pollen generative cell*: This cell is present in the pollen grains of gymnosperms like *Cycas*. It originates as a result of subsequent divisions of pollen nucleus and is destined to produce stalk- and body cell.
- *Pollen generative nucleus*: It is present in the pollen grains of angiosperm, originates from pollen nucleus and gives rise to two sperm nuclei – the male gametes.
- *Ray cells*: These are parenchyma cells, sometimes called vascular rays that are developed in the secondary vascular tissues by the division of ray initial of vascular cambium. They are set horizontally and may be uniseriate (one cell wide) or multiseriate (several cells wide). In the *Multilingual glossary of terms used in wood anatomy* compiled by International Association of Wood Anatomists (IAWA), 1964 describes ray as a "ribbon like aggregate of cells extending radially in the xylem and phloem".
- *Root cap cells*: These are parenchymatous cells, situated outside the root apical meristem. They are derived from root apical meristem (e.g. Dicots) and calyptrogen (e.g. Monocot). Because of their terminal position they protect the inner meristem of root apex.
- *Root hairs*: These are tubular outgrowths of some single cells present on the epidermis of root. They occur in greatest abundance at a little distance away from the apical meristem. They have thin cell walls and function in absorption of water and solutes, and in anchorage.
- *Secretory cells*: They have secretory function in the secretion of oils, nectars, terpenes etc. They occur in different parts of the plant and secrete particular substances. The secretory structure sometimes characterizes certain families.

- *Stomatal guard cells*: These are reniform epidermal cells, a pair of which forms a stoma by union at their concave ends (e.g. Dicot leaves). In grasses the guard cells are dumb-bell shaped. The cell walls are differentially thickened and the cells contain prominent nucleus and chloroplastids.
- *Stomatal subsidiary cells*: Also known as accessory cells which occur in association to stomatal guard cells; these cells are morphologically distinct from neighbouring epidermal cells and guard cells.
- *Synergids*: These cells occur in the embryo sacs of angiosperms, a pair of which in association with the egg forms the egg apparatus.
- *Tracheid*: It is one of the nonliving elements of xylem. It is an elongated cell with tapering, obtuse or oblique ends. The cell wall is thick, lignified with pits on its lateral wall. The cells lie parallel to the long axis of the plant and conduct water and solutes. These cells are formed from a single cell. They are mechanical cells, i.e. give strength and support to the organ in which they occur.
- *Transfer cells* (Fig. 2.1): These are specialized parenchyma cells characterized by wall ingrowths. The ingrowths are formed by the deposition of wall materials on the inner side of primary wall. The plasma membrane of the cell gets invaginated along with the outline of wall ingrowths and thus the surface area of plasma membrane is increased by many folds. These cells are concerned with absorption and secretion where the invaginated plasma membrane facilitates the process. They occur in association with xylem and phloem, glandular hair, hydathodes, nectaries, salt glands, synergids etc.

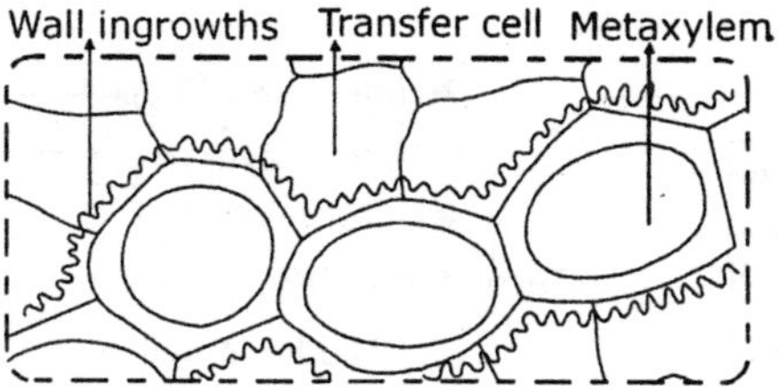


Figure 2.1

Xylem parenchyma transfer cell showing wall ingrowths. Adapted after Singh *et al.*, 1994.

- *Tapetum*: It is usually composed of a single layer of cells present within the anther of angiosperm. It is food rich layer, which surrounds the sporogenous tissues. The tapetal cells disintegrate to liberate the contents and thus they supply metabolites for the developing sporogenous cells.
- *Xylem vessel*: It is also known as trachea, which is one of the non-living elements of xylem. Vessels are elongated tube like thick walled lignified cells with perforated ends and pits on the side walls. They are set end to end in longitudinal series and lie parallel to the long axis of the plant. Conduction of water and solutes occur through it, which also gives mechanical support.

CHAPTER 3

# Cell Wall

Cell wall is a layer of structural materials (composed of polysaccharides, glycoproteins and phenolic compounds) found external to protoplast and the wall defines the shape and size of the cell, and fastens to its neighbours.

**Gross structure of cell wall**: A plant cell may consist of either primary wall only or the cell may contain both primary and secondary walls (Fig. 3.1).

• *Primary wall*: The first formed wall of a cell is the primary wall. It usually grows in surface area.

Primary wall is about 0.1 μm thick normally, but the walls of collenchyma may be 10 μm thick. The cuticularized epidermal cells may be thicker. Thick cellulosic primary cell walls are observed in the endosperm cells of *Phoenix dactylifera, Strychnos nux-vomica* etc. Thick primary wall, in contrast to secondary wall, is plastic and reversible. The meristem, cambium cell, parenchyma,

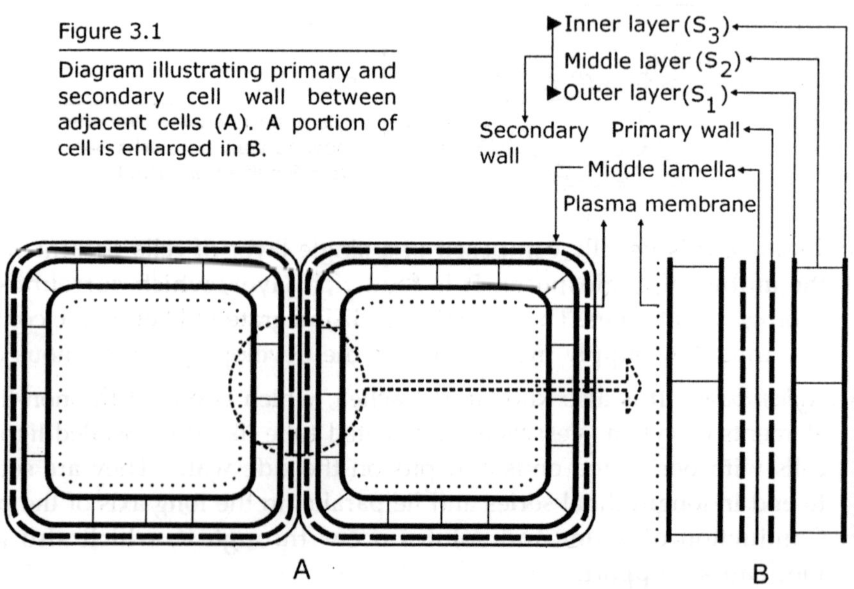


Figure 3.1

Diagram illustrating primary and secondary cell wall between adjacent cells (A). A portion of cell is enlarged in B.

collenchyma, root hairs etc. possess primary walls. The primary wall of cambium cells shows seasonal changes in thickness and the thick wall of endosperm cells of *Phoenix*, *Strychnos* etc. are digested during germination.

The components of primary wall are deposited on both the sides of middle lamella, which initially creates a boundary between two nuclei at the end of nuclear division. As more wall materials are deposited and growth occurs, the distinction between middle lamella and primary walls is obliterated. At maturity the middle lamella and the two primary walls of adjacent cells become confluent and it becomes difficult to define a precise boundary between them.

• *Secondary wall*: A secondary wall is one whose polysaccharide components are deposited over the primary wall. In contrast to primary wall, secondary wall is formed after the cessation of surface growth. Therefore, it causes growth in thickness only. Since the secondary wall materials are deposited outside the plasma membrane, it is present internal to primary wall surrounding the cell lumen. In addition to cellulose, hemicelluloses and other polysaccharides, the secondary wall contains lignin. When lignification occurs, it may begin at either the primary wall or middle lamella. In the deposition of secondary wall materials layering can be observed. The lignified tracheid and fibre show three layers in their secondary wall—the outer layer ($S_1$), the central layer ($S_2$) and the inner layer ($S_3$), among which the central ($S_2$) is the thickest. The $S_1$ and $S_3$ layers lie adjacent to primary wall and cell lumen respectively. Cells with secondary wall consist of five layers: a three layered secondary wall, the primary wall and the middle lamella. In some cells, such as primary xylem, the secondary thickening materials are laid down in such a way that various patterns are formed on the cell wall, e.g. annular, spiral, reticulate, scalariform and pitted. In contrast to primary wall, secondary walls are irreversible and provide mechanical strength. Usually secondary walls are present in those cells, which are devoid of active protoplast.

• *Middle lamella*: This layer lies in between the two primary walls of adjacent cells. It is the cementing material that fastens cells to their neighbours; therefore, the cells are held together by their primary walls. The middle lamella is very thin (less than 30nm thick) and thickest at the cell corners. Pectic substances, possibly in association with calcium and others compose the middle lamella.

> Various chemical reagents dissolve the pectic substances and this property is utilized in maceration technique where the middle lamella solubilize, the cells are separated and can be studied individually. In nature, the enzyme pectinase of pathogenic fungal origin dissolves the middle lamella and thus facilitates the entry of pathogen when it comes in contact to the host.

Middle lamella originates at the end of nuclear division and creates a new boundary between the two nuclei, which had previously shared a common cytoplasm. As growth continues the primary wall materials are laid on either side of middle lamella. At maturity, the layers are indistinguishable and it becomes difficult to define a precise boundary between them.

**Composition and chemical nature of cell wall:** The cell wall is mainly composed of carbohydrate rich materials. The major components of cell wall are cellulose, pectins, hemicelluloses, proteins and phenolics. The cell wall is a biphasic structure consisting of relatively rigid cellulosic microfibril embedded in gel-like non-cellulosic matrix. The microfibrillar phase consists of only cellulose (β1, 4-glucan) and the matrix is composed of non-cellulosic polysaccharides and other substances, which include pectins, hemicelluloses, proteins and phenolics. The microfibrillar phase is readily visible in Electron Microscope and is crystalline, i.e. its molecules are arranged in a definite way and the matrix is non-crystalline. Moreover, it is relatively homogeneous in chemical composition than matrix.

• *Cellulose*: It provides shape and strength to the cell wall. It composes 20-30% of the dry weight of primary wall and accounts 40-90% of the dry weight of secondary wall. Cellulose is an unbranched β1,4-glucan. The microfibrils are made up of cellulose molecules. It is clear from x-ray diffraction and chemical studies that the bulk of microfibril is made up of crystalline β1,4-glucan. The microfibrils are long, thin structure with oval or circular in cross section and have uniform width of about 10 nm (1 nm = $10^{-9}$m) in higher plants. The cellulose molecules consist of long chain of linked glucose residues. It is a polymeric carbohydrate with the general formula $(C_6H_{10}O_5)n$. The glucose residues $C_6H_{10}O_5$ are linked together with oxygen atoms. There are at least 8000 to 15,000 glucose monomers per cellulose molecule. The microfibrils are aggregated to form macrofibrils that are composed of about 500,000 cellulose molecules in transection. The macrofibrils are about 0.4 μm wide and can be visible under light microscope. Several macrofibrils are combined together to form the cell wall.

The non-crystalline and non-cellulosic phase of cell wall are the matrix. It is made up of polysaccharides, proteins and phenolic compounds whose composition differs in different species, in different types of cells, in different parts of cell wall and even at different stages of the cell growth.

The matrix polysaccharides are partitioned into two classes—pectin and hemicellulose. This traditional division is based on the extraction procedure. The cell wall, after the removal of lignin (if necessary), can be extracted either with a hot aqueous chelating agent or with hot dilute acid. This extracted fraction is called pectin. The remainder of the matrix polysaccharides, after depectination, can be extracted with alkaline solution. This extracted fraction is the hemicellulose. Growing primary wall contains about 65% water, almost of which is in the matrix. Water is an important structural component of the matrix.

• *Pectins*: These are a group of polysaccharides, which are rich in galacturonic acid, rhamnose, arabinose and galactose. The other pectin polysaccharides are arabinan, galactan, arabinogalactan I and homogalacturonan. Pectic polysaccharides may be linked covalently to cellulose, proteins and phenols. Pectins are present in much larger amount in primary walls than the secondary walls, suggesting a role in growth. They are highly hydrophilic polysaccharides, and the water that they introduce into the matrix may loosen the wall enabling the cellulose microfibrils to separate—necessary for wall expansion. Pectins are present in high concentration in the middle lamella where they presumably serve the function of cementing adjacent cells together.

• *Gum*: It is the collective term of the disintegration products of the cell wall carbohydrate especially of starch. Gums are colloidal plant products, which either dissolve or swell in water. They are related to pectic compounds and on hydrolysis give pentoses, hexoses and complex organic acids. Gums appear as a result of breakdown of walls, which are often caused by physiological or pathological disturbances, diseases, insect or mechanical injury.

• *Mucilages*: It is the polymer of galactan and related to pectic compounds. Mucilages are found in seed coats and outer cell layers of plant bodies of many aquatic plants. They apparently increase the water holding capacity of the wall.

• *Hemicelluloses*: These are matrix polysaccharides built up of a variety of different sugars. They differ in different species and in different cell types. The principal hemicelluloses are the followings.

Xylan: It typically makes up roughly 5% of primary cell wall and 20% of secondary wall in dicots. This hemicellulosic polysaccharide is linked with xylose and arabinose. Other sugars may also be present. The primary wall of dicotyledonous cells contains glucuronic acid and arabinose. In dicots, the secondary walls contain glucuronoxylan as hemicellulose with small amount of arabinose. The secondary walls of monocot cells also contain arabinoxylan with more glucuronic acid.

Glucomannan: This polysaccharide forms the major hemicellulose of secondary wall of gymnosperm and angiospermous cells. It contains glucose and mannose in a ratio of approximately 1 : 3.

Mannan and Galactomannan: These hemicellulosic polysaccharides are found in the walls of endosperm. These cells function as food reserves.

• *Glucuronomannan:* This polysaccharide is found in low proportion in the cell walls. It contains mannose, glucuronic acid, xylose and arabinose.

• *Xyloglucan:* It is storage polysaccharide. Xyloses are the major components of the thick storage walls of some seeds, e.g. *Nasturtium* and tamarind. They typically make up about 20% of the primary cell walls of dicots, and one to five percent of the primary walls of grasses. They appear to be absent from most secondary walls. Xyloglucans have a backbone identical with cellulose. Xyloglucans contain glucose and xylose residues. Some xyloses are substituted by fucose, galactose and arabinose. Fucose is absent from the storage polysaccharides of endosperm cell walls.

• *Callose*: It normally occurs in small amounts in healthy cell walls, where it has been suggested to form a thin layer between the plasma membrane and the wall proper. It is also found lining the sieve pores of sieve tubes, microfibrils of pollen tube walls and secondary walls of cotton fibres. Callose may be formed on the surface of protoplasts in response to wounding or on the surface of stigma in response to incompatible pollen. Callose is β1,3-linked glucan.

• *β1,3,β1,4-glucan*: This hemicellulose polysaccharide is found in the cell wall component of barley, cereals and other grasses.

Arabinogalactan II: This polysaccharide is found in the walls of gymnospermous cells.

• *Proteins*: Different varieties of protein are present in the cell wall, most of which are linked with carbohydrate forming glycoprotein. The cell wall glycoprotein extensin contains an unusual aminoacid hydroxyproline (about 40%), which is generally absent from the protoplast. In addition to hydroxyproline valine, tyrosine, serine and lysine are present in large amounts. Histidine is also found in some extensins. Extensins are present in the primary cell walls of dicots making up one to ten percent of the wall. They are present in epidermal cells in large quantities. They are present in the primary walls of grasses in considerably smaller amounts. Secondary walls are devoid of extensins. The control of cell extensibility is attributed to wall proteins and accordingly these proteins are named as extensin. It regulates the orientation of cellulose microfibrils, forms the skeletal construction and controls the growth of cell wall. It also resists the invading pathogens.

Arabinogalactan-proteins are another protein component of the cell wall. The protein portion contains a high proportion of hydroxyproline, serine, alanine and glycine.

In addition to above glycoproteins, the following enzyme proteins are present in the cell wall: invertase, peroxidase, pectinase, cellulase, pectin methylesterase, acid phosphatase and malate dehydrogenase.

• *Lignin and other phenolic compounds*: Lignin is the cell wall component of sclerenchyma, which includes fibres and sclereids. It is also found in the tracheids and vessels of xylem. In some cells, lignin deposits in response to attack of microorganisms.

Lignin is a phenolic polymer and contains three subunits namely—coumaryl, guiacyl and sinapyl-propane. Cell wall lignin of monocots contains all the three subunits more or less in equal amounts. The dicots contain about equal amount of guiacyl and sinapyl-propane subunits in their cell wall lignin while in gymnosperm guiacyl subunits predominate. Wood contains about 30% lignin. It is usually absent from growing walls or may be present in very little amount. Initially, it is deposited over the middle lamella and gradually extends to primary walls that have developed secondary walls.

Lignin deposits to fill up the spaces present between the macromolecules of cell walls. Thus, it cements the wall components and forms a very strong structure. As a result, the cell wall loses the capability of plastic extension and hence growth ceases. Moreover, lignified walls form a waterproof system of xylem and resist the pathogens. It forms the mechanical cells of plants.

Apart from lignin, other phenolic compounds-cutin and suberin are found as the component of cell wall.

• *Cutin*: It is present in the epidermal walls of roots, root hairs, stems and leaves. It may be found in growing walls also. It is indigestible polymer and composed of polyester of hydroxy-fatty acids. Cutinized walls and cuticle contain the phenol—ferulic acid.

• *Suberin:* It is found in the cell walls of seed coats, cork and casparian strips of endodermis. In contrast to cutin, it contains more phenolic materials.

• *Tannin*: It is present in condensed form. Tannins of the proanthocyanidin type seem to be linked to polysaccharides. It is not known whether the walls of living cells contain tannins.

• *Wax*: It is a fatty substance and occurs in and upon the cell wall. It is usually noted on the surfaces of leaves and fruits. It may occur either above or below the surface of cuticle. It may be present within cutin also.

• *Mineral substances*: Mineral substances like silica and calcium carbonate may deposit on cell wall. Silica is commonly noted in the epidermal walls of *Equisetum*, leaves of grasses etc.

In conclusion, it can be stated that the cell wall can be visualized as a network of polymer, which on superimposition forms the overall complex structure.

**Function and role of cell wall in the life of a plant**: The cell wall represents a kind of exoskeleton. It is the frontier zone where the cell is in contact with the environment. It performs a great diversity of functions:

• *Gives mechanical strength*: The nonlignified wall derives strength from the cellulose microfibrils. It is believed that the mechanical strength of plant cell walls is due to the presence of skeletal framework formed by cellulosic microfibrils. Cellulose is the major component of paper, cotton etc. Maximum strength is obtained from collenchyma and sclerenchyma. The walls of collenchyma cells are thickened with pectin, hemicellulose, protein and cellulose. Collenchyma cells are living and retain their protoplast even when mature. Therefore, they can regulate the deposition and orientation of wall materials according to the need of developing organs.

The cellulosic walls of wood are further strengthened by lignification. The large size and structural strength of a woody plant is achieved by the cell wall. In these plants the constituents of cell wall contributes 95% of the dry weight of wood. The rigidity of the whole plant and strength is due to cell wall.

In some species where cellulose is absent, other polysaccharides assume strengthening role and may form microfibril.

• *Maintains cell shape*: The various shapes of cell are attributable to cell wall. Isolated protoplasts are more or less spherical whose boundary is the plasma membrane. The plasma membrane is the most common site of cellulose synthesis and cellulose is one of the components of cell walls. Additional wall component may include pectin and/or lignin. These wall components surround the plasma membrane and thus the cell wall determines the shape of cells.

• *Controls cell expansion* (Fig. 3.2): Growth of plant cell depends on the elastic (i.e. reversible) and plastic (i.e. nonreversible) nature of cell walls.

In turgid cell turgor pressure presses the cell wall to expand and due to elastic nature the wall regains its original position at flaccid condition. In growing cell, when the critical turgor pressure exceeds, the wall expands in a plastic manner thus resulting in a permanent increase in cell volumes and growth. This is a very slow process. The elastic stretching of wall due to turgidity accelerates the

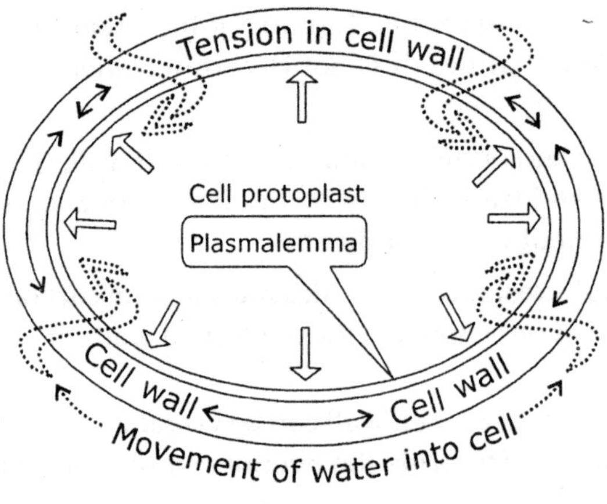


Figure 3.2

The movement of water into the cell causes to press the protoplast against the cell wall and thus the cell wall extends.

⇧ = turgor pressure against plasmalemma and cell wall (outward).

phenomenon. Thus, the turgor pressure is the driving force behind the extension. In multicellular plants the pressure, exerted by neighbouring cells, has much control in cell extension. Probably the covalent or non-covalent bonds within the matrix of wall are weakened by acid. The covalent bond affects the cell wall through the action of an enzyme – probably transglycosylase. This enzyme causes the transfer of polysaccharide fragments from one site of glycosidic attachment to another. The non-covalent bonds may be the calcium bridges between polygalacturonate molecules or glycoproteins or glucuronoxylan. Another non-covalent bonding is given by lectins in the cell wall. Lectins are proteins, which are involved in cross-linking polysaccharides or glycoproteins. Later the phenolic-cross-links between polymers are regarded to control wall extensibility. Phenolic links are found in lignin. A cell ceases to elongate when lignin is laid down over the whole area of the wall. In some primary tracheids and phloem fibres, where lignin is deposited to a part of wall only, elongation and growth continue in the non-lignified areas.

• *Controls intercellular transport*: The transport of substances between plant cells involves cell wall and it occurs in two ways: apoplastic and symplastic transport (Fig. 3.3). In apoplastic transport, the movement of substances between neighbouring cells occurs through the matrix of cell wall and across the plasma membrane whereas in symplastic transport the movement occurs via plasmodesmata present between cells. In apoplastic transport the size of the substance is very significant. The microfibrils and matrix polymer of the cell wall form sieve like structure, which inhibit the entry of large molecules and microorganisms. The smaller molecules, ions, small proteins and polysaccharides move through the aqueous channels present in the matrix.

The charges of molecules also have an effect on movement. The positively charged molecules tend to bind with negatively charged polymers present in the wall and thus their movement is retarded; the negatively charged, neutral or uncharged molecules move freely. The primary wall thus provides an ease transport.

The cuticle present on epidermis, the casparian strip of endodermis and lignified walls resist the apoplastic movement. The plasmodesmata, through

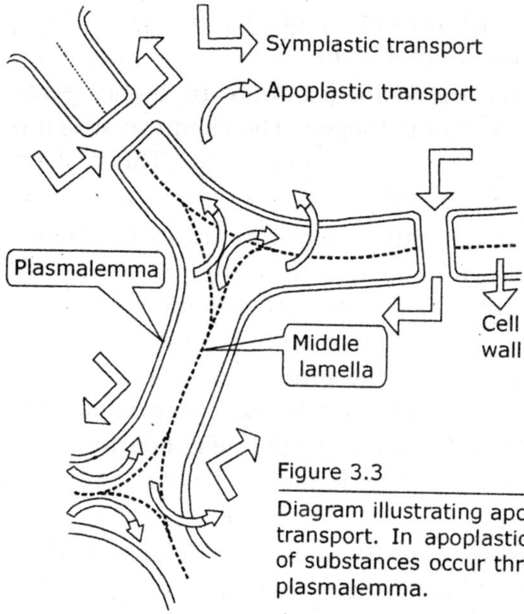


Figure 3.3

Diagram illustrating apoplastic and symplastic transport. In apoplastic transport movement of substances occur through the cell wall and plasmalemma.

which symplastic movement occurs, exist in pit pairs and in the sieve pores of sieve tubes.

• *Protects against infective organisms*: The epidermal cell walls of higher plants are always subjected to attack by pathogenic microorganisms like virus, bacteria and fungi. This pathogen penetrates the host either by enzymatic dissolution of part of cell wall or through the opening of stomata or opportunistic entry through breaks or wounds in the wall. So, a cell wall possesses passive and active defense mechanisms against attack by microorganisms.

The intact cell wall forms an effective and passive physical barrier. The pores on cell wall are too small to allow entry even the tiny microorganisms. The cuticle, lignin, cutin and in grasses – silica in the epidermal cell wall provide an effective barrier against fungi. To penetrate cutin and middle lamella, secretion of cutinase and pectinase respectively are required by fungi. Silicon forms a mechanical barrier and act as a toxin against fungi.

Intact cell wall also possesses active defense mechanisms. It forms an effective barrier around the point of parasitic infection by the deposition of lignin and sometime suberin thus sealing the path of penetration of the pathogen. The death of cells around infection also prevents further penetration. Sometimes callose deposits at the point of penetration of fungal haustorium adjacent to plasma membrane thus providing an effective barrier.

Evidences suggest that due to pathogenic attack cell wall polysaccharides degrade to oligosaccharides. Oligosaccharides from injured tissues diffuse out to the neighbouring cells and there it stimulates the active defensive mechanisms.

• *Cell wall has signalling device against attack of microorganism*: The cell wall can react in response to pathogenic attack. It triggers active defence mechanisms to prevent the entry and establishment of a pathogen. The most common example is the production of phytoalexin by higher plants.

Phytoalexins are nonspecific toxins produced by higher plants when they are stimulated by the attack of a pathogen. The chemical structure of it varies from plant genera (e.g. Pisatin, Phaseolin, hircinol etc.). This protective substance takes an active part in the defence mechanisms of plants.

Apart from phytoalexin formation, a cell produces inhibitors of microbial proteins in response to signal emanated from injured tissues. Proteinases of microbial origin degrade protein of the host and the inhibitors prevent the protein from degradation.There is evidence that pectic polysaccharides and oligosaccharides are the signals for the production of proteinase inhibitor and phytoalexin. Due to pathogenic attack cell wall polysaccharides degrade to oligosaccharides. The latter diffuses out to the neighbouring cells to stimulate the active defensive mechanism. Oligosaccharides have also role in cell extension and cell differentiation.

The cell wall also possesses passive and active defence mechanisms against larger predators. The passive mechanism includes—presence of lignin on cell wall rendering plant tissues tough and unpalatable to higher animals. Presence of silicon to the epidermal cells also decreases the palatability.

The active defence mechanism of plants is triggered when the cell wall is stimulated by the attack of insects. In response, the plant produces the same inhibitor of microbial proteinase.

• *Acts as a reservoir of food*: A substantial quantity of food is always present in a seed to provide nutrition to the young seedling until it can absorb and photosynthesize independently. The foods may remain stored within the cell or as a component of the cell wall. The polysaccharides (mannans, xyloglucans and galactans) of cell wall are the example of the latter type. Mannans are cell wall storage polysaccharides. Endosperm of Ericaceae and Apiaceae contain 90% mannose approximately. Glucomannans (i.e. glucose and mannose) are present in the endosperm of Liliaceae and Iridaceae. Galactomannans (mannose and galactose) are found in the endosperm of Fabaceae. Xyloglucans are found in those families where the food is stored in cotyledons, e.g. Legumes. Galactans are present as cell wall food reserve in the seed of *Lupinus*.

Cell wall also plays some important roles in the activities like secretion, transpiration, absorption, translocation etc.

**Ultrastructure of cell wall**: The cell wall is a biphasic structure consisting of cellulose microfibril embedded in gel-like non-cellulosic matrix. The microfibrillar phase consists of cellulose (β1,4-glucan) only and the ultrastructure of cell wall is based on it. The microfibrillar phase is readily visible in Electron microscope and is crystalline, i.e. its molecules are arranged in a definite way. Moreover, it is homogeneous in chemical composition.

The microfibrillar phase is composed of microfibrils which are long, thin structure with oval or circular in cross section and have uniform width of about

10nm (1 nm = $10^{-9}$m) in higher plants. The cambium cells have narrow width (± 3 nm). The microfibrils are made up of cellulose molecules, which are precisely defined polymer composed of purely glucose molecules linked to each other by β1,4 bond and are unbranched 1,4-glucan. The glucose residues $C_6H_{10}O_5$ are linked together with oxygen atoms (Fig. 3.4). There are at least 8000 to 15,000 glucose monomers per cellulose molecule and are 0.25 μm to 0.5 μm long. The molecules are flat and ribbon like, and lie parallel to each other. Hydrogen bonding occurs between the molecules, thus crystallizing and producing aggregates. These aggregates are called microfibril. Each microfibril contains 40 to 70 chains, which lie side by side, and these can be seen in Electron micrographs. Electron microscopic study reveals that about 40 cellulose molecules are grouped together to an elementary fibril of about 3.5 nm in diameter. Later the idea of elementary fibril is considered as misconception (Preston, 1974). The cellulose molecules form chains, which are at some regions of microfibrils, are arranged in parallel into 3-dimensional crystalline lattices termed micelles. The lattices are connected with each other by intra and inter molecular hydrogen bonds. The spaces between the

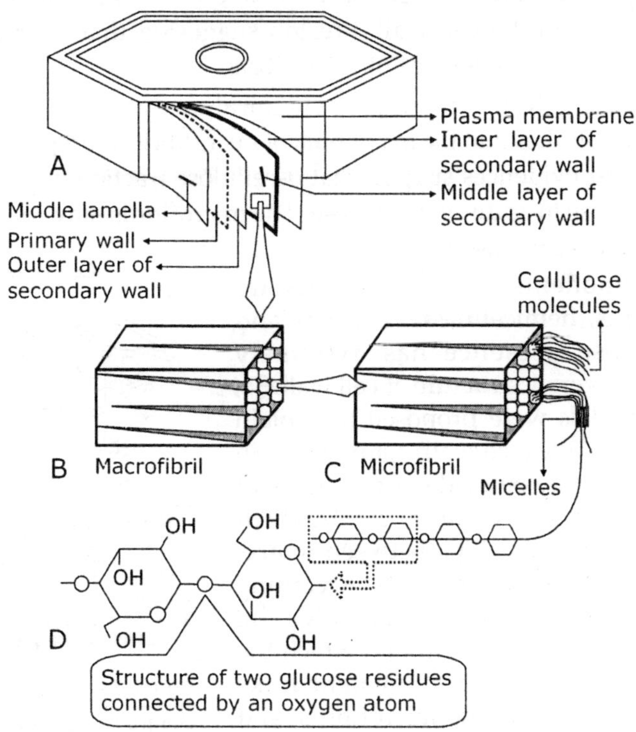


Figure 3.4

Diagram illustrating the structure of cell wall of a fibre. A. Diagrammatic representation of cross section of fibre in three dimensional view showing middle lamella, the primary wall and three layers of secondary wall. B. Macrofibril from a portion of the middle layer of secondary wall. C. Microfibril from a portion of macrofibril. D. Structure of chain-like molecules of cellulose. Adapted after Esau, 1961.

microfibrils are filled up with lignin, cutin, pectic substances, hemicellulose, water etc. (Stace, 1963). Thus, the microfibril gains considerable strength.

In primary cell wall, the orientation of microfibril is transverse to the long axis, and during growth the arrangement may be longitudinal. The orientation in secondary wall may differ from primary wall. Tracheids and fibres show three layers in their secondary wall—the outer layer ($S_1$), the central layer ($S_2$) and the inner layer ($S_3$), among which the central ($S_2$) is the thickest. The $S_1$ and $S_3$ layers lie adjacent to primary wall and cell lumen respectively. These layers $S_1$, $S_2$ and $S_3$ may be distinguished by their respective orientation of cellulose microfibrils. In $S_1$ and $S_3$, the microfibrils are in the form of a lax helix and in $S_2$, it is a steep one (Fig. 3.5).

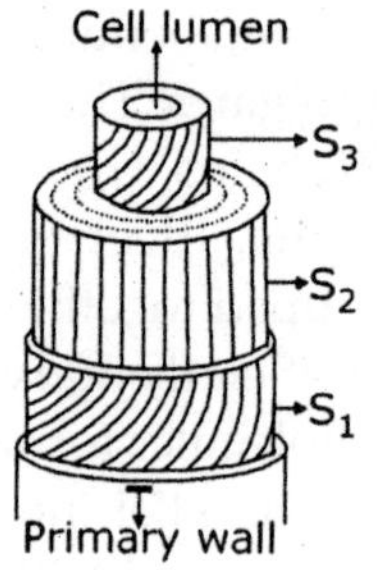


Figure 3.5

Diagrammatic illustration of the orientation of microfibrils in the secondary wall of a cell.

The microfibrils are aggregated to form macrofibrils, which are composed of about 5,00,000 cellulose molecules in transection. The macrofibrils are about 0.4 μm wide and can be visible under light microscope. Several macrofibrils are combined together to form the cell wall. Preston suggested that microtubule directs the arrangement of microfibrils.

It is certain from chemical analysis and X-ray diffraction studies that the major bulk of microfibril is composed of crystalline β1,4-Glucan. Later evidences suggest that α-cellulose fraction of cell wall contains mannose and xylose in addition to glucose. The microfibrils may consist of a central core of crystalline—cellulose micelle.

Major components of the cell walls are cellulose, pectins, hemicellulose, proteins and phenolics whose presence has extremely complicated the overall structure of cell wall. So, a number of models were proposed to explain the arrangement of the (cell wall) components in the wall. Mention may be made of the model of (Fig. 3.6) Lamport and Epstein, 1983, which explains the interrelationships between matrix molecules and cellulose microfibrils. According to this model the protein molecules lie perpendicular to cell surface through which the microfibril passes. There are covalent links between protein and cellulose microfibrils, and between the proteins.

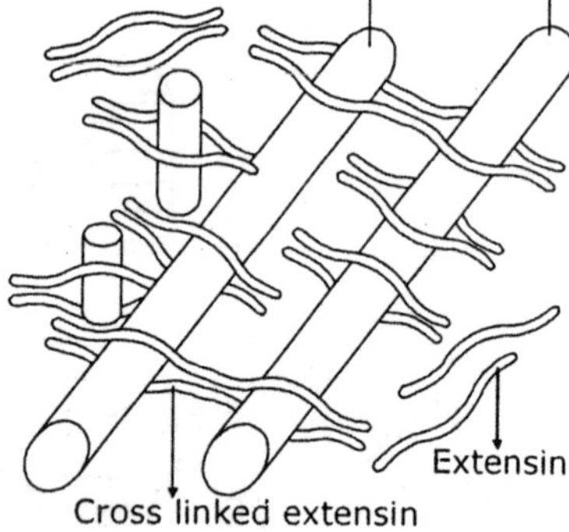


Figure 3.6

Diagram representing Lamport and Epstein model. Adapted after Brett and Waldron, 1996, p. 52.

## Plasmodesmata

Plasmodesma (plural: plasmodesmata) is thin irregular cylinder of cytoplasm lined by plasmalemma, passing through fine pores in the cell walls, thus forming a connection between the cytoplasm of adjacent cells.

They are found in higher plants and fluctuate widely in abundance and distribution. They are commonly present in primary pit fields and pit membranes of young and mature living cells respectively. They may be scattered over the entire wall or occur in groups when they are concentrated on primary fields (e.g. cambium of *Pinus strobus*, ray cells of *Sequoia sempervirens* etc). In a meristematic cell, the number of plasmodesma ranges from 1000 to 10,000 and their distribution may not be uniform. The frequency of distribution may vary even in different walls of a single cell. Plasmodesmata, at the intercellular canal between the common walls of living cells, are encircled by plasmalemma, which is continuous with that of the adjacent cells. At the centre of plasmodesma, there occurs a tube of membrane, termed desmotubule. The desmotubule is composed of protein sub-units and contains an axial central rod. Diameter of the lumen of plasmodesma is very narrow, 30 nm to 60 nm in diameter through which the cell organelles cannot move to the adjacent cells. The diameter of desmotubule ranges from 16 nm to 20 nm. A space is present in between the plasmalemma and desmotubule—termed cytoplasmic annulus. Sometimes plasmodesmata are branched; it is observed in the pit membranes of living fibres of *Tamarix* and between sieve tube and companion cell where it is branched towards the latter side. The above structure of plasmodesmata is revealed by electron microscopic study (Fig. 3.7).

Plasmodesma originates during cytokinesis when cell plate is formed. It is formed at those regions of the cell plate where the endoplasmic reticulum (ER) is present and prevents the fusion of vesicles. At this region, the cellulose microfibrils and pectic substances are not accumulated. As a result intercellular canal is

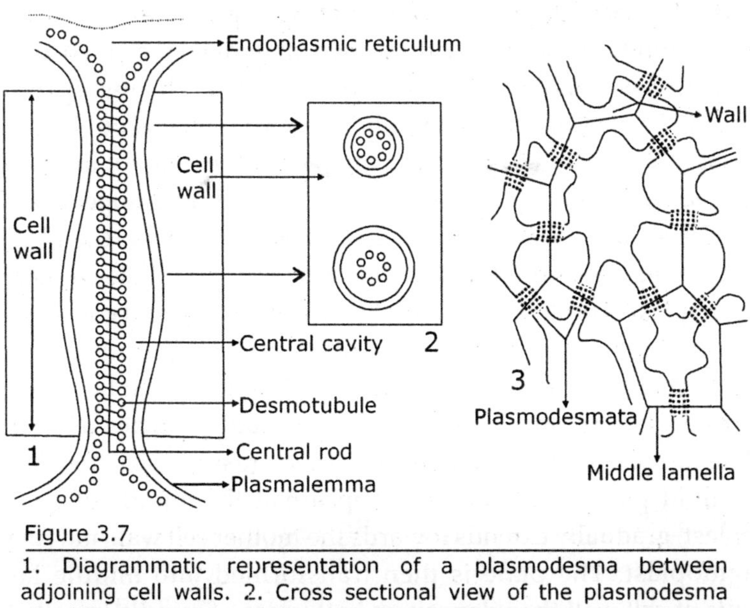


Figure 3.7

1. Diagrammatic representation of a plasmodesma between adjoining cell walls. 2. Cross sectional view of the plasmodesma at different levels. Adapted after Cutter, 1978. 3. Diagram showing plasmodesmata. Adapted after Holman *et al.*, 1939, p. 52.

formed. It is observed that the desmotubules are continuous with the ER of adjoining cells through the intercellular canals. Therefore, it is regarded that the desmotubules are derived from ER. There are reports of the formation of plasmodesmata between mature cells, between the host cell and haustorium of parasite, between tyloses, in grafts between the cells of stalk and scion etc. In these cases there are wall degrading enzymes and signaling devices, which perforate the common wall to form intercellular canal and align the plasmodesma situated at the adjoining cells.

Plasmodesmata exist in thick cell wall also, e.g. endosperm of the seeds of *Phoenix dactylifera*, *Coffea arabica* etc. They can be easily observed in the endosperm of seeds of *Aesculus*, *Diospyros* etc. It is best studied in plasmolysed cells where the protoplast shrinks from all the regions of cell wall, except the places where plasmodesmata occur.

• *Function*: (1) It helps in the short distance transport of materials; (2) the relay of stimulus occurs through it; (3) viruses can pass through plasmodesmata; (4) plant hormones move through plasmodesma; (5) the movement through plasmodesma is bi-directional. It is suggested that the desmotubules act as a valve and regulate the direction of flow; (6) small molecules and ions pass readily through plasmodesma.

**Origin of cell wall:** Initially middle lamella creates a new boundary between the two-daughter nuclei of a dividing cell. The components of cell wall deposit on either side of middle lamella to form the cell wall.

In a non-vacuolated cell, nucleus is usually present at the centre where it divides. At telophase, the two nuclei lie at opposite poles. At early telophase, there occurs spindle fibre between the two daughter nuclei. At the equatorial line, these fibres become stronger and form a fibrous spindle termed phragmoplast. During development, phragmoplast widens out between the two nuclei and becomes barrel shaped. Phragmoplast is the basis for the middle lamella formation and subsequently the cell wall. As the cell plate appears at the middle of phragmoplast, the fibrous spindle disappears. However, the fibrous spindle remains at the margins until the cell plate appears there. Later researches reveal that phragmoplast is proteinaceous in chemical structure and consists of elements of endoplasmic reticulum, vesicles, microtubules etc. During the formation of cell wall, cytoplasm at the equatorial line and phragmoplast become denser and fibrous termed kinoplasm. Small droplets of fluid particles appear along the median line of phragmoplast. These particles increase in size, fuse laterally thus forming a fluid plate. The plate first appears at the middle of median line of phragmoplast, gradually extends towards the mother cell wall and finally divides the phragmoplast. The plate is then transformed into middle lamella. The components of cell wall then deposit on both sides of middle lamella to form the cell wall (Fig. 3.8).

In a vacuolated cell, nucleus lies towards the cell wall due to the presence of central vacuole. Cell plate originates in the region that was previously occupied by vacuole. The nucleus moves to the centre along the cytoplasmic strands where

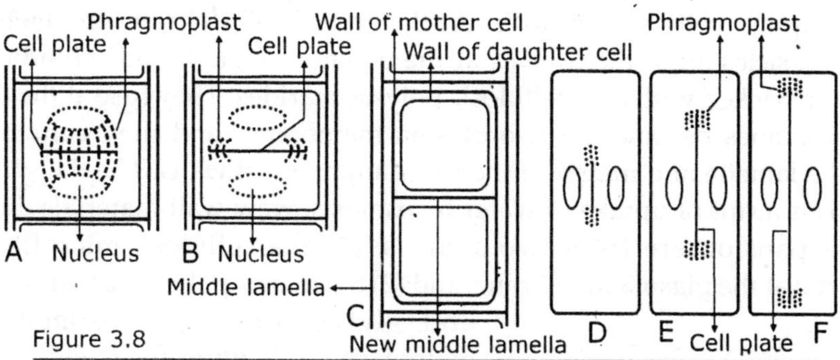


Figure 3.8

Diagrams illustrating cell wall formation in isodiametric (A, B & C) and elongated cells (D, E & F). Adapted after Esau, 1961 (A, B & C) and Eames and MacDaniels, 1947 (D, E & F). Vide text for details.

it divides. The cytoplasmic strands form a plate perpendicular to the plane of nuclear division. The term phragmosome was coined by Sinnot and Bloch (1941) to designate the cytoplasmic plate. Then in the phragmosome, phragmoplast and cell plate are formed in the usual way that is ultimately transformed to middle lamella.

Difference between phragmosome and phragmoplast.

| Phragmosome | Phragmoplast |
|---|---|
| 1. It originates before nuclear division. | 1. It originates at the end of nuclear division. |
| 2. It is formed in vacuolated cell. | 2. It is formed in both vacuolated and non-vacuolated cell. |
| 3. Cytoplasmic plate forms phragmosome. | 3. The fibrous spindle fibre between the two nuclei forms phragmoplast. |
| 4. It is formed in the vacuole of cell. | 4. It is formed in the phragmosome. |
| 5. It is the medium where phragmoplast develops. | 5. It is the medium where the cell plate is formed and thus a cell is divided into two. |
| 6. It is oriented at right angles to the future plane of nuclear division. | 6. It is oriented parallel to the phragmosome. |

In the very long cells like fusiform initial of cambium, when divison occurs tangentially, phragmoplast initially forms plate between the two daughter nuclei instead of wall to wall. The phragmoplast, which consist of short spindle fibres and dense cytoplasm, becomes barrel shaped and gradually expands towards the cell wall. The expanding bands, as they approach towards the wall, are broken into arcs—called kinoplasmasomes. The cell plate is formed along the progressing arcs which ultimately divide the cytoplasm completely. In face view, the expanding phragmoplast appears as a circular ring or 'halo' surrounding the nuclei. This is termed as phragmosphere, which is merely the circular band of kinoplasmasome.

Later studies reveal that plasma membranes appear first during the formation of two daughter cells after mitosis. At the end of cell division, the membrane-bounded vesicles align themselves at the equator of the mitotic spindle. These vesicles are packed with non-cellulosic polysaccharides. They fuse at the equator; their membranes become the two plasma membranes and their contents form the middle lamella and matrix of the two daughter cell walls. The daughter cells expand as a result of uptake of water by osmosis. New wall materials deposited during expansion are from two sources: (1) the cellulose microfibrils are synthesized at the plasma membrane, and (2) the matrix polysaccharides, proteins and glycoproteins are formed in the endoplasmic reticulum and Golgi apparatus. Lignin is formed within the cell wall. Preston suggested that some small ordered granules containing glucan synthetases are present on the outer face of plasmalemma. These granules are (Fig. 3.9) responsible for synthesis and orientation of cellulose microfibrils. The enzyme glucan synthetase is involved in the elongation of microfibrils by addition of glucose residues.

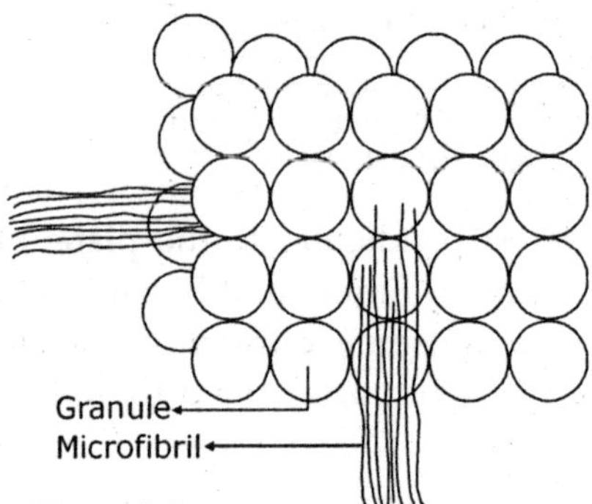


Figure 3.9

Diagram showing microfibrils with their ends in contact with granules. Adapted after Cutter, 1978.

Primary wall material is deposited over the middle lamella during the phase of cell growth. Secondary wall is subsequently laid in some cells once wall extension has ceased.

Later researches with isolated plant protoplast culture indicate that no cell plate is formed, instead, cell wall appears at the outer surface of plasmalemma. Initially an envelope originates surrounding the plasmalemma and then cellulose microfibrils are formed between them. Scan Electron microscopic study reveals that it takes 24 hours or more to form the cell wall of tobacco protoplast (Burgess, 1976).

**Growth of cell wall**: Growth of cell wall involves the growth in surface area as well as growth in thickness.

Normally, the entire cell wall grows. Localized growth occurs in fibres, root hairs, tracheids and pollen tubes.

A new primary wall grows in surface area. The wall expands due to the turgor pressure exerted on the walls by the protoplast. Turgor pressure is developed by the tendency of protoplast to absorb water by osmosis. It presses the protoplast outwards against the wall and the wall begins to stretch in a plastic (i.e. non-reversible) manner. New cell wall materials like matrix polysaccharides and cellulose microfibrils are deposited over the expanding wall.

The cellulose microfibrils are formed between plasmalemma and outer envelope. This is revealed by isolated protoplast culture where the initial cell wall formation is not associated with cytokinesis. Preston is of opinion that small granules occur on the outer face of plasmalemma. These are involved in cellulose synthesis and orientation of microfibrils. The granules contain glucan synthetases and when they come in contact with the ends of microfibril, the later is extended by condensation of glucose residues. Dictyosome, vesicles, endoplasmic reticulum or microtubules may be the site of precursors of cellulose.

Previously there were two theories to explain the growth of cell wall regarding the deposition of wall materials. In one view, growth occurs by intussusception where new microfibrils are laid down between the existing microfibrils of expanding wall in an interwoven manner. The other view, which is more common, states that growth occurs through apposition (Fig. 3.10). During this process, new microfibrils are deposited over the existing one like the pages of a book. In apposition, the deposition of microfibrils is usually centripetal, i.e. it occurs from periphery towards the lumen of cell. Cell walls of tissues grow by centripetal growth. Centrifugal growth, i.e. deposition of wall materials from cell lumen to periphery is also observed during apposition. It occurs during exine formation of the specialized cells like pollen grains and spores.

A different view, termed mosaic growth, was proposed by Frey-Wyssling and Stecher (1951) to describe the growth of primary cell wall. Their opinion was based on the observation with Electron microscope. At the time of growth, the cell wall expands due to turgor and as a result, the microfibrils are separated at certain regions. The gap, thus produced due to the separation, is then filled up by the deposition of new microfibrils by either apposition or intussusception. It is suggested that growth hormones and enzymes, present on the cell wall, are involved in the process. Another concept, regarding the longitudinal growth of cell wall, was put forward by Houwink and Roelofsen (1954) in their theory of

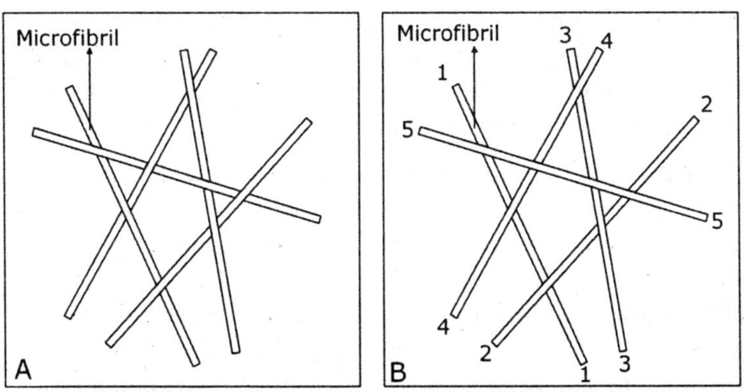


Figure 3.10

Diagrammatic representation of the growth of cell wall. A. Intussusception. B. Apposition. The microfibrils are numbered 1-1, 2-2, 3-3, 4-4 & 5-5 according to the sequence of formation. Adapted after Cutter, 1978.

multinet growth. According to this view, microfibrils are laid down perpendicular to long axis of the cell. New microfibrils are added centripetally by apposition and as a result, older microfibrils are gradually pushed towards the peripheral side. During longitudinal growth, the peripheral microfibrils are stretched, and as more and more elongation occurs, these microfibrils are gradually oriented in longitudinal plane. Electron microscopic observation on the cell wall formation of fibres and tracheids, and autoradiographic study confirm the multinet theory.

The deposition of secondary wall materials in fibres initially occurs at the middle and gradually extends towards the tip. Therefore, the fibres have thicker cell wall at the middle. In the velamen cells of the aerial roots of *Ansellia gigantea*, the deposition of wall materials begin near the ends of the cells. The localized growth at the tips of stellate parenchyma cells, pollen tubes, root hairs, fibres and tracheids are explained based on the localized type of multinet growth. It is regarded that cytoplasm determines the localized deposition of microfibrils (Roelofsen, 1965) Fig. 3.11.

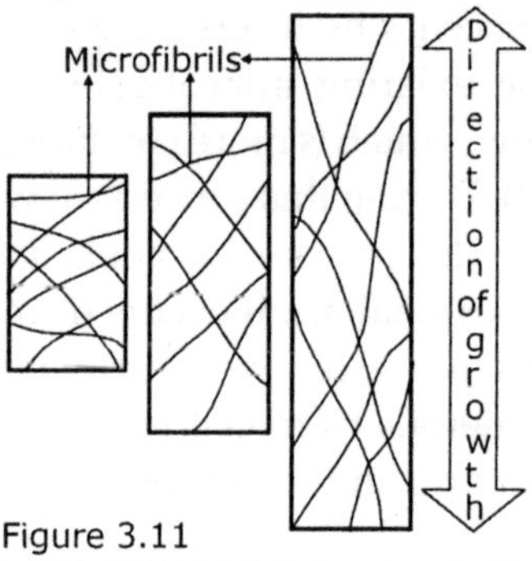


Figure 3.11

Diagram illustrating the change in orientation of microfibrils during elongation of primary walls. After Roelofsen, 1965.

As growth continues the cell wall expands along with the middle lamella. The components of the latter are synthesized within it or nearby and on incorporation cause the elongation. As a result the plasmalemma is stretched. The microfibrils are deposited on the stretched area and as a result the frequency of plasmodesma increases. When the incorporation of microfibril fails the stretched areas form the primary pit field. A pit is formed on secondary wall.

To sum up, cell growth involves: (i) stretching of cell wall due to turgor, which causes the loosening and separation of the fibrillar structure of cell wall, and (ii) mending the gaps by adding new microfibrils by either apposition or intussusception. The loosening of the fibrillar texture of cell wall is attributed to the hormone and enzyme present on the cell wall and the cytoplasm plays the role to fill up the gap with new microfibrils.

**Modification and chemical changes of the cell wall:** Cellulose is the basic constituent of cell walls. It is associated with various substances and may be impregnated with lignin, cutin, suberin, minerals or mucilage. The phenomenon of impregnation is accordingly referred to as lignification, cutinization, suberinization, mineralization and mucilaginous changes. The above phenomena are as follows.

• *Lignification*: In this phenomenon lignin is the substance of impregnation. Lignin is a phenolic polymer and contains three subunits namely—coumaryl, guiacyl and sinapyl-propane. It is found in sclerenchyma, which includes fibres and sclereids, and in the tracheids and vessels of xylem. It sometimes infiltrate in response to attack of microorganisms.

Wood contains about 30% lignin. It is usually absent from growing walls or may be present in very little amounts. It is deposited over the middle lamella and primary walls that have developed secondary walls. Lignin deposits to fill up the spaces present between the macromolecules of cell walls. Thus, it cements the wall components and forms a very strong structure. As a result, the cell wall loses the capability of plastic extension and hence growth ceases. Moreover, lignified walls form a waterproof system of xylem and resist the pathogens. It forms the mechanical cells of plants.

• *Cutinization*: In this phenomenon the impregnation of cell wall with cutin occurs. Cell wall impregnated with cutin is described as cutinised wall. Cutin is an indigestible polymer and composed of polyester of hydroxy-fatty acids. Cutin may deposit on outer surface of the cells as a separate layer. This is cuticle and the phenomenon of its formation is cuticularization. Cuticle forms a continuous layer over the aerial parts of higher plants interrupted only by stomata and lenticels. Cutinised walls and cuticle contain the phenol—ferulic acid.

Cutin is found in the outer epidermal walls of leaves, stems and fruits. The presence of cutin is also reported from roots and root hairs. Cutinised walls may be smooth or rough. In aerial parts, the chief function of cutin is to prevent the excessive water loss. It protects the cell against mechanical injury and from the attack of microorganisms. Cutin has high chemical stability and is preserved as such in fossil materials. This property is utilized in the identification of fossil species.

• *Suberinization*: It refers the phenomenon of impregnation of cell wall with suberin. It is found in certain specialized cell walls, e.g. cork, seed coats and the casparian strip of endodermis. It is also found in exodermis. It may occur in association with cellulose or cutin. It is polyester and contains more phenolic materials than cutin. Suberin impregnated wall layers are protective tissues. This layer is impervious to water and gases. In cork cells suberin occurs in between the middle lamella and the inner cellulose layer of cell walls. Thus, suberin makes cork an excellent protection for the stem against excessive loss of water and mechanical injury. It makes the cork resistant to acid also.

• *Mineralization*: It refers to the phenomenon of the deposition of inorganic substances in a cell wall. The substances are anhydrides of silica and calcium salts. Silica salt is a common wall component of grasses, horsetails (*Equisetum*), sedges, and many species of Cyperaceae, Arecaceae, Moraceae, Aristolochiaceae and Magnoliaceae. It is also noted inside the cells. Calcium oxalate crystals in the form needles (raphides) occur in *Pistia*, Arum etc. Calcium oxalate crystal in the form of star (sphaeraphides) occurs in *Pistia*. Calcium carbonate crystals occur in the epidermal cells of leaves of *Ficus* sp. (cystolith). Cystoliths are commonly noted in Moraceae and Urticaceae.

• *Mucilaginous change*: The cell wall may undergo changes due to deposition of mucilage. It is the polymer of galactan and related to pectic compounds. They are found in seed coats (ex. *Salvia* sp., *Ocimum* sp., *Plantago* sp. etc) and outer cell layers of plant bodies of many aquatic plants. The cell walls of the families Moraceae, Malvaceae, Rhamnaceae, Thymelaeaceae and Euphorbiaceae may become mucilaginous. In seeds, the outer walls of epidermal cells become mucilaginous, e.g. *Linum usitatissimum* and species of *Alyssum*. The epidermal cells of nectaries also become mucilaginous at the time of nectar secretion.

Mucilage apparently increases the water holding capacity of the wall. It also hinders the entry of oxygen into the cell. In *Sinapis arvensis*, the dormancy of seed is considered to be due to a specific growth inhibitor, which is produced at low oxygen concentration. Mucilage is present in the testa and it is thought that it hinders the entry of oxygen in the seed.

## Biochemical pathways of major wall component formation

• *Polysaccharides*: They are synthesized from sugar nucleotides by polysaccharide synthases (Glycosyltransferase). The monosaccharides glucose, galactose, mannose, xylose, arabinose, glucuronic acid, galacturonic acid and fucose etc. compose the wall polysaccharides. Each of these is one of the components of sugar-nucleotide substrate. In the sugar-nucleotide, also called nucleoside diphosphate sugar, the monosaccharide is attached through its glycosidic hydroxyl to the β-phosphate of a ribonucleoside diphosphate. In a nucleoside there occurs a ribose sugar and a nitrogen-containing base that is usually uracil or guanine (Fig. 3.12).

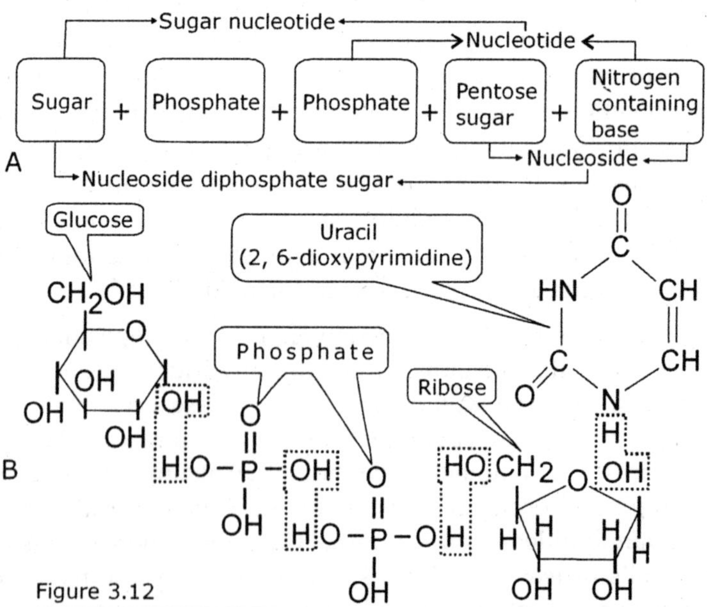


Figure 3.12

A. Diagram representing the general structure of sugar nucleotide. B. Schematic representation of the formation of uridine-diphosphate-glucose. The linkage sites are shown by dotted lines. The linkage occurs with the formation of $H_2O$.

The formation of different sugar-nucleotides is schematically represented in the following Fig. 3.13. The names of sugar-nucleotides are alluded in abbreviated form in the text and illustration, the full name being mentioned in the caption of illustration. UDP-Glc is synthesized either from Glc-1-P or from sucrose. Other uridine containing sugar like UDP-GlcA and UDP-Gal can be formed from UDP-Glc. UDP-GlcA may be the precursor of UDP-GalA and UDP-Xyl; the latter may form UDP-Ara. Glc-1-P is formed from Glc-6-P, which is derived from Fr-6-P. Fr-6-P is normally formed in Calvin pathway and Pentose-phosphate pathway. The guanosine-containing sugars are formed from Man-1-P, which is derived from Man-6-P. Fr-6-P gives rise to Man-6-P. Man-1-P gives rise to GDP-Man. GDP-Fuc and GDP-Glc can be formed from GDP-Man.

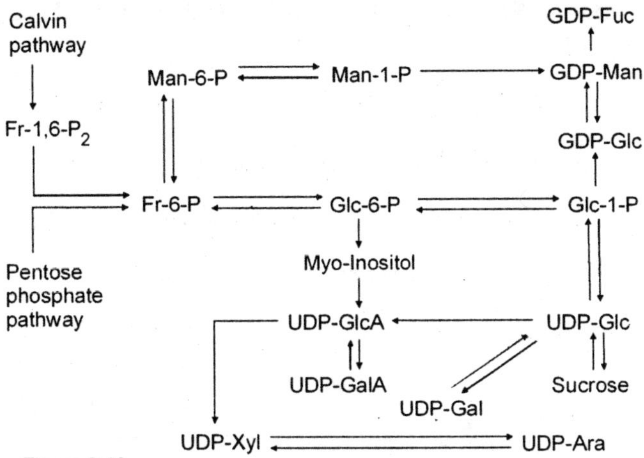


Figure 3.13

Diagram representing the pathways leading to formation of some nucleotide sugars. Abbreviations: UDP-Ara=Uridine-diphosphate-arabinose; UDP-Gal=Uridine-diphosphate galactose; UDP-GalA=Uridine-diphosphate-galacturonic acid; UDP-Glc = Uridine-diphosphate-glucose; UDP-GlcA=Uridine-diphosphate-glucuronic acid; UDP-Xyl=Uridine-diphosphate-xylose; Fr-6-P= Fructose-6-phosphate; Fr-1,6-$P_2$=Fructose-1,6-diphosphate; GDP-Fuc=Guanosine-diphosphate-fucose;GDP-Glc=Guanosine-diphosphate-glucose;GDP-Man=Guanosine-diphosphate-mannose; Glc-1-P=Glucose-1-phosphate; Glc-6-P=Glucose-6-phosphate; Man-1-P=Mannose-1-phosphate and Man-6-P=Mannose-6-phosphate. Adapted after Brett and Waldron,1990, p.63.

The different nucleoside diphosphate sugar, thus formed, transfers the monosaccharides to the growing polysaccharide chains. The membrane bound enzymes 'sugar-nucleotide-polysaccharide glycosyltransferase' help in the reactions. These enzymes are commonly called polysaccharide synthases or synthetases. The reaction is as follows:

$$\text{Nucleoside diphosphate sugar} + \text{Polysaccharide}_{(n)} \xrightarrow{+\ \text{synthetases}} \text{Nucleoside diphosphate} + \text{Polysaccharide}_{(n+1)}$$

The 'n' and 'n+1' indicate the number of monosaccharides in the polypeptide chains. The monosaccharide is attached to the non-reducing end of the polysaccharide.

The polysaccharides like β (1-4) glucan (cellulose), β (1-3) glucan (callose), [β (1-3), β (1-4)] glucan, xyloglucan etc. are formed from UDP-Glc. GDP-Glc forms glucomannan. Galactan is formed from UDP-galactose. GDP-mannose gives rise to glucomannan and galactomannan. Xyloglucan and β (1-4) xylan are formed from UDP-xylose. Arabinan is derived from UDP-Ara. UDP-GlcA, UDP-GalA and GDP-Fuc form glucuronoxylan, α (1-4) galacturonan and xyloglucan respectively.

• *Proteins*: They are polypeptides where the amino acids are joined by peptide linkages between —COOH group of one amino acid and the $NH_2$ group of another amino acid (Fig. 3.14).

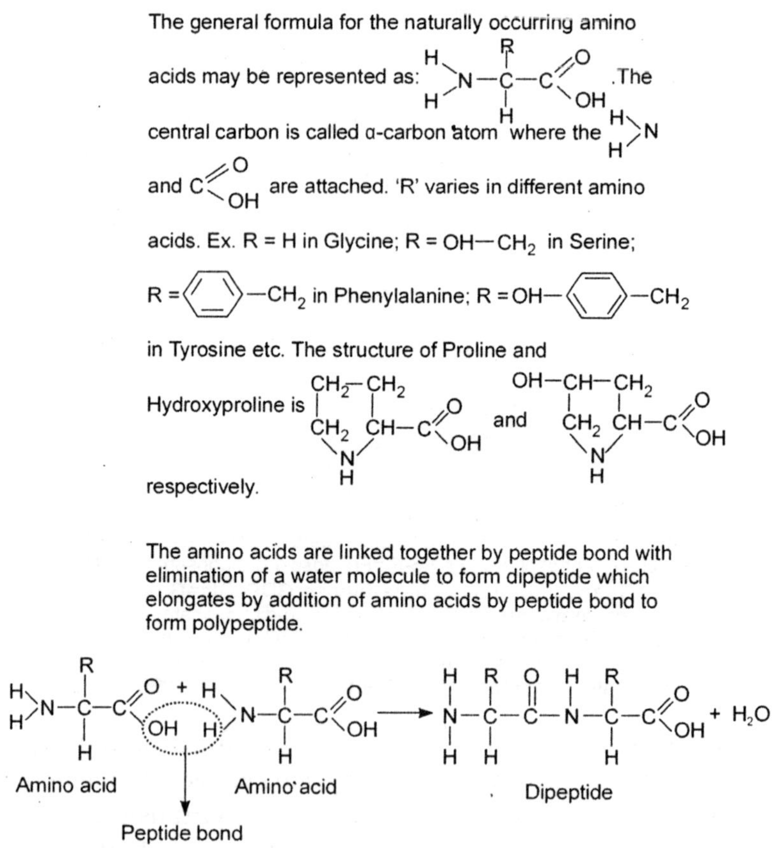


Figure 3.14

Diagram illustrating polypeptide formation.

Wall polypeptides are built up of the twenty amino acids specified by the genetic code. In addition two more amino acids—hydroxyproline and isodityrosine are produced as a result of post-translational modification. The polypeptides are formed by the ribosomes of rough endoplasmic reticulum and then exported to the cell wall via Golgi apparatus.

Wall proteins are glycoproteins, i.e. polypeptides with carbohydrate side-chains. The carbohydrates may be mono-, oligo- or polysaccharide. [Oligo- and polysaccharides are made up of monosaccharide units. There exists no sharp distinction between them. The former contains fewer in number of monosaccharides than the latter. The term polysaccharide is used when there are at least ten monosaccharide units]. The linkage between the polypeptide and the carbohydrate side-chain is often a normal glycosidic bond where one—OH group of the polypeptide is attached to the reducing terminus of carbohydrate.

The wall protein extensin is extremely rich in hydroxyproline in addition to serine, lysine, valine, tyrosine and histidine, the last one being absent from other glycoproteins. The proline residues are hydroxylated to hydroxyproline. This occurs after the completion of polypeptide chain. The enzyme prolyl hydrolase acts as catalyst in the reaction. Ascorbate and ferrous ions are the cofactors of prolyl hydrolase. The enzyme requires α-ketogluterate as a co-substrate and uses molecular oxygen as the oxidizing agent (Fig. 3.15). The site of hydroxylation of proline residues is the endoplasmic reticulum and the Golgi apparatus.

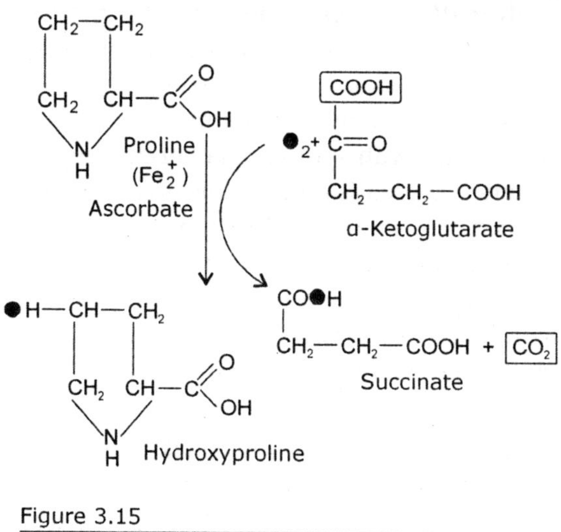


Figure 3.15

Diagram illustrating the hydroxylation of proline leading to the formation of hydroxyproline. Adapted after Brett *et al.*, 1990.

Uridine-diphosphate-arabinose is the donor of arabinose to hydroxyproline. Uridine-diphosphate-galactose transfers galactose to serine.

• *Lignin*: The precursor of lignin is the three aromatic alcohols—coumaryl, coniferyl and sinapyl alcohol. The aromatic amino acids tyrosine and phenylalanine form the aromatic alcohols via several intermediates (Fig. 3.16).

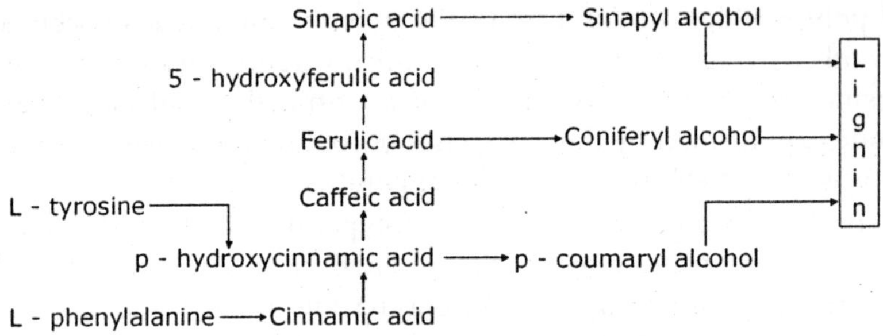


Figure 3.16

Diagram showing the major components formed in the pathway of lignin biosynthesis. Adapted after Brett and Waldron, 1990, p. 72.

Phenylalanine initiates the path of alcohol formation through the enzyme phenylalanine-ammonia lyase. The alcohols are oxidized within the wall by the enzyme peroxidasc. As a result phenoxy radicals are produced. Lignin is formed from these radicals. The radicals can react with some polysaccharides to form linkages between lignin and wall polysaccharides. Lignin deposition begins in the middle lamella and then spreads first to the primary wall. The deposition on secondary wall occurs after the primary wall. Lignin precursors are formed at the endoplasmic reticulum. It is then transported to the cell wall via Golgi apparatus and vesicle.

**Thickening of cell wall:** The cell plate separates the two protoplast of a dividing cell. The cell plate gives rise to the middle lamella upon which the wall materials are laid down to form primary wall. During cell growth, each daughter cell forms its own primary wall. Many cells, at maturity, contain primary walls only. In these cells, the primary wall synthesis generally stops and the growth is checked. However, in fibres, tracheids and vessels further wall materials are laid down on the inner surface of primary wall to form secondary wall. Secondary wall materials are deposited after the wall extension has ceased. In contrast to primary wall, secondary wall is thicker and more rigid due to lignin deposition. The deposition of secondary wall materials occurs in such a way that various patterns are formed on the cell wall. The following patterns are commonly noted (Fig. 3.17).

1. *Annular thickening*: These are ring like thickening present on the inner side of primary wall. The rest of the wall is thin. Ex. Protoxylem.
2. *Spiral* or *helical thickening*: These are spiral or helical thickening of secondary wall materials. There may be more thar one helix. Ex. Protoxylem.
3. *Scalariform thickening*: The thickenings appear as parallel transverse bands like the rungs of a ladder. Ex. Protoxylem.
4. *Reticulate thickening*: The secondary wall appears as a network as the meshes remain thin. Ex. Metaxylem.
5. *Pitted thickening*: These are more or less circular areas, called pit, where secondary wall materials are not deposited. Ex. Metaxylem.

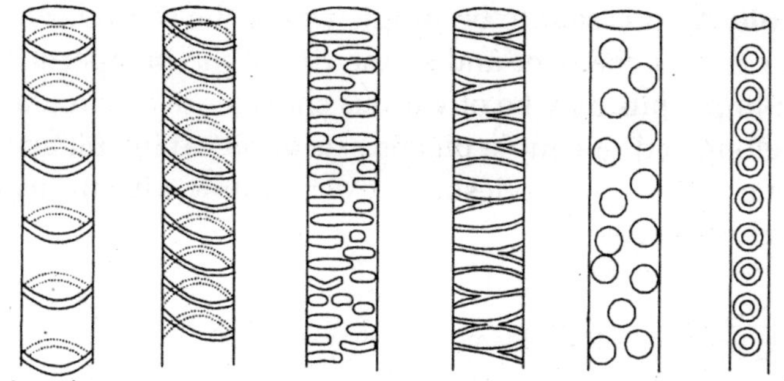


Fig 3.17

Diagrammatic representation of the different types of thickening of cell wall in longitudinal view.

Lignification decreases the water content of cell wall. Therefore, a cell with secondary wall is thought to be inextensible since water probably lubricates the extension process. However, in some phloem fibres and primary tracheids with annular and helical thickenings, the non-lignified regions of the wall continue to grow. As a result, the rings of annular tracheids are widely separated and in helical wall thickenings, the helix extends as in an extended spring. That is why the youngest elongating parts of roots, stems and leaves possess the characteristic annular and helical tracheids. Additional secondary wall materials are laid down between the spaces of rings and gyres of extended tracheids to form scalariform and reticulate thickenings. The former two and pitted thickening lead to cessation of cell wall elongation. The metaxylem elements may also have annular or spiral thickening and here the rings and helix are interconnected by uniform deposition of lignin that leads to the cessation of growth.

• *Primary pit fields*: The meristematic cells and their derivatives without secondary walls have numerous deeply sunken areas on their wall. These depressions in the primary wall are termed as primary pit fields. They are also called primary pit or primordial pit where plasmodesmata exist. In longitudinal sectional view, a primary wall with numerous sunken regions appears as beaded. In primary pit areas, the walls are thin and continuous. These primary pit field areas comprise the middle lamella and the primary wall of both adjacent cells (Fig. 3.18.1).

*Pit*: These are small sharply defined more or less circular areas on the cell wall and in surface view appear as depressions in wall. In these areas, the secondary wall materials are not deposited. In contrast to primary pit, the secondary wall layers are not continuous at the pit region, i.e. the primary wall is not covered. Pit may be developed over primary pit field or irrespective of it, i.e. over the primary wall. A pit consists of pit cavity, pit aperture and pit membrane. The pit cavity, also called pit chamber, is the interrupted region of secondary wall. The pit aperture is the mouth or opening of the pit chamber and it lies towards the cell lumen. The pit membrane, also termed as closing membrane, comprises the middle lamella and the two thin primary walls of two adjacent cells. The following types of pit are found on the cell wall with secondary deposition (Fig. 3.18).

i. **Simple pit:** A pit is said to be simple where the secondary wall does not arch over the pit chamber and so the rim of the pit aperture contains no border. Simple pits may be of various shapes—oval, circular, pentagonal or even elongated. A simple pit consists of pit cavity, pit aperture and pit membrane where plasmodesmata exist. Simple pits are usually present in parenchyma cells.

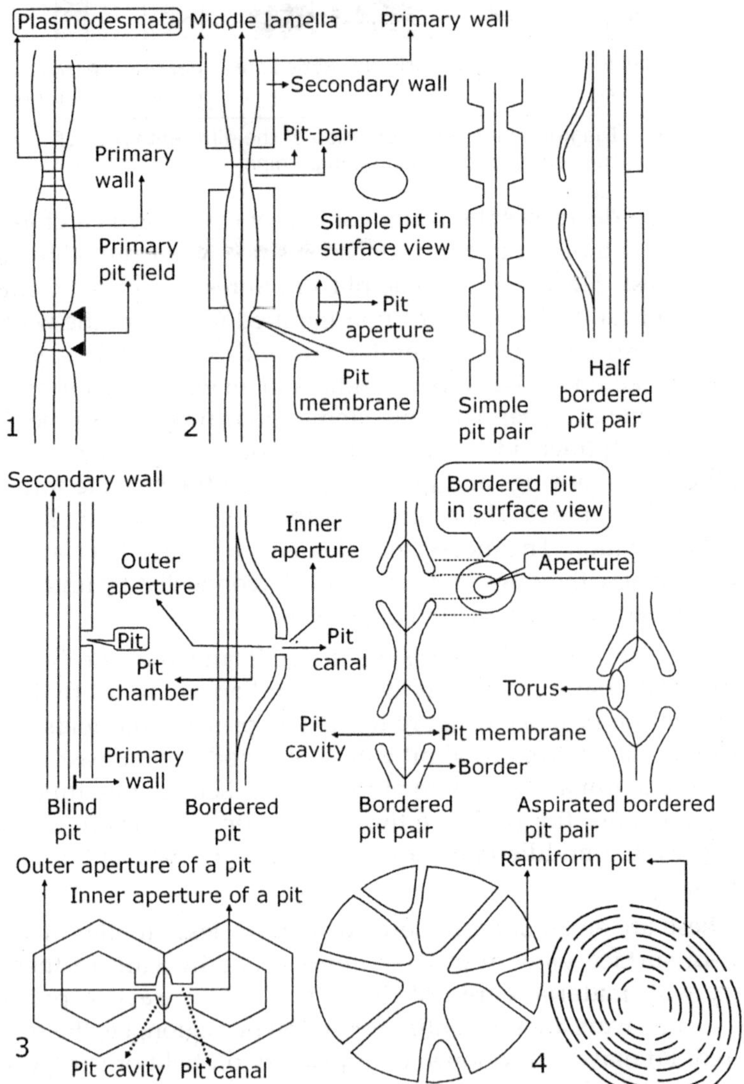


Figure 3.18

Diagrammatic representation of structure of pits as seen in longiudinal sections. 1. The two primary walls of adjacent cells with pits (primordial pits) and the middle lamella in between. 2. Composition of cell wall having secondary wall. At the region of pit the secondary wall is discontinuous. Adapted after Esau, 1961 and Fahn, 1987. 3. A pit pair between two fibers in transverse section. 4. Brachysclereids (diagrammatic).

ii. **Ramiform pit:** Ex. *Cydonia* fruit. In the transverse section of very thick walled brachysclereids, the simple pit appears as a canal on the cell wall. Some of them have the form of branches due to coalescence with the neighbouring canals. These branched shaped pits are ramiform pit (Fig. 3.18.4).

iii. **Bordered pit:** It is found in lignified fibres, tracheids and trachea. In these pits, the pit cavity is partly enclosed by over-arching of the secondary cell wall, which is revealed in longitudinal section. In surface view, bordered pits appear as two concentric rings where the inner circle is the pit aperture and the outer circle forms the border. A bordered pit is thus composed of (1) pit chamber, formed by the over-arching of secondary cell wall, and pit aperture, the shape of which may be circular, oval or linear; (2) the middle lamella or pit membrane or closing membrane. Sometimes the overarching of secondary wall may be very thick to form a canal - termed pit canal, which lies between the pit chamber and cell lumen. Therefore, the pit canal has two openings—outer aperture, which lies towards the cell wall and inner aperture, which is the other opposite opening towards the cell lumen (Fig. 3.18.3). The size and shape of pit chambers and inner apertures depend on the thickness of secondary wall. The pit chambers and outer apertures are always circular in outline in thick wall. The shape of inner aperture varies. it may be elliptical, narrowed and elongated, and slit like. The length of the narrow and elongated inner aperture may be less than the margin of pit cavity or equal to the limit of the cavity. The length may also be much elongated than the limit of pit chamber and in this case the pit canal is funnel shaped. In a pit pair, where both the inner apertures are elongated beyond the margin of pit cavity, the apertures are crossed in appearance in face view (Fig. 3.19). In the wood fibres

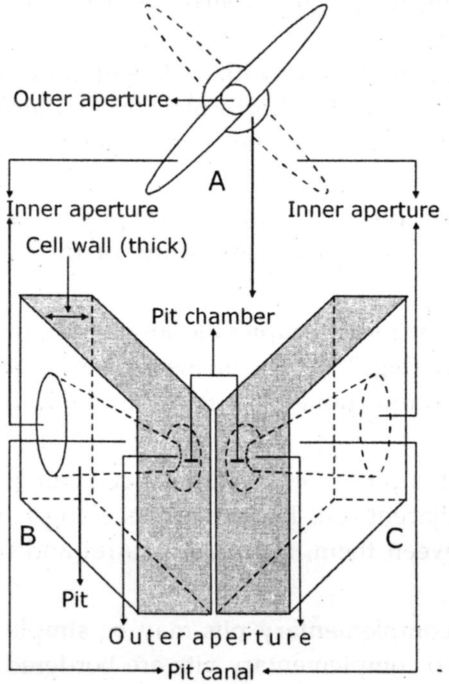


Figure 3.19

A. Diagrammatic representation of a crossed pit apertures in surface view. B. A portion of thick cell wall with a pit of a pit pair (three dimensional view). C. A portion of thick cell wall with the complementary pit of B of a pit pair. Adapted after Esau, 1965 & Eames and MacDaniels, 1947.

(e.g. *Magnolia, Populus*) these 'crossed pit apertures' are very conspicuous. In some gymnospermous wood like *Pinus, Cedrus* etc. the closing membrane between the bordered pit pairs is thickened at the middle. The thickened portion is termed torus (Fig. 3.18), which is more or less disc shaped and wider than the pit aperture. The torus remains suspended by microfibrils. The microfibrils are loosely arranged and radially oriented to form a raised border over the closing membrane where the tori are present. The raised border is known as margo that appears as a thick circular border around the torus. There are many perforations in the margo. Studies with Electron microscope (Liese, 1965) confirm the porous nature of margo though it was discovered as early as in 1913 by Bailey with the aid of a light microscope. In *Cedrus* the torus has fringed margin—a characteristic feature that distinguishes it from other genera.

In gymnosperm, where the vessels are usually absent, conduction of water occurs mainly through the bordered pits of tracheids. The torus is apparently impermeable. Usually the torus occupies the median position and so water can move freely. The pit membrane is flexible and during water movement, the torus swings. During swinging, torus may be at the lateral position where it blocks the pit aperture and thus prevents the flow of water. In conifers, bordered pits regulate the flow of water with the help of torus that acts as the valve.

The shape and arrangements of bordered pit in tracheary elements may vary and accordingly the following three types are noted (Fig. 10.6).

(a) **Scalariform pitting:** In this type, the bordered pits are either linear or elongated and arranged in regular transverse bands like a ladder (ex. *Magnolia*).

(b) **Opposite pitting:** Here the pits are either elongated or circular or elliptic in outline and arranged in horizontal rows; when such pits are crowded, the outline of pits become either rectangular or square (ex. *Liriodendron*).

(c) **Alternate pitting:** In this type the round or oval pits are arranged in diagonal rows and on crowding, the outline of the pit may be hexagonal (ex. *Salix*).

iv. **Aspirated pit:** In these pits, the torus occupies the lateral position. It blocks the outer aperture of pit canal (Fig. 3.18). The pit membrane loses flexibility and thus the pit becomes nonfunctional. It is found in the heartwood of *Pinus*.

v. **Pit pair:** In a functional pit, a pit on one wall must coincide in position with a pit in the wall of adjacent cell. These two pits along with the pit membrane, present in between them, form a structural and functional unit-termed pit pair.

In a pit pair, (1) the two complementary pits may be simple—termed simple pit pair; (2) the two complementary pits are bordered—termed bordered-pit-pair; (3) a bordered pit coincides with simple pit—termed half bordered pit (Fig. 3.18); (4) a large pit is complemented by two or more pits—known as unilateral compound pitting.

vi. **Blind pit:** When a pit is not complemented by any pit on the adjacent wall, it is termed as blind pit (Fig. 3.18). These pits are nonfunctional and may occur opposite to intercellular spaces.

vii. **Vestured pit:** These pits are characterized by the presence of some minute projections in the pit chamber (Fig. 3.20A). These outgrowths originate at the side wall of the pit cavity and pit aperture by the accumulation of cytoplasmic materials. The projections form sculpturing of various shapes.They are found in the secondary tracheary elements of some members of Leguminosae, Cruciferae, Myrtaceae and Caprifoliaceae. Vestured pits are considered to be advanced, as they are present in developed xylem.

6. *Trabeculae*: Trabeculae are bar or rod-like radial projections of the cell wall. They extend across the cell lumen in radial direction and so occur in radial rows (Fig. 3.20C). Sometimes these small bars may extend up to the phloem through cambium. They occur frequently in the tracheids of conifers (ex. *Abies*), rare elsewhere.

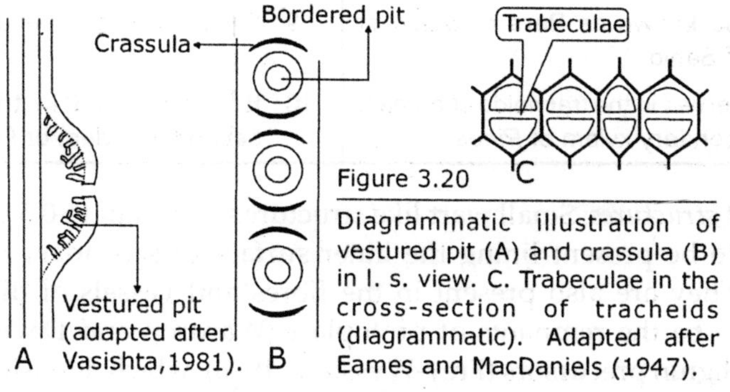


Figure 3.20 Diagrammatic illustration of vestured pit (A) and crassula (B) in l. s. view. C. Trabeculae in the cross-section of tracheids (diagrammatic). Adapted after Eames and MacDaniels (1947).

7. *Crassulae*: These are linear or crescent shaped thickening present along the upper and lower margins of individual or small group of bordered pits. The primary wall and middle lamella, at the rims of primary pit-field thickens to form crassulae. These thickenings may surround or partly encircle individual bordered pits. The deposition of secondary wall materials is interrupted at the regions of crassulae and they are visible when stained properly. Therefore, crassulae represent the elevated rim of primary pit-fields. Crassulae were also named as 'Bars of Sanio' or 'Rims of Sanio', but now a days these terms are of little use. They are commonly found in gymnospermous tracheids, ex. *Pinus* (Fig. 3.20B).

Difference between crassulae and trabeculae.

| Crassulae | Trabeculae |
|---|---|
| 1. It occurs on cell wall. | 1. It occurs within cell cavity. |
| 2. These are linear or crescent shaped thickenings. | 2. These are rod or bar-like radial projections. |
| 3. The thickenings occur along the upper and lower margins of individual or small group of bordered pits. | 3. It extends across the cell lumen. |
| 4. They may surround or partly encircle individual bordered pits. | 4. They are present in radial rows. |
| 5. It is formed in individual tracheids only. | 5. It may extend up to the phloem through cambium. |
| 6. The primary wall and middle lamella, at the rim of primary pit-fields, thickens to form crassulae. | 6. The secondary wall materials form trabeculae. |
| 7. The secondary wall materials do not deposit over the crassulae. | 7. It is formed by the deposition of secondary wall materials. |
| 8. It represents the elevated rim of primary pit fields. | 8. It represents the extension of secondary wall. |
| 9. It is also known as Bars of Sanio or Rims of Sanio. | 9. It is known as trabeculae only. |
| 10. It is observed in the tracheids of primary and secondary xylem of *Pinus*. | 10. It is observed in the tracheids of secondary xylem of *Pinus*. |

8. *Wart structures*: Small wart like structures of 0.1 μm to 0.5 μm in diameter are found to be present lining the inner surface of secondary wall of conifer tracheids. They are also present in the fibres and vessels of dicots. They are supposed to be the remnants of protoplast (Wardrop and Davis, 1962) or the remnant of lignin precursors (Frey-Wyssling, 1976), which deposited on the inner wall of tracheids, fibres and vessels to form warts.

**Intercellular space:** In mature tissues, spaces frequently occur between cells. These spaces are known as intercellular spaces. The spaces may be small, large (ex. *Canna*)—called chambers and elongated—called canals. The spaces may be considerably elongated in one plane—called duct.

The spaces are exceptionally well developed in aerenchyma of many aquatic plants. They form a connected system throughout the entire plant body. The spaces may be vertical and are filled with air and water, thus forming air spaces and air canals. Air spaces and air canals help in aeration and add buoyancy to plant. It has been suggested that, the honeycomb-like system of intercellular spaces in aquatic plants can withstand considerable mechanical stress.

In terrestrial plants, the enlarged intercellular space gives rise to secretory glands (ex. *Citrus* sp.) or ducts (ex. *Pinus*).

Parenchyma of the endosperm of most seeds is devoid, or almost so, of intercellular spaces. These cells are very tightly packed and are in close contact all round. Meristematic tissues are usually thought to be devoid of intercellular spaces. Observation with electron microscope, on the root tip of *Allium*, reveals the presence of intercellular space of 20 μm. Such a small space could not be readily observed with the aid of light microscope.

According to the mode of development, there are two types of space, schizogenous and lysigenous. The former is most common and results from the separation of cell walls from each other along certain areas of their contact. The word schizogenous is derived from the Greek word—schizo meaning split and genesis meaning beginning. Previously it was thought that these spaces result from the splitting of the middle lamella and so the space are called schizogenous. The schizogenous development of an intercellular space is described as follows (Martens, 1937). At the end of cell division a cell plate is formed separating the two sister protoplasts. The plate forms the middle lamella. The components of the cell wall then deposit on both the sides of middle lamella to form primary cell wall. The newly formed middle lamella makes a contact with the primary wall inside of the parent cell at the sides only and not with the middle lamella, that cements the parent cell with neighbours. At the point of contact between the new middle lamella and primary wall of parent cell a small space develops. This space enlarges due to the dissolution of primary wall of parent cell at opposite side of space. This space is more or less triangular in outline. The components of middle lamella deposit on the sides of triangular space. The new middle lamella becomes continuous with the recent deposits. Therefore, the unit structure appears to be forked. The forked portion gradually extends because of growth of daughter cells and ultimately joins with the middle lamella of parent. Thus, an intercellular space is formed within the forked middle lamella (Fig. 3.21). The parent cell may have similar space with the neighbour. This old space may join with the new one to form a large intercellular space. In schizogenous development of spaces, a portion

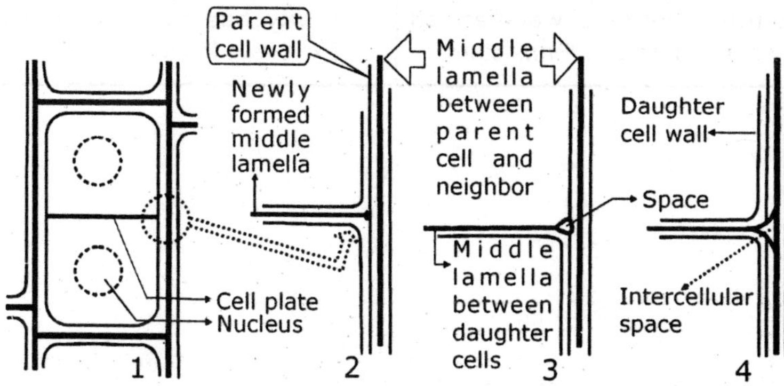


Figure 3.21

Diagrammatic representation of the stages in the formation of intercellular space. 1. Dividing parent cell at the end of telophase. 2. Newly formed middle lamella in contact with the primary wall of parent cell. 3. Appearance of space. 4. Newly formed middle lamella in contact with the middle lamella of parent cell. Adapted after Chandurkar, 1983.

of parent cell wall is dissolved only. These newly formed cells enlarge in all directions to assume the normal size. The cell walls separate and the schizogenous space is the result. Meristematic cells are tightly packed, and during differentiation to parenchyma, schizogenous spaces develop. Examples of this schizogenous cavities or ducts are: resin ducts of *Pinus* and secretory ducts of Asteraceae and Apiaceae, secretory cavities of *Eucalyptus* etc.

The lysigenous intercellular spaces are formed by the disintegration of entire cells. The term lysigenous is derived from the Greek word—lysis meaning loosening and genesis meaning beginning. The large air spaces of water plants, air spaces in *Zea* root, primary resin duct of *Mangifera indica* etc. are the examples of lysigenous spaces. The protoxylem lacunae of *Zea mays* is formed by both schizo- and lysigenous way. Such spaces are called schizolysigenous.

Comparison between resin duct and oil cavity.

| Resin duct | Oil cavity |
|---|---|
| 1. It is an internal cavity, which is considerably elongate in one plane. | 1. It is an internal cavity, which is more or less isodiametric. |
| 2. These are schizogenous duct. | 2. These are lysigenous or schizolysigenous glands. |
| 3. The ducts are formed by the dissolution of middle lamella. | 3. The cavity is formed both by the phenomenon schizogeny (separation of cell walls) and lysis (dissolution of entire cells), i.e. the middle lamella and some of the adjoining cells separate and disintegrate to form cavity. |
| 4. In the ducts, resin is secreted. | 4. In the cavity, oil is secreted. |
| 5. The duct remains surrounded by a ring of intact parenchyma cells, termed epithelium that forms a well-defined boundary of the duct. Ex. *Pinus* stem. | 5. The cavity remains surrounded by intact or broken parenchyma cells and the cavity has no clear–cut boundary. Ex. *Citrus* fruits. |

CHAPTER 4

# Pollen and Spore Wall

Pollen and spore wall define the shape of a pollen grain and spore, and are composed of sporopollenin—'the most extraordinarily resistant material known in the organic world' and other substances. They include all layers [exine/ exospore (the peripheral layer of wall), intine/endospore (the innermost layer of wall) and perine/perispore (the outermost layer) when present] of the wall. They are also known as sporoderm (L. *sporodermis*), spore coat, spore wall and the integument of spore.

Morphologically pollen and spore (microspore) are homologous units. Pollen grains are produced in phanerogams whereas spores are formed in cryptogams. Except some algae and fungi, pollen and spore are the products of meiosis that involves a reduction by half of the chromosome content. Both pollen and spore require dispersal in space in order to perform their functions.

## Spore (n., pl. –s)

Spore is the asexual reproductive unit. It is usually unicellular and haploid. It is formed as a result of reduction division in embryophytes. It is a propagative body that consists the gametophyte enclosed and protected by spore coat. The spore coat alone is distinguishable in the spores present in coal. The following common types of spore are noted in embryophytes.

• *Isospore* (n., pl. –s, syn, Homospore): Isospores (in Greek 'isos' means equal)/ homospores (homo-, a Greek prefix, meaning 'alike', 'similar') are characterized in having one kind of spore only. They all are of same size. They are generally small, morpholcgically similar and not distinguishable as to sex. Plants producing isospore/homospore are described as isosporous/homosporous (adj.). Homospores are produced in the sporangium borne on homosporous plants as a result of reduction division. The unicellular isospore/homospore develop into multicellular gametophytic plant. In homothallic gametophytes there originate both male and female sex organs and accordingly male and female gametes develop, e.g. homosporous pteridophytes like *Lycopodium*, *Equisetum*, *Ophioglossum*, *Polypodium* etc. In heterothallic gametophyte, e.g. *Equisetum arvense*, two types of gametophyte develop. In one type male sex organ develops while the other has female sex organ only.

• *Heterospore* (n., pl. –s; hetero– a Greek prefix meaning 'different', 'not normal'): Heterospores are characterized in having two kinds of spores—microspore and megaspore. Microspore and megaspore are distinguishable on the basis of sex and size. The megaspores are always larger than the microspores of same species. It is interpreted that the larger size of megaspore is due to an adaptation for the storage of food, which nourishes the embryo arising later. In heterosporous embryophytic plants megaspore germinates to produce multicellular female gametophyte, also called megagametophyte. The microspores are smaller in size than the megaspores of the same species. In heterosporous plants microspore germinates to produce multicellular male gametophyte, also called microgametophyte. Plants producing heterospores are described as heterosporous (adj.). Microspore and megaspore are the product of reduction division that occurs in the respective sporangia, accordingly termed microsporangium and megasporangium produced on heterosporous plants. Example: *Selaginella, Isoetes* etc.

Coal microscopists describe all small spores as microspores regardless of biological function. Traverse (1988) suggests avoiding this practice since it may tend towards confusion when more precise information becomes available. In palaeopalynology the term macrospore is used as a synonym of megaspore, though macrospore may not have similar biological function like megaspore. All spores including pollen grains that are greater than 200 micra in diameter are grouped under macrospore (Traverse, 1988).

• *Meiospore* (n., pl. –s): Spores produced as a result of reduction division or meiosis is termed as meiospore. Microspore, megaspore, isospore, pollen etc. all are meiospores as they develop respectively from their mother cells as a result of meiosis.

• *Miospore* (n., pl. –s; in Greek meion means 'less', 'smaller'): A pollen grain or spore having diameter less than 200 micra, regardless of biological function is termed as miospore (Traverse, 1988, p. 507). When the term miospore is used, pollen grains and spores having diameter more than 200 micra are referred to as megaspore irrespective of their biological function. In palaeopalynology sometimes it becomes difficult to distinguish spores from isospores / homospores of homosporous plants, microspore of heterosporous plants, small megaspores, pollen and pre-pollen when they all are observed individually. So palynologists adopted the term 'small spore' to describe the spores of relatively small size, regardless of their biological function. Later the term 'miospore' is proposed (Guennel, 1952) to denote isospores/homospores, true microspore of heterosporous plants, small megaspores, pollen grains and all fossil spores having diameter less than 0.20 mm. On the other hand spores having diameter greater than 0.20 mm are referred to as macrospores (Guennel). The 0.20 mm measurement is used as the demarcation line because of technical reasons only and bears no biological meanings. The standard sieves that are used in the preparation techniques of spore and pollen slides have meshes of that size (Kremp, 1965, p. 91, Meyen, 1987, p. 243).

[Prepollen is defined as a pollen grain characterized by the presence of proximal aperture, presumed proximal germination and distal aperture. The pollen tube that emerges through proximal aperture is meant for releasing fertilizing nuclei, presumably in antherozoids—thus exhibiting zoidogamy. The haustorial pollen tube emerges through distal aperture. Prepollen is haploid and is observed in some extinct (e.g. Cycadofilicales) and extant (e.g. *Cycas* and *Ginkgo*) seed plants. Prepollen grains have laesurae. In this sense one may exclude *Cycas* and *Ginkgo* because they lack proximal laesurae (Traverse, 2007, p. 694).]

## Pollen (n. pl., sing. a pollen grain)

A pollen grain is defined as male gametophyte/microgametophyte of seed plants (Gymnosperm and Angiosperm) that remains enclosed within pollen wall. 'The definition of pollen is not morphological but functional: the microspore wall of seed plants, plus the microgametophyte that develops within the microspore wall'—Traverse (2007; p. 11). Fossil pollen grains consist of exine only (Traverse, 1988). Intine and microgametophyte are destroyed during fossilization. An extant pollen grain represents a highly reduced male gametophyte in angiosperm. A pollen grain at an early stage of development is also termed as microspore because both pollen and microspore develop into male gametophyte in heterosporous plants. But the terms 'pollen' and 'microspore' are not homologous—the microspore is the immediate product of meiosis of microspore mother cell whereas a pollen grain contains the microgametophyte developed from microspore within its wall (Erdtman, 1943).

Grammatically the term pollen is a 'collective plural noun' and the term refers to many individuals. So the correct plural is 'pollen' not pollens (Traverse, 1988). It is incorrect to refer a single individual as 'a pollen'. A single individual is to be referred to as 'a pollen grain' (Traverse, 2007, p. 691)/'pollen grain' (Traverse, 1988, p. 511). 'Examples: Correct: The pollen of all sections of family Poaceae is monoporate. Incorrect: The pollens of euphorbiaceous plants are eurypalynous' —Traverse (2007, p. 691). The word pollen in classical Latin means 'fine meal'/ 'fine flower'. The first to use it was the great Linnaeus in the title *Philosophia Botanica* in 1751 to denote the dusts that are shed from the male sex organ (anther) of a flower to execute fertilization. Later botanical writers cited the word in their literatures from whose pages the term 'pollen' has become gradually 'diffused and internationalized' (Kremp, 1965, p. 119).

## Sporoderm stratification (Erdtman's concept)

Sporoderm (L. *sporodermis*), i.e. the spore and pollen wall is composed of distinct layers or subdivisions, the sequence of which is as follows from outside inwards in a fresh pollen grain and spore (Erdtman, 1952).

*Sclerine* (Gr. Scleros means hard—Fig. 4.1A & B): Sclerine is the peripheral layer of sporoderm excluding intine. It is entirely, or chiefly, hard (sclerodermic) and preserved in fossil spores. This layer has two well-defined subdivisions—perine and exine. The term can be used when there is doubt regarding the presence of perine, i.e. sporoderm excluding intine.

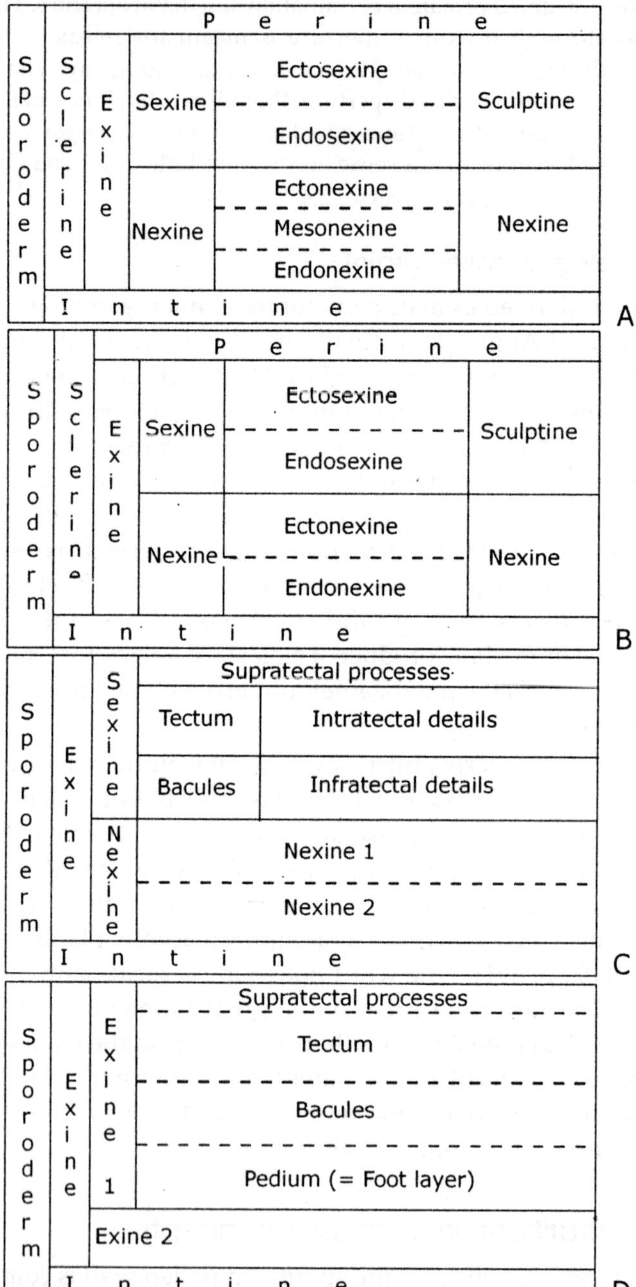


Figure 4.1

Diagrammatic comparison of sporoderm stratification according to A. Erdtman (1948), B. Erdtman (1952), C. & D. Erdtman (1969). The terms Exine 1 & Exine 2 in D are based on staining characteristics. Vide text for details.

Comparison between pollen and spore.

| Pollen | Spore |
|---|---|
| 1. It is a container that houses male gametophyte of phanerogams. | 1. It is the resting spore and dispersal phase of cryptogams. |
| 2. The target of a pollen grain after transportation is stigma. | 2. There is no such target. |
| 3. A pollen grain can perform its function when it reaches on a compatible stigma. | 3. Spore can perform its function when it lands on suitable substratum and environment. |
| 4. Aperturate pollen grains are colpate/ porate etc. | 4. Aperturate spores are monolete or trilete. |
| 5. Apertures occur on distal pole. | 5. Apertures occur on proximal pole. |
| 6. Perine is absent in phanerogams. | 6. Perine is present in the spores of Bryophyte and Pteridophyte. |
| 7. Division of exine into sexine and nexine is prominent. | 7. Division of exine into sexine and nexine is indistinct. |

*Perine* (perisporium, L. *perinium*, Syn. perispore): Perine is the outermost layer of sporoderm and more or less sculptured like sexine. It is chiefly present in spores of ferns and mosses. It is also reported from gymnosperm pollen, e.g. *Taxodium*. This layer is less resistant to certain chemicals and decay, e.g. *Psilotum*. This layer is usually wrinkled, folded in various ways and envelopes the mature

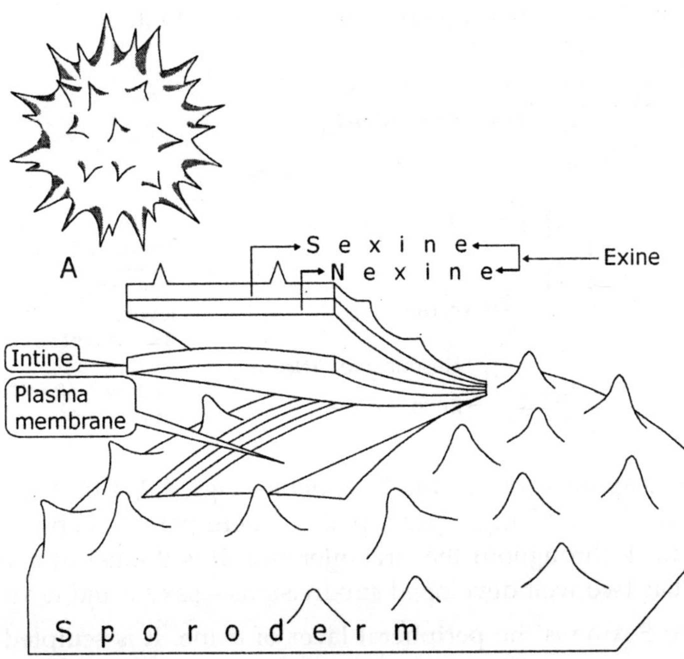


Figure 4.2

A. A pollen grain (diagrammatic). B. A portion of sporoderm and its stratification in three dimensional view.

spore in a loose manner. Perine often cannot be differentiated from the sexine in fossil material. In paleontological considerations, the exterior parts of exine that form a natural separate surface are considered as perispore. It is interpreted as an additional wall layer external to exine. Sometimes it is difficult to ascertain whether a certain stratum is a part of perine or sexine. In such cases the term sclerine is used as a comprehensive term for perine and exine. So the adjective 'sclerinous' can be applied to the layer which may be perinous, exinous or both of the two. Similarly sometimes it becomes difficult to decide whether the supratectal element is of perinous or sexinous nature. Such cases are described as sculptine. It is used as a neutral term and signifies that the sculptured element may be a part of either perine, sexine or both. The term sculptine is used to describe the sculptural layer of sclerine.

The term perine is usually used in conjunction with exine and intine whereas the synonym perispore is used with exospore and endospore. The term exospore and endospore are homologous to exine and intine respectively. Mycologists use the two terms to describe spores that are formed outside and inside respectively of a sporangium. Perine, in contrast to exine, is less resistant to decay and acetolysis.

[Acetolysis is a procedure followed by palynologists to stain pollen present in peat and anther. Erdtman introduced the practice of acetolysing pollen grains by treating the materials in a mixture of nine parts of acetic anhydride and one part of concentrated sulphuric acid. The latter is a catalyst and desiccating agent. After acetolysis pollen grains stain yellow to orange colour. Sporopollenin remains unscathed and exine sculpturing becomes clearer to study.]

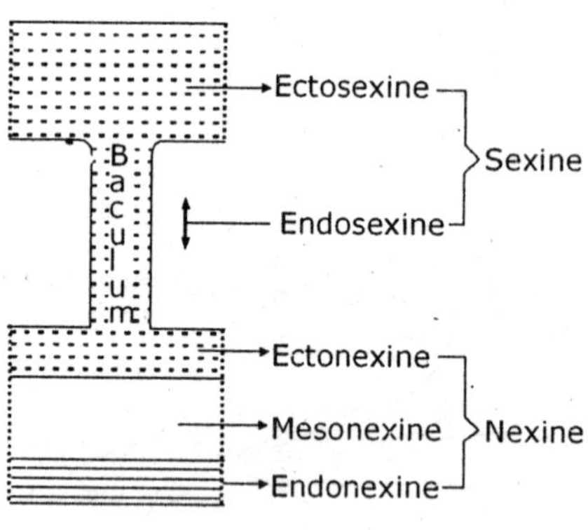


Figure 4.3

Diagram of exine layers except perine as illustrated by Erdtman, 1948. Redrawn after Tschudy *et al.* (1969), p. 24.

*Exine* (exospore; exosporium; L. *exinium*—Figs. 4.2, 4.1C & D, 4.3): Exine is the outermost layer of angiosperm pollen where perine is absent. Exine is not uniformly thick throughout the circumference. It is thinner at apertural region. This layer has two well-developed subdivisions - sexine and nexine.

• *Sexine*: Sexine is the peripheral layer of exine. It is sculpted according to taxon-specific manner. The term sexine is derived "from 's' in sculptured, and exine". This layer has two well-developed subdivisions—ectosexine (also spelt as ektosexine) and endosexine.

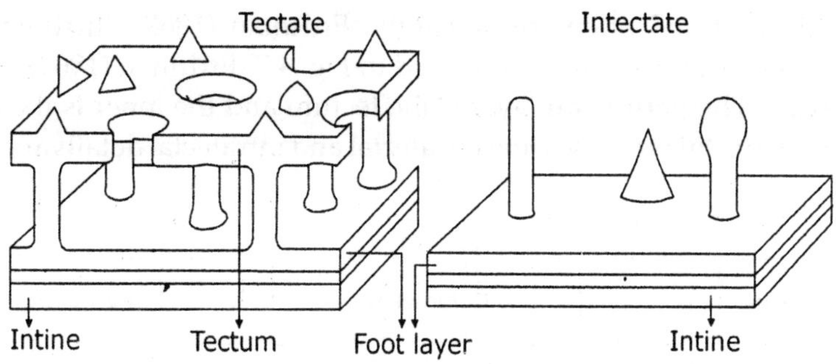


Figure 4.4

Diagrammatic representation of three dimensional view of a portion of sporoderm illustrating tectate and intectate pollen. Adapted after Hesse *et al.*, 2009, p. 20.

Ectosexine (Fig. 4.6A-H) forms a roof-like layer on the periphery of a pollen grain. This layer covers a pollen grain partially or wholly. This layer is termed as tectum (n., pl. tecta, adj. tectate; Syn. tegillum). Tectum may be absent when a pollen grain is described as intectate. When tectum is present a pollen grain is said to be tectate where the tectum appears to be as a continuous sheet without any perforations or covers 80 percent or more of the total surface of a pollen grain (Fig. 4.4). In a semitectate or subtectate pollen grain the tectum is discontinuous where perforations exist. When the perforation is more or less one micron in diameter, it is termed as punctum (pl. puncta). Each punctum is like a cylinder and occurs as perpendicular channel on the tectum (Fig. 4.6D). It has two openings—one on the supratectal (L. *supra* means over) and the other on the infratectal (L. *infra* means under) side. In many pollen grains the perforations on tectum are very large, more or less angular and regularly or irregularly polygonal. Such honeycomb-like perforations are described as reticulum (adj. reticulate). Each reticulum has a roofless space termed lumen (pl. lumina) surrounded by sexinous ridges termed muri (sing. murus). Muri are the walls of a lumen and they anastomose to form reticulum. The murus stands on nexine, which is also the floor of lumen. So a reticulum has no tegillum (Fig. 4.6E & F). Tegillum is present on negative reticulum. Negative reticulum is formed by grooves. Grooves on sexine anastomose and enclose sexine isles. A network of anastomosing grooves separates the sexine isles thus constituting an inverse or negative reticulum (Fig. 4.6G & H). Apart from punctum and reticulum small lumen occurs on tectum. Each lumen has very wide muri. This type of perforation cannot be regarded as reticulum. Such perforations are referred to as scrobiculi (sing. scrobiculus). The lumina of reticula when become very small and the spaces that separate the lumina become wider, then the lumina are called scrobiculi. The sculpture on the supratectal region of ectosexine is discussed later under surface ornamentation.

Endosexine consists of columella, baculum and void spaces separating the bacula from each other. The columella (pl. columellae, adj. columellate) is a column like element of sexine that supports the tectum at the top. A baculum is also a sexine element, which is slim, cylindrical rod like structure and stands on nexine. Each baculum is more than one micron in length and less than this in diameter.

Bacula are straight and often end abruptly. Erdtman (1969) illustrates different types of baculum and they are represented in Fig. 4.7. Erdtman (1969) regards two layers in sexine. The peripheral layer is the tectum and the inner is the bacules. In Fig. 4.8 the supratectal processes, the intratectal and infratectal details are illustrated.

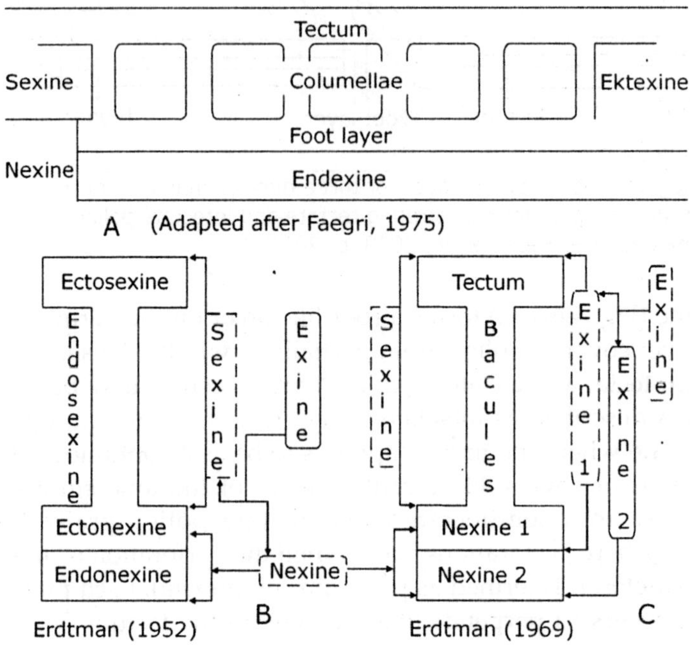


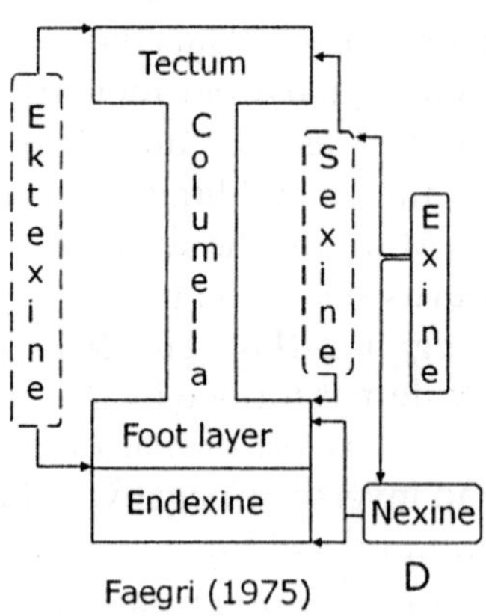


The spelling 'ectexine' was introduced instead of ektexine because of its latinized spelling. The former is the orthographical variant of the latter. The spelling ektexine is still in use (Faegri, 1975; Beug, 2004; Cutter, 1971 and Bhattacharya *et al.*, 2006).

Figure 4.5

A. Diagrammatic representation of exine stratification according to Faegri. B., C. & D. Diagrammatic comparison of exine stratification according to Erdtman & Faegri.

• *Nexine*: Nexine occurs below the sexine and it is nonsculptured part of exine. The term nexine is derived "from 'n' in nonsculptured and exine". Erdtman (1952) distinguished two subdivisions (ectonexine and endonexine) in this layer on the basis of thickness 'and refractive zone'. Ectonexine is the peripheral layer of nexine, exhibits more or less thick and 'not very refractive zone'. Endonexine is the inner layer of nexine, more or less thin and exhibits 'more refractive zone'. Erdtman (1948) reported the presence of a layer in between ectonexine and endonexine—termed mesonexine (Figs. 4.1A & 4.3).

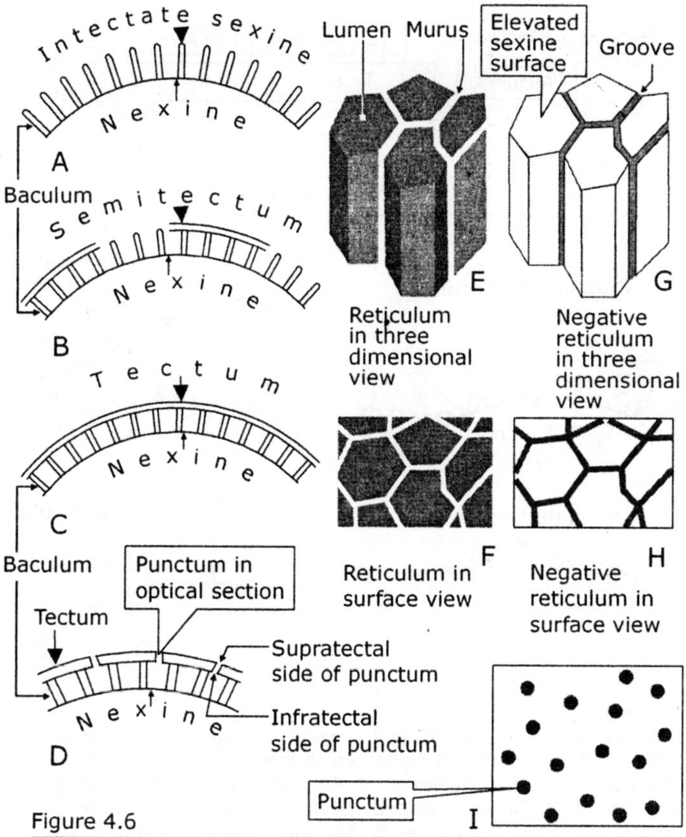


Figure 4.6

Diagrams illustrating different forms of ectosexine. A. Intectate exine. B. Semitectate exine. C. Tectate exine. D. Punctitegillate exine. E. & F. Reticulate exine. G. & H. Negatively reticulate exine. I. Puncta in surface view. F, H & I are observed at high adjustment of microscope.

Nexine seems to be homogeneous. In acetolysed pollen grains (ex. species of Malvaceae and Onagraceae) it is reddish brown while the sexine is yellowish. Erdtman (1969) differentiated two layers or subdivisions in nexine based on staining characteristics. The layers are nexine 1 and nexine 2. In acetolysed pollen grains nexine 1 stains red with alcoholic fuchsin and nexine 2 does not show any staining characteristic. Nexine 1 is also called pedium (the Latin word *pedium* means foot layer) as the bacules stand on this layer. Sexine also stains red with alcoholic fuchsin like nexine 1. Due to this similar staining characteristic Erdtman (1969) called sexine and nexine 1 together as exine 1. Nexine 2 is referred to as exine 2 (Figs. 4.1D & 4.8). Exine 1 and exine 2 are also referred to as ectexine and endexine respectively. Ectexine and endexine are distinguished on the basis of staining characteristics with basic fuchsin. The former stains positively with basic fuchsin while the latter remains relatively unstained.

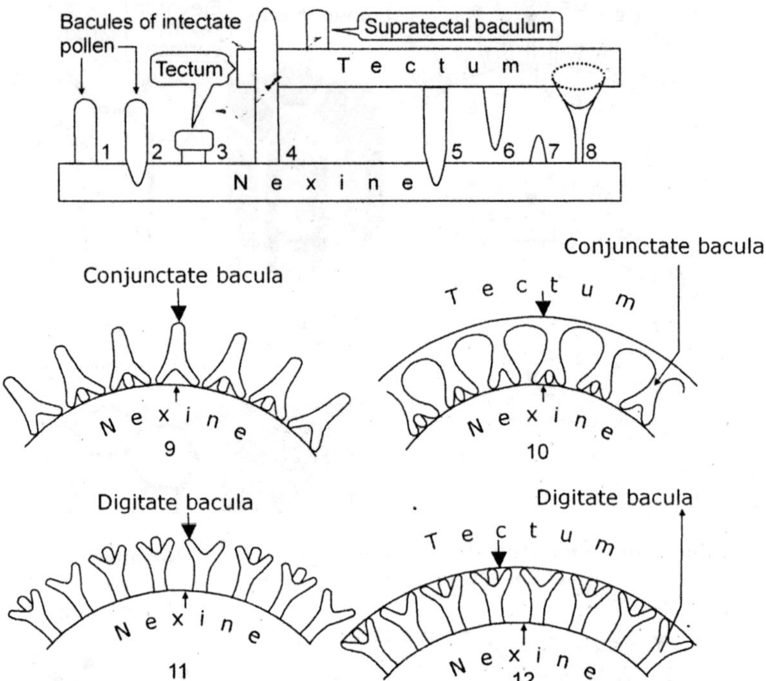


Figure 4.7

Diagram illustrating the different forms of bacules (after Erdtman, 1969). 1. Baculum having higher length than breadth. 2. The base of baculum is within the nexine. 3. Baculum having higher breadth than length. 4. Baculum extends beyond tectum. 5. The base of baculum is within the nexine. 6. Baculum hangs from tectum ex. Caryophyllaceae. 7. The top of baculum is below the tectum. 8. The top of baculum is funnel shaped. Figures 3-8: Bacula belong to tectate pollen. 9. Conjunctate bacula / columellae of intectate pollen. Bacula are branched into two or more parts on the proximal side (ex. Compositae). 10. Conjunctate bacula or columellae of tectate pollen. 11. Digitate bacula / columellae of intectate pollen. The bacula are branched into two or more parts distally (ex. *Polygonum bistorta*). 12. Digitate bacula of tectate pollen. Figures 9-12 are adapted after Punt *et al.*, 2007.

*Intine* (endosporium, L. *intinium*): Intine is the innermost layer of sporoderm. It is soft (malacodermic), fragile and pliable. It is colourless layer, encircles completely the protoplast and remains tightly attached to the plasma membrane. It is slightly resistant layer. It decomposes when boiled with acids and alkali. It is not well preserved in fossil spores and pollen. The thickness of intine is related to that of exine. In the place where the exine gets thinner, the layer of intine becomes thicker. So thickened portions of intine occur under pores. Exceptions are observed in *Canna*, *Musa*, and *Crocus* etc. where a thick layer of intine envelops the protoplast uniformly. During germination of a pollen grain the intine emerges through the apertures forming the membrane of pollen tube.

Rowley (1959) recognized a layer in between exine and intine and termed it as mesine. This layer is faintly lamellate, acetolysis resistant and electron dense. Saad (1963) called this layer as medine.

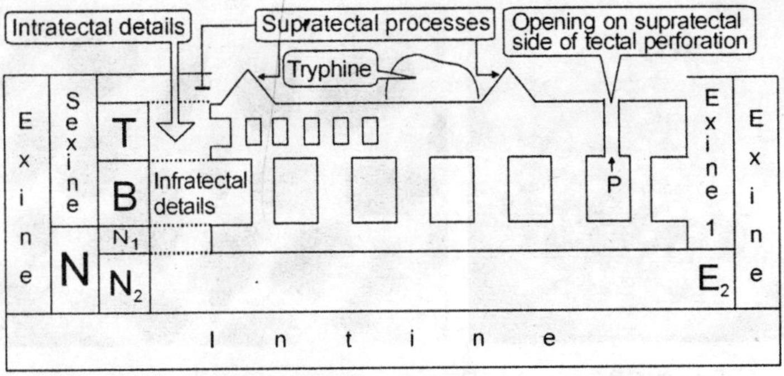


Figure 4.8

Sporoderm stratification according to Erdtman (1969) in a typical angiosperm pollen grain. N: nexine, $N_1$: nexine-1, $N_2$: nexine-2 (syn. Exine-2), T: tectum, B: bacules, $E_2$: exine-2 (syn. Nexine-2), P: opening on infratectal side of tectal perforation.

[Tryphine: It is the deposition of structural materials (derived from break down products of tapetum) on the exine and contains the membranous components like mitochondria, endoplasmic reticulum etc. Pollenkitt, the other break down product of tapetum, differs from tryphine in lacking the membranous components. It is sticky and cements pollen together during entomophilous dispersal.]

In this context it is worthy to mention the concept of Faegri (1975) regarding sporoderm stratification and its difference from that of Erdtman that are illustrated in Fig. 4.5A-D.

## Surface ornamentation

The surface of a pollen grain and spore may be smooth or sculptured due to the occurrence of supratectal processes of various types. Elements/processes that stand on the surface of exine can be called sculpture. The sculpture elements are more or less regularly or irregularly arranged on tectum. The sculpture of a pollen grain is genetically labeled, fairly constant and an excellent means of recognition of pollen. Followings are few sculpture types that ornament both pollen and spore (Figs. 4.10A-H & 4.11A-L).

• *Chagrenate* (Turkish 'gaghri' means untanned leather – Fig. 4.11A): The term describes the surface of a pollen grain and spore that are smooth, shiny, structureless and translucent (Tschudy *et al.*, 1969). Traverse (1988) considers the term as 'a variant of psilate'. Individual structure particles are not recognizable.

• *Psilate* (adj., in Greek psilos means smooth – Fig. 4.11B): The term describes a pollen grain and spore with surfaces completely smooth. The tectum is devoid of any processes. The surface appears as a mat that may have small depressions or pits that are less than one micron in diameter. Example: *Aconitum*, *Betula*.

• *Spinula* (pl. spinules, adj. spinulose – Fig. 4.9A): This supratectal process is pointed towards apex and the length of a spinula never exceeds three micra.

• *Spina* (pl. spines, adj. spinose): This supratectal element has pointed apex and the length of it always exceeds three micra. A spina has a broad base that narrows slowly towards apex.

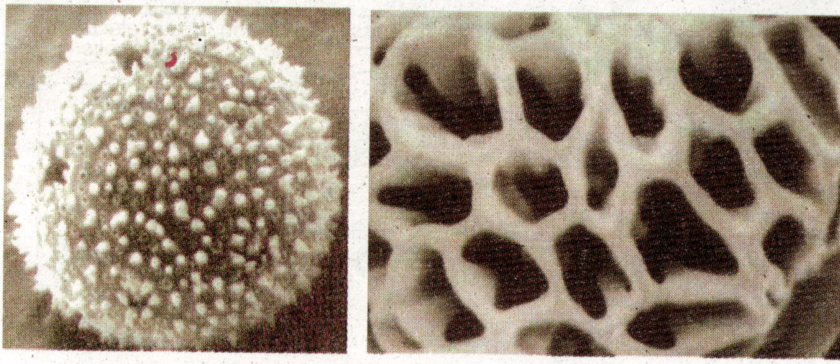

Figure 4.9 A B

A. Scan electron micrograph of *Boerhaavia punarnava* pollen showing puncta and spinules. B. Scan electron micrograph of a pollen grain of *Basella rubra* showing reticulate exine. After Roy & Chanda, 1992.

• *Pilum* (pl. pila, adj. pilate – Fig. 4.10L, M & N): Pilum is blunt and drumstick-shaped element. It has a swollen more or less round or elongated top called caput and a rod like part that holds the caput – termed baculum (Erdtman, 1952). A pilum is taller than one micron. Caput and baculum together, apart from supratectal projections, may form sexine where the capita and bacula represent respectively ectosexine and endosexine. Example: *Mercuriales*, some elements of *Ilex* and *Nymphaea*.

• *Verruca* (pl. verrucae, adj. verrucate; in Latin *verruca* means a wart – Fig. 4.10D & E): Verruca is a rounded projection on the tectum where the breadth, i.e. diameter of radial projection at the lower part is greater than length. The base of verruca is not constricted. The surface of verrucae can be round, flat or irregular. In a grain the verrucae differ in shape and size. Thus in surface views round and less round verrucae are observed. The diameter of a verruca is more than one micron. Example: *Plantago, Fumaria* and *Aristolochia arborea* etc.

• *Gemma* (pl. gemmae, adj. gemmate; in Latin *gemma* means a bud – Fig. 4.10F & G): Gemma is a spherical sculptural element whose greatest diameter of radial projection is greater than height. It may be more than one micron in diameter. The lower part is constricted. Example: *Juniperus, Linum* and some elements of *Nymphaea* etc.

• *Clava* (pl. clavae, adj. clavate; in Latin *clava* means a club – Fig. 4.10H & I): Clava is a club like supratectal process where the length is greater than breadth. It is more than one micron in length. The apex has greater diameter than base. Clava gradually narrows towards the base. It is like a baculum where the apex is thicker than base. Example: Some elements of *Ilex*.

• *Baculum* (n., adj. baculate; in Latin *baculum* means a rod – Fig. 4.10B & C): Baculum is rod like the length of which is much greater than the diameter. The rod is of equal thickness throughout apex and base. Each baculum is more than one micron high. It is straight and often ends abruptly. It may occur as supratectal

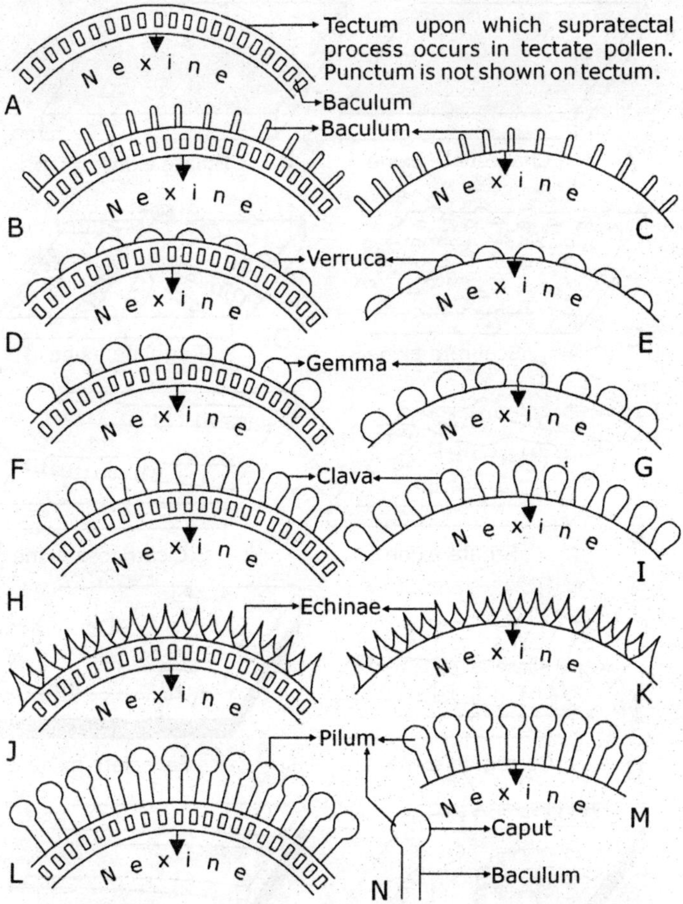


Figure 4.10

Diagrammatic illustrations of a few common pollen sculpture types as observed in optical section. (A) Plan view of sexine illustrating the position of supratectal processes in tectate pollen of the following diagrams. Baculate exine in tectate (B) and intectate (C) pollen. Verrucate exine in tectate (D) and intectate (E) pollen. Gemmate exine in tectate (F) and intectate (G) pollen. Clavate exine in tectate (H) and intectate (I) pollen. Echinate exine in tectate (J) and intectate (K) pollen. Pilate exine in tectate (L) and intectate (M) pollen. N. A single pilum. The processes are not drawn proportionately for clarity.

process in tectate pollen grains. In intectate or semitectate pollen grains it may form a part of sexine, where bacula stand at right angles to nexine. Example: Some elements of *Nymphaea* and *Viscum* etc.

• *Scabrate* (L. *scaber* means flecked – Fig. 4.11C): These are small sculpturing elements on the surface of pollen. The projections are radial and less than or equal to one micron in diameter. The shape may vary. This term is also used synonymously with granulate. It is a general term for all elements that are less than one micron in diameter. Example: *Artemisia campestris*, *Quercus*, *Populus*, *Urtica* etc.

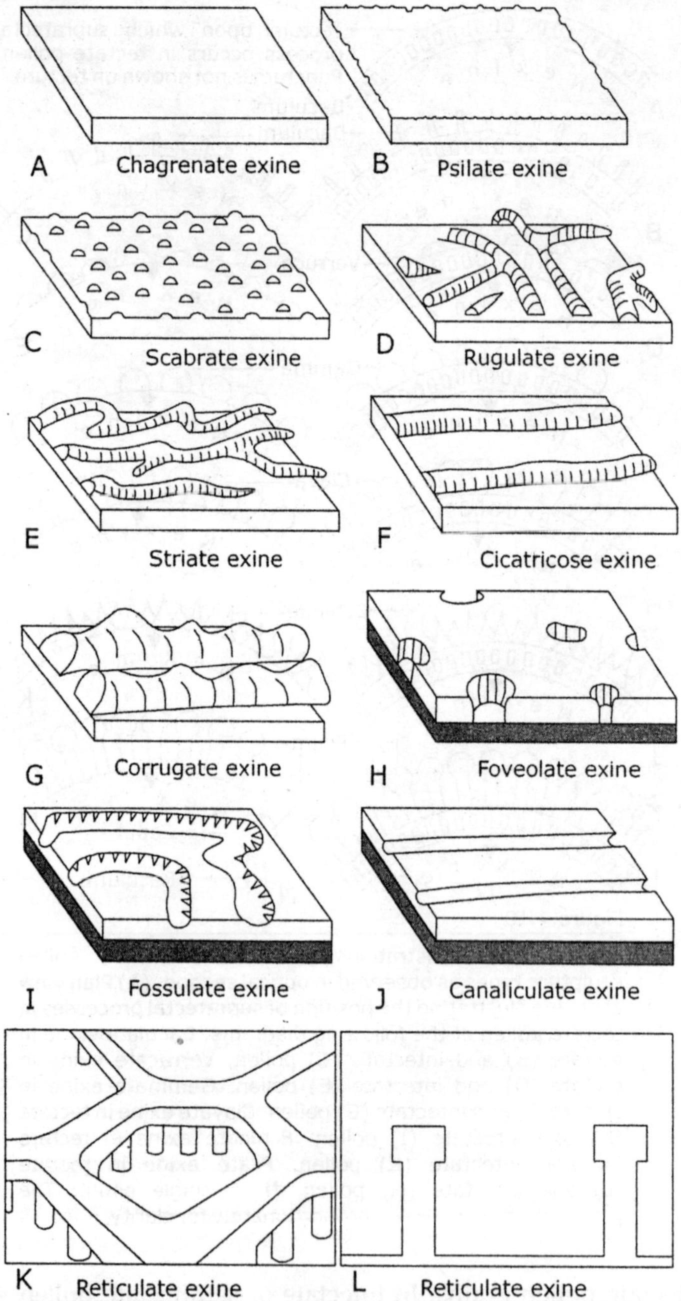


**Figure 4.11**

Diagrammatic representation of ornamentation in a portion of exine in three dimensional view (A to K). L. Reticulum as seen in optical section. Figures A-J and K & L are adapted after Tschudy *et al.* (1969), p. 24 and Erdtman (1969) respectively.

• *Rugulate* (L. *ruga* means a wrinkle, L. *rugolosus* means 'covered with creases'; L. *rugulae*, sing. *rugula*; Erdtman, 1969—Figs. 4.11D & 4.12): The surface of a pollen grain and spore is provided with radial projections, the length of which is at least twice the breath. The ridges are irregularly distributed over the surface. Rugula appears long and narrow in surface view. Example: *Nymphoides peltatum*, megaspore of *Selaginella densa* (Tschudy *et al.*), *Botrychium australe*, *Sedum*, *Ulmus*, *Apium* and *Erodium* etc.

• *Striate* (L. *stria* means striped—Figs. 4.11E & 4.12): The surface of a pollen grain is provided with ridges and grooves. The ridges and grooves are numerous, more or less parallel to each other in the exine. The sculpture elements form linear or 'fingerprint' like pattern and thus exhibit parallelism. The striate ornamentation may also be referred to as muri and grooves. The sculpturing elements are elongated sideways and the length is at least two times greater than breadth. In striate pattern muri/ridges are sometimes called valla (sing. vallum) or lirae (sing. lira). The inter-ridge spaces are 'streak-like furrows' termed striae (sing. stria, streak; Erdtman, 1969). Grains having striae are described as striate pollen. Example: *Menyanthes*, *Saxifraga*, *Potentilla*, *Gentiana* and *Helianthemum* etc.

• *Cicatricose* (L. *cicatrix* means a scar—Fig. 4.11F): The surface of a pollen grain is provided with stripes or ridges. The ridges run more or less parallel in the manner of a fingerprint over the surface of grain for a long distance. The stripes are not arranged in a net. The ridges are narrower than the spaces that separate them. Example: *Anemia phyllitidis*, *Cicatricosisporites*.

• *Reticulate* (L. *rete* means net—Figs. 4.6E, F, 4.11K & L, 4.9B): The surface of a pollen grain is provided with ridges. The ridges anastomose and form reticulum that may be irregularly polygonal or regularly hexagonal. The ridges, also termed muri, are narrower or equal than the spaces they enclose. The enclosed space is termed as lumen (pl. lumina). Lumina are more than one micron in diameter. Example: *Fritillaria*, *Potamogeton*, *Hedera* and Cruciferae.

• *Corrugate* (L. *corrugatus* means wrinkled—Fig. 4.11G): The surface of spores is provided with projections or ridges. The ridges are with regular or irregular bulges or radial humps. Example: *Riccia natans* (Tschudy *et al.*).

• *Foveolate* (L. *fovea* means a pit—Figs. 4.11H & 4.12): The surface of a spore is provided with more or less circular pits. The diameter of pit is equal to or more than one micron. Sometimes the lumina or cavities may be up to two micra in diameter, or if larger, are too widely separated to form a reticulum. The spaces, i.e. the ridges that separate the pits are wider than the diameter of pit. Foveolae

Figure 4.12

Diagrams illustrating the sculpturing types visible in surface view. The raised areas are shown light, lower areas and the holes are shown dark. Adapted after Moore *et al.*, 1991, p. 76.

are larger than scrobiculae. Example: *Lycopodium phyllanthum*, *L. selago*, *Tilia* and *Fagopyrum* etc.

• *Fossulate* (L. *fossa* means a ditch – Fig. 4.11I): The surface of spore is provided with grooves. The grooves are irregular on the surface and do not anastomose. They may be straight or sinuous. The grain has a reticulate appearance. The boundary of the reticulum is formed by grooves. The spaces between the boundary are elevated sexine elements. Such reticulum is described as negative reticulum. A true reticulum differs in having ridged muri and the depressed lumina. Example: *Pteris tripartia*, *Ledum palustre* and *Huperzia selago*.

• *Canaliculate* (L. *canalis* means a channel – Fig. 4.11J): The surface of a spore and pollen grain is provided with channels or cavities. Raised stripes occur between the channels. The cavities or channels are narrower than the raised stripes that separate them. Example: *Anemia simii* (Tschudy *et al.*).

• *Echinate* (L. *echinatus* means prickly – Fig. 4.10J & K): This term describes a pollen grain with echinae. Echinae are processes that have broad base that tapers towards apex to a sharp point. They may vary from cylindrical to cone-shaped. The processes are more than one micron high. The term is used as a synonym of spina. Example: *Taraxacum*, *Malva*, *Cirsium* and *Lonicera* etc.

In describing sculpture, it is customary to record the size range of sculptural elements. This includes height and breadth of spinae, verrucae etc. In describing reticulum, diameter of lumina and the width of muri, shapes of lumina (polygonal/ rounded/hexagonal etc.) together with the nature of sides (converging/ diverging /parallel etc.) are to be recorded.

The ornamentation can be grouped into two broad categories, namely Excrescence type (e.g. ehinate, baculate and gemmate etc.) and Depression type (e.g. foveolate, scrobiculate and fossulate etc.). Nair (1970) discussed about the relative primitiveness of the two types and concluded that 'the two categories represent two parallel lines of morphological evolution' – p. 42.

Faegri *et al.* (1989, p. 238) formulated a key to identify some common sculpturing types, which are reproduced in Box 4.1.

It is regarded that the sculpturing of exine of a pollen grain has functional significance. It is expected that the conspicuous processes that occur on exine are associated with adhesion to pollination vectors. It is assumed that pollen grains with psilate exine are anemophilous whereas the grains with other sculptured exine are entomophilous. But this is not applicable in all cases. As for example, pollen grains of Araceae have psilate, echinate, verrucate, striate and reticulate exine. It is observed that the psilate grains are pollinated by slow moving beetles. The psilate grains get stuck to the smooth surfaces of pollinators with the aid of sticky secretions from the stigmas. A few entomophilous pollen grains of Araceae have their own adhesive materials, referred to as pollenkitt that is not evident in acetolysed pollen. The echinate grains are largely pollinated by flies and it is regarded that such grains are better adapted for attachment to the hairy bodies of fast-flying flies. For more details consult Grayum (1986).

Box 4.1 depicting key of sculpturing types. Reproduced from Faegri *et al.* (1989, p. 238).

$A_2$ No distinct sculpturing elements sensu stricto present.
- $B_3$ Surface pitted, diameter of pits less than 1μm ................................psilate.
- $B_2$ Surface pitted, diameter of pits equal to or more than 1μm .......foveolate.
- $B_1$ Surface with grooves ......................................................................fossulate.

$A_1$ With sculpturing elements sensu stricto
- $C_2$ Radial projection of sculpturing elements ± isodiametric.
  - $D_2$ No dimension equal to or larger than 1μm.........................scabrate.
  - $D_1$ At least one dimension equal to or larger than 1μm.
    - $E_2$ Sculpturing elements not pointed.
      - $F_2$ Lower part of elements constricted.
        - $G_2$ Greatest diameter of radial projection equal to or greater than height of element...........gemmate.
        - $G_1$ Height of element greater than greatest diameter of projection..............................................clavate.
      - $F_1$ Lower part of element not constricted.
        - $H_2$ Greatest diameter of radial projection equal to or greater than height of element..........verrucate.
        - $H_1$ Height of element greater than greatest diameter of projection.............................................baculate.
    - $E_1$ Sculpturing elements pointed................................echinate.
- $C_1$ Radial projections of sculpturing elongated (length at least twice the breadth).
  - $I_3$ Elements irregularly distributed........................................rugulate.
  - $I_2$ Elements ± parallel...........................................................................striate.
  - $I_1$ Elements form a reticular pattern....................................reticulate.

## LO-analysis

LO-analysis is a means to elucidate the details of sexine pattern on the basis of light microscopic observations of a pollen grain and spore at different adjustments of the microscope. The ultra thin section (0.1-0.5 micron thick) and optical section of a pollen grain provide details of sexine. The term optical section refers to the view of a pollen grain and spore under light microscope when the focal plane is halfway through the grain. In optical section thickness, structure, layers and

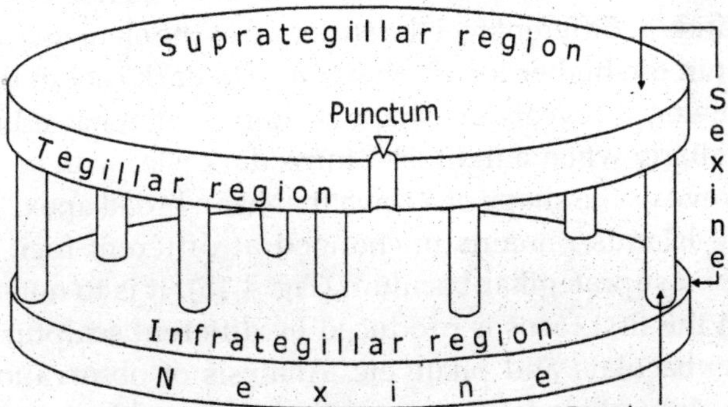


Figure 4.13

Schematic diagram showing the different regions of sexine in three dimensional view.

ornamentation etc. of a grain wall are visible. Sometimes optical section does not make the fine structures of exine as clear as is expected. The advantage of LO-analysis is that the details of sexine can be established with full clarity in unsectioned pollen and spore.

Erdtman (1952) illustrated LO-analysis where 'L' is derived from the Latin word *lux* meaning light and 'O' is derived from the Latin word *obscuritas* meaning darkness. In sexine of angiosperm pollen, three regions are encountered – suprategillar, tegillar and infrategillar (Fig. 4.13). The term tegillum (syn. tectum) denotes the layer of sexine that forms a roof over columella and baculum etc.

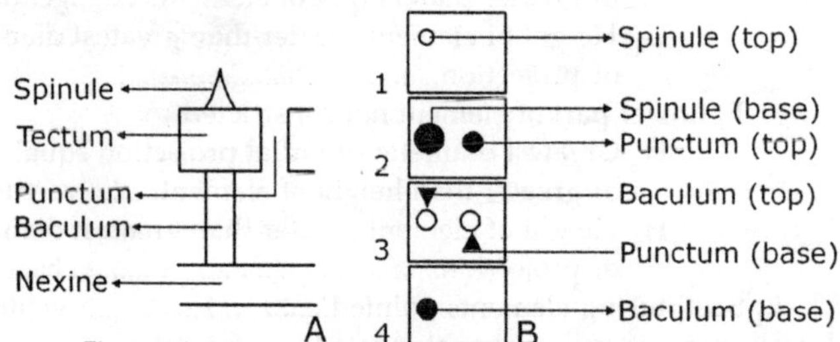


Figure 4.14

Diagram illustrating LO-analysis. A. Layers of exine. B. Light and dark islands that appear from high adjustment to low adjustment of light microscope. 1, 2, 3 & 4 represent first, second, third and fourth focus of microscope from high to low adjustments.

For LO-analysis (Fig. 4.14) a pollen grain is focused under light microscope. The objective is raised until the pollen grain is not in the focus. Then the objective is slowly moved down until details are visible. At this high adjustment the suprategillar elements are visible only. The elements appear as light islands. The objective is further racked down and the light islands turn to dark islands. So the suprategillar elements exhibit LO-pattern, i.e. light islands at first focus become dark when focused at a slightly lower level (Figs. 4.14A & B, 4.16A & B). The light and dark islands reveal the cross sectional view of elements. On analyzing the light and dark islands at different foci the structure of the element can be deduced. A small light islet in the higher focus when turns to dark islet in the lower focus with greater diameter, it reveals the shape of spinula that has a broad base and pointed apex. Similarly when a light islet turns dark islet with smaller diameter at lower focus, it reveals the shape of a clava that has a broad apex. If the diameter of light and dark islands remains unchanged at different foci, it reveals the columnar shape of a suprategillar baculum (Fig. 4.15). It is to note that the same surface pattern in the first focus is produced by different sculpturing types, e.g. gemmate, clavate, baculate and pilate etc. Analysis of observations of second, third and if necessary other lower foci will elucidate the shape of projecting processes. When the objective is at the focusing level of the base of suprategillar elements, it also focuses other structure present on tegillar surface. The

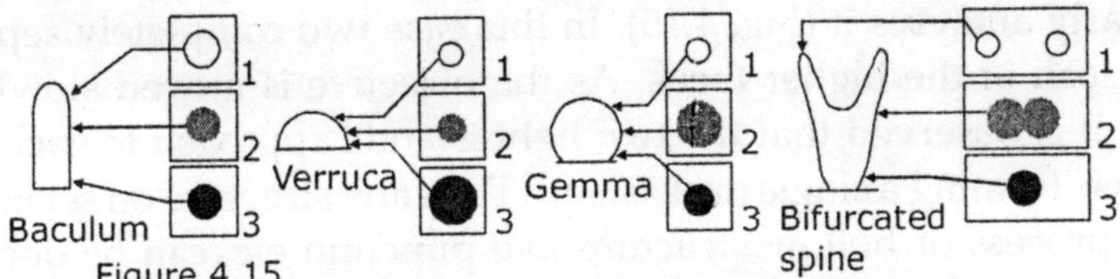


Figure 4.15

Diagrams illustrating LO-analysis of different types of supratectal processes. 1., 2. & 3. represent first focus, second focus and third focus respectively of LO-analysis.

tegillum may be punctate, scrobiculate, reticulate etc. When the objective is focused on the suprategillar side of punctum, scrobiculus, reticulum etc. all appear as dark islands. In lower focus the dark island turns bright. So the hollow structures like punctum, reticulum and scrobiculus etc. exhibit OL-pattern, i.e. dark islands at suprategillar side turn bright on infrategillar side (Figs. 4.16C & D). The size, shape and diameter etc. of dark and bright islands differentiate between punctum, scrobiculus and reticulum etc. When the objective is moved further down and adjusted at the top level of infrategillar region small white islands appear. These are caused by the bacula standing on nexine. They later become dark at further low adjustments. In intectate pollen bacula, clava, pila etc. directly stand on nexine. The infrategillar elements exhibit LO-pattern. The

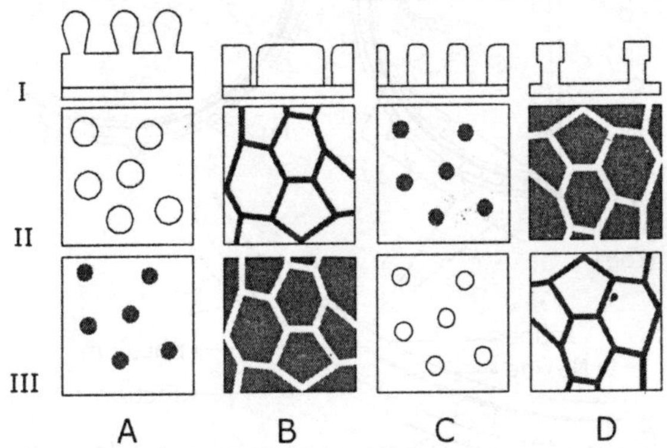


Figure 4.16

Diagrammatic illustrations of LO-pattern (A & B) and OL-pattern (C & D) in LO-analysis. I. Types of exine ornamentation in optical section. II. Ornamentation in surface view at high adjustment of microscope. III. Ornamentation in surface view at low adjustment. A. Clavate exine. B. Exine with negative reticulum. C. Punctate / scrobiculate exine. D. Reticulate exine.

patterns exhibited on the different layers of sexine are referred to as S-, T- and I-pattern where S = suprategillar, T = tegillar and I = infrategillar. It is to note that solid objects like spine, clava and baculum etc. exhibit LO-pattern whereas the hollow structures like punctum, scrobiculus and reticulum etc. reveal OL-pattern. Transitional forms are observed when the objective is moved to lower focus slowly. Sometimes a supratectal process may be bifurcated at the tip. LO-analysis clearly analyses it (Fig. 4.15). In this case two completely separate light islands are seen at the higher focus. As the objective is moved slowly towards lower focus it is observed that the two light islands approach to each other and gradually fuse forming a single dark island. Thus any structure on sexine, whether it is a solid process or hollow structure like punctum etc. can be demonstrated with full clarity by LO-analysis.

The S-, T- and I-patterns are depicted in diagrams. This diagram elucidates the sexine pattern of a pollen grain and spore on the basis of LO-analysis and becomes a part of palynogram. A palynogram (Fig. 4.17) is a diagrammatic representation of a pollen grain and spore illustrating the shape, size, aperture, sporoderm stratification, sexine patterns etc. (Erdtman, 1952).

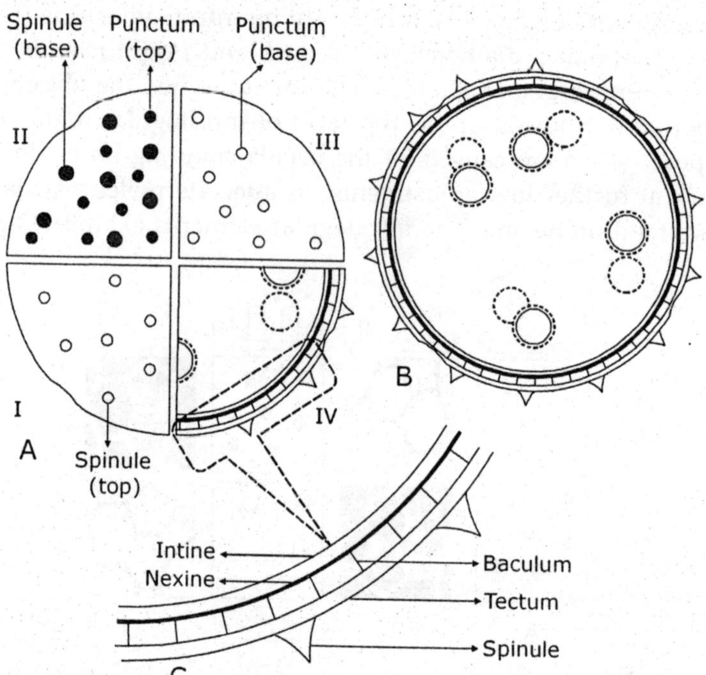


Figure 4.17

Diagram illustrating palynogram. *Drypis spinosa*. A. Pollen in four different foci. I. First focus showing top of spinule. II. Second focus showing base of spinule and top of punctum. III. Third focus showing base of punctum. IV. Portion of pollen in optical section that is magnified in C. B. General view of pollen showing the distribution of apertures and optical section. C. A portion of pollen showing layers of exine and intine. Redrawn after Roy and Chanda, 1992.

Pollen grains and spores are acetolysed for LO-analysis. After acetolysis the grains are mounted in glycerine jelly that has refractive index 1.474 (Chanda, 1966, p. 357). This value is lower than the refractive index of sporopollenin in exine (Moore *et al.*, 1991, p. 77). A difference in refractive index eases to detect a solid structure and its surroundings not only under phase contrast microscope but also in light microscope. For observation in light microscope oil immersion objective greatly aids in the visualization of exine stratification. The observations are clearer in a binocular microscope. A flat or slightly curved surface on apocolpium/mesocolpium/apoporium/mesoporium is convenient for observation (Fig. 4.18).

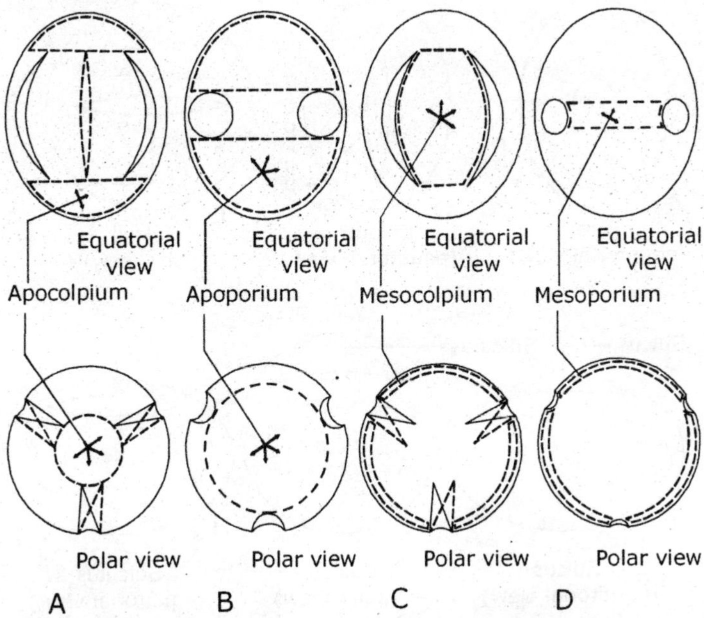


Figure 4.18

Diagrams showing how apertures partition the surface of a pollen grain into polar and non-polar areas. The areas are termed for descriptive purposes which are as follows : A. Apocolpium: The term denotes the area on the poles (distal or proximal) of a zonocolpate pollen grain. The lines joining the apices of colpi delimit the area. B. Apoporium: The term denotes the area on the poles (distal or proximal) of a zonoporate pollen grain. The area is pole-ward part that begins from the pole-ward limits of pores. C. Mesocolpium (pl. Mesocolpia): The term denotes the non-polar area on pollen surface that occurs between two adjacent colpi of a zonocolpate pollen grain. Transverse lines drawn connecting the ends of colpi delimit the area. D. Mesoporium (pl. Mesoporia): The term denotes the non-polar area on pollen surface that occurs between two adjacent pores of a zonoporate pollen grain. Transverse lines drawn through the polar edges of pori delimit the area.

It is to note that a punctum and a baculum both may appear as dark islands but at different focal planes. So the structure of exine is to be deduced by concentrating attention on a small area of exine and carefully observing the appearance of different images at different focal planes in that area.

Now a days Scan Electron Microscope (SEM) is introduced to analyze the sexine surface pattern and to assemble three-dimensional structures of certain

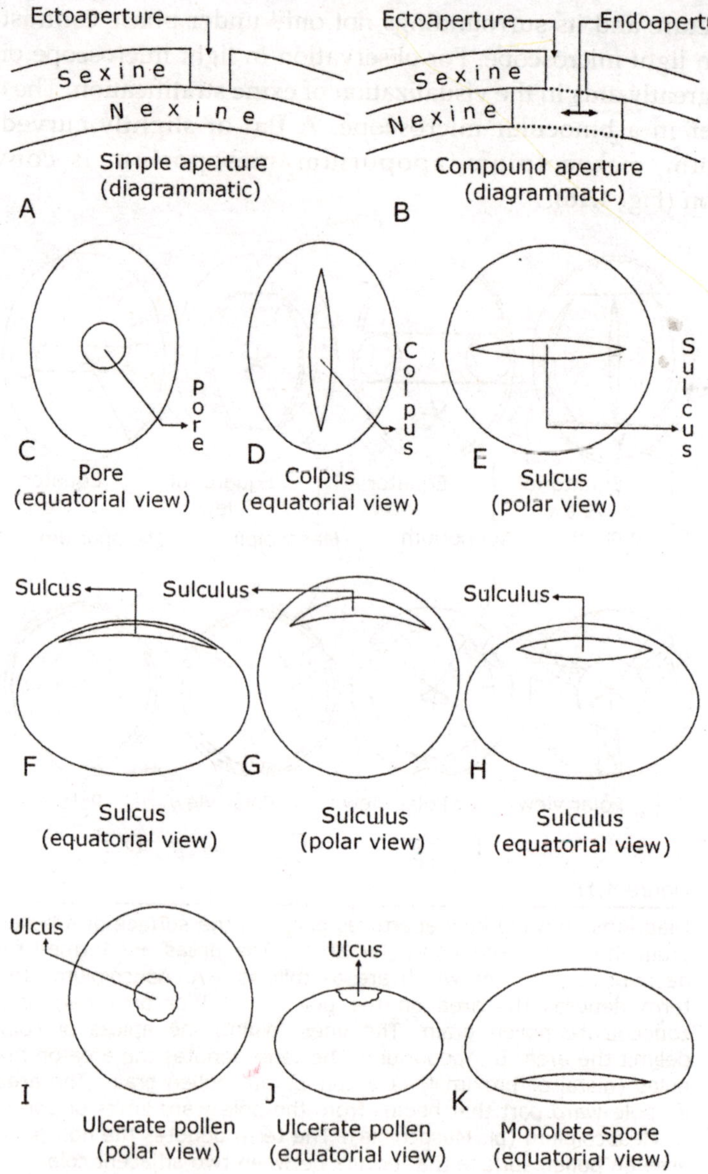


Figure 4.19

Diagrammatic representation of different types of aperture. Adapted after Punt *et al.*, 2007.

pollen types. Still the light microscopic observations on the basis of LO-analysis are an indispensable tool because SEM images do not provide clearly the structure of endoaperture. The structure can be established with full clarity on the basis of LO-analysis.

## Aperture

It is a weak, preformed locus on the exine of spore and pollen through which the intra-exinous substance exits and pollen tube emerges during pollen germination (Erdtman, 1952).

Apertures may be absent from a pollen grain or spore when they are termed as inaperturate/atreme/non-aperturate (ex. *Potamogeton*). Some non-aperturate pollen grains belonging to the family Euphorbiaceae are omniaperturate. The sexine of omniaperturate pollen is either absent or very thin. The intine is thicker than exine. It is regarded that the whole surface of pollen wall is apertural in nature (Thanikaimoni, 1984).

In aperturate pollen at the region of aperture the exine is either missing or greatly reduced. When reduced the exine represents a thin layer over aperture. The intine layer is thicker under apertures. The intine, in the aperture, may have a covering on the nexine side – termed aperture membrane. Sometimes granular elements may occur on the membrane forming granulated aperture membrane. An aperture may be formed on sexine only or on both sexine and nexine. In the former the aperture is described as simple. In the latter the aperture is compound or composite. The aperture formed on sexine is called ectoaperture (= ectopore). In contrast the aperture formed on nexine is termed as endoaperture (= endopore). Ecto- and endoaperture together constitute a compound or composite aperture (Fig. 4.19A & B).

## Simple aperture

• *Pore / porus* (n., pl. –s / pori, adj. porate – Fig. 4.19C): Pore is rounded or elliptic (ratio between length and breadth is less than two), more or less distinct aperture of a pollen grain. As a rule they are equally distributed on the equatorial plane or over the general surface of a pollen grain. The number of apertures is indicated by the prefixes mono-, di-, tri-, tetra-, penta-, hexa- and poly-. A pore is formed on sexine only. At poral region continuity of sexine is interrupted, the nexine being continuous. This is simple aperture and ectoaperture as it is present only in one layer, i.e. sexine. Through the pore pollen tube emerges during pollen germination and intra-exinous material exits.

• *Colpus* (n., pl. colpi, adj. colpate – Figs. 4.19D & 4.21M): Colpus is an elongated aperture (the ratio between length and breadth is greater than two) of a pollen grain. Colpi are equally spaced on the equatorial plane or over the general surface of a pollen grain. In dicotyledon a colpus is a longitudinally elongated meridonial furrow that crosses the equatorial plane at right angle. Exceptions are noted in some tetra-, hexa- and polycolpate grains where colpi do not always cross the equator. In palynology the term 'meridonial' is used to describe those longitudinal features of pollen surface that occur perpendicular to equatorial axis. The number of colpi is indicated by the prefixes mono-, di-, tri-, tetra-, penta-, hexa- and poly-. A colpus is an ectoaperture and formed on sexine only. It is a simple aperture as it is present only in one layer, i.e. sexine. The sexine is interrupted at the region of colpus and the nexine is continuous. Colpus is the germinal apparatus of extant pollen and may function during the emergence of pollen tube. In extant pollen the colpus is the normal exit of intra-exinous substance only. It is interpreted that during evolution a colpus, by shortening and contraction developed to a pore.

• *Sulcus* (n., pl. sulci, adj. sulcate – Fig. 4.19E & F): Sulcus is an elongated latitudinal ectoaperture present on the surface of most monocotyledon and primitive dicotyledonous pollen crossing the polar axis at right angle. It has the shape of a colpus, i.e. the ratio between length and breadth is greater than two. It differs from colpus on the basis of orientation. It occurs as a latitudinal aperture in contrast to colpus, which is longitudinal aperture. In palynology the term

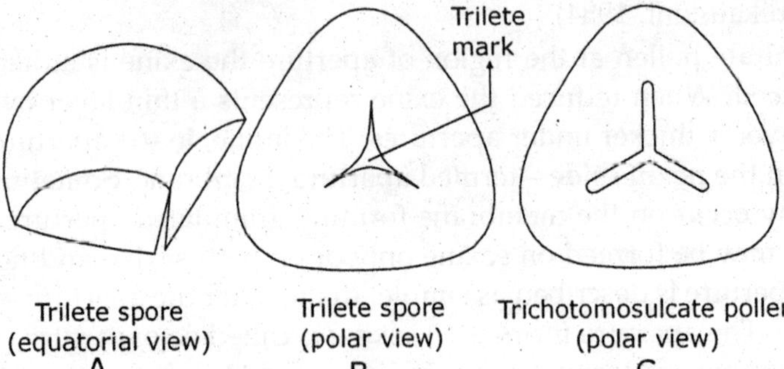


Figure 4.20

Diagrammatic representation of different types of aperture. Adapted after Punt *et al.*, 2007.

'latitudinal' is used to describe the features that run parallel to equatorial axis of a pollen grain. In most monocotyledons the term sulcus is restricted to distal aperture of a pollen grain only. Sulcus with its centre may occur at the distal pole or on proximal pole of a pollen grain and accordingly termed as anasulcate and catasulcate. A pollen grain with one sulcus is called monosulcate. A pollen grain with two sulci is termed as disulcate. Example: *Metroxylon*, *Chamaerops* (Palmae) etc. In some pollen the sulcus may be in the form of three or four-slit like opening. Such pollen grains are respectively termed as trichotomosulcate (Fig. 4.20 C) and tetratomosulcate. It is simple aperture as it is present only in one layer, i.e. sexine.

• *Sulculus* (n., pl. sulculi – Fig. 4.19G & H): The term sulculus describes a latitudinal aperture of a pollen grain. The aperture is an elongated furrow and situated more or less parallel to equator near a pole, generally distal. It is not centered at the distal pole like sulcus. Sulculi may unite encircling a grain thus forming a ring. Example: *Eupomatia*.

• *Lete* (Erdtman, 1969): Lete is always used as a suffix to denote the absence or presence of laesura(e). Laesura is the scar mark of a spore. The mark represents the original contact of spores at their tetrad stage. So laesura is always proximal. A laesura comprises a commissure – the line of dehiscence of spore along which the spore germinates. A spore is termed as monolete (Fig. 4.19K) when there is one un-branched straight tetrad scar on the proximal side (ex. *Psilotum*, *Dryopteris*). A spore is called trilete (Fig. 4.20A & B) when there is a three-pronged laesura resembling the English letter capital 'Y' on the proximal side (ex. *Lygodium, Pteridium* etc.). Palynologists regard that trilete spores have a single tri-radiate or three-branched laesura. A spore is termed as alete when there is no laesura. An aletė spore has completely smooth surface that is devoid of dehiscence marks and tetrad scar. The spore shows no recognizable characters after separation from tetrad. Dilete spores are rare (Traverse, 1988). Pollen grains of *Simethis* (Family: Liliaceae) exhibit a three-slit aperture resembling a trilete scar, which is very rare in angiosperm.

• *Pantoporate* (adj., syn. periporate—Fig. 4.21A): The term pantoporate describes a pollen grain with pores spread over the surface. Panto- (syn. pan-, peri-) is a prefix that illustrates global distribution. Pores are more or less circular where the ratio between length and breadth is less than two. The number of pores is always more than four and may be as high as 80. The pores remain evenly distributed on the surface of a pantoporate pollen grain (ex. Chenopodiaceae). Though the terms polyporate and periporate are synonymous, the former is favoured (Traverse, 1988). The term pantoporate is more widely used than other synonyms.

• *Ulcus* (pl. ulci, adj. ulcerate—Fig. 4.19I & J): Ulcus is an ectoaperture and a thin place on exine. The aperture is more or less like a circular pore, the ratio between length and breadth is less than two. The outline of a pore is irregular and often torn open widely (Traverse, 1988 and Chanda, 1966). Example: Some members of Restionaceae. Ulcus occurs either on distal or proximal pole. Accordingly a pollen grain is described as ana-ulcerate and cata-ulcerate.

• *Pantocolpate* (adj., syn. pericolpate—Fig. 4.21B): The term pantocolpate describes a pollen grain with colpi spread over the surface. Panto- (syn. peri-) is a prefix that illustrates global distribution. Colpus is an elongated furrow where the ratio between length and breadth is greater than two. The number of colpi is more than three and they remain evenly distributed over the surface in a pantocolpate pollen grain. Example: *Portulaca, Drymaria* etc.

• *Trichotomosulcate* (Greek tricha = in three parts, tome = cut—Fig. 4.20C): The term trichotomosulcate describes a monsulcate pollen grain in which the sulcus is three-branched. It appears more or less triangular and simulates a trilete laesura (Traverse, 1988). It is an ectoapertue and latitudinal aperture. In palynology the term 'latitudinal' is used to describe the features that run parallel to equatorial axis of a pollen grain. Example: *Elaeis guineensis* (Palmae).

• *Syncolpate* (adj.—Fig. 4.21C & D): The term syncolpate describes a pollen grain where the colpi run together at the poles (Erdtman, 1969). It appears that the ends of colpi join/anastomose normally near distal and proximal poles. Syncolpate pollen have two or more simple ectoapertures that are elongated furrows (the ratio between length and breadth is greater than two). The prefix 'syn-' is used to indicate the fusion/anastomosis of features. Example: *Illicium, Pedicularis* etc.

• *Parasyncolpate* (adj.—Fig. 4.21E & F): The term parasyncolpate describes a pollen grain where the apices of colpus divide into two branches and the branches anastomose / join with that of other colpus near the proximal and distal pole.

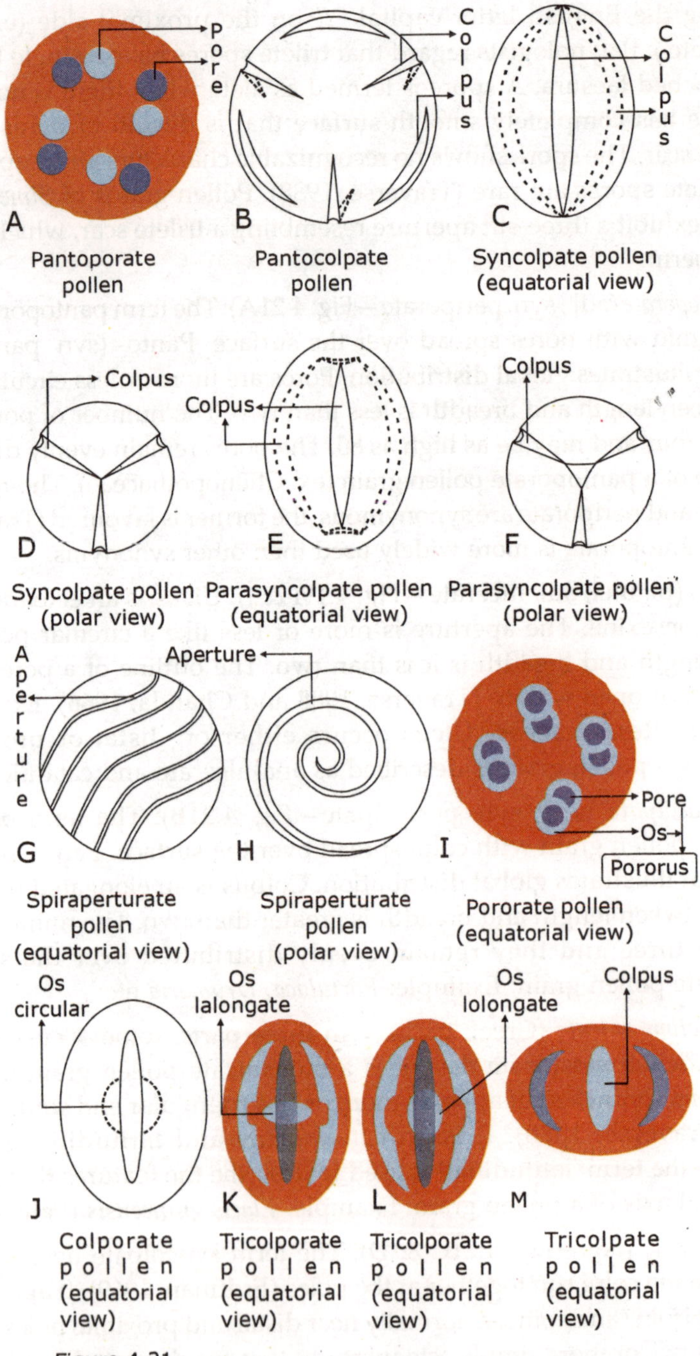


Figure 4.21

Diagrammatic representation of different types of aperture. Adapted after Punt *et al.*, 2007.

The branches join in such a way that they delimit an isolated area called apocolpial field. The apocolpial field normally has a definite shape that is usually triangular. Example: *Eugenia* (Myrtaceae), *Nymphoides* (Menyanthaceae) etc.

• *Brevicolpate* (adj.): The term brevicolpate describes a pollen grain where the colpus is short in length. The prefix 'brevi-' is used to indicate a short feature. The distance between proximal pole and adjacent apex of colpus together with the distance between distal pole and adjacent apex of colpus is greater than the length of the colpus in a brevicolpate pollen grain. Example: *Dillenia* (Bhattacharya *et al.*, 2006).

• *Leptoma*: Leptoma is a thin region on exine present on the distal pole of a pollen grain through which pollen tube emerges (Traverse, 1988). Example: *Corollina*.

• *Pseudoaperture* (n., adj. pseudoaperturate): Pseudoaperture is an area on the exine, which resembles an aperture. In this area the exine is thin and intine is also thin. Normally the intine is thicker where exine is thin. It is presumed that pseudoaperture is not the site of emergence of pollen tube (Thanikaimoni, 1980). Pseudoaperture may be in the form of an elongated furrow and a pore. Accordingly they are termed as pseudocolpus and pseudopore. In pseudocolpus (pl. pseudocolpi, adj. pseudocolpate) the pseudoaperture is colpus like, i.e. the ratio between length and breadth is more than two. Example: *Myosotis* (Boraginaceae). In pseudopore (pl. pseudopores, adj. pseudoporate) the ratio between length and breadth of aperture is less than two. Traverse (1988) defines the thin area present in leptoma as pseudopore. Example: Pollen of Cupressaceae and Taxaceae , and the fossil family Cheirolepidiaceae.

• *Spiraperturate* (adj.—Fig. 4.21G & H): Spiraperturate pollen grains have one to several colpi. The colpi are fused giving the appearance of a set of spirals surrounding the whole pollen grain. The apertures may be irregular in outline, irregularly spaced and spiral around the pollen grain. Example: *Eriocaulon aquaticum* (Eriocaulaceae) and *Crocus* etc.

## Compound/Composite aperture

• *Pororate* (adj., n. pororus Fig. 4.21I): The term pororate is used to describe a pollen grain that has compound aperture consisting of ectoaperture and endoaperture where both ecto- and endoaperture are pores and they are not congruent. Ectoaperture is the pore (the ratio between length and breadth is less than two), which is formed on sexine. The endoaperture is also a pore that is formed on nexine. The endoaperture is also called os (pl. ora, adj. orate). Pore with os forms the pororate pollen. This term is commonly used in adjective form. Example: *Casuarina equisetifolia*.

• *Colporate* (adj., sing. colporus, pl. colpori—Fig. 4.21J, K & L): The term colporate is used to describe a pollen grain that has compound aperture consisting ectoaperture and endoaperture where ectoaperture is colpus and the endoaperture is os. Colpus is formed on sexine and os is formed on nexine. Colpus with os forms the colporate pollen grain. Example: Pollen grains of Compositae and Solanaceae etc.

The term is commonly used in adjective form. In a colporate pollen grain the colpus is meridonial aperture, i.e. it occurs parallel to polar axis. The endoaperture os may be longitudinally elongated (Fig. 4.21L) termed lolongate (ex. *Cassia sofera*, *Mangifera indica*). The endoaperture may be also latitudinally elongated

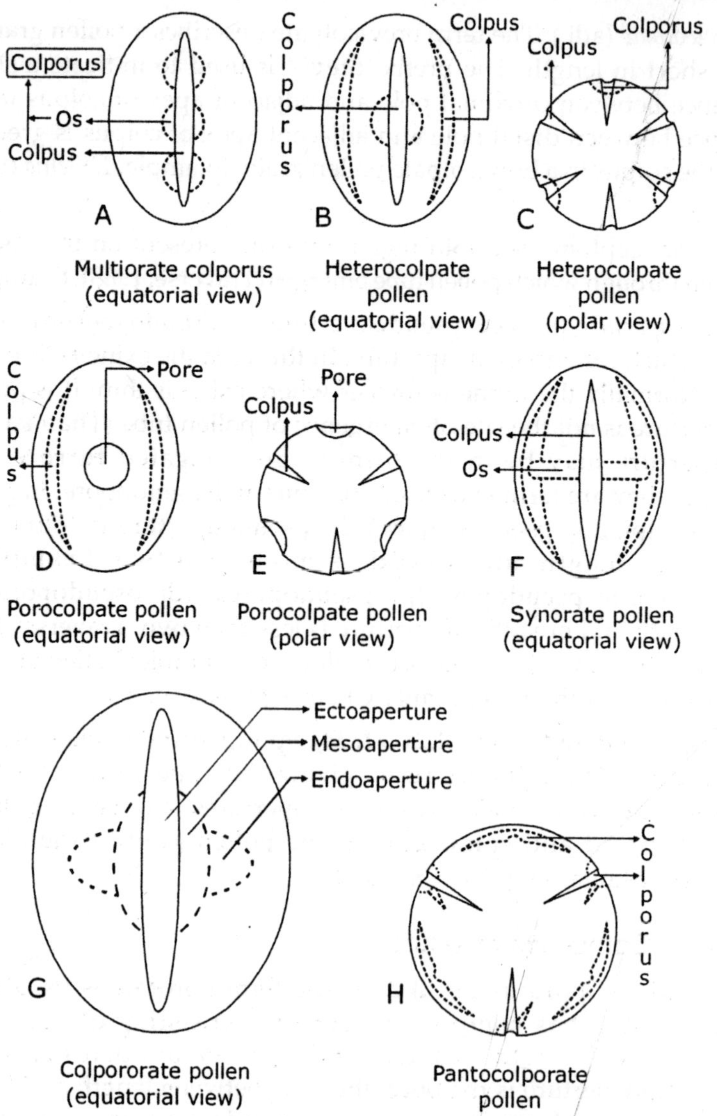


Figure 4.22

Diagrammatic representation of different types of aperture. Adapted after Punt *et al.*, 2007.

(Fig. 4.21K), i.e. elongation occurs perpendicular to polar axis – termed lalongate (ex. *Datura metel*, *Lippia alba*), Banik *et al.* (1986). Apart from meridonial position colpori may spread over the surface of a pollen grain. Such grains are described as pantocolporate pollen (Fig. 4.22H).

• *Multiorate* (adj. – Fig. 4.22A): The term multiorate describes a colporate pollen grain where the number of ora is more than two in a colpus. Example: *Congea* (Bhattacharya *et al.*).

• *Colporoidate* (adj.): The term colporoidate describes a colporate pollen grain where the os is indistinct in a distinct colpus, i.e. a colporate pollen grain with a weakly developed pore. In a colporoidate pollen grain the pore is either absent or not clearly recognizable (Traverse, 1988). Example: *Scoparia dulcis, Lindernia crustacea* etc.

• *Colpoidorate* (adj.): The term colpoidorate describes a colporate pollen grain where the colpus is indistinct and the os is distinct. Example: *Alangium villosum* (Bhattacharya *et al.*).

• *Heterocolpate* (adj.—Fig. 4.22B & C): The term heterocolpate describes a pollen grain where both simple aperture colpus and compound aperture colporus are present. Example: *Lythrum, Peplis* etc.

• *Porocolpate* (adj.—Fig. 4.22D & E): The term porocolpate describes a pollen grain where the simple apertures both colpus and pore occur, and their arrangement is alternate to each other round the equator. Example: *Pardoglossum* (Boraginaceae).

• *Colpororate* (adj.—Fig. 4.22G): The term colpororate describes a compound aperture of a pollen grain consisting an ecto-, endo- and mesoaperture. The ectoaperture is an elongated furrow and simulates a colpus. The endoaperture is lalongate. The mesoaperture is lolongate and short. Thanikaimoni (1980) defines mesoaperture as the middle part of a composite aperture. In this region there exist also an ecto- and endoaperture. Example: *Sonchus* (Compositae).

• *Synorate/synclinorate* (Fig. 4.22F): The terms synorate/synclinorate describe an os of a colporate pollen grain. The os is latitudinally elongated and joins with other of similar os. Example: Polygalaceae, *Krameria*.

In many pollen grains the margin of aperture is distinct from the remainder of exine. The margin may be thin or thick, or may have different ornamentation. The followings are a few common examples.

• *Annulus* (n., pl. annuli, adj. annulate): Annulus is an area on sexine that occurs encircling a pore (Fig. 4.23A). The area can be sharply differentiated from the rest of sexine. The area may be elevated (Fig. 4.23B) or depressed (Fig. 4.23C), or may have different sculpturing than the remaining part of sexine. The area appears as a halo surrounding a pore. Example: Poaceae. Some members of Caryophyllaceae have annulus with thinned sexine.

• *Margo* (n., pl. margines, adj. marginate—Fig. 4.23D): Margo is an area on sexine that occurs encircling a colpus. The area can be sharply differentiated from the rest of sexine. The area may be elevated or depressed, or may have different sculpturing than the remaining part of sexine. The area appears as a halo surrounding a colpus. Example: *Cassia sophera* (Banik *et al.*).

• *Costa* (n., pl. costae, adj. costate, costal [Traverse, 1988]—Fig. 4.23E, F & G): Costa is a rib-like thickening that occurs on endonexine. It is associated with colpi. Costae are usually meridonial and border colpi. The thickening may also occur following the outline of an ectoaperture. Example: *Melia.* Costae are best observed in equatorial view of a pollen grain (Traverse, 1988). Punt *et al.* (2007) reported the presence of costa in pore also.

• *Operculum* (n., pl. opercula, adj. operculate—Fig. 14.23H, I & J): Operculum is a thickening of pore membrane that can be observed distinctly and measured. It is an isolated part of sexine and separated from the rest by a narrow zone where sexine is absent. So the structure of operculum is similar to that of the

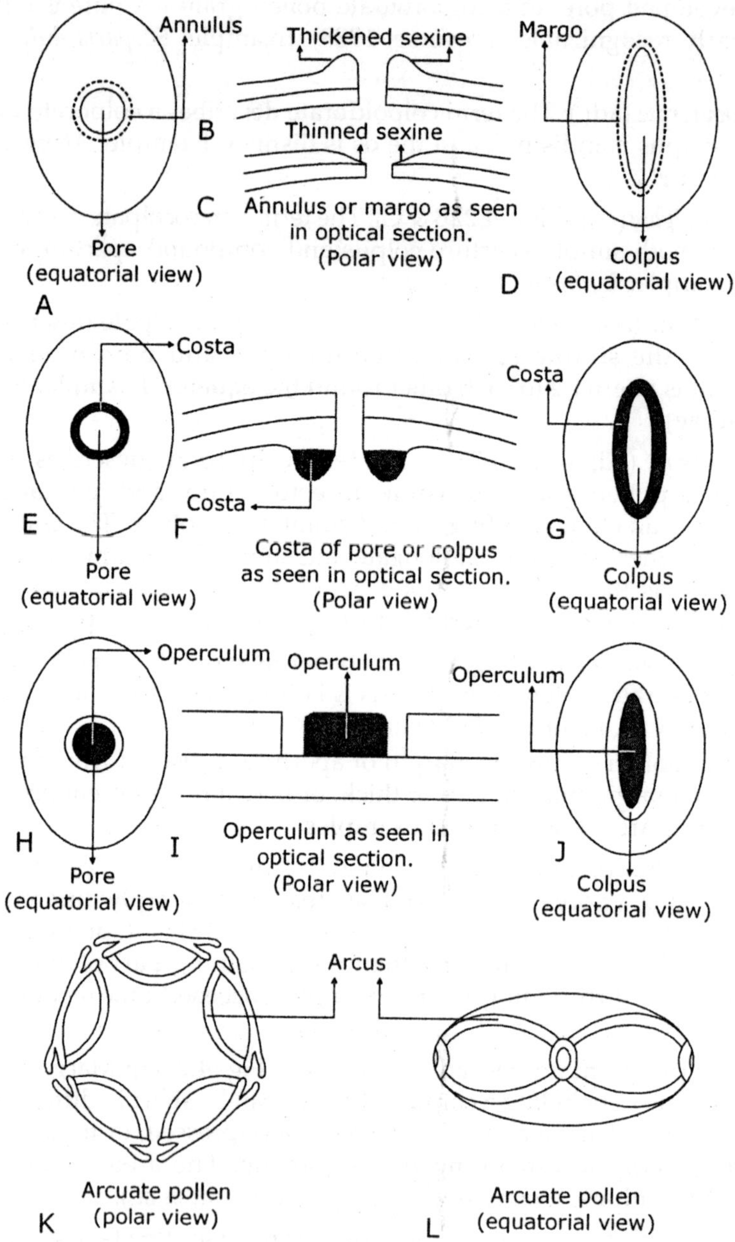


Figure 4.23

Diagram illustrating different types of margin of apertures (adapted after Punt *et al.*, 2007).

remaining exine. Operculum covers aperture partially or wholly. It remains surrounded by aperture. So opercula are more or less circular in pore and elongate in colpi. The zone of operculum is less strong than the rest of exine. A pollen grain ruptures along operculum when pressure is exerted from inside. In fossil pollen pore-opercula are frequently lost.

• *Arcus* (adj. arcuate—Fig. 4.23K & L): The term arcus describes a kind of thickening of sexine. The thickening is in the form of a band that occurs between the apertures of a pollen grain. The bands are curved or bend like a bow. Example: *Alnus* (Betulaceae), Punt *et al.*.

• *Oncus* (pl. onci—Fig. 4.24A): The term oncus describes a structure that occurs beneath the aperture of certain unacetolysed pollen grains. The structure is lens-shaped and not resistant to acetolysis. Example: *Corylus* (Corylaceae), Punt *et al.*

• *Aspis* (adj. aspidate/aspidote—an orthographical variant of aspidate—Fig. 4.24B & C): The term aspis describes a kind of thickening around a pore. The thickening is in the form of a dome-like protrussion where pore occurs. It differs from annulus in not having any thinning or depression of sexine. Example: *Betula* (Betulaceae), *Dorstenia* (Moraceae), Punt *et al.*

• *Atrium* (Fig. 4.24D): The term atrium describes the shape of a space that occurs within the aperture of a complex pore. The endopore is much wider than

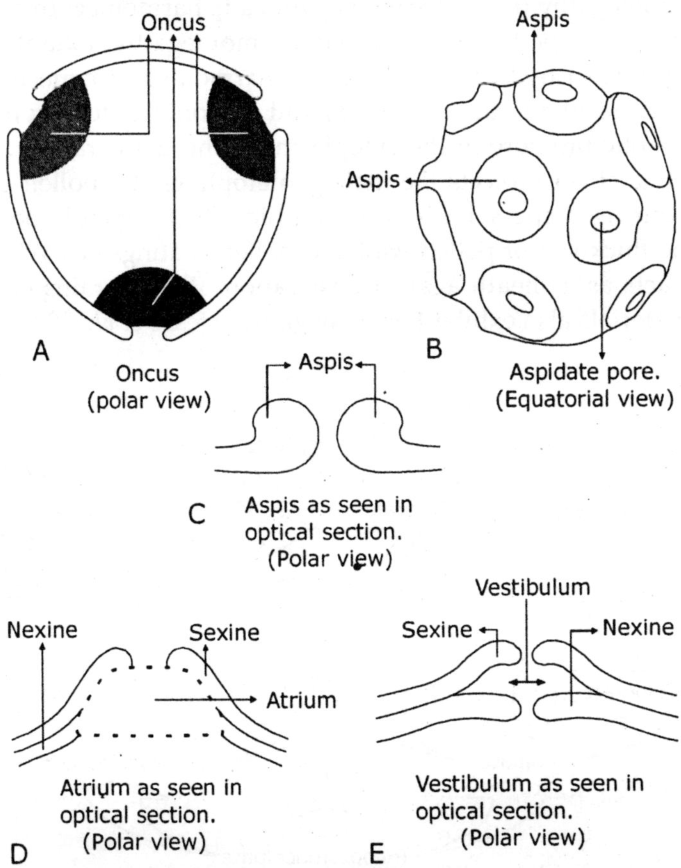


Figure 4.24

Diagram illustrating different types of margin of apertures (adapted after Punt *et al.*, 2007).

the ectopore. As a result the pore canal is narrow towards sexine and wider towards nexine. Example: *Myrica* (Myricaceae), Punt *et al.*

• *Vestibulum* (Fig. 4.24E): The term vestibulum describes a small chamber formed by the sexine and nexine of a pollen grain. Sexine and nexine split apart near a porus thus forming a chamber. The chamber is the site of communication between outside and inside of a pollen grain. Example: *Betula, Alnus.*

The main function of an aperture is to allow passage for the emergence of pollen tube. Polyaperturate grains have better opportunity for the pollen tube to emerge close to stigma surface than monoaperturate one. The apertures also provide exits of proteins that recognize the compatible stigma.

The number of porus, colpus and colporus may be indicated by prefixes di-, tri-, tetra-, penta-, hexa- and so on. It is also of common use to write numbers instead of prefixes, e.g. 3-colpate instead of tricolpate and 4-porate instead of tetraporate etc. Many palynologists prefer to use prefixes. The term pantoaperturate is used to indicate the globally distributed apertures of a pollen grain.

The aperture membrane (e.g. colpus, porus and leptoma etc. of a pollen grain) exhibits harmomegathy (n.; pl. harmomegathi; adj. harmomegathic). Apart from aperture membrane, pollen wall exhibits harmomegathy. Extant pollen grain exists either in dry or hydrated condition. In harmomegathic mechanism pollen wall and aperture membrane either expand or contract to accommodate the change of osmotic pressure in the cytoplasm during dehydration or hydration. Harmomegathic effect protects the male gametophyte of a pollen grain against desiccation during dispersal. Harmomegathic effect depends on pollen size, apertures, the thickness of pollen wall and pollen coatings etc. When coating is abundant it acts as a sheath against desiccation. Pollen grains collected from herbarium materials also exhibit harmomegathy (Hesse *et al.*, 2009, p. 23).

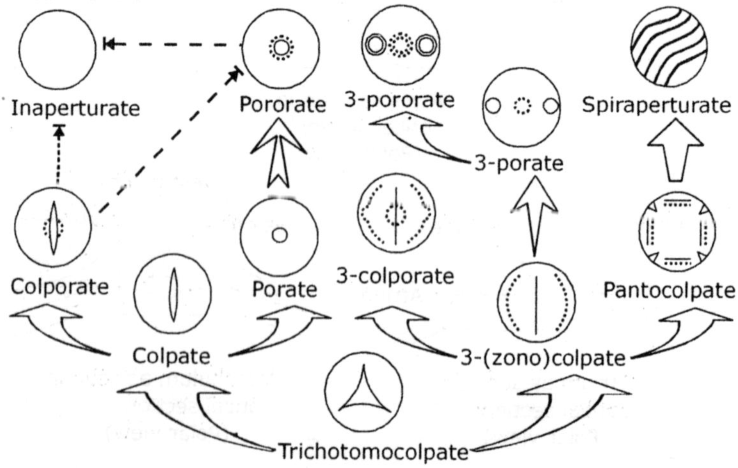


Figure 4.25

Diagram illustrating some of the lines of apertural evolution in pollen and spore. Adapted after Nair, 1970, p. 40.

Box 4.2 illustrating relative primitiveness among apertural forms. Adapted after Nair, 1970, p. 43.

| Primitive | Advanced |
|---|---|
| Trilete<br>Trichotomocolpate<br>Colpate<br>Zonocolpate<br>Pantocolpate<br>Nomotreme (aperture present) | Monolete<br>Monocolpate<br>Porate and colporate<br>Panto-colpate or porate<br>Spiraperturate<br>Atreme (aperture absent) |

Regarding apertural evolution trichotomocolpate (= trilete) form is regarded as the basic type from which other apertural forms (e.g. colpate, porate, colporate, pororate, pantocolpate, pantoporate and spiraperturate etc.) have evolved (Nair, 1970, p. 40). The lines of evolution are illustrated in Fig. 4.25. The relative primitive and advanced apertural forms are depicted in the Box 4.2.

## NPC classification

NPC is an artificial system of classification of pollen and spore based on the three features of aperture only, i.e. number, position and character. Erdtman and Straka (1961) proposed NPC classification and palynologists all over the world accepted it. According to NPC system each pollen grain has an arithmetic cardinal number consisting of three digits. The first digit reveals the absence or presence of aperture, and when present it mentions the total number of aperture(s) present in a pollen grain. The second digit illustrates the position of aperture(s), i.e. distal, proximal, latitudinal, meridonial, equatorial etc. The microspores reveal the position of aperture(s) with full clarity when they are in tetrad. The third digit explains the character of an aperture, i.e. circular / oval or elongated, simple or compound etc. 'N' from number, 'P' form position and 'C' from character of aperture compose the NPC-classification.

• *Classification of aperture based on number(s)*: In NPC system 'N' denotes the number (N; L. *numerus*) of aperture(s) present in a pollen grain. Aperturate pollen, i.e. pollen having apertures are divided into seven groups. The groups are mentioned as $N_1$ to $N_7$. Each group has characteristic number of aperture, i.e. $N_1$ has one aperture and $N_2$ has two apertures and so on. The $N_7$ group has seven or more apertures. $N_1$ to $N_7$ groups are also referred to respectively as monotreme, ditreme, tritreme, tetratreme, pentatreme, hexatreme, and polytreme (Greek trema means hole, opening, aperture; pl. tremata). There are pollen grains where apertures are absent. Such pollen grains are termed as inaperturate or atreme and they are placed in $N_0$ group. Another special group $N_8$—termed anomotreme is created where the pollen grains and spores have one or several irregular or irregularly spaced apertures.

• *Classification of aperture based on position* (Fig. 4.26): In NPC system 'P' denotes the position (P; L. *positio*) of aperture in a pollen grain and spore. The position may be proximal, distal and equatorial. There are seven groups of aperture based on position namely—$P_0$ to $P_6$. Pollen grains having $P_0$ group have uncertain or unknown position of aperture. $P_1$ groups of pollen and spores are catatreme (Gr. Kata = down; -treme is suffix used as a synonym of aperture). Catatreme pollen grains have one aperture that occurs on the proximal part of a grain. The proximal (L. *proximus,* nearest) part is the face of a pollen grain or spore that faces inward/nearest or toward the centre of tetrad (Fig. 4.31). $P_2$ groups of pollen

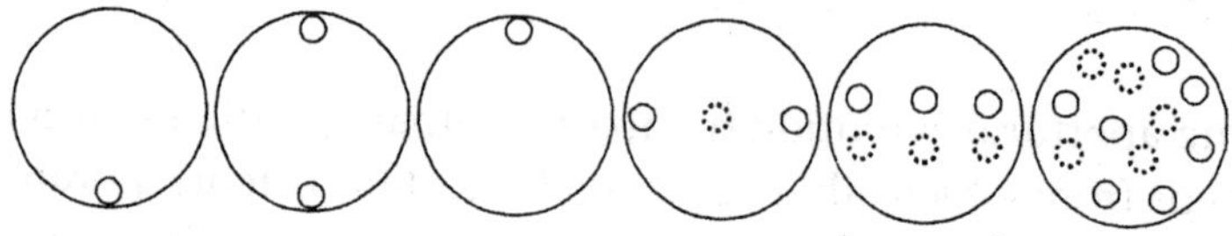


Figure 4.26

Schematic representation of position (P) of aperture(s) only in pollen and its number in NPC.

and spores are anacatatreme (Greek ana = up). Anacatatreme pollen and spores have two apertures. One aperture with its centre occurs at the proximal pole. The other aperture with its centre occurs on the distal pole. The distal (L. *distalis,* remote, outer) part is the face of a pollen grain and spore that faces outward, i.e. away from the centre of tetrad and opposite the proximal part (Fig. 4.31). $P_3$ groups of pollen and spores are anatreme, i.e. the aperture is distal in position. $P_4$ groups of pollen and spore are zonotreme. A zonotreme (zono—a prefix used to indicate the equatorial/subequatorial region) pollen grain is characterized in having apertures on equator or sub-equator. The equator is the part of a pollen grain or spore that runs midway between the proximal and distal poles and perpendicular to polar axis. $P_5$ groups of pollen and spore are dizonotreme. Dizonotreme pollen grains have apertures arranged in two or more zones. The apertures occur parallel to equator. $P_6$ groups of pollen and spore are pantotreme (Greek pan, gen. Pantos, all, wholly). Pantotreme pollen grains have apertures scattered over the whole surface uniformly. As a rule, pantotreme pollen grains are spheroidal.

• *Classification of apertures based on character* (Fig. 4.27): In NPC-system 'C' denotes the character (C; L. *character*) of an aperture in a pollen grain and spore. The character groups of pollen and spore are seven and they are mentioned as $C_0$ to $C_6$. $C_0$ groups have apertures whose character cannot be established with certainty. $C_1$ groups of pollen and spore have leptoma (Greek leptoma means thin place). Leptoma is a thin area, aperture like and functions like an aperture. Pollen grains having one leptoma are termed as monlept. The leptoma may occur on either proximal—or distal face of a pollen grain and spore and accordingly termed as catalept and analept. $C_2$ groups are trichotomocolpate (Gr. Tricha, in

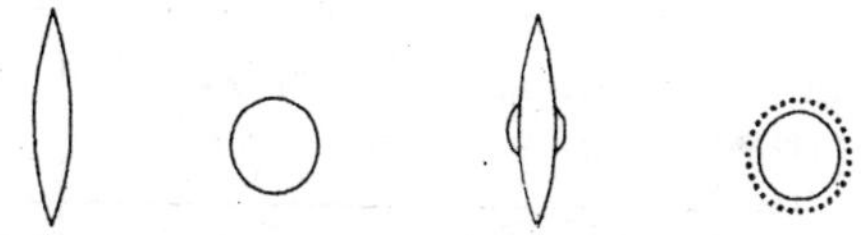


Figure 4.27

Schematic representation of character (C) of some apertures present in a pollen grain and its number in NPC. Dotted line indicates different plane of focus.

three parts; tome, cut; kolpos, depression, furrow). Trichotomocolpate is a three-branched aperture, the branches of which are more than two times longer than breadth. Trichotomocolpate pollen and spores having aperture on porximal face are termed as trilete. The group $C_3$ has colpate grains. The group $C_4$ comprises porate pollen grains. The group $C_5$ comprises colporate pollen. The group $C_6$ comprises pororate pollen. $C_3$, $C_4$, $C_5$ and $C_6$ groups of aperture are previously discussed under aperture.

In NPC classification a grain is mentioned in three-digit number (Figs. 4.28, 4.29 & 4.30), e.g. 343 instead of $N_3P_4C_3$. Pollen grains having NPC 343 are tritreme zonocolpate, which is also described as tricolpate pollen. NPC 764 characterizes those pollen grains that are polytreme pantoporate, which are also described as pantoporate or polyporate. Pollen grains of Amaranthaceae, Chenopodiaceae etc. have NPC 764. Examples of tricolpate pollen grain, i.e. NPC 343 are *Rumex*, *Vitex*, *Tectona*, *Argemone* etc.

**Merits of NPC classification**

1. It is a simple system of classification and illustrates the apertures of a pollen grain and spore.
2. NPC makes the description of apertures precise.
3. With the aid of NPC pollen grains and spores of pteridophyta, monocotyledon and dicotyledon, to some extent, can be differentiated. Most of the spores of pteridophyta are monolete or trilete. Monocots are characterized by inaperturate, monosulcate and monoporate pollen grains. Dicots, with a few exceptions, have pollen grains that are mostly with three meridonial furrows and polyaperturate. Thus NPC narrows the search list of identification of unknown sporomorphs.
4. NPC is supposed to be of primary classificatory character because apertures are most conservative. It is supplemented by surface ornamentation, size and shape etc. of a pollen grain. Sometimes it becomes possible to identify the family or genus or even species of a pollen grain with the aid of NPC in combination with other morphological characters.
5. Palynologists all over the world accepted NPC-classification as it is basically simple and consistent where pollen grains and spores could be arranged easily. This helps to identify unknown sporomorphs.

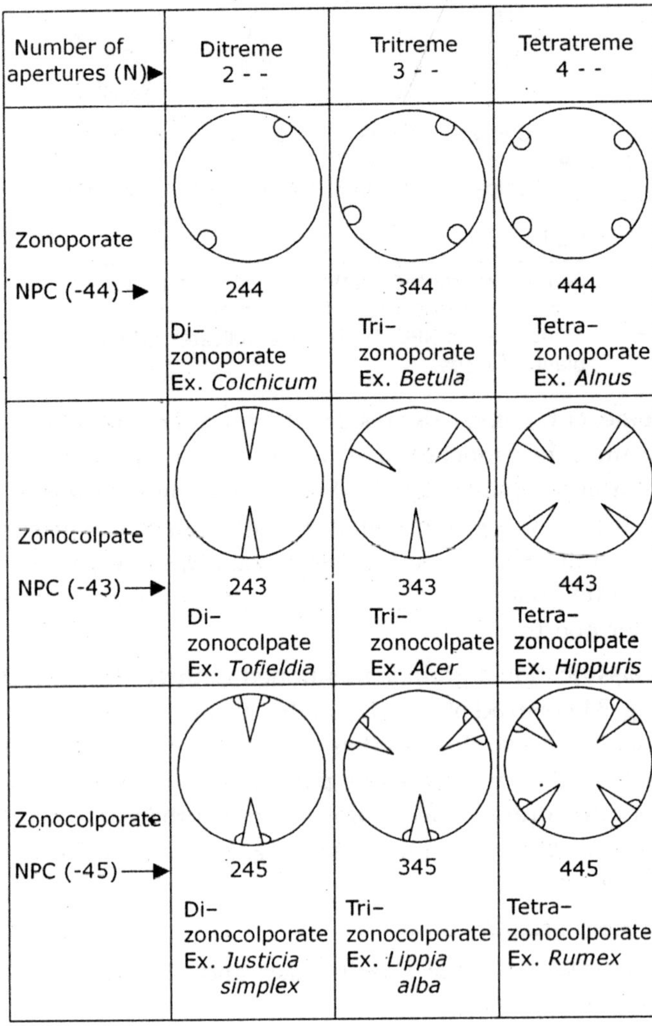


Figure 4.28

Diagrammatic representation of NPC of some pollen in polar view.

6. NPC, sporoderm stratifications, exine patterns, size and shape etc. of a pollen grain are genetically stable. This property is utilized for various purposes and the followings are a few illustrations. With the aid of NPC and other characters a key can be formulated that helps to identify unknown pollen and spores. Identification of pollen and spores is the essential prerequisite in the applied aspects of palynology, i.e. aeropalynology, melissopalynology, forensic palynology and palaeopalynology etc. The interfamily and intrafamily affinities of taxa, to some extent, can be determined with the aid of NPC. As for example the family Gramineae seems to be closely related to Restionaceae, Centrolepidaceae and Flagellariaceae, because pollen grains of above taxa are monoporate.

| Number of apertures (N) ▶ | Pentatreme 5 - - | Hexatreme 6 - - | Polytreme 7 - - |
|---|---|---|---|
| Zonoporate<br>NPC (-44) ⟶ | 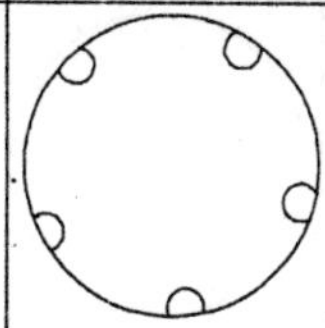<br>544<br>Penta-zonoporate<br>Ex. *Alnus* | 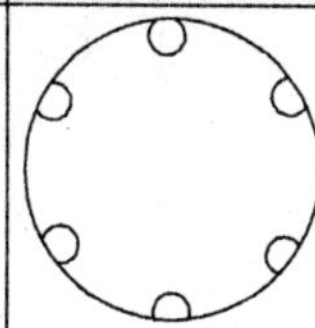<br>644<br>Hexa-zonoporate<br>Ex. *Ulmus* | 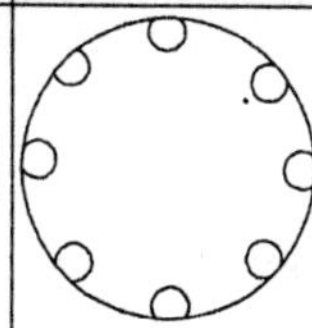<br>744<br>Poly-zonoporate |
| Zonocolpate<br>NPC (-43) ⟶ | 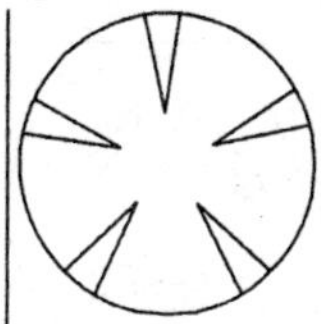<br>543<br>Penta-Zonocolpate | 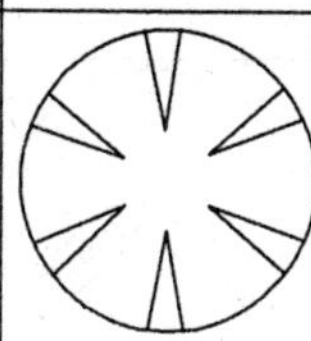<br>643<br>Hexa-Zonocolpate | <br>743<br>Poly-Zonocolpate |
| | ⟵ Ex. Labiatae, Rubiaceae ⟶ | | |
| Zonocolporate<br>NPC (-45) ⟶ | 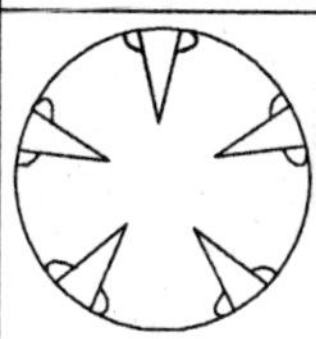<br>545<br>Penta-zonocolporate<br>Ex. *Viola* | 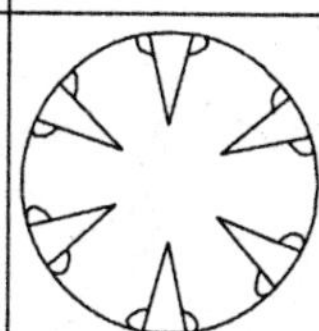<br>645<br>Hexa-zonocolporate<br>Ex. *Sanguisorba officinalis* | 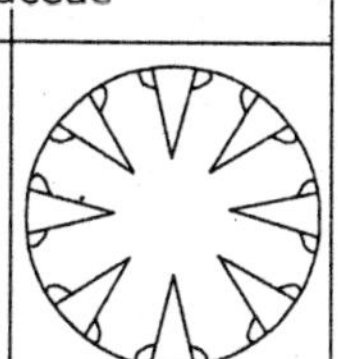<br>745<br>Poly-zonocolporate<br>Ex. *Utricularia* |

Figure 4.29

Diagrammatic representation of NPC of some pollen in polar view.

7. NPC and the various types of exine patterns and ornamentation provide characters of taxonomic significance and thus become one of the sources of alpha taxonomy. Example: Bombacaceae is segregated from Malvaceae; Zingiberaceae, Cannaceae and Musaceae are amalgamated into Scitamineae.

**Demerits of NPC classification**

1. It is an artificial system of classification.
2. Syncolpate and parasyncolpate pollen grains do not fit neatly in NPC system.

| Number of apertures (N)▶ | Pentatreme 5 - - | Hexatreme 6 - - | Polytreme 7 - - |
|---|---|---|---|
| Pantoporate<br>NPC (-64)→ | 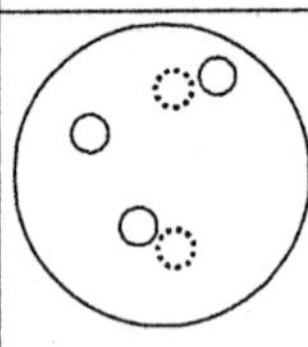<br>564<br>Penta-pantoporate<br>Ex. *Plantago* | 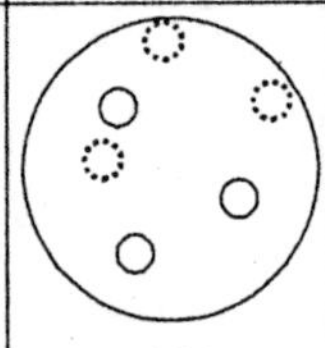<br>664<br>Hexa-pantoporate<br>Ex. *Plantago* | 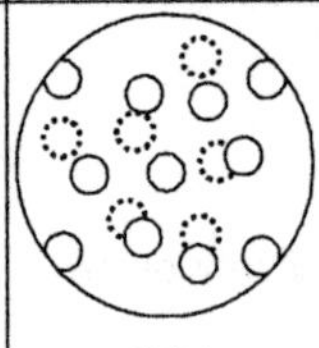<br>764<br>Poly-pantoporate<br>Ex. *Chenopodium* |
| Pantocolpate<br>NPC (-63)→ | 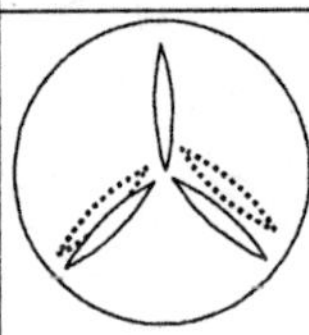<br>563<br>Penta-pantocolpate | 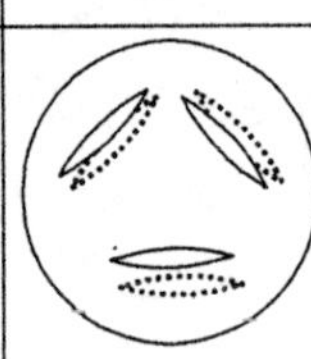<br>663<br>Hexa-pantocolpate | 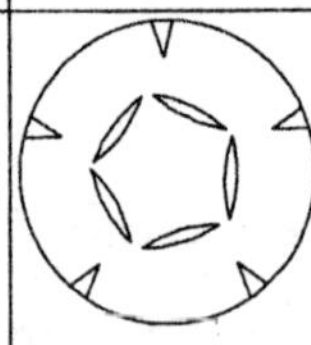<br>763<br>Poly-pantocolpate<br>Ex. *Polygonum amphibium* |
| Pantocolporate<br>NPC (-65)→ | 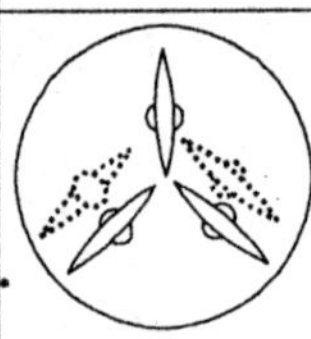<br>565<br>Penta-pantocolporate | <br>665<br>Hexa-pantocolporate<br>Ex. *Polygonum oxyspermum* | 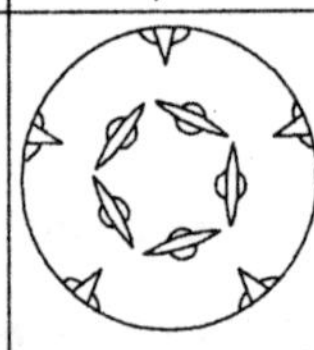<br>765<br>Poly-pantocolporate |

Figure 4.30

Diagrammatic representation of NPC of some pollen in equatorial view. Dotted lines indicate different plane of focus.

3. Pollen grains that are characteristically present as aggregates in tetrads, e.g. Ericaceae, Typhaceae and polyads, e.g. Orchidaceae, *Mimosa* etc. are not grouped in NPC system.

NPC-system of classification is always compared with Linnaeus's system of classification, because the latter is also an artificial system of classification. The characters of stamen were the basis of classification. 'Linnaeus accepted the weakness of his classification but claimed that it was propounded mainly as an aid to identification'—Chanda, 1992, p. 40. This is also applicable to Professor Erdtman. Palynologists from every discipline of palynology utilize Erdtman's NPC classification and other characters related to pollen morphology as an aid to the identification of unknown sporomorphs.

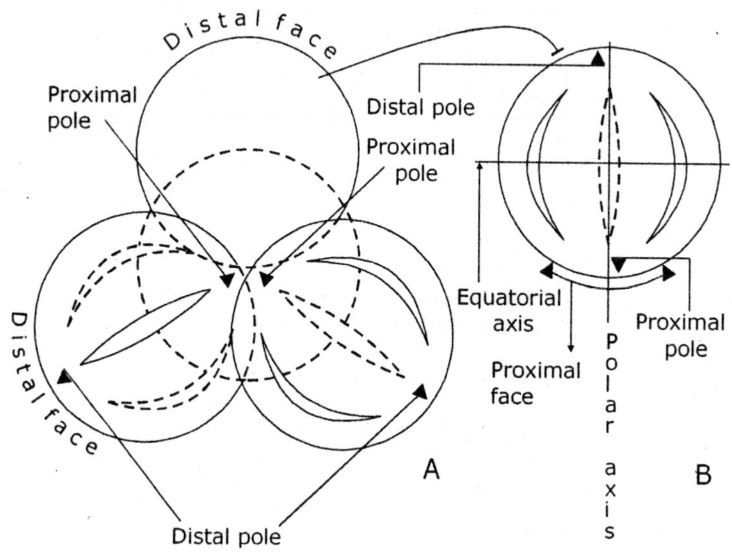


Figure 4.31

A. Diagrammatic representation of tetrahedral tetrad showing the orientation of 3-colpate pollen grains within the tetrad and determining the poles of a pollen grain. B. Diagram illustrating how the axes of a pollen grain are named. The measurement of a pollen grain along the equatorial axis is referred to as equatorial diameter. Dotted lines in A and B indicate a different focal plane. Adapted after Moore *et al.*, 1991, p. 70.

## Polarity

Polarity is defined as the condition of having distinct poles in a pollen grain. Jackson in 1928 (Kremp, 1965, p. 117) assumed that a pollen grain points to the poles like *Silphium*. *Silphium laciniatum,* also known as compass plant, has leaves that 'turn their edges north and south and avoid the mid-day radiation'—Willis (1973, p. 1067).

Pollen mother cell after meiosis forms four pollen grains that commonly remain in tetrads and later separate to form individual pollen grain. At tetrad stage the polarity of a grain is best detected.

There are two poles in a pollen grain—proximal and distal (Fig. 4.31). The proximal face is the part of a pollen grain and spore that faces inward/nearest or toward the centre of tetrad. The distal face is the part of a pollen grain and spore that faces distal, i.e. away from the centre of tetrad and opposite the proximal part. An imaginary straight line labeled as polar axis connects the proximal pole (more precisely the proximal pole at the centre of the proximal face) to the distal pole (more precisely the distal pole at the centre of outer face) of a pollen grain. Equatorial axis/diameter (= equator) is also an imaginary straight line that occurs perpendicular to polar axis of a pollen grain midway between the poles. The term meridonial is used to indicate the surface of a pollen grain that is oriented 'in a pole-to-pole direction at right angles to the equator'—Erdtman (1969, p. 26). The following common terms are used in connection with polarity.

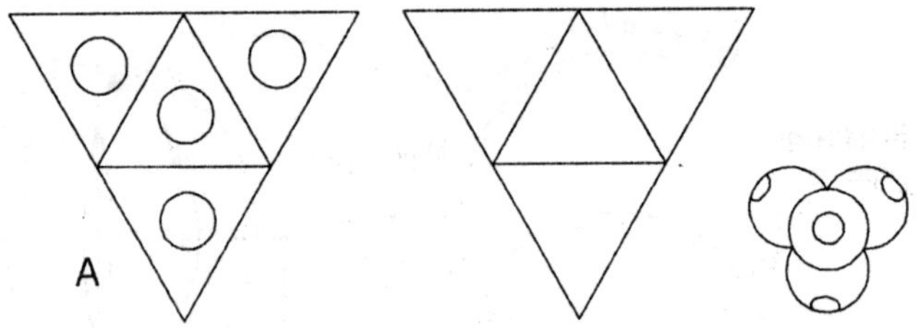


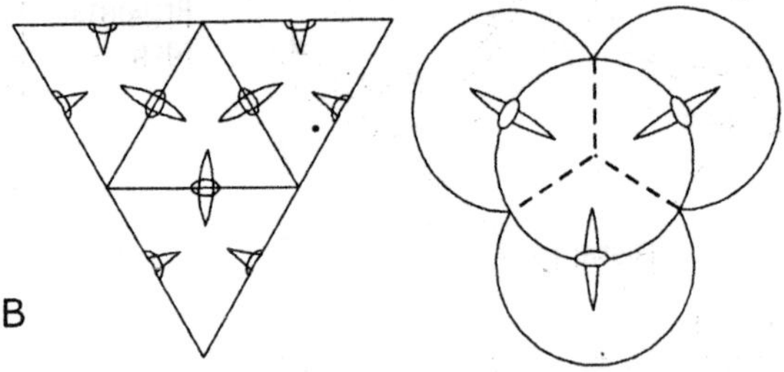


Figure 4.32

Diagrammatic illustrations of polarity in pollen tetrad. A. *Drimys* with heteropolar pollen. B. *Calluna* with isopolar pollen. The central pollen grain has apertures at '2, 6 & 10 o'clock' position. Adapted after Erdtman, 1969, p. 23.

• *Apolar.* A pollen grain is said to be apolar in which the proximal pole, distal pole, proximal region and distal region cannot be distinguished. Pollen grains that occur in linear tetrads are apolar because they exhibit no polarity. Example: *Periploca ephedriformis* and *Typha* etc.

• *Cryptopolar*: Pollen and spores are said to be cryptopolar spore that appears to be apolar, i.e. poles and polar region cannot be distinguished, 'but on closer examination they reveal a more or less distinct polarity (*Larix, Pseudotsuga, Equisetum* and others)' – Erdtman (1952, p. 11).

*Polar.* A pollen grain is said to be polar in which proximal and distal pole can be distinguished in individual state. Isopolar, heteropolar and subisopolar/ paraisopolar belong to this category.

• *Isopolar*: A pollen grain is said to be isopolar (Fig. 4.32B) in which the proximal – and distal half have similar appearance. A uniform plane that occurs halfway between the poles and at right angles to polar axis divides a pollen grain into two halves – proximal and distal. The half that faces proximal pole is labeled as proximal half and the half that faces distal face is referred to as distal half. The two halves of pollen are equal with regard to apertures and surface ornamentation etc. ex. *Calluna* (Ericaceae).

• *Heteropolar*: A pollen grain is said to be heteropolar (Fig. 4.32A) in which the proximal—and distal half is essentially dissimilar to each other with regard to apertures. As for example *Lilium* exhibits distal colpus and *Elaeis* has three-lobed distal furrow etc.

• *Subisopolar* or *paraisopolar*: A pollen grain is said to be subisopolar/paraisopolar that is ± intermediate between isopolar and heteropolar pollen grains. As for example *Diastella, Nivenia* and *Ulmas* etc. have small differences in surface details between proximal - and distal half.

The polarity of a pollen grain forms the basis of aperture terminology and the followings are few examples. A circular aperture that occurs equatorially or globally is called porus. The circular aperture that occurs distally is termed as ulcus. An elongated aperture that occurs equatorially or globally is called colpus. When an elongated aperture occurs distally it is called sulcus. The polarity gives a pollen grain two views: a proximal polar and a distal polar. It is the polarity that distinguishes a spore from pollen. Emergence of pollen tube from distal pole is the characteristic of a pollen grain. Proximal germination is restricted to spores that germinate at tetrad scar termed laesura.

It is to note that polarity exists throughout megasporogenesis. But during microsporogenesis the time of initiation of polarity varies. Usually microspore mother cell is non-polar (Shivanna *et al.*, 1975). Microspore mother cell undergoes reduction division. It is observed that polarity is established during the appearance of spindle at the time of first meiotic division. The aperture sites (in *Lilium longiflorum*) develop in each microspore tetrad in the face away from the centre. Following the first meiotic division all events in the development of microspore are polarized and occur in pre-determined way.

Polarity can be disturbed by colchicine treatment or by centrifugation. Both treatments disturb spindle formation. By colchicine treatment, in *Lilium longiflorum*, the spindles of first and second meiotic division can be blocked. In this case microspore mother cell, as a whole, develops with regular exine patterning but in it the colpus may not be developed at all or if develops, it may be with random distribution and irregular form. Vide Shivanna *et al.* (1975, p. 20-21) for details.

## Shape

Sporoderm defines the shape of a pollen grain and spore. Erdtman (1952) distinguishes nine common types of shape (Fig. 4.33). The shape classes are based on the ratio between the length of polar axis (P) and equatorial diameter (E). P and E are measured from the equatorial view of a pollen grain and spore. Measurements are taken from the optical section of a pollen grain and spore. The imaginary straight line (Fig. 4.31A & B) that occurs between the proximal pole and distal pole of a pollen grain is called polar axis. Equatorial axis/diameter are also an imaginary straight line that occur halfway perpendicular to polar axis. In measuring the P and E axes longest side of pollen and spore are considered in optical section. The followings are the common shapes according to Erdtman

(1952): peroblate (ratio of P to E < 0.50), oblate (ratio of P to E between 0.50 and 0.75), subspheroidal (ratio of P to E between 0.75 and 1.33), suboblate (ratio of P to E between 0.75 and 0.88), oblate spheroidal (ratio of P to E between 0.88 and 1), prolate spheroidal (ratio of P to E between 1 and 1.14), subprolate (ratio of P to E between 1.14 and 1.33), prolate (ratio of P to E between 1.33 and 2) and perprolate (ratio of P to E > 2). Erdtman in 1943 (Kremp, p. 140) mentioned another shape class termed spheroidal (ratio of P to E between 1.14 and 0.88).

Shape of a pollen grain has taxonomic significance. It is used in the formulation of pollen keys. As for example 3-colpate grains of *Rumex dentatus* are differentiated from the 3-colpate grains of *Leonurus sibiricus, Tectona grandis* and *Vitex negundo* on the basis of shape. *Rumex dentatus* has subspheroidal to spheroidal pollen and the rests have subprolate to prolate pollen (Banik *et al.*, 1986).

## Size

There exist much variations in sizes of pollen and spores. The size may be 5 μ x 2.4 μ as observed in pollen of forget-me-not (*Myosotis*) or as high as 200 μ or more as is noted in certain genera of Cucurbitaceae and Nyctaginaceae etc. (Erdtman, 1969, p. 34). The diameter of megaspores of *Selaginella exaltata* is about 1.5 mm, i.e. 1500 μ. In radially symmetrical pollen grains the size is expressed with regards to the length of polar axis (PA) and equatorial diameter (ED). The axes are measured from the equatorial view of pollen. In bilateral pollen and spores also the size is expressed with regards to length of polar axis and equatorial diameter. But it is a rule (Erdtman, 1969, p. 36) to measure the equatorial breadth (= equatorial diameter) of bilateral pollen and spores from the polar view also. In polar view the diameter of amb is measured. Amb is defined as the outline of a spore or pollen when viewed from directly above with one of the poles. In isopolar, not equatorially constricted spore and pollen, the diameter of amb is same as that of equator.

Erdtman in 1945 (Kremp, 1965, p. 141) and 1969, p. 36 suggested the following pollen and spore-size classes. The measurements were based on the length of the longest pollen and spore axis.

1. Very small grains or spores (*pollina perminuta; sporae perminutae*) ...< 10 μ. Example: *Lycoperdon, Myosotis palustris.*
2. Small grains and spores (*pollina minuta; sporae minutae*) ... 10 μ-25 μ. Example: *Daphne mezereum.*
3. Medium size grains or spores (*pollina media; sporae mediae*) ... 25 μ-50 μ. Example: *Stellaria.*
4. Large grains or spores (*pollina magna; sporae permagnae*) ... 50 μ-100 μ. Example: *Osmunda regalis.*
5. Very large grains or spores (*pollina permagna; sporae magnae*) ... 100 μ-200 μ. Example: Spruce.
6. Gigantic grains or spores (*pollina gigantea; sporae giganteae*) ... > 200 μ.

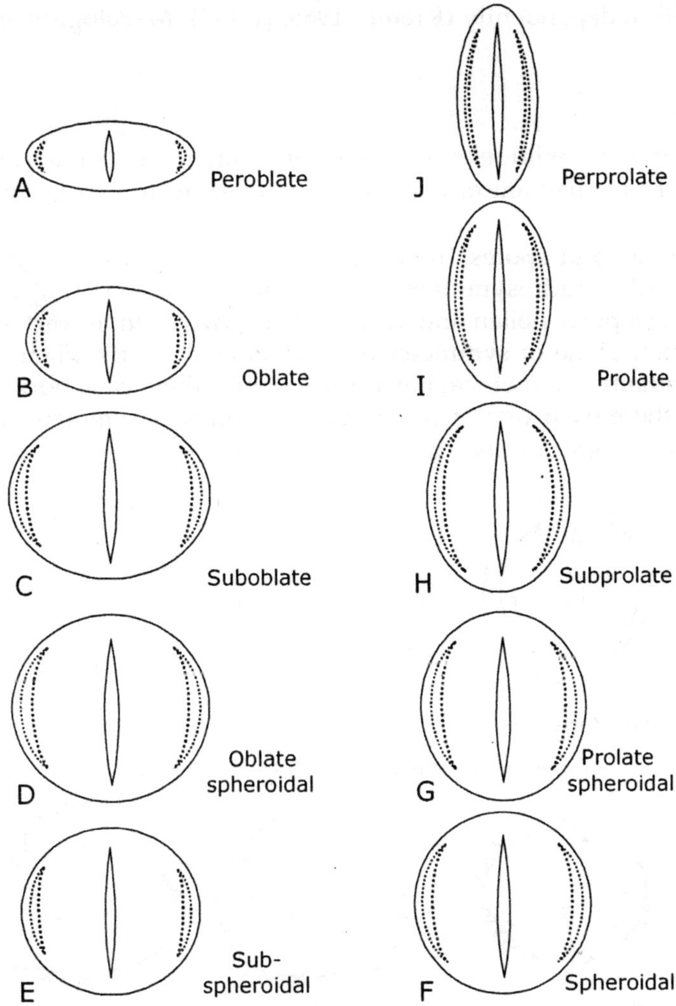


Figure 4.33

Diagrammatic illustrations of the different shape classes of pollen. All pollen grains are tricolpate with equatorial view. Dotted lines indicate different plane of focus.

The size of a pollen grain is taxonomically significant. It is frequently used in the formulation of pollen keys that are used to differentiate the sporomorphs. As for example *Melandrium noctiflorum* (diameter greater than 37 μ) is differentiated from *M. rubrum* (diameter 37 μ or less), *Dianthus deltoides* with ± 40 μ in diameter is differentiated from *Stellaria media* that has diameter less than 40 μ (Chanda, 1963, p. 87), *Heliotropium indicum* (PA × ED = 45.0 μ × 30.0 μ) is differentiated from *Euphorbia hirta* (PA × ED = ± 28.0 μ × 17.0 μ) and *Cestrum nocturnum* (PA × ED = ± 39.0 μ x 32.0 μ) is differentiated from *Dalbergia sissoo* (PA × ED = ± 28.0 μ × 24.0 μ) etc. Banik *et al.*, p. 17 [PA = polar axis; ED = equatorial diameter].

It is to note that in the general dimensions of pollen and spores the sculptural elements like spines etc. should not be included. The sculptural elements are to be dealt with independently (Kremp, 1965, p. 141). Mycologists also adopt this view.

## Symmetry

The term symmetry with relation to a pollen grain implies similarity of halves on either side of a median line or plane so far as aperture and ornamentation etc. are concerned.

Most pollen and spores are symmetrical. As a rule, symmetric pollen and spores are either radiosymmetric (Fig. 4.34) or bilateral (Fig.4.35). Radially symmetrical isopolar pollen and spores exhibit two or more vertical planes and one horizontal plane of symmetry (ex. *Centaurea*). All the planes are of equal length. In radially symmetrical heteropolar spores there is no horizontal plane of symmetry; there exist two or more vertical planes of symmetry (ex. *Osmunda regalis, Ophioglossum vulgatum* and *Pteridium aquilinum* etc.).

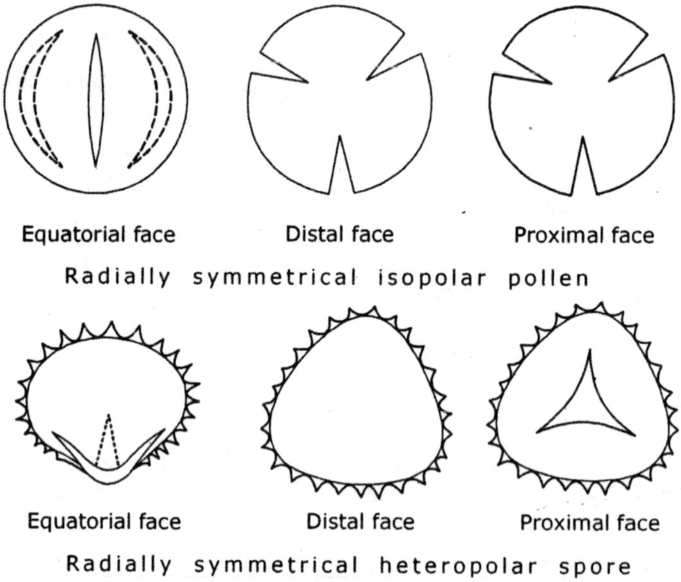


Figure 4.34

Diagram illustrating radially symmetrical pollen and spore. Adapted after Erdtman, 1969, p. 24 & 25.

Bilateral isopolar spores exhibit two vertical planes and one horizontal plane of symmetry. All the planes are not of equal length (ex. *Rungia grandis*). In bilateral heteropolar spores, there exist two vertical planes of symmetry. The planes are unequal in length and intersect each other at right angles (ex. *Ephemerum serratum, Picea abies* and *Cycas revoluta* etc.).

There exists no plane of symmetry in asymmetrical pollen and spores (ex. *Berberis darwinii, Papaver argemone* and *Myriophyllum alterniflorum* etc.). Asymmetric pollen and spores may be with fixed shape, termed fixiform or may be without any definite shape called nonfixiform. Asymmetric pollen and spores are rare.

## Composition of sporopollenin

Sporopollenin is the major component of exine of spores and pollen. 'It is most extraordinarily resistant material known in the organic world'. It is extremely resistant to non-oxidative chemicals, enzymes and strong chemical reagents. It is chemically stable and remains unchanged in pollen grains that are 500 million years old. It is well preserved in sediments. Sporopollenin remains unchanged when extant pollen grains are heated to 300°C or treated with concentrated acids and bases. Because of this resistance, the definite chemical structure of sporopollenin is inadequately known.

At present, it is assumed that sporopollenin is a polymer. Monomers of sporopollenin skeleton are linked by ether-linkage thus composing a complex polymer. The sporopollenin polymer has a uniform composition, i.e. relatively small monomer moieties of similar primary structure are present (Henning Bubertb *et al.*, 2002). Kawase *et al.* (1995) report that the main structure of sporopollenin is a simple aliphatic polymer containing aromatic conjugated side chains. The elemental contents of Carbon, Hydrogen and Nitrogen as determined from CHN analysis indicate that sporopollenins are N-free substances. Zetsche

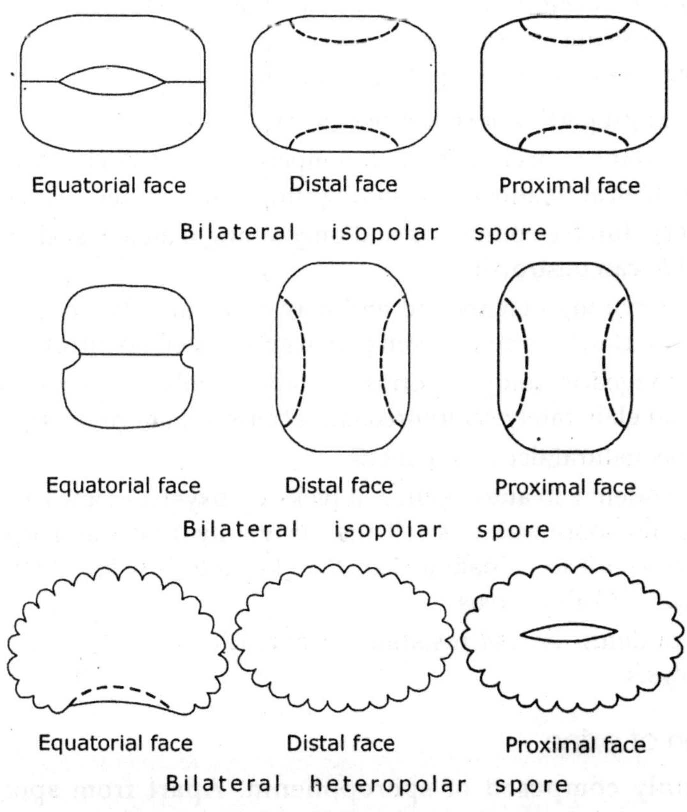


Figure 4.35

Diagram illustrating bilateral spores. Adapted after Erdtman, 1969, p. 24.

and Vicari (1931) estimated the presence of 90 Carbon, 142 Hydrogen and 27 Oxygen atoms in a sporopollenin molecule of *Lycopodium.* Sporopollenins differ specifically in composition and quantity in different pollen species. It is to note that the very resistant *Pinus* pollen grains possess the greatest quantity of sporopollenin (Faegri *et al.*, 1950). Chemical analysis reveals that approximately half of the sporopollenin carbons are bonded as CH and the other half as $CH_2$. The greater part of oxygen found in sporopollenin originates from hydroxyl groups and is derived from aliphatic and not from aromatics. It is regarded that aromatics, long unbranched aliphatic components compose the complex structure of polymer. Hesse *et al.* (1995) report the occurrence of benzoic acid moieties in the sporopollenin of *Typha angustifolia.* Ahlers *et al.* (2000) found α, β unsaturated carbonyl group in the sporopollenin of *Typha angustifolia.* Chemical analysis reveals that sporopollenin is a biopolymer containing mainly long chain fatty acids, phenylpropanoids, phenolics, carotenoids and xanthophyll. The presence of phenylalanine was detected by tracer experiments. Sporopollenin is close in structure to the potassium salt of glucuronic acid. Infrared absorption spectra of the sporopollenin of *Beta vulgaris* reveal it (Traverse, 1988, p. 37). In the biosynthesis of sporopollenin the chain-elongating enzyme systems of lipid metabolism are involved (Wilmesmeir & Wieramann, 1995). Biochemical studies indicate this by using thiocarbamate herbicide—the inhibitors of the chain-elongating enzyme systems of lipid metabolism.

The various properties of sporopollenin are:

- It is resistant to acetolysis, high temperature, many chemicals, physical and biological treatments including fungal and bacterial attack.
- It is very durable compound and only strong oxidants such as hydrogen peroxide can destroy it.
- It is a very tough compound and may occur in pollen even 500 million years old. Due to sporopollenin pollen gets preserved during fossilization.
- It is arranged in complex forms in walls of pollen in such a way that it forms an elaborately sculptured characteristic pattern on sexine.
- It checks natural decay of pollen.
- It has tendency to auto-oxidize. It picks up oxygen atom. This causes the pollen and spore wall to swell and eventually dissolute. The endexine is more resistant to oxidation than the ektexine (Rowley, 2001). It cannot survive in alkaline areas.
- Sporopollenin is less resistant to oxidation. It can be degraded by ozonolysis.

## Composition of exine

Exine is mainly composed of sporopollenin. Apart from sporopollenin glycoproteins are present in the chambers formed by bacula in tectate pollen. It is important in pollen-stigma interaction. The presence of glycocalyces (sing. glycocalyx) is reported from exine (El-Ghazaly and Rowley, 1999 and Meir-

Melikyan *et al.*, 2003). Glycocalyx forms a fine layer on the outer surface of many cells. It is composed of polysaccharides, polypeptides or both. In some cells it forms a coating consisting of glycoproteins and proteoglycans. It enables a cell to adhere, protects the cell from dehydration, and helps in absorption and cell individual identity. It is reported from Asteraceae (*Calendula officinalis*), Lauraceae (*Persea americana*), Boraginaceae (*Borago officinalis*), Alismataceae (*Echinodorus cordifolium*), and Nymphaeceae etc. In Lauraceae and Boraginaceae glycocalyx is found to consist of radially arranged helical cylindrical units. These units are the receptors of sporopollenin deposition and this is revealed with the study by means of electron microscopy.

The intine contains cellulose at the inner part and pectin on the outer part. It also contains callose at the apertural region. Various proteins are present in it (Brett and Waldron, 1990). The intine also contains polyuronids or a mixture of polyuronids and polysaccharides. The pollen tube wall is made up principally of callose [β (1-3) glucan] and α (1-3) arabinan.

**Importance of exine**: (i) It protects the protoplasm from desiccation and abrasion during pollination. Harder exine than other parts resists deterioration more successfully; (ii) it is involved in cell-to-cell recognition and carries the chemicals (e.g. glycoproteins) that recognize the compatible receptive stigma. The chemicals are highly specific and also help in adhesion between pollen and stigmatic surface; (iii) it helps to adhere to a pollinator; (iv) through it the pollen tube emerges during fertilization; (v) it allows for the expansion and reduction of pollen size with changing humidity; (vi) the surface details of exine are remarkably variable and it yields the most important characters for the identification of pollen and spores; (vii) the physical strength of a pollen grain is obtained from it. It can withstand shock without permanent deformation. It is very flexible and resilient (Rowley and Skvarla, 2000). Generally the shape of pollen does not change after long-term sediment underground, mainly due to the existence of exine.

Light microscopy was the only means to study exine. After 1950s with the introduction of scan electron microscope and transmission electron microscope the exine structure is analyzed and tested with the light microscopic observation. After 1990s atomic force microscopy seems to be the ideal method to investigate the exine substructure. The morphological subunit of sporopollenin and their arrangement patterns in the exine of angiosperm and gymnosperm can be well studied from atomic force microscope (AFM) images. A few reports are available so far concerning the substructure of exine (Rowley, J. R. *et al.*, 1995, XING Shuping *et al.*, 2000). From the AFM images the exine is regarded as a complex unit where the morphological subunits are shaped as spheroid and tubule. They are arranged in helical pattern (ex. *Nuphar luteum*). The subunits may also be in the form of granules that form 3-dimensional lattice pattern in the exine (ex. *Cedrus deodara*).

Sporopollenin makes the exine tough and renders pollen grains and spores resistant to decay. As a result pollen and spores survive in ancient sediments for at least 500 million years. Shape, sculpturing and apertures etc. of a grain remain unchanged after long term sediment underground. These properties of a pollen grain form the basis of palynology.

**Palynology** (n. Gr. paluno = to strew or sprinkle; L. *pollen* = flour, dust; Gr. logia = combining form denoting theory or science) can be defined as the collective study of spores of embryophytic plants and pollen of seed plants. In a broad sense the study also includes a number of other acid-resistant microorganisms and palynomorphs and other biological materials that are studied by palynological techniques. Hyde and William in 1944 advocated the term palynology.

[The term palynology replaced the name 'pollen analysis'. In *Pollen Analysis Circular* (No. 8, 1944) it was expressed to have a better name other than pollen analysis. The *Circular* was a cyclostyled research bulletin. The Editors were Professor Paul B. Sears of Oberlin College, Ohio, Mr. H. A. Hyde of the Natural Museum of Wales and Mr. D. A. Williams of Llandough Hospital, Cardiff. In the October issue of the *Circular* the term 'palynology' was suggested. 'It is hoped that the sequence of consonants p-l-n (suggesting pollen, but with a difference) and the general euphony of the new word will commend it.'—Kremp, 1965, p. 110; http://www.geo.arizona.edu/palynology/riteword.html.]

## Major branches of palynology and importance of studying them

Palynology is divided into two broad fields—palaeopalynology and neopalynology. The former is the study of fossil pollen and spores. Neopalynology is the study of extant pollen and spores. It includes the following important disciplines: palynotaxonomy, aeropalynology, melissopalynology, copropalynology, forensic palynology and entomopalynology etc. Brief accounts of the disciplines with their applications are mentioned below. It is to be noted that definite identification of a pollen grain is the essential prerequisite to study any discipline of palynology.

• *Palynotaxonomy*: Palynotaxonomy is the study of spore morphology of embryophytic plants and pollen morphology of seed plants.

The contributions of Professor Erdtman on pollen morphology marked the beginning of a new phase on plant taxonomy. Erdtman and many palynologists all over the world made available pollen descriptions of all angiosperm families including most of the genera and species to taxonomists (Erdtman, 1969, p. 96-112; Faegri *et al.*, 1989, p. 241-288; Moore *et al.*, 1991, p. 86-162 and Kapp, 1969). Taxonomists employ pollen morphological attributes in systematic works, i.e. in solving controversial taxonomical and phylogenetical problems, identification and determining affinity etc.

On the basis of pollen morphological data plant families are divided into two groups—stenopalynous and eurypalynous. Stenopalynous families do not exhibit marked variations of pollen types with regard to apertures and sporoderm stratification etc. ex. Capparidaceae, Chenopodiaceae, Casuarinaceae and Gramineae etc. Eurypalynous families exhibit marked variation of pollen types with regard to apertures and sporoderm stratification etc. ex. Bignoniaceae, Boraginaceae, Cucurbitaceae and Euphorbiaceae etc. 'Based on stenopalyny or eurypalyny the inter or intrafamily affinity of taxa, to some extent, can be determined, e.g. the pollen grains of grasses are all monoporate and phylogenetically the Gramineae seem to be closely related to Restionaceae, Centrolepidaceae and Flagellariaceae group.'—Chanda, 1970-71, p. 35. Morphological characters of pollen have been used to substantiate many taxonomical revisions. As for example 'segregation of Trapaceae from Onagraceae, Bombacaceae from Malvaceae, the union of Zingiberaceae, Cannaceae and Musaceae into Scitamineae, and Moraceae and Cannabinaceae into Urticaceae ...'—Chanda, 1970-71, p. 35.

• *Aeropalynology*: Aeropalynology is the study of palynomorphs found in the atmosphere. The term palynomorph encompasses pollen, spores and other bioparticles that are acid resistant.

Pollen grains have impact on human systems. They cause allergy on sensitive individuals. Hay fever or allergic rhinitis is the best known allergy of all. Hay fever is caused by pollen grains of *Betula, Populus, Salix* and *Chrysanthemum* etc. Allergy is also caused by the pollen grains of *Cynodon dactylon, Amaranthus spinosa, Chenopodium album, Cannabis sativa* and *Cassia occidentalis* etc. The symptoms of hay fever are sneezing, clogged nose, itching nose, throat and eye, conjunctivitis and watering eyes. Aeropalynology comes under the purview of aerobiology that includes the study of bioparticles present in the atmosphere in addition to pollen and spore. In aerobiological research palynologists in collaboration with clinicians search the causative agents present in the atmosphere that have impact on human health.

• *Melissopalynology* (= *melittopalynology*): Melissopalynology is the study of pollen and spore present in honey.

Pollen analysis of honey samples reveals the followings:

- It provides the plant species foraged by honeybees.
- The identified plants provide the fingerprint of environment and pollen spectrum from where the honey comes.
- It provides the botanical and geographical origin of honey.
- It is an indispensable tool to authenticate honey—unifloral honey or multifloral honey. If unifloral whether the honey sample contains pollen grains of the same plant as is declared. As for example—*Brassica*-honey, *Acacia*-honey and *Citrus*-honey etc.
- It detects whether honey sample contains allergenic pollen grains.
- It detects whether honey sample contains poisonous pollen, e.g. *Datura* and *Rhododendron* etc.
- It identifies pollen that are toxic to bees and cause bee poisoning, e.g. *Kalmia, Cuscuta, Tilia* and *Solanum* etc.

• *Forensic palynology*: Forensic palynology is the study of palynomorphs that are used as evidence in criminal cases and resolve other legal problems. Forensic palynology also includes identifying and linking the suspect to the scene of the crime.

The following attributes of pollen and spore form the basis of forensic palynology:

- Pollen grains are abundant in the atmosphere.
- Pollen and spores are microscopic and may be unknowingly taken away or left at a crime scene. Such pollen and spores can be used as evidence.
- Pollen and spores cannot be removed from a crime scene.
- Sporopollenin is the major component of pollen wall and it forms 'species-specific-pattern' that persists for thousands of years without any change in shape, aperture and sculpturing etc.
- A pollen grain can be identified to specific plant type, site, region or country.

The study of palynomorphs collected from crime scene and the belongings of suspect enables forensic palynologists to decide whether the pollen analyses will be of any probative value. Forensic palynology has been utilized effectively in many cases, e.g. assault, rape, forgery, drug dealing, fraud and murder etc. In many cases mud, dust or dirt are collected from shoes, vehicles etc. for pollen and spore analyses. There are several court cases where pollen evidence becomes important clues that helped to obtain conviction. A few cases are mentioned in the chapter 5.

Sometimes 'the pollen grains themselves can be the villains'– Erdtman (1969, p. 134). There are reports of several cases of death after consuming poisonous honey in the state of São Paulo in Brazil. The honey contained poisonous pollen grains of sapindaceous *Serjania lethalis*. Pollen analysis of the stomach content of a deceased boy revealed the presence of the lethal pollen.

• *Palaeopalynology*: Palaeopalynology 'deals with the study of fossil micro-organisms such as pollen grains, spores, marine and fresh water algae, foraminifera and other planktonic forms' – Agashe (1995, p. 290). The study of chitinozoans (animal remains), fungal spores, dinoflagellates, acritarchs and other organisms (except diatoms) is also included in palaeopalynology.

Palaeopalynology depends on the following attributes of a pollen grain:

- Pollen wall is tough and it survives better and longer than other biological materials.
- Sporopollenin composes exine and it forms a 'species-specific-pattern' of ornamentation/sculpturing on pollen wall. A pollen grain and its characteristic pattern fossilize well.
- Due to presence of sporopollenin the shape, size, aperture(s) and sculpturing etc. remain unchanged after long term sedimentation underground. These properties of a pollen grain provide means of identifying an unknown sporomorph up to the respective family or genera or even species. Correct identification of a pollen grain and spore is an essential prerequisite to study any discipline of palynology.
- A little quantity of peat or sediment sample contains a huge number of pollen grains and thus the vegetation that provided pollen can be studied.

Palaeopalynology is an applicative discipline of palynology. It is of immense value in a wide range of scientific studies. With palaeopalynological data stratigraphical correlation and dating of ages etc. can be established. The presence of fossil fuels—coal and oil can be predicted through fossil pollen morphological studies. Pollen grains present in Quaternary deposits can be referred to extant genera and species. Thus the past vegetation can be ascertained. From the identified plants the climate and ecology etc. can be determined. By studying the vegetation chronologically of the different strata the past history of vegetation, plant evolution, climatic and ecological change, geographical distribution, affinity between groups of plant, appearance and decline of some plants etc. can be interpreted.

• *Copropalynology* (Greek 'kopros' means dung): Copropalynology is the study of palynomorphs present in coprolite (= fossilized excrement) of animals (Erdtman, 1969). Human coprolite is preserved in dry caves. Pollen and spore analyses of human coprolite provide information about the diet of prehistoric humans. Pollen analysis of herbivore coprolite provides data to reconstruct the composition of pastures of that time.

The term copropalynology also includes the pollen analysis of recent excreta. Forensic palynologists determine the last diet of a murdered person by pollen analyses of stomach, gut and feces.

• *Entomopalynology*: Entomopalynology is the study of pollen grains that are associated with insects. This study encompasses melissopalynology, researches on honeybees, foraging distances of insects and pollination biology etc.

The purpose of studying entomopalynology is to find the relation of pollinators and crop yield. It is observed that in apples, melons and almonds etc. the yield increases with the increase of pollinators. Some pollinators are pests on crops. By studying pollinators and pests efficient management practices can be developed to increase crop yield. The studies include determining habit, habitat and life cycle of pollinators, the frequency of visits of pollinators, whether the pollinators are after nectar, pollen, or both, whether genetically modified plants affect pollinators, the future of plant species in absence of pollinators and to save pollinators from being endangered etc.

• *Palynodebris*: "Palynodebris is the organic 'junk' = organic mater ("OM") found in palynological preparations along with palynomorphs"—Traverse (2007, p. 3). It has the size of palynomorph and does not contain sporopollenin and chitin. It consists of charcoal or cellulosic tissue fragments like wood fragments that escape acetolysis. Palynodebris is studied for its 'relevance to palynofacies/environment relationship'. The analysis of charcoal reveals much about the environment at the time of deposition. Palynofacies—this term refers to the total assemblage consisting of palynodebris and palynomorph present in sediment. The assemblage characterizes the sediment and reveals the environment at the time of deposition. Palynodebris is now studied along with palaeopalynology.

• *Iatropalynology*: It is the 'study of spores and pollen as applied to human health problems'—Traverse (1988, p. 504). It is known that pollen and spore have medicinal properties. As for example Arabian physicians use *Lycopodium* spores—commercially called 'vegetable sulphur' for the treatment of stomach disorders. The spores of *Adiantum phillippinense* are used to cure coughs. The pollen grains of *Cycas circinalis* have narcotic properties. For more details vide Nair (1985, p. 11).

CHAPTER **5**

# Basic Concept on Applied Aspects of Palynology

Palynological science is broadly divided into basic palynology and applied palynology. The former is related to pollen morphology and the latter deals with palaeopalynology, aeropalynology, copropalynology, melittopalynology etc. Later Crop Palynology, Archaeopalynology, Cytopalynology, Aerobiology, Palynopathology, Palynotaxonomy and Phylogeny, Pollination Biology, Palynoenergetics, Pollen Biotechnology, Iatropalynology etc. have come under the purview of palynological science (Nair, 1998, p. 256). The materials associated with the dispersed pollen grains and spores, called 'Palynodebris' are also included in the study of palynological science. Palynology is now not only a fundamental science but also an applicative discipline.

The following four—aeropalynology, melissopalynology, forensic palynology and palaeopalynology are the modern applicative disciplines of palynology. They are discussed in numerous articles, Internet Websites and Literatures and will be mentioned briefly in the following essays.

**Aeropalynology**: Erdtman (1969) defines aeropalynology as the study of palynomorphs found in the atmosphere. The term palynomorph encompasses pollen grains, spores and other bioparticles that can not be dissolved in hydrofluoric acid. Aeropalynology is a branch of palynology and related to the study of pollen and spores that are dispersed in the atmosphere. The study also includes their eventual dissemination, deposition and impact on human systems.

Pollen grains are dispersed more than 400 miles away from the source plants. They can be found more than two miles above the surface. The airborne pollen grains originate from anemophilous plants and so they are small, light, smooth walled, colourless, produced in large numbers, dry and lack nectar. In other words they are custom made for wind transport. The first and foremost requirement in the study of airspora is to identify and detect the airborne palynomorphs and sporomorphs. A sound knowledge of ground flora is essential for this purpose. The study of ground flora of a particular area will reveal the genera and species present in that locality and their flowering season. Pollen grains from each species are to be studied and the slides of them are to be deposited in herbarium for future references. The airborne pollen are trapped for sampling by two major principles – (1) simple gravimetric method where pollen and spores are deposited by normal gravitational force on slides and (2) suction method where the atmospheric air is sucked in with the help of instruments – called sampler. There are many types of sampler that can suck certain volume of air according to a known velocity for a definite time. Along with the air pollen and spore are sucked in and they get stuck either on a plastic band or sticky cello-tape (fitted according to sampler), which were previously placed there for the purpose of trapping pollen. Mention may be made of Burkard seven days sampler, Burkard personal slide sampler, Burkard petriplate sampler, Anderson sampler and Hirst spore trap (Singh *et al.*, 2004). The next task is to analyze the trapped grains. Analysis includes identification and quantification of sporomorph and palynomorphs. Microscopic identification is dependable in most cases. Certain fungal spores cannot be identified under the microscope. If the spores are culturable, they are isolated and cultured in cultural media. Colonies are formed and thus they are identified. Pollen and spores are identified on the basis of sporoderm stratification, shape, size and apertural characters. The morphological characters of trapped pollen and spore are compared with those of ground flora and thus the airborne grains get identified. The frequency of the identified grains is also calculated. Trapping, identification and quantification are carried out throughout the year and the results are recorded. The data help to prepare the pollen calendar of a particular locality. A pollen calendar reveals the name of plant species that release pollen and spore in particular month(s). It also reveals the approximate amount of grains released in that season. The term 'pollen count' is frequently used to represent the concentration of pollen in atmosphere. It is a measure of how much amount of pollen and spore is present in the atmosphere in a particular area at specific time. The count may consist of a particular type of pollen or spore, or all the pollen grains and spores present in the air. Pollen count is expressed as the number of pollen present in a cubic metre or other standard volume of air over a twenty-four hour period at a particular place. The samplers collect pollen grains. A sampler has a drum or rod coated with silicone grease or sticky cello-tape. During trapping the rod/drum is rotated one turn only in an entire 24 hours of a day with the help of a motor fitted with the sampler. The trapped pollen grains are then analyzed for identification and quantification.

Anemophilous plants of local vegetation determine the frequency of airspora and so the pollen counts. Meteorological factors also influence the pollen count. The change in climatic condition has a profound effect on the concentration of air spora. Pollen count increases in dry, warm and breezy days. It decreases in chilly with high humidity and rainy days. Moreover pollen count can be changed due to pollution, industrialization, population growth, tree plantings and cuttings, land use etc.

Singh *et al.* (2004) published a detailed list of airspora of the various parts of India along with their sources in literature. Chanda and Sarkar (1972) reported the existence of the following pollen at the atmosphere of greater Kolkata. The grass pollen contributed 39% to the total pollen present in air. The dominant pollen types are from *Amaranthus, Argemone, Azadirachta indica, Caesalpinia, Carica papaya, Chenopodium, Mangifera indica, Xanthium strumarium* etc. In West Bengal 59 types of pollen in air were reported (Singh *et al.*, 2004). In the month of May the concentration of pollen was maximum. The maximum concentration of pollen grains belonging to the family Asteraceae and Chenopodiaceae was found in the month of June.

Pollen grains have impact on human systems. The air, without which life cannot exist, contains an array of bioparticles among which pollen grains form one of the dominant concentrations. People inhale the free-floating pollen grains and spores without knowing them. Pollen enter and move through nasal passages to the throat. People, who are not sensitive to pollen, swallow or cough out the grains. But different conditions arise in pollen sensitive individuals. These people manifest hypersensitive reactions that are called allergy. von Pirquet in 1906 introduced the term allergy. The term allergy is defined as "an altered or accelerated reaction of a person to a second or subsequent exposures to a substance, usually harmless to the general population, to which he/she has been sensitized during the first exposure" —Shivanna (2003, p. 41) and Agarwal *et al.* (1998, p. 358). The sensitizing agents are known as allergens and it is defined as a substance to which a person is allergic (e.g. pollen and spores, food protein, house dust mites, furry animals etc.). Sometimes the term allergen is used as a synonym for antigen or immunogen. It is protein in nature. The symptoms of pollen allergy are as follows: in case of asthma coughing, wheezing and difficulty in breathing due to narrowing of the bronchial passages and excess mucous production; in case of eczema rash and itchy skin; in case of hay fever also known as seasonal rhinitis sneezing accompanied by a runny or clogged nose, itching nose, throat and eye, persistent upward rubbing of nose by a child, conjunctivitis, watering eyes; the lips and mouth show symptoms in case of gastrointestinal allergy. The upper tract respiratory allergy, naso-bronchial allergy and seasonal rhinitis are very common. The symptoms are manifested after respiratory tract infection. If asthma symptoms accompany the symptoms of hay fever, it indicates that the bronchial tubes have become involved indicating the need for immediate medical attention.

Hay fever or allergic rhinitis also called pollenosis (also spelt as pollinosis) is the best-known allergy of all. *Ambrosia* pollen grain is the dominant pollen type to cause allergic rhinitis and pollen asthma in North America. The allergy problems are caused by *A. artemisiifolia* and *A. trifida* although 21 species of *Ambrosia* occur in North America. *Ambrosia* pollen grain is also reported from the atmosphere of Tulsa, Oklahoma (Levetin, 1998). The other pollen types that cause hay fever throughout the world are *Alnus, Betula, Chrysanthemum, Populus, Salix* etc. In India Singh *et al.* (2000) published a detailed list of allergenic species during spring, autumn and winter. Mention may be made of grasses (*Cynodon dactylon,*

*Eragrostris tenella, Poa annua* etc.), weeds (*Cannabis sativa, Amaranthus spinosa, Chenopodium album, C. murale, Cassia occidentalis, Xanthium strumarium, Ricinus communis, Suaeda fruticulosa* etc.) and trees (*Alnus nitida, Cedrus deodara, Eucalyptus* sp., *Putranjiva roxburghii* etc.).

Free-floating allergenic pollen grains enter in the human system through nostrils during inhalation. Most of them are swallowed and others are either coughed off or remain in the nose. Within a few minutes after entrance a chain of reactions occur that leads to the release of chemical mediator like histamine. Histamine induces allergic symptoms in pollen sensitive individuals. When pollen and spore come in contact with the mucous membranes present in nose, respiratory tracts, stomach, oral cavity and eyes, the mucous secretion hydrates them. The mucous liquid helps to dissolve the pollen allergens. The allergens then diffuse out of pollen and penetrate the mucous tissue. The mucous tissue contains mast cells that are present in the upper respiratory tract, intestinal tract, lungs and skin. The allergens induce the formation of antibodies. The antibodies are largely of Immunoglobulin E class or in short IgE. It is glycoprotein in nature. The mast cells of allergic person have a high concentration of IgE. IgE occurs on the surface of mast cells. As more and more pollen allergens are released they bind and cross-link with specific IgE. Thus a complex is formed between antigen, e.g. pollen allergen and IgE antibody. The antigen-antibody complex activates the enzyme that causes the release of chemical mediator like histamine. Histamine induces edema of the mucous membrane thus causing narrowing of nasal or bronchial passages, contraction of nasal or bronchial muscles, the dilation of blood capillaries, excess mucous production and secretion of watery nasal fluid, sneezing and itching (Chanda, 1994).

People with hay fever may at first suspect that they have a summer cold. It is important to consult doctor if the allergic symptoms persist more than a week or two. Doctor will study the medical history of the patient. If it is noted that the symptoms recur at the same time each year, the doctor will proceed to diagnose the causative agent of allergy under the hypothesis that a seasonal allergen is involved.

Respiratory allergy is a seasonal and local problem. It is related to the appearance and the concentration of specific airborne allergens in that particular area. It is to correlate among the presence and abundance of airspora type, meteorological conditions and respiratory symptom incidence.

Blood and skin test are performed to detect the causative agents of respiratory allergy. Among the blood tests mention may be made of Enzyme Linked Immuno Sorbent Assay (ELISA), Paper Radio Immuno Sorbent Test (PRIST) and Radio Allergo Sorbent Test (RAST). These tests indicate the amount of IgE present in the serum of patients. Consult Shivanna (2003) and Bhattacharya *et al.* (2006) for details of these tests. Although these tests offer advantages over skin test, they are time consuming, expensive and somewhat less sensitive. These tests become compulsory for those patients who have certain disorders in the skin. The skin test is most commonly employed as it is simple, convenient, and highly specific and less expensive. To perform the test the antigenic extract of each type of pollen found in the local area of the patient is applied to a scratch made on the arm of patient or injected under the patients' skin. In positive reaction at the test site of skin (scratch/prick) an elevated reddened area with a surrounding flush, called wheal, appears. The size of the wheal provides the diagnostic clue.

Medication is the best treatment to relieve symptoms of allergy due to pollen and spore. Though antihistamine and corticosteroids are very effective in controlling allergic disorders, they have serious side effects after prolonged usage. Avoidance of allergen as a primary mode of therapy might give relief to some extent. Avoidance includes the followings: (1) reduction in exposure to the particular species in which a patient is allergic, (2) moving to a place where the allergenic plant does not grow. Allergy specialists strongly discourage this approach because patient may develop allergies to other plants. (3) It is advisable to remain in indoors when the pollen count is highest especially in the early morning and late afternoon. (4) To wear Face Mask, which is designed to filter pollen out of the air during inhalation. (5) After returning from outdoors it is recommended to take bath and use fresh cloth. (6) Use of air cleaners like air conditioners and air purifiers inside the rooms can reduce pollen level inside. (7) To dry wet clothes in dryer instead of hanging outside, where pollen may get trapped on the wet clothes. (8) Pollen grains generally settle down in the evening and so it is advisable to close all windows in the evening. (9) To avoid planting the allergenic plants around house. (10) The house compound must be free from weeds and grasses. (11) To install Electronic/electrostatic precipitator (Singh *et al.*, 2004).

Respiratory allergy is a very common disease among all populations all over the world. In India 20 to 30% of the population suffer from allergic rhinitis and 15% suffer from asthma (Singh *et al.*, 2004). So it is necessary to develop a network of centres in various biozones of India to collect data of the dispersed palynomorphs and sporomorphs. These centres will prepare pollen calendars from the collected data and broadcast the pollen count. This will be very useful for clinicians and as well as patients. The centres will also perform the clinical and immunological tests based on local allergens to local patients. In this respect some work has been done. Singh *et al.* (1992) published the pollen calendars of 12 different states of India.

In conclusion it is to mention that aeropalynology comes under the purview of Aerobiology. Aerobiology includes the study of bioparticles present in the atmosphere in addition to pollen and spores. These bioparticles might affect human and animal health, crops etc. Aerobiology also includes the identification of bioparticles and their geographical origin. The study of morphology, physiology, concentration, dispersion and seasonal patterns of bioparticles are also included in the study of aerobiology. In aerobiological research palynologists in collaboration with clinicians search the causative agents present in the atmosphere that have impact on human health.

**Melissopalynology**: Melissopalynology or melittopalynology [respectively also spelt as mellissopalynology or mellittopalynology. International Commission of Bee Botany (ICCB), Louveaux *et al.*, (1978) spelt as melissopalynology. So in this essay the spelling of ICCB is followed] is the study of pollen and spores present in honey. By extension, it also includes honeydew elements (HDE), i.e. fungal spores, hyphae and microscopical algae and other bioparticles. 'Melissa' and 'melitta' mean a bee according to Paxton's Botanical Dictionary, 1868 (Bryant, 2001). *Apis mellifera* is the scientific name of common honeybee. In *Latin* the word *mellifera* means honey. In addition to *Apis mellifera* L. the other Indian honeybees are *A. indica* F., *A. florea* F. and *A. dorsata* F.

Honeybee requires nectar, pollen, resin and water for their survival. The bees are adopted for feeding on nectar and pollen. Resin reinforces the hive and water cools the hive. Moreover water dilutes the honey that is fed to larvae. In a hive there are three types of bee: the single queen bee, a variable number of male bee or drone and 20000 to 40000 or more worker bees. The worker bees travel long distances to collect nectar. They suck nectar from nectary present in the flower. The nectar is stored in the crop of bee. Regurgitation of nectar starts before the return of bees. On return to the hive the other worker bees remove the nectar. The foraging bees and other worker bees regurgitate the nectar. Regurgitation occurs a number of times till the nectar is transformed into honey. During the process the natural complex sugar (sucrose) of the nectar is converted to simple sugar (fructose, glucose etc.) by the enzymes. The sugary fluid is then stored in honeycomb, which is left unsealed. The honeycomb is a group of hexagonal cells composed of beeswax and propolis. The sugary fluid is high in water content. The formation of ripe honey is accomplished by evaporating the excess water. This happens when the bees inside the hive fan their wings thus circulating air across the honeycomb. The reduction in water content causes the increase of sugar concentration. The honeycomb containing the ripe honey is then capped (sealed) with beeswax and propolis.

Honey is thick, sweet syrupy, white or yellowish or brown or other colours, viscid fluid. Honeybees from the nectar of flowers produce it. Honey is consumed primarily as an energy source and is stored in the honeycomb for unfavourable situation like winter etc. Pollen grains are used as the primary source of protein and other nutrients. Honey is composed of 14-18% of water, 76-80% of glucose, fructose, levulose, dextrose etc., pollen, mineral salts (compounds of calcium, potassium, sodium, magnesium, copper, silica, phosphorus, manganese, silicon, iron etc.), acids (malic, tartaric, citric, succinic etc.), plant pigments like carotene, xanthophyll and chlorophyll; enzymes like invertase, diastase, catalase etc., vitamins ($B_1$, $B_2$, K, folic acid, biotin, pyridoxine, pantothenic acid etc.) and amino acids (cystine, lysine, glycine, aspartic acid, glutamic acid, alanine, valine etc.). The composition varies and it is dependent upon the flower that provided the nectar. The variation in colour is also dependent on the plant that foraged the bee, i.e. alfalfa and clover produce white honey and *Acacia* produces straw colour honey. The honeys produced from cotton, *Plectranthus* and *Eucalyptus* are also white. Golden or light yellow coloured honeys are obtained from litchi, mustard and rubber. Honey is significantly sweeter than table sugar. It was almost the

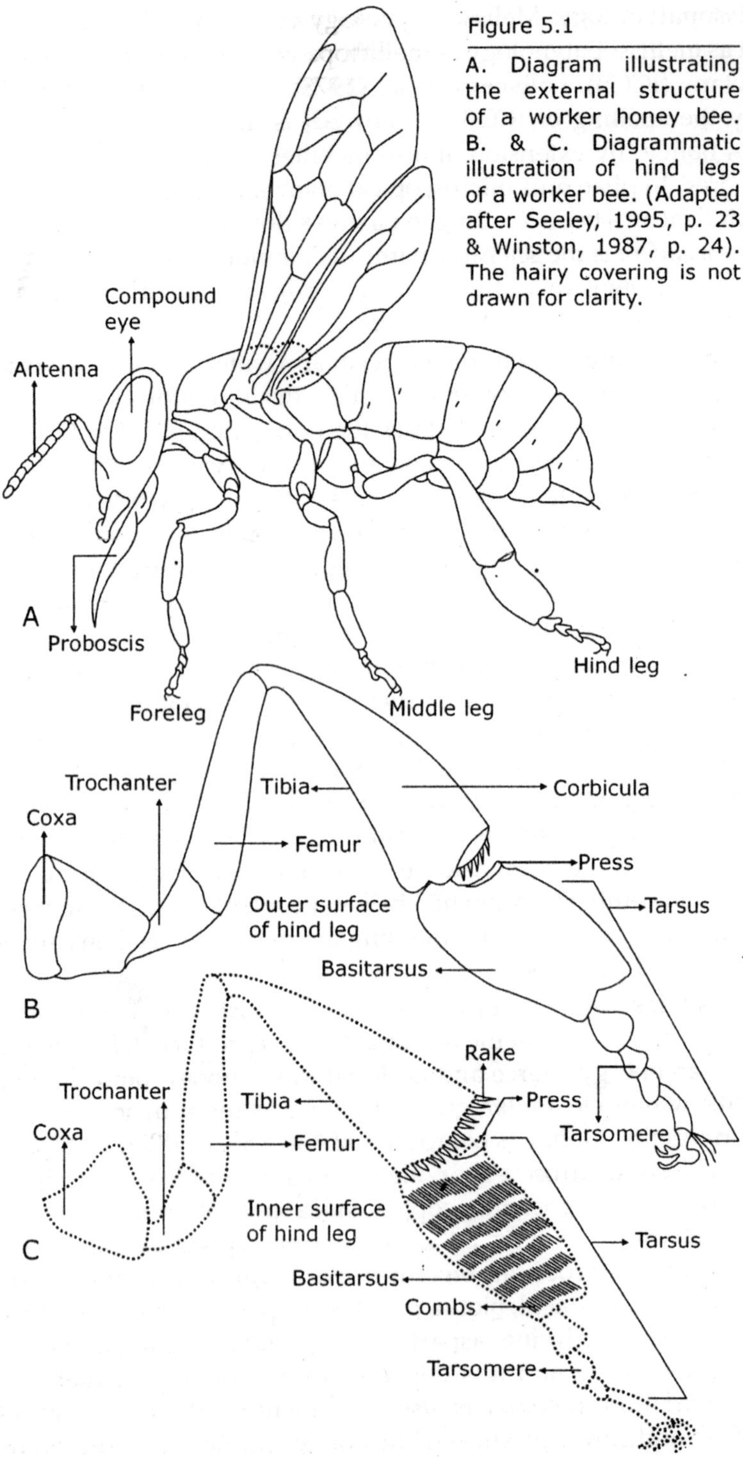


Figure 5.1

A. Diagram illustrating the external structure of a worker honey bee. B. & C. Diagrammatic illustration of hind legs of a worker bee. (Adapted after Seeley, 1995, p. 23 & Winston, 1987, p. 24). The hairy covering is not drawn for clarity.

sole sweetening agent from ancient times until cane sugar became commercially important. Honey being a complete natural and biological product and free from any type of impurities, it is often preferred over sugar. It is one of the best health food consumed by man. Honey has become a common ingredient in many types of ayurvedic medicines as it is found to contain certain types of medical benefits. Apart from food, honey has always been held in high regard. In language, literature and culture honey is frequently used as a symbol that is supposed to bring good luck or to keep away evil. It is the symbol of sweetness of every kind.

The honey does not spoil easily. Due to the presence of low water and high sugar concentration, it does not allow to grow any airborne yeast or bacteria. Yeast/bacteria get plasmolyzed when they come in contact with concentrated sugary honey. So long the moisture content remains below 18% almost no microorganisms can multiply in honey.

Pollen grains are collected during the visit of plants by bees for nectar. The legs of bee help in this process. The legs are not simple structure to walk on. They are highly specialized organs that enable the bee to collect pollen. The legs are present on the thorax of a bee. The bee thorax consists of three body segments and six legs, one pair of legs being present per thoracic segment (Fig. 5.1A). Each pair of legs has the same construction, i.e. consists of coxa, trochanter, femur, tibia and tarsus. The tarsus consists of five sub-segments—one elongated basitarsus and four smaller tarsomeres. The basitarsus has pollen brush on the inner side. The hind legs of worker bees are highly modified to carry pollen and propolis. The tibia of worker bees has corbicula, commonly called pollen basket, on the outer surface of tibia (Fig. 5.1B). The pollen basket is an expanded, slightly concave region of tibia. Its surface is smooth. A stiff row of bristles borders the inner edge of tibia, called pollen rake or rastellum. The basitarsus of worker bees has pollen combs on the inner side that consist of a regular series of stiff bristles (Fig. 5.1C). The base of basitarsus has a flattened area called pollen press. The above mentioned structures are meant for manipulating pollen into the pollen basket. During foraging pollen grains stick to head, abdomen and other body parts. The bee moistens the forelegs with the protruding tongue and collects pollen grains from the front body. Pollen grains from the forelegs are then transferred to middle legs with the help of pollen brushes. The pollen grains now present on the middle legs are passed to the pollen combs present on the inner basitarsus of inner hind legs. This happens when pollen combs scrape pollen from the middle legs and abdomen. Then pollen grains pass to pollen basket of opposite hind legs. This occurs in a very ingenious way. The pollen rake scrapes the inner surface of each opposing pollen comb (Fig. 5.2). As a result pollen grains come to lie on pollen press of opposite leg. Pollen grains thus accumulated over the press are pushed into the pollen basket by straightening the hind leg. As a result pollen grains are pressed, compacted and squeezed into the pollen basket. Pollen-collecting-bees can be recognized instantly by conspicuous balls of bright-coloured pollen packed onto their hind legs. All the above manipulations regarding the collection of pollen and deposition of pollen in pollen basket are carried out at brief intervals. Thus the pollen basket is gradually filled with pollen. The basket

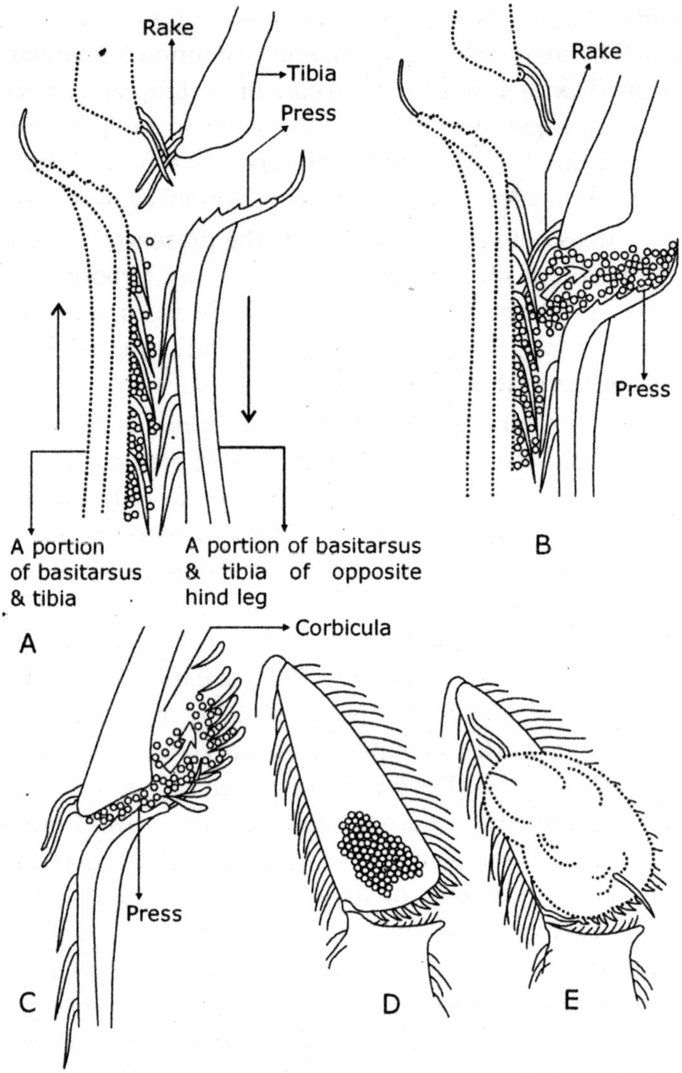


Figure 5.2

Diagram illustrating the transfer of pollen grains from combs of one hind leg to the pollen basket of the opposing hind leg. A. Rubbing of hind legs against each other during transfer of pollen. A portion of basitarsus and corbicula are shown. Arrows indicate the direction of scraping. B. Pollen rake scrapes pollen grains from combs of opposing hind leg and deposits them on press. C. Pushing of pollen grains into pollen basket by pumping the leg. D. Pollen aggregate within corbicula. E. A corbicula (surface view) filled with pollen grains. Adapted after Winston, 1987, p. 27.

is also filled with pollen mixed with nectar and regurgitated nectar. The compressed pollen content in the pollen basket is known as pollen pellets. A microscopical study of pollen grains present in pollen pellets will reveal the plant species that the bee visited during foraging.

Honeybees also collect honeydew instead of nectar. Honeydew is a sugary substance secreted by aphids and other plant-sap-sucking insects. Honey is also produced from honeydew by bees and is known as honeydew honey. It consists water (16%), fructose (38%), glucose (27%), sucrose (7%), dextrose (9%), acids (7%) and minerals etc. The following plants are the hosts of aphids and other insects that produce honeydew—*Abies alba, Acer, Corylus, Fagus, Fraxinus, Carya, Larix, Malus, Pinus, Prunus, Quercus, Salix alba, Sorbus aucuparia, Ulmus americana* etc.

Melissopalynology is an early branch of palynology (the study of pollen and spore). Pollen analysis of honey or melissopalynology has great importance because it reveals the plant species foraged by honeybees. The identified plants provide the fingerprint of the environment and pollen spectrum from where the honey comes. Pollen analysis also provides the botanical and geographical origin of honey, information about honey extraction, presence of contaminants like rust and smut spores, kinds of adulterant etc. Today pollen analysis has become an indispensable tool to authenticate honey.

Microscopical analysis is the prerequisite to determine the pollen spectrum of a honey sample. For microscopical examination the honey sample is diluted with water and the microscopical elements are concentrated by centrifuging. The elements are studied under microscope either after acetolysis (it is a technique used in the preparation of slides of pollen grains and spores to study the morphological details specially the sporoderm stratification) or without further treatment. The procedure of acetolysis is usually omitted to determine HDE, thin-wall-pollen grains and other bioparticles as they get dissolved during the process. For the identification of pollen grains acetolysis is a must because palynological literatures provide the identifying characters from acetolysed materials.

Slides are prepared for microscopical examination. Louveaux *et al.* (1978) and von Der Ohe *et al.* (2004) published the details regarding the method of preparation of slide and counting methods of pollen. In a nutshell, the sediments after centrifugation (acetolysed/without any treatment) of a honey sample are spread evenly with a micro spatula over an area of 22mm x 22mm on the slide. For this purpose a square of 22mm x 22mm was drawn on the back of microscopical slide with a permanent marker to prevent unintentional dyeing of pollen and discolouration of the markings. The spread was mounted in glycerine jelly and covered with cover slip of 22mm x 22mm or other appropriate size.

The microscopic analysis of the honey sample includes the identification of pollen grains and other microscopic constituents (e.g. HDE), quantification and determination of frequency classes and their percentages. A collection of reference slides is indispensable for the identification. The reference slides may be acetolysed or unstained. In the latter case the pollen may be degreased or without the removal of fat. It is recommended to collect mature buds instead of open flower and letting them to open in the laboratory for preparation of slides to avoid contamination from anemophilous pollen.

In a honey sample pollen of different plant species may be present as the worker bees visit different plants. In addition anemophilous pollen may be incorporated into honey. Airborne pollen grains (ex. Gramineae, Cyperaceae, *Rumex, Cannabis, Quercus, Amaranthus, Abies, Pinus, Picea, Larix, Alnus, Betula* etc.) enter the hive on wind currents. Moreover airborne pollen grains deposit when bee-keeper removes honey from a hive. A honey sample may also contain honeydew elements like spores of Uredinaceae, Ustilaginaceae, Peronosporaceae etc. Pollen from non-nectariferous and more or less entomophilous (ex. *Papaver, Plantago, Thalictrum,* Chenopodiaceae, *Artemisia* etc.) may also be present in a honey sample. Sometimes bees visit these plants for pollen collection.

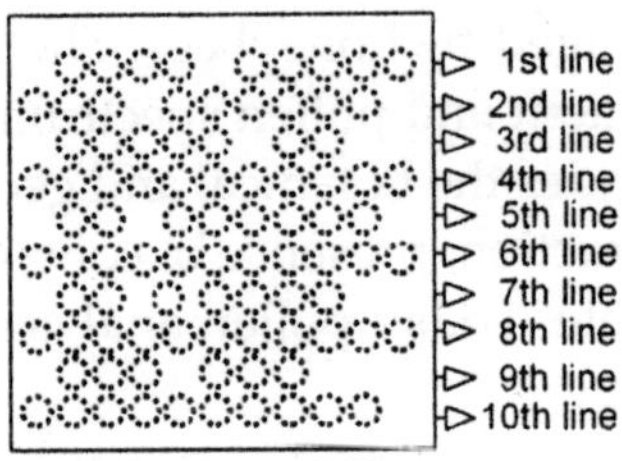


Cover slip 22 mm X 22 mm

Figure 5.3

Diagram illustrating the procedure used in counting 1000 pollen grains for homogenous examination of the slide. From each line 100 grains are to be counted. ⬚ represents the whole microscopic field of view. (Adapted after Der Ohe *et al.*, 2004, p. S20).

After the identification of pollen grains in a honey sample, it is to estimate the relative frequencies of pollen type. It is necessary to count 1000 pollen grains for the determination of relative frequencies. The sequence of counting is illustrated in Fig. 5.3. Pollen of wind-pollinated or nectarless plant, pollen of nectarless but more or less entomophilous plant, HDE consisting of fungal spores and hyphae, algae and plant-pathogenic fungal elements, e.g. spores of Uredinaceae and Ustilaginaceae, *Peronospora* etc. are to count separately. The terms used in interpreting the counts and estimating the pollen grain frequencies and their frequency classes are illustrated in Table 5.1.

| % of pollen in the total count | Terms used to express the frequency of entomophilous pollen | Terms used to express the frequency classes of entomophilous pollen | Terms used to express the frequency of anemophilous and other nectarless plant |
|---|---|---|---|
| > 45% | Very frequent | Predominant pollen | Very frequent |
| 16 - 45% | Frequent | Secondary pollen | Frequent |
| 3 - 15% | Rare | Important minor pollen | Rare |
| < 3% | Sporadic | Minor pollen | Sporadic. |

Table 5.1

Table representing the terms used to express the different pollen frequency classes in relation to % of pollen in the total count. Adapted after Louveaux *et al.*,1978, p. 144.

After analyzing the pollen and HDE contents it is possible to deduce the pollen spectrum of a honey sample. This can be verified with pollen spectrum of the declared botanical and geographical origin of honey. For the determination of botanical origin the frequencies of pollen type are calculated on the nectariferous species only. The pollen grains from the anemophilous and nectarless plant are excluded from the count. The determination of botanical origin is possible for those honeys that are extracted centrifugally. Sometimes honey is filtered through diatomaceous earth or similar materials to remove pollen and other particles. In this case due to lack of pollen the botanical origin is to be verified by other method. The honeydew honey contains higher percentage of pollen from nectarless plant than that of flower-honey. The honeydew honey also contains more HDE than pollen (P), the ratio being HDE/P > 3. The geographical origin of honey can be verified by comparing with pollen spectrum of that region. In some instances the geographical origin can be determined by the presence of characteristic pollen of that region, e.g. the presence of *Banksia* pollen indicates Australian origin; the presence of *Roystonea* pollen is the geographical marker of Cuban honey etc. The pollen spectrum of honey depends on the local vegetation where it was produced. Political boundaries are rarely associated with it. So microscopical analysis and interpretation indicate the geographical provenance of honey rather than the country of origin.

Honeys are of two types – unifloral honey and multifloral honey. In the former the bees collect nectar from one particular source plant, e.g. heather honey, citrus honey etc. In multifloral honey the bees collect nectar from different source plants e.g. honey sample obtained from greater Kolkata contains the pollen grains of *Cassia, Cocus, Solanum, Tridax, Borassus, Croton, Polygonum,* Chenopodiaceae etc. The important honey plants of Jammu & Kashmir are *Mesembryanthemum, Adhatoda vasica, Celosia, Impatiens, Tecoma, Opuntia, Cassia, Helianthus* etc. In Kashmir region the following honey plants are present: *Salix, Prunus, Pyrus, Juglans, Rosa macrophylla, Cucumis, Frageria, Iris, Robinia* etc.

In unifloral honey it is frequently observed that the pollen grains are over-represented, i.e. the percentage of pollen in honey is greater than the percentage of corresponding nectar. Ex. *Castanea* (>80%), *Brassica* (>60%) etc. The reverse situation is also observed where the percentage of pollen in nectar is greater than the percentage of the corresponding honey; in this case the pollen grains are under-represented. Ex. *Citrus* (10-20%), *Medicago* (20-30%) etc. In unifloral honey the percentage of single pollen type is 45% or more. A single pollen type with 45% concentration is the minimum requirement for a honey to be classified as unifloral. The over-represented type fulfils the above criterion only. There is no value to determine the under-represented pollen type. Later the idea of absolute pollen concentration (APC) of a particular flower type was developed. In this concept a pollen coefficient class expressed as an average number for a particular flower type was deduced to determine the unifloral honey sample (under - represented or over - represented) regardless of data revealed by calculating the percentage of relative pollen concentration. The average number of pollen coefficient class was obtained in the following way. A small number of caged honeybees were allowed to feed on flowers of a particular plant species. The honey thus produced contains the pollen grains of that particular plant species. It is expected that this honey sample is the valid representation of absolute pollen concentration (APC) for that particular flower type. Different plant species of unifloral honeys were used as experimental material and in each case an average number of APC is deduced. These data were illustrated in a tabular form called pollen coefficient table. These data are used to determine the unifloral nature of honey in question.

Unifloral honey is always preferred over multifloral honey. The former is considered as the premium type and has high demand. Microscopical analysis determines the botanical and geographical origin of the premium honey. But the verification of the source plant often becomes difficult because many samples have pollen grains that are traditionally under-represented. So pollen coefficient values are currently being used throughout the honey industry to verify the premium honeys.

From ancient times beekeeping was traditionally practiced for bees' honey harvest. The other major hive products are beeswax, royal jelly, propolis and pollen.

Beeswax is the byproduct of the honey making process. The young worker bees secrete it. The bees have eight wax-producing glands on inner side of the abdomen. These glands secrete wax in the form of thin scales. Bees use the wax to build the honeycomb and to seal the opening of the filled honeycomb cells. The honeycomb is the site of storage of pollen and honey, and of raising the young bees. The colour of beeswax varies from nearly white to brownish. It is usually yellowish and the colour is dependent upon the plant species that the bees visit. Beeswax is used for the production of candles, furniture polish, art materials, cosmetics, drugs etc.

Royal jelly (RJ) is the secretion of hypopharyngeal glands of bees. It is a creamy, milky white, acidic with high nitrogenous substances. It tastes bitter and has slight pungent smell. It is to feed the queen bees throughout their larval and adult lives. Though it is not derived directly from plants RJ contains pollen grains coming from the foraging activity of honeybees. So pollen analysis of RJ will reveal its geographical origin and the environment where the beehive is located.

Royal jelly contains B-complex vitamin, water, several amino acids, simple sugars, fatty acids, trace minerals, enzymes, antibacterial and antibiotic components, and trace amount of vitamin C. Bees use RJ because it has the potential utility as food for rearing queen bees. People use RJ as dietary supplement as it has medical benefits.

Propolis is a sticky, gummy and resinous substance. The bees collect it from the buds and barks of trees. In the hive it is mixed with beeswax, enzyme and pollen. Bees use porpolis (i) as cementing material to cement the honeycomb cracks thus reinforcing the structural stability of the hive, (ii) to seal the alternate entrance thus making the hive more defensible, (iii) as a disinfectant, (iv) to protect colonies against bacteria, putrefying corpuses, and unpleasant odour etc.

The colour of propolis varies from green to brown and reddish. It depends on its botanical source. The composition of propolis also varies between hives, seasons and places of origin. In general it consists of resins and vegetable balsam, beeswax, essential oils, pollen grains, amino acids, vitamins, salts of calcium, potassium, zinc, iron etc. It is sticky at room temperature. It becomes hard and brittle at lower temperature. Propolis has medical benefits and is mainly used as disinfectant.

Pollen grains are the source of vitamins and proteins to the bees. It is highly nutritious and is used as dietary supplement. To collect it, the beekeepers force the bees to fly through meshes (also called pollen trap) where the openings scrape the pollen from the legs of bees. The pollen grains are dropped in a container from where they are collected.

There are certain plants where pollen are toxic to bees and the cause of bee poisoning, i.e. *Rhododendron, Kalmia, Corynocarpus, Zygadenus, Cuscuta, Aesculus, Tilia, Solanum* etc. (Abrol, 1997). In case of pollen poisoning dead bees are found near the plants that the bees visited while in pesticide poisoning dead bees are found near the hives.

Honey is not always edible. It may contain allergenic pollen and poisonous pollen grains, e.g. pollen of *Datura, Rhododendron* etc. Moreover the bacterium *Clostridium botulinum* can survive on honey and cause botulism in infants when fed.

Lastly it is to mention that the correctness of pollen identification, their quantification and determination of relative frequencies, and the subsequent interpretation of results of a honey sample form the basis of melissopalynological analysis.

**Forensic palynology**: Forensic palynology is the study of palynomorphs trapped in materials associated with criminal or civil investigations to solve criminal cases and resolve other legal problems. It also includes identifying and linking the suspect to the scene of the crime. According to Mildenhall, 1992 "The term Forensic palynology refers to the use of pollen and spore evidence in legal cases" – Bryant *et al.* (1998). By extension the study of dinoflagellates, acritarchs and chitinozoans that are present in fresh water and marine environments (Faegri *et al.*, 1989) are also included in forensic palynology. Most of these microorganisms are restricted to fossil deposits and sometimes provide legal information. Forensic palynologists seldom encounter these microorganisms.

Forensic investigators have lately employed the implication of pollen morphological characters in solving criminal cases. Most pollen grains are aperturate and each pollen grain and spore have definite shape, size, aperture and sporoderm stratification. The apertures have definite number, position and character (NPC). The exine often bears processes of various types, which form patterns and ornamentations. With the aid of NPC, sporoderm stratification, size and shape it becomes possible to identify pollen. Identification includes family, genus and sometimes species or even smaller taxonomical units. Palynologists have sometimes called pollen and spore 'the fingerprint' of plants, as they are distinctive, easily recognizable and identifiable.

Most pollen grains fall on any exposed surface to the air. They may end up as components of soil and dust on objects. The types of pollen present at any one location, or on any object and their relative frequency will reveal the particular vegetation type of a site or a region. This is unique for a location, a larger geographical area or even a country. In other words 'pollen fingerprint' of a region can be obtained from the pollen assemblage of that region. The pollen fingerprint is used as the control sample as it reflects the regional vegetation. It is this aspect that enables the forensic palynologists to compare pollen samples between a suspect, victim, object and localities. It also enables to prove and disprove a relationship between them.

The value of forensic palynology stands on the following attributes of pollen and spore. First, pollen and spores are produced in large numbers and so they are abundant in the atmosphere. Second, they are microscopic in size. They may be unknowingly taken away or left at a crime scene as evidence in large numbers. They remain unnoticed, as they are microscopic. They cannot be removed from a crime scene. Third, sporopollenin is present on the walls of pollen where it forms species-specific-pattern. The pattern persists for thousands of years. Moreover due to the presence of sporopollenin pollen grains are very much resistant to damage and decay. They get well preserved for many years in nature without specialized storage. The identifying characters like size, shape, aperture and sporoderm stratification remains unchanged after long periods of time and thus enables to identify a pollen grain. Fourth, pollen grains and spores can be identified to specific plant type, site, region or country. To identify pollen and spores palynologists compare forensic pollen samples with references. The references may be in the form of slides, drawings and photographs published in literatures. Moreover a large number of Internet Websites provide images and description of pollen of regional vegetation.

A basic knowledge regarding the production and dispersal pattern of pollen of regional vegetation is extremely useful to the crime scene personnel. It enables the personnel to decide whether pollen analysis will be of any probative value at the crime scene in question. If the pattern of pollen dispersal and pollen productivity of vegetation are known, then the expected pollen content of a forensic sample from that particular region can easily be predicted. If atypical species are present in the forensic sample, research is needed to find the reason behind the presence of atypical species.

Each plant has its own method of pollen dispersal. So the flowers may be hydrogamous, autogamous, cleistogamous, zoogamous and anemophilous. Accordingly the production of pollen per anther varies. Pollen, whose exine contains sporopollenin, are only preserved in long term sediments. Pollen of hydrogamous plant lack sporopollenin and so rarely preserved. All the above criteria, i.e. methods of pollen dispersal, production of pollen per anther and presence or absence of sporopollenin are essential in determining the forensic potential of different plants. The forensic potential of plants with different methods of pollen dispersal is depicted in Table 5.2.

| Pollen dispersal method | Pollen produced per anther | Occurrence of pollen in forensic sample | Comments on forensic potential |
|---|---|---|---|
| Water in hydrogamous flower | 1000 or more | Sporopollenin absent and so pollen are rarely preserved | Generally of little value in forensic investigation |
| Pollen of same flower pollinate the stigma in autogamous flower | < 100 | Rarely dispersed in the atmosphere and so pollen are rare in sediments of nearby plants | Though rare, it has high value when present in forensic investigation |
| Cleistogamous flower never open and so pollen of same flower pollinate the stigma | < 100 | Very rarely dispersed in atmosphere and so pollen are rarely present in the sediments near the plants | Though rare, it has high forensic value when present |
| Insect, bird, bat and snail are the agents of pollination in Zoogamous flower | 100 - 1000 | The pollen may be few to many in the sediments of nearby plants | The forensic value is excellent as the assemblage may be free from atmospheric contaminate |
| Wind is the agent of pollen dispersal in Anemophilous flower | 1000 - > 100,000 | Pollen are abundant in the sediments near the plants and neighbouring areas | The forensic value may be poor or excellent and it depends on the plant association of plant types of a region |

Table 5.2

Table summarizing the different types of flower, their approximate number of pollen production per anther and their forensic potential (After Lynne *et al.*, 2005, p. 226).

The ability of pollen to travel on air currents or to remain aloft is also extremely useful in forensic palynology. The 'sinking rate' or speed of a pollen grain determines it. The sinking speed is the rate of movement of a pollen grain while sinking to the ground and is usually measured as the distance travelled in unit time. As for example, the sinking speed of *Alnus, Betula* pollen is about 2 cm per second (Bryant *et al.*, 1998). Several factors influence the sinking speed, e.g. the mass and size of pollen and their aerodynamic shapes, wind currents, humidity, obstacles in the movement and the height at which they are released from anther. Researches reveal that 95% anemophilous pollen travel 25m to 100km and few may be carried thousands of kilometres. The small and light pollen move longer distances than the large and heavy pollen, because the fall or sinking speed of the former is much lower than the latter. So the potential distribution area of large and heavy pollen is small and restricted. To explain the potential distribution area, Bryant *et al.* (1998) cited the examples of *Zea mays* and *Abies* pollen. Pollen of the above plants are large and heavy. They fall to the ground or on any object fifteen times faster than the small and light pollen found in *Alnus, Betula, Cannabis* and *Juniperus*. Studies show that the dispersion area covered by *Zea* and *Abies* pollen is smaller and more restricted than those of *Alnus*, *Betula*, *Cannabis* and *Juniperus* pollen. The presence of large and heavy pollen like *Zea* and *Abies* in a forensic sample will indicate the small dispersion area. This will enable to identify the source plant and accordingly the region. So, the forensic potential of anemophilous pollen may be poor or excellent and it depends on the association of plant types of that region from where the anemophilous pollen originated (Lynne *et al.*, 2005).

The value of pollen evidence in forensic investigation depends on careful collection, documentation and preservation for later forensic analysis of the collected sample (Fig. 5.4A). To obtain accurate results the forensic samples are collected and processed with great care. Forensic palynologists collect, store and analyze the samples with great precision. Precautions are taken throughout all the steps to avoid contamination from atmosphere or other sources. The source material of forensic pollen may be almost anything that are exposed to or in contact with air. Even the packaged food and dry food like prunes, sultanas etc. contain pollen from where the food originated. The source material of forensic pollen is different for each case. Lynne *et al.* (2005) published a detailed list of materials from where pollen in forensic cases were collected. Mention may be made of air filters, antique furniture, apricots, breads, *Cannabis*, carpet, car, clothes, colons, dirt and dust, dried fruits, fur, hair, honey, intestines, leaves, mud present on victim and suspect's belongings, soil, wood, wool, weapons etc. Each item has an interesting case history and Lynne *et al.* (2005) illustrated many of them.

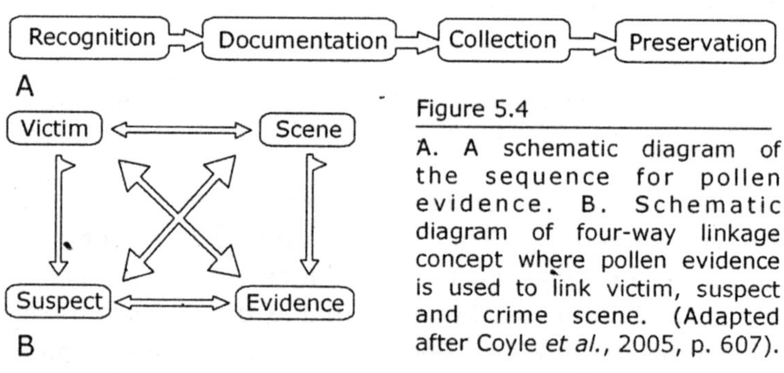


Figure 5.4

A. A schematic diagram of the sequence for pollen evidence. B. Schematic diagram of four-way linkage concept where pollen evidence is used to link victim, suspect and crime scene. (Adapted after Coyle *et al.*, 2005, p. 607).

The collected pollen samples are stored in secure location to avoid any intentional switching or contamination. The result of forensic pollen analysis will guide the investigator to establish link between victim, suspect and crime scene using four-way linkage concept (Fig. 5.4B). Crime scene documentation is necessary to establish the relationship of evidence to overall scene and other evidences. The documentations are usually in the form of photography, sketches, diagrams, general note taking and videotaping. Documentation is important for any type of evidence and it is meant to complement the other.

Forensic palynology has been utilized in many cases, e.g. assault, rape, forgery, drug dealing, armed robbery, fraud etc. In many cases mud, dust or dirt collected from shoes, vehicles etc. for forensic evidence contain pollen from more than one locality. The pollen types of forensic samples are compared with the control samples of pollen assemblage from different region. Thus it becomes possible to trace the region where the shoes or vehicles were present and to determine the travel history of the same items. Similarly pollen analysis from soil present on fingernail or clothing enables to reconstruct the recent movements of a person or animal. The travel history of a new car can be determined by pollen analysis. Pollen, collected from a car radiator and their subsequent analysis, reveal the area and path the car traveled. The presence of similar type of pollen between the forensic samples collected from shoe prints and the shoes that made the prints indicate a match correlation. Pollen evidence is very much effective in crime scenes that are restricted to a few square metres, such as rape or burglary. A comparison between forensic pollen samples collected from the entry point of burglary and suspect's shoe could provide link in a case. In many cases pollen evidence prompted the suspect to confess to their crimes. In a crime scene it happens that the body was shifted to the present location from other locality where the crime occurred. Forensic pollen analysis from victim's clothing reveals the pollen assemblage from both the localities, i.e. the locality where the body was present and the locality of the crime. Analysis of pollen type in the stomach and intestine of a victim revealed what the victim ate in the last meal. Accordingly the time of death or location can also be determined. Identification of pollen in stomach contents can be useful in verifying a person's alibi. Pollen analysis of a honey sample has been used to authenticate it as labeled. Forensic palynology can also be used to prove or disprove alibis, narrow down a list of suspects, determine the travel history of items and drugs etc., determine the geographical source of drugs, fruits and various other things.

There are several actual court cases where pollen evidence became important clues that helped to obtain conviction. Erdtman (1969) describes an Austrian case where pollen evidence was the basis to discover the murdered victim's body and convict the suspect. A man disappeared during a vacation at Danube, near Vienna. During investigation the police suspected one with a motive for killing the missing person, who was the male colleague of the suspect. The suspect was detained while on a boat journey along the Danube River. But the police could not progress further without the missing person's body or confession from the suspect. Later search was made on suspect's room and a pair of boots with mud attached to the soles was taken for forensic analysis. Geologist Wilhelm Klaus along with Austrian Geological Survey analyzed the mud sample attached to the boots. The sample revealed the presence of pollen both from a fossil and living plants. The living pollen grains were from pine and alder. The only fossil pollen grain was from miocene-age 20-million-year-old hickory plant. Fortunately one area was known where pine and alder pollen grew together on tertiary strata. Klaus pinpointed a small area along Danube valley where such a unique assemblage of living and a fossil pollen grain occur in the mud. This is due to the fact that in that area an exposed tertiary stratum was present and that contributed the fossil pollen grain. Pollen from pine and alder mixed up with hickory pollen.

The suspect must have walked there and thus attached the mud on the boots. The identification of the area and the establishment of link between the mud from the area and from the boots of suspect on the basis of similar pollen composition prompted the suspect to confess to the crime.

In another case in Australia a man was physically assaulted near a river in southern Queensland (Lynne *et al.*, 2005). The victim and assailant rolled into the river. When questioned, assailant denied presence anywhere near the river. The forensic samples were collected from the river as control sample. An analysis of the control sample revealed the presence of algal cysts and other palynomorphs. Forensic samples were also collected from the clothing of victim and suspect and the results of analysis of samples were compared with the control sample. All the samples, i.e. samples from river, victim and assailant had the same assemblage of algal cysts and other palynomorphs. This evidence prompted the assailant to confess to the crime.

Forensic palynology is also used to catch thieves. In New Zealand a rustler stole 300 sheep from a farm (Bryant *et al.*, 1998). Later a rancher who lived further north took 350 sheep to an auction for sale. The auctioneer knew the rancher. He became suspicious because the rancher's farm was not large enough to accommodate 350 sheep. The pasture can support about 200 sheep. The 350 sheep were impounded and the rancher was unable to show the bill of sale. He also failed to point the farm from where the sheep were bought. The rancher, who lost sheep, was asked to identify the impounded sheep. He inspected the sheep and believed they might be his sheep. He did not brand the sheep by putting any identifying mark. So he could not prove his claim due to lack of any mark of identity. The rancher, who put the sheep up for sale, insisted that he had bought the sheep a week earlier though he could not produce any evidence of purchase. Police collected the forensic sample from the pastureland from where the sheep were stolen and from where the sheep were put to auction. The pollen assemblage of these controls was analyzed. Forensic samples were also collected from the impounded sheep. Forensic palynologist sheared wool from small areas from the top of the backs and heads of impounded sheep and analyzed the pollen spectra of the wool. The result revealed that the pollen assemblage matched the type of plants that are present in the region from where the sheep were stolen. The pollen evidence confirmed that the impounded sheep were not from the farm from where they were put to auction. Thus, the pollen evidence helped to support the claim of the rancher who lost sheep to the impounded sheep.

The following is a good example where pollen analysis has been used to establish the time of death (Coyle *et al.*, 2005). A mass grave containing 32 male skeletons were discovered from Magdeburg, Germany in the month of February of 1994. The identity of victims was not known. It was assumed that they died in a military interaction as all the skeletons were of males. Two hypotheses were proposed: First, Gestapo at the end of World War II killed the victims in 1945 sometimes in the spring; second, the victims were Soviet soldiers. The Soviet secret police killed them in June of 1953 after the revolt of German Democratic

Republic. It was very critical to determine whether the soldiers were killed in spring or in summer. The skeletons were dug out of earth. Forensic control samples were collected and analyzed from the soil from where the skeletons were recovered. Forensic pollen samples were also collected from nasal cavities of exhumed remains to determine pollen that were inhaled before death. The nasal cavities were ringed with saline solution and thus pollen were collected. The forensic sample collected from the nasal cavity contained high amounts of pollen from plantain, rye and lime tree. All these plants flower in summer. Pollen analysis supports the second hypothesis. The skeletons were of Soviet soldiers. The Soviet secret police killed them after June 1953 revolt.

The following is the case example where pollen evidence narrowed the search list. A European company exported some machineries by sea to a location in Asia (Lynne *et al.*, 2005). The machinery was packed in wooden crates. On the way the ship stopped at a number of ports. At each port various cargo was loaded and unloaded. On reaching the destination it was found that the wooden crates contained soil bags only. Pollen types present on the soil were analyzed. Pollen analysis revealed the plants that are present in South Africa. The results of pollen analysis narrowed the search to Cape Town, the only port in South Africa where the ship stopped. Though the ship stopped at many other ports but they were away from South Africa. After several months the machineries were recovered from a warehouse in South Africa.

The above are a few illustrations where pollen evidence is used to solve crimes and resolve other legal problems. The forensic applications of palynology are almost limitless. Literatures and a large number of Internet Websites provide many case histories throughout the world. Pollen evidence is now accepted in courts of law. Pollen evidence has helped in solving several cases of murder, rape etc. in Sweden, New Zealand and many other countries. New Zealand leads the world so far as the forensic application of palynology is concerned. In India forensic palynology has not received much attention. If due attention is given many cases of rape, murder, kidnapping, adulteration of food and other eatables such as honey, milk etc can be solved easily (Dev, 2005).

## Palaeopalynology

Palaeopalynology (palaeo-/paleo- is the prefix used in a combining form to denote relation with fossils) is one of the disciplines of palynology and concerned with the study of fossil pollen grains and spores. By extension, the study also includes chitinozoans (animal remains), fungal spores, dinoflagellates, acritarchs and other organisms (except diatom) that are preserved in sedimentary rocks, composed of sporopollenin and chitin etc., resistant to acids and the size being 5 μ to 500 μ (Traverse, 1988, p. 509).

Palaeopalynology has become an applicative discipline of palynology due to the following features possessed by pollen grains and spores.

1. Pollen and spores survive better and longer than other biological materials due to the presence of tough exine.

2. The exine is mainly composed of sporopollenin that renders the pollen grains and spores resistant to decay. Grains survive well where microbial activity is depressed due to drought, low availability of oxygen and presence of toxic salts in soil.
3. The resistant exine is always ornamented/sculptured. The ornamentation occurs in a 'species-specific-pattern'.
4. Pollen grains always have definite type and number of aperture(s) through which pollen tube emerges.
5. The type of ornamentation and the number, position and character of aperture(s) are genetically stable. Variations in ornamentation of exine and properties of aperture(s) provide means of identifying an unknown sporomorph up to its respective family or genera or even species. Definite identification of a pollen grain and spore is the essential prerequisite to study any discipline of palynology.
6. The size of a pollen grain and spore is small and this facilitates an ease of aerial transport to long distances. This property helps to trace the place of migration of a plant.
7. Pollen analysis, in comparison to megafossil studies is more advantageous by virtue of the fact that a little quantity of sample unfolds the vegetation of that area from where the samples are collected. Numerical abundance of pollen and possibility for uniform scattering help in statistical calculation and enable to determine the vegetation of the area. Uniform scattering of pollen and spore makes their fossils available in continental sediments and marine rocks. This is in contrast to marine fossils that are available only with marine rocks.
8. Pollen and spores are always produced in very large numbers. This is due to unreliability of particular grain finding its target. To be effective pollen and spores are produced in far large numbers than would be needed. This (excess pollen and spore) forms the basis upon which palaeopalynology depends. Pollen grains from atmosphere settle to soils from where they are washed out by rain water into lakes, streams, seashores, mires etc. Pollen grains lie there for very long periods of time. Humus collects and thus pollen grains get fossilized. But the characteristics of surface pattern and aperture are still retained, and are of great diagnostic value. These deposits become a storehouse of information from where ancient vegetation can be reconstructed. In such deposits types of pollen and their abundance become means of statistical analysis that leads to reconstruct palaeoenvironment. Changes of types of pollen and spores in the different strata reveal the change of vegetation and accordingly the changing pattern of past climate can be interpreted. It is to note that in the study of palaeopalynology most of pollen found come from anemophilous plants. Entomophilous pollen grains are under-represented.

Palaeopalynology is of immense value in a wide range of scientific studies, a few of which are mentioned below.

• *To trace the history of vegetation*: Palaeopalynology is most effectively exploited in the investigation of vegetation history. It is fruitfully employed with regard to the pollen grains of Quaternary deposits, because most of pollen in these deposits is referable to extant genera and species. Pollen and their chronological abundance in sediments provide a means of tracing the history of vegetation. On the basis of palynological data Bryant *et al.* (1985) reconstructed the vegetation changes that occurred in Texas during late Quaternary. The data indicate that the flora of United States, including *Picea glauca,* invaded Texas during late Quaternary. Coniferous forests were present in Texas during the full-glacial period but those were not much widespread. During post-glacial and part of the late-glacial period grassland savannas were present in North Texas. Chronological abundance of pollen grains provides information regarding arrival, expansion and decline of major trees like pine, birch and oak etc. Pollen stratigraphic studies enabled to calculate the population doubling time for forest trees during their phase of exponential population growth. This type of study was carried out in early Holocene of Eastern England (Moore *et al.*, 1991, p.7). The exponential population growth ranges from 58-73 years for *Pinus sylvestris* and about 100 years of *Tilia cordata.* Apart from origin and expansion the decline of plant taxa can also be detected through pollen stratigraphic studies, e.g. the decline of *Castanea dentata* and *Tsuga canadensis* in Northeastern United States. Pollen analysis of peat profiles collected from Calcutta Metro Railway Excavation Project reveals the existence of *Heritiera, Exocaria, Phoenix paludosa* and *Barringtonia* etc., that form the present-day mangrove vegetation of Sundarban. Pollen contents of peat samples were investigated at the laboratory of Division of Palynology and Environmental Biology, Bose Institute, Kolkata. With the aid of pollen stratigraphic studies the past vegetation of an area can be reconstructed. This becomes a tool in interpreting palaeoclimate, palaeoecology etc. of the region under study.

• *To study plant assemblage at a specific stratigraphic horizon*: Each stratum of a particular region has definite types of fossil pollen and spores. The abundance of grains plus the durability of grains due to the presence of sporopollenin ensure pollen analysis from each stratum. The grains are most often preserved in bogs and lakes. Samples collected from these sources are treated chemically to stain pollen and spores thus allowing individual grains to be identified and counted. The distinctive number, position and aperture plus the surface ornamentation of an individual pollen grain and spore allow them to be used to identify the plants that produced them. Light microscopic study of the grains allows developing pollen spectrum. A pollen spectrum indicates the types of plant present in a specific stratigraphic horizon. Thus it reflects the plant assemblage that existed at the time of pollen and spore production.

Generally sediments have distinct groups of pollen and spore. These sporomorphs are called index fossil for that layer. The terms guide fossil, type fossil, zonal fossil, key fossil and diagnostic fossil are used as a synonym for index fossil. Index fossils provide information about the sediment from where they are found. From index fossil the age of sediment and the environment in which the sediments were deposited can be determined. The relationship between sediments in distant locations can be established with the aid of index fossil. If the sediments from distant locations contain same index fossil, they are likely part of the same sediment and thus from the same time period. Difference in the composition of index fossil in sediments from distant locations will indicate the sediments are distinct and not related to each other. With the aid of index fossil floral succession can be determined. The followings are the characteristics of ideal index fossils: (1) they must be abundant because it must be easy to find them in the sediments, (2) they must be easy to identify, (3) they must be short lived and (4) they are to be widely distributed. Mention may made of the following pollen genera that are important index fossils in North Tethyan floral belt: *Mansoniaepollis*, *Reevesiapollis* and *Tahitiaoipollis* (Krutzsch, 2004).

• *To correlate deposits and assigning dates*: Each stratum of a particular region during a particular geologic time has characteristic pollen assemblage. With the aid of pollen assemblage palynologists determine a relative chronology or a sequence of events that occurred during plant succession. The prominent changes in pollen proportions being correlated with a particular timing form the basis of using pollen analysis as dating tool. But the absolute age of sediment cannot be determined using pollen assemblage alone. The absolute age is derived from absolute dating obtained by radiocarbon methods. A well-authenticated datum horizon can be of value on a particular region. A distinctive pollen horizon and sampling of adjacent horizon extend the records of the plant communities that occupied the site. A number of distinctive pollen horizons over wide areas are of broader value. As for example the decline of elm at about 5000 years ago in North Europe and the decline of hemlock in Eastern North America were revealed by comparing a number of pollen horizons over wide areas (Moore *et al.*, 1991, p. 8). Now a days radiocarbon dating is favoured over pollen dating. Yet pollen analysis remains a major stratigraphic tool in older strata (Traverse, 1988).

• *To study climatic change*: Climate influences vegetation patterns. Determination of past vegetation types through pollen analysis is a useful tool for the study of past climates. Relative pollen frequencies are used exclusively in the reconstruction of past climate.

Pollen grains are produced in large numbers and most of them after dissemination settle to the ground throughout the year. In the following year another fresh layer of pollen grains deposits over the preexisting layer(s). This process repeats year after year. Thus, the vegetation that provides pollen is recorded in each layer. Any change of vegetation is reflected to the change in pollen assemblage. Pollen analysis of each layer provides data to reconstruct vegetation of that layer. The climate can be determined by studying the nature of vegetation. A comparison between pollen assemblages of two successive layers reveals the change of vegetation and accordingly the change of climate. Pollen analyses of soil, especially the soil of bogs and lakes provide data to interpret the change of climate pattern.

• *To study extinct genera*: Ex. *Ephedripites* and *Vittatina.* Wilson in 1959 (Andrews, 1961, p. 460) reported pollen grains of the above two genera from the Middle Permian of Oklahoma. The grains are believed to be the ancestral forms of *Ephedra* and *Welwitschia*. It is regarded that these grains are the most ancient records of Gnetales.

• *To study evolution of plants and establish affinity*: It is regarded that monocolpate pollen grains are most primitive. Monocolpate pollen grains are found in extinct Bennettitales and extant Magnoliaceae of angiosperms. So there exists an affinity between Bennettitales and Magnoliaceae so far as pollen morphology is concerned. Trichotomous aperture of bryophytic and pteridophytic spore is regarded as most primitive from which other morphotypes have originated. All known primitive vascular plants, both living and fossil, have trichotomous aperture (Chanda, 1970-71).

• *To study past distribution of flora*: Gnetales has extremely scanty macrofossil record (Andrews, p. 460). But their pollen grains have been reported in fossil forms. Pollen grains attributed to *Ephedra* have been reported from early Tertiary horizons in Tasmania, South Australia etc. Pollen grains of *Welwitschia* have been found in Russia, United States and Australia. Pollen grains of *Ephedra* are polycolpate whereas *Welwitschia* has monocolpate pollen. *Gnetum* has no recognizable germinal opening.

• *To study palaeoecology*: Pollen spectrum reflects the plant assemblage of a specific stratigraphic horizon. From the type of plants the past vegetation of an area can be reconstructed. This helps to interpret the palaeoecology of that area. It has been proved that pollen grains are of immense value in determining changes in climate associated with changes in the extent of Pleistocene ice. Most plant genera of the Mesozoic and Paleozoic are extinct (Andrews, 1961). In this case palaeoecology is determined on the basis of physical character of the deposit and correlative evidence from associated fossils.

• *To determine coal-bearing strata*: Coal beds have characteristic abundance of certain pollen and spore genera on the basis of which they are identified (Fig. 5.5). The seams can be traced laterally by comparing the pollen assemblage of the adjacent sections. The pollen assemblage is expressed in the shape of tables and diagrams, and separate horizons that have definite pollen genera. Such horizons are compared with horizons of adjacent sections to correlate and extend the coal seam. This correlation is successfully utilized in the Paleozoic, especially in the Pennsylvanian coal seams. The correlation tool has been proved to be of immense value in coal investigation of North America and Europe. In India Birbal Sahani Institute, Lucknow, has successfully employed such correlation tool in the investigations of coal basins of Assam, West Bengal, Bihar, Madhyapradesh and other states.

• *To define ancient shoreline*: The fossil fuel—oil accumulates in sediments that occur parallel to seashore. So in the exploration of oil, the defining of ancient shoreline bears special significance. In this respect palaeopalynology has immense value. With the aid of palaeopalynology it is possible to determine strandlines, because pollen assemblages are limited near the shore. In modern marine sediments various oceanic plankton including dinoflagellates, diatoms and hystricosphaerids occur in association with pollen and spores (Fig. 5.5). By segregations the quantities of marine forms from pollen and spore the distance and direction of ancient

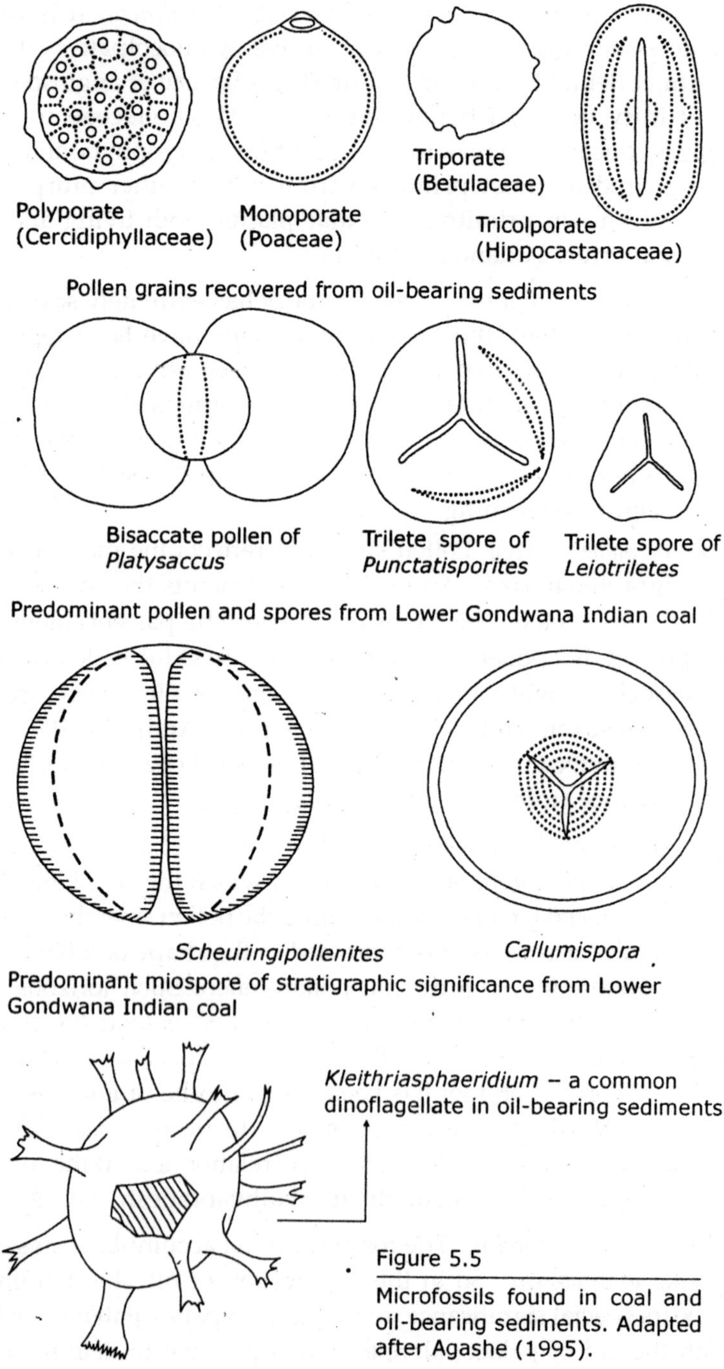


Figure 5.5

Microfossils found in coal and oil-bearing sediments. Adapted after Agashe (1995).

shorelines can be determined. In seaward direction the quantity of marine forms increases with a corresponding decrease of pollen and spore.

• *To study archaeological palynology*: Archaeological palynology is regarded as an information bank of the past including climatic conditions in a particular time bracket. Sediments along with pollen are collected for analysis from the archaeological sites. These are dated with radiocarbon dating to measure the changing pollen presence over time. Individual pollen grain from archaeological sites is identified, counted and estimated the types of pollen present in relation to time scale. These data are used to trace the history of the domestication and cultivation of plants. Moreover pollen analysis of human coprolites (= feces fossils—these are typically preserved in dry caves) provides the dietary preferences. Pollen analysis of sheep excrements found in Switzerland over several thousand years provides data regarding the reconstruction of the composition of pastures of that time. With the aid of palynological data the land use, forest clearance, cultivation and abandonment all can be identified. These data when collaged with other archaeological evidences become extremely informative. *Zea mays* is the most commonly found pollen of cultigens. Other cultigens are cotton, *Agave, Cleome, Opuntia* and various weeds etc. Pollen data from the excavation sites of San Andres, Southeast Mexico confirm the use of cultigens like maize and manioc (*Manihot*), Bryant (2003). With the aid of radiocarbon dating the archaeological deposits containing pollen of cultigens were dated. and they were found to be of 5800 to 6200 years old. In the early Holocene humans first selected and cultivated the food and fibre plants upon which they are to depend. Pollen data obtained from Pueblo dwelling in Eastern Arizona revealed that Pueblo cultures (AD 900-1250) collected and ate a wide variety of plants like *Cleome, Chenopodium, Helianthus* and *Amaranthus* etc. Previously many archaeologists assumed that the Pueblo groups subsisted on agriculturally grown foods (Bryant *et al.*, 1996). In old Egyptian tombs honeys are found. Pollen analysis of these honey samples helps to trace back over four to five thousand years the vegetation of Nile Valley. In India Birbal Sahani Institute of Paleobotany, Lucknow, investigated the sites of the excavation at Lahuradewa District, Sant Kabir Nagar, U. P. and Porbandar. The excavation near Porbandar has brought to light a late Bronze Age settlement. Pollen analysis from archaeological samples reveals shallow river depositional environment. The relative rise and fall of sea level corresponding to dry and warm climatic periods influenced the depositional environments.

CHAPTER **6**

# Classification of Tissue

A tissue can be defined as a group of cells having it or diverse morphological structures, common origin, function and occupy definite position in the plant.

Classification of tissues and the cells included in them: Tissues are generally classified on the basis of development and the types of constituent cell—a brief account of which is as follows.

(A) *Classification of tissues based on development*

(a) **Meristematic tissue:** Meristematic tissues can be defined as a group of cells which have the power of division, usually situated at the apices of roots, stems and leaves. These cells are always in a state of division and their derivatives either compose or supplement the dermal, vascular and ground tissue system of a plant.

(b) **Permanent tissue:** Permanent tissues can be defined as a group of cells, which do not divide and are the differentiated products of meristematic tissues. Sometimes these cells may regain the power of division—a phenomenon termed dedifferentiation. In some plants dedifferentiation occurs at definite regions at certain stages of development to supplement the conducting and protective tissue system.

The meristematic and permanent tissues are discussed in chapter 7, 8 and 9.

Haberlandt (1914) reviewed 'the various anatomico-physiological systems' in plants and recognized twelve types of tissue systems. The types are grouped into four tissue systems, namely— meristematic tissue system, protective tissue system, tissue system concerned with metabolic activity and tissue system comprising 'the execution of movements for the perception or transmission of stimuli'. A brief account of the various tissue systems is given below.

i. Meristematic system: The tissues of this system are always in a state of division and compose or supplement the dermal, conducting and ground tissue system of a plant. Primordial, primary and secondary meristem belong to this category. Meristematic tissue system is placed at the top of all tissue systems due to the fact that (a) it consists of embryonic condition of all permanent tissues and (b) it is an independent tissue system and provides cells to increase the respective permanent tissues.

Protective system: The tissues, belonging to this category, protect the inner tissues from the adverse natural calamities and mechanical injury; they also add mechanical strength to the plant. These tissues are grouped into two systems, namely:

ii. The tegumentary or dermal system which includes epidermis and periderm.

iii. The mechanical system which is composed of collenchyma, sclerenchyma, wood-fibres and bast-fibres.

System concerned in metabolic activity: In this category the following six systems are recognized.

iv. The absorbing system: The tissues of this system have absorptive function. Example: Rhizoids, roots, haustoria, velamen etc.

v. The photosynthetic system: The tissues of this system can photosynthesize. Example: Mesophyll cells of leaves, peripheral chlorophyllous cells of stem etc.

vi. The conducting system: Xylem and phloem compose this system and they are concerned with upward conduction of water and downward translocation of solutes. Xylem parenchyma, medullary rays and latex-tubes are also included in this system.

vii. The storage system: This system consists of water tissues and storage tissues of seeds, tubers and bulbs.

viii. The ventilating system: The tissues of this system are meant for gaseous exchange between internal tissues and external atmosphere. Example: Stomata, lenticels etc.

ix. Secretory organs and excretory reservoirs: The nectaries, digestive glands, hydathodes etc. compose this system and they are concerned with secretion and excretion. Mucilage glands, oil glands and resin glands etc. are also included in this tissue system.

Systems concerned with the execution of movements for the perception or transmission of stimuli: The followings belong to this system.

x. Motor tissues: The insectivorous plant *Dionaea* possesses these tissues on their leaves, which trap the insects. When the insects rest on leaves, the motor tissues direct the leaves to fold.

xi. Sense-organs: The hairs, statolith and ocelli (Fig. 6.1) etc., which perceive the external stimuli, compose this system.

xii. Stimulus-transmitting organs and tissues: The tissues of this system transmit the external stimulus. Example: Leaves of *Mimosa pudica* droop down when touched. Protoplasmic connecting threads are also included in this system.

(B) *Classification of tissues based on types of cell:* There are two types of tissue – simple and complex. Simple tissues are homogeneous, i.e. they are composed of one type of cells only, which have similar morphological structure, perform same function and originate from the same mother cell. Parenchyma, collenchyma and sclerenchyma are the examples of simple tissues. Complex tissues are heterogeneous, i.e. they are composed of dissimilar types of cells, which have different morphological structures. Xylem and phloem are the complex tissue through which upward conduction of water and downward translocation of solutes occur. The constituent cells of simple and complex tissue are mentioned in the following table.

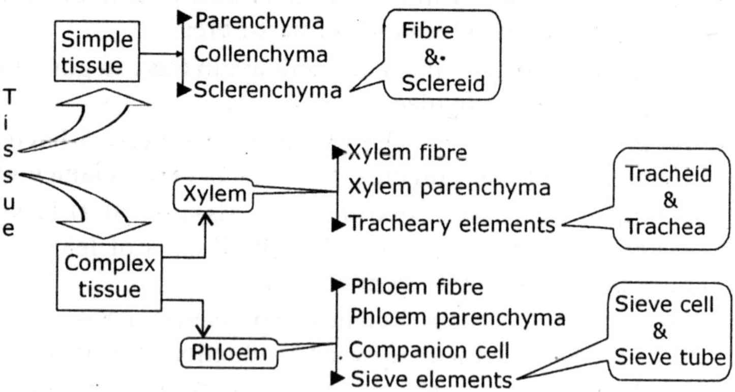


Simple and complex tissue with their constituent cells are discussed in chapter 8 and 9.

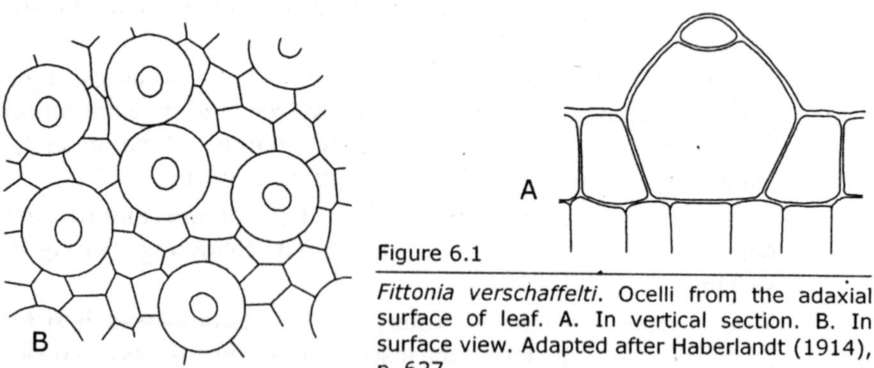


Figure 6.1

*Fittonia verschaffelti*. Ocelli from the adaxial surface of leaf. A. In vertical section. B. In surface view. Adapted after Haberlandt (1914), p. 627.

[The localized organs of light-perception differ anatomically from the other epidermal cells. They are distinguished by the name ocelli. The cells composing ocelli have protruding outer wall, flat inner wall and sloping lateral walls. A minute biconvex lens-cell occurs at the protruding wall.]

CHAPTER 7

# Meristem

Meristem can be defined as mitotically active centre that is restricted to specific portion of higher plants and the cells of meristem divide indefinitely by multiple mitoses thus resulting in continuous addition of new cells to the developing plant organs.

The zygote, in all sexually reproducing higher plants, divides and subdivides to produce an undifferentiated packet of cells. After differentiation these cells form root, stem, leaves and other organs. But among these a group of cells always remains undifferentiated and retains the capability of cell division. These cells undergo frequent divisions by multiple mitoses and add new cells to the developing organs. Such cells are collectively termed as meristem or meristematic tissue. The term meristem is derived from the Greek word 'merizein' meaning to divide (http://en.wikipedia.org/wiki/Meristem). A cell of meristem is regarded as totipotent; that is the cell has the capability to develop the different types of cells that compose a plant body.

Meristems occupy the growing regions of organs concerned. The regions include stem tip, leaf tip, leaf bases and root tip. Meristem in the form of cambium occurs as fascicular cambium (present within vascular bundle), interfascicular cambium (present in between vascular bundles), cork cambium (= phellogen) and accessory cambium (originates within stele). Meristem also occurs in reproductive apex and flower buds.

The cells of meristem are isodiametric in shape. They are tightly packed living cells with little intercellular spaces. The cell wall is thin and composed of cellulose only. Cells have large nucleus, rich cytoplasm with less ergastic inclusions, crystals and reserve food; vacuoles if present are very small and scattered throughout the protoplast; plastids occur in rudimentary form (proplastid); cells have smaller amount of endoplasmic reticulum and less elaborate structure of mitochondria. Though meristematic cells exhibit the above mentioned criteria, exceptions are observed in cambial cells. The cells of fusiform initial of vascular cambium are long and narrow. Cork cambium or phellogen contains ergastic substances, reserve food materials and chloroplasts. The ray initials of vascular cambium may contain starch and tannin. Vacuoles occur in the cells of apical meristem of *Ginkgo*. Fusiform—and ray initial cells of vascular cambium may be highly vacuolated (Fig. 7.1). In *Ginkgo* the cells of the apical meristem have thick primary cell walls with prominent primary pit fields. The cambial initials also exhibit thick cell walls with depressed primary pit fields.

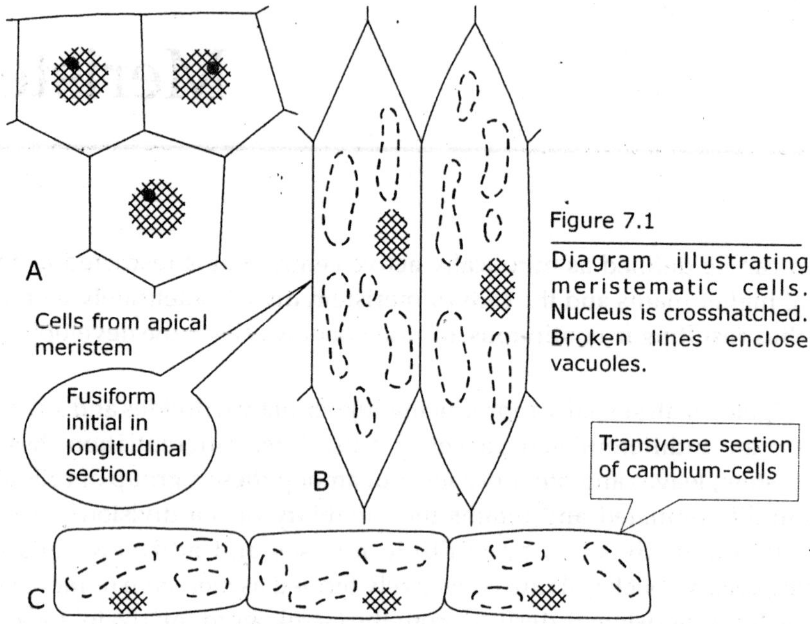


Figure 7.1

Diagram illustrating meristematic cells. Nucleus is crosshatched. Broken lines enclose vacuoles.

There exists much variation in the characteristics of meristem. It appears from morphological and physiological point of view that there exists no typical meristematic cell. So Kaplan in 1937 (Esau, 1965, p. 75) suggested the term eumeristem (= true meristem) to designate those meristems that have only the following characteristics: (a) small and isodiametric in shape; (b) cell wall is thin and (c) the cell has rich cytoplasm. The term was coined for descriptive purposes only.

The derivative cells of meristem transform into permanent tissues of plants. Permanent tissue includes epidermis and periderm that compose the dermal tissue system, xylem and phloem that compose the conducting tissue system and ground tissue system that consists of all the remaining tissues apart from epidermis, periderm and vascular tissues.

Meristem consists of a group of cells that continuously divide and the derivative cells differentiate into respective tissues. The undifferentiated group of cells is termed as promeristem. Promeristem is also referred to as primordial meristem or embryonic meristem. Promeristem occupies the apical portions of root, stem and leaf. When differentiation starts, the meristem is not regarded as promeristem.

## Classification of meristem

Meristems differ in their mode of origin, function, position and cell division. So for the sake of convenience and for effective description they are classified—brief account of which are as follows:

- *Classification of meristem based on origin*

Primary meristem and secondary meristem belong to this category.

(a) **Primary meristem**: Primary meristem can be defined as a meristem that owes its origin from promeristem, differentiates into permanent tissues that compose the primary body of a plant and persists for the whole life of a plant.

The primary body of stem and root, as seen in cross-section, consists of epidermis, cortex, phloem, xylem and pith that compose the dermal, vascular and ground tissue system of a plant. All tissues of these tissue systems are the derivatives of primary meristem. Primary meristem composes the apical meristem of root and shoot. Primary meristem also occurs at the leaf tip and forms the adaxial and abaxial epidermises of a leaf that enclose mesophyll tissues and vascular tissues. Primary meristems are also found to occur at the basal region of spine, at the basal region of internode, at the basal region of leaves of many monocotyledons and at the apex of axillary bud that develops into branches, inflorescences, thorns and tendrils etc. It is to note that as soon as differentiation starts in the promeristem it passes to primary meristem stage, the cells of which are also capable of division. After complete differentiation certain cells of permanent tissue may regain the ability to divide and this phenomenon is referred to as dedifferentiation.

(b) **Secondary meristem**: Secondary meristem can be defined as the meristem that owes its origin from primary permanent tissues by dedifferentiation and adds new tissues to dermal and vascular tissue systems for effective protection, conduction and mechanical strength.

Cork cambium also called phellogen, vascular cambium of dicotyledonous root and interfascicular cambium of dicotyledonous stem are examples of secondary meristem. Phellogen occurs at the extrastelar region of root and stem. It originates either adjacent to or within the epidermis. It also originates in the peripheral or deeper region of the cortex. Phellogen donates cells on the peripheral and inner side. The peripheral derivatives are phellem and the inner derivatives are phelloderm. Phellem, phellogen and phelloderm together constitute periderm. Periderm forms protective tissue system in plant. Vascular cambium of root occurs at the intrastelar region. The derivatives of vascular cambium supplement the vascular tissue—and mechanical tissue system of roots. Interfascicular cambium of stem originates from the primary permanent medullary ray cells present in between the two vascular bundles of stem by dedifferentiation. The interfascicular cambium along with fascicular cambium (the derivative of primary meristem) forms a cambium ring. The inner and peripheral derivatives are respectfully secondary xylem and secondary phloem and they compose the secondary vascular tissue or conducting tissue system in plants.

- *Classification of meristem based on function*

Haberlandt (1914) distinguished the following three types in primary meristem:

i. **Protoderm**: The formation of dermal tissue system in plant is attributable to protoderm. Dermal tissue system consists of epidermis/multiple epidermises and all other structures that occur on epidermis, e.g. stomata, hairs, glandular hairs etc.

ii. **Procambium**: The formation of vascular tissue system in plants is attributable to procambium. Procambium consists of cells that are elongated in the longitudinal plane of axis. The cells are densely cytoplasmic and contain large nucleus, proplastids and exhibit the usual features of meristematic cells. The cells are highly meristematic and undergo frequent mitoses. The peripheral derivatives of procambium are primary phloem and the inner derivatives are primary xylem. Primary phloem and primary xylem collectively form the collateral closed vascular bundle of monocot stem. In dicotyledonous stem procambium is not completely used up in the formation of primary xylem and primary phloem. A portion of procambium remains in between primary phloem and xylem and this portion of procambium later differentiate into cambium. Cambium with peripheral phloem and inner xylem forms the collateral open vascular bundle of dicotyledonous stem.

iii. **Ground meristem**: Ground meristem represents the undifferentiated meristematic tissue of primary meristem from where 'protoderm and primary procambial strands have been segregated'—Haberlandt (1914). Ground meristem divides frequently and the derivatives differentiate into cortex, pith and other tissues that are associated with them. Ground meristem is regarded as the precursor of fundamental or ground tissue system. Mesophyll cells of leaf and additional procambial bundles also originate from ground meristem.

The cells of an apical meristem give rise to protoderm, procambium and ground meristem that in turn give rise to particular tissues. The tissues mature and compose the primary body of a plant. In most terrestrial dicotyledons at the end of primary growth, additional secondary tissues are formed. The formation of primary and secondary tissues is diagramed in Fig. 7.2 and the cell lineages of their formation are illustrated in box 7.1.

- *Classification of meristem based on position*

According to the position in plant body meristems are divided into apical meristem, intercalary meristem and lateral meristem (Fig. 7.3).

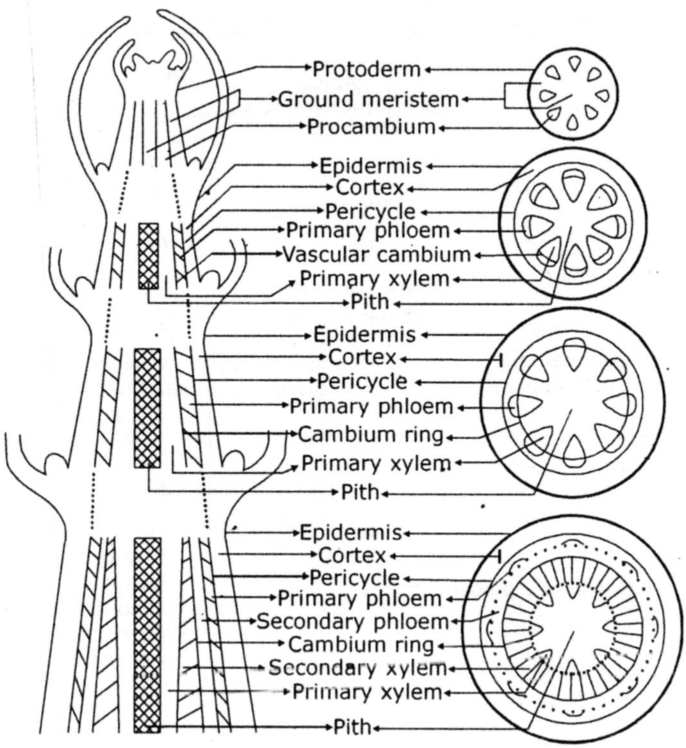


Figure 7.2

Part of a stem with apical meristem diagramed in longitudinal section (left) and in four cross sections (right) with their basipetal derivatives. At intervals from the tip, note vascular cambium develops into a complete ring producing secondary xylem and phloem. The vertical dotted lines in longitudinal section indicate unspecified number of nodes and internodes. Adapted after Meyer *et al.*,1965, p. 328.

**Apical meristem:** It is located at the apices of roots and at the apices of main and lateral shoots. It is also located in trichomes, glands and in other structures that they produce. Primordial meristem (= promeristem) composes the apical meristem and consists of undifferentiated homogeneous cells. Promeristem exhibits zonations in the apical meristem. The zones can be distinguished from each other by their relative thickness of cell wall, cell size, nuclear size and relative frequency of mitoses.

Apical meristem represents the region of initiation of primary tissue organization in roots and shoots. Promeristem transforms into primary meristem that consists of protoderm, procambium and ground meristem. Apical meristem is the growing point of shoot and forms leaves and branches in genetically predetermined sequences. As a result shoot elongates. Protoderm, procambium and ground meristem respectfully initiate the dermal, vascular and fundamental tissue system to form primary body of shoot and root. Flowers also differentiate from apical meristems.

Box 7.1

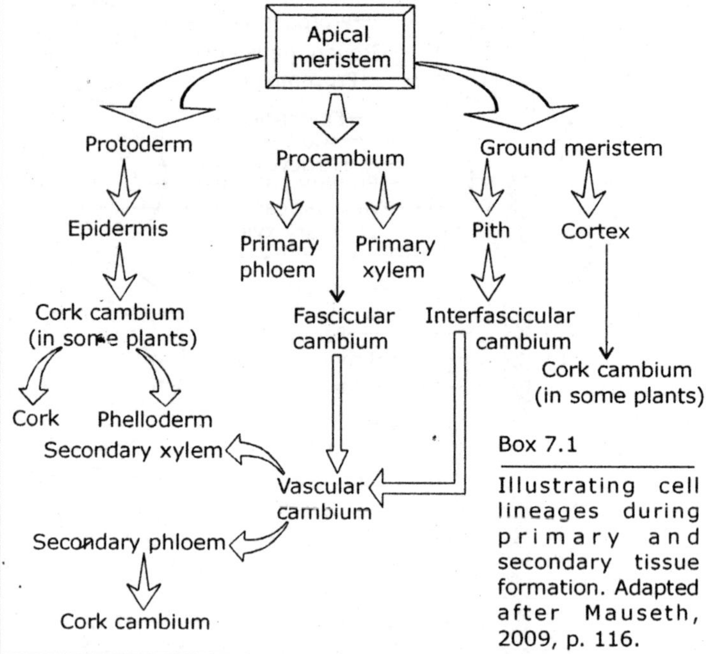


Box 7.1

Illustrating cell lineages during primary and secondary tissue formation. Adapted after Mauseth, 2009, p. 116.

**Intercalary meristem**: It is located at the base of internode. Intercalary meristem is short lived and divides frequently. It donates cells above and below. Vascular tissues are also differentiated from this meristem. The derivative cells differentiate into permanent tissues in basipetal succession; as a result stem elongates. Intercalary meristem is an isolated region in the internode and has permanent tissues above and below it. It remains widely separated from apical meristem. Ex. grasses (Fig. 7.4) and some plants belonging to the families chenopodiaceae, caryophyllaceae and polygonaceae etc. In monocots elongation of shoot axis occurs by random cell divisions and differentiation throughout the youngest internodes. This region of diffuse cell division is termed as uninterrupted meristem. Uninterrupted meristem is continuous with apical meristem and the derivative cells differentiate into permanent tissues in acropetal succession.

**Lateral meristem:** It is located parallel to the long axis of root and shoot. Examples of lateral meristems include vascular cambium, cork cambium, primary thickening meristem (PTM) and secondary thickening meristem (STM). Pericycle of root also could be referred to as lateral meristem because adventitious root originate from pericycle (Rudall, 2007, p. 22).

In dicotyledonous stem vascular cambium consists of fascicular and interfascicular cambium. The cambium divides tangentially and donates cells on the peripheral and inner side. The derivative cells differentiate into vascular tissues. Thus new cells are added to primary vascular tissues. As a result a stem increases in girth. Root also increases in diameter by the activity of cambium. These new tissues supplement the vascular and mechanical tissue system of root

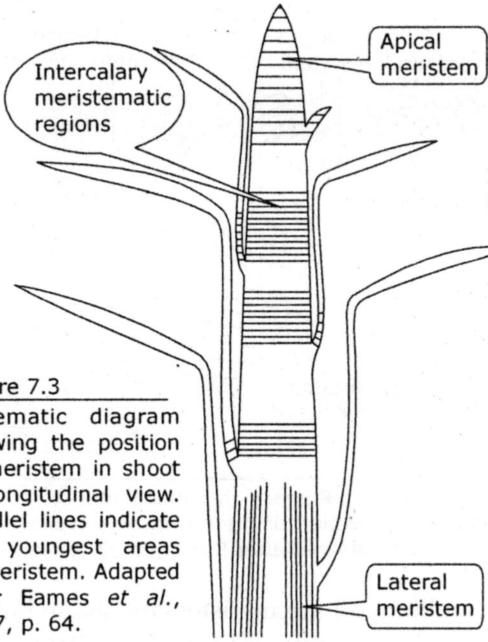


Figure 7.3
Schematic diagram showing the position of meristem in shoot in longitudinal view. Parallel lines indicate the youngest areas of meristem. Adapted after Eames *et al.*, 1947, p. 64.

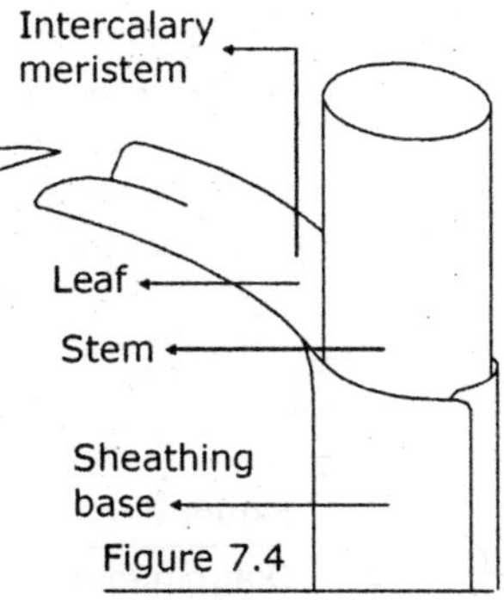


Figure 7.4
Diagram showing the position of intercalary meristem in monocot leaf. Adapted after Mauseth, 1988, p. 84.

and stem. Cork cambium (= phellogen) is located either in the epidermis or adjacent to epidermis or in the peripheral or deeper layer of cortex. Phellogen divides tangentially and donates tissues on the peripheral and inner side. The derivative cells form periderm that supplements the protective tissue system in stems and roots.

Primary thickening meristem (Fig. 7.5) and STM are present in monocotyledonous stems. Monocots lack vascular cambium. PTM and STM donate cells that cause the increase in diameter of monocot stems.

Primary thickening meristem (PTM) is located near the vegetative shoot apex. PTM is a narrow multiseriate zone having a concave form in young plants. The zone consists of meristematic cells that donate cells both centrifugally and centripetally. The centrifugal (peripheral) derivatives are differentiated into parenchyma cells only. The centripetal (inside) derivatives are differentiated into both parenchyma and discrete vascular bundles. PTM, in addition to stem thickening, also forms vascular tissues that link between the vascular tissues of root, stem and leaf. PTM is also responsible for the production of adventitious roots in some species.

In *Dracaena*, *Yucca* and *Aloe* etc. further increase in stem thickness occurs by the activity of secondary thickening meristem (STM). STM is located in the pericyclic region and is distant from the primary apex. STM divides tangentially and the inner derivatives are differentiated into parenchyma and vascular tissues. The vascular tissues are secondary vascular bundles that are leptocentric (= amphivsal) and radially elongated. In *Yucca whipplei* PTM and STM are axially continuous whereas in *Beaucarnea recurvata* they are axially discontinuous.

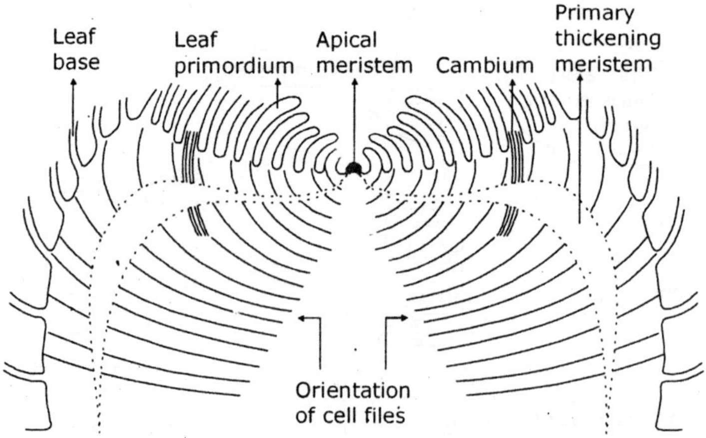


Figure 7.5

Diagrammatic illustration of Primary thickening meristem (PTM) in the longitudinal section of the crown of a typical thick-stemmed monocot. Adapted after Rudall, 2007, p. 39.

PTM, STM and vascular cambium all are meristematic tissue but PTM and STM are not homologous to vascular cambium. There exist differences between thickening meristems (PTM and STM) and vascular cambium. Vascular cambium exhibits the following characteristics: (i) it is present in dicotyledons; (ii) it is uniseriate; (iii) it originates within vascular tissue; (iv) it is distantly situated from apical meristem and (v) cambium donates cells both on peripheral and inner sides and the derivatives differentiate respectively into secondary phloem and secondary xylem. In contrast PTM and STM have the following characteristics: (i) they are present in monocots; (ii) they are multiseriate; (iii) they are present below the apical meristem; (iv) PTM occurs near the apical meristem and (v) PTM and STM donate cells mostly centripetally. The inner derivative cells differentiate into ground parenchyma and discrete amphivasal vascular bundles.

- *Classification of meristem based on the plane of cell division*

Mass meristem, plate meristem and rib meristem belong to this category.

**Mass meristem (or block meristem):** Ex. young embryos, endosperm, reproductive organs during the formation of spores and sperms etc. The cells of this meristem divide in all planes thus the tissue increases in volume. The derivative cells are isodiametric or spheroidal or have no definite shape.

**Rib meristem (or file meristem):** Ex. cortex and pith of root and cortex of stem. The cells of this meristem divide at right angles to the longitudinal axis of the plant organ. As a result parallel files of cells (= ribs) are produced. Rib meristem is the characteristic of organs that have cylindrical form.

**Plate meristem:** Ex. flat blades of angiosperm leaves. The cells of this meristem divide by anticlinal division. Anticlinal division increases the number of cells in one layer only. This division causes the meristem to grow as a sheet but not in thickness. Plate meristem is the characteristic of plant organs that have flat forms. As a result a flat organ like leaf is produced.

Plate—and rib meristem are the two growth forms exhibited by ground meristem. Plate meristem gives rise to the leaf like organs and rib meristem forms the elongated cylindrical structures like root, stem and petiole etc.

## Apical meristem

The growth centres of plants are root apical meristem and shoot apical meristem. They persist as long as plants live and donate cells to form new structures. The growing region of shoot and root has modular construction (Fig. 7.6). The shoot module includes leaf with axillary bud and stem with node and internode. The root module consists of elongating root bearing lateral roots. The growth of a plant is due to addition of modules. The cells of apical meristem divide and the derivative cells form the module behind the tip. The modules are formed in repetitive manner and in trees the module of shoot may be formed for centuries. When vegetative shoot apex is transformed to reproductive apex the basic modular construction persists and the pattern ends after the formation of reproductive organs. From the modular construction of growing region the positions of new organs can be predicted in relation to existing organs.

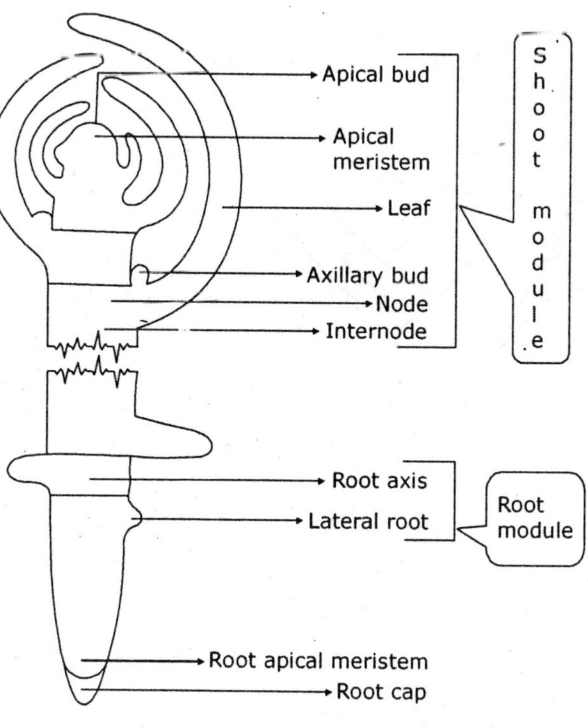


Figure 7.6

Diagram showing modular construction of a plant. Root apical meristem and shoot apical meristem produce respectively the basic units of structure - lateral root and axillary bud. The units are produced again and again, and when the units grow out the pattern is repeated. The internode elongates to form stem. Adapted after Lyndon, 1990, p. 20.

## Theories of shoot apical meristem

Several anatomists illustrated the shoot apical meristem on the basis of number, arrangement and the activity of initiating cells and accordingly proposed the following theories—brief account of which are discussed below.

**Apical cell theory:** Nägeli in 1944 (Cutter, 1971, p. 53) advocated this theory. According to this theory the apical meristem consists of a single apical cell (also called apical initial) and this cell is interpreted as the 'structural and functional unit of apical meristem'. The cell is very large and is shaped like an inverted pyramid. The apical cell is tetrahedral in shape and has three or four cutting faces among which single face is directed upward and the rest points downward. The side of apical cell that is directed upward is triangular or square in shape and forms a part of the outer surface of the shoot apex (Fig. 7.7). The cutting faces of apical cell divide in an orderly fashion, that is in helical succession (Beck, 2005, p. 84). The cell divides by an asymmetric division; as a result a narrow and flat cell is formed. The next division of the apical cell is also asymmetric. This type of asymmetric division is repeated in the downwardly pointed faces of the apical cell. As a result all cutting faces have their daughter cells. The daughter cells also divide and form large packet of cells. The packet of cells differentiates and forms different segments of shoot. So the apical cell is regarded as 'a reserve of one genetically sound cell'.

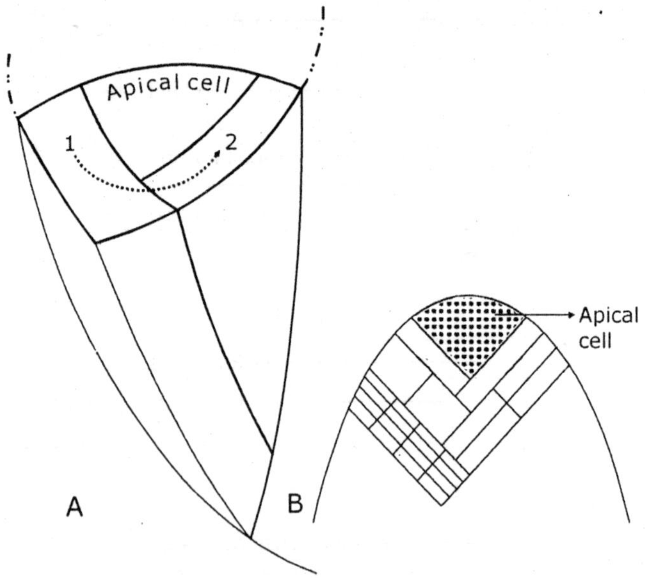


Figure 7.7

A. Diagram showing an apical cell of leptosporangiate fern and its derivatives that are formed in helical succession. The new cells are numbered as 1 & 2. B. Diagram showing a packet of cells formed by an apical cell by division and subdivision. B adapted after Mauseth, 1988, p. 86.

A single apical cell composing an apical meristem is present in vascular cryptogams. After the discovery of solitary apical cell in vascular cryptogams, it was supposed that such apical meristem might exist in higher plants as well. Later extensive investigations refuted the universal occurrence of solitary apical cell in a meristem. In higher plants the apical cell theory was replaced by the concept that the different parts of a plant body have independent origin. So the apical cell theory was later superseded by histogen theory (Esau, 1965).

**Histogen theory:** Hanstein in 1868 (Cutter, 1971, p. 53) put forward histogen theory (histogen means tissue builder). According to this theory the tissues of a plant body originate from a mass of meristem where the following three (histogens) can be distinguished (Fig. 7.8):

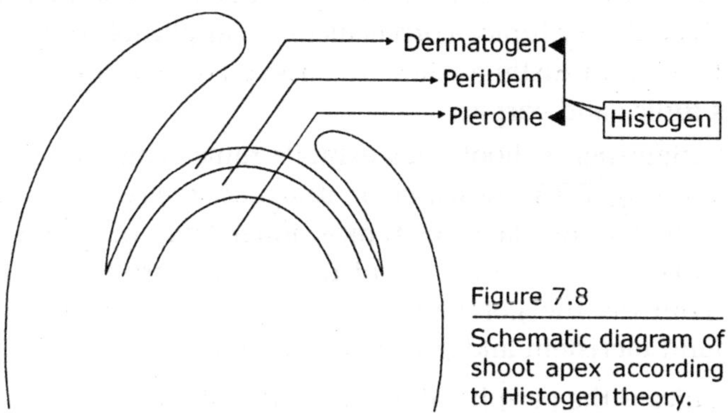


Figure 7.8

Schematic diagram of shoot apex according to Histogen theory.

(a) *Dermatogen*: (in Greek meaning skin). It is the outermost layer of the meristem. It gives rise to epidermises of root and stem.

(b) *Periblem*: (in Greek meaning clothing). This region occurs internal to dermatogen but peripheral to plerome. This histogen is destined to form cortex of root and shoot and inner tissues of leaves. It surrounds plerome.

(c) *Plerome*: (in Greek meaning that this fills). This region gives rise to vascular cylinder of stem and root including pith. It is the central core of stem and root and the cells composing this zone are very irregular. This region is enveloped by a variable number of mantle-like layers which are represented by dermatogen and periblem.

According to Hanstein dermatogen, periblem and plerome arise from independent initials of the apical meristem. Later investigations reveal that the sub-divisions—dermatogen, periblem and plerome have no universal application due to the following two reasons: (i) in gymnosperm and angiosperm there exists no clear distinction between periblem and plerome and (ii) the respective roles of the three histogens cannot be demonstrated (Foster, 1967, p. 33). The main weakness of Hanstein's concept was to assign specific destinies of histogens. The histogens—dermatogen, periblem and plerome are committal and respectfully give rise to epidermis, cortex and stele. Later this theory was superseded by tunica-corpus theory because the zones are noncommittal. Though histogen theory is rejected it is still regarded as 'classic' due to the fact that Hanstein regarded shoot and root apex as a composite unit consisting of different groups of histogens and the histogens differed fundamentally from each other producing different tissues. It is to note that Hanstein visualized this long before when the concept of gene and DNA was established. The present-day concept is that the different zones in the shoot apex are fundamentally same and capable of producing all tissues. It is the position in the meristem that determines the destiny of derivatives. This is due to the fact that in zones some genes are activated and others are repressed resulting in the production of different tissues.

**Tunica-corpus theory:** Schmidt in 1924 (Cutter, 1971, p. 55) postulated tunica-corpus theory on the basis of studies of shoot apices of angiosperm. This theory is concerned with planes of cell division in the apex. In contrast to apical cell theory and histogen theory tunica-corpus theory is applicable only to shoot apex and not to root. Schmidt distinguishes two tissue zones in the shoot apex and termed them as tunica and corpus.

Majority of angiosperm shoot apex exhibits tunica consisting of two layers of cells and corpus (Fig. 7.9A). Researchers designate the layers as L1, L2 and L3 to denote respectively outer layer of tunica, inner layer of tunica and corpus. Plasmodesmata exist between the cells of tunica and corpus. It is thought that plasmodesma controls the gene expression that leads to the formation of protoderm, ground meristem and provascular tissue.

- *Tunica*: Tunica is the peripheral tissue zone of shoot apex. It consists of one or more peripheral layers of cells. Dicotyledons exhibit one to five layers of cells in tunica; two layers of cells are represented by largest number of species. Monocotyledons have one to four layers of cells in tunica; one and two layered

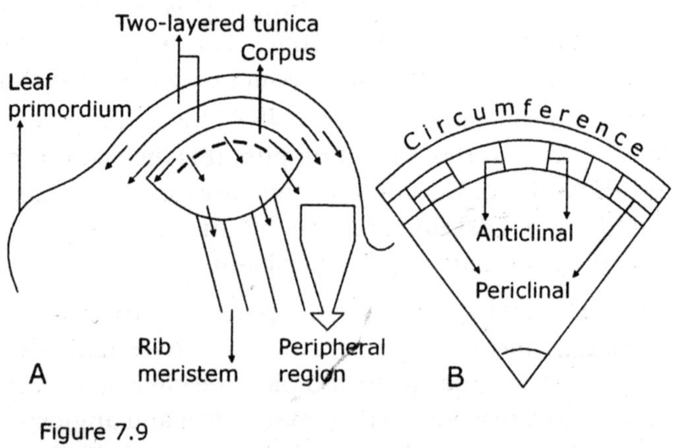


Figure 7.9

A. Diagram illustrating the tunica-corpus organization in dicotyledonous shoot apex. Arrows indicate the direction of cell formation in apical meristem. Adapted after Esau, 1961, p. 219. B. Schematic representation of anticlinal (= division wall perpendicular to surface) and periclinal (= division wall parallel to circumference) division.

tunica predominate in it. One single layered tunica is termed as monostratose. Many layered tunica is termed as multistratose. *Xanthorrhoea media* shows eighteen layered tunica.

Tunica is characterized in having anticlinal plane of cell division, i.e. the division wall is laid down perpendicular to the surface. This division reflects to the surface growth in the apex. As a result tunica grows as a sheet but not in thickness. To increase in thickness a tunica cell will have to undergo cell division with periclinal wall, i.e. the cell wall will have to be laid down parallel to surface (Fig. 7.9B). Normally, periclinal division does not occur in tunica where anticlinal division predominates except at the point of origin of leaf primordium and axillary bud. In multistratose tunica each layer arises from separate initials present in the shoot apex. The independence regarding origin among different layers of tunica and between tunica and corpus is strongly supported with the results of investigation by the use of periclinal chimeras. Such chimera in a plant body exhibits layers with different chromosome numbers. Such differences can be traced from promeristem through continuous cell lineages to similar differences in the derivatives of promeristem.

Promeristem may be induced to develop cytochimeras by treating the shoot apex with colchicine. Colchicine induces polyploidy in the cell(s) or layer(s) of promeristem and it is possible to relate the polyploid cell(s)/layer(s) to their progeny in the mature plant body. Using this technique it was determined that outermost layer of tunica gives rise to epidermis whereas the inner layers donate cells to form other tissues.

[Cytochimera: Chimaera, also spelt as chimera, (Fahn 1997), refers a plant or plant organ having tissues of varied genetic origin lying adjacent to one another. In some plant the meristematic region consists of at least two populations of cells that differ in their chromosome numbers. The chimera of this plant is recognizable only at the cytological level and termed as cytochimera. Usually two types of chimera are recognized: (1) Sectoral or mericlinal chimera—where the two populations of cells are arranged in sectors on the circumference around the axis of the meristem and (2) periclinal chimera—where the two populations of cells are arranged in different layers parallel to the surface of apical meristem. L1, that is the outer layer of tunica, is commonly destined to form the epidermis of all the leaves and stem formed by the shoot and it is revealed on analyzing the periclinal chimera.

Shoot apices are induced to form cytochimeras to study the function of L1, L2 and L3 layers. Treating the shoot apex with colchicine does this. The alkaloid colchicine, a chemical mutagen, is obtained from the plant *Colchicum autumnale*. Colchicine has the property to disrupt the spindle microtubules formed during mitosis. The drug colchicine is applied to the shoot tips by slow trickling or as paint. The treatment temporarily blocks mitosis. The cells that are in metaphase stage cannot complete chromosome movement due to the failure of spindle mechanism but the other events of mitosis do continue. As a result nuclei are formed with tetraploid chromosome numbers. Then colchicine was washed out and the meristems were allowed to continue cell division. Staining properties of nucleus can recognize the descendents of tetraploid cell. Using this technique it becomes possible to find that the L1 layer, i.e. the outermost layer gives rise to the epidermis of a leaf and the inner layers donate cells to form mesophyll and vascular tissues. In stems L1 gives rise to epidermis and the rest forms the cortex and vascular tissue. In *Datura* the shoot apices were experimentally induced to develop cytochimera. The apices exhibited that the cells of L1, L2 and L3 layers had nuclei with chromosome number respectfully 8n, 2n and 2n; 4n, 2n and 4n, and 2n, 8n and 4n (Cutter, 1971, p. 60)].

It is the anticlinal plane of cell division that characterizes tunica. In maize, certain grasses like *Agropyron repens* and *Chlorogalum pomeridianum* (Liliaceae) etc. periclinal divisions are observed in tunica. This led anatomists to modify the 'strictness of definition of the tunica'. One view regards that the tunica should include only those layer(s) that exhibit anticlinal plane of division. If any inner layer(s) of tunica show periclinal division they are to be assigned to corpus. Other workers treat tunica in a very loose sense and regard that the inner layer(s) may divide periclinally. So to avoid discrepancies the term mantle was proposed in a loose sense instead of tunica. Corpus was replaced by the term core. Mantle overarches core. Popham and Chan (1950) advocated the mantle and core concept in the shoot apex and they did not take into account to the planes of cell division.

Cytologically two zones are recognized in the tunica though all cells exhibit the same anticlinal cell division. The first zone is central apical zone and the second occurs between the central apical zone and leaf primordium. The central apical zone consists of one or few initial cells that are larger and contain large nuclei and vacuoles than the cells of the second zone. The cells of the second zone are small and more darkly staining than the cells of the first zone. In contrast to first zone periclinal division may occur in the second zone close to the leaf primordium in addition to anticlinal division.

The main function of tunica is to give rise epidermis. Sometimes the inner layers of tunica form cortex and vascular tissue.

• *Corpus*: Corpus is the inner tissue zone of shoot apex. It consists of cells that are several cell layers deep. Tunica overarches corpus. Meristematic tissues composing this zone are larger than tunica. The initial cells of corpus occur below the tunica. They are orderly arranged in contrast to haphazardly arranged cells in the mass of corpus. So the initials of corpus are difficult to differentiate from the initials of tunica (Esau, 1961, p. 221). The initials arise independently and not related to tunica. The initial cells divide periclinally and the derivatives divide to form the core of the shoot apex. In the division of derivative cells there is no definite orientation of cell wall formation, i.e. cells divide in all planes. As a result the shoot apex increases in volume. Generally corpus is destined to give rise cortex and vascular tissue.

Usually corpus is not homogeneous. It consists of several zones. The zones are illustrated by Popham (1952) in the usual angiosperm type. In this type the meristem is composed of tunica and corpus. The tunica consists of one or more layers of cells that overlie the corpus. The corpus is composed of three zones: (a) central mother cells – the uppermost zone of corpus, (b) pith-rib meristem that occurs below the central mother cell zone and (c) flank meristem (also called peripheral meristem) that surrounds both central mother cell zone and pith-rib meristem. The peripheral zone is shaped like a truncated hollow cone (Mauseth, 1988, p. 88).

The component cells of tunica and corpus differ in size, shape, plane of cell division and topography. Ultrastructurally each zone is composed of cells that have characteristic architectures as is revealed by quantitative techniques (Mauseth, 1988, p. 89).

*Merits of tunica-corpus theory:*

- It deals with one thing, i.e. planes of cell division. As a result the description of meristem becomes precise.
- It has topographical value in the studies of development of different tissue system in plants.
- The destiny of derivatives of corpus is not predetermined.
- The derivatives of the zones are not rigid like histogen theory.
- It explains clearly the growth pattern in the shoot apex of angiosperm.
- It enables to understand the development of leaves as they arise close to apex.
- The specific variation in the number of tunica layers may be of taxonomic significance, e.g. grasses (Brown *et al.*, 1957).

Tunica-corpus organization of shoot apical meristem is observed in angiosperm only. Cryptogams and majority of gymnosperm do not exhibit tunica-corpus organization. They do not have any stable surface layer that divides only by anticlinal division. *Araucaria* and *Ephedra* have a single stable surface layer that divides only by anticlinal division. Thus they exhibit tunica-corpus pattern. Tunica-corpus theory is noncommittal regarding the ultimate destiny of cell-lineages. In present-day concept all cells in the shoot apex with tunica-corpus organization are fundamentally same and have equal potentiality. In corpus the cells of peripheral zone and pith-rib meristem are also same as they owe their origin from central mother cell zone. But during differentiation some genes are repressed and others are activated resulting in the production of different tissues. It is thought that if some how or means the cells of peripheral meristem zone of corpus is transferred to the location of pith-rib meristem, the cells will produce pith cells.

In conclusion it is to mention that tunica-corpus theory provides a flexible description of cell arrangement in the shoot apex and this is not rigid like histogen theory. It has served well in understanding the structure of shoot apical meristem and the origin and development of leaf.

The shoot apices of angiosperm exhibit cytohistological zonation in addition to tunica corpus zonation (Fig. 7.10). Four zones are recognized namely (a) tunica

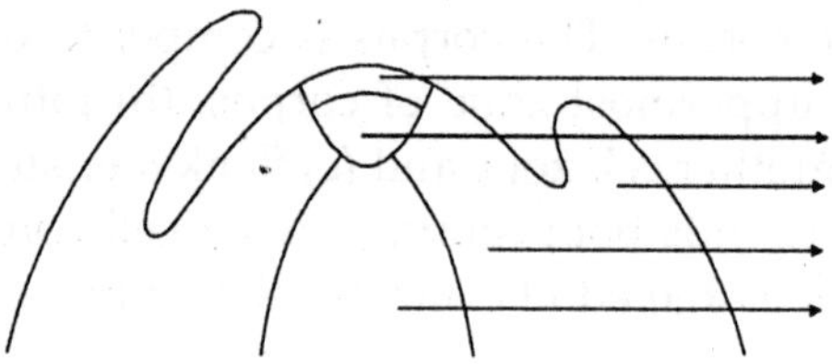


Figure 7.10

Diagram illustrating the cytohistological zonation of apical meristem. Adapted after Beck, 2005, p. 86.

initials—that consist of an apical group of cells, (2) corpus initials—that occur below apical initials and are similar to central mother cells, (3) a peripheral zone and (4) a rib meristem. Tunica initials contribute cells to central mother cell zone and to peripheral meristem. The central mother cell zone donates cells to the rib meristem and pith. The peripheral meristem is highly meristematic. Leaf primordia originate from this layer.

## Vegetative shoot apex

The terminal portion of shoot that occurs just above the youngest leaf primordium is designated as shoot apex. The shoot apex is the region of the primary organization of shoots and exhibits indeterminate or unlimited growth. It continuously produces leaf primordia at definite plastochron in a genetically predetermined manner. Sometimes shoot apex loses the capacity of unlimited growth and forms tendrils or thorns (ex. *Ulex europaeus* and *Bougainvillea* etc.). The meristem of apical shoot with indeterminate growth becomes determinate when it forms an inflorescence or flower.

Shoot apex exhibits great variation in size and diameter as seen in longitudinal section and in scan electron micrograph. Commonly the shoot apex is in the form of mound. It may be in the form of a slender with blunt-tipped cone (ex. grasses, *Elodea* and *Hippuris* etc.). The shoot apex of *Hibiscus syriacus* and *Drimys* is slightly concave (Fahn, 1997). The diameter of shoot apex exhibits rhythmic fluctuation during plastochron (vide the chapter *Plastochron* for a few more details). The diameter may be about 40 microns (*Syringa*), 500 microns (*Phoenix canariensis*) and 700-900 microns (*Trichocereus spachianus*). *Cycas revoluta* exhibits shoot apex with 3.5 millimeters in diameter. Generally large shoot apices are observed in ferns, cycads and cacti (Cutter, 1971).

**Shoot apex of pteridophyta:** In *Psilotum* and *Equisetum* etc. the apical meristem exhibits a single apical cell (Fig. 7.11). The cell is three sided (ex. *Salvinia* and *Selaginella*) or four sided and occurs on the surface layer of meristem. One side of apical cell is directed to the surface of apex and forms the outer surface of the shoot apex. This side of apical cell is triangular or square in outline and never divides. The other sides are cutting faces, pointed downward and divide in an orderly fashion. The three sided apical cell has two whereas the four sided apical cell has three cutting faces. All cutting faces divide and form a large packet of

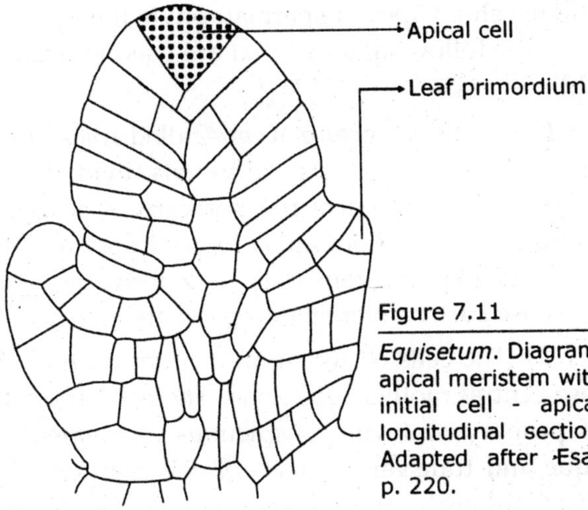


Figure 7.11

*Equisetum*. Diagram showing apical meristem with a single initial cell - apical cell in longitudinal sectional view. Adapted after Esau, 1961, p. 220.

cells. These cells on differentiation form the different segments of shoot. So this apical cell is regarded as 'structural and functional unit of apical meristem' in pteridophyte. The apical cell is genetically sound as its derivatives form the primary organization of shoot apex. The apical meristem of *Selaginella* exhibits 2-5 apical initials (Fig. 7.13).

In spermatophyte, according to the classical concept, the shoot apex is composed of promeristem/primordial meristem. Promeristem consists of undifferentiated mass of cells and the cells have equal morphology. Later investigations reveal that there exist complex zonations in the promeristem. The zones are distinct and can be distinguished on the basis of (1) cell size, (2) thickness of cell wall, (3) nuclear size, (4) staining properties, (5) relative frequency of mitoses, and (6) planes of cell division (Foster, 1949, Fahn, 1997).

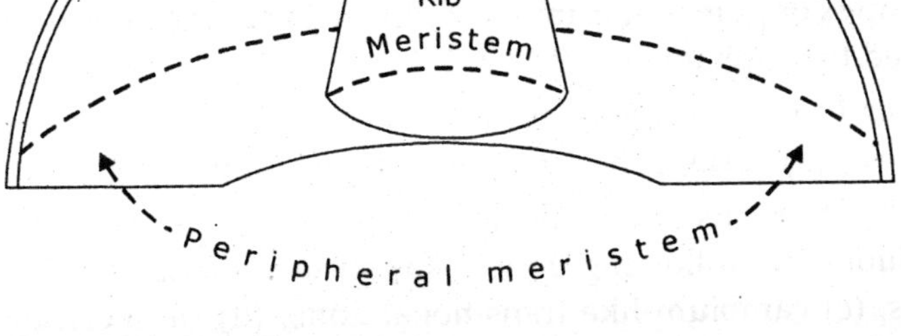


Figure 7.12

Diagram illustrating the different zones of apical meristem present in spermatophyte. Central mother cell-zone is cross-hatched.

Before illustrating the shoot apex of spermatophyte it may be useful to have a precise description of the following terms that are used to designate the zones of meristem (Fig. 7.12).

- *Surface meristem*: It consists of a group of cells called apical initials that occur on the outermost region of shoot apex. Apical initials divide by both anticlinal and periclinal division. In gymnosperm the derivatives of apical initials are regarded as the initiating zone of shoot apex. In angiosperm surface meristem commonly divides by anticlinal divisions only. The derivatives form epidermis of shoot apex. In this sense surface meristem represents tunica of angiosperm.

- *Central mother cells*: The cells occurring in median position just below the surface layer compose central mother cells zone. The cells are distinguished by their large-size, thick primary wall with conspicuous vacuolated cytoplasm. The cells divide by oblique and transverse planes of cell division. The derivatives produce the lateral peripheral zone consisting peripheral meristem/flank meristem and basal zone consisting of rib meristem.

- *Rib meristem*: The cells composing rib meristem zone may occur at the terminal region of apical meristem (ex. *Cycas*) or at the lower region below the central mother cells of apical meristem (*Ginkgo, Opuntia* etc.). Rib meristem is composed of vertical series of transversely dividing meristematic cells. The derivatives of this meristem differentiate into pith cells of a stem.

- *Flank meristem*: It is also termed as peripheral meristem and appears as a cylinder enclosing the zone of rib meristem. The cells of flank meristem contain dense cytoplasm. Surface meristem occurs on the peripheral side of flank meristem. Flank meristem produces cortex, procambium and leaf primordium. Flank meristem increases by its own derivatives. In gymnosperm surface meristem donates cells to flank meristem. In *Ginkgo* and *Opuntia* flank meristem is supplemented by cambium-like transitional zone.

- *Cambium-like transitional zone*: This zone is cup-shaped and occurs in between central mother cell-zone and rib—and flank meristem. The cells of this zone divide frequently usually by periclinal division and donate cells to rib—and flank meristem.

**Shoot apex of gymnosperm**: The followings are the three types of shoot apex as illustrated by Popham. Fig. 7.13 & 7.14

i. *Cycas* type: Example: *Cycas revoluta*. This type includes the following zones—(a) surface meristem, (b) rib meristem and (c) flank meristem.

ii. *Ginkgo* type: Example: *Ginkgo biloba, Sequoia sempervirens* etc. This type includes the following zones—(a) surface meristem, (b) central mother cells, (c) cambium-like transitional zone, (d) rib meristem and (e) flank meristem.

iii. *Cryptomeria-Abies* type: Example: *Cryptomeria japonica, Abies concolor* and *Taxus baccata* etc. This type includes the following zones—(a) surface meristem, (b) central mother cells, (c) rib meristem and (d) flank meristem.

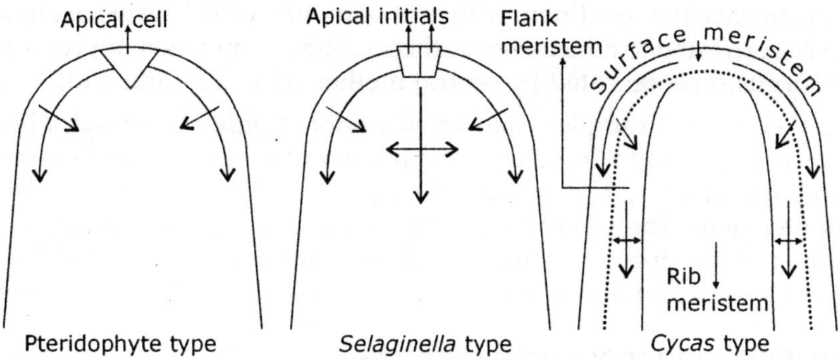


Figure 7.13

Diagram illustrating vegetative shoot apices. Arrows within zones indicate the lines of new cell formation. Adapted after Fahn, 1997

**Shoot apex of angiosperm:** The shoot apex of angiosperm was illustrated previously in histogen theory and tunica-corpus theory. Popham (1952) regards the following two types of shoot apex in angiosperm (Fig. 7.14):

- *Usual angiosperm* type: It includes the following zones – (a) surface meristem, (b) central mother cells, (c) rib meristem and (d) flank meristem. Rib – and flank

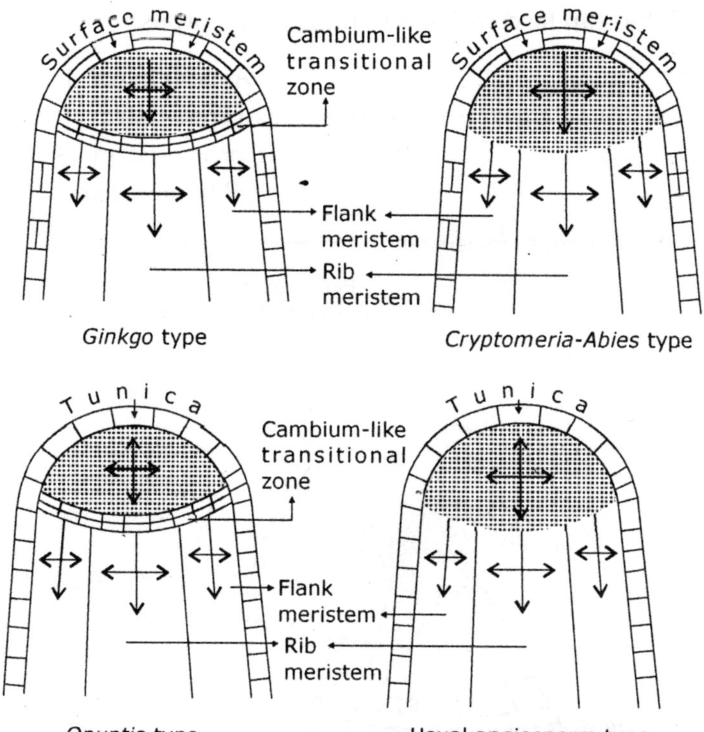


Figure 7.14

Diagram illustrating vegetative shoot apices. Central mother cell-zone stippled. Arrows within zones indicate the lines of new cell formation. Adapted after Fahn, 1997.

meristem appears to be confluent with central mother cells. The usual angiosperm type can be correlated to tunica-corpus theory. Surface meristem represents tunica and the corpus is represented by central mother cells, rib—and flank meristem.

• *Opuntia* type: Example: *Opuntia cylindrica, Xanthium pennsylvanicum* and *Liriodendron tulipifera* etc. It includes the following zones—(a) surface meristem, (b) central mother cells, (c) cambium-like transitional zone, (d) rib meristem and (e) flank meristem. The *Opuntia* type can be correlated to tunica-corpus theory. Surface meristem represents tunica and the corpus is represented by central mother cell-zone, cambium-like transitional zone, rib—and flank meristem.

## Anneau initial and méristème d'attente

Buvat (1955) put forward another interpretation regarding the apical organization of shoot apex (Fig. 7.15). According to this view the cells of central apical zone are mitotically inactive during the vegetative development of shoot apex. The cells become active when the shoot apex is transformed to reproductive, that is, when the shoot apex forms inflorescence and flower. This central apical zone is referred to as *méristème d'attente* (=waiting meristem) or *zone axiale.* The lateral and sub-terminal meristems are mitotically active and form the leaf primordium. This active meristem is called anneau initial or initiating ring. Buvat attributed the histogenic role to the initiating ring and recognized the following four zones in shoot apex that can be correlated to tunica and corpus.

*Anneau initial* ] — Tunica
*méristème d'attente* — [ Sporogenous promeristem ] — Tunica
*méristème d'attente* — [ Receptacular promeristem ] — Corpus
Pith meristem ] — Corpus

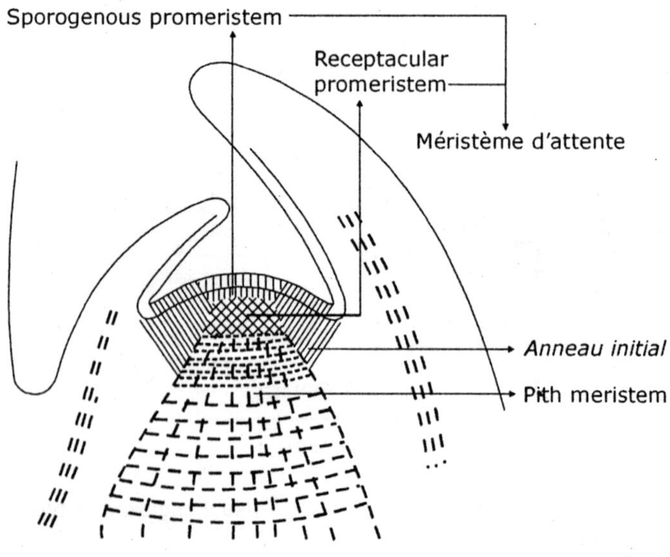


Figure 7.15

Diagram illustrating the concept of Buvat in the shoot apex of *Cheiranthus cheiri*. Adapted after Cutter, 1971, p. 54.

The bases of distinguishing the apical zonations are nuclear and nucleolar size, degree of vacuolation, distribution of RNA and the frequency of mitoses in the apical meristem (Cutter, 1971, p. 57). Later extensive investigations using colchine-induced cytochimeras reveal that cell divisions do occur in the central apical cell-zone. This view became controversial because the denial of mitotic division in the centrally situated apical cells that were once regarded as apical initials. Though this view gained little acceptance, it is widely discussed in literatures (Cutter, 1971, Fahn, 1997).

## Types of vegetative shoot apex

Newman (1965) illustrated three types of vegetative shoot apex based on continuing meristematic residue (c.m.r.). The term 'continuing meristematic residue (i.e., apical cell or cells) is proposed for a line (or lines) of cells inheriting the office of initial' – Newman (1965, p. 210). The derivatives of c.m.r. are the general meristem from where primary tissues elaborate. Newman regards that

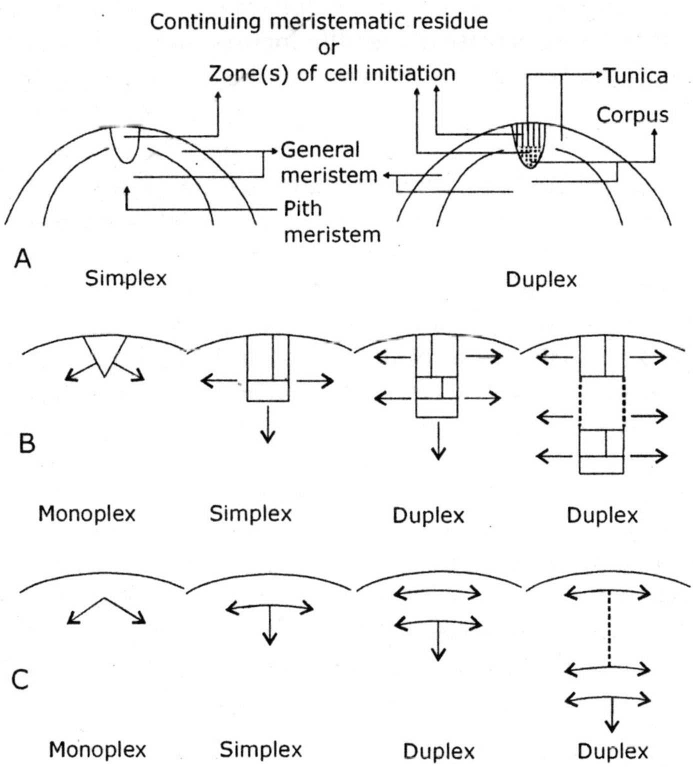


Figure 7.16

A. Diagram correlating the concept of Newman with Schmidt and others. B. & C. Diagram illustrating the basic types of continuing meristematic residue. Arrows indicate the line of new cell formation. Dotted lines in duplex represents unspecified number of layers that divide with anticlinal plane. Adapted after Newman, 1965, p. 192.

no cell being a permanent cell in c.m.r. and meristem. The following three types of shoot apex are recognized in vascular plants (Fig. 7.16):

• *Monoplex* apex: Ex. cryptogamic vascular plants. Continuing meristematic residue occurs in the superficial layer only. It may consist of more than one cell provided a common wall is present between the two cells. It may also consist of 3 or more cells and the cells have a common angle. During mitosis division walls lie parallel to the inclined walls. With inclined walls any one division in c.m.r. contributes to increase in both length and breadth, that is, one division provides for increase in bulk growth.

• *Simplex* apex: Ex. *Abies balsamea* (gymnosperm). In simplex apex continuing meristematic residue is in the superficial layer alone. Continuing meristematic residue may consist of more than one cell provided a common wall is present between the two cells. It may also consist of 3 or more cells and the cells have a common angle. Continuing meristematic residue is parallel-sided as viewed in longitudinal section. Cell division is restricted to one layer. Continuing meristematic residue divides by both anticlinal and periclinal divisions. Anticlinal division contributes to increase in breadth. Increase in length occurs as a result of periclinal divisions. So for bulk growth both periclinal and anticlinal divisions are needed.

• *Duplex* apex: Ex. angiosperms. In duplex apex c.m.r. consists of more than one layer and the cells composing them are parallel-sided as viewed in longitudinal section. The peripheral layer of c.m.r. forms the surface layer. Continuing meristematic residue may consist of more than one cell per layer provided a common wall is present between the two cells. It may also consist of 3 or more cells that have a common angle. The surface layer of c.m.r. divides by anticlinal divisions only. As a result increase in breadth or surface occurs. The inner layer of c.m.r. divides by both anticlinal and periclinal division. As a result bulk growth; i.e. increase in both length and breadth occurs. Duplex apex thus exhibits two contrasted mode of cell division in different layers— (i) peripheral layer of c.m.r. has anticlinal plane of cell division and (ii) inner layer of c.m.r. has both anticlinal and periclinal divisions. Hence the prefix 'duplex' was introduced to recognize two modes of activity in c.m.r. For bulk growth anticlinal division in peripheral layer of c.m.r. and both anticlinal and periclinal divisions in the inner layer(s) of c.m.r. are necessary.

Previously the shoot apex was neatly labeled by Hanstein, Schmidt, Buvat and Popham. The concept of continuing meristematic residue of Newman and its classification into monoplex, simplex and duplex appear to be simplest in comparison to previous views due to the followings:

- The classification is applicable to all groups of vascular plant and it includes stem and root.
- This classification is not rigid like histogen theory regarding the differentiation of derivatives.
- The classification does not take into account whether the shoot apex possesses tunica-corpus – or dermatogen-periblem-plerome organization.

- The concept of méristème d'attente of Buvat has not received universal acceptance and it is not as prominent as quiescent centre of root.
- The classification of shoot apex according to Popham into seven types has limited usefulness. The meristem fluctuates in structure during ontogeny at different stages of plastochron.

## Root apex

Root apex is situated opposite to the shoot apex. The seedlings of most dicotyledons have one main root, i.e. tap root that is soon replaced by adventitious and lateral roots. The roots have root cap at their apices and the apical meristem occurs below the apices. Lateral and adventitious roots have promeristem with similar composition. Promeristem gives rise to all the tissues of the root and it may consist of an apical cell or a group of initials. The structure varies according to group of plants—a brief account of which is discussed below.

• *Root apex of pteridophyte*: Most of the vascular cryptogams have a promeristem with a single apical cell. But some members of Marattiaceae exhibit a few initials in the root apex. The initials are arranged in one tier. The root apex of pteridophyte is illustrated below taking the example of small aquatic fern *Azolla* (Fig. 7.17) because it has been studied in details due to the presence of a conspicuous apical cell, prominent cell wall of derivative(s) of apical cell and the lineage of any cell, except the root cap, can be traced to apical cell. Moreover the meristem of *Azolla* root has genetically determined limit to its growth, i.e. root is determinate organ and this condition is in contrast to most roots where the activity of meristem is indeterminate—growth persisting as long as plant lives.

*Azolla pinnata* shows a large apical cell. The apical cell of the genus *Azolla* is four-sided. The side facing the root cap, i.e. the distal face is curved. The other three sides face the base of root and they are approximately triangular. The walls of basal three sides show numerous plasmodesmata. The apical cell is sparsely vacuolated in young roots. With the increase of age of the root, the vacuolation also increases, but the size of the apical cell remains unchanged. The root cap is generated towards the distal curved face. The apical cell divides and the cell plate is formed parallel to the curved face. The derivative cell again divides in the same plane resulting in two cell-layers. These two cell-layers then undergo numerous anticlinal divisions to produce the surface of the root cap. The root cap throughout its surface remains two cells thick only.

After the initial division to produce root cap, the apical cell divides sequentially from its three basal faces. The derivatives also divide and form a group of cells analogous to 'packet' of cells of shoot meristem. The initial divisions at the basal faces are parallel to surface. The divisions in the derivatives are also parallel to surface. As a result a series of successive segments are formed. These segments are referred to as merophyte (Lyndon, 1990, p. 22; Burgess, 1985, p. 159). A merophyte consists of derivative(s) of apical cell. The derivatives can be recognized as a group of cells by their thick walls and pattern of divisions.

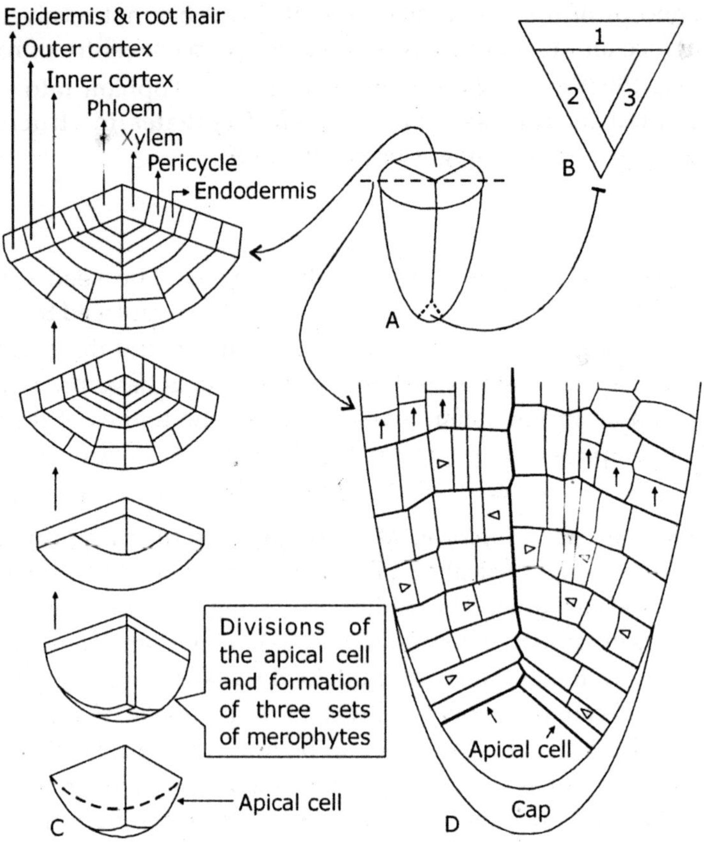


Figure 7.17

*Azolla*. A. A portion of root tip showing the position of apical cell (B), merophyte (C) and the plane of longitudinal section (D). B. Apical cell illustrating the sequential transverse divisions that are indicated by 1, 2 & 3. C. Diagram illustrating the formation of merophytes, their divisions and differentiation. D. Longitudinal section of root. The merophytes are bounded by thick lines. Horizontal arrows indicate

The initial division to produce merophyte is parallel according to figure 7.17B. The divisions occur in precise sequences in a clockwise or counter clockwise manner. Later divisions may be at right angles to the original parallel divisions. The divisions to produce merophytes are distinct, species-specific and the patterns of cell divisions are predictable. The figure 7.17C illustrates the production of merophytes and the formation of layers of root from them. The root cap is generated from the curved side of the four-sided apical cell and the three basal sides form epidermis, cortex and stele.

• *Root apex of gymnosperm*: Allen in 1947 (Fahn, 1997) illustrated the root apex of *Pseudotsuga* (Fig. 7.18B). The root apex consists of a central group of permanent initials with three groups of temporary initials. The temporary initials are generated by the central group of permanent initials on the peripheral side. The

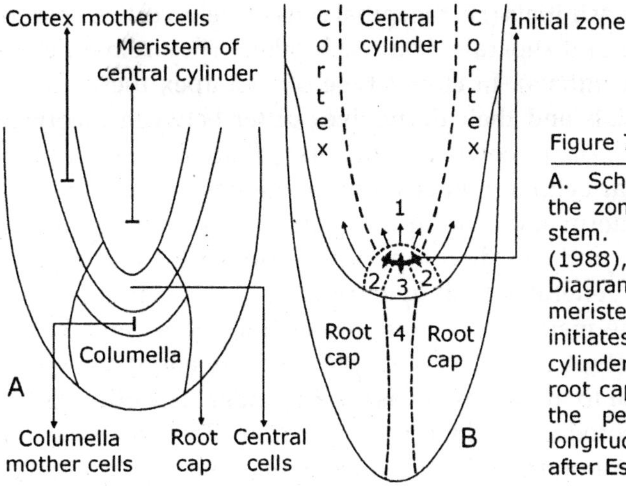


Figure 7.18

A. Schematic diagram showing the zones of a root apical meristem. Adapted after Mauseth (1988), p. 103. B. *Pseudotsuga*. Diagram illustrating the apical meristem of root. The initial zone initiates mother cells of 1. Central cylinder; 2. Cortex; 3. Column of root cap (4) that donates cells to the periphery of root cap by longitudinal divisions. Adapted after Esau, 1965, p. 119.

temporary initials are the mother cells that are destined to produce vascular cylinder, cortex and columella. The first group of temporary initials is the meristem of vascular cylinder from which stele develops. The second group of temporary initials is meristem of cortex. As the name implies cortex is developed from this meristem. The peripheral layer of cortex is transformed into protoderm that gives rise to epidermis. The third group of temporary initials is the mother cells of columella (Fig. 7.18A) that produce columella. In roots the term columella is used to designate the central part of a root cap where the cells are arranged in longitudinal files. Columella donates cells on the periphery of root cap by periclinal divisions. Wilcox in 1954 (Fahn, 1997) reports the presence of two groups of temporary initials in *Abies procera*. Among the two groups one group forms the central cylinder and the other forms the columella.

- *Root apex of angiosperm*: The root apex organization exhibits a central group of permanent initials with three groups of temporary initials. The temporary initials are the mother cells of protoderm, cortex, vascular cylinder and root cap, and they occur on the periphery of permanent initials. Guttenberg in 1960 (Fahn, 1997) regarded two types of root apex namely—closed and open (Fig. 7.19).

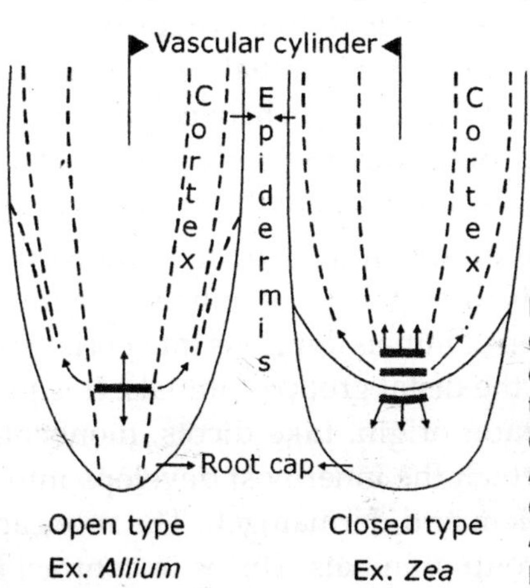


Figure 7.19

Diagram illustrating types of root apex. Vide text for details. Broad lines indicate the initials of protoderm, cortex, stele and root cap. Arrows indicate the direction of cell formation. Adapted after Tayal, 1994, p. 54.

In closed type the temporary initials (i.e. the initial cells of stele, cortex, epidermis and root cap) are distinct and separate from each other. They remain close to central cells (=permanent initials). In closed type of root apex there are three groups of temporary initials and their derivatives differ between dicotyledon and monocotyledon. In *Brassica* (dicotyledon) the first group of temporary initial forms meristem of vascular cylinder; other group of temporary initial produces meristem of cortex. Protoderm and root cap are generated from the third group of temporary initial. Protoderm develops into epidermis. The meristem (=temporary initial) that generates both epidermis and root cap is termed as dermatocalyptrogen. Some members of Compositae, Solanaceae and Rosaceae etc. also exhibit dermatocalyptrogen. In *Zea mays* (monocotyledon) one group of temporary initial gives rise to meristem of vascular cylinder. The other group of temporary initial forms protoderm and meristem of cortex. Root cap is generated from the third group of temporary initial. The meristem (=temporary initial) that generates root cap only is termed as calyptrogen by Janczewski in 1874 (Esau, 1965). Calyptrogen is also exhibited in some members of Zingiberaceae and Arecaceae etc. In open type of root apex the initials of epidermis, cortex and root cap are not distinctly delimited. All tissues of root (e.g. *Allium*) or all except the central cylinder (e.g. *Helianthus*) appear to be generated from a common meristematic group of cells that are situated on the periphery of central group of cells. Later the meristems of different tissue systems become discrete and occur some distance away from the central cells.

Eames and MacDaniels illustrated the root apices of vascular plants (Fig. 7.20) in the following ways. They interpreted the root apex according to the terms of histogen theory (dermatogen, periblem and plerome) though the theory was abandoned in case of shoot apex.

Vascular cryptogams have a solitary apical cell in the root apex. Root cap, epidermis, cortex and vascular cylinder are generated by the single apical cell, e.g. some species of *Selaginella*, ferns and *Equisetum* etc. In the above species the apical cells are similar and large. The root cap is structurally distinct from the other tissues of root and its origin can be traced back to the apical cell.

The root apex of gymnosperm consists of two groups of initials. Plerome is developed from the inner initials. The outer initials give rise to periblem and root cap. There exists no clear line of demarcation between plerome and periblem. The distal region of periblem proliferates to produce root cap. Dermatogen originates from periblem behind the apex where the periblem and cap adjoin.

The root apex of angiosperm differs in dicotyledon and monocotyledon. In dicotyledon the root apex exhibits three groups of initials. The innermost initials are the plerome. The median group of initials develops into periblem. Dermatogen and root cap are generated from the distal group of initials. It is to note that root cap and dermatogen have common origin. Like dicots, monocots also exhibit three groups of initials among which the innermost develops into plerome. The median initials give rise to periblem and dermatogen. The root cap is generated independently from the distal group of initials. This is in contrast to the dicots.

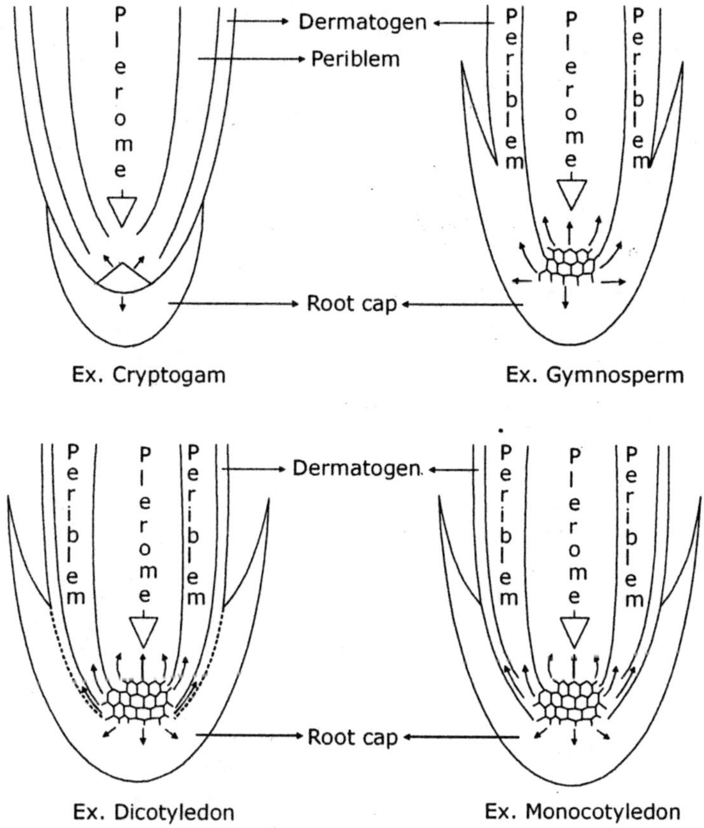


Figure 7.20

Diagrams illustrating the types of root apex of vascular plants according to the concept of Eames and MacDaniels (1947).

Newman (1965) illustrated the following five types of root apex (Fig. 7.21) based on the concept of Eames and MacDaniels (1947) and Esau (1965) with modifications. Newman did not propose any terminology and the types are numbered for convenience only without any significance.

*Type 1*: It has one initiating layer consisting one solitary apical cell. Root cap and all tissues of root are generated from the single initiating layer. Ex. Ranales, Amentiferae, *Pisum* and *Allium* etc.

*Type 2*: It has two initiating layers. The first or distal initiating layer forms epidermis, cortex and root cap. Vascular cylinder is generated by the second or inner initiating layer. Ex. gymnosperm, dicot and monocot.

*Type 3*: It has two initiating layers. The first or distal initiating layer forms epidermis, outer cortex and root cap. Inner cortex and vascular cylinder are generated by the second or inner initiating layer. Ex. Tiliaceae, Umbellales etc.

*Type 4*: It has three initiating layers. The first or distal initiating layer produces epidermis and root cap. The second or median initiating layer forms cortex. Vascular cylinder is generated from the third or innermost initiating layer. Ex. Rosaceae, Solanaceae and Compositae etc.

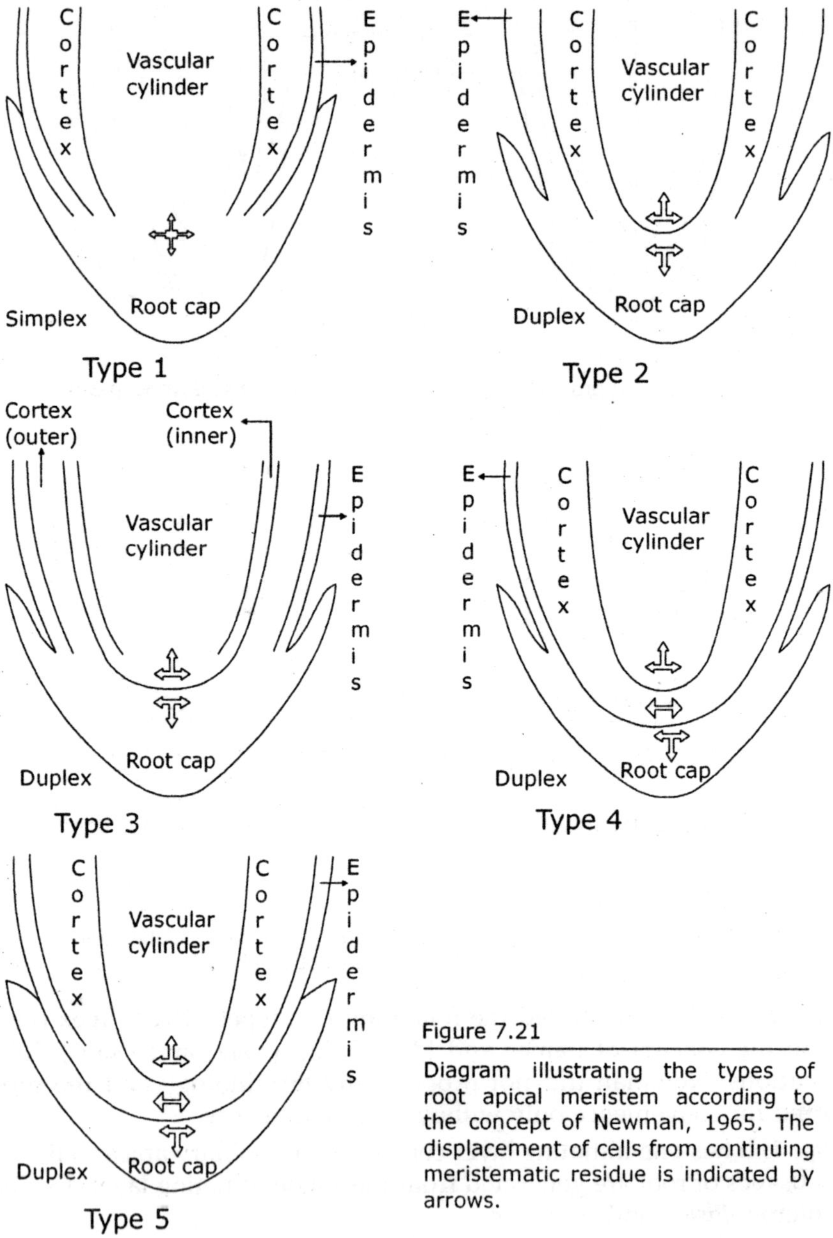


Figure 7.21

Diagram illustrating the types of root apical meristem according to the concept of Newman, 1965. The displacement of cells from continuing meristematic residue is indicated by arrows.

*Type 5*: It has three initiating layers. The first or distal initiating layer produces root cap only. The second or median initiating layer gives rise to cortex and epidermis. Vascular cylinder is generated from the third or innermost initiating layer. Example: Monocotyledon.

Type 1 represents the simplex apex and the types 2, 3, 4 and 5 exhibit duplex apexes.

## Theories of root apical meristem

There are mainly three theories to explain the root apices of vascular plant. The apical cell theory is applicable to vascular cryptogams only. Histogen theory and körper-kappe theory explain the root apices of flowering plants where more than one apical cell compose the meristem.

- *Apical cell theory*: This theory was proposed by Nägeli (Cutter, 1971) who drew the attention to the occurrence of a single apical cell or apical initial that composes the root meristem. A single apical cell is present only in vascular cryptogams, e.g. *Equisetum, Adiantum* and *Polypodium* etc. The apical initial is tetrahedral in shape and generates root cap from one side. The other three sides donate cells to form epidermis, cortex and vascular cylinder. In other words all tissues that compose a mature root including root cap are the derivatives of a single apical cell. Apical cell theory is confined to vascular cryptogams only as the root apical meristem of flowering plants does not have a single apical cell.

- *Histogen theory*: Hanstein in 1868 (Cutter, 1971, p. 8) advocated the theory. According to Hanstein root apical meristem consists of three cell-initiating regions called histogens (Fig. 7.22). The histogens are called dermatogen, periblem and plerome that respectfully form epidermis, cortex and vascular cylinder that are present in a mature root. The derivatives of dermatogen vary. In *Zea mays* (monocot) dermatogen generates root cap only and this histogen is referred to as calyptrogen. In *Brassica* (dicot) dermatogen generates both protoderm and root cap and this histogen is referred to as dermatocalyptrogen.

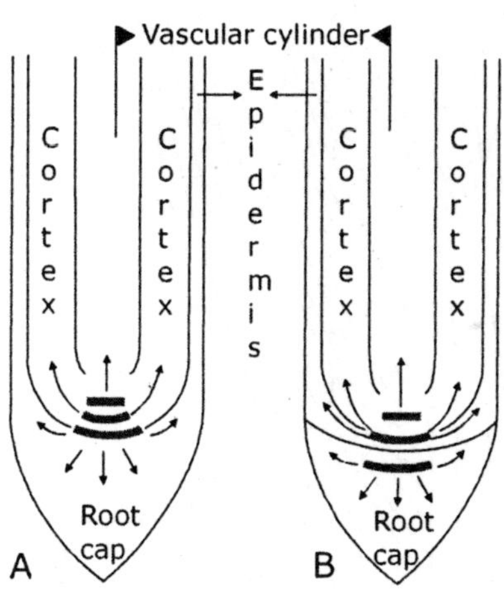


Figure 7.22

Schematic diagram of root apex based on Histogen theory. A. Root apex of *Raphanus sativa*. B. Root apex of *Zea mays* where root cap has its own histogen, i.e. Calyptrogen. Broad lines indicate histogens. Arrows indicate the direction of cell formation. Vide text for details. Adapted after Burgess, 1989, p. 157.

Histogen theory explains both root and shoot apical meristem. This theory attributes specific destinies to the derivatives of the three histogens. Though histogen theory is abandoned to explain shoot apex, Eames and MacDaniels illustrated the root apical meristem on the basis of histogen concept.

- *Körper-kappe theory*: This theory of root meristem was proposed in 1917 by Schüepp (Fahn, 1997) who regarded the occurrence of two systems of cell seriation that characterize the root apex with reference to planes of cell division in its parts. Körper-kappe concept is also referred to as body-cap concept (Körper = body and kappe = cap) and the concept illustrates distinct type of cell wall pattern formation during cell division. The body-cap concept is illustrated below on analyzing the divisions in the derivatives of apical cell (Fig. 7.23).

The root meristem exhibits multicellular structure. It consists of conspicuous longitudinal files of cells. During growth the root changes in diameter. This happens due to cell divisions that occur in such a way that a single longitudinal file of cells becomes double files. The initial cell divides transversely. The two cells thus formed one has the capability of cell division. This cell divides longitudinally and both the daughter cells inherit the property of cell division. The daughter cells are parallel in arrangement, share a common wall and divide by transverse partition followed by longitudinal partition in one cell. The sequences of wall formation when viewed together appear to form a configuration resembling the letter 'T' or 'Y'. Such divisions are described as T-divisions. Continuous T-divisions result in the formation of double-rowed region over a single rowed region.

It is the T-division that characterizes körper and kappe. In the kappe the initial cell first divides transversely and forms two cells. The daughter cell that faces the root apex inherits the initial function. It divides longitudinally. The two

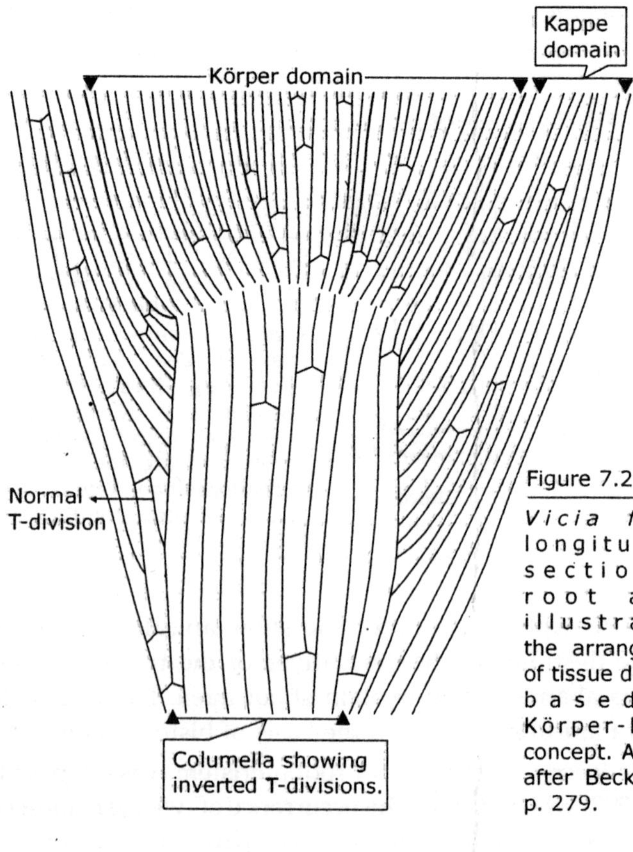


Figure 7.23

*Vicia faba*: longitudinal section of root apex illustrating the arrangement of tissue domains based on Körper-kappe concept. Adapted after Beck, 2005 p. 279.

cells thus formed have the capability of cell division. When transverse and longitudinal partition are viewed together the combined cell walls appear as 'T' that is right-way-up. When such division continues it is observed that a single rowed region is left behind over the double-rowed region. This occurs in downwardly pointed roots.

In the körper the initial cell first divides by transverse partition and forms two cells. The daughter cell that faces the base of root, i.e. away from the apex inherits the initial function. It divides longitudinally and the two daughter cells thus formed have the potentiality of cell division. The daughter cells divide by transverse partitions followed by longitudinal partitions. When transverse and longitudinal partitions are viewed together the cell walls form a configuration resembling an inverted 'T'.

Körper and kappe—these two zones of root are delimited by planes of cell division. The zones exhibit clear boundary when they originate from separate initials, e.g. root with calyptrogen. The zones do not exhibit sharp demarcation line when they are the derivatives of same apical cell. In root with dermatocalyptrogen the cap extends into protoderm.

The central part of root cap is the columella where the cells are arranged in longitudinal files. These cells seldom divide. When division occurs the partition walls form the configuration of an inverted 'T' that is observed in the körper. The 'T' has normal configuration in the peripheral region of root cap.

The körper-kappe theory of root apex is comparable with tunica-corpus theory of shoot apex. The body-cap concept and tunica-corpus concept both are based solely on the planes of cell division. Anticlinal division is the characteristic of tunica whereas corpus exhibits both anticlinal and periclinal division. On the other hand the inverted 'T'—and normal 'T' pattern of cell wall formation are the characteristic of körper and kappe respectively. The boundaries between körper and kappe, and between tunica and corpus are not always sharply demarcated.

## Quiescent centre

It is observed that in the root-promeristem of higher plants the initials divide much less frequently than other cells. This part of root-promeristem, which contains the initials of epidermis, cortex and stele and divides less frequently, is referred to as quiescent centre (QC). Quiescent centre is situated at the pole of cortex and stele and consists of four cells (e.g. *Arabidopsis*) or as high as 600 cells (ex. *Zea mays*, Fahn, 1997, p. 70) that have very low mitotic activity (Fig. 7.24). In closed meristem QC appears more or less hemispherical whereas in open meristem it is disc-shaped. The cells composing QC exhibit smallest dictyosomes, nuclei and nucleoli. The cells also exhibit little endoplasmic reticulum and fewer mitochondria. The cells contain lower concentration of RNA, DNA and protein than other cells of root apex.

Quiescent centre is unusual in shoots, but it is universal in roots of higher plants. Though the occurrence of QC was reported in vascular cryptogams (e.g. *Equisetum arvense*) later investigations have established that the entire root apex

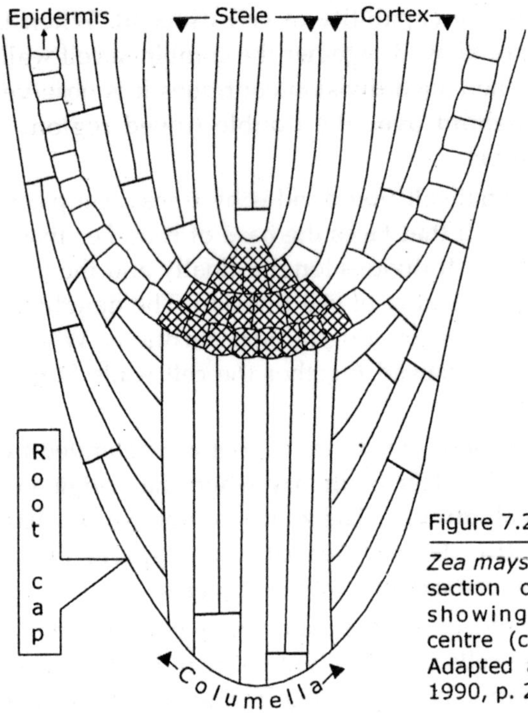


Figure 7.24

*Zea mays*. Longitudinal section of root apex showing quiescent centre (crosshatched). Adapted after Lyndon, 1990, p. 23.

of vascular cryptogams is mitotically active. Quiescent centre maintains a strict boundary in closed meristem. As a result the derivatives of QC also maintain a strict boundary in the apical zone. Quiescent centre consists of central cells only or central cell and part of its peripheral zones. It does not include the mother cells of columella. A separate cap meristem exists and it forms root cap. In open meristem the size and position of QC change. They fluctuate as the root ages and grow bigger. The volume and number of constituent cells of QC vary with the change of width of root. In growing roots QC is small. The columella mother cells are generated from the cells situated ahead of the central cell.

It needs special techniques to identify QC. It can be identified in autoradiograph of root sections. Supplying root with solutions of thymidine labeled with radioactive isotope hydrogen, tritium ($^3H$), the nuclei of root apical meristem can be labeled. Radioactive thymidine incorporates specifically into DNA of DNA synthesizing cells. All the nuclei except those of QC become radioactive because QC has low rate of DNA synthesis and so lack the incorporation of radioactive thymidine. To prepare autoradiograph roots were treated with radioactive thymidine and then fixed with fixative. The fixed roots were microtomed and the sections were placed in contact with autoradiographic stripping film. All were kept in dark for several days. The film was later developed and fixed. The film is sensitive to radiations emitted by radioactive substances and produces images of the locations from where the radiations were emitted. As the film is in direct contact with the sections of root the precise sites of radioactivity can be identified in autoradiograph. Mitotic figures and mitotic indices are also used to identify the QC.

Quiescent centre has the capability to generate the entire root including root cap. This is established in QC culture. Quiescent centre can be isolated and when the isolated QC is grown in suitable medium, it forms root from its proximal face and root cap from distal face. Normally root cap initials are generated outside the QC. But the isolated QC generates root cap initials. In an experiment the root cap of maize is carefully removed so that the root proper is not otherwise damaged. It is observed that the cell division in QC is stimulated and the root grows faster. The cell divisions slow down as the new root cap is regenerated. Later studies reveal that direct contact of root cap and differentiated cells with QC inhibit the differentiation of initial cells.

Quiescent centre is regarded as central mother cells that form promeristem of root apex. The followings are suggested regarding the function of QC.

- It provides a reserve block of diploid cells when the root tip is damaged at the time of pushing through the soil.
- It is the site of hormone synthesis.
- As it is the source of some growth substances, at certain concentration of hormone it inhibits cell division and growth.
- Quiescence is caused due to the presence of root cap and the pressure exerted by the dividing initial cells. Cells of QC maintain a long period in the presynthesis ($G_1$) phase of mitotic cycle and thus they may escape injury.

Earlier literatures emphasized regarding the presence and functioning of initiating layers in the root tips of seed plants. Researches during 1950s demonstrate the occurrence of QC in mature roots. Later studies reveal that in *Arabidopsis thaliana* QC and columella of root cap originate during embryogenesis. In this species genetic studies of root meristem development demonstrate the presence of *hbt* (*hobbit*) gene, which is responsible for the development of the normal initiating layers in roots, i.e. dermatogen, periblem and plerome. *Arabidopsis* with mutant *hbt* (*hobbit*) gene lacks QC. Moreover in *hbt* mutant calyptrogen develops abnormally that generates root cap where columella is lacking. Researches in molecular biology at the end of twentieth century definitely establish the fact that QC has a significant role in the development of root. Modern literatures describe QC 'as an architectural template in the root apical meristem of all angiosperm and gymnosperm root tips ......' – Beck (2005, p. 281).

Knowledge of meristem is an essential prerequisite in meristem culture. Meristems are cultured for vegetative propagation and micropropagation that are practiced in forestry, horticulture, floriculture and conservation of endangered species etc. – a brief account of which is mentioned in the chapter 32. Shoot apical meristem culture is also used successfully to remove viruses, bacteria and fungi from plants. This becomes possible due to: (i) the occurrence of viruses in the meristem is much less than other cells, (ii) the cells of meristem divide very fast and (iii) the viruses cannot move quickly through plasmodesmata and thus unable to spread within the actively growing meristem tip.

Comparison between root apical meristem and shoot apical meristem.

| Root apical meristem | Shoot apical meristem |
|---|---|
| 1. It occurs below the root cap and so it is not superficial in position. | 1. It occurs at the tip of shoot and so it is superficial in position. |
| 2. It forms lateral root only. | 2. It forms lateral organs like leaf and floral organs. |
| 3. It exhibits no rhythmic change in size and form during elongation and lateral root formation. | 3. It exhibits rhythmic changes during leaf primordium formation. |
| 4. Root cap encloses the apical meristem. | 4. Leaf primordium overarches the apical meristem. |
| 5. Differentiation of tissues occurs at some distance away from apex. | 5. Differentiation of tissue occurs just below the apex. |
| 6. The position of formation of new lateral root in relation to those already present is not predictable. | 6. The position of formation of new lateral organ in relation to those already present is predictable. |

Comparison of the derivatives of initiating layers according to Newman (1965) in different types of root apex.

| | Derivative(s) of distal initiating layer | Derivative(s) of median initiating layer | Derivative(s) of inner initiating layer |
|---|---|---|---|
| Type 2 | Epidermis, cortex and root cap | Median layer absent | Vascular cylinder |
| Type 3 | Epidermis, outer cortex and root cap | Median layer absent | Inner crotex and vascular cylinder |
| Type 4 | Epidermis and root cap | Cortex | Vascular cylinder |
| Type 5 | Root cap | Cortex and epidermis | Vascular cylinder |

Difference between tunica and corpus.

| Tunica | Corpus |
|---|---|
| 1. The chief mode of cell division is anticlinal resulting in surface growth of apex. | 1. The cells of corpus divide in any plane including anticlinal and periclinal resulting in volume growth of apex. |
| 2. It is the peripheral zone of shoot apical meristem. It overarches corpus. | 2. It is the inner zone of shoot apex and occurs below tunica. |
| 3. It may consist of one or more layers of cell. | 3. It is several cell layers deep. |
| 4. The cells of tunica are more or less rectangular in transverse section and smaller than cells of corpus. | 4. The cells are more or less round in transverse section and larger than the cells of tunica. |
| 5. It forms epidermis. | 5. It gives rise to ground tissue and vascular tissue. |
| 6. Ontogenetically it is not related to corpus. | 6. Ontogenetically it is not related to tunica. |

Difference between Tunica-corpus theory and Histogen theory.

| Histogen theory | Tunica-corpus theory |
|---|---|
| 1. Hanstein postulated the theory and it is applicable to root and shoot apex. | 1. Schmidt postulated the theory and it is applicable to shoot apex only. |
| 2. Three zones are present in shoot apex namely dermatogen, periblem and plerome. | 2. Two zones are present in the shoot apex namely tunica and corpus. |
| 3. Dermatogen gives rise to epidermis, periblem forms cortex and plerome gives rise to stele. | 3. Tunica gives rise to epidermis and corpus forms the rest of tissues. |
| 4. There is no definite types of cell division in histogens. | 4. Tunica divides by anticlinal division only. Corpus divides in all planes. |
| 5. The histogens are rigid in the formation of respective tissues in a plant body. | 5. There is no rigidity in the formation of tissues. The inner layers of tunica may form tissues other than epidermis. |
| 6. It is observed in gymnosperm and angiosperm shoot apex. | 6. It is observed mainly in angiosperm shoot apex. |

Comparison between Tunica-cropus theory and Körper-kappe theory.

| Tunica-corpus theory | Körper-kappe theory |
|---|---|
| 1. This theory is applicable to shoot apex only. | 1. This theory is applicable to root apex only. |
| 2. Shoot apex exhibits tunica-corpus organization. | 2. Root apex exhibits körper (body) and kappe (cap) organization. |
| 3. This theory was proposed by Schmidt in 1924. | 3. This theory was proposed by Schüepp in 1917. |
| 4. Tunica-corpus organization is mainly observed in angiosperm shoot apex. | 4. Körper and kappe organization is observed in vascular plants. |
| 5. This theory is based on anticlinal division only on peripheral layer and periclinal and anticlinal division in inner layer. | 5. This theory is based on the T-pattern of cell division only. |
| 6. Tunica exhibits anticlinal division only. | 6. Kappe exhibits T-division where 'T' is straight-way-up, i.e. the top stroke of 'T' faces the base of root (away the root apex). |
| 7. Corpus exhibits both anticlinal and periclinal divisions. | 7. Körper exhibits T-division where 'T' is inverted, i.e. the top stroke of 'T' faces the apex of root. |

CHAPTER **8**

# Simple Tissue

The tissues, which are composed of one type of cells are termed as simple tissue. It is homogeneous, i.e. consists of one type of cells only, has similar morphological structure, performs same function and originates from the same mother cell. Parenchyma, collenchyma and sclerenchyma are the examples of simple tissue.

## Parenchyma

Parenchyma (Figs. 8.1, 8.6) is a cell and tissue type in which the cells have only thin primary walls; the cells are unspecialized, lack the characteristic wall of collenchyma and the secondary walls of sclerenchyma; the cells have live nucleate protoplast concerned with various physiological activities in plants; the cells are meristematic, or permanent, simple homogeneous (i.e. composed of one type of cells only), fundamental or ground tissue upon which other simple and conducting tissues appear to be embedded.

• *Distribution*: Large portion of all plant organs are occupied by parenchyma. They occur in pith, cortex and pericycle of root and stem. They are also found in leaves as mesophyll tissue. They are also present in the fleshy parts of fruits and endosperm of seed. They also occur in the conducting tissues as xylem and phloem parenchyma. The innermost layer of periderm—phelloderm is also composed of parenchyma.

• *Origin*: Parenchyma cells present in the primary plant body, i.e. pith and cortex, originates from the ground meristem. Procambium and cambium give rise to xylem and phloem parenchyma of the primary and secondary conducting tissues respectively. Phelloderm originates from phellogen. In leaves, protoderm and ground meristem give rise to parenchymatous epidermis and mesophyll respectively.

• *Shape*: Isolated parenchyma cells grown in culture media are spherical. In cell aggregates, they are polygonal due to contact with other cells. Fourteen sided polyhedral cells are most common, although cells with 12, 13, and 15, 16 or fewer are found. The other most common shapes are : elongate (e.g. mesophyll tissue of *Lilium* leaf), stellate (e.g. stem of *Scirpus* and *Juncus*), inner wall protuberated (e.g. phloem transfer cell of *Sherardia* leaf) etc.

• *Arrangement*: Mature parenchyma tissues may be compactly set without any intercellular spaces. Ex. endosperm tissue of seeds. They may also be arranged loosely so that well developed intercellular spaces are present between them. Ex. stems and leaves of hydrophytes.

• *Structure and content*: Parenchyma cells usually have primary walls (e.g. storage and chlorophyllous parenchyma). The storage parenchyma cells of endosperm of *Phoenix*, *Asparagus* have very thick walls. The thickness of wall is due to the deposition of hemicellulose.

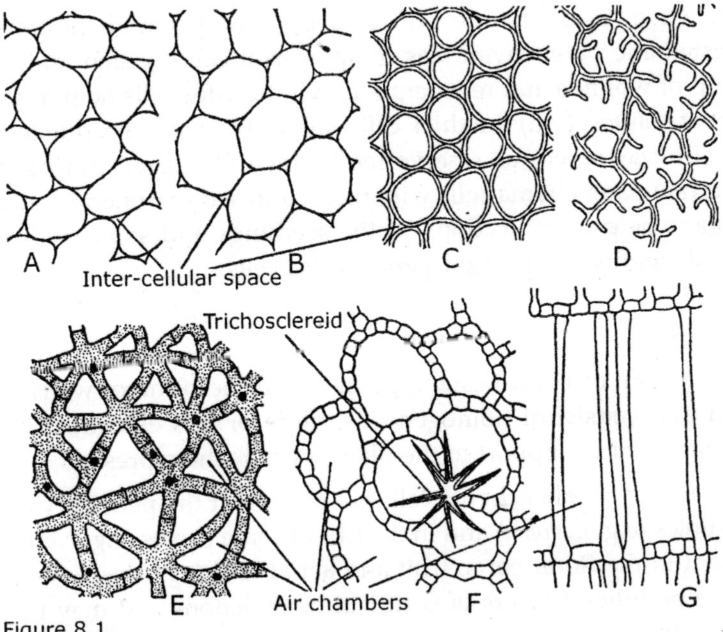


Figure 8.1

Parenchyma. A. Thin-walled parenchyma from cortex of *Helianthus*. B. Same from pith. C. Thick-walled parenchyma from pith of *Clematis*. D. From leaf of *Pinus*. E. Aerenchyma from petiole of *Musa*. F. Same from petiole of *Nymphaea* with trichosclereid. G. Same from stem of *Jussiaea*.

Usually parenchyma cells contain living protoplast with single or numerous vacuoles. They may also contain leucoplasts, chloroplasts etc. (chloroplast containing parenchyma cells are termed as chlorenchyma). Various carbohydrates, nitrogenous and fatty substances are found in the cell sap of parenchyma. The parenchyma cells of the roots of sugar beet and the bulb scales of onion contain amides, proteins, sugars etc. in their cell sap. The cotyledon of many leguminous plants contains protein and starch in their parenchyma cells. The parenchymatous endosperm of *Ricinus communis* contains protein and oils. Protein and starch are present in the parenchyma of potato tuber. Starch is present in the parenchyma of endosperm, tubers, cortex, fruits, xylem and phloem etc. In the parenchyma of storage organs and succulent, water is present. Tannins are also found in many parenchyma cells. Some parenchyma cells, termed idioblasts, which markedly differ in size, content and function than the neighbouring cells, may contain resinous substances (e.g. Rubiaceae, Rutaceae etc.), mucilaginous substances (e.g. Tiliaceae, Portulacaceae etc.), oily substances (e.g. Lauraceae) and the enzyme myrosinase (e.g. Cruciferae).

• *Function*: (i) Living parenchyma cells are the site of all metabolic activities; (ii) chlorenchyma, i.e. parenchyma cells containing chloroplastids, can photosynthesize; (iii) storage parenchyma stores different reserve materials; (iv) parenchyma cells present in conducting tissues, i.e. xylem and phloem parenchyma help in the movement of water and solutes; (v) aerenchyma, i.e. parenchyma cells present surrounding the air spaces in aquatic plants help in aeration, gaseous exchange and add buoyancy which helps in floating; (vi) meristematic parenchyma has the power of cell division and thus helps in healing up of wounds and regeneration; (vii) transfer cells help in short distance transport of solutes; (viii) the thick cell wall in *Asparagus* etc. due to the presence of hemicellulose, serves as reserve food which is utilized during germination; (ix) epidermal parenchyma cells with their cutinised peripheral walls of different plant organs help in protection; (x) the seedlings and some hydrophytes gain mechanical rigidity from turgid parenchyma cells.

## Collenchyma

Collenchyma is a cell and tissue type in which the primary walls are unevenly thickened and consist of homogeneous, more or less elongated living cells; it provides mechanical support to the plant organs where present.

• *Distribution* : Collenchyma cells are present at the periphery of herbaceous stems, petioles (e.g. *Begonia*) and in the ribs of some leaves (e.g. *Nerium*). They are also present in the floral parts, fruit and aerial root (ex. *Monstera*). They usually occupy the peripheral layers of cortex in dicotyledons and may be present just beneath the epidermis or below a few peripheral layer of parenchyma. They may be present as a continuous layer to form hypodermis. Sometimes the cells form distinct ribs or ridges at the outer edges of stems (e.g. *Leonurus*, *Cucurbita* etc.) and petioles (e.g. *Apium*). They are uncommon in roots, monocotyledonous leaves and stems. Subterranean roots of *Vitis* and *Diapensia* contain collenchyma. The rind of fruits is collenchyatous in *Vitis* and *Cassia tora*.

• *Origin*: Collenchyma develops from the ground meristem or from procambium (e.g. *Apium graveolens*). They may also originate from elongated cells, which resemble procambium.

• *Shape*: Isolated collenchyma ranges from round, elliptical to elongated cell. Transitional forms are also present. The round or elliptical cells resemble ground parenchyma. The cells may assume the shape of a short prism. The elongated cells look like fibre with both ends tapered and may attain a length of 2 mm, though rarely.

• *Arrangement*: Collenchyma tissue appears to be more or less compactly arranged as the thickening materials deposit more heavily at the corners and on the radial walls of cell in addition to normal uniform thickening. The nature of wall thickenings varies and accordingly the following three types of collenchyma are distinguished.

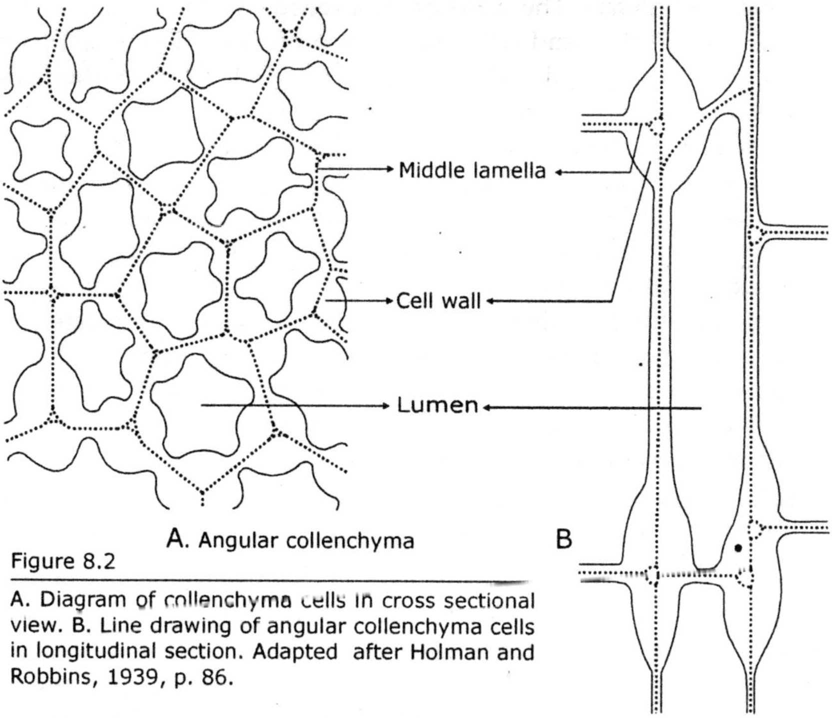


Figure 8.2

A. Diagram of collenchyma cells in cross sectional view. B. Line drawing of angular collenchyma cells in longitudinal section. Adapted after Holman and Robbins, 1939, p. 86.

- *Types of collenchyma*: (Figs. 8.2, 8.3) (a) Angular: The thickening materials deposit at the angles or corners of the cells. In cross sectional view the thickenings occur at those places where several cells meet. Example: petioles of *Cucurbita, Beta* etc. (b) Lamellar: The thickening materials deposit heavily on the tangential walls of the cell than the radial walls, ex. stem of *Sambucus*. (c) Lacunar: The thickening materials deposit at those places of the cell wall, which are in direct contact with the intercellular spaces. Example: petioles of *Salvia, Malva* etc.

Duchaigne (1955) reported another type, called annular collenchyma in the petiole of *Nerium* (Fig. 8.3C). In this type the cell lumen appears to be more or less circular in cross sectional view. Fahn (1987) opined that this might be the case

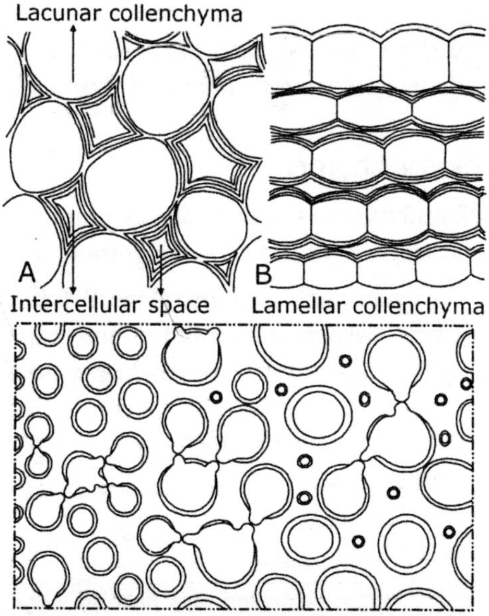


Figure 8.3

Collenchyma cells in cross-sectional view. C. Adapted after Metcalfe & Chalk, 1979, p. 57.

of mature collenchyma. It is observed that due to continued and heavy deposition of wall materials the angular appearance of the lumen may be lost.

• *Structure and content*: The walls are thickened by high amounts of pectin and hemicellulose. Protein and cellulose are also present. Primary pit fields can be distinguished in the walls. The cellulose microfibrils are either oriented transversely or alternately transverse and longitudinal. The longitudinal microfibrils may also alternate with noncellulosic materials, as is revealed by electron microscopic study with the petiole of *Apium graveolens*. The composition of these alternating layers varies. One layer is rich in cellulose and poor in pectin; the alternate layer is pectin-rich and cellulose-poor. It is now suggested that the visible lamellation of the cell wall of collenchyma is due to composition and orientation of microfibrils at alternate layers. Usually lignin is completely absent. But in some cases the cellulose-rich layer may be impregnated with lignin (e.g. stem and petioles of *Salvia officinalis*, *Viscum album*, *Medicago sativa* etc.). Lignified lamellae may also be deposited on the wall. Thus the pecto-cellulosic wall of collenchyma may become sclerified.

Collenchyma contains living protoplast. Chloroplastids may also be present in some cells. The cells may also contain tannins.

• *Function*: (1) The cells are extensible with a considerable degree of plasticity and so support the organs in which they occur. (2) Chloroplast containing collenchyma can carry out photosynthesis. (3) In some cases the peripheral thick walled collenchyma becomes thin and regains the meristematic activity, e.g. phellogen, the cork cambium, which divides to form the periderm. (4) The sclerified collenchyma is the mechanical cell of mature plants. (5) Collenchyma can serve as storage sites for defensive antibacterial compounds. The compounds are hydroxyproline-rich bacterial agglutinins found in *Solanum tuberosum*.

## Sclerenchyma

Sclerenchyma is a kind of mechanical tissue consisting of oval or elongated cells with thick lignified or nonlignified walls and small lumen.

• *Types of sclerenchyma* (Figs. 8.4, 8.5, 8.6): There are two kinds of sclerenchyma: (i) fibre—which is elongated cell, tapering to a wedge-shape at both ends and (ii) sclereids—which are oval or rounded cells.

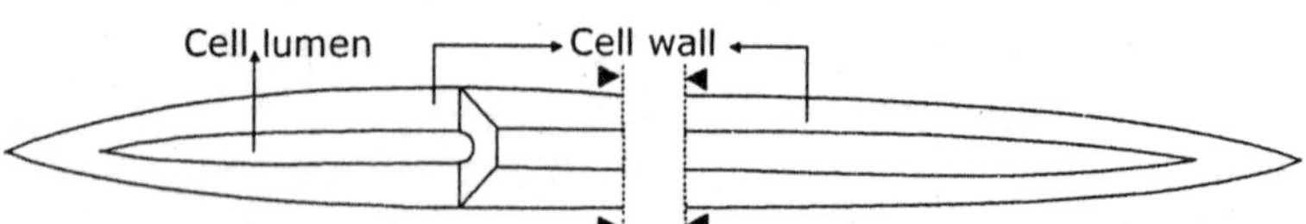


Figure 8.4

Line drawing of a fibre in longitudinal section. The arrow heads indicate the distance that may be of several millimeters. Adapted after Brett and Waldron, 1996, p.69.

## Fibre

Fibre is narrow, elongated cell with thick lignified wall and narrow lumen tapering to a wedge-shape at both ends.

• *Distribution* and *arrangement*: Fibre occurs in the ground and vascular tissues of all plant organs like leaves, stems, roots, fruits etc. in association with other tissues. It occurs either as bands or an uninterrupted cylinder (e.g. *Linum usitatissimum*) or single as idioblasts (ex. *Cycas* leaflet). They are most commonly associated with xylem and phloem. They are found in groups above the vascular bundle as bundle cap (ex. sunflower stem) and around the vascular bundle as bundle sheath (ex. maize stem). They are also located encircling the vascular cylinder—termed pericyclic fibre. They also occur between the vascular bundles (ex. leaf of *Agave*) extending between upper and lower epidermis. They are also associated with the parenchyma cells of pith and cortex. The fibres usually lie parallel to the long axis of organ in which they occur. They may be interlocked at the ends (Fig. 8.6).

• *Shape, structure* and *content*: They are much-elongated cells with many times longer than the breadth, tapering to a wedge-shape at both ends. These cells have very thick secondary walls made up of lignin or cellulose (ex. *Linum usitatissimum*). Sometimes the deposition of lignin is so thick as to obliterate the central lumen. The fibres appear more or less angular in cross sectional view with little intercellular spaces (Fig. 8.6). The cell wall is pitted and may show slight indentation (ex. *Zingiber*). The pits may be cross-pit, i.e. pits of a pit pair have crossed apertures (Cutter, 1978). Usually fibres are considered as dead cells and contain no living protoplast at maturity. But the works of Fahn and Leshem (1963) reveal that the wood fibre can retain living protoplast for several years. The fibres of *Vitis*, *Zingiber* etc. are septate and the transverse wall is very thin. Septate fibres contain starch, oils, calcium oxalate crystals etc.

• *Types of fibre*: Fibres are classified into two groups according to the tissues in which they are associated: (i) xylem or xylary fibre and (ii) extraxylary fibre.

**(i) Xylary fibre:** This fibre is associated with the xylem as xylem fibre.

• *Origin of xylary fibre*: These fibres are present in the primary and secondary xylem and accordingly their origin differs. They are developed from the procambium when associated with primary xylem and originate from cambium when associated with secondary xylem. The fusiform initials of cambium give rise to fibres.

The xylary fibres are classified into (1) libriform fibre and (2) fibre-tracheid on the basis of pit and thickness of the wall.

1. **Libriforn fibre** (Fig. 9.1F): These fibres are longer in length, have thicker cell wall with simple pit in contrast to tracheids of the plant in which they occur. The pit canal is longer than that of tracheids. The inner aperture of a pit-pair appears slit-like due to wall thickening and is usually formed at right angles to outer aperture.

2. **Fibre-tracheid** (Fig. 9.1E): This fibre is longer in length, has thicker cell wall with bordered pit than tracheids but shorter in length, has thinner cell wall than libriform fibre of the plant in which they occur. The border of the bordered pit is much reduced than those of tracheids. As a result the pit chamber becomes smaller than tracheids. The inner aperture appears as slit due to thickening and is usually formed at right angles to outer aperture. Thus the fibre-tracheids are intermediate in all respects between tracheids and libriform fibres.

**Gelatinous or mucilaginous fibre:** These fibres are characterized by either absence or presence of lignin in little amount on the cell wall. Thus gelatinous or mucilaginous fibres differ from other xylary fibres in having less lignin and much α-cellulose on their cell wall. α-cellulose molecules are deposited on inner secondary wall of these fibres instead of lignin. Gelatinous fibres are found in tension wood of dicotyledon. (Tension wood is one type of reaction wood [vide chapter 20] that is formed in leaning branches or trunk of dicots. The tension wood, i.e. secondary xylem is formed on the upper side of the branch or trunk when they lean due to gravity. These woods usually show eccentric growth rings).

**(ii) Extraxylary fibre:** These fibres occur in the cortex, pith and in association with phloem as phloem or bast fibre. They never associate with xylem elements. In monocotyledonous stems the extraxylary fibre may surround the individual bundles–termed bundle-sheath; it may form an uninterrupted cylinder on the ground tissue. These fibres originate partly from ground meristem and the rest from the procambium. In some dicotyledons (e.g. *Aristolochia*, *Cucurbita* etc.) fibres occur as bands or uninterrupted cylinder on the peripheral side of vascular cylinder and at the innermost layers of cortex. These fibres originate from the ground meristem and termed as pericyclic fibres. The structure, content and shape of these fibres are more or less like xylary fibres.

• *Function*: (i) fibres are mechanical cells that support the organs in which they occur; (ii) they protect the inner tissues due to the presence of thick wall; (iii) septate fibres are the storage tissue where starch and oils are present; (iv) fibres exhibit elastic properties and so can withstand tension.

• *Source and use of commercial fibres*: Fibres are used in textile industry, as filling fibres and in the preparation of ropes, brushes, brooms etc. The textile fibres are soft fibres that may or may not contain lignin on the cell wall. They are obtained from the dicotyledonous plants like *Gossypium* sp. (cotton), *Corchorus* sp. (jute), *Boehmeria nivea* (ramie), *Linum usitatissimum* (flax), *Cannabis sativa* (hemp) etc. Cordage fibres are hard fibres due to the presence of lignin on the cell wall. They are used in the preparation of ropes and binder twines and are obtained from the monocotyledonous plants like *Musa textiles* (abaca), *Agave sisalana* (sisal) etc. *Agave* fibres are used in the preparation of brushes and brooms. Cotton, jute etc. are used as filling fibres.

## Factors influencing differentiation of fibres

• *Stress*: Haberlandt (1914) pointed out that physical stress could stimulate the differentiation of fibres in *Cyclanthera* tendrils. An estimate regarding the amount of fibres present was carried out between the free tendrils and the tendrils of same age attached to a support. It was observed that the latter had tendrils with more fibres than the former. The fibres were thick-walled.

• *Water*: It was observed in *Cannabis sativa* that the plants, which grew in well-watered soil, had fibres that were much stronger than others (Cutter, 1978).

• *Gibberellic acid*: *Corchorus capsularis* treated with gibberellic acid (GA) produced more amounts of fibres than the untreated one (Sircar and Chakraverty, 1960). The fibres of *Cannabis* and *Corchorus* treated with GA have thick-walled, wider and larger fibres. GA alone has little effect on the differentiation of fibres. In an experiment with *Phaseolus vulgaris*, the leaves were excised and those were replaced by auxin. It was observed that xylem differentiated and in it the proportion of vessels were high in comparison to fibres. But when auxin and GA replaced the excised leaves, xylem differentiated. In it the proportion of fibres were high in comparison to vessels.

**Sclereids:** Sclereids are either oval, rounded or elongated cells with thick lignified walls that occur singly or in groups in plants.

• *Distribution*: Sclereids occur singly or in groups and in association with xylem and phloem in all plant organs. They occur in the aerial roots of *Monstera* and in the roots of *Nymphaea*. The stems and petioles of *Hoya*, *Nymphaea* etc. have sclereids in their parenchymatous pith and cortex. They are also present in the leaves of *Camellia*, *Trochodendron*, *Nymphaea*, and *Monstera* etc. The flesh of fruits of *Pyrus*, the seed coats of *Phaseolus*, *Pisum* etc. contain sclereids. In the seed coats they may be arranged in a complete layer or occur as idioblasts, i.e. as cells that are readily distinguishable from the neighbours by their shape, cell contents and wall thickness.

The distribution of sclereids in the mesophyll tissues of leaves varies. They are apparently random or irregular in distribution (termed diffuse sclereid) in the leaves of *Trochodendron*, *Nymphaea*, and *Olea* etc. They may also occur at the veinlet endings (termed terminal sclereids) as observed in the leaves of *Capparis spinosa*, *Arthrocnemum* etc. Sclereids also occur near the margin of leaves, e.g. *Camellia*.

• *Origin*: The origin of sclereids varies and it depends on the position in which they occur. The diffuse sclereids, which are irregularly distributed, originate from parenchyma cells, whose walls are secondarily thickened. The cortical parenchyma gives rise to sclereids that occur in the cortex (e.g. *Nymphaea*, *Monstera* etc.). The terminal sclereids, which occur at the veinlet endings, have common origin with the veinlet. The same meristem gives rise to terminal sclereids and the procambium that gives rise to veinlet. Astrosclereids, present in the leaves (e.g. *Nuphar*) develop from mesophyll cells.

• *Structure and content*: Sclereids are thick-walled lignified cells and thickening may not be uniform. The sclereids in the bark of *Cinnamomum* are strongly thickened on their tangential walls. The cell wall of sclereids may have numerous simple pits with round apertures and narrow lumen. Ramiform pits are also present.

Usually mature sclereids are described as dead cells. But it has been observed that sclereids can retain living protoplast. The stone cells of pear also retain living protoplast for several years.

• *Shape*: Sclereids are of various shapes and accordingly Tschirch in 1889 distinguished four types (brachysclereid, macrosclereid, osteosclereid and astrosclereid) and Bloch (1946) added one more type—the trichosclereid.

• *Types of sclereids* (Fig. 8.5): (a) Brachysclereid: These sclereids are more or less isodiametric and resemble parenchyma cells. These are also commonly known as stone cell due to hard walls. They are also called grit cells and the gritty texture of *Pyrus* fruit is attributable to these cells, e.g. flesh of fruits of *Pyrus*, phloem of *Hoya*, *Cinnamomum* stem, barks of *Alstonia* etc. (b) Macrosclereid: These are elongated, rod-like or columnar in shape. The testa of many leguminous seeds is built entirely of macrosclereids, e.g. seed coat of *Pisum* (pea), *Phaseolus* (bean) etc. (c) Osteosclereid: The shapes of osteosclereids are columnar—the end of which may be lobed or branched or simply enlarged like a narrow bone, e.g. the seed coats of *Pisum*, the leaves of *Hakea* etc. (d) Astrosclereid: It is stellate cell, i.e. the cell is deeply lobed or branched in such a manner that it resembles stars, e.g. leaves of *Thea* (tea), *Olea*, and petiole of *Nymphaea* etc. (e) Trichosclereid: It is hair like, very much elongated cell with branches, which extends into the intercellular spaces, e.g. leaves of *Olea*, *Nymphaea*, aerial root of *Monstera* etc.

The sclereids present in the leaf of *Olea europaea* are very much elongated, fibre-like and about one millimetre in length. Arzee (1953) termed these as filiform sclereids.

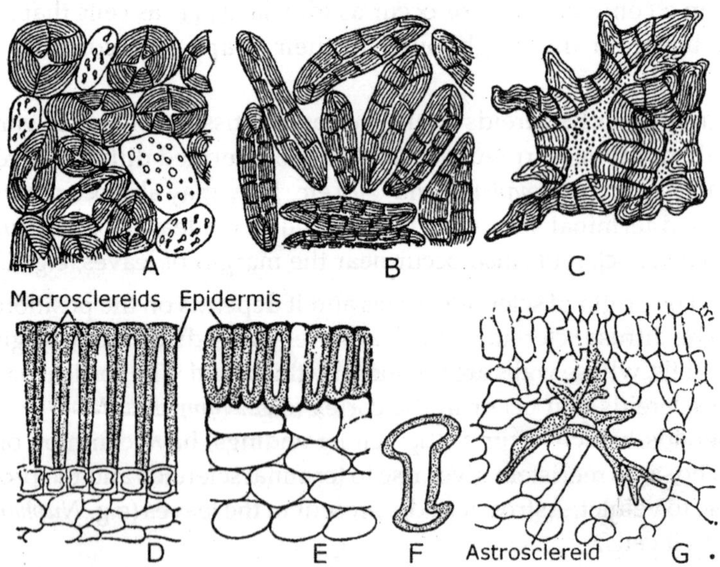


Figure 8.5

Sclereids. A. Brachysclereids from flesh of *Pyrus*. B. Same from *Cocos*. C. Irregular sclereids from *Tsuga*. D. Macrosclereids from epidermis of *Phaseolus* and E. from epidermis of *Allium sativum*. F. Osteosclereids from seed coat of *Pisum*. G. Astrosclereid from a leaf.

• *Function*: (i) Sclereids are mechanical cells and support the tissues in which they occur; (ii) sclereids, which form a continuous layer at the periphery, protect the inner tissues.

## Factors influencing differentiation of sclereids

• *Ageing*: In the barks many parenchyma cells change to brachysclereids. It is regarded that the sclereid differentiation is due to ageing of the tissues.

• *Wounding*: In *Monstera deliciosa* leaves and aerial roots stone cells appear close to wound tissues. Bloch (1944) opines that the appearance of stone cells is in response to physiological disturbances. Later observations reveal that wounds have inhibitory influence on the leaves of *Fagraea fragrans* (Rao, 1969) in the differentiation of brachysclereids. In this species brachysclereids did not appear in the tissues present around the injury. These tissues normally produced sclereids.

• *Position*: Sclereids differentiate along the margins of the developing leaves of *Trochodendron aralioides* (Cutter, 1978). In an experiment the buds of the above species were excised and grown in sterile culture media. It was observed that astrosclereids appeared along the margin of leaf. Brachysclereids in the air roots of *Monstera deliciosa* and astrosclereids in the leaf of *Camellia japonica* differentiate in close proximity to the surface. In the leaf of *Camellia* (Foard, 1959) additional sclereids could be induced to differentiate by experimental incisions on the leaf surface. Sclereids appeared along new surfaces in incised leaf.

• *Osmotic pressure*: Osmotic pressure influences sclereid development. In an experiment the leaves of *Camellia japonica* (Foard, 1958) were detached and cultured in a medium where sucrose concentration was high. It was noticed that fewer sclereids were developed in the leaf. A similar result was observed when the detached leaves were cultured in a medium with control level of sucrose and mannitol was added to it.

• *Enzyme*: The initials that are destined to differentiate into sclereids show more activity of enzyme than their neighbours. So it is regarded that enzyme influences sclerenchyma differentiation. In *Rauwolfia* the initials showed intensified activity of the enzyme cytochrome oxidase (Mia *et al.*, 1965). The initials have higher degree of metabolic activity than the other cells and as a result more enzyme activity is observed in them.

• *Hormone*: The following experiments indicate that hormone influences sclereid differentiation. In *Fagraea fragrans* sclereids developed in the leaves when they were treated with low concentration of auxin (Rao *et al.*, 1968). But the cell walls of sclereids were non-lignified and so thin-walled. In another experiment the buds of *Pseudotsuga taxifolia* were grown in culture medium. Few sclereids were differentiated in the leaves. In the culture medium various auxins with higher concentration were added. It was noticed that the number of sclereids was further reduced than the previous experiment (Al-Talib *et al.*, 1959). It indicates that auxin at certain concentration inhibits sclereid formation. In the other experiment with *Pseudotsuga* (Al-Talib *et al.*, 1961) the branches were defoliated and treated with indoleacetic acid in lanolin paste. It was observed that in the developing leaves sclereid differentiation was inhibited.

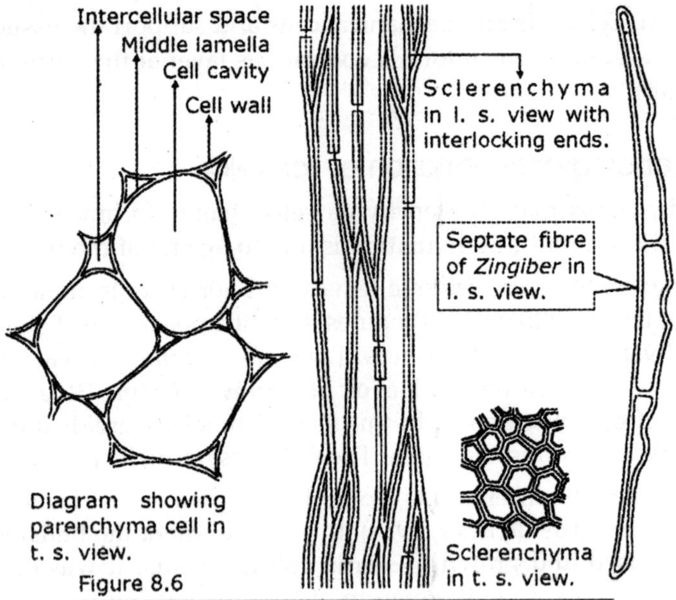


Figure 8.6
Diagram illustrating parenchyma and sclerenchyma.

Difference between sclereid and fibre.

| Sclereid | Fibre |
|---|---|
| 1. Parenchyma cells give rise to sclereid by secondary thickening and lignin deposition on their walls. | 1. Originates from meristematic cells; the cell walls are composed of lignin. |
| 2. They are usually isodiametric and may be of various shapes. | 2. They are elongated with pointed and usually interlocking ends. |
| 3. According to shapes, they may be brachy-, macro-, astro-, trichosclereids etc. | 3. They may be septate termed septate fibre. |
| 4. It has long numerous pits with rounded pit aperture. | 4. They have pits with slit like pit aperture. |
| 5. Ramiform pits are present on brachysclereids. | 5. Ramiform pits are absent from cell wall. |
| 6. They are present in fruits, in the bark, in pith and cortex, mesophyll cells, in seed coat etc. and rarely in association with xylem and phloem. | 6. They occur in leaves, stems, roots and fruits. They are present in cortex, pith, in association with xylem and pholem, bundle caps, pericycle etc. |
| 7. They add mechanical strength. | 7. They give mechanical rigidity and elasticity. |

Difference between parenchyma, collenchyma and sclerenchyma.

| Parenchyma | Collenchyma | Sclerenchyma |
|---|---|---|
| 1. Protoderm and ground meristem give rise to parenchyma. | In the ground meristem, procambium like cells give rise to collenchyma. | Protoderm, procambium & ground meristem give rise to sclerenchyma. |
| 2. Cells contain living protoplast. | Cells contain living protoplast. | Cells are devoid of living protoplast. |
| 3. Nucleus, dense cytoplasm and small vacuoles are present. | Nucleus, cytoplasm and larger vacuoles are present. | Nucleus, cytoplasm and vacuoles are absent. |
| 4. Cells are of various shapes and usually isodiametric. | Cells are generally elongated. | Cells are isodiametric (sclereids) and elongated (fibres). |
| 5. The cell wall is thin and uniformly thickened. | The cell wall is thick and unevenly thickened; usually more thickened at the corners. | The cell walls are thick and more or less uniformly thickened. |
| 6. The cells have primary cell wall. | The cells have primary cell wall. | The cells have secondary cell wall |
| 7. Sculpturing is absent from cell wall. | Sculpturing is absent from cell wall. | Various sculpturing like annular, spiral, reticulate etc. are present on cell wall. |
| 8. Cell walls are usually plastic. | Cell walls are usually plastic. | Cell walls are elastic. |
| 9. Primary pit fields are present on cell wall. | Pits are rare on cell wall. | Simple and bordered pits are present. |
| 10. Profuse intercellular spaces are present in this tissue and the ends of cell walls are not interlocked. | Little intercellular spaces are present in this tissue and the ends of cell walls may be interlocked. | Intercellular spaces are absent and the ends of fibers may be interlocked. |
| 11. Collenchyma and sclerenchyma are embedded upon this ground tissue. | It remains embedded on parenchyma. | It remains embedded on parenchyma. |
| 12. The main function is storage. Turgid parenchyma gives mechanical strength. | In addition to storage, it gives mechanical rigidity to the region where present. | It is mechanical tissue. |

CHAPTER 9

# Complex Tissue

Complex tissue can be defined as a special type of heterogeneous tissue that is composed of different types of cell having specialized functions like conduction of water and mineral salts, translocation of elaborate food matters and mechanical strength.

Xylem and phloem are the examples of complex tissue and together they are referred to as vascular tissue/conducting tissue, and accordingly labeled as composing vascular tissue system/conducting tissue system in plants. They always remain in close association with each other and compose the different types of vascular bundles that are observed in the roots and stems.

## Xylem

Xylem can be defined as a complex tissue that is composed of four basic types of cell (tracheids, trachea, xylem fibre and xylem parenchyma), remains in close association with phloem and has specialized functions like conduction of water and solutes, and mechanical strength.

Xylem and phloem together constitute the conducting tissues in plants. They occur both in primary—and secondary vascular tissues. Accordingly their origins differ. The procambial strands of apical meristem develop into primary xylem and primary phloem. Secondary vascular tissues originate from cambium—the fascicular part of which originates from procambium strands. Xylem is heterogeneous tissue and the four basic cell types (Fig. 9.1 & 9.2) composing it are xylem parenchyma, xylem fibre, tracheids and trachea.

- Xylem parenchyma: Parenchyma that forms one of the constituent of xylem is referred to as xylem parenchyma.

Xylem parenchyma is present in primary—and secondary xylem and respectively originates from procambium and cambium. In secondary xylem xylem-parenchyma is also termed as wood parenchyma and is present as axial—and radial parenchyma that accordingly occur as parallel and perpendicular to the long axis of organ where they lie.

Xylem parenchyma has no definite shape. It may be irregular, rectangular, round, oval and elongated etc. (9.1A, B & C). The cell wall is usually thin when it is composed of cellulose only. At later stage lignin may deposit and as a result cell wall becomes thick. Pits frequently occur on cell walls. It is simple when occurs between parenchyma cells. It may be simple, bordered or half bordered when occurs between parenchyma and tracheary elements (tracheids and tracheae).

Xylem parenchyma possesses living protoplast and contains starch, fat, crystals and tannins, and sometimes chlorophyll also.

Xylem parenchyma is mainly concerned with storage of starch, fat and ergastic substances etc. Transportation of minerals, solutes and water etc. occur through it. It gives mechanical support when turgid. The thick-walled parenchyma gives mechanical strength to the plant.

• Xylem fibre: Fibre that occurs as element of xylem is called xylem fibre. It is also referred to as xylary fibre and wood fibre.

Xylary fibres are elongated cells the length of which is many times longer than breadth. The two ends of a fibre are tapering to a wedge shape. The cell wall is usually thick. Due to lignin deposition the lumen of fibre becomes narrow. Pits are present on walls and they may be simple and bordered. Fibres may be septate (Fig. 9.1G). Usually it is regarded that fibres lack living contents at maturity, but there are reports that wood fibres may retain living protoplasts as long as twenty years, e.g. *Tamarix aphylla*.

Fibres may occur in primary—and secondary xylem, but proportionately their occurrences are much less in primary xylem. In primary xylem fibre originates from procambium. In secondary xylem fibres originate from fusiform initials of cambium.

The following three types of xylary fibres are noted:

- Libriform fibre (Fig. 9.1F): These fibres are characterized in having elongated cell with thick cell wall, small number of simple pits on the wall and very narrow cell lumen in comparison to tracheids in which they are associated in the same plant. Bordered pit rarely occurs. The inner pit aperture usually is slit like and pit canal is more elongated than that of tracheids. The inner apertures of the pit-pairs in libriform fibre are crossed with each other. Libriform fibre (liber = phloem) simulates phloem fibre, hence, the name libriform.
- Fibre-tracheid (Fig. 9.1E): This fibre has the characteristics that are intermediate between tracheids and libriform fibres. In contrast to tracheids fibre-tracheid is much elongated; the cell wall is thicker with reduced size of bordered pits. In comparison to libriform fibre fibre-tracheid is shorter in length and the cell wall is thinner with bordered pits. The inner apertures of a pit pair in fibre-tracheids are crossed with each other. The inner apertures are usually slit like.
- Gelatinous or mucilaginous fibre: This fibre is either libriform fibre or fibre-tracheid where the secondary wall lacks lignin or contains it in very small amount. The secondary wall has α-cellulose in its innermost layer termed G-layer. As a result the wall becomes highly hygroscopic. Gelatinous fibres are characteristic of the tension wood of angiosperm.

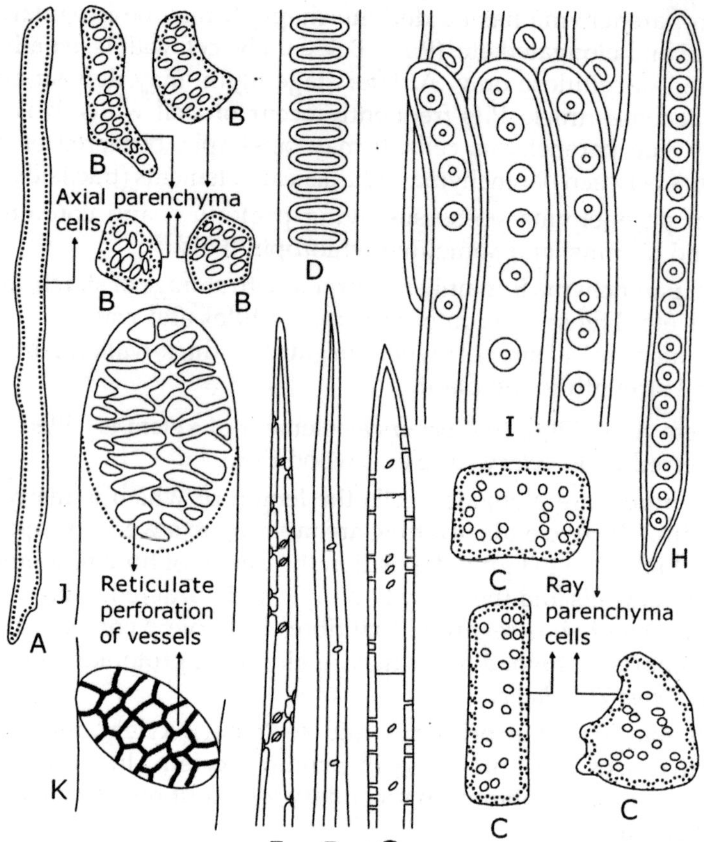


Figure 9.1

Diagram illustrating isolated elements of xylem. A. & C. *Ephedra californica*. B. *Aristolochia brasiliensis* (adapted after Esau, 1965, p. 240 & 241). D. Scalariform pitted tracheid of fern (adapted after Chandurkar, 1983 p. 72). E. Fibre-tracheid of *Quercus ithaburensis*. F. Libriform fibre of *Quercus ithaburensis*. G. Septate fibre of *Vitis*. H. Tracheid of *Pinus*. (Adapted after Fahn, 1987, p. 89 & 104). I. Conifer tracheids in longitudinal section showing overlapping ends and bordered pits on radial walls (adapted after Beck, 2005, p. 29). J. *Hymenocallis caribaea* (Adapted after Fahn, 1987, p. 109). K. Adapted after Cutler *et al.*, 2008, p.264).

Tension wood is formed in response to gravity that causes lateral stress. Branches of angiosperm dicotyledonous plants exhibit tension wood that is formed to counteract the lateral stress caused due to gravity.

• Tracheids (Fig. 9.1H): Tracheid is long, single-celled, non-living water conducting element of xylem with hard, thick (due to lignin deposition) and pitted wall, and the two ends of cell are chisel-like, oblique, imperforate and obtuse or tapering.

Tracheids occur in all groups of vascular plants. It is present exclusively in the xylem of pteridophytes and gymnosperms. Primitive angiosperms like *Trochodendron, Tetracentron* and *Drimys* etc. also have it. It occurs in primary—and secondary xylem. In primary xylem it originates from procambium whereas in secondary xylem it develops from cambium. Tracheids originate from a single cell.

Tracheid occurs parallel to long axis of an organ where it lies. It overlaps each other and communicates with neighbours by means of pits (Fig. 9.1I). Pits are present on lateral and oblique end walls. In xylem the tracheids are situated one above the other and the end walls are in contact with that of others. Water diffuses through the pits (Fig. 9.3A). Pits may be simple, bordered or half bordered. Bordered pit is observed in the pit pairs present between tracheids. The pit pair that occurs between tracheids and parenchyma exhibits simple pit on the parenchyma-cell wall and bordered pit on tracheid-cell wall. In ferns the bordered pit is transversely elongated termed scalariform pitting (Fig. 9.1D). Tracheids are angular in cross-sectional view.

The cell wall of tracheids is thick due to the deposition of lignin. Lignin does not deposit uniformly and as a result different types of pit, annular and helical sculpturing are formed. Annular and spiral thickening are observed in protoxylem. Metaxylem exhibits bordered pit and scalariform pitting.

The function of tracheids is conduction of water and minerals in solution (Fig. 9.3A). It gives mechanical support to the organ where it lies. There is report that tracheids store water.

• Vessel: Vessels (also termed tracheae; sing. trachea) are long, non-living thick walled (due to lignin deposition) elements of xylem and composed of vertical rows of single-cell units the end walls of which are perforated thus forming a continuous water-conducting tube.

Vessels are exclusively present in angiosperm. They also occur in gymnosperm, e.g. *Ephedra, Welwitschia* and *Gnetum*. In pteridophyta vessels are present in *Selaginella, Equisetum* and in the following four ferns: *Actiniopteris, Pteridium, Regnellidium,* and *Marsilea*. The occurrence of vessel in pteridophyte and gymnosperm is regarded as anomalous. In angiosperm, vessels occur both in primary—and secondary xylem and respectfully originate from procambium and cambium.

Each vessel unit is long and cylindrical in shape and lack living protoplast. Sometimes the length of a vessel unit may be shorter than breadth when it appears as drum shaped. The cell wall is thick due to deposition of lignin. Lignin does not deposit uniformly and as a result the following types of pitting (Fig. 10.6) are formed on the lateral walls: scalariform pitting, opposite pitting and alternate pitting.

The two ends of a vessel unit are perforated. Several perforate-units are aggregated longitudinally to form a chain of cells. The chain of cells forms a tube-like system and usually referred to as trachea or vessel. The unit cells that make up the tube-like system are referred to as vessel members, vessel elements or vessel units. Vessel member that terminates in the tube-like system has tracheid-like terminating end, i.e. the terminating end is not perforated. Apart from terminal position perforations may be sub-terminal or lateral. Part of a vessel member bearing perforation is known as perforation plate. A perforation plate may consist of a single or more than one perforation and accordingly the following perforation plates are noted (Fig. 9.2).

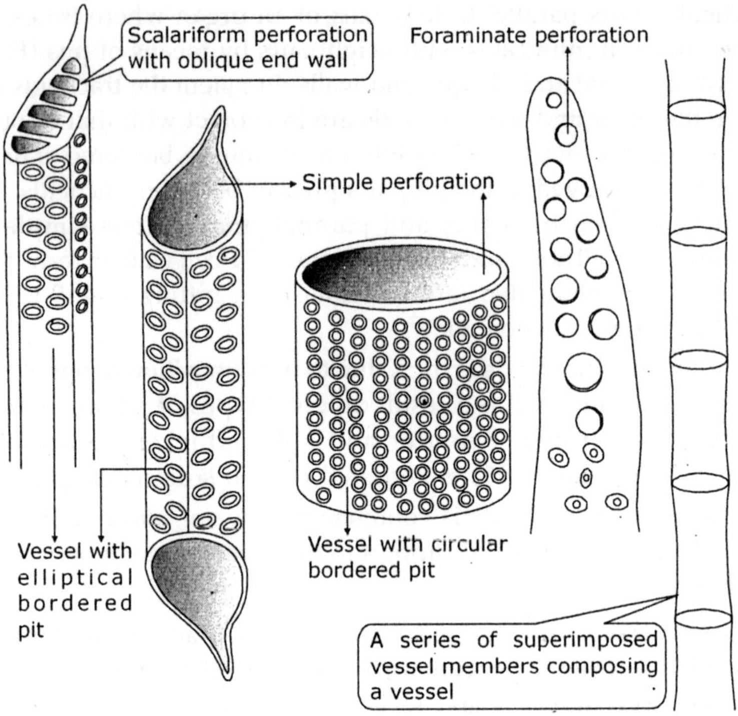


Figure 9.2

Diagram of vessel members showing perforations. Adapted after Beck, 2005, p. 32. & Fahn (1997).

❒ Simple: Ex. *Quercus*. The perforation plate is composed of a single perforation that is present at each end of a vessel member. Usually vessel with transverse end wall has simple perforation plate.

Multiple: More than one pore compose multiple perforation plate. The pores are arranged in various ways and the following types are observed:

❒ Scalariform: Ex. *Liriodendron* and *Phoenix dactylifera* etc. The pores are laterally elongated and occur parallel to each other in the perforation plate. The intervening bars of thickenings appear ladder-like. Commonly vessel with oblique end walls has scalariform perforation plate.

❒ Foraminate: Ex. *Ephedra*. Several and almost circular pores are grouped together to compose the perforation plate. It is also referred to as ephedroid perforation plate.

❒ Reticulate (Fig. 9.1J & K): Ex. *Rhoeo discolor, Hymenocallis*. Many small pores compose the perforation plate. The arrangement of pores makes a mesh-like structure. The intervening regions between pores are in the form of network formed by secondary wall thickening materials. Reticulate perforation occurs less frequently.

Usually a particular type of perforation plate occurs in a plant. But sometimes the two ends of a vessel member may have different types of perforation, e.g.

simple and scalariform. The neighbouring vessel elements may also possess different types of perforation plates.

• Function: (a) conduction of water and dissolved minerals occur from root to stem and leaves through vessels (Fig. 9.3B) and (b) vessels give mechanical rigidity to the organs where they lie due to the presence of hard lignified cell wall.

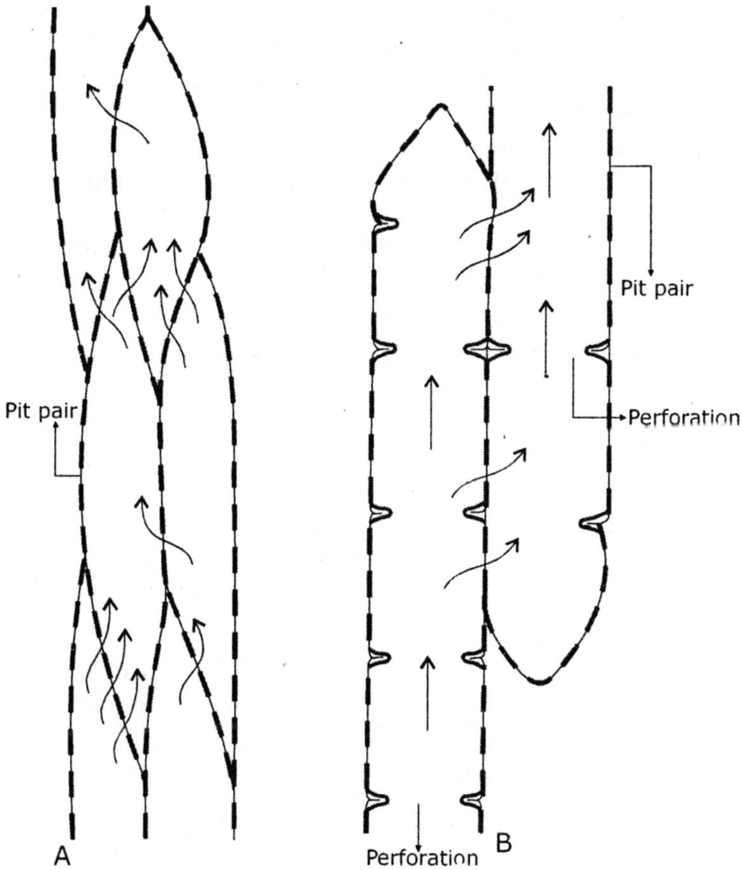


Figure 9.3

Diagram illustrating the path of movement of water and minerals: (A) in tracheid water passes between them only through pit pairs; (B) in vessel water passes between vessel members through perforation and pit pairs. Adapted after Mauseth, 2009, p. 107.

The tracheary elements, i.e. the tracheids and vessels are like capillary tubes. The ability of these tubes to allow water transport through them is known as conductance. Conductance is related to vessel diametre. The conductance is greatly increased in the vessels with large diametre in comparison to those with small diametre. One of the properties of water is that it tends to stick to many substances by adhesive force. Water sticks to inner sides of tracheary elements thus decreasing the diametre of them. So the movement of water through tracheary element is not uniform. Water present near the wall is stuck to the wall and so its movement is very slow. The water in centre is free form adhesion and so its movement is faster. The vessels with wide diametre, are advantageous because the friction caused by adhesion is minimized here. Though adhesion hinders water movement it is advantageous in the fact that it strengthens the water column present within tracheary elements. It is important for the upward movement of water. The water column in tracheary elements is continuous from root to leaf via stems. All the water molecules are held together by cohesive force. As a result the water column tends to act as a unit. The continuity and erectness of the column are maintained by cohesion and adhesion. There is also continuity of water among the mesophyll cells, guard cells and tracheary elements of a leaf. The leaves lose water by cuticular and stomatal transpiration. As a result a suction force is developed among the mesophyll cells, which are in direct contact with tracheary

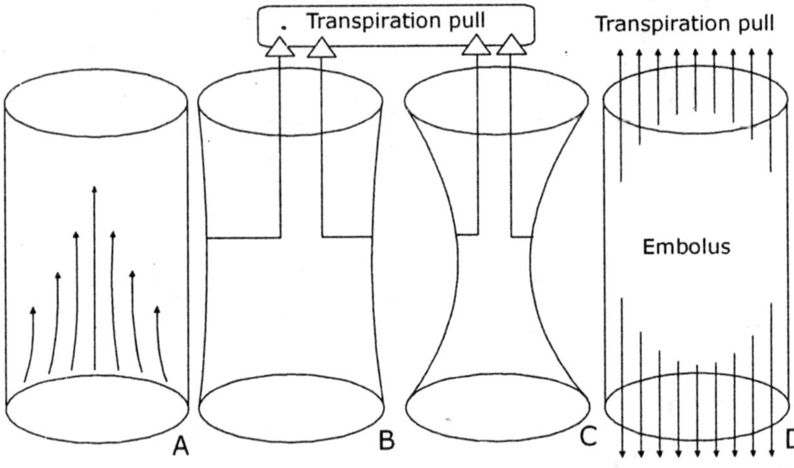


Figure 9.4

A. Diagram representing the flow of water in a vessel. The water in centre moves faster than those present near the wall. B. The vessel wall is pulled inward by the transpiration pull. C. The vessel wall buckles when the wall is not rigid. D. The embolus or air bubble expands due to transpiration pull and gravitational pull.

elements. This causes a tension in the water and a pull is established, and it is transmitted to the water present in root. This happens, as the column is continuous. This pull, also known as transpiration pull, drags water column and the water moves upward. During movement water column is under tension, which is also transmitted to cell wall of the tracheary elements. This happens because water molecules and cell wall act as a unit due to their bonding by cohesive and adhesive forces. During transpiration pull (Fig. 9.4B) the cell wall of tracheary element is also pulled inwardly. If the cell wall has not enough rigidity to resist it may buckle and in extreme case it may collapse. So for effective conductance the cell walls of tracheary elements must have enough strength to avoid collapse.

The other prerequisite of conductance is adhesion and cohesion. If somehow the adhesive and cohesive forces are overcome embolus (Fig. 9.4D) may appear in the water column—a phenomenon called embolism thus stopping water transport. Embolus expands and stops when it comes in contact to solid surface only. Embolus cannot expand in tracheids, as they have no perforation plate. In vessels embolus spreads from one to other through perforations thus causing vessels functionless.

The movement of water is dependent upon its chemical potential. 'The chemical potential of a substance in a system is a measure of the capacity of that substance to do work' – Kramer, 1969, p. 16. The chemical potential of pure water at one atmosphere of pressure or one bar of pressure is defined as water potential and its value is zero. Addition of solutes dilutes water and reduces its activity. The potential of this system is less than that of pure water. Hence the water potential has negative value and is expressed as a negative number. Water moves from regions of higher to regions of lower potential. Transpiration, i.e. loss of water from leaves, lowers the water potential. This is advantageous for transpiration pull and upward water transport. Water in the tracheary element has high cohesive force. It remains confined within the wettable cell walls where adhesive forces operate. During upward movement tension develops, which may be as high as 300 or more atmospheres. This is also advantageous for upward movement of water and to keep water column erect.

## Phloem

Phloem is a complex tissue or heterogeneous vascular tissue that stores and conducts principally the products of photosynthesis in vascular plants and sometimes adds mechanical strength.

Phloem in association with xylem constitutes the vascular bundle and forms the conducting tissue system in plants. In roots phloem occurs as isolated patches alternating with xylem (radial vascular bundle). Phloem in the stems is usually external to xylem. A cylinder of phloem may surround a central core of xylem (e.g. haplostele) or discrete strands of xylem are surrounded by phloem (ex. mixed protostele). Two cylinders of phloem may occur on the external and inner side of xylem (ex. amphiphloic siphonostele). In amphivasal vascular bundle a central strand of phloem is encircled by xylem (ex. *Dracaena*). In most dicotyledonous stem phloem strand occurs external to xylem (e.g. collateral vascular bundle). In some species of the families Cucurbitaceae, Asclepiadaceae, Apocynaceae, Solanaceae etc. in addition to outer phloem, it occurs on the inner side of xylem. These two parts of phloem, according to position, are designated as outer or external phloem and inner or internal phloem (Fig. 31.2B). Esau (1965) suggested the terms abaxial and adaxial phloem to designate outer and inner phloem respectively. The internal phloem is also termed as intraxylary phloem. Sometimes phloem strands or layers are formed embedded in the secondary xylem. This phloem is termed as included or interxylary phloem. The interxylary phloem is called concentric when it arises as layers alternating with xylem layers, foraminate when it arises as strands encircled by xylem.

Phloem is composed of (A) parenchyma termed phloem parenchyma, (B) specialized parenchyma cells known as companion cell and albuminous cell, (C) phloem fibre and (D) sieve cell and sieve tube. In addition to the above elements, sclereids, laticifers and resin ducts are also present in phloem tissue of some species. In dicots phloem parenchyma, sieve tubes, companion cells and phloem fibres are present. Monocots and a few ranunculaceous genera do not possess phloem parenchyma in phloem elements. Sieve cells and albuminous cells are present in gymnosperm and vascular cryptogams.

**(A) Phloem parenchyma:** The parenchyma cells, other than albuminous and companion cells, which occur in association with phloem are referred to as phloem parenchyma.

Distribution: Phloem parenchyma occurs in both primary and secondary phloem. In the latter case they are present in two systems, the axial and the ray system, and accordingly they are termed as axial phloem parenchyma and phloem rays.

Shape, structure, contents and arrangement: Phloem parenchyma is more or less rectangular or rounded in cross section. In longitudinal section it appears as elongated cell with both ends rounded or pointed; it may also appear as rectangular or cylindrical.

Active parenchyma cells have thin walls that are primary and not lignified. The wall of inactive parenchyma, in some cases, becomes sclerified and thickened. Numerous pit fields occur on the cell wall. Through these pit fields protoplasmic connections are established between the axial and ray parenchyma, between the parenchyma and companion cells or sieve elements. In some plants crystal-containing parenchyma cells occur. The phloem parenchyma cells of *Sherardia arvensis* leaf may develop wall ingrowths and these cells are known as phloem transfer cell. Phloem parenchyma cells possess living contents. These cells have nucleus, cytoplasm and may store starch, fats, resins tannins etc.

The parenchyma cells of primary phloem are oriented parallel to the long axis of xylem. In the secondary phloem, the axial and ray parenchyma lie parallel and perpendicular respectively to the long axis of xylem in which they are associated. The ray parenchyma cells are radially elongated. The phloem parenchyma cells that commonly occur at ray margins may be vertically elongated.

• Function: (1) Active phloem parenchyma stores fat, starch etc.; (2) accumulates resins, tannins etc.; (3) phloem transfer cells help in short distance transport of solutes; (4) in many plants phloem parenchyma gives rise to phellogen that forms the protective tissue – periderm; (5) the cell walls of some inactive phloem become thickened by lignin deposition and add mechanical strength of the organ in which they occur.

**(B) Companion cell:** The densely cytoplasmic nucleated parenchyma associated with sieve tube with which it has common origin from the same mother cell and plays some role in the functioning of sieve tube in angiosperms, is referred to as companion cell.

Shape, structure, content and arrangement: The companion cells are vertically elongated and somewhat angular in cross section. In length they may be as long as the associated sieve tube or may be shorter.

The cell wall of companion cell is uniformly thick and possesses many depressed areas. These areas may be either sieve areas that occur on the side of sieve tube or primary pit field that occurs on the side of other companion cell or parenchyma. In the sieve areas and primary pit fields there exist the plasmodesmata, which are branched on the companion cell side. In some companion cells, wall materials deposit on the inner side of the primary wall to form transfer cell.

In contrast to sieve tubes, companion cells have prominent nuclei at maturity. The nucleus may be elongated or lobed and bounded by a normal double membrane. The cytoplasm is very dense due to the presence of abundant organelles that are dictyosomes, endoplasmic reticulum, mitochondria with well-marked cristae, ribosomes, plastids like leucoplasts or chloroplasts etc. In some companion cells P-proteins are found.

A single meristematic cell gives rise to companion cell and sieve tube. They remain strongly attached to each other. They are so tightly appressed that they cannot be separated by the usual maceration technique. Companion cells vary in number in relation to a single sieve tube. Usually the number is one or two and occasionally up to five (Calycanthaceae) or several. These cells may be developed on one side of sieve tube only or formed on all sides. Companion cells are present in most dicots and monocots. They are absent in some primitive woody dicotyledons and primary phloem (protophloem). They are also absent in gymnosperms and pteridophytes. In gymnosperm the associated parenchyma with sieve cells are termed as albuminous cell.

• Function: (1) Companion cell and sieve tube are ontogenetically related and so it is thought that there exists a physiological and functional relationship between them. It is observed that in nonfunctional sieve tube the associated companion cell dies; (2) they are the active site of protein synthesis; (3) the endoplasmic reticulum, plastids and plasmodesmata form a route through which sucrose is transferred to neighbouring cells; (4) Cutter (1978) is of opinion that there exists a complex functional relationship in between the sieve tube and companion cells or other nucleated cells for effective transport of solutes.

❐ Albuminous cell (Fig. 9.5): The densely cytoplasmic nucleated parenchyma, which is associated with the sieve cells of gymnosperm and plays some functioning role of it, is referred to as albuminous cell.

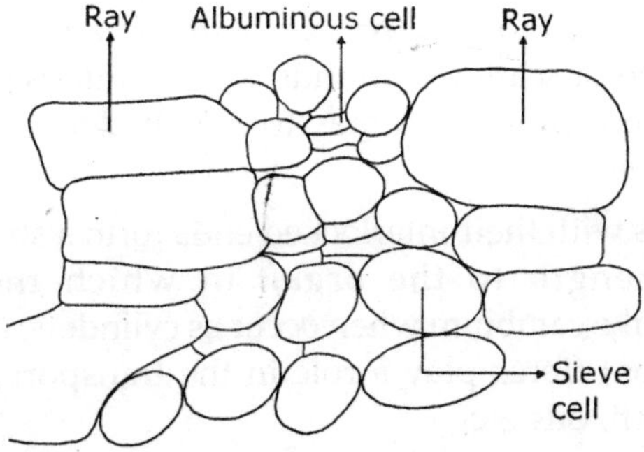


Figure 9.5

*Acmopyle pancheri*: Diagram illustrating albuminous cells of gymnosperm phloem in cross sectional view. Adapted after Cutler, 1978, p. 50.

*Shape, structure, content and arrangement:* Albuminous cells are vertically elongated and may be of same length of the sieve cells or shorter. The end walls may be oblique or tapered. The cell wall is thin and there is connection with the associated sieve cells. Sieve cells have sieve areas on the walls facing the albuminous cell. The cells contain protein rich cytoplasm and stain deeply with cytoplasmic stains, and so these are designated as albuminous cells. In these cells starch is usually absent. These cells contain nucleus and slime body. Albuminous cells occur in rays and among the axial parenchyma cells. They can be differentiated from neighbouring parenchyma cells by their usual connection with the sieve cells and absence of starch. Albuminous cells occur at the margin of rays and these cells form erect ray cells.

Origin: Albuminous cell is present in primary and secondary phloem; accordingly its origin differs. In contrast to companion cell it is not ontogenetically related to sieve cells, i.e. albuminous cell originates from individual mother cell. These cells develop either from phloem rays or from phloem parenchyma, which are the derivatives of procambium. In the secondary phloem of *Ephedra* albuminous cells originate from the fusiform initials of vascular cambium.

• Function: (1) Though there is no ontogenetic relation between albuminous cells and sieve cells, there exist morphological and functional relationship between them. It is evident from the fact that in the nonfunctional sieve cells the associated albuminous cells die. Moreover the attachment between the sieve cells and albuminous cells is such that they remain attached even after maceration; (2) albuminous cells possibly helps in the conduction of protein.

**(C) Phloem fibre:** The fibre, which occurs in association with phloem, is referred to as phloem fibre.

Sometimes the terms bast fibre or bass fibre or basswood or bast wood fibre are synonymously used to mean phloem fibre.

*Shape, structure, content and arrangement:* The fibres are elongated cells and may be very long. The two ends of a fibre are usually tapering to a wedge-shape and interlocked with other fibres. The cell wall is thick and lignified. But the wall of *Linum* phloem fibre is made up of cellulose. Gelatinous fibre also occurs in phloem. The cell wall contains simple pits with linear or round apertures. Sometimes slightly bordered pits occur. The phloem fibre of *Vitis* is septate. Phloem fibres are considered as dead cells and contain no living protoplast at maturity. The septate fibres contain starch, oils, resins, calcium oxalate crystals etc. The fibres are arranged in parallel to the long axis of the organ in which they occur. They may occur as isolated or scattered strands, as continuous or irregular bands, as clusters over the phloem strand and may form cylinders of tangential sheets encircling the inner tissues.

• Function: (1) Phloem fibres with their interlocked ends form a strong strand and provide mechanical strength to the organ in which they occur; (2) they protect the inner tissues like cambium when occur as cylinders; (3) Mitchell and Worley (1964) suggested that fibres play a role in the transport of solutes; (4) septate fibres may store starch, oils etc.

**(D) Sieve elements:** The conducting elements of phloem are referred to as sieve elements that are characterized by the presence of sieve areas and absence of nuclei from mature protoplasts.

Sieve element is the collective term of sieve cell and sieve tube (or sieve tube member or sieve tube element), which are distinguished on the basis of sieve areas and sieve plates. The sieve cells do not contain sieve plates. They characterize Pteridophyta and Gymnosperm. Exceptions are noted in the four species of *Equisetum* (*E. arvense*, *E. giganteum*, *E. hyemale* , *E. telmateia*) and the fern *Cyathea gigantea*. Long sieve element in secondary phloem is considered as primitive. It is to be noted that short sieve tube element does not always indicate an advanced condition because sieve tube element is frequently formed from cambium initial following transverse division.

❑ Sieve tube (Figs. 9.6 & 9.7): The thin walled, living, enucleate, longitudinally arranged conducting elements of angiospermic phloem with sieve plates and sieve areas on their transverse end walls are designated as sieve tube.

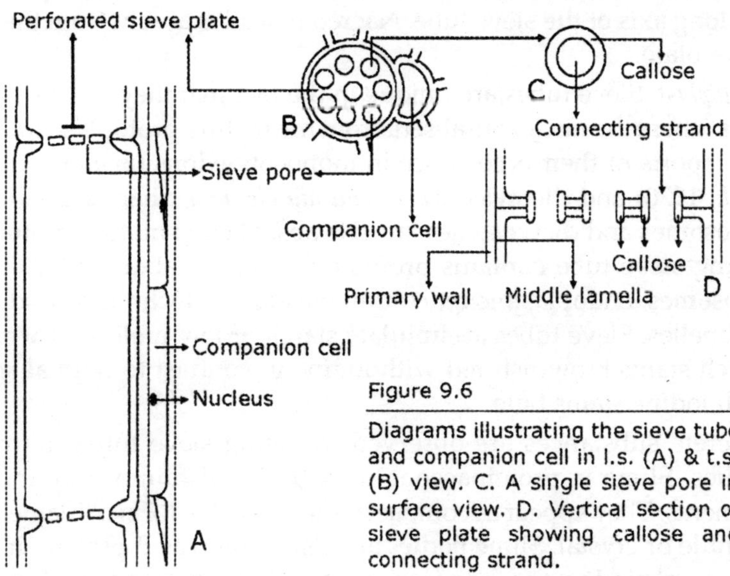


Figure 9.6

Diagrams illustrating the sieve tube and companion cell in l.s. (A) & t.s. (B) view. C. A single sieve pore in surface view. D. Vertical section of sieve plate showing callose and connecting strand.

The sieve tube consists of longitudinal files of cells that are connected with each other through sieve areas on their transverse end walls. Each cell is a sieve tube member and is composed of (i) cell wall, (ii) protoplast and (iii) sieve plate.

i. *Cell wall:* The cell wall of sieve tube may be thin or thick and is usually primary. The wall is composed of mainly cellulose and pectin. The thin walls are one micron thick. The thick wall may almost fill the cell lumen. The thick wall consists of two layers—the outer thin and inner thick layer. The outer thin layer lies towards the middle lamella and the inner layer is towards the cytoplasm. The inner layers may have glistening properties and therefore the thick wall is termed as nacreous wall (Ex. *Magnolia*,

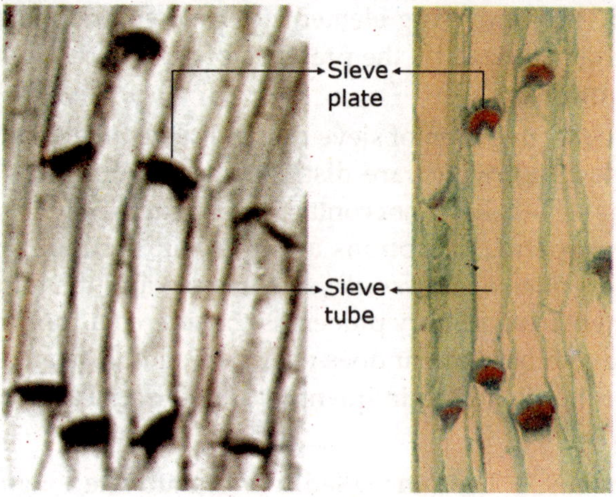


Figure 9.7

Microphotograph of L. S. of sieve tube of *Cucurbita* stem.

*Cucurbita* etc.). The nacreous wall is polylamellate in *Cucurbita* and is composed of microfibrils. The microfibrils are oriented perpendicular to the long axis of the sieve tube. Nacreous wall is absent from the region of sieve plate.

ii. *Protoplast:* Sieve tubes are unique in the fact that they are the only living cells where the nuclei are absent from the mature protoplast, though there are reports of their occurrence in monocotyledons (*Smilax hispida,* Ervin *et al.,* 1970) and dicotyledons (*Neptunia oleracea,* Shah and James, 1968). Ribosomes and dictyosomes are also absent from mature protoplast. The young sieve tube contains prominent nucleus, abundant dictyosomes, ribosomes, endoplasmic reticulum, plastids, mitochondria and other cell organelles. Sieve tubes accumulate starch of low molecular weight. This starch stains brownish red with iodine in contrast to normal starch that with iodine stains blue.

Discrete substances are observed in young sieve tubes, termed slime bodies. Slime is proteinaceous in nature and found only in the sieve elements. They appear as compact structure in the form of filament, tubule, granule or crystal. Slime bodies are also known as P-protein. It occurs in all dicotyledonous species so far investigated and is rare in monocotyledons. They are absent in gymnosperms (except *Ephedra*) and pteridophytes. P-proteins are synthesized in the cytoplasm and occupy the peripheral position. In the stained preparation of sieve tubes P-proteins accumulate at the transverse end walls of tubes and plug sieve plate pores. This plug is termed as slime plug.

Plastids occur in the protoplast of sieve tubes. Two major types of plastid are distinguished on the basis of accumulation of protein and/or starch. Sieve element plastids accumulating only starch are defined as S-type plastids while those containing protein accumulation are called P-type plastids. S-type plastids are found in Bataceae (*Batis maritima*), Polygonaceae (*Polygonum bistortum, Rumex patientia*), Plumbaginaceae

(*Plumbago europaea*), Gyrostemonaceae etc. P-type plastids can accumulate starch in addition to protein. In P-type plastids a central crystalloid remains surrounded by a ring shaped bundle of protein filament. This type is specific for Caryophyllales (Behnke and Turner, 1971). In some cases crystalloids may be absent, only the ring shaped bundle of filaments are present. The crystalloids may be globular or polygonal. Therefore, three types of P-type plastids are represented (Behnke, 1976), i.e. with globular crystalloid, with polygonal crystalloid and without crystalloid (Fig. 9.8). P-type plastids with globular crystalloid occur in Aizoaceae, Basellaceae, Cactaceae, Didiereaceae, Halophytaceae, Hectorellaceae, Molluginaceae, Nyctaginaceae, Tetragoniaceae and most of the genera of Phytolaccaceae. *Stegnosperma* (Stegnospermataceae), *Limeum* (Phytolaccaceae) and Caryophyllaceae have P-type plastid with polygonal crystalloid, whereas Chenopodiaceae and Amaranthaceae are characterized by not having any crystalloid. These ultrastructural details of sieve-element-plastids are, now a days, applied to characterize some higher taxa like Magnoliophyta, Caryophyllidae etc.

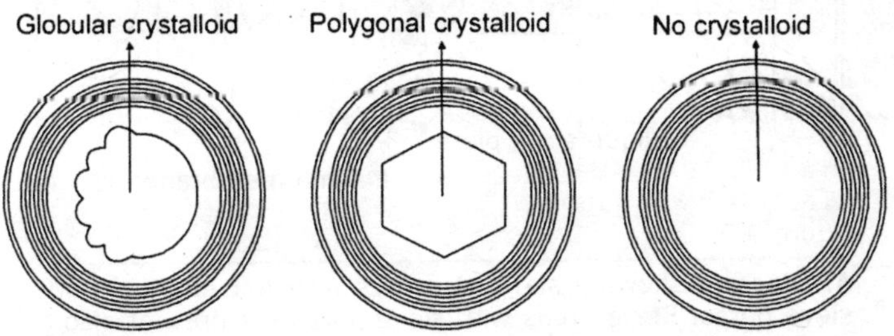


Figure 9.8

Diagram representing three types of P-type plastid with ring shaped bundle of filaments found in Caryophyllales. Adapted after Behnke, 1976, p. 48.

iii. *Sieve plate:* Sieve plate is the region where sieve areas occur. The plate lies at the end walls of sieve tube member and is usually horizontal or oblique to the longitudinal axis of them.

Sieve areas appear as depressed region in the wall where pores occur. In the pores there exist the connecting strands that connect the protoplast of one sieve tube member to the neighbouring member. Thus there is continuity between the sieve tube members, which form the sieve tube. There may be one or several sieve areas in each sieve plate (Fig. 9.9) and accordingly they are termed as: (i) simple sieve plate—where there is one sieve area only (Fig. 9.9A) on the plate (ex. *Cucurbita*) and (ii) compound sieve plate—where there are more than one sieve areas (Fig. 9.9D) in a plate (ex. *Vitis, Pyrus* etc.). The pores in sieve areas vary in size. They may be less than one micron (e.g. *Spiraea*) or may be more than ten microns (e.g. *Cucurbita, Ailanthus* etc.) in diameter.

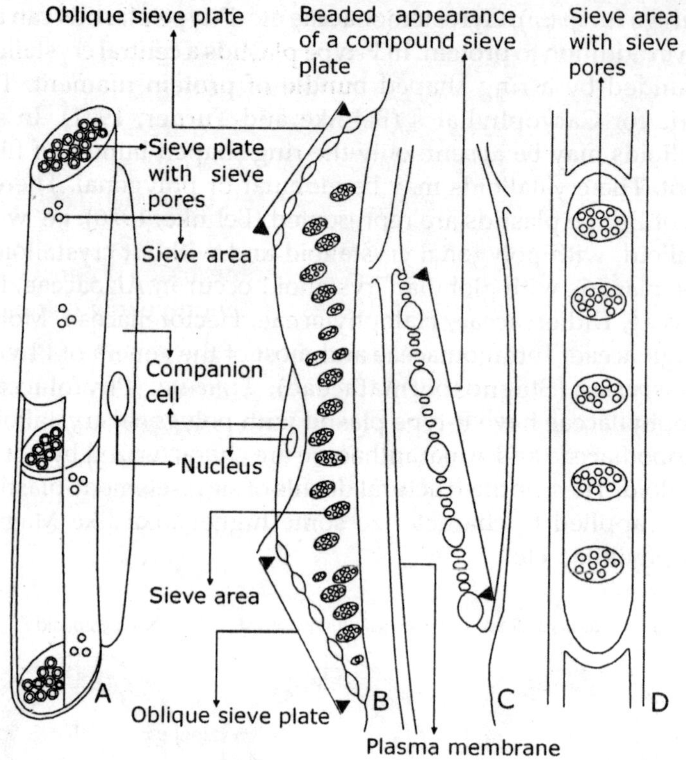


Figure 9.9

A. Diagram showing sieve elements with flat end walls with sieve pores. Sieve areas with sieve pores are present also on lateral wall. B. L. S. view of a sieve element with oblique compound sieve plate. Sieve areas are shown on side wall as seen in surface view. C. Enlarged L. S. view of a compound sieve plate illustrating its beaded appearance. D. Enlarged face view of a compound sieve plate showing sieve areas with sieve pores. A. Adapted after Mauseth, 2009, p. 111. B, C & D adapted after Bowes & Mauseth, 2008, p. 87.

Sieve pores contain callose (Fig. 9.6C). Callose is carbohydrate and is composed of β-1,3-linked glucan. Callose forms a sheath around the connecting strand in a pore. In mature sieve areas callose also deposits on the surfaces present between the pores. As more and more callose deposit the connecting strands in the sieve pore gradually become thin and ultimately disappear. The callose masses of the neighbouring sieve areas may coalesce to a single mass that forms callose pad. As a result the sieve pores are plugged when the sieve tube becomes nonfunctional. The mature sieve areas appear as raised portions instead of depressions, which occur in young sieve tube member. Callose may not be present in the pores and may be apparently absent in some palms.

Callose may deposit seasonally or permanently. Seasonal accumulation of callose on the sieve areas is usually referred to as dormancy callose. In *Vitis* the callose accumulates in the autumn when sieve tubes become nonfunctional. In spring callose gets dissolved and the sieve tubes start to function again. Permanent accumulation of large masses of callose on the sieve areas causing the cessation of function of sieve tubes is designated as definitive callose. Usually sieve tubes remain active during a single growing season, but in *Suaeda, Tilia, etc.* sieve tube may remain active for two or more years.

Callose is studied by staining it with aniline blue. When viewed with a microscope using ultraviolet light it fluoresces lemon yellow colour.

❑ Sieve cell (Fig. 9.10): It is the phloem element of pteridophyta and gymnosperm. It is arranged longitudinally, but the cells are not arranged one above the other in a series. The cells are single, elongate and tapered at both ends. They often do not have a distinct end wall. When present the end walls are either tapered or very oblique. The end walls may overlap one another. The cell wall is thin and made up of cellulose. In *Pinus* the sieve cells are thick walled. Perforations are present on the walls and these regions are called sieve areas.

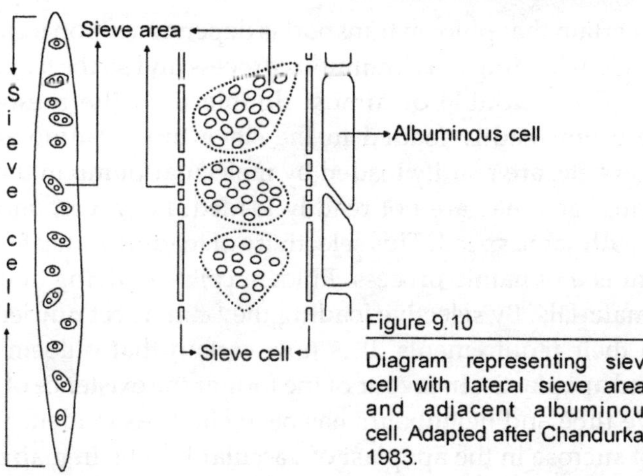


Figure 9.10

Diagram representing sieve cell with lateral sieve areas and adjacent albuminous cell. Adapted after Chandurkar, 1983.

Sieve areas are less specialized in sieve cells in comparison to those of sieve tubes. In contrast to sieve plate no wall parts can be distinguished in sieve areas. Sieve plates are reported in four species of *Equisetum* (*Equisetum arvense, E. hyemale, E. giganteum, E. telmateia*) and the fern *Cyathea gigantea*. Sieve areas are present on lateral walls and sometimes also occur on terminal wall. They are more numerous in those positions where the sieve cells overlap each other. Callose deposits in the perforations of sieve areas.

Sieve cells are living cell where nucleus is absent at maturity. But the nucleus is present in the mature sieve cells of *Pinus strobus* and in the family Taxaceae. A large central vacuole is present. Cytoplasm is present surrounding the peripheral layer of vacuole. Mitochondria, plastids and slime bodies are present. Starch grains are absent in sieve cells.

In contrast to sieve tube, sieve cells are devoid of companion cell. They are associated with albuminous cell and they are not ontogenetically related.

**Functions of sieve elements:** (1) Translocation of photosynthetic products and other organic solutes occur through sieve elements. It is suggested that P-protein is involved in the process. (2) It stores carbohydrates, proteins etc.

Phloem forms a tubular network from leaf to root. In the leaf the mesophyll cells are interconnected through plasmodesmata. There exists connection between mesophyll cells and sieve tubes. Sugars and other metabolites are produced in the mesophyll, move from cell to cell and finally dumped into sieve tube. The leaf is the source from where the sieve tubes are loaded. At sink, i.e. at root the sugars and other metabolites are unloaded and here they are utilized.

When loading occurs at source osmotic potential and water potential in the mesophyll cells become more negative. As a result water enters in the mesophyll and ultimately in the sieve tube. Sieve tube becomes turgid and it pushes the solution to the next sieve tube. This process continues and as a result the solution moves from source to sink. Mesophyll cells continuously produce and dump sugar and other metabolites to the sieve tube. Accordingly the loading at source and unloading at sinks continue.

It is now certain that phloem transport is dependent upon its loading at source (Fig. 9.11). Phloem loading is a continuous process and so the solute concentration in the sieve tube is double or almost double than the mesophyll cells. All substances are not readily loaded in the sieve tube. Amino acids, ions, non-reducing sugars etc. are readily loaded by molecular pump in the plasmalemma. Malic acid, citric acid etc. are not readily loaded. However, once loaded all are translocated with same speed. This selectivity of loading indicates that movement in the phloem is a dynamic process. Phloem cells are living and they can select the loading materials. By selective loading they can direct nutrients to the organs according to their requirements. It is now certain that phloem loading is both symplastic and apoplastic. In favour of the former the existence of plasmodesmata between sieve tube and neighbours has been cited. As evidence of the latter the occurrence of sucrose in the apoplast of vascular bundle in maize has been cited. The unloading may be apoplastic or symplastic. The molecular pumps of plasmalemma, which help in loading, also help in unloading. In this context it is to be mentioned that apoplastic loading is advantageous due to the fact that it occurs in response to drought, high temperature etc.

Although the source is constant, i.e. the photosynthesizing cells, but the sink, i.e. the utilizing cells are different. The upper leaves load assimilates for the shoot apex while the lower leaves for the roots. The middle leaves supply to both leaves and roots. Loading at source and unloading at sink cause a difference in water potential, which must be maintained for effective translocation. The unloaded materials must be utilized to maintain the difference in potential. If the sink fails to utilize, the rate of loading will be reduced. So the functioning of sink will determine the rate of loading at source.

E. Münch hypothesized Source—Sink concept and Phloem Loading and Unloading as early as 1930. This hypothesis is also known as Mass flow hypothesis, Pressure flow hypothesis or Münch's hypothesis. Though there are strong

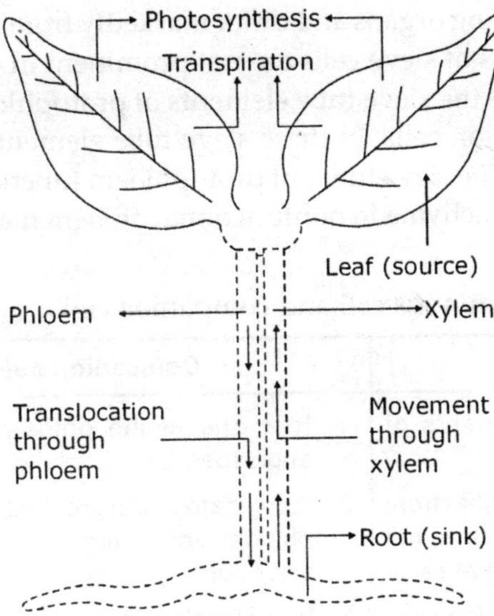


Figure 9.11
Diagram representing Münch's hypothesis.

arguments against this hypothesis still this is dominant in elucidating the concept regarding phloem loading and unloading.

**Phylogenetic specialization of sieve element:** Sieve elements originated from parenchyma cells, which modified themselves as to the function. The modification and specialization occurred in the following way: (i) Sieve cells are considered as primitive and it is assumed that they appeared in pteridophytes and gymnosperms though they may be present in angiosperm (*Austrobaileya scandens* and *Sorbus aucuparia* of Rosaceae). They are not arranged in axial files; (ii) sieve tube members originated in angiosperms; (iii) disappearance of nuclei from sieve elements; (iv) association with nucleated parenchyma like albuminous cells in pteridophyta and gymnosperm, and companion cells in angiosperm for metabolic activities; (v) reduction of sieve areas on the side walls; (vi) localization of sieve areas on the end walls; (vii) orientation of end walls from oblique to transverse; (viii) development of simple sieve plate from compound sieve plate; (ix) development of thick from thin connecting strands that are present in pteridophyta and gymnosperm; (x) the diametre of pores increased and thus the percentage of pore-area occupied in sieve areas was also increased; (xi) in monocots the development and specialization of sieve tube members occurred from leaves to roots (Esau, 1965); (xii) in angiosperm sieve tube members became shorter in length during evolution.

**Protophloem and metaphloem:** Protophloem and metaphloem together constitute the primary phloem. The first elements of primary phloem formed from the procambium strands are designated as protophloem. The lately differentiated elements of primary phloem are known as metaphloem.

Metaphloem is the completely developed primary phloem and consists of sieve elements, phloem fibre and phloem parenchyma. They usually occur in the mature parts of a plant and may remain active for several years. In contrast, protophloem occurs in developing organs and differ markedly from metaphloem. In gymnosperm, the sieve areas of sieve cells are not prominent in protophloem elements. In some angiosperms the sieve tube elements of protophloem are long, narrow and devoid of companion cells. In these sieve tube elements sieve areas are not easily distinguishable. The sieve tubes of protophloem function for a short period and disappear. The parenchyma in nonfunctional phloem may form fibres or collenchyma.

Comparison between albuminous cell and companion cell.

| Albuminous cell | Companion cell |
|---|---|
| 1. It is one of the phloem elements of conifers and *Ginkgo*. | 1. It is one of the phloem elements of angiosperm. |
| 2. In primary phloem it originates either from phloem parenchyma or phloem rays and ontogenetically not related to sieve cells. | 2. It originates from procambium in primary phloem and ontogenetically related to sieve tube. |
| 3. It is closely associated morphologically and physiologically with sieve cells. | 3. It is closely associated morphologically and physiologically with sieve tubes. |
| 4. There is no report of labyrinthine wall in it. | 4. It may develop wall labyrinths and form labyrinthine wall of transfer cell. Ex. *Hieracium*. |

Comparison between sieve tube and sieve cell.

| Sieve tube | Sieve cell |
|---|---|
| 1. It is present in angiosperm. | 1. It is present in pteridophyta and gymnosperm. |
| 2. They are arranged in series. | 2. They are not arranged in a series and the cells are single. |
| 3. They have end walls. | 3. They often do not have end walls. |
| 4. Sieve areas are present on end walls. | 4. Sieve areas occur on lateral and sometimes on terminal wall. |
| 5. They have sieve plates. | 5. They do not have sieve plates. |
| 6. The sieve areas are specialized. | 6. The sieve areas are relatively less specialized. |
| 7. They remain associated with companion cell. | 7. They are associated with albuminous cell. |
| 8. Sieve tube and companion cell are ontogenetically related. | 8. Sieve cell and albuminous cell are not ontogenetically related. |

Difference between protophloem and metaphloem.

| Protophloem | Metaphloem |
|---|---|
| 1. It is the first element of primary phloem formed from procambium strands. | 1. It is lately differentiated element of primary phloem. |
| 2. It occurs in the elongating portions of plant organs. | 2. It occurs in mature portion of plant organs. |
| 3. The sieve elements may be shorter and narrower than metaphloem. | 3. The sieve elements may be longer and wider than protophloem. |
| 4. The sieve areas of sieve cells are not easily distinguishable. | 4. The sieve areas are prominent in sieve cells. |
| 5. There is no companion cell in many plants. | 5. There is always companion cell. |
| 6. The sieve tube remains functional for a short period only. | 6. The sieve tube remains functional for several years. |
| 7. Phloem fibre is present. | 7. Usually phloem fibre is absent. |
| 8. Parenchyma of non-functional protophloem gives rise to fibres. | 8. Parenchyma of old metaphloem may become sclerified. |
| 9. Less number of transfer cells is present in comparison to metaphloem. | 9. More number of transfer cells is present in comparison to protophloem. |

Difference between perforation plate and sieve plate.

| Perforation plate | Sieve plate |
|---|---|
| 1. It is the region of vessel where perforations occur. | 1. It is the region of sieve tube where sieve areas occur. |
| 2. The position is usually terminal but sub-terminal and lateral plates are also noted. | 2. The position of sieve plates may be horizontal or oblique. |
| 3. The perforation plates may be simple, scalariform, foraminate or reticulate. | 3. The sieve plates may be simple or compound. |
| 4. The pores of the perforation plate are lined by lignin. | 4. The sieve plate pores contain callose. |
| 5. Water and minerals exist in the perforations. | 5. Plasmodesmata exist in the sieve pores. |
| 6. It is composed of secondary wall materials. | 6. It is composed of primary wall materials. |
| 7. The perforations always remain open. | 7. Sometimes callose deposits encircling the sieve pores to form callose pad and covers the sieve areas. |

Difference between sieve plate and sieve field.

| Sieve plate | Sieve field |
|---|---|
| 1. Sieve plates have well developed sieve areas. | 1. Sieve fields are vestigial sieve plates with lattice like appearance. |
| 2. They are definite in outline. | 2. They are not definite in outline. |
| 3. The sieve pores are larger in diametre than perforations in sieve fields. | 3. The perforations in the sieve fields are exceedingly minute or often lacking. |
| 4. The sieve pores are larger than plasmodesmata. | 4. The perforations in the sieve field are not larger than plasmodesmata. |
| 5. Sieve plates are also present in herbs. | 5. They are absent in herbs. |

Difference between tracheid and trachea.

| Tracheid | Trachea |
|---|---|
| 1. It predominates in xylem elements of pteridophytes and gymnosperm and rarely occurs in angiosperm. | 1. It is also known as vessel and predominates in xylem elements of angiosperm and rarely occurs in gymnosperm (Ex. Gnetales) and pteridophyta (Ex. *Selaginella*). |
| 2. The cells are elongated with narrower diametre than trachea. | 2. They are elongated tubular and wider in diametre than tracheids. |
| 3. The two end walls may be pointed, obtuse or chisel like. | 3. The end walls are transversely of obliquely flat. |
| 4. The end walls are not perforated. | 4. The end walls are perforated and the perforation may be simple, scalariform, foraminate or reticulate. |
| 5. Communication with adjacent tracheids occurs through the pits present on the lateral walls. | 5. Communication occurs through perforated end walls and lateral walls where pits are present. |
| 6. It is situated parallel to the long axis of the organ with overlapping and interlocking ends. It never forms a continuous tube. | 6. Vessels form a parallel tube-like series of cells arranged end to end and their end walls are perforated and jointed. |
| 7. The cell walls have annular, spiral, reticulate, scalariform and pitted thickenings. | 7. The cell walls commonly have bordered pits. |
| 8. It forms the supporting and conducting tissue mainly in pteridophyte and gymnosperm. | 8. It forms the supporting and conducting tissue in angiosperms in association with tracheids. |

Difference between sieve tube and vessel (trachea).

| Sieve tube | Vessel (Trachea) |
|---|---|
| 1. It is one of the components of phloem. | 1. It is one of the components of xylem, also known as trachea. |
| 2. This is a living cell without nucleus. | 2. The cells are nonliving. |
| 3. Sieve tube contains cytoplasm and food matters. | 3. Vessels contain water and minerals. |
| 4. The end walls termed sieve plates may have many sieve areas. | 4. The end walls are called perforation plates, which have many perforations. |
| 5. The sieve plate may be simple or compound. | 5. The perforation plate may be simple, scalariform, foraminate etc. |
| 6. In sieve areas plasmodesmata exist. | 6. In perforations water and minerals exist. |
| 7. Each sieve tube is always associated with companion cell which contains nucleus, phloem fibre and phloem parenchyma (nucleated). | 7. Vessels are associated with fibres, tracheids and xylem parenchyma among which the former two are devoid of nucleus. |
| 8. The cell wall is thin and mainly composed of cellulose. It is primary in origin. | 8. The cell wall is thick and contains lignin. It is secondary in origin. |
| 9. The lateral walls contain primary pit fields only. | 9. The lateral walls may be annular, spiral, reticulate and pitted etc. |
| 10. Downward translocation of food occurs through sieve tubes. | 10. Ascent of sap occurs through xylem. |
| 11. These are not mechanical cells. | 11. They are mechanical cells. |
| 12. Sieve tube may become inactive by the formation of callose pad. | 12. Vessels become inactive by the formation of tylosis. |

Difference between primary xylem and primary phloem.

| Primary xylem | Primary phloem |
|---|---|
| 1. In stem, it is situated towards the center side. | 1. In stem, it lies towards the peripheral side. |
| 2. In root it lies alternating with phloem. | 2. In root, it lies alternating with xylem. |
| 3. Proto- and metaxylem can be easily differentiated. | 3. Proto- and metaphloem cannot be easily differentiated. |
| 4. Xylem consists of tracheids, trachea, xylem fibre and xylem parenchyma. | 4. Phloem consists of sieve tube, companion cell, phloem fibre and phloem parenchyma. |
| 5. All elements except xylem parenchyma are dead cells. Living protoplast is absent in xylem except xylem parenchyma. | 5. Phloem fibre is dead cell only. |
| 6. The cell wall of tracheary elements and fibre is thick and lignified. | 6. The cell wall of sieve element, companion cell and phloem parenchyma is thin. |
| 7. The thickening of cell wall may be annular, spiral, scalariform, reticulate and pitted etc. | 7. Primary pit field and pits are present on cell wall only. |
| 8. Ascent of sap occurs through xylem. | 8. Downward translocation of prepared food occurs through phloem. |
| 9. Xylem gives mechanical rigidity to the plant. | 9. Phloem fibre gives little mechanical strength to the plant. |

CHAPTER **10**

# Ontogeny and Phylogeny of Xylem and Phloem

The complex tissue xylem consists of xylem parenchyma, fibre, tracheids and vessel whereas phloem is composed of sieve cell, phloem parenchyma, sieve tube with companion cell and phloem fibre. Their ontogeny and phylogeny are discussed separately.

## Xylem

**Xylem parenchyma:**

• *Origin*: Xylem parenchyma cells are present both in primary and secondary xylem; accordingly their origin also differs. In primary xylem they originate from procambium. In secondary xylem ray parenchyma cells originate from the ray initials of cambium. The fusiform initial of cambium gives rise to axial parenchyma along with tracheary element and fibres.

Fusiform initials of cambium normally divide vertically in the longitudinal plane. Transverse divisions do occasionally occur during the formation of additional initials, but these are much less frequent than vertical division. The vertical divisions are mainly periclinal. In these cells anticlinal division also occur to keep pace with the growth of stem in girth. During longitudinal division cell wall first originates between the two newly formed nuclei and gradually extends towards the ends of cell. The cell wall formation may not be completed for some time after mitosis.

• *Phylogeny of ray parenchyma*: In radial longitudinal section it is observed that the ray parenchyma is composed of square (isodiametric), erect (upright or vertically elongate) and procumbent (radially elongate) cells. It is considered that the square cells are morphologically equivalent to erect cells. The ray cells may be homocellular and heterocellular. The homocellular rays are composed either of square cells or of erect cells, or of procumbent cells, or of erect and square cells. The heterocellular rays consist of both square and procumbent cells or of erect and procumbent cells. Rays may be uniseriate when they are composed of one cell in width, biseriate (ray is two cells in width) or multiseriate where the rays consist of a portion more than two cells wide. The latter may have uniseriate wing similar to uniseriate rays. Multiseriate ray, as seen in tangential longitudinal section (TLS), is tapered towards both upper and lower margins. Tapered ends are described as wings, which is commonly uniseriate. The outline of rays appears to be fusiform in TLS. Primitive and advanced wood exhibit the following types of rays.

| Primitive wood | Advanced wood |
|---|---|
| 1. In primitive woods both multiseriate and uniseriate rays are present. | 1. In advanced wood either multiseriate or uniseriate ray is present, i.e. one is lost during evolution. |
| 2. In primitive wood the uniseriate and multiseriate rays were of marked vertical length, i.e. the rays appear to be very high. | 2. In advanced forms the multiseriate rays are reduced in size and number. |
| 3. The primitive woods exhibit heterocellular uni- and multiseriate rays. | 3. The advanced woods exhibit homocellular rays. During evolution the erect cells are lost and so there is a tendency for loss of heterogeneity. |
| 4. In primitive woods uniseriate wings are present in multiseriate rays. | 4. The advanced woods exhibit extremely short uniseriate wing on multiseriate rays. |

Carlquist (1961) showed some trends of evolution of ray types in dicots based on the works of Kribs (1935) with modifications and recognized the following type of rays.

**Heterogeneous I:** It consists of uniseriate ray and multiseriate ray with uniseriate wings. The cells of wings are more or less similar to the cells of uniseriate rays. Both rays are of marked vertical length, i.e. very high. The cells are square, erect and procumbent.

**Heterogeneous II(A):** This type is composed of multiseriate rays with short height and short wings. The cells are square, erect and procumbent. This type also consists of uniseriate ray with square and erect cells.

**Heterogeneous II(B):** The rays are uniseriate and the cells are square, erect and procumbent.

**Heterogeneous III:** The rays are uniseriate and the cells are square, erect and procumbent.

**Homogeneous I:** The rays are uniseriate and multiseriate with very short wings. The multiseriate rays are mostly fusiform. The cells of multiseriate portion together may be oval, round or radially elongated in outline. The cells of the uniseriate tips are identical with the cells of multiseriate portions of rays. The cells are procumbent.

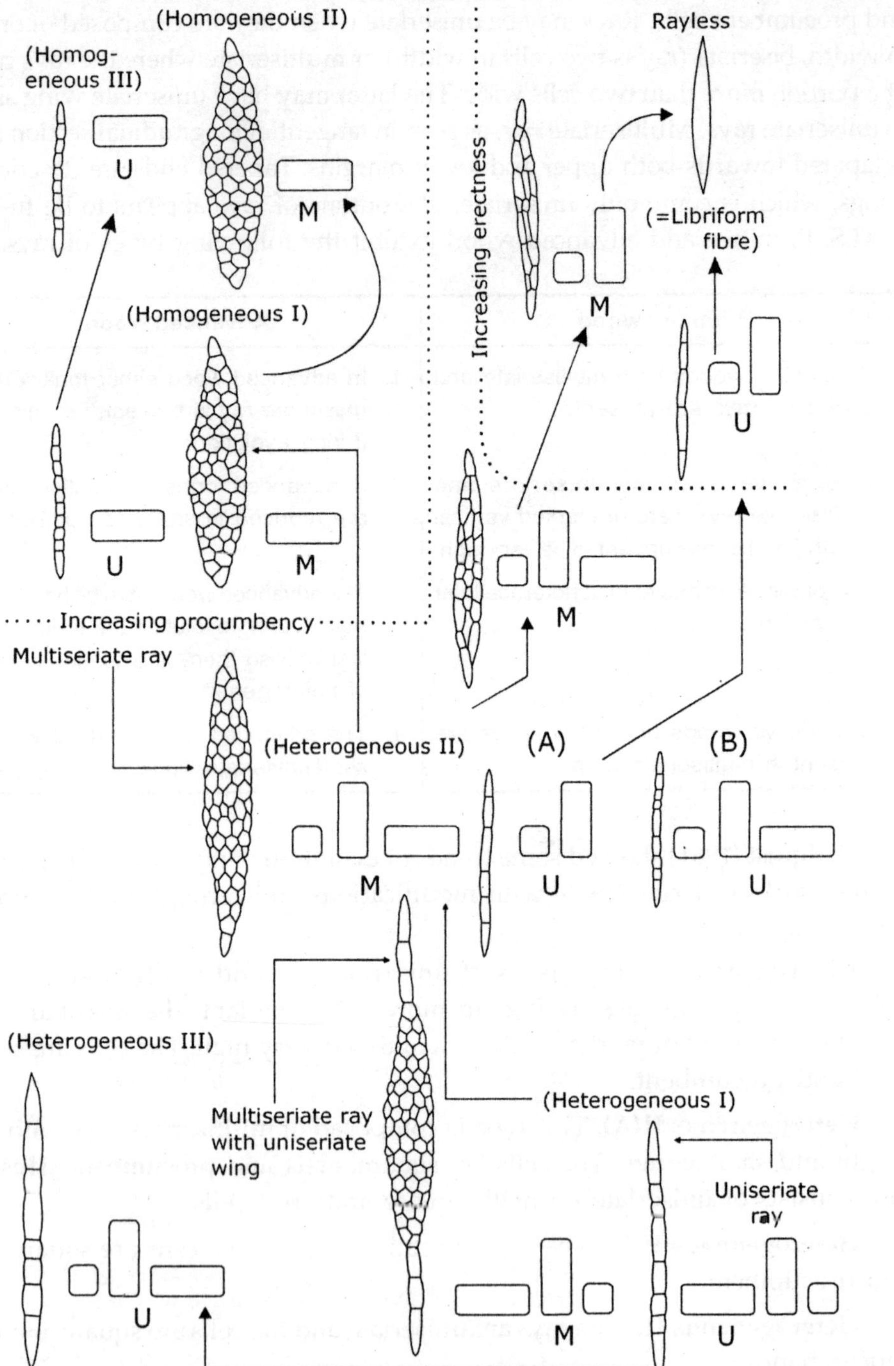


Figure 10.1

Schematic representation of evolution of ray types in dicots. Rays are in tangential longitudinal sectional view. The square, erect and procumbent cells are in radial longitudinal view for multiseriate (M) and uniseriate (U) ray. The arrows indicate the trends of evolution (after Carlquist, 1961, p. 59).

**Homogeneous II :** The rays are multiseriate with extremely short wings and the cells are procumbent. The shape of multiseriate rays is fusiform. The cells may be round or radially elongated.

**Homogeneous III :** This type consists of uniseriate rays with procumbent cells.

Carlquist showed the trends of evolution of rays in dicots which is presented in Fig. 10.1 in diagrammatic figures.

Woods having heterogeneous I type of ray is considered as primitive. The advanced form of rays is either uniseriate or multiseriate with procumbent cells (Homogeneous I, II and III). It is regarded that the rays with erect cells increased in length and continued elongation resulted the conversion of ray initial into fusiform initial. The ultimate result is raylessness.

• *Phylogeny of axial parenchyma*: The fusiform initial of cambium gives rise to axial (vertical) parenchyma. The distributions of axial parenchyma are studied in transverse sections. Axial parenchyma may lie independent or associated with vessels. Accordingly timbers may be apotracheal (parenchyma are not associated with vessel) and paratracheal (parenchyma are distinctly associated with vessels). The common apotracheal forms are (i) diffuse (axial parenchyma occurs as isolated strand), (ii) diffuse-in-aggregates (axial parenchyma occurs as aggregates), (iii) banded (axial parenchyma appears as bands; the bands may be narrow or wide), (iv) marginal [parenchyma occurs either at the beginning of growth ring (initial) or at the end of growth ring (terminal)]. The common paratracheal forms are (i) scanty (the parenchyma cells do not form a continuous sheath surrounding a vessel), (ii) vasicentric (parenchyma cells encircle the vessels), (iii) abaxial (vasicentric parenchyma occurs more in width on abaxial side of vessel), (iv) adaxial (the parenchymatous sheath is more in width on adaxial side of vessel), (v) aliform (vasicentric parenchyma extends laterally in the form of wings) and (vi) confluent (vasicentric parenchyma extends and coalesces with other forming a continuous band) (Fig. 10.2).

The phylogenetic sequences among the distributional type of axial parenchyma are discussed below (Fig. 10.3).

1. The primitive wood may exhibit no parenchyma; ex. vesselless wood of Winteraceae.
2. The primitive wood exhibits diffuse parenchyma. The cells of fusiform cambium differentiate into parenchyma.
3. The advanced form of wood exhibits diffuse-in-aggregate, where the tendency towards grouping of axial parenchyma is noticed.
4. The advanced form of wood exhibits apotracheal-banded parenchyma. The wide band of apotracheal parenchyma is considered to be more advanced than narrow band of parenchyma. It is assumed that diffuse-in-aggregate gives rise to narrow banded apotracheal type that ultimately forms wide band apotracheal type.
5. Marginal parenchyma (collective term of terminal and initial parenchyma) is formed as a result of changes in climatic conditions. So it is assumed that marginal parenchyma arose independently.

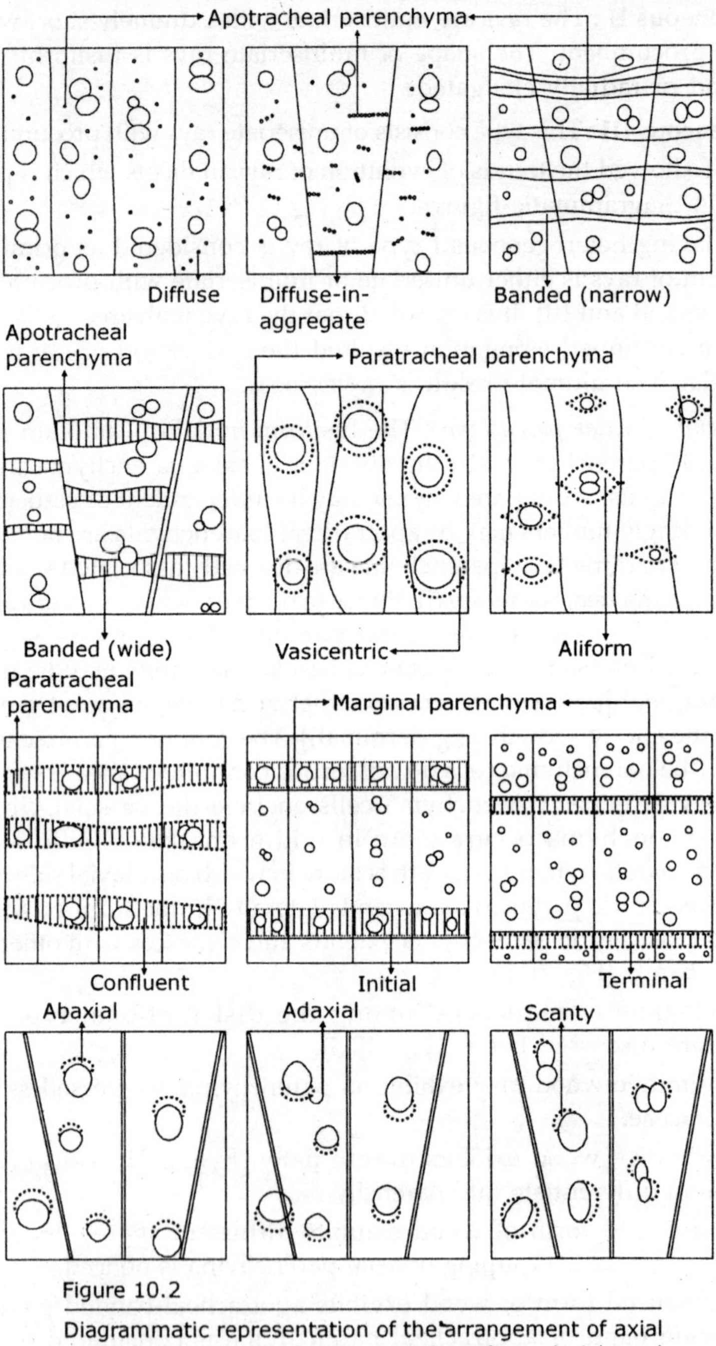


Figure 10.2

Diagrammatic representation of the arrangement of axial parenchyma as seen in cross-section of wood. Adapted after Cutter, 1971, p. 87 & Carlquist, 1961, p. 62.

6. Scanty parenchyma seems to be less specialized than vasicentric.
7. The banded types gave rise to vasicentric, aliform and confluent types.

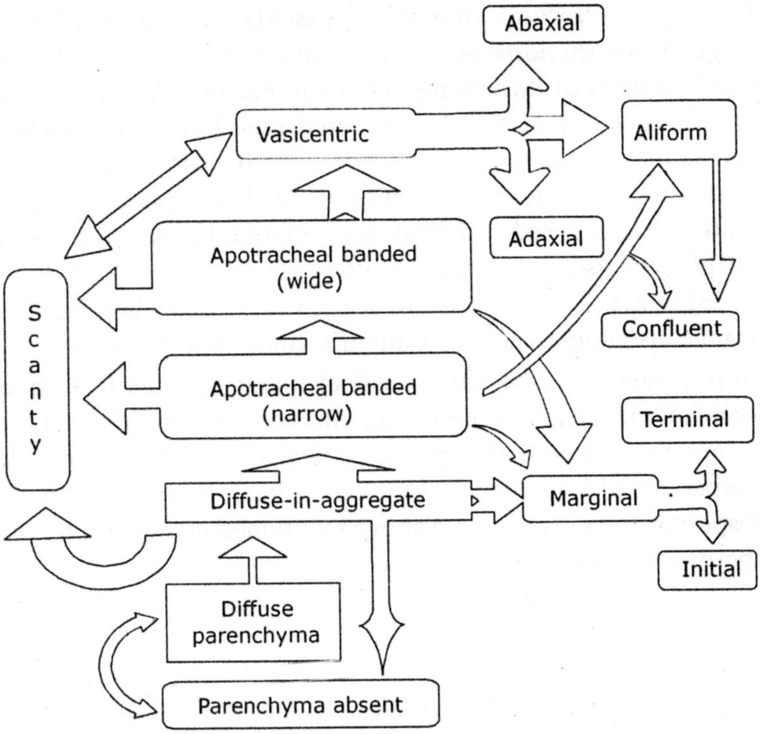


Figure 10.3

Diagram illustrating the trends of evolution in axial parenchyma distribution. Adapted after Carlquist, 1961, p. 62.

## Xylem fibre

• *Origin of xylem fibre*: In primary xylem fibres originate from procambium whereas they are developed from fusiform initial of cambium in case of secondary xylem.

• *Phylogeny of xylem fibre*: Fibres and tracheids are phylogenetically related and it is suggested that the former evolved from the latter. During evolution the length of fibre decreased, the bordered pits reduced in size and the thickness of cell wall increased. In secondary xylem of dicots, libriform fibre evolved in the sequence of tracheid, fibre-tracheid, and libriform fibre. Bordered pits are prominent in tracheids. The pit border is diminished in fibre-tracheid where the bordered pits have less developed border. Disappearance of border on pit occurs in libriform fibre. Here the pits are simple or nearly so. The cell wall is thick in fibre-tracheid and thicker in libriform fibre. The diameter of fibre becomes narrower.

## Tracheid

• *Origin of tracheid*: Tracheids of primary xylem originate from procambium. In secondary xylem they originate from cambium ring. A single fusiform initial forms a tracheid.

• *Phylogeny of tracheid*: In general, vascular cryptogams have very long tracheids; gymnosperm possesses tracheids of intermediate length. The following types of secondary wall thickening may be found in tracheids: annular, helical, scalariform, reticulate and circular bordered pits. From ontogenetic viewpoint annular elements precede helical elements. The helical bands became joined at certain areas thus forming scalariform elements. In the phylogenetic scale pitted elements are most advanced. Tracheids with annular thickening are reported from one of the oldest fossils *Cooksonia*, a leafless vascular plant from upper Silurian deposits (Stewart, 1983). Another very early vascular plant—the leafy *Baragwanathia* also contained the annularly thickened tracheids.

From phylogenetic viewpoint vessel elements evolved from tracheids and are polyphyletic in origin. Vessel elements occur in pteridophyta (*Selaginella*, *Equisetum* and four ferns—*Actiniopteris*, *Pteridium*, *Regnellidium* and *Marsilea*) and in gymnosperm (*Ephedra*, *Welwitschia* and *Gnetum*). It is to note that aside from the angiosperms, the vessels of the above mentioned genera of pteridophyta and gymnosperm are considered as anomalies.

## Vessel

• *Origin and development of vessel*: Trachea (or vessel) originates from meristematic cells (Fig. 10.4) that are grouped together in a longitudinal file. The meristematic cells are known as xylem mother cell that develops from procambial cells in the primary xylem. In secondary xylem cambial derivatives give rise to xylem mother cell. At the last stages of development the fusion of xylem mother cells end to end results in the formation of trachea. Due to fusion and subsequent loss of end walls the vessels are grouped together to form a continuous long tube.

Each xylem mother cell forms a vessel member. The primordial vessel member may or may not elongate in length during differentiation. But it usually widens laterally. Secondary wall materials deposit after the completion of longitudinal elongation and lateral expansion. The deposition does not occur uniformly and forms specific pattern characteristic for the particular vessel element. The secondary wall materials do not cover certain portions of the primary wall. These portions are the site of future pits and perforations. The cell wall at the future perforation region becomes thicker than the rest of primary walls. The thickening occurs due to the swelling of intercellular pectic substances and no secondary wall material is deposited on it. The swollen regions of primary wall later disintegrate to form perforations. This happens after the completion of deposition of secondary wall substances.

In longitudinal section the end walls of primordial vessel element appear as lens-shaped (e.g. Celery) or plate like (e.g. *Robinia*) after swelling. As many as three layers can be distinguished in this region. Middle lamella and the two primary walls of the two adjacent cells compose the three layers. The middle lamella consists of pectic substances whereas non-cellulosic polysaccharides compose the two adjacent primary walls.

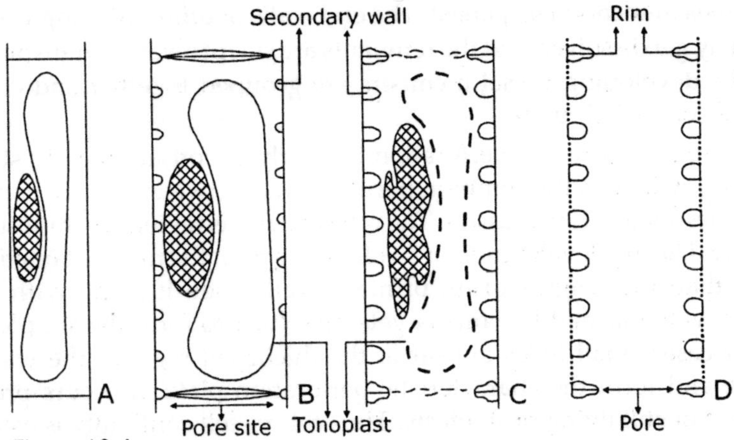


Figure 10.4

Diagrammatic illustration of development of a vessel member with helical thickening. A. Cell with primary wall. B. Enlarged cell and enlarged nucleus. Secondary wall material has begun to be deposited over primary wall. C. Tonoplast ruptures and nucleus is deformed. D. Vessel member with open pore at both ends. Protoplast degenerated and primary wall partly hydrolyzed between secondary thickenings. Adapted after Esau, 1977, p. 116. Nucleus = crosshatched.

Xylem mother cells are thin walled, densely cytoplasmic, uninucleate and vacuolated. Cytoplasm remains active throughout the development of vessel elements. It slowly disintegrates as the vessel members mature. As development progresses the nucleus becomes small and flattened, and lie either at the centre, on the side of lateral wall or against the end wall—the site of future perforation. Differentiating metaxylem becomes polyploid and contains more DNA than neighbouring cells. In some members of Euphorbiaceae the developing vessel members become multinucleate. Commonly the cells remain uninucleate with large endopolyploid nuclei. At maturity nucleus disintegrates. The tonoplast-bound vacuolar sap contains hydrolytic enzymes. After the rupture of tonoplast cytoplasm and nucleus are exposed to hydrolytic enzymes. As a result autolysis of cytoplasm occurs.

After the introduction of electron microscope it is observed that xylem mother cell contains microtubules, endoplasmic reticulum, dictyosomes and mitochondria etc. Evidences suggest that microtubules direct the deposition of new secondary wall materials. It is observed that the concentration of microtubules increases at the region of developing thickening of wall. Cytoplasm present between the thickenings is devoid of microtubules. In this region endoplasmic reticulum is present in close association with the plasmamembrane. In the cytoplasm active dictyosomes are also present. It is thought that vesicles produced from dictyosomes contribute wall components at the site of wall thickenings. The role of endoplasmic reticulum is to inhibit the process of wall thickenings. As the wall reaches its final state of maturity, the vacuoles present in the differentiating vessel member coalesce. As a result a single large vacuole is formed. Later vacuolar membrane ruptures and this causes the degeneration of cytoplasmic contents.

Sometimes mitochondria persist. The end walls of differentiating vessel members ultimately get dissolved and the remnants are swept away by transpiration stream. Thus the developing vessel members are grouped together and converted to a series of connected tubes.

The region of future perforation is clearly set off at an early stage from the secondary wall. It is the common partition wall between the two contiguous cells that are destined to form vessels. The cell wall at this region is composed of primary wall only. During development it swells along with the middle lamella. It becomes thicker than the neighbouring primary wall. Then it along with the adjacent primary wall and middle lamella gets dissolved to form the simple perforation. In other cases small holes appear at the thickened regions, the walls gradually dissolve and form the characteristic perforation plate. It occurs presumably by the action of still living protoplasts. Thus cell-to-cell continuity is established and a continuous tube is formed. Now the vessels become functional in conduction. Around the simple perforation the secondary wall appears as a rim.

• *Phylogeny of vessel*: The evolutionary specialization of vessel is discussed within the vessel itself along with other features related to it. These include length, end wall pitting, thickening of cell wall and the outline in transverse section of a vessel while the other features comprise the abundance, groupings and ring porousness of vessels. The following discussion is based on Metcalfe and Chalk (1950), and Carlquist.

- *Length*: Vessels have arisen from no other cells than tracheids that are usually long and narrow spindle shaped. So it is assumed that long vessels are primitive. The advanced forms are short and wide. As specialization increases the length of fusiform initial that forms vessels also reduces. The length of a vessel may be as short as a few centimetres or as long as several metres. The size is not random. It is to be noted that certain organs have only long vessels and others have only short ones.

- *End wall*: Vessel with long slanted end is considered as primitive as it is present in tracheid like vessels. Evolution proceeded towards the formation of transverse end walls.

- *Perforation plate*: Vessels with scalariform perforation plate with many bars in a very slanted end wall is considered as primitive. This type of perforation is associated to long vessel, which is also considered as primitive. During evolutionary specialization the number of bars reduced. Accordingly the number of pores also reduced. The final stage in specialization is the loss of all bars and simple perforation is the result at the end walls. Simple perforation with circular rim is more advanced than long oval rim (Fig. 10.5).

  The specialized vessels are short with large diameter. The perforation plate is simple with circular rim in an almost transverse end-wall. This type of vessel is advantageous for good conduction of water. In a wide vessel the adhesive interaction between the molecules of water and cellulose of cell wall is decreased and this minimizes the contact between water and cell

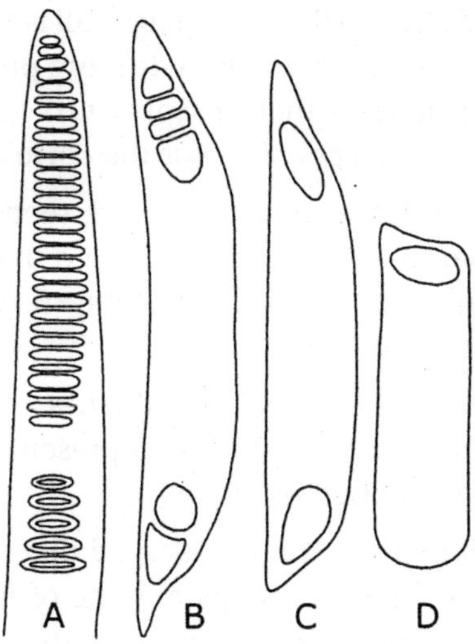


Figure 10.5

Diagrammatic representation of different types of perforation plate. A. A portion of long tracheid-like vessel with scalariform perforation plate with many bars in a very slanted end wall. B. Reduced number of bars and pores. C. & D. Simple perforation with tendency toward less oblique end wall. Adapted after Carlquist, 1961, p. 41.

wall. As a result water flows easily. Moreover the simple perforation plate with circular rim allows easy flow of water. In scalariform perforation plates the bars may offer resistance to water flow.

The long vessels require extra reinforcement in addition to that provided by the secondary wall. Scalariform bars and the rim of perforation plate provide the extra means of strengthening the long vessel. In short vessels with simple perforation the mechanical strength is obtained from the rim of perforation plates in addition to the secondary wall.

- *Lateral wall pitting*: It is suggested that scalariform pitting is the most primitive. Evolutionary sequence advanced through transitional types to opposite type. The final stage in the evolution is alternate pitting. Scalariform pitting is associated with long vessel, whereas alternate pitting is associated with short vessel. Vessels with medium length are associated with transitional and opposite pitting (Fig. 10.6).

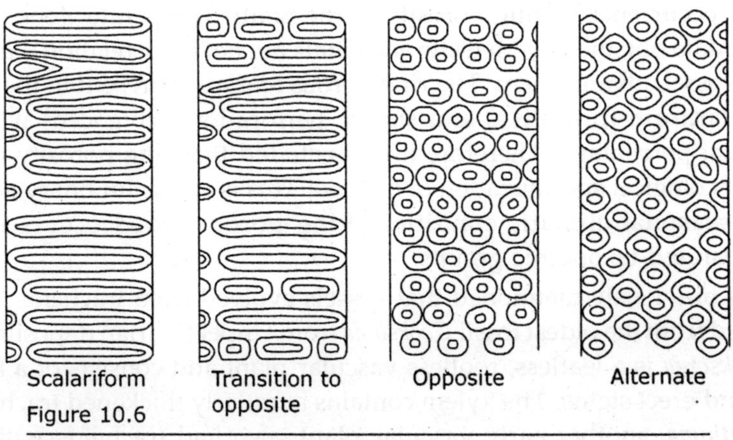


Figure 10.6

Diagram illustrating different types of lateral wall pitting of vessel member. Adapted after Fahn, 1997.

- *Thickening of cell wall*: Spiral thickening on the secondary wall is an advanced character. This type of thickening is closely linked with ring porousness of wood. Metcalfe and Chalk is of opinion that probably ecological factors associated with ring-porousness are responsible for the correlation.
- *Outline in transverse section*: Vessel, as seen in transverse sections, ranges from angular in outline to circular in outline. It is suggested that transition occurred from angular to circular in outline. This type of transition is observed both in dicots and monocots.
- *Abundance*: The abundance of vessel offers characteristics of taxonomic significance. It is measured as the number of vessels present per square millimetre of areas in a transverse section of wood.
- *Grouping*: In the transverse section of wood it is observed that vessels may be present as single or in groups. The groupings may be of multiples of two or three. In a group the vessels may be arranged in radial, oblique, or tangential lines. This type of distribution and character are considered as most valuable taxonomic character. These characters can be used to trace the trends of evolutionary specialization and to indicate affinity. It is suggested that solitary vessel represents the primitive condition. During specialization vessels are arranged in groups.
- *Ring porousness*: Ring porous wood has two marked zones based upon differential distribution of large and small vessels. In this wood numerous large vessels occur at the beginning of growth ring and fewer small vessels are situated at the end. This type of wood is formed due to either different seasonal climatic variation, cold winters or very dry season followed by wet climate. In contrast, diffuse porous wood has no marked zones of small and large vessels and their distribution is at random throughout the growth ring. Gilbert (1940), after an extensive study on North Temperate woods opines that ring porosity is an advanced feature. But Carlquist does not consider ring porosity as one of the trend of vessel evolution and showed the occurrence of ring porosity in the South Temperate Zone. Moreover the data obtained by Metcalfe and Chalk are not in correlation with features of vessel advancement. The ring porous feature is present in the vessels of scalariform (primitive) and simple perforation (advanced) with almost equal percentage. Therefore it is concluded that ring porosity represents an ecological specialization of a wood. This is an evolutionary adjustment to seasonal climatic condition. Ring porous wood occurs in widely separated taxonomic groups.

It was previously mentioned that vessels evolved from tracheids. Tracheids are reported from the oldest fossil *Cooksonia* from Upper Silurian deposits (Stewart, 1983). *Cooksonia* is a leafless, rootless vascular plant and consists of a horizontal rhizome and erect stems. The xylem contains annularly thickened tracheids only. *Baragwanathwia*, another early vascular plant, also had tracheids with annular thickenings in the xylem, but in contrast to *Cooksonia* it had leaves.

Vessels are polyphyletic in origin. Vessels occur in Pteridophyta (*Selaginella, Equisetum* and the four ferns—*Actiniopteris, Pteridium, Marsilea* and *Regnellidium*), Gymnosperm (*Ephedra, Gnetum* and *Welwitschia*) and angiosperms. Aside from angiosperms, vessels of the above mentioned genera of Pteridophyta and Gymnosperm are considered as anomalies. Vessels appeared first in the secondary xylem of dicots and later in the primary xylem. In monocots vessels first arose in roots and then later in the stems and leaves.

## Phloem

*Ontogeny and phylogeny of phloem*: Phloem consists of sieve cell, sieve tube, companion cell, phloem parenchyma and fibre.

- *Sieve cell*-Origin: The mother cells of sieve cells vary in shape. They are slender, short cylindrical to elongate with both ends tapering. Numerous primary pit fields occur on lateral walls. During differentiation the mother cells elongate. Vacuole appears in the cytoplasm that streams actively. The cell wall increases in thickness. Sieve areas develop in the position of primary pit-fields. Cytoplasmic strands appear in the sieve areas and they gradually become prominent and increase in size. At later stage callose develops surrounding the cytoplasmic strands.

Sieve cell predominates in pteridophyte and gymnosperm. It is reported among angiosperms in *Austrobaileya scandens* and *Sorbus aucuparia* of Rosaceae. Sieve cells are considered as primitive. They are not arranged in axial files. Moreover the structure of end walls is similar to the lateral walls.

- *Sieve tube*: Sieve tubes are arranged in axial files. The end wall of it is the sieve plate. Sieve plate is a specialized structure and differs from the sieve areas present on the lateral wall of a sieve tube and all walls of a sieve cell.

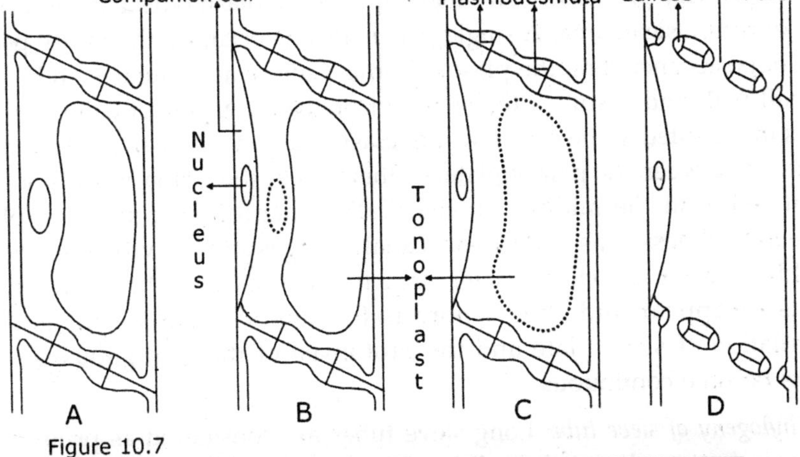


Figure 10.7

Diagram representing the development of sieve tube. A. Sieve tube mother cell that divides longitudinally to form companion cell (B). C. The disappearing tonoplast. D. Mature sieve tube. Adapted after Esau, 1961.

❑ *Origin and development of sieve tube*: Sieve tubes are present in primary and secondary phloem and accordingly they originate from procambium and cambium respectively (Fig. 10.7). The procambium and cambium are arranged in a longitudinal file. They contain living protoplast with prominent nuclei, endoplasmic reticulum, dictyosomes, tonoplast bound vacuoles, ribosomes, mitochondria, chloroplast etc. The mother cells divide longitudinally to form two daughter cells. One of the daughter cells forms the companion cell and the other cell develops into sieve tube member. During development the companion cell may divide transversely to form more companion cells. The differentiating sieve tube member increases in length and its protoplast undergoes a profound change. In mature sieve tube mitochondria, plastids, P-protein, plasmalemma and some endoplasmic reticulum persist. The other cellular components degenerate during differentiation. The nucleus disappears or may persist as collapsed body. Nucleus does not degenerate in *Taxus*, *Neptunia oleracea* etc. Ribosomes and dictyosomes, though abundant in differentiating sieve tube member, are absent from them at later stage. At early stages of differentiation the endoplasmic reticulum (ER) is of rough type due to the presence of ribosomes. Later ER becomes smooth by loosing the attached ribosomes and aggregation in parallel stacks to form cisternae. The cisternae are arranged in parallel or perpendicular to long axis of the cell wall of the sieve tube member and occupy the parietal position. As development progresses ER diminishes in amount. Some changes also occur in mitochondria where the inner membrane disorganizes and so lacks cristae. The vacuolar membrane ruptures and as a result there lies no delimitation between cytoplasm and vacuole. After the disappearance of tonoplast the P-protein occupies the parietal position of sieve tube member and sieve plate pores.

Sieve areas originate at the sieve plate, which is the common transverse wall of the sieve tube elements. The future sieve plate, at early stages, is smooth with no sign of primary pit fields. Later there appear the plasmodesmata, which mark the site of future sieve area. A single plasmodesma occupies the future pore site. Callose deposits encircling the plasmodesma. Callose accumulates on both sides of the cell wall and assumes the shape of platelets. A pair of callose platelets, which is interrupted at the centre where plasmodesma exists, occupies the future pore site. In between the pair of platelets there occur the primary walls of the two adjacent cells and the middle lamella. Callose deposition continues, platelets increase in thickness, plasmodesmatal canal enlarges and a cavity is developed at the middle lamella region. At a later stage the two cell walls between the paired platelets disappear and thus a pore is formed. The end result is the full differentiation of sieve plate and the protoplast of the contiguous sieve tube elements become continuous.

❑ *Phylogeny of sieve tube*: Long sieve tubes are considered as primitive. The short one indicates advanced condition. The cambial initials, during evolutionary specialization, tend to become shortened and this caused the formation of short sieve tubes. It is frequently noted that cambial initial undergoes transverse septation before the differentiation of sieve tube. As a result short sieve tubes are formed. So, Carlquist opined that the studies on the length of sieve tube must note whether it was directly derived from the cambial initial or formed after the septation of initial.

The sieve tube that has a long end wall with numerous sieve areas is considered as primitive. Small pores in the sieve plate are regarded as primitive by Zahur (1959). Hemenway (1913) regards that the lateral sieve areas, during evolutionary specialization, become smaller and less conspicuous. The terminal sieve areas become more specialized and form the sieve plate. The sieve plate may be compound or simple and the latter is an advanced feature. The position of a sieve plate also changed from oblique to almost transverse during phylogenetic advance. So the primitive sieve tube has compound sieve plate with oblique position, whereas the advanced one has simple sieve plate with horizontal orientation. A significant statistical correlation was found between the compound sieve plate and plate with oblique orientation. A similar correlation was also found between simple plate and those that are horizontal in position.

Zahur noted that there exists a correlation between the length of a sieve tube and length of a sieve plate. Zahur grouped the sieve tubes into three categories based on the length of it, position of sieve plate and number of sieve areas present per sieve plate. These types were compared to other features in a statistical fashion to indicate phylogenetic advance. The three categories are (1) sieve tube long, sieve plate is oblique in position and the number of sieve areas is more than ten per plate, (2) sieve tube medium in length, sieve plate is oblique in position and the number of sieve areas per plate ranges between two to ten and (3) sieve tube is short, sieve plate is almost transverse to transverse in position and the number of sieve area is single per plate. Zahur concludes that in course of evolution (1) the length of sieve tube is decreased and this is due to either reduction in length of cambial initial or septation of the initial; (2) the length of end walls decreased with the reduction in number of sieve areas per sieve plate; (3) in the secondary phloem of dicots, the size of sieve areas increased and (4) in the functional sieve tube the presence or absence of nucleoli has no phylogenetic significance and this feature may be regarded as taxonomic criteria.

The sequence of specialization in the metaphloem of monocots seems to have occurred in the following way: roots to stems and terminating in rhizomes, corms and inflorescence axes. The least advanced sieve tubes are noted in roots and the most advanced forms are found in rhizomes, corms and inflorescence axes, whereas the intermediate forms are present in the aerial stems (Cheadle and Whitford, 1941).

• *Companion cell*: The companion cell and sieve tube originate from a common parent cell. It is present in the primary and secondary phloem and accordingly the parent cell originates either from procambium or cambium. The parent cell divides by unequal longitudinal division and the smaller cell develops into companion cell.

Companion cells remain associated with sieve tubes. Zahur, in the secondary phloem of dicots, distinguished three types of companion cell to indicate phylogenetic specialization. The basis of distinction is length and septation of companion cell. The followings are the three types: (1) companion cell is much shorter than the accompanying sieve tube; (2) companion cell is almost as long as the accompanying sieve tube and (3) companion cell is as long as the accompanying sieve tube, but the companion cell is septate. Zahur considers that the above three types form a natural group. Zahur regards that the first type, where the companion cell is much shorter than the accompanying sieve tube, is primitive. The other two types are advanced. The decrease in the number of companion cell indicates evolutionary advance (Cheadle and Esau, 1958). In course of evolution the length of sieve tube is shortened. This shortening may be correlated with the increasing contact between the companion cell and sieve tube (Carlquist).

• *Phloem parenchyma and fibre*: Procambium gives rise to phloem parenchyma of primary phloem. In secondary phloem the axial phloem parenchyma and phloem rays are developed from fusiform initial and ray initial of cambium respectively. Some parenchyma cells originate from a common mother cell of sieve elements. Fibres occur in both primary and secondary phloem and accordingly their origin differs. The primary phloem fibre originates from procambium where as the secondary phloem fibre originates from cambium. The fusiform initial of cambium gives rise to fibre, which composes the axial system of the organ in which it occurs. In many dicotyledonous stems the parenchyma of protophloem often differentiates into fibre in later stages of development, i.e. when the protophloem elements become functionless.

The phloem parenchyma and fibre of secondary phloem bear no phylogenetic trend in phloem evolution. The distribution and morphology of them may be of comparative value (Zahur).

Sieve elements are the most labile cells of a plant. The fossils do not provide any useful details of phloem structure though other tissues show excellent preservation. An idea of fossil phloem can be obtained from *Palmoxylon* (Bonde and Biradar, 1981) and *Rhynia* (Satterthwait and Schopf, 1972). Therefore the phylogeny of phloem is inferred from extant plants.

CHAPTER **11**

# Secretory Structures in Plants

The group of cells concerned in secretion of cutin, wax, suberin etc. is generally termed as secretory structures. The term secretion includes excretion and recretion also.

The term secretion implies the act of separation of by-products of metabolism from protoplast. These substances may be stored in insoluble forms within the cell or exuded from it, and have either a special physiological function or no use to the plant and regarded as waste. The removal of waste products of metabolism is defined as excretion. The process, which eliminates the salt from a cell and thus regulates the ion content of it, is defined as recretion (Fahn, 1979, p. 1). In the strict sense, the term secretion refers to those end products that take part in the metabolism process (ex. hormones and enzymes). The process of cell wall formation, cutinization, cuticularization, wax deposition, suberization etc. are also the examples of secretion. Examples of excretion are terpenes, saponins, rubber, tannins, crystals etc. Later findings reveal that the borderline between secretion and excretion is ill defined. As for example, the terpenes are regarded as excretions, though they play important roles in (i) attracting pollinators, (ii) repelling animal foes, (iii) blocking wounds in plant organs and (iv) having antiseptic properties etc. Therefore, the present day authors use the term secretion to denote excretion and recretion as well. Many of the end products have enormous commercial value, e.g. rubber, opium, gutta-percha etc.

The secretory structures vary greatly in structure and position. They may be either simple glandular trichomes or multicellular glands with vascular tissues. They may be external when originate from epidermis, or deep seated or internal such as laticifers and resin ducts. The various structures, both external and internal, involved in secretion are discussed below.

**External secretory structure:** These include glandular trichomes, nectary, osmophores, hydathodes and salt glands, of which hydathodes and salt glands are discussed in the chapter 12.

**Glandular trichomes:** These trichomes consist of a stalk with a head above. The stalk may be unicellular or multicellular and in the latter case the cells may be arranged in several rows. The head is the secretory part and may be composed of single cell (ex. *Pelargonium*) or many cells (e.g. *Callitriche*). The head is covered with a cuticle. The secretion is accumulated beneath the cuticle. This type is exampled by the volatile oils like camphors, balsams, peppermint oil, resins etc. Trichomes may also secrete nectar (ex. stipules of *Vicia sepium*) and water (ex. leaves of *Hygrophila*). Several of the secretions of the glandular trichomes have role in plant defense. Some are repellent to insects. The glandular hairs of tomato and wild potato (*Solanum berthaultii*) provide resistance to aphids. These trichomes secrete sticky exudates that trap the aphids. Plant breeders are now trying to raise a hybrid of cultivated potato that possesses these useful aphid-trapping glandular hairs. The glandular trichomes of insectivorous plants have specialized functions: (i) they secrete mucilage that is effective in trapping insects; (ii) they contain proteolytic enzymes; (iii) they have absorptive functions, i.e. the products of digestion move into the leaf through them.

**Nectary:** Nectary can be defined as a gland or part of a flower that secretes nectar to the exterior of plants. They are divided into floral and extrafloral nectaries. The former is situated within the flower and is directly involved in pollination; the latter occurs on the vegetative organs and is not directly associated with pollination. The nectaries are present on the epidermis. They may be deeply sunken, or raised above over an outgrowth, or at the level of the organ that bears them. The tissues that compose the nectary are commonly known as nectariferous tissues. These tissues may not form any anatomically differentiated structure – termed non-structural nectaries (e.g. on leaves of *Dracaena reflexa*, in bracts of *Sansevieria zeylanica*, on tepals of *Cattleya percivaliana* etc.). In some cases it is difficult to distinguish the non-structural nectaries from the surrounding tissues. The histochemical test reveals that the parenchyma cells composing the non-structural nectaries have a high acid phosphatase activity like the anatomically differentiated ones. The nectariferous tissues may form anatomically differentiated structure termed structural nectaries, which are macroscopically recognizable also. It consists of stalk cells and secretory cells with cuticle. The cells of the nectary are specialized parenchyma cells that are small, thin walled and contain dense cytoplasm, dictyosomes, endoplasmic reticulum, small vacuoles and large nuclei (Fig. 11.1). The nectaries may be provided with special vascular tissues. The nectar is supplied by phloem and it accumulates between the cuticle and secretory cells.

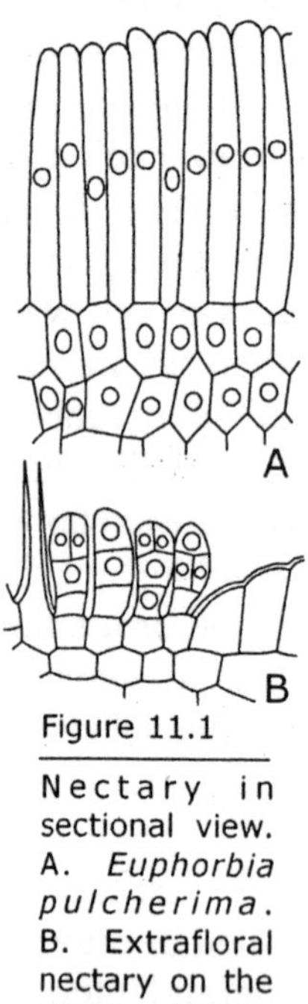


Figure 11.1

Nectary in sectional view. A. *Euphorbia pulcherima*. B. Extrafloral nectary on the stipule of *Vicia*.

The nectar of non-structural nectaries is exuded through stomata. Nectar, in structural nectaries, is secreted by epidermal cells or trichomes directly to the outside. The nectar contains glucose, sucrose and fructose as major components. The followings are also recorded in the nectar of different plant species: maltose, melobiose, mucilage, proteins, phosphates, mineral ions, organic acids, oxidases, sucrose, vitamins etc. The non-enzymatic proteins are detected in the floral nectary of *Erica, Bergenia* etc. The lipids are reported from *Jacaranda, Trichocereus* etc. The nectar also contains all the essential aminoacids, which with their smell attract some of the anthophilous insects. It is suggested that nectar provides the aminoacid requirements for the insects. The cells of nectaries may absorb the sugary fluid. So they possess both secretory and absorptive properties.

The extra floral nectaries are very common in dicotyledonous and much less in monocotyledonous plants. It rarely occurs in Poaceae (e.g. *Andropogon* and *Eragrostis*). In dicots it is present on all organs, namely—in the cotyledons (e.g. *Ricinus communis*), on leaf margins (e.g. *Rosa, Populus* etc.), on phyllodes (e.g. *Acacia longifolia*), on pulvinus (e.g. *Thunbergia grandiflora*), on stipules (ex. *Vicia fava)* etc.

**Osmophores:** Osmophores can be defined as certain special areas on floral organs, which differ in structure from the neighbouring cells and have the fragrance producing properties to attract pollinators.

The fragrance of a flower is due to volatile low molecular weight terpenes. The fragrant materials are not exuded by the glandular trichome. This substance occurs as droplets in the cytoplasm of epidermal cells. The oil droplets of terpenes, at an appropriate temperature diffuse out of the cell in gaseous form through the cell wall and cuticles. As a result the fragrance is produced. In many plants certain localized cells are only involved in producing the fragrant materials. The cells differ markedly from the other normal cells situated near by and are termed as osmophores. The osmophores secrete terpenes as the main fragrant materials. In some species belonging to Araceae the fragrant substance may contain amines and ammonia in addition to terpenes. The osmophores appear as flaps, cilia or brushes; they can be stained with neutral red and thus can be identified from the neighbouring cells. Osmophores are present in Asclepiadaceae, Araceae, Orchidaceae, Aristolochiaceae and Liliaceae. The osmophores of *Ceropegia* (Fig. 11.2) consists of an epidermal layer and two rows of isodiametric cells situated below the epidermis. The epidermis contains dense cytoplasm and the isodiametric cells are filled with starch grains. The starch disappears after the emission of fragrant materials.

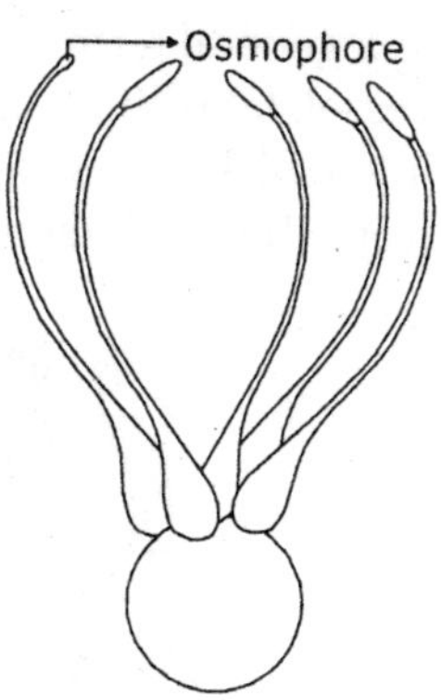


Figure 11.2

Osmophore in the flower of *Ceropegia loranthiflora* (After Fahn, 1979, p. 159).

**Internal secretory structure:** The internal secretory structures may be composed of a single cell or groups of cells. They may occur throughout a tissue (e.g. oils or enzymes) or may be localized in distribution. Example: castor oil obtained from the endosperm of *Ricinus;* ground nut oil extracted from the cotyledons of *Arachis;* the source of palm oil is the mesocarp of the fruit of *Elaeis guineensis;* the seed of *Carthamus tinctorius* yields safflower oil etc. Sometimes resin or oil—secreting idioblasts are formed, e.g. oleo-resin cells are present in the ground tissue of the rhizome of *Zingiber officinale*, the oil cells secreting the aromatic oil occur in the phloem of *Cinnamomum zeylanicum*.

**Glands and ducts:** They comprise a group of cells or sometimes a single cell that is readily distinguishable from the neighbouring cells and secretes a specific substance. These cells are thin walled with dense protoplasm and sometimes occur as layer surrounding a cavity, known as secretory cavity. The secretion is discharged and accumulates within the internal cavity. These cavities may be more or less spherical or much elongated like tubes and respectively termed as glands or ducts. The internal cavities originate by three ways and accordingly the following three types of glands are recognized: (i) Schizogenous glands: these are formed by the dissolution of middle lamella, thus separating apart the cells to form cavity. Example: oil glands of *Eucalyptus*, the secretory ducts of *Rhus glabra*, resin duct of *Pinus* etc. This cavity remains surrounded by a ring of intact parenchyma cells, termed epithelium, which forms a well-defined boundary of the gland. (ii) Lysigenous glands: these glands originate by lysis of a few cells thus forming the cavity (ex. glands present in the leaves and fruits of *Citrus* sp., that are also formed schizogenously) and (iii) Schizolysigenous glands: these glands arise by both phenomenon schizogeny and lysis, i.e. the middle lamella and some of the adjoining cells disintegrate to form cavity. Example: *Eugenia caryophyllata* where the glands are present in the floral parts that are the source of oil of clove. In contrast to schizogenous gland the lysigenous – and schizolysigenous gland have no clear-cut boundary.

**Laticifers** (Fig. 11.3): Laticifers can be defined as a specialized cell or a row of such cells that secrete the milky fluid termed latex. The word laticifer is used as a general term to denote the various latex-secreting structures – latex cell, latex vessel, latex duct, latex tube and laticiferous duct. The laticiferous duct is a cavity into which latex is secreted. The latex cell may be simple or branched and is derived from the enlargement of a single cell. The latex ducts are elongated, branched

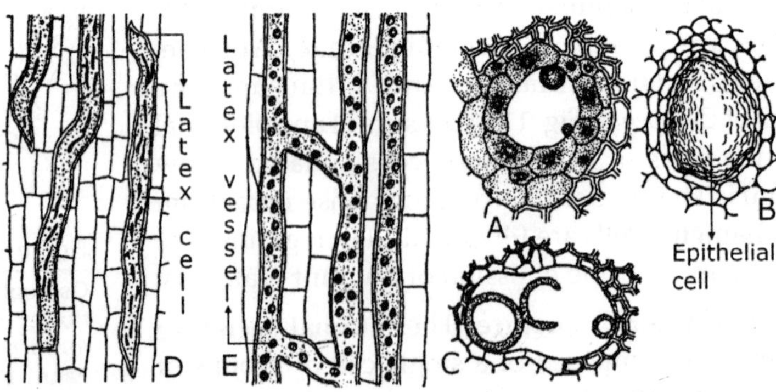


Figure 11.3

Diagrams showing D. latex cell of *Euphorbia pilulifera* (non-articulate), E. latex vessel of *Carica papaya* (articulate) and intercellular spaces : A. Resin duct of *Pinus* (schizogenous), B. oil-cavity of *Citrus* fruit (Lysigenous) and C. Protoxylem lacuna of maize (schizo-lysigenous).

and aseptate. The latex vessel is simple or branched tube that usually anastomoses with similar tubes; it is formed as a result of enlargement and union of chain of cells. The latex tube usually means either a latex cell or the latex vessel. The borderline between the different terms of latex secreting structures (i.e. latex cell, -duct, -vessel, -tube etc.) is ill defined and so the word laticifer is introduced as a general term. The laticifers may be simple or compound on the basis of origin. The simple laticifer is derived from a single cell whereas the compound laticifer originates from a longitudinal file of cells.

Difference between latex tube and latex vessel.

| Latex tube | Latex vessel |
|---|---|
| 1. It is also known as non-articulate laticifers. | 1. It is also known as articulate laticifers. |
| 2. It originates from a single cell. | 2. It is compound in origin. |
| 3. It is capable of unlimited growth. | 3. It has limited growth. |
| 4. They never anastomose with each other. | 4. They frequently anastomose with each other. |
| 5. The end walls of the tube remain intact. | 5. The end walls are usually broken, and the lateral walls are perforated and thus a continuous vessel is formed. |
| 6. The enzyme cellulase is absent. | 6. The enzyme cellulase is present. |

Latex is fluid produced in the latex vessels or cells. It is usually white and milky (ex. *Euphorbia, Asclepias, Lactuca* etc.), yellow and brown (e.g. *Cannabis*), orange and sometimes colourless and clear (e.g. *Morus, Nerium* etc.). It contains a number of substances including sugars, proteins, alkaloids, oils, mineral salts, organic acids, terpenes, resins, rubber etc. The latex of *Euphorbia milii* contains starch grains that are dumb-bell shaped. The proteolytic enzyme papain is present in the latex of *Carica papaya*. The latex of *Asclepias syriaca* contains the enzyme pectinase. The latex of some *Euphorbia* species is rich in vitamin $B_1$.

The cell wall of laticifers is thick and may be thicker than the adjacent cells. They are not lignified. The growth of cell wall occurs through apposition process. The tip of latex cell is thin walled. The walls are composed of cellulose, hemicellulose and pectin.

The laticifers may occur throughout the plant body or their distribution is restricted to certain tissues. Laticifers are grouped into two: non-articulate and articulate. The former is derived from the enlargement of a single cell. This cell has the potentiality of unlimited and rapid growth, and elongates to form long latex tubes. The tubes may remain unbranched termed non-articulate unbranched laticifer (e.g. *Vinca, Cannabis, Urtica* etc.). In some plants (e.g. *Euphorbia, Nerium* etc.) the tubes may branch called non-articulate branched laticifers that seldom anastomose with similar types. The non-articulate laticifers are coenocytic and multinucleated, and also termed as laticiferous cell. There is continuity of laticifers between the shoots and branches. Laticifers grow through the intercellular spaces and the enzyme pectinase helps in the process. Pectinase is secreted by the growing tip of laticifers and dissolves the middle lamella.

The articulate laticifers, also termed laticiferous vessel, consist of longitudinal files of cells. The transverse end walls of the individual cell either remain intact or break down partly or wholly to form a continuous tube – the latex vessel. So the articulated laticifers are always compound in origin. They occur in primary or secondary phloem and may be present in cortical parenchyma. They may remain either as a single chain of cells without any anastomosis termed articulated non-anastomosing laticifer (e.g. *Convolvulus, Allium, Musa* etc.), or they may anastomose with similar file of cells to form a complex anastomose system called articulated anastomosing laticifer (e.g. *Lactuca, Papaver, Carica papaya* etc.). The enzyme cellulase is found in the latex of articulated laticifers and it is absent from non-articulated laticifers. So it is suggested that cellulase is involved in the lysis of common transverse walls during the formation of articulated laticifers.

Latex occurs in 900 genera distributed in 20 families (Fahn, 1979) of mostly in dicotyledons ( e.g. Apocynaceae, Asclepiadaceae, Compositae, Euphorbiaceae, Papaveraceae etc.) and in a few families of monocotyledons (e.g. Araceae, Musaceae and Liliaceae). They are very much economically important, a few of which are mentioned below: The opium, a medicinally important alkaloid is obtained from *Papaver somniferum*. The most important latex is rubber whose principal source is *Hevea brasiliensis*. The species of *Palaquium gutta* yields gutta-percha. The latex of *Achras sapota* yields chicle, from which chewing gum is made.

• Function of laticifers: Regarding the function of laticifers different views were expressed since 1877, a brief account of which is mentioned below.

(1) It is a vital sap vessel and similar to the blood vessels of animals; (2) it takes part in translocation of assimilates as it is associated with phloem; (3) it stores food materials; (4) it is now considered as secretory tissues where the secretory substances do not re-enter the plant metabolism; (5) Sen and Chawan (1972) suggested that laticifiers regulate the water balance in plants; (6) it has role in the transport of oxygen; (7) it has role in healing up of wounds; (8) it acts as a defense against herbivores and microorganisms. [The human ocular tissues are damaged by the latex of *Calotropis procera*]; (9) McCay and Mahlberg (1973) reported the absence of bacterial activity from laticifers of *Asclepias* in vivo.

**Myrosin cell:** It is a special type of secretory cell where the enzyme is produced and stored. Myrosin cells are idioblasts, which contain the enzyme myrosinase (β-thioglucosidase). They are reported from Cruciferae, Capparidaceae, Moringaceae, Tropaeolaceae etc. They are more or less similar in shape to the neighbouring cells and may be elongated or assume various shapes. They occur in the integument of seeds, in the exocarp and endocarp of fruits, in the cotyledons, in the cortex of stem, in the leaf lamina, spongy parenchyma etc.

The enzyme myrosinase hydrolyses thioglucosides to produce mustard oil and other substances. The enzyme and substrate occur in different cells, i.e. the enzyme is produced in myrosin cells and thioglucosides are present in the other parenchyma cells. The reaction between the enzyme and substrate is brought about only when the tissues are damaged and thus mustard oil is released. When insects feed on such plants mustard oil is released in the insect gut. It is demonstrated that mustard oil is toxic to a number of insects.

Chapter **12**

# Epidermis

The outermost layer or layers of cell covering all plant organs are the epidermis. It is in direct contact with the environment and so it modifies itself to cope up with the natural surroundings. It thus protects the inner tissues from any adverse natural calamities like high temperature, desiccation, mechanical injury, external infection etc. In some plants the epidermis may persist throughout the life, while in others it is replaced by periderm when the epidermis is sloughed off along with underlying tissues.

**Origin:** The epidermis of all organs originates from the outermost layer of apical meristem. Haberlandt, Hanstein and Schmidt called this surface layer of meristem as protoderm, dermatogen and tunica respectively. In cryptogams epidermis originates from a single initial cell that also forms cortex and stele. The epidermis of gymnospermous root originates, in association with root cap, from periblem. In dicotyledonous root, epidermis develops from initials of dermatogen, which are not distinct from those of root cap. The epidermis of monocotyledonous root owes its origin from the periblem along with the cortex.

**Structure and contents:** Usually the epidermis consists of one layer of cells. Several-layered epidermis, termed multiple epidermis, is found in the leaves of *Ficus*, *Nerium* and in the aerial roots of orchid. The initials of epidermis divide periclinally to form multiple epidermis. The multiple epidermis of orchid root has the special name—velamen.

The epidermis of aerial parts of a plant consists of living parenchyma cells whose shape, size and arrangement may differ. The epidermal cells are more or less tabular (=horizontally flattened) in cross sectional view. In leaves, the epidermal cell walls appear as sinuous in dicots and in monocots they appear as straight or sinuous in surface view. Usually the cells of epidermis are compactly set with none or few intercellular spaces (e.g. flower petals).

The epidermal cells are devoid of chloroplasts. The guard cells of stomata that are specialized epidermal cells contain chloroplastids. Other pigment like anthocyanin may occur in epidermal cells. In some plants silicon may be deposited in the epidermal cells either in the lumen or wall. The wall of trichome may be silicified. Silicon containing cell can be differentiated from the adjacent epidermal cells by its shape and size. This cell is solitary and may be either scattered over the leaf surface or situated over the veins in longitudinal rows. Silicon is deposited in the bracts of rice, in the marginal trichomes of oat, in the leaves of *Cyperus*, *Avena* etc. In the internode of *Avena sativa*, the epidermal cells at the intercostal position form cork-silica cell pairs, i.e. cork and silica containing cells are in close contact with each other. In the leaf of *Ficus*, some of the epidermal cells contain crystals of calcium carbonate, known as cystolith. These cells are easily distinguishable from the other epidermal cells by their large size and these specialized epidermal cells containing cystolith are called lithocyst.

Usually the walls of epidermal cells are thin. Thick walled lignified epidermal cells occur in some gymnosperms. Cutin, a fatty substance, is very often deposited on the outer surface of the epidermal cell wall to form cuticle over which wax may also be deposited. The cuticle is resistant to decay and is well preserved in fossils. The cuticle often preserves the characteristic features of the epidermal surfaces such as the types and distribution of hairs and stomata. Thus the fossil plants may be identified by cuticular studies. Palmer (1976) used the fossil grass cuticle instead of grass pollen, as a new palaeoecological tool to reconstruct the nature of past vegetation of East African lake sediments. Now a days cuticular pattern is used in recognizing small fragments of plants, which are necessary in forensic medicine, animal nutrition, pharmacognosy etc. Cuticular pattern is also taxonomically useful to characterize genus and species. The cuticle is impervious to water but in grapes water diffuses out when it is transformed to sultana. It has protective function. Cutin is resistance to microorganisms and prevents the entry of the pathogen.

In many plants wax is deposited on the surface of the cuticle. This forms a powdery coating on various fruits, e.g. plum, grapes etc. and on leaves. This gives a glossy appearance to the surface of leaves and fruits (e.g. grapes). Wax is deposited either in the form of granules, rods or tubes, which form various specific patterns on the surface. The deposition may also be in the form of projections and folds. Wax is also deposited on the inside of the pitcher of *Nepenthes* in the form of overlapping scales. The scales adhere to the feet of insects, which fall inside the pitcher. So the insects cannot climb out of the pitcher. The morphological form of the deposition of wax is typical for the species. With the aid of scanning electron microscope the wax and cuticular pattern can be observed directly. The study of wax pattern on the epidermal surface is extremely useful in agricultural practices. Waxy epidermis is not wetted. So the effectiveness of fungicide and herbicide can be obtained by studying the extent of wax deposition. Wax obtained from the wax palm *Copernicia cerifera* is commercially used in making polishes and phonographs records.

**Hairs:** Hairs are present all over the surfaces of plant organs namely – roots, stems, leaves, floral parts, seeds (e.g. *Gossypium*) and stamens (ex. *Tradescantia*).

**Root hair:** They are present at a short distance behind the root tip of most monocots and dicots. Root hairs in some species are formed from distinct epidermal cells termed trichoblast. Trichoblasts may be morphologically similar to other epidermal cells or they are distinguishable by their smaller size with dense cytoplasm. Trichoblast prolongs to form unicellular root hairs. The cell wall is generally thin and is covered by a thin layer of cuticle; mucilage may also occur on the surface (Fig. 12.1).

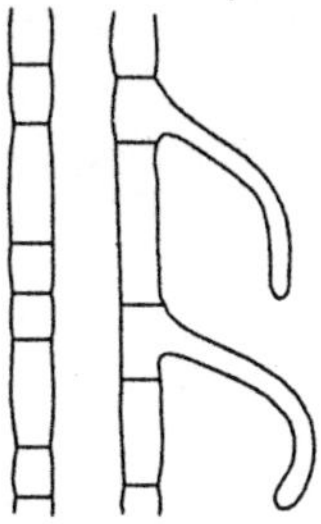

Figure 12.1

Root-hairs from short cells in *Cyperus*.

Apart from anchorage the other main function of root hairs is absorption. It has been found that the rate of absorption in the epidermal surface with and without root hairs is same. Moreover the short root hairs are more efficient in absorption than longer ones (Rosene, 1954). So it appears that the longer hairs absorb the distantly situated water. The root hairs synthesize cellulose at their tips.

**Hairs other than root hairs:** Hairs or trichomes are the outgrowths of epidermal cells. They are either unicellular or multicellular. Multicellular hairs may be composed of a single linear row of cells or several rows. Trichomes, either unicellular or multicellular, are classified into glandular and non-glandular hairs. The former is secretory in function and the latter is the covering hair and does not secrete. It is believed that non-glandular trichomes are protective in function and may prevent undue water loss (Fig. 12.2). The covering trichomes may have a star like appearance (stellate hair) or a miniature tree (dendroid hair, e.g. *Verbascum*). In *Hamamelis* the covering hairs occur in tufts. Hair like projections are present in may flower petals termed papillae. The glandular hairs consist of a stalk and a head that may be unicellular or multicellular. A cuticle like structure covers the head. The secretory substances accumulate in the sac formed between the cuticle and the cell of head. Example: Oils, resins, camphor, peppermint (e.g. *Mentha*) etc. The leaves of *Olea* have scales, which are composed of a short stalk and head, consisting of discoid plate of cells. Some of the epidermal cells of *Mesembryanthemum* may enlarge where water accumulates. These specialized epidermal cells are termed as vesicles or bladders. Most trichomes have thin and cellulosic cell wall; lignified cell walls also occur (ex. seed coat of *Strychnos nux-vomica*).

Trichomes with their different types may be of taxonomic significance (Carlquist, 1961). The types of trichome can identify the several species of Oleaceae and *Rhododendron* to some extent.

**Bulliform cell:** It is a group of outer epidermal cells that can be easily distinguished from the typical epidermal cells by their fan like appearance in cross section and larger sizes. The median cell of bulliform cell is the tallest and the size of the other cells, present on the two sides, diminishes gradually. The cells are thin walled, hyaline and have large vacuole. The cells contain much water and are devoid of chloroplastids. The cell walls are composed of cellulose and pectic substances; cutin occurs on the outermost wall that is covered by cuticle. They usually form isolated strips that are situated on parallel between the veins. They are present on the outer epidermis of the leaves of Poaceae and other monocotyledons except Helobiae (Fig. 12.3).

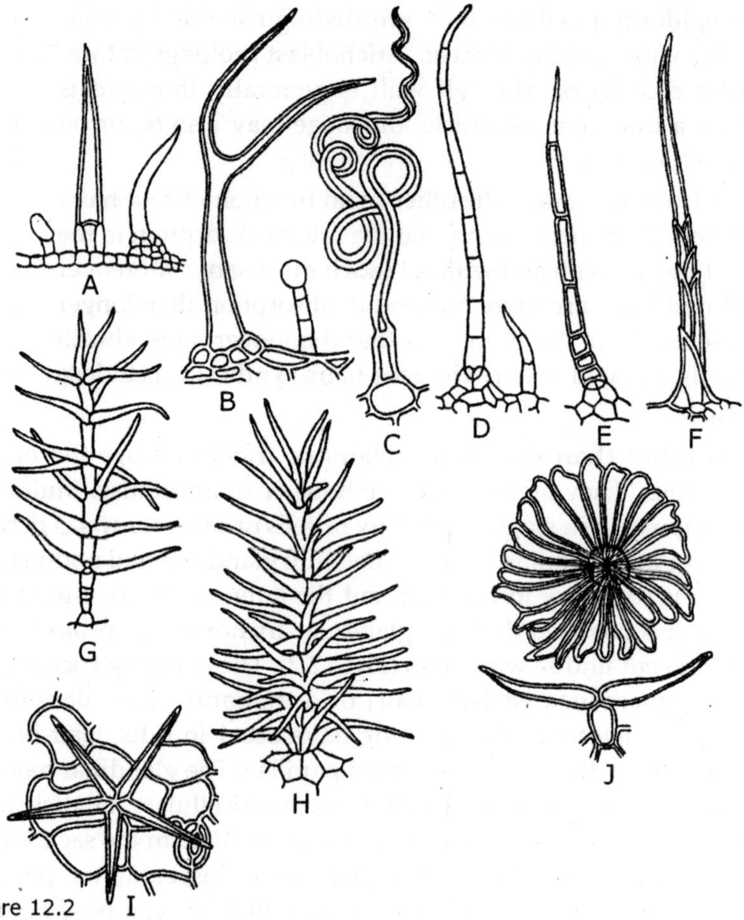


Figure 12.2

Epidermal hairs - different types. A. Sharp hairs of *Lantana*. B. Lobed hair of *Amaranthus*. C. Wooly hair of *Banksia*. D. *Lycopersicum*. E. *Helianthus*. F. *Mimosa*. G. Dendroid hair of *Platanus*. H. Dendroid hair of *Mimosa*. I. Stellate hair of *Althaea*. J. Peltate hair in surface and side view of *Olea*.

Opinion varies regarding the functions of bulliform cell, which are: (i) the rolled leaf in bud unrolls with the help of bulliform cell; (ii) the turgidity and flaccidity of bulliform cell due to water uptake and loss respectively cause the closing and opening of mature leaves; (iii) these cells act as water reservoir and (iv) these cells are often filled with silica and their outer walls become thick and cuticularized thus providing mechanical rigidity to leaves.

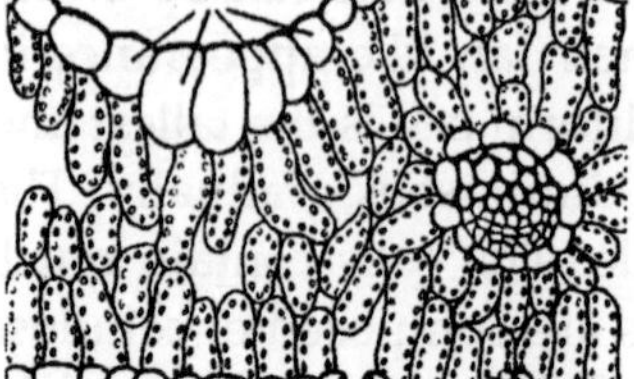


Figure 12.3

Bulliform cells in the section of *Triticum* leaf.

**Stomata:** Stoma (pl. stomata, physiologists usually call stomate) that occurs predominantly on leaves and young stems can be defined as a pore enclosed by two specialized cells—the guard cells that move to open and close the pore, and thus control gaseous exchange during transpiration, respiration and photosynthesis.

In Greek the word stoma means mouth. 'Link and de Candole in 1827 jointly claimed to be the first to have called the pores by that name'—Willmer, 1983, p. 1. Stoma is reported as early as lower Devonian period (about 390 million years ago) from the extinct genera *Rhynia, Asteroxylon, Zosterophyllum* and *Drepanophycus.*

## Occurrence and position of stoma

Stoma occurs in the epidermises of any of the aerial parts of plants. It occurs in Bryophytes, Pteridophytes and Spermatophytes. In Bryophyta stoma is found on the capsules of Anthocerotopsida and Bryopsida. In Marchantiopsida stomata are absent; instead pores are present on the surface of thalli. The pores remain open permanently.

The phylum Pteridophyta has stomata in all of its four divisions. Large stoma occurs in Psilopsida, Lycopsida and Pteropsida. Psilopsida (ex. *Psilotum*) and Sphenopsida (ex. *Equisetum*) possess stomata that are confined to longitudinal grooves of the aerial stem ridges. In Lycopsida stomata are present on leaves and stems. In *Equisetum* each scale leaf has two rows of stomata. In Pteropsida stomata are present in leaves, predominantly on abaxial surface. In the phylum Spermatophyta the epidermal layers of most aerial parts have stomata. They are found in sepals, petals, stamens and carpel of flowers. Awn, palea and lemma of cereal inflorescences also possess stomata. Stomata are also present in the inner and outer surfaces of fruits, in the surface of seeds, inside pods and in the skins of banana. The presence of stoma is also recorded from rhizomes. The chlorophyll-less parasitic plants like *Monotropa* and *Neottia* lack stomata. Another chlorophyll-

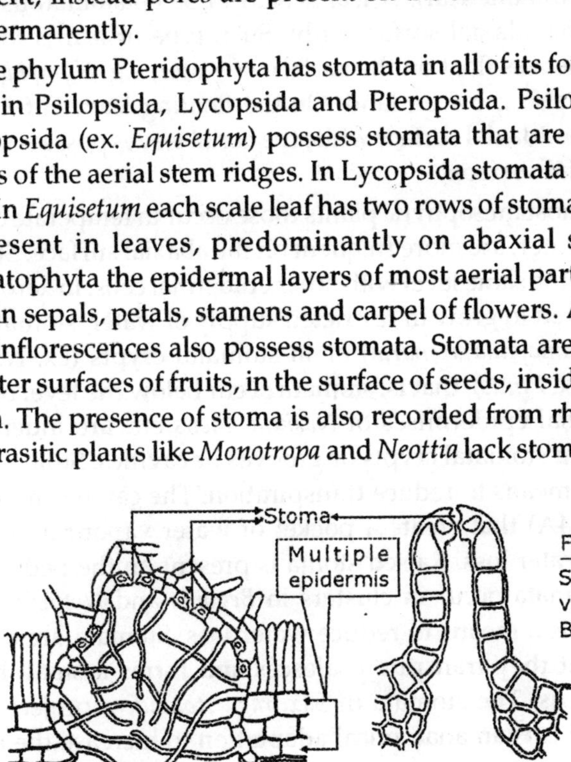


A. *Nerium* leaf B. *Cucurbita* peduncle

Figure 12.4
Stomata in sectional view. A. Sunken stoma. B. Raised stoma.

free plant *Orobanche* possesses stomata on the stems. Plants belonging to Balanophoraceae, Monotropaceae and Rafflesiaceae are chlorophyll-free and there are reports that these plants lack stomata. The submerged aquatic plants do not possess stomata. Usually stomata are absent in roots. But in seedling roots of *Ceratonia siliqua* and *Pisum arvense* stoma is observed (Fahn, 1997, p. 164). Non-functional stoma is found in albino plants, petals, staminal filaments (ex. *Colchicum*), carpel and seeds.

Stoma predominantly occurs in young stems and leaves. Stoma is found in both adaxial (upper) and abaxial (lower) surfaces of a leaf. Stomata may be present on both surfaces, on abaxial surface and on adaxial surface only of a leaf. Accordingly the following terminologies are proposed to describe a leaf on the basis of surface(s) where stoma occurs.

- *Amphistomatous* (= amphistomatic leaf): A leaf is said to be amphistomatous where stomata are present on both adaxial and abaxial surfaces. The abaxial surface contains more number of stoma in comparison to adaxial surface. Most plants of herbaceous dicotyledons possess such leaves. Ex. *Basella*.
- *Hypostomatous* (= hypostomatic leaf): A leaf is said to be hypostomatous where stomata are present on abaxial surface only. Most tree species of dicotyledons possess such leaves. Ex. *Ficus, Nerium* etc.
- *Epistomatous* (= epistomatic leaf): A leaf is said to be epistomatous where stomata are present on the adaxial surface only. Such type of leaf is found in aquatic plants where the leaves float on the surface of water, e.g. *Nymphaea*. The aquatic fern *Marsilea* can also grow on land. On the floating leaves of *Marsilea* stomata are restricted to the adaxial surface while the aerial leaves possess stomata on both adaxial and abaxial surfaces.

Stomata, in leaves of most mesophytic plants those occur in temperate climate with adequate supply of water, are more frequent on the abaxial surface. Stomata are situated more or less at the same level with other epidermal cells. In xerophytes where the plants are adapted to grow in restricted supply of water, stomata may be located in pits (ex. *Ammophila arenaria*) or in stomatal crypts (ex. *Nerium*). Stomata may be sunken (ex. *Pinus*), that is, stomata occur below the level of other epidermal cells. The abaxial epidermises of *Nerium, Ficus* etc. are indented at several places thus forming stomatal crypts or grooves or cavities. Stoma occurs in these cavities. This is a means to reduce transpiration. The cavities may have hairs (ex. *Nerium,* Fig. 12.4A) that create a pocket of water vapour and this is means to reduce further water loss. Raised stoma is present on the peduncle of *Cucurbita* (Fig. 12.4B). Stomata occur in clusters in *Begonia* and *Nothofagus* etc. The clustering of stomata is a means to reduce water loss. Stomata are close to each other. The water that they transpire coalesces and forms a small humid zone that reduces water loss. The stomata of *Saxifraga stolonifera* project above the level of leaf surface. This is an anatomical adaptation to increase the rate of transpiration.

## Distribution of stomata on leaves

Stomata most frequently occur on regions over the chlorenchymatous mesophyll cells of leaf. Stomata do not occur over the fibre bundles of veins. They also do not occur over the sclerenchyma present on the leaf margins.

In majority of monocotyledons where the veins are parallel, e.g. Gramineae, the stomata are arranged in rows parallel to the long axis of leaf. In dicotyledons where the venation is reticulate there is no particular order of arrangement of stomata. Here the stomata are irregularly distributed over the leaf surface, but they are roughly equidistant apart from each other. This is due to the fact that the stomata are mutually inhibitory, that is, existing stoma inhibits the formation of new one. It is regarded that there occurs an inhibitory zone around a developing stoma that prevents to develop another stoma. It is suggested that there may be hormonal control behind this phenomenon. It is observed that in mustard cotyledon stomata do not fully develop in dark. But when the mustard cotyledons are exposed to far-red illumination fully developed stomata result. So it is thought that the hormone phytochrome is involved in the development of stomata.

There are reports of unusual distribution stomata over leaf surface. In *Begonia semperflorens* stomata occur in clusters of two to four each. In *Nothofagus nitida* the clusters of stomata are much larger than *Begonia*. In *Saxifraga stolonifera* (formerly *sarmentosa*) stomata are located on raised patches. Unusual pattern of distribution is also observed in the different species of *Saxifraga*. In certain species stomata occur near leaf margin while in other species stomata only occur near leaf tip. In *Daphne petraea* (Mauseth, 1988) stomata occur only in two bands immediately adjacent to the midrib. *Mimosa cruenta* has stomata distributed throughout the upper surface. On the lower surface stomata are restricted to just one of the two longitudinal halves of the leaflet (Mauseth).

## Stomatal frequency and stomatal index

Stomatal frequency can be defined as the number of stomata present per unit area of a leaf. Determination of stomatal frequency and the total area of stomata covered in a leaf are the essential prerequisite to assess the rate of water loss through stomata. Several environmental and genetical factors affect stomatal frequency. The following environmental factors change the stomatal frequency— water availability, carbon dioxide concentration, temperature and light intensity. Water stress results in a greater stomatal frequency. Plants growing in wet soil with high humidity have lower frequency than the plants growing in dry soil with low humidity. Stomatal frequency is reduced in polluted atmosphere, e.g. *Trifolium partense* (Sharma *et al.*, 1975). Light intensity has marked effect on stomatal frequency. In low light intensity frequency reduces. Stomatal frequency becomes higher when a plant grows in full sunlight. In an experiment it is observed that tomato plants grown in controlled condition have hypostomatic leaf and when grown at high light intensity the leaves are amphistomatous. The degree of ploidy also affects stomatal frequency. Polyploid plants have less stomatal frequency and larger stomata. Generally stomatal frequency decreases where the stomata are large.

Stomatal frequency is not constant within a plant. Highest frequencies are often found in those leaves that occur on the top. Even within a leaf there exists variation in frequency. As for example the grass leaf has lowest frequency at the tip. The highest frequency is observed at the point of insertion of the leaf where the cells are still developing and smallest (Willmer, p. 12).

Because the stomatal frequencies often vary Salisbury (1928) proposed the term 'stomatal index'. Stomatal index relates the number of stomata to number of epidermal cells. Stomatal index is calculated in the following way:

$$\text{Stomatal index} = \frac{\text{Number of stomata present per unit area of leaf}}{\text{No. of stomata + no. of epidermal cells within a unit area of leaf}} \times 100$$

This value is found to be reasonably constant for any particular species.

## Structure of stoma

A stoma consists of two guard cells that surround an aperture. In the extinct Devonian Pteridophyte *Zosterophyllum myretonianum* and *Z. ilanoveranum* (Meyen, 1987, p. 268) had one guard cell with an aperture in its centre that simulates the stoma of moss sporophyte. In the extant Pteridophyte *Azolla pinnata* the stoma consists of a unicelled guard cell with a pore. The guard cell is binucleate and there is no wall separating the nuclei (Sen and De, 1992b, p. 53). The aperture leads to a large intercellular space below. This space is termed as sub-stomatal chamber and it is continuous with the intercellular spaces of the mesophyll. Differential expansion between guard cell mother cell and the developing underlying mesophyll cells result in the formation of sub-stomatal cavity or chamber. There are certain epidermal cells that adjoin the guard cells and differ from the rest of the epidermal cells in size, arrangement and content. Such cells are termed as subsidiary cells. Conventionally a stoma means a pair of guard cells and the pore. A stoma along with neighbouring subsidiary cells is referred to as stomatal complex/stomatal apparatus (Fig. 12.21). The term stomatal pore indicates the intercellular space only (Mauseth).

• *Guard cell*: There are two basic shapes of guard cells. One is crescent-shaped with blunt ends as seen in surface view; a pair of such cells is attached to each other by their concave ends thus enclosing a pore. A pair of guard cells and the pore together constitute an elliptical stoma that appears in surface view. Such type of stoma is the characteristic of most dicotyledons (Fig. 12.5 B). The other is dumb-bell shaped. The two ends of such cells are bulbous and they are continuous

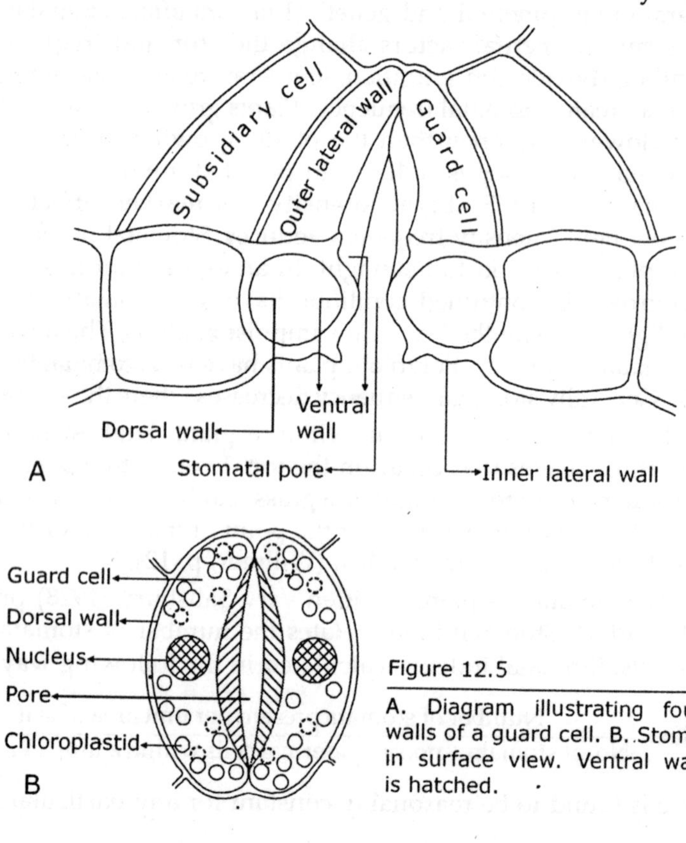


Figure 12.5

A. Diagram illustrating four walls of a guard cell. B. Stoma in surface view. Ventral wall is hatched.

with each other through a narrow middle section. Two such guard cells are attached to each other at the bulbous ends. An aperture is formed at the narrow middle portion of the two dumb-bell-shaped guard cells (Fig. 12.9 Gramirieous). Such type of stoma is the characteristic of Gramineae and also reported to be present in Cyperaceae, Flagellariaceae, Marantaceae, Anarthriaceae and Rapateaceae etc.

There exist considerable size variations in a guard cell. Depending on the species the length of a guard cell ranges from 10 μ to 80 μ. The width may vary from a few microns to 50 microns. The variations of width are dependent on the condition of a stomatal aperture.

Conventionally, anatomists call the four sides of a guard cell (Fig. 12.5A) as ventral wall (= wall that occurs towards pore), dorsal wall (= wall that occurs opposite to ventral wall and juxtaposed to neighbouring epidermal cells), outer lateral wall (= wall that occurs parallel to leaf surface toward the atmosphere) and inner lateral wall (=wall that occurs toward the sub-stomatal chamber).

There exist variations in the guard-cell-wall thickenings. In kidney-shaped guard cells (=also called reniform guard cells) the dorsal wall is thin. The ventral wall is heavily thickened and may be ornamented. Localized ledges or projections of thick wall or cuticle often occur on the upper and lower edges of ventral wall of guard cells. The ledges appear as horns in cross-sectional view. Ledges may be absent, occur only on upper side or occur both on upper and lower sides. When present on both sides, the upper ledge delimits the front cavity above the stomatal pore; the lower ledge is situated below the back cavity of stomatal pore (Fig. 12.6) and above the sub-stomatal chamber. In dumb-bell-shaped guard cells the bulbous ends are thin walled. The narrow middle section that connects the bulbous

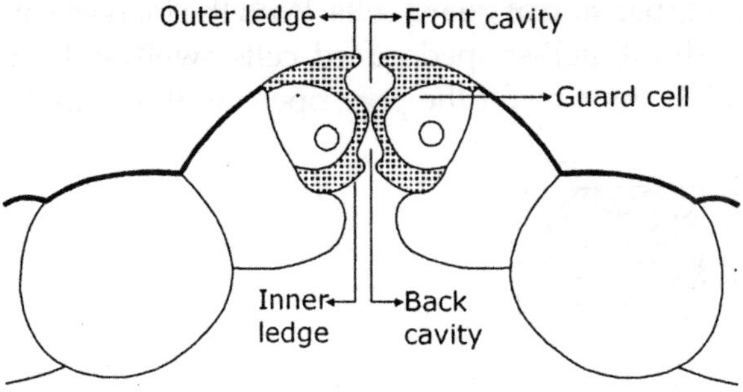


Figure 12.6

Diagram illustrating ledge of a stoma. Ex. *Pastinaca*. Guard cells are in cross-sectional view. Adapted after Esau, 1965.

ends is thick walled. The middle portion has very thick upper and lower lateral walls and thin ventral and dorsal walls. Electron microscopic study indicates that the bulbous ends of guard cells of *Zea mays* have thick walls (Srivastava *et al.*, 1972). The guard cells are attached to subsidiary cells or other epidermal cells by their dorsal wall only.

Guard-cell walls are rich in pectins that impregnate the cellulose microfibrils. Electron microscopic study reveals that in kidney-shaped guard cells the cellulose microfibrils/micellae have radial orientations, i.e. microfibrils radiate out from the pore around the guard cell (Fig. 12.7). In dumb-bell-shaped guard cells the

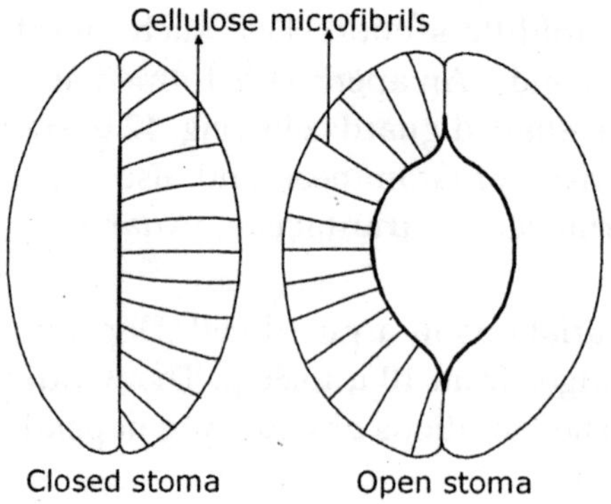


Figure 12.7

Diagrammatic representation of radial direction of cellulose microfibrils in one guard cell of each open and closed stomata. Adapted after Willmer, 1983, p. 59.

microfibrils radiate out from the pore and they exhibit axial arrangements. The guard-cell walls contain lignin, e.g. many vascular cryptogams, gymnosperms (ex. *Abies pinsapo* and *Juniperus chinensis*) and some angiosperms (Fahn, 1997, p. 166). In the guard cell wall of *Ophioglossum* β-1,3 glucan, a callose, has been detected next to plasmalemma (Willmer, p. 25). Silicon, in the form of silicon dioxide is located below the cuticle of guard cells of sugarcane and *Equisetum*.

It is thought that the thick-walled region of guard cells and the radial micellation (= radial orientation of microfibrils) play a major role in the mechanism of opening and closing of pore. Stomatal movement is attributed to changes in turgor of guard cells. With the increase of turgor the thin dorsal wall expands more than the thick ventral wall. As a result the shape of guard cells is deformed in such a way that the pore of stoma opens. There exists a relationship between potassium accumulation in guard cell and opening of stoma. Guard cells have greater concentration of potassium than the neighbouring cells. So it is regarded that greater concentration of potassium in the guard cells leads to the opening of stoma and loss of potassium brings about closure of stoma. The radial micellation of guard-cell walls allows the expansion of guard cells. With the increase of turgor the thin-walled ends of the dumb-bell-shaped guard cells swell and push the middle portion apart (Fig. 12.8). As a result the pore opens with parallel sides.

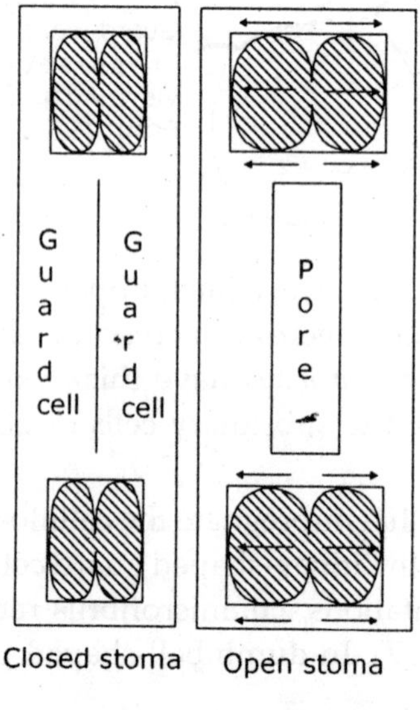


Figure 12.8

Diagram illustrating turgid bulbous ends of guard cells of dumb-bell-shaped stoma pushing guard cells apart in the direction of arrows. As a result stoma opens. Adapted after Willmer, 1983, p. 61.

All the usual cell organelles are present in the guard cells, e.g. mitochondria, microbodies, peroxisomes, spherosomes (ex. *Campanula persicifolia*), microtubules, rough endoplasmic reticulum, vesicles, plastid and nucleus etc., among which the last two are in brief discussed below.

❑ Plastids: All functional stomata have chloroplasts in the guard cells. There is report that the functional guard cells lack chloroplasts, e.g. *Paphiopedilum* and *Pelargonium zonale.*

The number of chloroplasts varies depending on the species. *Anthoceros* has two chloroplasts per guard cell, *Selaginella* has three to six chloroplasts per guard cell and *Polypodium vulgare* exhibits as many as 100 chloroplasts per guard cell.

The size of chloroplast varies. In some species of *Allium* it is too small to be observed under light microscope. It is large in ferns. In contrast to the mesophyll chloroplasts, guard-cell-chloroplasts are poorly developed. Guard-cell-chloroplasts have little thylakoid structure and scant granal stacking (Willmer, 1983, p. 37).

A common feature of guard cell chloroplasts is the abundance of starch. Exceptions are noted in the families Iridaceae, Amaryllidaceae and Liliaceae.

❑ Nucleus: The kidney-shaped-guard cells have prominent nucleus similar to other cells of leaf. The nucleus is usually centrally located close to ventral wall of guard cell. The nucleus exhibits change in shape during stomatal closure and opening. In *Vicia faba* the nucleus is oval when stomata are closed and it becomes rounded when stomata are open.

In dumb-bell-shaped guard cells a mass of nuclear material is present instead of nucleus. The mass of nuclear material is located at each bulbous end of a guard cell. A strand of nucleoplasm that occurs at the narrow middle portion of guard cell connects the masses of nuclear materials present at the two bulbous ends.

Two dumb-bell-shaped guard cells are attached to each other by their bulbous ends thus forming a stoma. In contrast to crescent-shaped-guard cell the bulbous ends are fused. In the common walls of guard cell pair large pores of about one micron or more in diameter exist. Through the pores exchange of organelles between each guard cell pair occurs. So the guard-cell-pair is regarded as a single binucleate unit.

The guard cells are distinct within the epidermal layers both anatomically and biochemically. They have adapted a specialized set of metabolic pathways that cause rapid change in osmotic potential. Minor changes in the external environment stimulate the metabolic pathways. As for example, a change in carbon dioxide concentration in atmosphere will stimulate a series of reactions that result in opening and closing of stoma. The stomatal movement is associated with rapid fluxes of $K^+$ and $H^+$ between guard cells and neighbouring cells. Guard cells possess high metabolic activity due to the presence of abundant mitochondria and protein synthesizing machinery (Willmer, p. 18). Calvin cycle is not detected in guard cells. It is to note that functional plasmodesmata do not exist between guard cells and neighbouring cells though plasmodesma exists in developing stoma.

• *Subsidiary cell*: Subsidiary cell (also called accessory cell) can be defined as specialized epidermal cell that differs in size, arrangement from the rest of epidermal cells and lies close to guard cells. In *Pinus* and *Equisetum* subsidiary cell envelops the guard cells. They are smaller than the epidermal cells. They contain dense cytoplasmic content and greater frequency of cell organelles. Normally subsidiary cells are devoid of chloroplasts, anthocyanins and crystalline inclusions. Subsidiary cell plays a major role in the movement of stoma composed of dumb-bell-shaped guard cells.

## Types of stoma

Stomata are classified (1) on the basis of number and arrangement of subsidiary cells/epidermal cells in relation to guard cell [Metcalfe and Chalk (1950) and Stebbins and Khush (1961) described the stomatal complexes on the basis of observation under light microscope] and (2) on the basis of development of guard cells and subsidiary cells [Pant (1965), Florin (1933) and Payne (1979)].

Metcalfe and Chalk illustrated in the Introduction of their influential *Title* (p. xv) five types of stoma in dicotyledons and the sole Gramineous stoma in monocotyledons. They classified dicotyledonous stomata on the basis of number and arrangement of subsidiary cell/epidermal cell in relation to guard cells. The followings are the different types of stoma according to Metcalfe and Chalk.

• *Anomocytic* (also called ranunculaceous or irregular-celled type): Anomocytic stoma remains surrounded by cells that are not different from other epidermal cells so far as size and shape are concerned. There exist no definite number and arrangement of cells that surround a stoma. A stoma appears to be embedded in epidermal cells. Ex. Apocynaceae, Boraginaceae, Chenopodiaceae, *Cucurbita* etc. Metcalfe and Chalk mentioned 142 families where anomocytic stoma is present.

• *Anisocytic* (also called Cruciferous or unequal-celled type): Anisocytic stoma remains surrounded by three unequally sized subsidiary cells, among which one is distinctly smaller in size than the other two. Ex. Cruciferae, *Solanum*, *Petunia*, *Sedum* and *Nicotiana* etc. Metcalfe and Chalk mentioned 37 families where anisocytic stoma occurs.

• *Paracytic* (also called rubiaceous or parallel-celled type): Paracytic stoma is accompanied on either side by one or more subsidiary cells. The longitudinal axes of subsidiary cells lie parallel to that of aperture and guard cells. The subsidiary cells may or may not meet over the poles. In the family Rubiaceae the subsidiary cells usually meet over the poles. The subsidiary cells of *Drimys* and *Linum* etc. usually fall short of the poles. Ex. Rubiaceae, Convolvulaceae, *Phaseolus*, *Arachis* and *Psoralea* etc. Metcalfe and Chalk cited 105 families where paracytic stoma is found.

• *Diacytic* (also called caryophyllaceous or cross-walled type): Diacytic stoma remains surrounded by a pair of subsidiary cells. The common wall of the subsidiary cells is at right angles to guard cells. Ex. Caryophyllaceae, Acanthaceae, *Hygrophila*, *Dianthus* etc. Metcalfe and Chalk mentioned eleven families to have diacytic stoma.

• *Actinocytic*: Actinocytic stoma remains surrounded by a circle of radiating cells (Metcalfe and Chalk, p. xv). Ex. *Ancistrocladus* and *Euclea pseudebenus* (Ebenaceae). It is to note that Metcalfe and Chalk in the original definition of actinocytic did not state the number of subsidiary cells that enclose a stoma. Van Cotthem (1970, p. 244) described actinocytic stoma where at least five subsidiary cells surround a stoma (actinocytic = star-celled). Stace (1965, p. 48) described the subsidiary cells of actinocytic stoma as somewhat 'radially elongated' cells.

• *Gramineous*: Gramineous stoma possesses two guard cells that are shaped like dumb-bells. Each guard cell has a narrow middle portion and two bulbous ends. The narrow middle portion is strongly thickened. The subsidiary cells occur parallel to the long axis of pore. Ex. Gramineae and Cyperaceae.

Metcalfe and Chalk (p. xv) in the Introduction of their book illustrated anomocytic, anisocytic, paracytic and diacytic as Type A, Type B, Type C and Type D respectively. They thought that more terms would be necessary and so added another term—actinocytic.

The terms ranunculaceous, cruciferous, rubiaceous and labiatous or caryophyllaceous stoma were coined by Vesque in 1889. Later studies revealed that these types are not confined to respective families only. Identical types occur in very distantly related families. Therefore Metcalfe and Chalk proposed new terms respectively anomocytic, anisocytic, paracytic and diacytic instead of ranunculaceous, cruciferous, rubiaceous and labiatous or caryophyllaceous. They added the fifth type—actinocytic to the four 'classic' ones of Vesque. They did not employ the new terms in the text and commented that 'the older and more familiar terms have been mentioned in this book to allow time for the new ones to become assimilated'—p. xv.

In addition to gramineous stoma Metcalfe (1961) described a new stomatal type in monocotyledons. It is termed tetracytic where the guard cells are surrounded by four subsidiary cells—two laterals and two polar, each being present on the four sides. The two laterals lie parallel to guard cell. The two polar subsidiary cells are often smaller. Tetracytic stoma characterizes numerous monocotyledonous families. It is also reported from dicotyledons, e.g. *Tilia* and some members of Asclepiadaceae. The seven types of stoma (five from dicotyledons and two from monocotyledons) according to Metcalfe and Chalk and Metcalfe are shown in Fig. 12.9.

In addition to the above five types of stoma in dicotyledons Van Cotthem illustrated two more types of stoma, which are as follows (Fig. 12.10):

• *Hemiparacytic*: The stoma is accompanied by a single subsidiary cell, which is placed parallel to the long axis of the pore and this cell may be long or short in length in contrast to the guard cells. Example: *Glinus latioides* and *Trianthema lancastrum* etc.

• *Hexacytic*: The stoma is surrounded by six subsidiary cells among which two are situated on the two polar sides and rest two pairs occur on the two lateral sides being parallel to the long axis of the guard cells. The size of subsidiary cells may be of two different types: (a) the two polar cells may be as broad as the

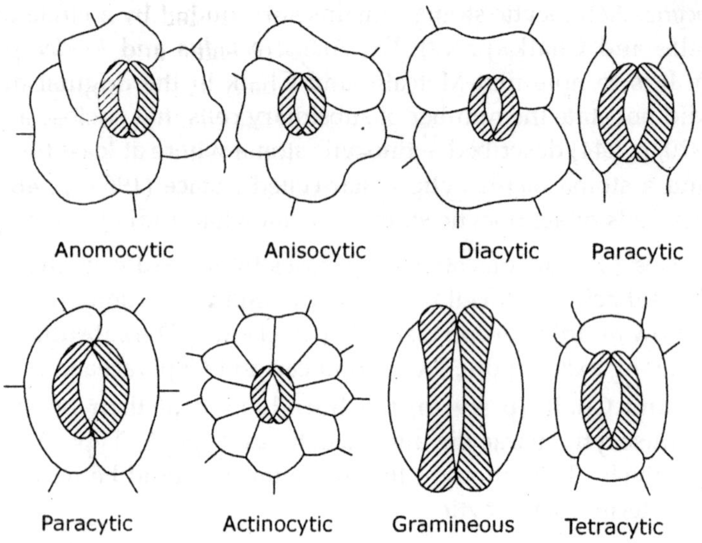


Figure 12.9

Diagrammatic representation of different types of stoma according to Metcalfe and Chalk (1950) and Metcalfe (1961) in dicotyledons and monocotyledons. Guard cells are hatched. The others are subsidiary cells / epidermal cells.

stomatal complex (e.g. *Geogenanthus*) and (b) the length of the outer most lateral subsidiary cells may as long as stomatal complex (e.g. *Commelina*).

Stace (1965) illustrated the cyclocytic (Fig. 12.10) stoma in dicotyledons. The stoma is surrounded by four or more subsidiary cells, which form one or two narrow rings around the guard cells. Example: *Lumnitzera, Laguncularia* etc.

The above discussed stomatal types are tabulated in table 12.1.

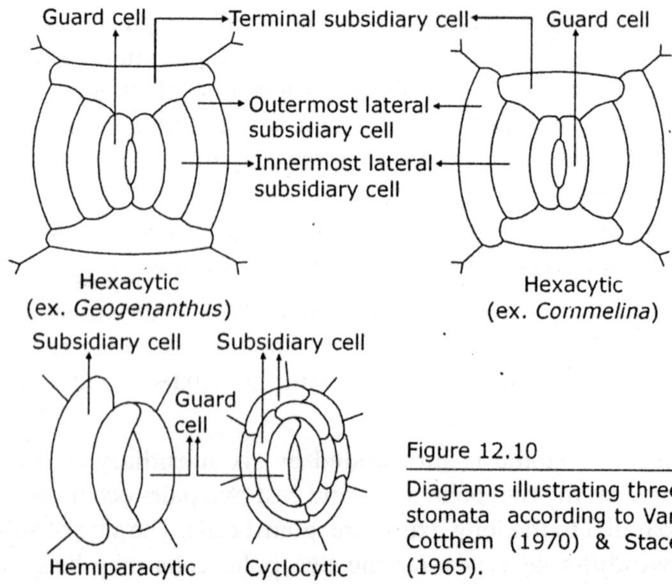


Figure 12.10

Diagrams illustrating three stomata according to Van Cotthem (1970) & Stace (1965).

Stebbins and Khush (1961) recognized four categories of stomatal complex in monocotyledons. The basis of classification is the presence or absence of subsidiary cells and when present the number, shape and arrangement in relation to guard cells form the basis. Stebbins and Khush did not propose any terminology to aid their concepts and distinguished them as First type, Second type, Third type and Fourth type which are as follows (Fig. 12.11).

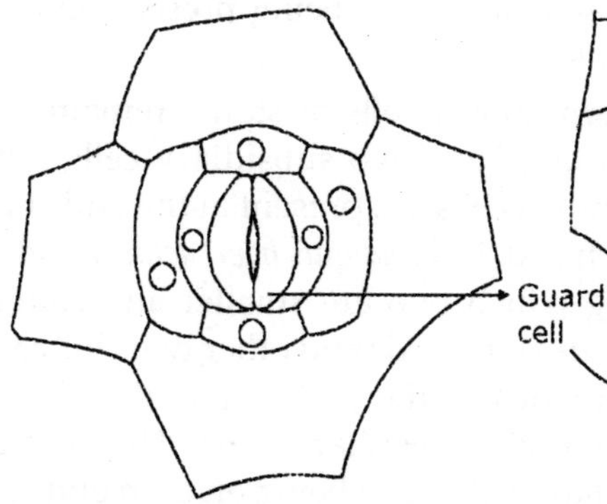


First type
with 6 subsidiary cells
(*Commelina communis*)

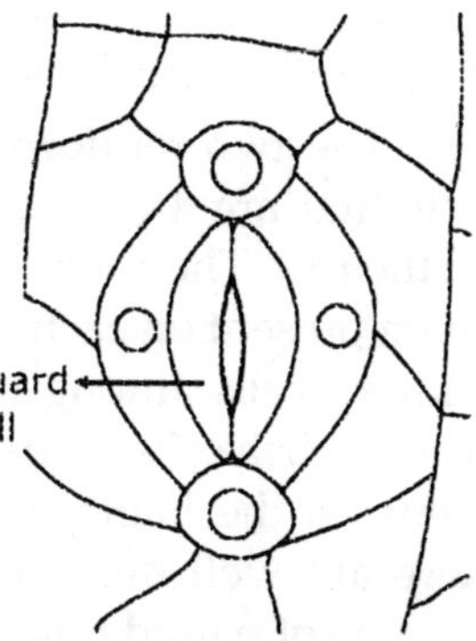

Second type
with 4 subsidiary cells
(*Pandanus haerbachii*)

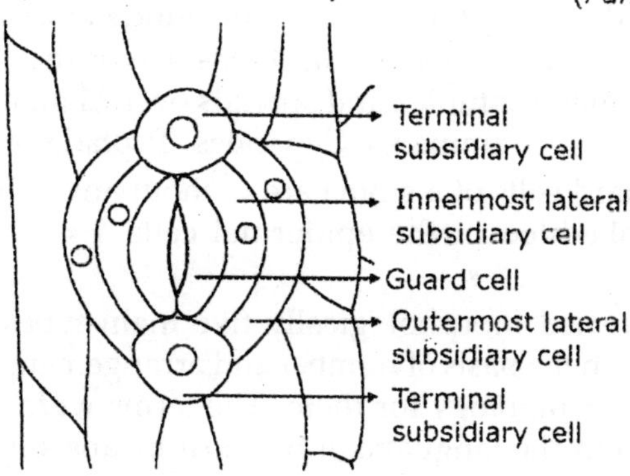


Second type
with 6 subsidiary cells
(*Phytelephas microcarpa*)

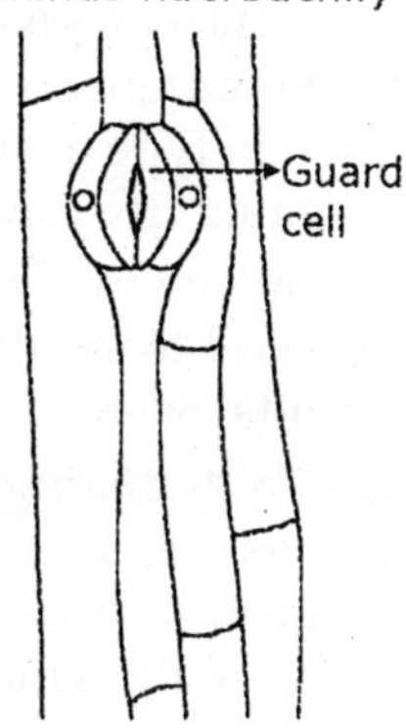


Third type
with two subsidiary cells
(*Calamus oxleyanus*)

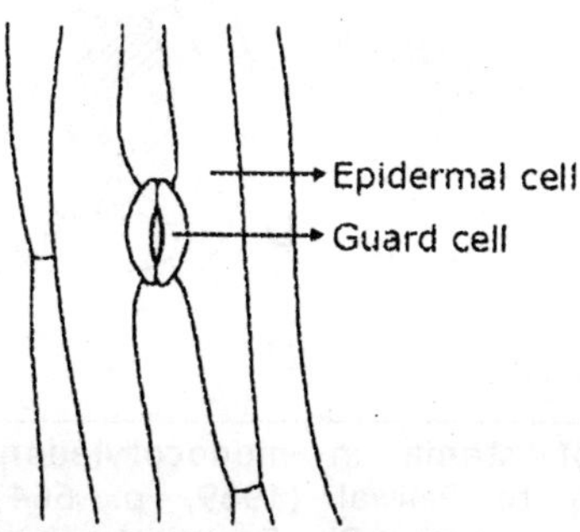


Fourth type
with no subsidiary cells
(*Nothoscordum inodorum*)

Figure 12.11

Diagrams illustrating the four types of stoma in monocotyledons according to Stebbins and Khush (1961). The subsidiary cells are with spherical nucleus.

*First type*: The two reniform guard cells of stoma remain surrounded by subsidiary cells that are 4-6 in number. In *Rhoeo, Tradescantia* and *Zebrina* of Commelinaceae there are four subsidiary cells. Each subsidiary cell occurs on each of the four sides of paired guard cells. As a result the stomatal complex has square appearance in surface view. *Commelina* of the same family has six subsidiary cells. The additional subsidiary cells occur on the lateral sides of stoma, each being present on each of the two lateral sides. In contrast to *Zebrina, Commelina* has four subsidiary cells on lateral sides, two being present on each of the two lateral sides of paired guard cells.

*Second type:* The two reniform guard cells of stoma remain surrounded by subsidiary cells that are 4-6 in number. Two subsidiary cells are smaller and roundish than the rest. The roundish cells are present at the ends of paired guard cells—each being present on each end. In *Pandanus haerbachii* (Pandanaceae) there are four subsidiary cells among which two are smaller and roundish than the other two. The roundish cells occur at the ends and other two elongated subsidiary cells are present on the lateral sides of the paired guard cells. In *Phytelephas microcarpa* there are well defined six subsidiary cells. The roundish cells are situated at the ends of paired guard cells, each being on each end. The rest occurs on lateral sides, two being on each side. Such type of stomatal complex with four subsidiary cells occurs in Calycanthaceae.

*Third type:* The two guard cells of stoma remain surrounded by two subsidiary cells each being present on each lateral side of guard cells. This is the most common and predominant type of stomatal complex and spreads over 24 monocot families so far investigated. Ex Juncales, Graminales, Cyperales, Typhales etc.

*Fourth type:* The two guard cells of a stoma are without any subsidiary cells. A stoma appears to be embedded in the epidermal cells. Ex. Amaryllidales, Iridales, Orchidales etc.

Paliwal (1969) distinguished morphologically five main types of stoma in monocotyledons (Fig. 12.12) on the basis of number and arrangement of subsidiary cells and suggested special terminology for them. The terms were derived from the word 'sahkoshik' which in the language of Sanskrit means subsidiary cells. The stomatal complexes are as follows:

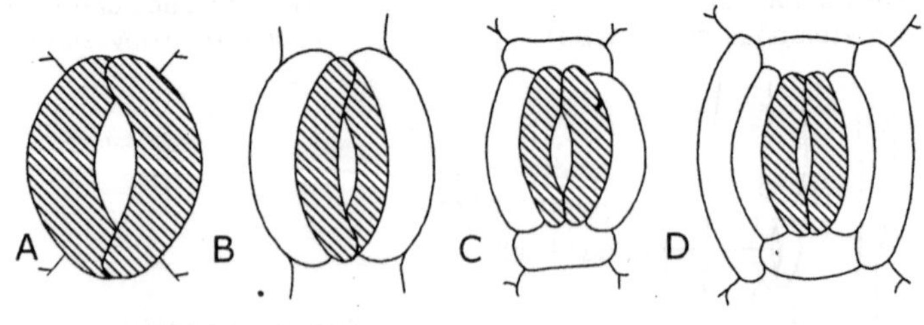


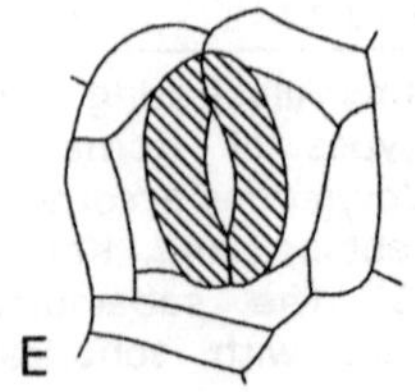


Figure 12.12

Types of stoma in monocotyledons according to Paliwal (1969, p. 664). A. Asahkoshik. B. Dwisahkoshik. C. Chatushsahkoshik. D. Shatsahkoshik. E. Bahusahkoshik. Guard cells are hatched.

1. Asahkoshik (Sans: 'A'= without + Sahkoshik): There are no subsidiary cells in this type of stoma and the stomatal complex consists of two guard cells only. Ex. Members of Liliaceae, Orchidaceae etc.
2. Dwisahkoshik (Sans: 'Dwi' = two + Sahkoshik): In this type of stoma there are two subsidiary cells each of which is situated laterally on either side of the pair of guard cells. Ex. Cyperaceae, Palmae etc.
3. Chatushsahkoshik (Sans: 'Chatus' = four + Sahkoshik): In this type of stoma there are four subsidiary cells, which are arranged in two different ways: (a) a pair of subsidiary cell is placed laterally on two sides of the guard cells. Ex. Members of Zingiberaceae and (b) each of the four subsidiary cells is situated on the four sides of guard cells, i.e. two polar and two laterals. Ex. *Rhoeo.*
4. Shatsahkoshik (Sans: 'Shat' = six + Sahkoshik): In this type of stoma there are six subsidiary cells of which two are placed on the two polar sides and a pair on two lateral sides of the guard cells. Ex. Members of Commelinaceae, Palmae, Musaceae etc.
5. Bahusahkoshik (Sans: 'Bahu' = many + Sahkoshik): In this type of stoma there are more than six subsidiary cells, which are either arranged irregularly or form a ring around the guard cells. Ex. Agavaceae, Araceae etc.

## Ontogeny of stoma

Before illustrating the ontogeny of stomatal complex it may be useful to have a precise definition of the following terms which are used in this essay. The definitions are in accordance with Pant (1965).

- Stoma: It includes the pore and two guard cells.
- Stomatal complex/stomatal apparatus: It includes a stoma and the adjoining subsidiary cells. It also includes a stoma and the neighbouring epidermal cells when subsidiary cells are not distinguishable.
- Stomatal meristemoid: It refers a protoderm cell whose first or consecutive divisions lead to the formation of guard cells of a stoma. Marginal meristems produce marginal initials that in turn form the abaxial and adaxial leaf epidermal layers. The marginal initials undergo anticlinal divisions and subdivisions. The derivative cells form protoderm cells. Some specialized protodermal cells transform into stomatal meristemoid.
- Meristemoid: It refers the cells that are the derivatives of stomatal meristemoid and retain the capacity for further divisions.
- Guard cell initial/guard cell mother cell (gcmc): It refers 'a meristemoid whose bisection results in the formation of two sister guard cells'.
- Mesogene: It refers to the subsidiary cells that have a common origin with guard cells, that is, the subsidiary cells and the two guard cells are the derivative cells of same meristemoid.
- Perigene: It refers to the subsidiary cells that have a different origin from guard cells. The cells that form subsidiary cells occur around the meristemoid that forms guard cells.

Stomatal meristemoid, by several divisions, forms a stoma. The meristemoid is usually isodiametric or polygonal in shape and possesses a single conspicuous nucleus with denser cytoplasm. Usually the stomatal meristemoids are initiated between the mature stomata (e.g. gymnosperms and angiosperms). Sometimes their origin is acropetal (e.g. stem of *Psilotum*); basipetal origin is observed in Lycopodiales, Convolvulaceae, Rubiaceae etc. while simultaneous differentiation is observed in *Erythrina*, a member of Leguminosae. The differentiation of a stomatal meristemoid and its subsequent divisions leading to the formation of a stoma vary from group of plants.

The ontogeny of elliptical stoma is illustrated below taking the example of *Allium cepa* (Fig. 12.13) because the developmental process is not complicated by the formation of subsidiary cells.

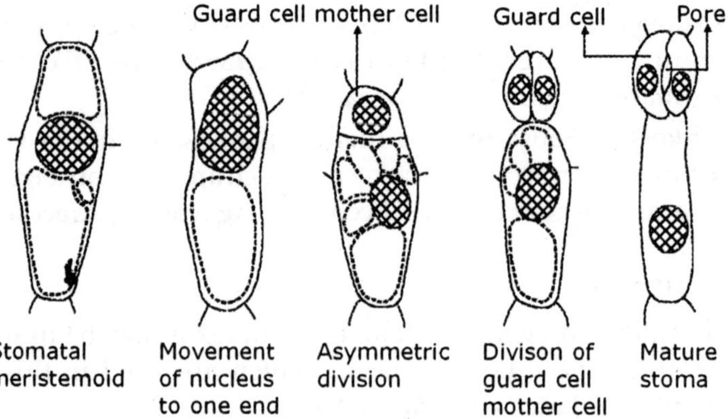


Figure 12.13

*Allium cepa.* Diagram illustrating ontogeny of stoma. Adapted after Fahn, 1997, p. 171. Nucleus = crosshatched. Dotted lines enclose vacuoles.

In *Allium* the stoma consists of a pair of guard cells and a pore. The guard cells are crescent shaped. The stoma is devoid of any subsidiary cell. Stoma develops from stomatal meristemoid which is a transformed protodermal cell. The protodermal cell is elongated and contains dense cytoplasm with a conspicuous nucleus and vacuoles. The nucleus occupies central position. Prior to division the nucleus becomes displaced to one end of the stomatal meristemoid and the vacuoles occupy the opposite end. A differential distribution of cytoplasm occurs. The distal end accumulates most of the cytoplasm and organelles including the nucleus. The stomatal meristemoid undergoes asymmetrical division resulting a distal small cell and a large proximal one with less activity. The smaller distal cell with dense cytoplasmic contents becomes the guard cell mother cell (also called guard mother cell). The large proximal cell has large vacuolated cytoplasm and large nucleus. It becomes a less specialized cell. The guard cell mother cell (gcmc) contains proplastids, microbodies, dictyosomes, starch and all the usual cell organelles. A band of microtubules becomes evident surrounding the nucleus. Subsequently the gcmc divides symmetrically by a wall parallel to the long axis of leaf to give rise to two daughter guard cells.

The two daughter guard cells separate to form a pore. The formation of stomatal pore in *Allium cepa* is not completely understood. In other species it is observed that the daughter guard cells secrete the enzyme pectinase. The enzyme dissolves the pectinaceous middle lamella that is present between the daughter guard cells resulting in the formation of pore. It is believed that increased pressure within the guard cells causes the mechanical separation thus forming a pore. The two daughter guard cells gradually assume the characteristic crescent shape thus forming an elliptical stoma.

In contrast to *Allium cepa* the stomatal complex of *Cyperus alternifolius* includes a pair of guard cells and one lateral subsidiary cell per guard cell. The subsidiary cells originate from the adjoining epidermal cells. The gcmc like *Allium cepa* is formed as a result of unequal division of protodermal cell. The gcmc induces divisions in the adjacent epidermal cells. The epidermal cell forms two cells by asymmetric division. The smaller cell constitutes the subsidiary cell. So *Cyperus alternifolius* exhibits perigene stomatal complex.

The classifications proposed by Metcalfe and Chalk and Stebbins and Khush take no account of the ontogeny of stomata. It is now evident that mature stomatal pattern does not reveal the ontogenetic relationship between guard cells and subsidiary cells. Subsidiary cells may be perigene, mesogene or may be mesogene and perigene in a stomatal complex. Accordingly a new classification is proposed by Pant based on the ontogenetic relationship of subsidiary cells and guard cells. Pant classified stomata into following types:

**Mesogenous:** A stomatal complex is said to be mesogenous where all subsidiary cells or a single ring-like subsidiary cell have a common origin with the guard cells, that is, the same meristemoid gives rise to subsidiary cells and guard cells.

Pant recognizes the following four subtypes in mesogenous:

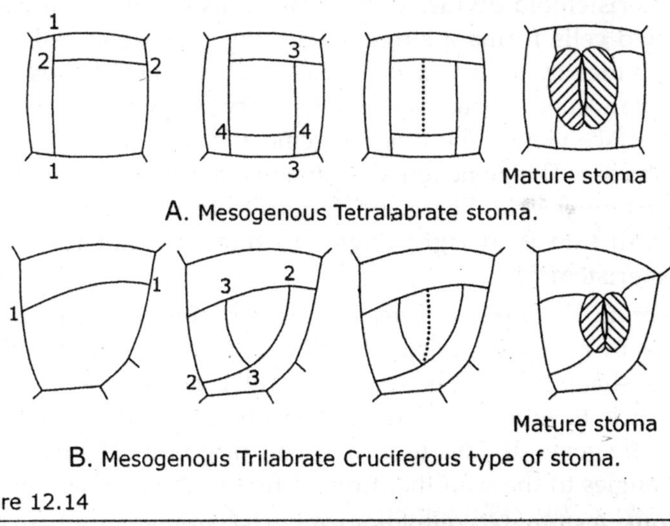


A. Mesogenous Tetralabrate stoma.

B. Mesogenous Trilabrate Cruciferous type of stoma.

Figure 12.14

Diagram illustrating main ontogenetic types of stomata. Adapted after Pant (1965), p. 6. Guard cells are hatched. 1-1, 2-2 and 3-3 indicate the sequence and plane of cell divisions in the stomatal meristemoid to form guard cell mother cell. Dotted line indicates the plane of division of guard cell mother cell.

(a) Tetralabrate: Example: *Lactuca viscosa* (Fig. 12.14A). The meristemoid is more or less rectangular in shape and has four cutting facės. The four sides of meristemoid cut off four mesogene subsidiary cells in a spiral sequence depicted in diagram as 1-1, 2-2, 3-3 and 4-4. All cells enclose a central guard cell initial. The initial divides and forms two guard cells.

(b) Trilabrate (= cruciferous): Example: *Beta vulgaris, Sedum spurium* and *Datura stramonium* etc. (Fig. 12.14B). The meristemoid has three cutting faces. Each side divides by anticlinal division in a spiral sequence that is depicted in diagram in numericals. All cells enclose a central guard cell initial that divides to form two guard cells.

(c) Dolabrate: The meristemoid has two parallel cutting faces. Pant sub-classified dolabrate into the following three on the basis of division of guard cell mother cell in relation to the plane of divisions of meristemoid that are as follows:

  i. Mesoparacytic (= rubiaceous): Example: *Ricinus communis, Phaseolus vulgaris* etc. (Fig. 12.15 C). The meristemoid divides by two successive parallel divisions and the resulting cells enclose the central guard cell mother cell (gcmc). The gcmc divides to form a pair of guard cells by a wall that is parallel to the wall formed earlier.

  ii. Mesodiacytic: Example: *Ocimum basilicum* an *Asteracantha longifolia* (Fig. 12.15D). The meristemoid divides by two successive parallel divisions and the resulting cells enclose the gcme. The gcmc divides to form a pair of guard cells by a wall that is at right angles to earlier division.

  iii. *Pyrrosia* type: Example: *Pyrrosia lingua* (Fig. 12.15E). The stomatal meristemoid divides to form two cells by anticlinal wall. One of the two cells forms a single encircling mesogene cell. The other cell divides by a curved periclinal wall. Two cells thus formed, one is 'planoconcave mesogene subsidiary cell which appears annular in surface view'. The other is gcmc and it is hemispherical and dome shaped. The gcmc remains attached to the outer periclinal wall by its flat outer side. The gcmc divides to form a pair of guard cells by a wall that is at right angles to the wall that formed first in the meristemoid.

(d) Unilabrate: Example: *Anemia* and *Schizaea* (Fig. 12.15F). The meristemoid divides by a curved periclinal wall to form a subsidiary cell that encloses a dome shaped gcmc. The subsidiary cell is annular as observed in surface view. It is to note that there occurs no anticlinal division in unilabrate type. The gcmc divides to form a pair of guard cells by a wall that is at right angles to the wall that formed first in the meristemoid.

**Perigenous:** A stomatal complex is said to be perigenous where the subsidiary cells and guard cells do not have a common origin. The subsidiary cells originate from cells that lie around the meristemoid. Meristemoid divides symmetrically to form two daughter guard cells (Fig. 12.16G). Example: *Psilotum, Selaginella, Isoetes, Ginkgo,* Cucurbitaceae, Nymphaeaceae and Lemnaceae etc.

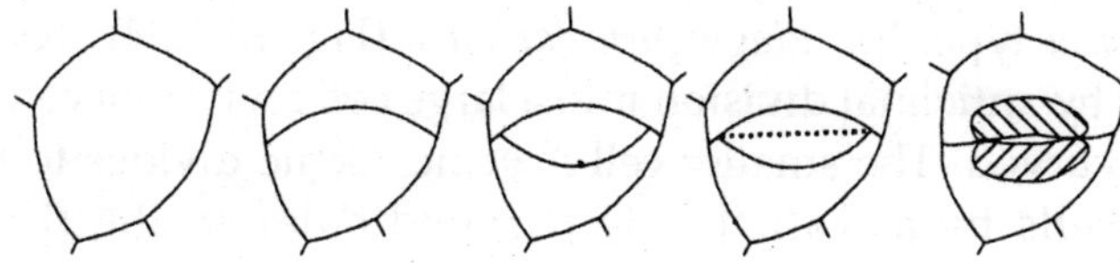


C. Mesogenous Dolabrate mesoparacytic rubiaceous type of stoma.

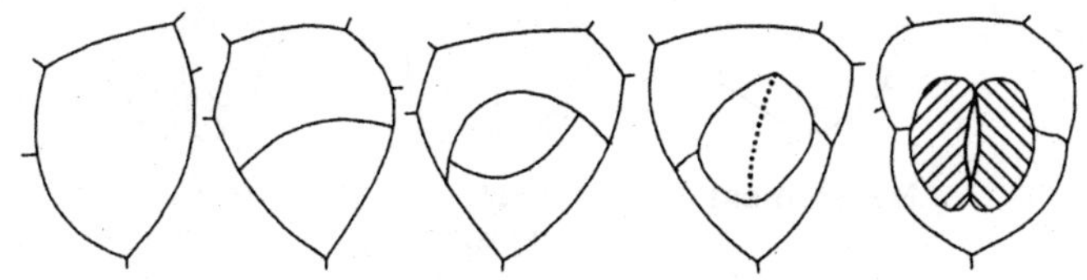


D. Mesogenous Dolabrate mesodiacytic stoma.

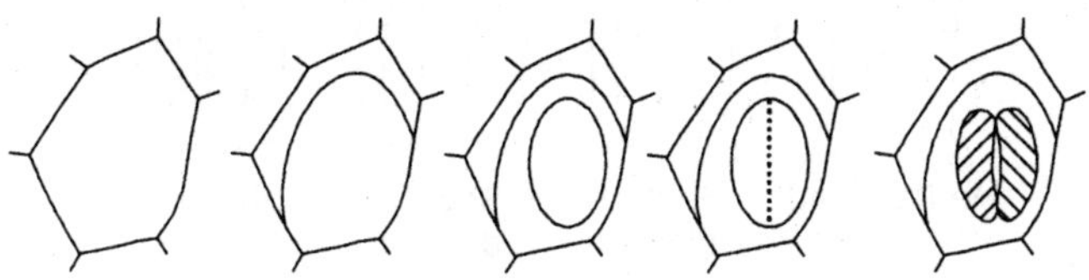


E. Mesogenous Dolabrate *Pyrrosia* type of stoma.

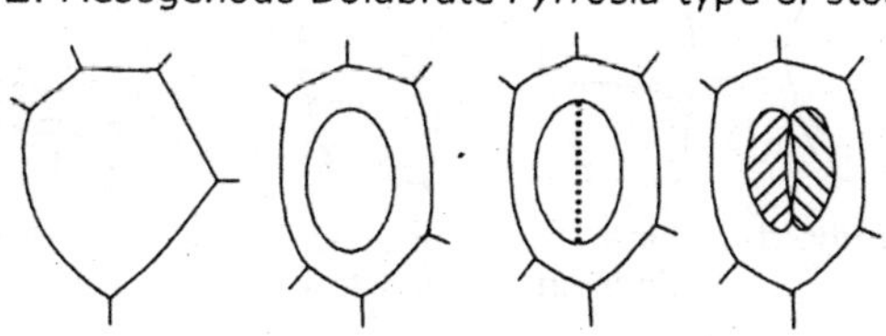


F. Mesogenous Unilabrate *Anemia* type of stoma.

Figure 12.15

Diagram illustrating main ontogenetic types of stomata. Adapted after Pant (1965), p. 6. Guard cells are hatched. Dotted line indicates the plane of division of guard cell mother cell.

**Mesoperigenous:** A stomatal complex is said to be mesoperigenous where one stomatal meristemoid gives rise to one mesogene subsidiary cell and gcmc that divides symmetrically resulting two daughter guard cells. The rest of subsidiary cells are perigene and originate from cells lying around the meristemoid.

Mesoperigenous is sub-classified by Pant into following three types on the basis of the division wall of gcmc in relation to the wall formed first in the meristemoid.

i. *Plagiogyria* type: Ex. *Plagiogyria japonica* (Fig. 12.16H). The meristemoid divides by anticlinal division into a large cell and small cell. The partition wall is curved. The smaller cell is gcmc. Gcmc divides to form a pair of guard cells by a wall that is perpendicular to the first division of meristemoid. A single mesogene subsidiary cell girdles the three sides of the stoma. Perigene subsidiary cell occurs on the fourth side of stoma.

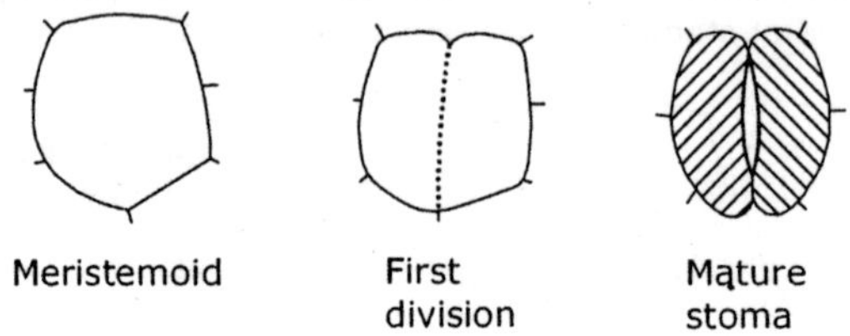


G. Perigenous stoma

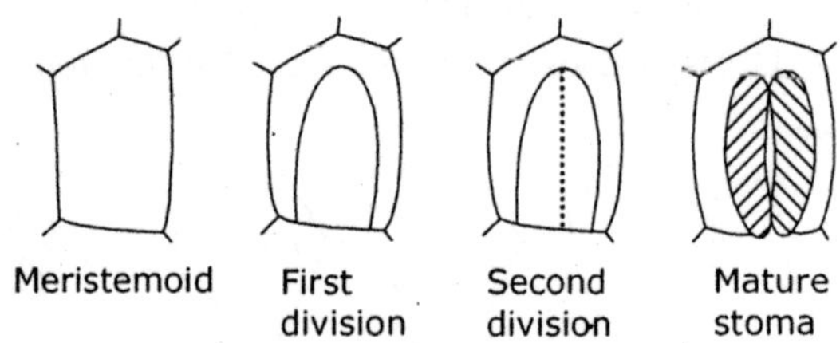


H. Mesoperigenous *Plagiogyria* type of stoma.

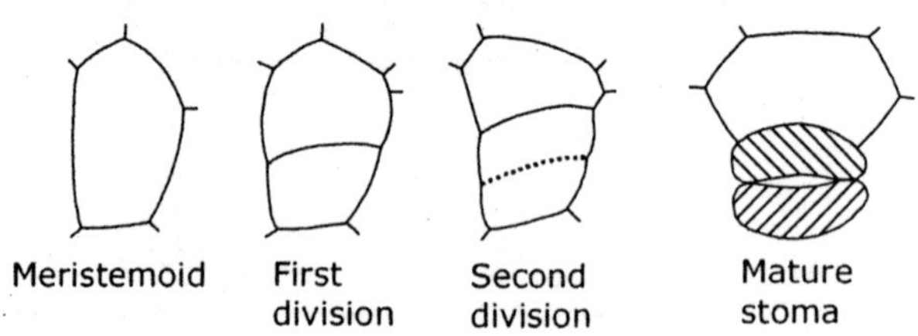


I. Mesoperigenous *Tetracentron* type of stoma.

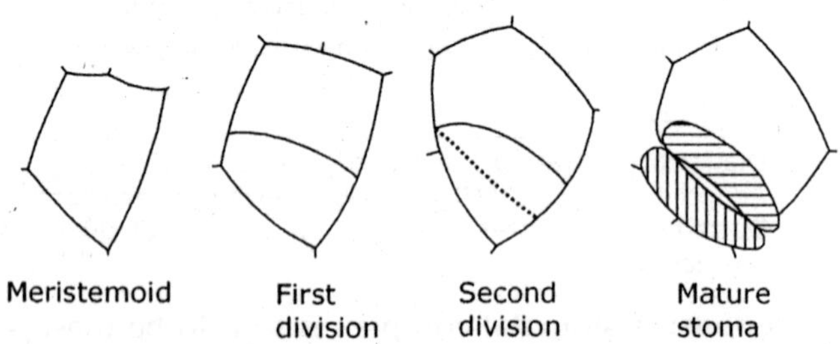


J. Mesoperigenous *Ranunculus* type of stoma.

Figure 12.16

Diagram illustrating the main ontogenetic types of stomata. Adapted after Pant (1965), p. 6. Guard cells are hatched. Dotted line indicates the plane of division of guard cell mother cell.

ii. *Tetracentron* type: Example: *Tetracentron sinensis* and *Trochodendron aralioides* (Fig. 12.16I). The meristemoid divides by anticlinal wall into two cells among which one is gcmc. Gcmc divides to form a pair of guard cells by a wall that is parallel to the wall formed in the first division of meristemoid.

iii. *Ranunculus* type: Example: *Ranunculus, Delphinium* and *Nigella* (Fig. 12.16J). The meristemoid divides to form two daughter cells among which one is the gcmc. Gcmc divides to form a pair of guard cells. The partition wall in gcmc has no relation to the first wall of meristemoid. The wall may be at any angle other than perpendicular and parallel.

Rudall (2007, p. 15) reported another stomatal type – agenous. Agenous stoma lacks subsidiary cells. The meristemoid (= guard cell mother cell) divides to form two sister guard cells. In contrast perigenous stoma also develops from guard cell mother cell, but it has subsidiary cells and the subsidiary cells develop from the neighbouring cells of guard cell mother cell.

In photosynthesizing organs stomatal meristemoids appear between older ones and mature stomata. They may appear gradually in a basipetal sequence, in an acropetal sequence and simultaneously. In leaves with reticulate venation there is no regularity in the appearance of stomata (Fahn, 1997).

Earlier, Florin (1933) proposed the following two stomatal patterns based on origin:

(a) *Haplocheilic*: In haplocheilic stoma two guard cells originate from a common meristemoid. The meristemoid functions directly as guard mother cell (GMC). There occurs no division in the meristemoid before its transformation to GMC. GMC and guard cells influence the neighbouring protodermal cells to divide and form subsidiary cells. So in this type of stomatal complex the subsidiary cells are perigene. Ex. *Ephedra*. Haplocheilic stoma is similar to perigenous stoma of Pant.

(b) *Syndetocheilic*: In syndetocheilic stoma the stomatal meristemoid undergoes a number of divisions resulting in the formation of GMC and subsidiary cells that adjoin the guard cells. So in this type of stomatal complex the subsidiary cells are mesogene. Ex. *Welwitschia*. Syndetocheilic stoma is similar to mesogenous stoma of Pant.

Florin and Pant recognized perigenous stomata as distinct type. But Payne (1979) regards that there exists no perigenous stoma. It is evident that a protodermal cell undergoes either equal or unequal division to form daughter cells. In equal division the number of protodermal cells increases. A protodermal cell when undergoes unequal division, is destined to form meristemoid. This finds support due to the fact that unequal cells resulted from unequal division exhibit differential staining. There occurs no truly perigenous stoma in which a protodermal cell directly serves as GMC. In this sense a stoma is either mesogenous or mesoperigenous. Considering the above facts Payne proposed the following classification of stomata based on 'the manner of production and division of the GMC' – p. 127.

• *Diameristic*: Example: mesogenous stoma of Lamiaceae (Fig. 12.17A) and mesoperigenous stomata of monocots and Caryophyllaceae. A stoma is said to be diameristic where GMC divides with a wall formed at right angles relative to the wall that cuts it off from its meristemoid. Meristemoid divides repeatedly resulting two or more subsidiary cells which are parallel to each other. They enclose a central GMC. GMC divides to form two guard cells and the cell wall is formed at right angle to long axes of subsidiary cells. At maturity the stomatal complex consists of a pair of guard cells and a pair of subsidiary cells, the long axes of which are at right angles to each other.

• *Parameristic*: Example: mesogenous stoma of Rubiaceae (Fig. 12.17B) and mesoperigenous stoma of *Liriodendron*. A stoma is said to be parameristic where GMC divides with a wall parallel to the wall that cuts it off from its meristemoid. Meristemoid divides thus forming two cells. One of the daughter cells becomes the first lateral subsidiary cell. The other daughter cell again divides in such a way that the resulting cell wall curves to meet both ends of the cell wall that is formed in the first division. Among the two cells thus formed one becomes the second lateral subsidiary cell. These two divisions enclose GMC. GMC divides parallel to the long axes of subsidiary cells. Thus a mature stoma exhibits a pair of guard cells enclosed by a pair of lateral subsidiary cells. The long axes of guard cells and subsidiary cells are parallel to each other.

• *Anomomeristic*: Example: mesoperigenous stoma of *Ranunculus* (Fig. 12.17C). A stoma is said to be anomomeristic where GMC divides with a wall formed at any angle relative to the wall that cuts it off from its meristemoid. Meristemoid divides to form GMC. GMC in turn divides to form a pair of guard cells the partition wall of which has no definite orientation in relation to the wall that formed GMC.

The contributions of Florin on the investigation of the development of leaves in gymnosperm marked the beginning of studying ontogeny of stoma. Florin's work revealed two types of development: (1) perigenous and the adult appearance of stoma is termed as haplocheilic and (2) mesogenous and the adult appearance is termed as syndetocheilic. Later in 1965 and onwards, Pant, followed by other workers, revised ontogenetic terminology. Pant did not subdivide the perigenous group. Mesogenous group is subdivided into four subgroups and mesoperigenous group is subdivided into three types in Pant's classification. Later van Cotthem followed by other workers, reviewed the whole subject thoroughly and proposed additional terminology. They subdivided the perigenous group. The study of stomatal ontogeny gained much prominence due to the fact that mature stomata which appear alike may have different developmental pathway. This aspect is to be considered while assessing the phylogenetic position of a species based on stoma.

## Importance of studying stoma

Followings are a few reasons for the great interest in stoma.

Stoma provides characters of taxonomic significance. This is due to the fact that genera and even families show great constancy for their possession of a

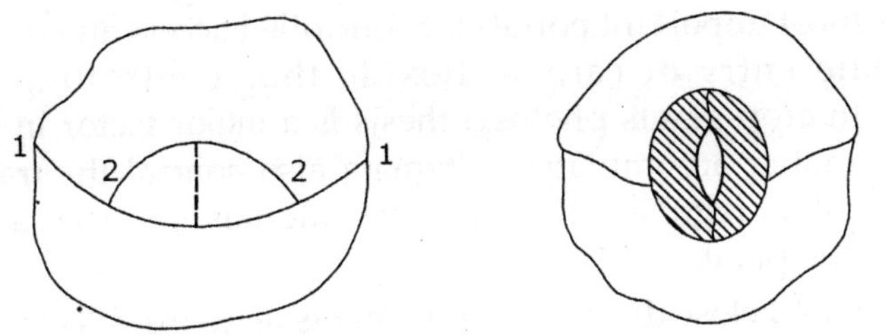


Division in meristemoid          Mature stoma

A. Ontogeny of diameristic mesogenous lamiaceous stoma.

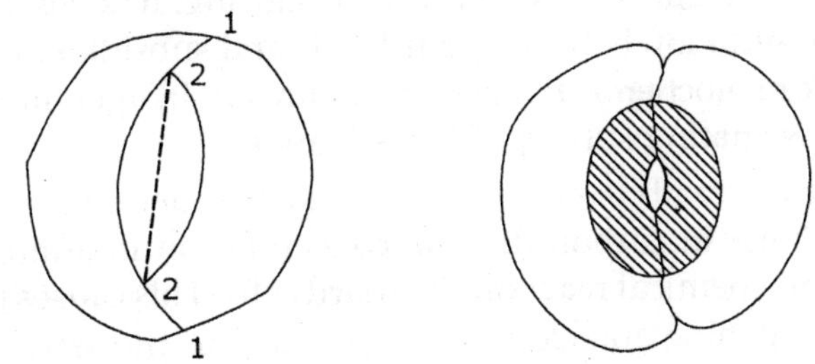


Division in meristemoid          Mature stoma

B. Ontogeny of parameristic mesogenous rubiaceous stoma.

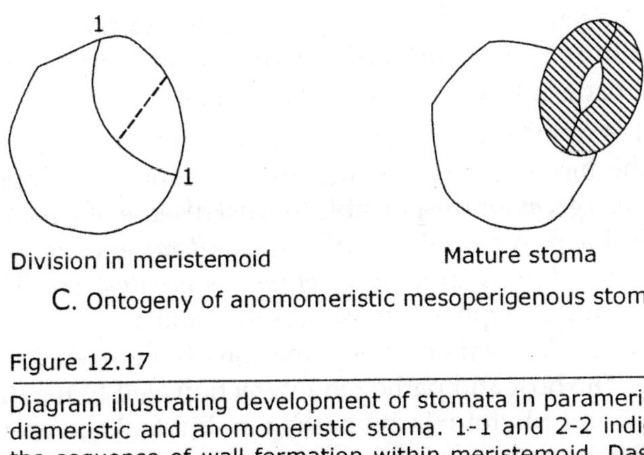


Division in meristemoid          Mature stoma

C. Ontogeny of anomomeristic mesoperigenous stoma.

Figure 12.17

Diagram illustrating development of stomata in parameristic, diameristic and anomomeristic stoma. 1-1 and 2-2 indicate the sequence of wall formation within meristemoid. Dashed lines indicate the plane of division of guard cell mother cell. Adapted after Payne, 1979, p. 128. Guard cells are hatched.

particular stomatal complex. The study of stomatal complexes becomes very useful to understand the evolutionary relationship of plants when the classical methods of cytology and genetics etc. cannot be applied. Apart from stomatal complex surface view of guard cells and the aperture have taxonomic importance. The aperture is unique in *Azolla* where it is oriented at $90^{o}$ to guard cells (Rajagopal and Ramayya, 1977, p. 57). The frequency and distribution of stoma (epistomatic, hypostomatic or amphistomatic) form different taxonomic groups. Appearance of stoma in surface view and their ontogeny provide characters not only to understand the evolution of stomata but also form the bases of different classificatory systems. Vide the chapter 32 for a few more details.

Stomata are most important portals for controlled access and egress of gases. Stomata regulate entry of carbon dioxide thus controlling the rate of photosynthesis. In crop plants photosynthesis is a major factor in determining the rates of dry matter accumulation. Stomata also control the transpirational water loss and it thus regulates the entry of materials into root and transport materials within the plant.

Stomata open and close due to the movements of guard cells thus regulating gas exchange. Stomata appear to open and close in response to accumulation of potassium ion. During opening potassium ion accumulates in guard cells. Potassium ion moves to subsidiary cells during closing. It is observed that there is a shuttle of potassium ion between guard cells and subsidiary cells in grasses. There are many more biochemical reactions that result in opening and closing of stoma and it is a fascinating subject of investigation.

Stomata can sense small changes in carbon dioxide concentration and respond accordingly. A change in carbon dioxide concentration in atmosphere sets in motion a series of biochemical reactions in guard cells. This causes the movement of guard cells thus opening and closing of stomata. This property of stomata can be utilized in determining the air quality in the atmosphere. Sen and De (1992b) states that 'the attributes of stomata of certain fern taxa can be used as biomonitors of air-pollution'.

The study of ontogeny of stoma and stomatal complex has greatly improved our knowledge about the establishment of cell polarity, development, differentiation and pattern formation over leaf surface.

Another aspect of stoma is proving to be a fascinating subject for investigation. The stomata are the mode of ingress of some economically important plant pathogens. Ex. *Puccinia graminis* (responsible for rust disease of wheat), *Plasmopara viticola* (responsible for downy mildew of *Vitis vinifera*) and *Pseudoperonospora humuli* (responsible for downy mildew of *Humulus lupulus*) etc. The nature of stimulus that attracts fungal hyphae / zoospores to stomatal aperture has received considerable attention. A number of explanations have been forwarded but scarcely conclusive. The host and pathogen interaction lead to the production of phytoalexin by host and phytotoxin by pathogen. Phytoalexin and phytotoxin both affect stomatal movement. As for example, pisatin – a phytoalexin produced by peas inhibits stomatal opening in peas. Victorin—a phytotoxin produced by *Helminthosporium victoriae* causes stomatal closure.

Stomata are also proving to be the subject of investigation of antitranspirants. Antitranspirants are used to close stomata enabling a crop plant to withstand a period of drought. Due to closure of stomata photosynthesis is prevented thus reducing yield. Transpirational water loss is also prevented. Perennial crops are subjected to such treatment for a particular season when their survival becomes more important than good yields. Abscisic acid, aspirin etc. are used as antitranspirants.

Stomata are intimately linked to physiological processes that are critical for survival in terrestrial environments. They have adapted many ecological features that are vital for survival—a few of which are discussed in the chapter 29.

• Hydathodes: Haberlandt (1914) introduced the term hydathode to designate the structure that secretes water. Haberlandt distinguished two kinds of hydathode—epithem hydathode and epidermal hydathode. The epidermal hydathode, in addition to water, secretes ions and minerals. They are now referred to as salt glands (Metcalfe and Chalk).

**Epithem hydathode:** Generally referred to as hydathode or water stoma.

□ *Distribution*: Hydathode occurs in many plant families (e.g. Poaceae, Araceae, Ranunculaceae, Papaveraceae etc.) including pteridophytes (*Equisetum*). They may occur over the entire surface of the leaf or may be restricted to margins or tips. They are also observed at the tip of tendrils (e.g. *Vitis vinifera*). They are reported in plants of various life forms and of various ecological habits. Ex. *Lathraea* (a total parasitic plant), *Crassula argentata* (succulent leaf), *Ranunculus fluitans* (submerged leaves) etc.

□ *Structure* : A typical hydathode consists of (a) water pore, (b) epithem and (c) tracheid (Fig. 12.18).

(a) **Water pore:** The continuity of the epidermis at the hydathode region is interrupted by means of opening - termed water pore. These water pores are generally regarded as modified stomata, which are formed by guard cells. The guard cells lack chloroplasts and contain large nuclei and mitochondria, which are distributed throughout the cytoplasm. A large part of the protoplast is occupied by vacuum. The guard cells are unable to perform the closure movements and so the water pores always remain open. In this context, it may be mentioned that the water pores of *Impatiens* and *Tropaeolum* are active. The opening of the water pores of *Impatiens* is observed in presence of light.

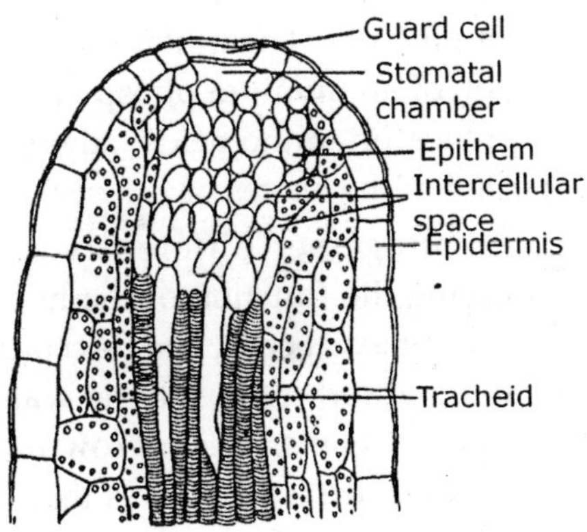


Figure 12.18

L. s. view of hydathode from *Primula*.

The water pores lead to a chamber termed sub-stomatal chamber. One or two layers of cells bound the chamber. Each cell is more or less isodiametric and contains large nucleus, large vacuole and mitochondria, which are localized mainly around the nuclei. The cells are larger than the neighbouring epithem cells.

(b) **Epithem:** It consists of parenchyma tissues, which are situated below the sub-stomatal chamber. The cell wall is thin and the cell lacks chloroplastids. In some cells fragmented and reduced vacuole is present. Electron microscopic study with the epithem cells of *Taraxacum officinale*, *Papaver rhoeas* reveals the presence of highly developed wall protuberances in them. These cells are known as 'transfer cell'. These cells contain dense cytoplasm with mitochondria, endoplasmic reticulum, ribosomes and polysomes.

A sheath of cells completely or incompletely surrounds the epithem of many dicots. The sheath of *Saxifraga* consists of two layers and the cell walls are suberized. The cells may contain tanniniferous substances.

Abundant intercellular spaces occur in the epithem except *Crassula argentata.* These spaces are sometimes equal to the size of a cell.

(c) **Tracheid:** The tracheids often abut on the intercellular spaces of the epithem. The epithem may have single, double or even triple vein supply. The veins fuse with each other before reaching the epithem. The terminal tracheids may extend through the epithem up to the sub-stomatal chamber (e.g. *Primula vulgaris*). In *Apomogeton distachyus* the vascular bundles extend up to the opening of the hydathode, as a result the surface layer may be torn. The terminal tracheids usually havė spiral, scalariform or annular thickenings.

**Function:** The main function of hydathode is guttation, i.e. exudation of water in liquid form. A single leaf of *Colocasia* may guttate 10ml—100ml of water in a single night (Kramer, 1956). It is generally agreed that guttation occurs as a result of root pressure. The guttation fluid may be composed of pure water or a dilute solution of ammonium phosphate, magnesium chloride, calcium nitrate etc. The occurrence of amino acids, sugars and water-soluble vitamins in the fluid are reported in the leaves of *Zea mays* and *Colocasia antiquorum.*

An altogether different view was expressed by Voronin *et al.*. (1976) concerning the function of the hydathodes of *Crassula,* though no experimental evidences have been cited. According to them the hydathodes, in addition to guttation, absorb atmospheric water from the dew. They observed that, in the species of *Crassula,* which grow in dry habitats, hydathodes generally occur on the exposed surface of leaves that led them to suggest this interesting view.

**Salt gland:** Salt glands occur in several angiospermic families (Plumbaginaceae, Tamaricaceae etc.) and in the some species of Acanthaceae,

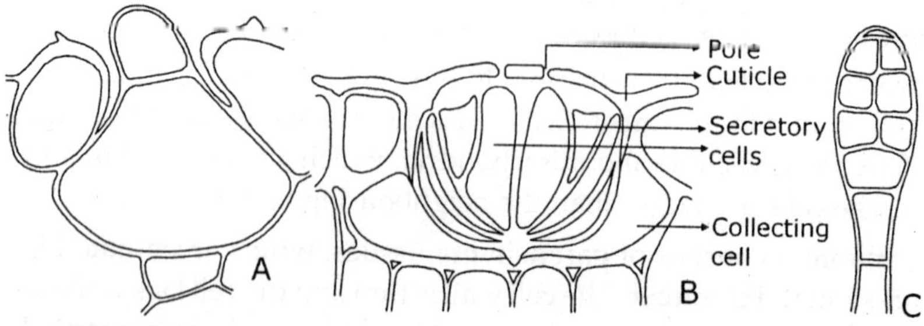


Figure 12.19

A. Salt gland of *Spartina*. B. Salt gland of *Limonium*. C. Salt secreting trichome (= trichome-hydathode) in *Cicer arietinum*. Adapted after Fahn, 1997 & 1979.

Chenopodiaceae, Primulaceae, and Gramineae etc. They also occur in mangrove plants (e.g. *Avicennia, Aegiceras* etc.). In contrast to epithem hydathodes these glands have no direct connection with the vascular elements and do not possess water pores.

Salt glands may be composed of two cells (e.g. *Spartina*) or many cells (e.g. *Acanthus, Aegiceras, Limonium* etc.) Fig. 12.19. The salt glands of *Atriplex* have hairs that consist of a distended head bladder cell with one celled or 2–3 celled uniseriate stalk (Fig. 12.20).

Salts are stored in these glands and may be exuded by guttation flow. These glands play an important role in the regulation of mineral contents of plants.

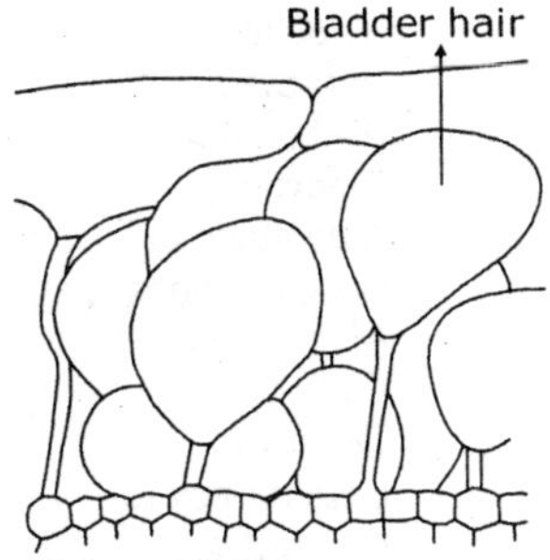


Figure 12.20

Cross section of the leaf of *Atriplex halimus* showing bladder hairs which are the salt glands. (After Fahn, 1979, p. 25).

**Function of epidermis:** (i) The thick cuticularized epidermal cells protect the inner tissues from adverse natural calamities, resist attack of insects and prevent the entry of pathogen; (ii) cuticularized epidermis prevents cuticular transpiration; (iii) guard cells of stomata, the chloroplast containing modified epidermal cells can photosynthesize; (iv) the specialized epidermal cells—the bulliform cells help in water storage, unrolling of developing leaves and the opening and closing movements of mature leaves; (v) velamen of orchid root and epiblema with root hairs help in absorption; (vi) the epidermis is the storage site for water, unwanted material (salt glands) and food; (vii) the watery vacuoles, present in the epidermal cells, absorb ultraviolet and infrared radiation and thus protect the plants from their harmful effects; (viii) the wax deposition, suberization, cutinization etc. represent the process of secretion, which in certain cases occur in the epidermis and (ix) the epidermal cells can perceive the external stimuli and can react accordingly.

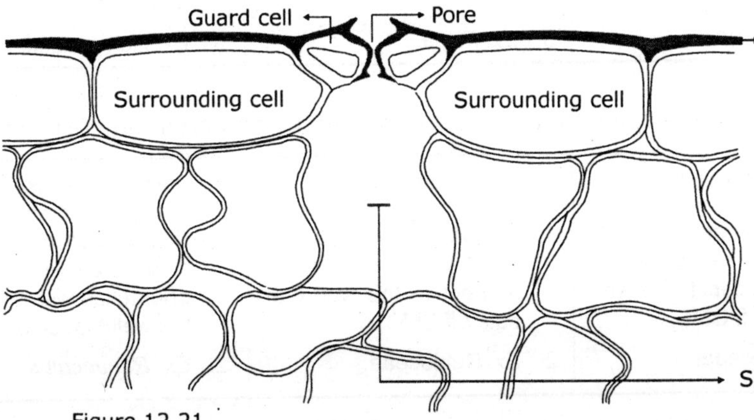


Figure 12.21

Diagram illustrating stomatal apparatus in transverse section. Adapted after Metcalfe & Chalk, 1979.

Table 12.1

Table depicting the types of stoma, number of subsidiary cell(s) and sources in literature. The older terms are mentioned in brackets.

| Types of stoma | No. of subsidiary cells | Sources in literature |
|---|---|---|
| Anomocytic | 0 | Metcalfe and Chalk (1950) |
| (Ranunculaceous) | 0 | Vesque (1889) |
| Anisocytic | 3 | Metcalfe and Chalk (1950) |
| (Cruciferous) | 3 | Vesque (1889) |
| Paracytic | 2, 4, 6 or more | Metcalfe and Chalk (1950) |
| (Rubiaceous) | 2 | Vesque (1889) |
| Diacytic | 2 | Metcalfe and Chalk (1950) |
| (Labiatous or Caryophyllaceous) | 2 | Vesque (1889) |
| Actinocytic | more than 4 | Metcalfe and Chalk (1950) |
| Cyclocytic | more than 4 | Stace (1965) |
| Tetracytic | 4 | Metcalfe (1961) |
| Hemiparacytic | 1 | Van Cotthem (1970) |
| Hexacytic | 6 | Van Cotthem (1970) |
| Gramineous | 2 | Metcalfe and Chalk (1950) |

Comparison between mesogenous, perigenous and mesoperigenous stomata.

| Mesogenous | Perigenous | Mesoperigenous |
|---|---|---|
| 1. The stomatal meristemoid forms guard cell mother cell (gcmc) and initials of subsidiary cells. | 1. The stomatal meristemoid directly acts as gcmc only. | 1. The stomatal meristemoid forms gcmc and at least one initial of subsidiary cell. |
| 2. The guard cells and subsidiary cells are formed from a single stomatal meristemoid. | 2. The guard cells and subsidiary cells are formed from separate meristemoid. | 2. The guard cells and at least one subsidiary cell are formed from a single meristemoid. The other subsidiary cells originate from separate meristemoid. |
| 3. Ex. Rubiaceae. | 3. Ex. *Pistia*. | 3. Ex. *Ranunculus*. |

Comparison between diameristic, parameristic and anomomeristic stoma.

| Diameristic | Parameristic | Anomomeristic |
|---|---|---|
| 1. The division of GMC is by a wall that is perpendicular with respect to the wall that cuts off GMC. | 1. The division of GMC is by a wall that is parallel with respect to the wall that cuts off GMC. | 1. The division of GMC is by a wall that has no special orientation with respect to the wall that cuts off GMC. |
| 2. Ex. Lamiaceae. | 2. Ex. Rubiaceae. | 2. Ex. *Ranunculus*. |

CHAPTER **13**

# Mechanical Tissue

The tissue that supports a plant and their growing organs against any deformation and provides mechanical strength is termed as mechanical tissue. Haberlandt (1914) called the mechanical tissue as stereome. Schwendener in 1874 (Foster, 1949, p. 93) termed the mechanical cells (e.g. collenchyma, bast fibres and libriform fibres) as stereids. Schwendener is of opinion that stereids collectively constitute the stereome or the mechanical tissue system of plants.

Plant organs are to withstand various strains like stretching due to presence of large fruits, bending due to natural calamities like high wind and passing animals and downpour, heavy snow etc. The stems are compressed due to presence of large number of branches and leaves at their top; branches are to withstand bending as they lie either in oblique or horizontal position; the fruit stalks tend to be extended due to weight of fruits; the roots are also subjected to extension when the stem bends due to high wind. In high wind the leaves lacerate. To withstand these strains the cell wall provides mechanical strength in all cells–unthickened or thickened, lignified or nonlignified. The cell walls of parenchyma, collenchyma and sclerenchyma provide mechanical rigidity to the plant.

The large size and structural strength of a woody plant are achieved by the cell wall. In these plants the constituents of cell wall contribute 95% of the dry weight of the wood. The strength and rigidity of the entire plant are due to cell wall.

The nonlignified wall derives strength from the cellulose microfibrils. It is believed that the mechanical strength of plant cell walls is due to the presence of skeletal framework formed by cellulosic microfibrils. Cellulose is the major component of paper, cotton etc. The cellulosic wall of wood is further strengthened by lignifications. In some species where cellulose is absent, other polysaccharides assume the strengthening role and may form microfibril. In algae the microfibril are composed of either xylose or mannose. Chitin is the strengthening materials of fungi where it forms microfibril. In some higher plants, the hemicellulosic xylans form long strands, which lie parallel to microfibrils. The lignin contributes strength in lignified walls. The strength is greatest in the direction parallel to the microfibrils in both types of cell walls. There are several layers in a cell wall and the orientation of microfibrils is different in each layer. The construction of the wall can be compared to plywood where the orientation of grains is analogous to the direction of orientation of microfibrils in the cell walls. So, a cell wall is enough strong to resist forces from any direction.

In unthickened cells, the wall gives limited support but the strength is greatly increased when the outward pressure of the turgid cytoplasm supplements it. The hydrophytes, some herbs and the growing seedlings gain the mechanical rigidity from turgid parenchyma cells. The orientation of microfibrils in the wall is of great importance to withstand the major mechanical stresses imposed by environment. In the stem parenchyma, the microfibrils are oriented transversely on the vertical walls so that the cells bend without breaking. In roots the microfibrils have steeply-pitched and helical orientation to resist extension forces.

Maximum strength is obtained from collenchyma and sclerenchyma. The wall of collenchyma cells is thickened by pectin, hemicellulose, protein and cellulose. Lignin is completely absent. The thickening materials are deposited mainly at the corners or at the tangential walls. In growing cells also, deposition occurs and the microfibrils show transverse and longitudinal orientation at different alternate layers. Collenchyma is living cell and retains its protoplast even when mature. So, it can regulate the deposition and orientation of wall materials according to the need of developing organs. The collenchyma cells, in addition to mechanical strength, also provide elasticity to the cell due to the presence of hydrated pectin on the wall.

Collenchyma (Figs. 13.1, 13.6) is one of the important mechanical cells of the growing organs and the mature organs of herbaceous plants. In stems it usually occurs just beneath the epidermis (e.g. *Cucurbita*). In *Tilia* stem it is separated from epidermis by parenchyma cells that may be one or two layered. It occurs as a continuous cylinder in *Sambucus, Helianthus* etc. It is present as individual bundle in *Cucurbita, Pastinaca* etc. Collenchyma cells are very conspicuous below the ridges in the petioles and stems with projecting ribs (e.g. *Chenopodium*). Collenchyma may occur as bundle cap (e.g. *Apium graveolens*). It may form a sheath

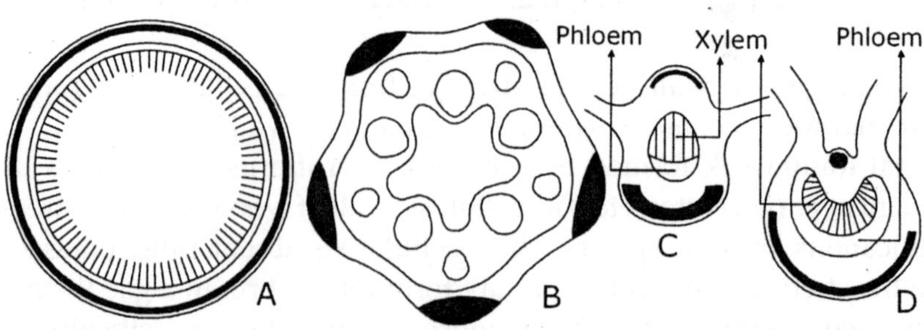


Figure 13.1

Diagram illustrating the distribution of collenchyma (shaded) in transverse section of stems and a leaf (C). A. *Sambucus* stem. B. *Cucurbita* stem. D. Midvein of *Humulus* leaf. Adapted after Esau, 1961 & 1965.

around the entire vascular bundle of many plants (Esau, 1965). In leaves collenchyma occurs in the petiole and blade. In the latter it may be present in one or two sides of the vascular bundles. In the leaf of *Sambucus* it occurs below the phloem of large vascular bundles and forms a prominent cap under the phloem. These caps are formed in large vascular bundles that protrude on the underside of the leaf as ribs. Collenchyma cells also occur along the margins of a leaf thus making it tough and resistant to tearing. The rinds of some fruit also gain mechanical rigidity from collenchyma, e.g. *Vitis, Cassia* etc.

The other important mechanical cell is sclerenchyma. It may be non-conducting sclerenchyma and conducting sclerenchyma. The former includes sclereids and fibres.

Sclereids give strength, resistance and inflexible protection in the organs where they occur (Fig. 13.2). In stems they may occur just below the epidermis (e.g. *Trichocereus chilensis*), on the periphery of vascular region (e.g. *Hoya carnosa*) and in the pith of *Hoya* and *Podocarpus*..In *Trichocereus chilensis* the epidermis is thin walled. Below the epidermis there occur six or seven layers of sclereids. They are very thick walled. Astrosclereids are present in the cortex of *Trochodendron, Pseudotsuga* etc. Sclereids are also present in the leaves of *Trochodendron, Olea, Pseudotsuga, Camellia, Mouriria* etc. In *Mouriria* the sclereids are branched or fibre like and are restricted to the ends of veinlet. Foster (1947) called these as terminal sclereids. Sclereids are also present in fruits (*Pyrus, Cydonia* etc.) where they are present either singly or in clusters. The epidermis of seed coat of *Phaseolus, Pisum, Glycine* etc. contains sclereids.

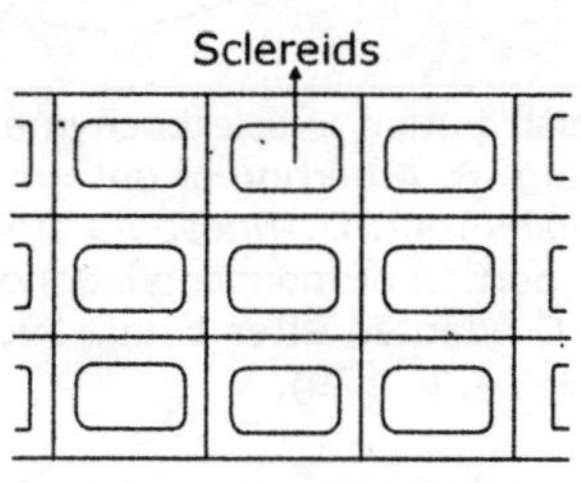


Figure 13.2

Diagrammatic representation of a group of sclereids that resists force from all directions because of the presence of cross wall (Adapted after Mauseth, 1988.).

Fibres (Fig. 13.3) add mechanical rigidity to the organ where they occur. The fibres may be extraxylary and xylary. The former fibres may occur uninterrupted just beneath the epidermis (e.g. *Zea mays* stem) or may be present in the ground tissue (e.g. *Asparagus* stem). They may be present in groups as isolated patches in the cortex (e.g. *Pandanus* root). A group of fibres may be present above the vascular bundles as bundle cap (e.g. *Xanthium* stem) or they may surround each vascular bundle forming bundle sheath (e.g. Maize stem). They may occur above and below of each vascular bundle (e.g. *Canna* stem). Fibres may be present beneath the endodermis in the form of an uninterrupted wavy cylinder (e.g. *Aristolochia* and *Tinospora* stem). Phloem fibres occur in the stem of *Sambucus, Tilia, Robinia* etc. These fibres may also occur in the secondary phloem (e.g. *Sequoia, Thuja, Tilia* etc.). Fibres are conspicuous in monocotyledonous leaves. They may form bundle sheath and connect the vascular bundles with the upper and lower epidermis,

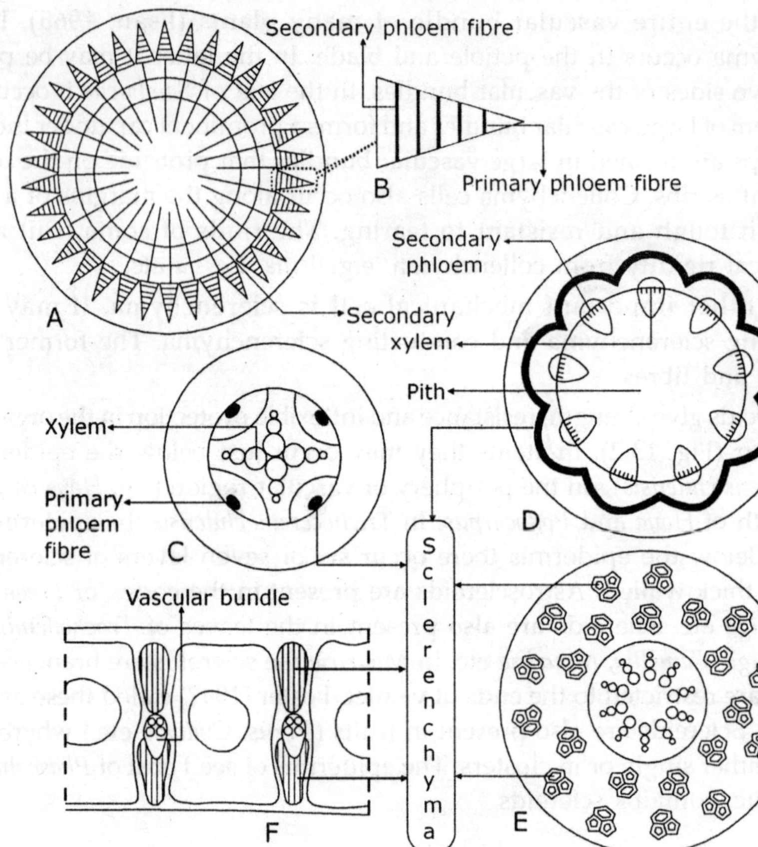


Figure 13.3

Diagrammatic illustration of distribution of sclerenchyma in transverse section. A. *Tilia* stem. B. A portion of phloem of *Tilia* stem enlarged. C. *Phaseolus* root. D. *Tinospora* stem. E. Stilt root of *Pandanus*. F. A portion of monocotyledonous leaf (*Phormium tenax*). A & C adapted after Esau, 1965. F. Adapted after Haberlandt (1914, p. 179).

e.g. Datepalm leaf. In this leaf sub-epidermal strands of fibres are very common and they are not associated with the vascular bundles.

The xylary fibres and the conducting sclerenchyma, i.e. tracheary elements also add mechanical rigidity to the organ where they are present. They are the constituents of vascular tissue and discussed in the chapter 9.

Some tracheary elements that are present in branches and stems show special adaptation to resist the force of gravity. This "reaction wood" differs in structure and location from ordinary wood. In conifers, this wood is located at the lower side of branches to resist compression while in dicots it is present on the upper side to resist tension. In conifer reaction wood, the innermost layer of the secondary wall is usually absent.

## Principles governing the construction and distribution of mechanical cells

The principle regarding the construction of mechanical tissue is to obtain maximum mechanical rigidity and elasticity with minimum expenditure of materials. This principle is observed in roots, stems, fruit stalks, branches etc., but the arrangement of mechanical cells varies greatly in them.

The organs of plants are subjected to different strains and stresses and accordingly the mechanical cells are developed to resist them. The resistances are designated as Inflexibility (resistance to lateral bending), Inextensibility (resistance to stretching), Incompressibility (resistance to compression) and Shearing stress (resistance to shearing action). The mechanical cells are developed accordingly, and the principle of construction and distribution of them are discussed below.

**Inflexibility:** The terrestrial plants with arboreal habit are mechanically rigid due to formation of heartwood at the centre. The herbs are the sufferers and so they form the mechanical cells to resist the forces. The aerial organs like stem, stilt root, branches etc. are the inflexible organs as they are subjected to lateral bending strain due to application of lateral forces like high wind. The distribution of mechanical cells in inflexible organs was compared with the railway line or girder, which is used in the construction of bridges, buildings etc. by Schwendener. The girder or railway line consists of two horizontal plates, upper and lower, which are connected by a vertical plate. In cross sectional view it resembles the English alphabet capital 'I' or two capital 'T's joined by vertical strands and so these are called I—girder. The horizontal plates and the connecting vertical plate are termed as flanges and web respectively (Fig. 13.4K). If a load is hanged at the middle of I-girder, it is subjected to bending strain. The upper flange tends to be compressed and the lower flange is subjected to extension (Fig. 13.4C). It is the two flanges that suffer the strains and the resistance to these strains depends on the flanges only. At the middle of the web, neither the compression nor the extension strain is operative. This line is termed as null-line as it is neutral regarding the strains.

A single I-girder can resist the lateral force from a single direction only. Forces may operate from any directions. To resist these forces several I-girders are constructed perpendicular to long axis in a circular manner. The flanges are arranged in such a way that the middle of the web or null-line intercepts a common point forming a composite I-girder that can resist forces from all directions (Fig. 13.4A). This simple I-girder formation is observed in plants and they operate either singly or in suitable combination of parallel or circular arrangements. The simple I-girder construction is observed in fibrovascular bundles (Fig. 13.4L). In these bundles bands of sclerenchyma are present above and below the vascular strand. These bands represent the upper and lower flanges and the vascular strand represents the web. The fibrovascular bundles are present in monocot leaves, e.g. *Cyperus*, where they are arranged in parallel. In monocot stem, the vascular bundles are completely encircled by sclerenchyma termed bundle sheath.

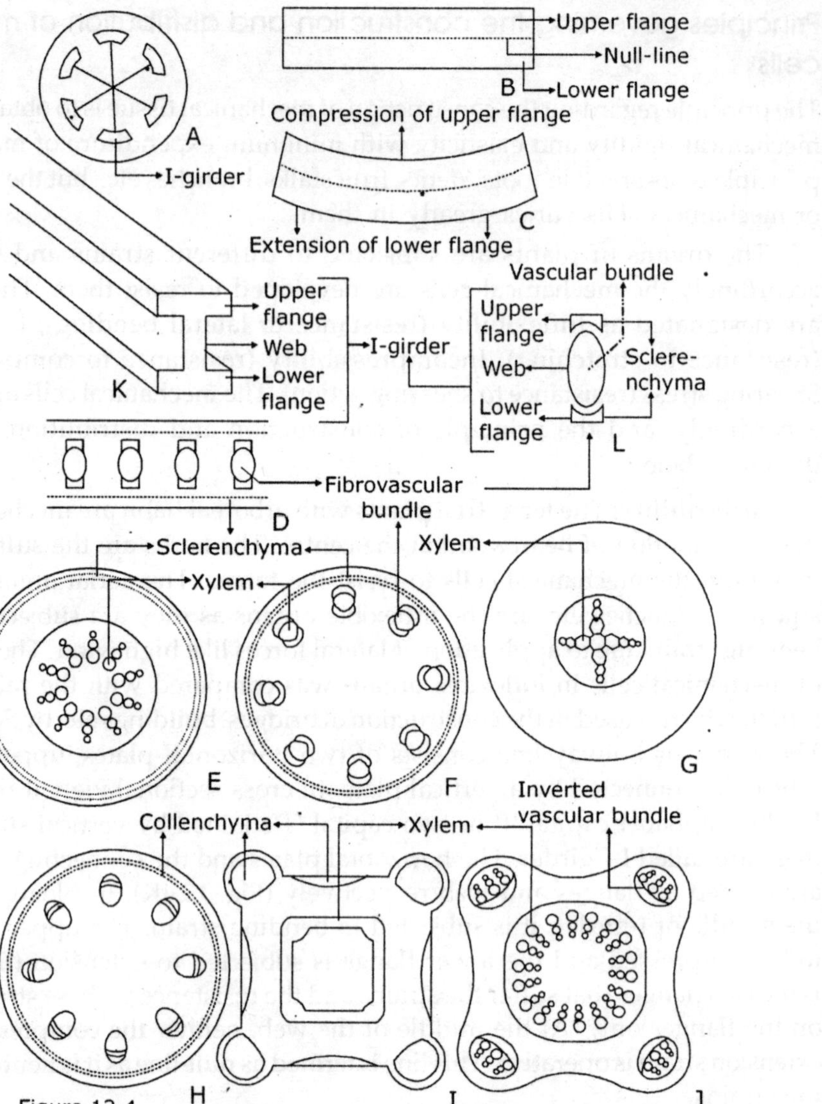


Figure 13.4

A. Composite I-girder formation. B. I-girder before bending. C. I-girder after bending. D. Distribution of mechanical cells and fibrovascular bundle in a leaf (diagrammatic). Diagrammatic representation showing the distribution of mechanical tissues in stilt root of maize (E), stem of *Cyperus* (F), dicot root (G), stem of *Helianthus* (H), stem of *Leonurus* (I) and stem of *Nyctanthes* (J). K. A portion of girder or railway line. L. A fibrovascular bundle.

Schwendener is of opinion that the same I-girder principle is applicable to cylinders also. When a cylinder is subjected to lateral bending strain, the side of the cylinder present towards the direction of force is extended and the side opposite to force is compressed. So the peripheral sides of a cylinder are mechanically rigid and they form the flanges and the web may be made up of some weak materials; in extreme case the centre region of a cylinder may be hollow.

In the cylindrical organs like stems etc. the mechanical cells are present at the periphery. So in the hypodermis, collenchyma and sclerenchyma cells are found in dicot and monocot stems respectively. They are present either continuously or as isolated patches. They form the flanges of composite I-girder. In some members of Compositae sclerenchymatous bundle caps are present. In *Zea mays* stem sclerenchymatous bundle sheath is present and the stele is atactostele. In rectangular or square stems (e.g. *Leonurus, Leucas* etc.) the mechanical cells (collenchyma) are present at the four corners (Fig. 13.4I). They form the flanges and the pith is the web. Thus diagonally opposite I-girders are formed. In *Nyctanthes* (Fig. 13.4J) inverted vascular bundles are present at the four corners that are considered as flanges.

Sclerenchymatous sheet or patches are present at the hypodermis of some monocot leaves and the fibrovascular bundles are arranged in parallel to resist the lateral forces and shearing stress.

**Inextensibility:** The rhizomes, roots and the other anchoring organs are sometimes subjected to longitudinal tension. The fruit stalks, the climbers and lianes are to withstand the longitudinal tension, as they are to bear the weight of fruits and their own weight respectively when they hang over the supporting object. Many rooted hydrophytes are longitudinally stretched due to water current. All these inextensible organs develop centralized mechanical cells to withstand longitudinal tension that can be compared with electrical cable. In the cable, mechanical rigidity is obtained from the central axile strand made up of metallic wire. This principle is observed in roots and other inextensible organs. In roots the stele is small as compared with the entire transverse section and the mechanical tissue like xylem and sclerenchyma are condensed within the stele. The formation of mechanical cells at the centre is also observed in hydrophytes (e.g. *Potamogeton lanceolatus*), in the fruit stalks (e.g. *Cucurbita*) and in the lianes (e.g. *Dioscorea*) etc.

**Incompressibility:** The cylindrical stems are to withstand the longitudinal compression as they are weighted at their upper end by the branches and leaves. If the line of action of the load coincides exactly with the longitudinal axis of the stem, mechanical tissues are condensed at the centre, and it acts like a pillar. But in nature this ideal condition is seldom found where the stem is as straight as a pillar. A slight asymmetry of construction is always present and the mechanical cells are distributed like inflexible organs.

The subterranean and submerged organs of plants are subjected to radial compression, as they remain surrounded by the soil and water respectively. In these organs mechanical cells are present at the peripheral sides to withstand radial compression.

The nodal roots of *Zea mays* and *Pandanus* exhibit a remarkable combination of incompressibility and inextensibility in the distribution of mechanical cells. The stilt roots occurring on the windward are to withstand longitudinal stretching and those present on the leeward suffer from longitudinal compression. Thus the same root is sometimes compressed and extended. So, in addition to centralized mechanical cells to withstand longitudinal extension, sclerenchyma is present at the periphery to ensure incompressibility. In *Zea mays* roots sheet of sclerenchyma is present at the periphery of cortex whereas in *Pandanus* isolated patches of sclerenchyma are present.

**Shearing stress:** The leaves and similar organs of terrestrial and aquatic plants are to withstand the shearing stress when they lacerate and move violently by the action of wind and water current respectively. Shearing stress is the action of force causing two contacting parts or layers to move apart in opposite direction. Dicot leaves are mechanically rigid, as they possess reticulate venation. In monocot leaves parallel I-girders formed by fibrovascular bundles are present. Moreover sclerotic strands are present at the hypodermal region and leaf margins. The leaf may be cuticularized to withstand shearing stress (Fig. 13.5).

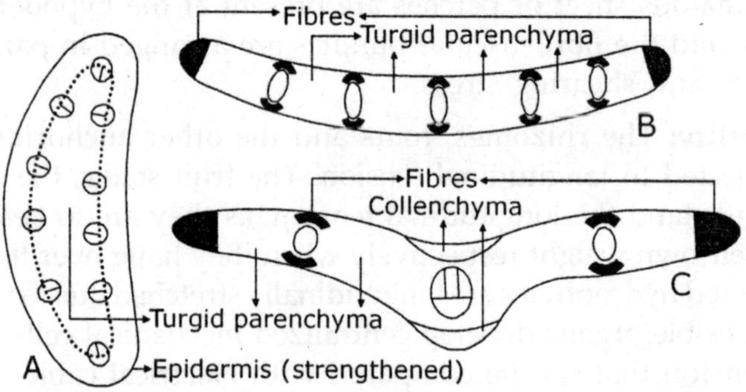


Figure 13.5

Diagrammatic representation of distribution of mechanical cells (fibres, collenchyma and turgid parenchyma) in A. succulent leaf of *Gasteria* (Liliaceae), B. monocotyledonous leaf and C. in a dicotyledonous leaf. Adapted after Cutler, 1978, p. 11 & 12.

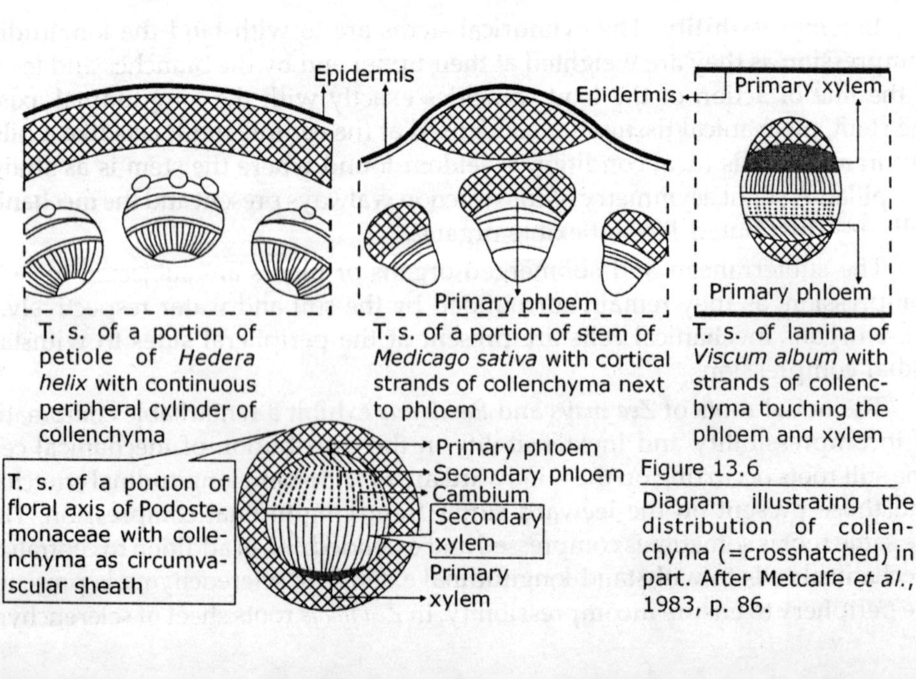


Figure 13.6

Diagram illustrating the distribution of collenchyma (=crosshatched) in part. After Metcalfe *et al.*, 1983, p. 86.

# CHAPTER 14

# Vascular Bundle

Vascular bundle is the isolated unit of the longitudinal strands of conducting tissues consisting essentially of xylem and phloem, frequently with a sheath of thick walled cells or other interspersed cells.

**Occurrence:** Vascular bundles are present in all plant organs, mainly in stem (including rhizomes or underground stems), leaves and roots (including aerial). In stem they are either scattered or located around the stem axis. Bundles may also be present in the cortex or pith and respectively they are referred to as cortical bundle (ex. members of Cactaceae, Oleaceae) and medullary bundle (ex. *Dahlia*, *Piper* etc.). The location of these bundles is considered as anomalous. In roots they are situated at the centre.

The differences between cortical and medullary bundle are as follows.

| Cortical bundle | Medullary bundle |
|---|---|
| 1. These vascular bundles occur in the cortex. | 1. The vascular bundles occur in the pith. |
| 2. These bundles are usually leaf traces that run independent of stelar bundles. | 2. These bundles are rarely leaf traces (e.g. *Dahlia*) that run independent of stelar bundles. |
| 3. The bundles may be normal, inversely oriented or centric—usually hadrocentric (ex. *Rochea*, *Aeonium* etc.). | 3. The bundles may be normal, inversely oriented or centric—usually leptocentric (e.g. *Phyllanthus*, *Ricinus* etc.). |
| 4. The bundles may be arranged in a ring (ex. *Casuarina*), in two rows (e.g. *Rhipsalis*) or restricted at the four corners of the stem (ex. *Nyctanthes*). | 4. The bundles may be two (e.g. *Achyranthes*) or several that may be scattered (e.g. *Amaranthus*), arranged in one (ex. *Piper*) or two (ex. *Boerhaavia*) rings. |
| 5. The bundles may be closed or open without any secondary vascular tissue formation. | 5. The bundles are usually open with little secondary vascular tissue formation. |
| 6. These bundles are concerned in translocation between stem and leaf. | 6. They are regarded as supporting tissue to stelar bundles. |

**Function:** Xylem and phloem compose vascular tissues, which transport water and dissolved minerals to the photosynthetic tissues through xylem. The elaborate food and other substances are translocated to all parts of the plant through phloem. Ground tissues between bundles are the medullary rays that assist in lateral transport of minerals, water and other substances. Vascular bundles also provide mechanical support. The fascicular cambium within bundles forms secondary tissues during secondary growth.

**Origin:** The vascular bundles originate from primary meristem. According to Hanstein and Schmidt plerome and corpus respectively give rise to vascular bundles along with other non-vascular tissues. Haberlandt (1914) called the meristem of vascular tissues as procambium that divides longitudinally. The peripheral derivatives form phloem while the inner forms xylem. The first cells of xylem to mature are termed as protoxylem whose development may be centripetal or centrifugal or both in relation to the centre of axis of the organ and respectively termed as exarch, endarch and mesarch (Fig. 14.1). The former two are the characteristic of root and stem respectively while mesarch arrangement is found

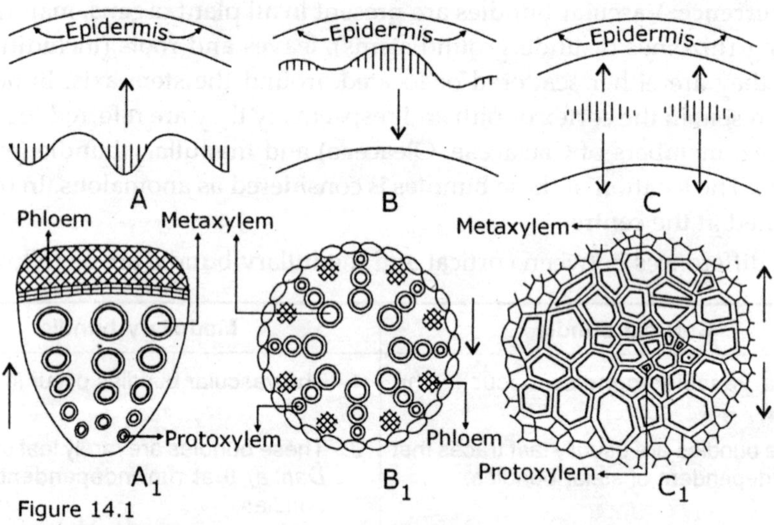


Figure 14.1

Diagrams showing order of development of primary vascular tissues in transverse views. A. & $A_1$. Centrifugal with endarch xylem. B. & $B_1$. Centripetal with exarch xylem. C. & $C_1$. Both centrifugal and centripetal with mesarch xylem.

in the stem bundles of *Tmesipteris* and ferns. The lately formed xylem is termed as metaxylem. Protophloem is the first cell of phloem to mature and their development is acropetal in relation to the centre of axis of organ. The lately formed phloem is termed as metaphloem.

Differences between protoxylem and metaxylem are given below.

| Protoxylem | Metaxylem |
|---|---|
| 1. It develops at the beginning of xylem differentiation. | 1. It develops late after protoxylem formation. |
| 2. It appears and matures in elongating organs. | 2. It appears in elongating organs but matures after the elongation is completed. |
| 3. In root it is situated towards the cortex, i.e. exarch and in stem it occurs towards the centre, i.e. endarch. | 3. In root it lies towards the centre and in stem it is situated towards cortex. |
| 4. Usually it has annular, spiral and rarely reticulate thickening. | 4. It has spiral, reticulate and pitted secondary wall. |
| 5. The vessels of it have narrower diametre than that of metaxylem. | 5. The vessels of metaxylem have wider diametre than that of protoxylem. |
| 6. The elements of it are usually destroyed in elongating organs. | 6. The elements of it are not destroyed as they mature after the organs completed their growth. |
| 7. No tylosis is formed. | 7. Sometimes tylosis is formed. |
| 8. Usually it contains few tracheids, vessels and so it is less compact than metaxylem. | 8. It has high proportion of lignified tracheids, vessels and fibres and so appears more compact that protoxylem. |

**Types of vascular bundle:** The two vascular tissues, i.e. xylem and phloem never occur alone. They are arranged variously and accordingly different types of vascular bundles are formed. Their arrangement may be side by side, i.e. appressed to each other or one surrounds or is surrounded by the other, or independent to each other, i.e. they are set apart (Eames and MacDaniels, 1947) and accordingly the following types are recognized (Figs. 14.2, 14.3).

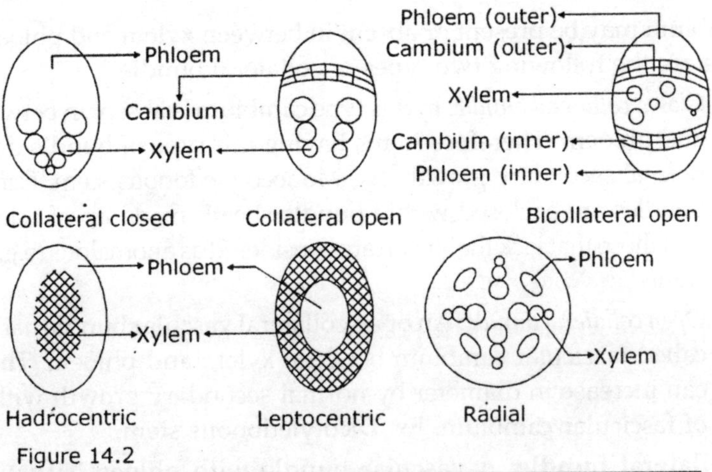


Figure 14.2
Diagram showing types of vascular bundle in t.s. (diagrammatic).

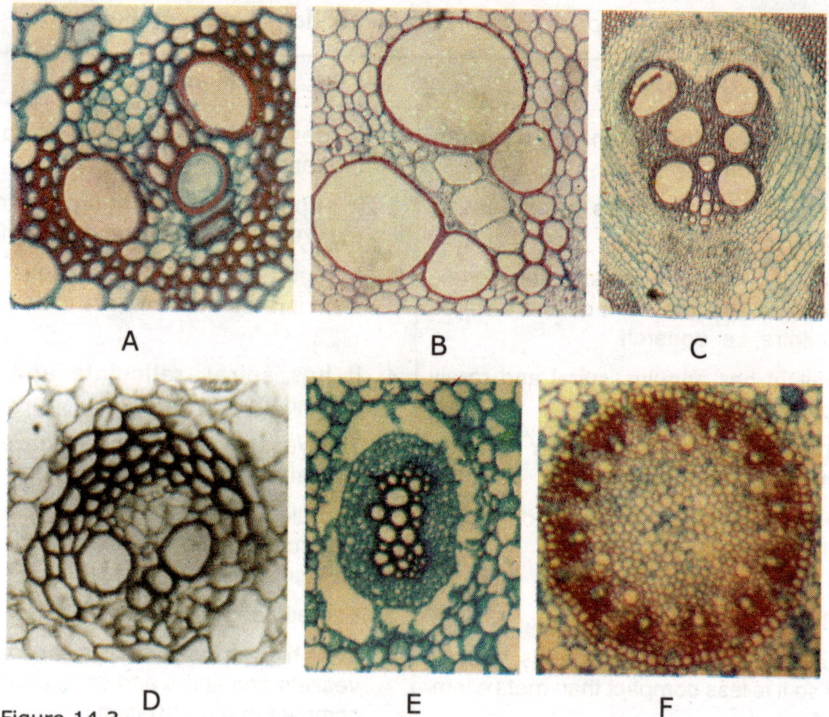


Figure 14.3

Microphotographs showing different types of vascular bundles. A. Collateral closed vascular bundle of *Zea mays* stem. B. Collateral closed vascular bundle of *Asparagus* stem without bundle sheath. C. Bicollateral open vascular bundle of *Cucurbita* stem. D. Leptocentric vascular bundle of *Dracaena* stem. E. Hadrocentric vascular bundle of *Selaginella* stem and F. Radial vascular bundle of *Vanda* root.

1. **Collateral bundle:** A vascular bundle in which a strand of phloem is present external to the strand of xylem on the same radius side by side is known as collateral bundle.

   Cambium may be present or absent in between xylem and phloem, and so there are the following two types of collateral bundle:

   (a) *Closed collateral bundle*: In this type cambium is absent in between xylem and phloem. Therefore stems having this type of bundle do not have normal secondary growth. Ex. Monocotyledonous stem. Usually these bundles are enclosed within bundle sheath made up of sclerenchyma and those that lack the sheath are considered as anomalous (e.g. *Asparagus* stem).

   (b) *Open collateral bundle*: An open collateral vascular bundle has cambium called fascicular cambium between xylem and phloem. The bundles can increase in diameter by normal secondary growth with the help of fascicular cambium. Ex. Dicotyledonous stem.

2. **Bicollateral bundle:** A vascular bundle with phloem situated on the peripheral and inner side of xylem is known as bicollateral bundle. A strip

of cambium termed outer cambium, is present between the peripheral phloem and xylem; another strip of cambium, termed inner cambium, is also present between inner phloem and xylem. The peripheral or external phloem is termed as outer phloem whereas the inner or internal phloem is called inner phloem. The sequence of vascular tissues in the bicollateral bundles from periphery toward centre is outer phloem, outer cambium, xylem, inner cambium and inner phloem. These bundles are open type as strips of cambia are present but the secondary thickening occurs only by the outer cambium, i.e. cambium present between the outer phloem and xylem. Ex. *Cucurbita* stem.

3. **Concentric bundle:** A vascular bundle in which one type of vascular tissue surrounds the other is known as concentric bundle. In this bundle xylem either encircles or is encircled by phloem and accordingly the following two types are recognized:
   (a) *Amphivasal* bundle: A vascular bundle in which xylem encircles the central strand of phloem is known as amphivasal bundle, also called leptocentric bundle. Ex. *Dracaena, Yucca.*
   (b) *Amphicribral* bundle: A vascular bundle in which phloem encircles the central strand of xylem is called as amphicribral bundle, also known as hadrocentric bundle. Ex. *Selaginella.*

   The concentric bundles, either amphivasal or amphicribral, are closed as there is no cambium in between xylem and phloem.
4. **Radial vascular bundle:** A vascular bundle, in which the primary xylem and primary phloem strands are separated from each other by nonvascular tissues and they are situated on alternate radii of an axis, is known as radial vascular bundle or radial bundle. These bundles are the characteristic of roots. There is no primary cambium in this bundle and the secondary thickening occurs by the secondary cambium that originates at the time of secondary growth in dicotyledonous root only. The dicot roots usually have four to six number of protoxylem poles in contrast to monocot root where many poles of xylem (more than six) are present. The number of protoxylem poles in a root may be 1, 2, 3, 4, 5, 6 or more. Accordingly they are called monarch, diarch, triarch, tetrarch and so on. The term polyarch is used when the number of protoxylem poles are more than six (Fig. 14.4).

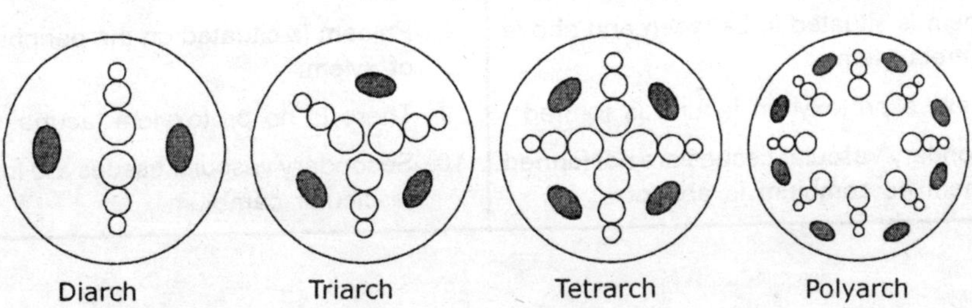


Figure 14.4

Diagram illustrating the different arrangements of the number of protoxylem groups in roots in cross-sectional view.

**Vascular bundle**

- → Collateral
  - → Collateral closed (ex. stem of *Zea mays*)
  - → Collateral open (ex. stem of *Helianthus*)
- → Bicollateral open (ex. *Cucurbita* stem)
- → Concentric
  - → Amphicribral (= Hadrocentric) (ex. *Selaginella* stem)
  - → Amphivasal (= leptocentric) (ex. *Dracaena* stem)
- → Radial (ex. root of dicotyledon and monocotyledon)

Chart summarizing the different types of vascular bundle.

Differences between the vascular bundle of monocot and dicot stem are given below.

| Vascular bundle of monocot stem | Vascular bundle of dicot stem |
|---|---|
| 1. The vascular bundles are scattered in the ground tissue. | 1. The vascular bundles are arranged more or less in a ring surrounding central pith. |
| 2. They are of different sizes, the smaller being restricted to peripheral sides. | 2. They are more or less uniform in size. |
| 3. They appear to be embedded in ground tissue. | 3. On the sides of vascular bundles medullary parenchyma occurs. |
| 4. Each vascular bundle is encircled by bundle sheath made up of sclerenchyma. | 4. Bundle sheath is lacking. |
| 5. There is no bundle cap. | 5. In some plants sclerenchymatous bundle cap is present. |
| 6. Each vascular bundle is conjoint, collateral or leptocentric and closed, i.e. cambium is lacking. | 6. Each vascular bundle is conjoint, collateral or bicollateral and open, i.e. there exists the cambium in between xylem and phloem. |
| 7. The protoxylem and metaxylem are arranged in the form of capital 'Y'. | 7. The proto- and metaxylem are arranged in rows. |
| 8. Phloem is situated in between and above the metaxylem. | 8. Phloem is situated on the peripheral side of xylem. |
| 9. Usually a protoxylem lacuna is formed. | 9. There is no protoxylem lacuna. |
| 10. Secondary vascular tissues are not formed in them as cambium is absent. | 10. Secondary vascular tissues are formed by fascicular cambium. |

CHAPTER 15

# Stele

Stele (= Greek word meaning a column) can be defined as the unit of vascular system that is made up of xylem, phloem, interfascicular tissues, medullary rays, pericycle and pith (if present).

The term stele refers to the central core of plant axis and is restricted only to primary tissues. The term is used in case of pteridophytes and is seldom applied in case of angiosperms and gymnosperms. Endodermis delimits a stele on the peripheral side. There are two basic types of stele – protostele and siphonostele.

I. **Protostele:** A protostele is composed of a solid core of xylem mass surrounded by phloem, which in turn remains encircled by pericycle. Endodermis delimits protostele on the peripheral side. In protostele pith is absent and the protoxylem is exarch. There is no leaf gap. Some dicotyledonous roots have radial stele where pith is completely absent. Such radial stele is also referred to as protostele. There are three types of protostele (Fig. 15.1) which are as follows.

(A) Haplostele: Haplostele has a cylinder of phloem that surrounds a smooth core of xylem. The xylem mass appears circular or oval in outline as seen in cross-section. Protoxylem is exarch. Ex. *Selaginella*, *Lygodium*, the extinct psilophytes *Rhynia* and *Horneophyton* etc.

(B) Actinostele (Figs. 15.1 & 15.3A): Actinostele has a cylinder of phloem that surrounds a star-like mass of xylem. As seen in cross-section, the xylem mass has radiating ribs of varying number. Protoxylem occurs at the tip of radiating ribs. Phloem also occupies the position between the xylem lobes and furrows. Ex. *Psilotum* and the extinct psilophyte *Asteroxylon*.

(C) Plectostele: Plectostele has masses of xylem that are in the form of plate-like lobes. As seen in cross-section the plates are of different sizes and some of the plates are united at one end. Cylinder of phloem surrounds xylem masses and phloem also occurs between xylem plates. Ex. *Lycopodium clavatum*.

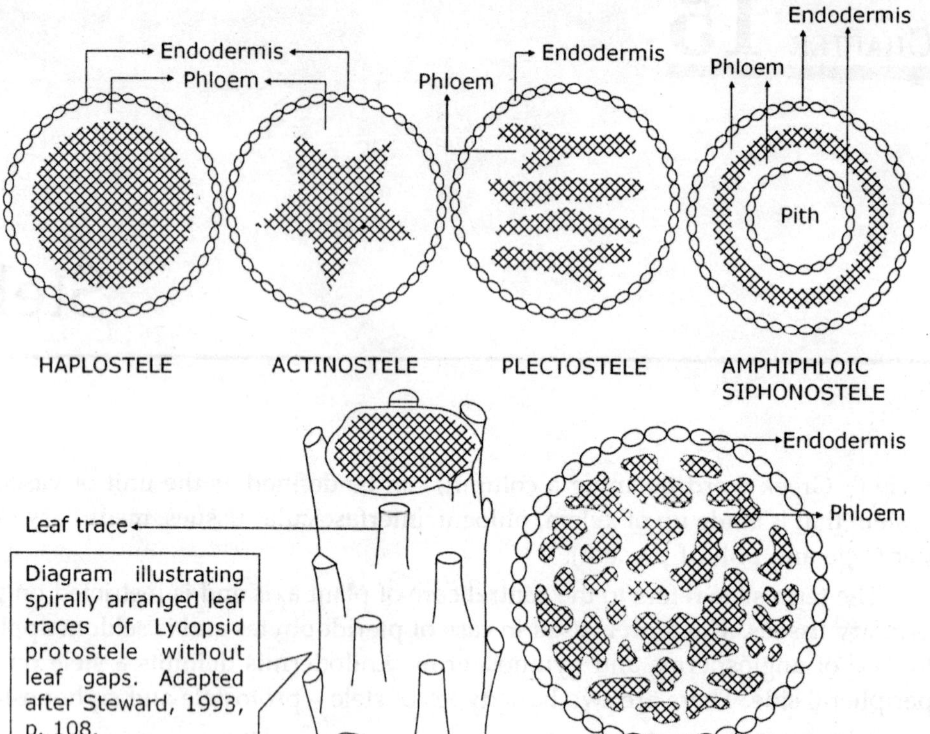


Figure 15.1

Diagrammatic illustration of different types of stele in t.s. view. Xylem = crosshatched.

*Lycopodium cernuum* has mixed protostele (Figs. 15.1 & 15.3B). In this type of stele as seen in cross-section the xylem is mesh-like mass that is uniformly distributed and appears to be embedded in the ground mass of phloem.

**II. Siphonostele:** Siphonostele, where xylem is in the form of a hollow cylinder, has parenchymatous pith at the central region of xylem. The xylem is surrounded by phloem that in turn remains encircled by pericycle. The whole stele is limited outside by a continuous endodermis. In siphonostele xylem and phloem are in the form of a continuous or split vascular cylinder. The following two types are recognized (Fig. 15.1) on the basis of position(s) of phloem in relation to xylem in siphonostele.

(a) Ectophloic siphonostele: Ectophloic siphonostele has a continuous cylinder of phloem surrounding the peripheral side of xylem. Parenchymatous pith occurs at the central region of xylem. The whole stele is delimited outside by a continuous endodermis. Sporne (1976) defines ectophloic siphonostele as 'medullated protostele'. Leaf gap is absent in ectophloic siphonostele. Ex. ferns like *Osmunda, Schizaea* etc. and dicotyledonous angiosperm like *Phlox, Lindenbergia* etc.

(b) Amphiphloic siphonostele (Figs. 15.1 & 15.3C): Amphiphloic siphonostele has cylinders of phloem on the peripheral and inner side of xylem. The peripheral phloem is termed as outer phloem and the other as inner phloem. Pericycle and endodermis appear both outside and inside of vascular tissues. To distinguish them the terms outer and inner pericycle, and outer and inner endodermis are used. The outer pericycle occurs surrounding the peripheral side of outer phloem whereas the inner pericycle is situated on the inner side of inner phloem. Outer endodermis delimits the whole stele and occurs between cortex and outer pericycle. Inner endodermis occurs between inner pericycle and pith. Ex. *Marsilea* and *Adiantum* etc.

Ectophloic and amphiphloic siphonostele may be cladosiphonic or phyllosiphonic. In cladosiphonic siphonostele the vascular tissues, in cross section, appear as continuous cylinder as the leaf traces are without gaps (e.g. *Selaginella*). The traces of phyllosiphonic siphonostele are with gaps and so the vascular tissues appear as isolated bundles in transverse section (e.g. *Polypodium*).

• Solenostele: Solenostele can be defined as a type of amphiphloic siphonostele with non-overlapping leaf gap. The leaf gaps are distantly spaced. Solenostele consists of two vascular strands—the small leaf trace and the large principal vascular strand as seen in a cross-section of stem at node. The principal vascular strand appears horse-shoe-shaped due to the presence of parenchymatous leaf gap. In the vascular strands phloem appears both outside and inside of xylem. The two vascular strands have individual continuous endodermis. The vascular strand is in the form of a continuous cylinder between leaf gaps. The vascular cylinder is interrupted at the places corresponding to the origin of leaf traces. Ex. *Anemia, Adiantum pedatum, Davallia* etc. It is to note that ectophloic siphonostele with non-overlapping leaf gap is also referred to as solenostele.

• Dictyostele: Dictyostele can be defined as a type of amphiphloic siphonostele with overlapping leaf gaps. The upper part of a leaf gap overlaps the lower part of the upper adjacent leaf gap. The gaps are not distantly spaced from each other and occur in parallel manner. As a result a longitudinal cylindrical network of interconnected vascular strands (Fig. 17.1C) is formed when viewed as three-dimensional object. The vascular strand is perforated as seen in cross-section. The vascular strands are arranged in a ring-like manner and parenchyma occurs in between the vascular strands. Each vascular strand is composed of xylem surrounded by phloem. This amphicribral vascular strand is surrounded by a pericycle and the whole being bounded on the outside by a continuous endodermis. Ex. *Mohria, Polypodium falcatum, Ophioglossum, Dryopteris* etc. Each vascular strand of dictyostele (Fig. 15.2) is referred to as meristele.

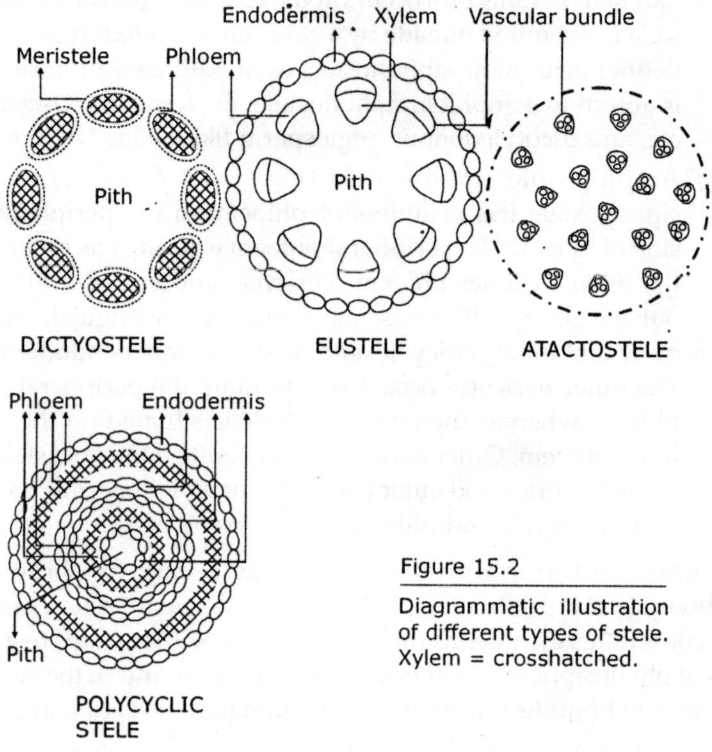


Figure 15.2

Diagrammatic illustration of different types of stele. Xylem = crosshatched.

• Eustele: Eustele can be defined as a type of ectophloic siphonostele with overlapping leaf gaps. The leaf gaps occur parallel to each other and are not distantly spaced. The upper part of a gap overlaps the basal part of the upper adjacent gap. When viewed as three-dimensional object the vascular strands form an interconnected network. The vascular strands are separate as seen in cross-section. Each vascular strand, also called vascular bundle, is conjoint and collateral. Parenchyma occurs at interfascicular region. All vascular bundles are arranged in a ring like manner. Pericycle surrounds the vascular bundles on the peripheral side, the whole being bounded by a continuous endodermis. Eustele is the characteristic of gymnosperm and dicotyledonous stem. Ex. *Helianthus, Xanthium* etc. Eustele with bicollateral vascular bundle is observed in the families Cucurbitaceae, Solanaceae etc.

• Atactostele: Atactostele can be defined as a type of eustele where collateral vascular bundles are arranged in an irregular manner (Fig. 15.2). It is the characteristic of monocotyledonous stem where there is no distinction between pith and cortex. Parenchyma bounded by epidermis is designated as ground tissue in monocotyledons. The vascular bundles are scattered on the ground tissue as seen in cross-section of stem. The vascular strands form an interconnected network when viewed as three-dimensional object. Ex. stem of *Zea mays* , *Asparagus* (Fig. 15.3E) etc.

• Polystele (Fig. 15.3D): Polystele can be defined as having more than one protostele as observed in the cross-section of a stem. In *Selaginella willdenowii* (Fig. 15.1) three protosteles occur. Each protostele has xylem surrounded by phloem, the whole being bounded by endodermis. In angiosperm polysteles occur in the families like Acanthaceae, Nymphaeaceae, Palmae etc. In cross-section polysteles appear to be scattered or organized into a ring. In longitudinal section it is revealed that individual steles, by anastomosis among themselves, form a network.

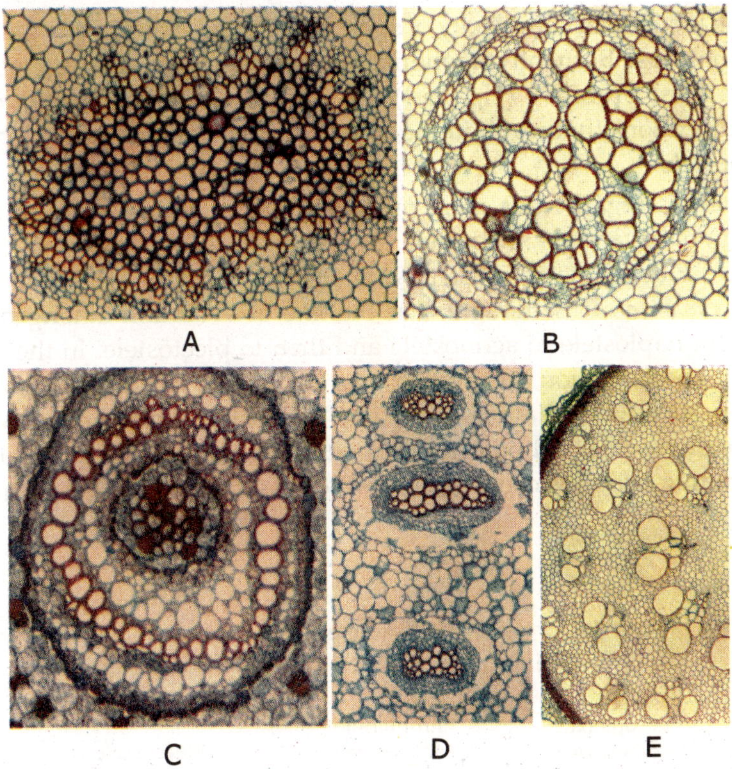

Figure 15.3

Microphotograph of different types of stele. A. Actinostele of *Psilotum* stem. B. Mixed protostele of *Lycopodium* stem. C. Amphiphloic siphonostele of *Marsilea* stem. D. Polystele of *Selaginella* stem and E. Atactostele of *Asparagus* stem.

• Polycyclic stele: Polycyclic stele can be defined as having two or more co-axial cylinders of vascular strands as observed in the cross-section of a stem. The individual cylinders are interconnected at the base of inner stele. Polycyclic steles are also referred to as polycyclic siphonostele where the innermost vascular cylinder is amphiphloic siphonostele (Fig. 15.2). The other cylinders remain separated by parenchyma. Ex. *Matonia pectinata.*

## Evolution of stelar forms

The well known genus *Rhynia* represents the simplest kind of vascular plant. It is rootless, leafless and the axis is dichotomously branched. A stele runs through the centre of stem. The cells in the very centre of stele consist of tracheids only and this xylem is among the simplest kind of wood cells known. A zone of phloem tissue surrounds the xylem core. Phloem is conspicuous among living plants, but not especially evident in extinct plants. This very simple kind of cylindrical stele where phloem surrounds a solid core of xylem is termed as protostele. Protostele is found not only in ancient plants like *Rhynia* and *Psilophyton*, but also in stems of *Lycopodium*, some ferns and in many kinds of dicotyledonous roots.

.With the increase in complexity of sporophyte of vascular plants the vascular system elaborates. Both phyolgenetically and ontogenetically protostele is regarded as most primitive (Fig. 15.4). During evolution protostele undergoes certain modifications that result in the formation of different forms of protostele and siphonostele.

Haplostele is regarded as most primitive among protostele. Actinostele is somewhat more advanced kind of stele, e.g. extinct *Asteroxylon*, extant *Lycopodium serratum* etc. Plectostele is most advanced kind of stele in protostele, e.g. *Lycopodium clavatum*. It is thought that the line of evolution among protostele proceeds from haplostele to actinostele and then to plectostele. In the course of evolutionary specialization the circular vascular strand of haplostele, as seen in cross-section, becomes stellate thus forming actinostele. This happens when the smooth core of vascular cylinder folds up at different places. Plectostele results when the folding becomes complete followed by separation of xylem masses.

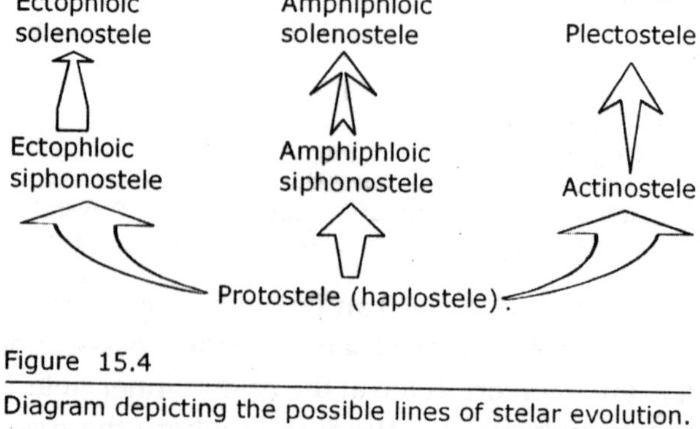


Figure 15.4

Diagram depicting the possible lines of stelar evolution.

It is always observed that non-living conducting cells like tracheids etc. are in close association with living cells of some kind. There is no known physiological basis of this association. Anatomists believe that 'because of the very close association between non-living conducting cells and living cells, there must exist some unexplained relationship between them' (Delevoryas, 1966, p. 66).

In haplostele, when the stele is small, each tracheid is not very far from a living cell, that is, phloem that surrounds the xylem. With the growth of sporophyte the stele also increases by the formation of living cells. Many tracheids may be of considerable distance from living cells. So to have a close association between non-living and living cells the xylem mass of haplostele becomes ridged and furrowed to form actinostele. During the course of evolutionary specialization plectostele arises where the living cells surround the xylem plates and intersperse between xylem masses.

The living-nonliving cell relationship is also maintained by the formation of pith. Pith is composed of living cells. Pith occupies the centre of stele and is composed of parenchyma cells. Most of the higher vascular plants exhibit steles with pith in the centre. Pith remains surrounded by xylem that in turn is surrounded by phloem on the peripheral side (ectophloic siphonostele) only or on both outside and inside (amphiphloic siphonostele). Such steles are referred to as siphonostele.

Ectophloic siphonostele, in its primitive form, has a continuous cylinder of vascular tissues. The advanced form consists of a network of bundles. In cross-section the bundles appear as separate vascular strands that are arranged more or less in a ring. Parenchyma exists at interfascicular regions. These regions are usually the position from where leaf traces originate from the vascular cylinder of stem. This interfascicular region is referred to as leaf gap. The leaf gaps may be considerably distant from each other. In the other case the upper part of one gap is at the level of the lower part of another gap. The gaps are said to be non-overlapping when they are distant from each other. In overlapping gaps the base or apex of one gap is at the level of apex or base of another gap. In the course of evolutionary specialization non-overlapping leaf gap arises and this results in the formation of siphono-eustele (also termed as ectophloic solenostele). In this stele the vascular tissue is interrupted at the place where leaf trace appears. In other places the vascular tissue is in the form of a continuous cylinder. Overlapping leaf gaps dissect the stele and thus eustele originates. In course of evolution atactostele arises where there is no definite arrangement of vascular bundles. Siphono-eustele, eustele and atactostele have collateral vascular bundles. The evolutionary specialization proceeds from ectophloic siphonostele to siphono-eustele and then to eustele. This line of evolutionary sequence is exhibited in seed plants.

Amphiphloic siphonostele, in its primitive form, is in the form of a continuous cylinder. Non-overlapping leaf gap appears in course of evolution and dissects the vascular cylinder to form solenostele (also termed amphiphloic solenostele). Ultimately overlapping leaf gaps appear and a dictyostele results. This line of evolutionary specialization proceeds from amphiphloic siphonostele to amphiphloic solenostele and then to dictyostele. This line of evolutionary sequence is exhibited in cryptogams.

The polycyclic stelar structure as seen in cross-section of *Matonia* etc. is regarded to have arisen from protostele by elaboration. In favour of this view the stelar structure of *Matonia pectinata* has been cited. The basal portion of this plant is protostele. In the higher region the stele is solenostele and ultimately the stele is polycyclic.

From phylogenetic point of view protostele is regarded as primitive than siphonostele. In the course of evolutionary specialization siphonostele originated from protostele. Protostele became large with the addition of living cells. The living cells at the centre never became modified into nonliving tracheids. They remained living and are referred to as pith. Thus siphonostele arises and the followings are the two schools of thought regarding the origin of pith namely—expansion theory and invasion theory.

• *Expansion theory*: Proponents of this theory believe that pith has originated from stelar tissues. During differentiation of xylem certain living cells at the centre never modified into non-living xylem. In protostele the solid central core of xylem mainly consists of tracheids. The siphonostele (of pteridophyta) possesses parenchyma cells at the centre that is pith, surrounded by tracheids. In the transitional forms some amount of parenchyma remains mingled with tracheids—termed mixed pith. So the evolution occurred in the following way: stele without pith—stele with mixed pith—stele with pith (siphonostele). In favour of this view Gewirtz and Fahn (1960) cited the stelar structure of *Ophioglossum lusitanicum*. The basal portion of this species is protostele, the upper portion shows siphonostele dictyostele and in the transition zone, below the level of the first leaf gap, some parenchyma cells are found at the centre in association with tracheids. The number of parenchyma cells gradually increases in the more upper level. It appears that during transition from protostele to siphonostele, some of the initials of tracheids are transformed to the initials of parenchyma that formed pith. This view also finds support due the presence of isolated tracheids in the pith of *Botrychium lunaria* and *Osmunda regalis*.

• *Invasion theory*: According to this view pith is regarded as extrastelar in origin. Jeffrey (1917), who proposed the invasion theory, is of opinion that the cortical parenchyma cells invaded the central core of xylem of protostele where they established as pith. According to Jeffrey the invasion occurred through the leaf—and branch gaps. As evidence Jeffrey cited the occurrence of inner endodermis between pith and vascular cylinder as endodermis is regarded as the integral part of cortex. Jeffrey opined that endodermis together with cortical parenchyma penetrated through the leaf—and branch gaps towards the centre of protostele. This view has been a subject of much debate because in *Selaginella*, *Pteridium* etc. the endodermis is stelar in origin. This view gained little support.

In this context it is worthy to mention the researches of Namboodiri and Beck (1968) who establish that leaf gaps do not exist in seed plants like ferns. The origin of eustele in seed plant is different from that of dictyostele. According to

Namboodiri and Beck haplostele is the original protostele. During evolution the xylem protrudes in the areas from where leaf trace originates. The rest of the areas of xylem strand become indented. Thus the margin of xylem mass becomes undulated and this results in the formation of actinostele. This stage is exampled by *Aneurophyton* (Fig.15.5). During evolutionary specialization the protruding areas of actinostele become isolated. As a result a stele is formed that consists of interconnected vascular strands. Parenchyma surrounds the vascular strands. This parenchyma is cortical parenchyma and it extends between the regions of vascular strands. This type of stele is represented by the extinct plant *Stenomyelon* (Fig. 15.5). In *Archaeopteris* (Fig. 15.5) the number of individual vascular strands increases and the vascular strands are arranged more or less in a ring surrounding a well-defined region of parenchyma. This parenchymatous region is termed as pith and it is also present in conifers (Fig. 15.5) and in other seed plants. The above mentioned evolutionary sequence regarding the appearance of pith reveals that the seed-plant-eustele is not a dissected siphonostele. Previously the node of seed plants was interpreted on the basis of nodal anatomy of ferns and to describe it the terms leaf gap, leaf trace, cauline bundle and common bundles were used. But the information and ideas presented by Namboodiri and Beck reveal that leaf gaps do not exist in seed plants as they do in ferns.

Whatever might be the reason, pith originated in protostele and that resulted in the formation of siphonostele. Siphonostele may be cladosiphonic or

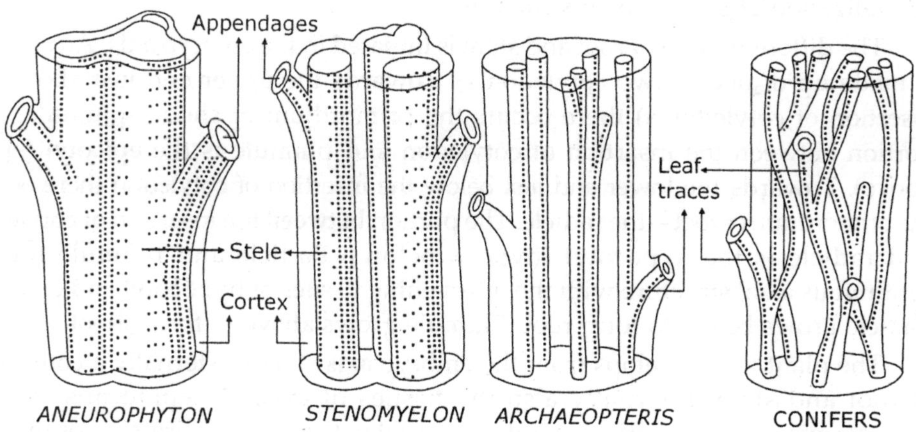


Figure 15.5

Diagrams illustrating the stelar organization of plants that Namboodiri and Beck, 1968 presented to hypothesize about the origin of pith in seed plants. Vide text for details. Adapted after Esau, 1977, p. 272.

phyllosiphonic. In cladosiphonic siphonostele the vascular cylinder is interrupted at the region of branch gap. Branch gap is associated with branch trace. Branch trace originates from central vascular cylinder and enters into lateral branches. Cladosiphonic siphonostele is observed in the microphyllous plants present in Psilopsida, Lycopsida and Sphenopsida. In phyllosiphonic siphonostele the vascular cylinder is interrupted due to the presence of leaf gaps. It is observed in plants present in Pteropsida.

CHAPTER **16**

# Root Stem Transition

The change in arrangement of vascular tissues of roots having separate strands of phloem and xylem with exarch protoxylem to collaterally placed phloem and xylem with endarch protoxylem of stem is commonly referred to as root-stem transition.

In higher plants root and shoot form a continuous axial structure. The simple form of axial structure is exhibited in Psilophytales. In higher plants the differentiated plant body into root, stem and leaf is regarded as an evolutionary specialization of simple axial structure.

The differentiation of root and stem is initiated in the embryo axis. An embryo is made up of one or two cotyledon(s). Towards the upper end just above the insertion of cotyledon(s) there occurs the primordium of shoot—plumule. The portion between the insertion of cotyledon and plumule is the epicotyl (epi = above). Towards the lower end just below the insertion of cotyledon there occurs the primordium of root—the radicle. The portion between the insertion of cotyledon and radicle is the hypocotyl (hypo = below). Radicle and plumule form a continuous axial structure with the intervening tissues of hypocotyl and epicotyl. Radicle grows below to form root. Plumule grows above to form shoot.

The plant that develops from the embryo axis shows externally a continuity of root and stem. Internally also the tissues of axis, i.e. epidermis, cortex, endodermis, pericycle and secondary vascular tissues are directly continuous from root to stem. Continuity is also present in primary vascular tissues, but they do not merge directly because the orientation of primary vascular tissues is markedly different in root and stem. In root the primary vascular tissues have independent strands of phloem and xylem that show alternate arrangement, the xylem being exarch. In contrast the stem shows collateral arrangement of primary vascular tissues with endarch protoxylem. So the primary vascular tissues of root and stem though apparently continuous merge with each other by changing one type of arrangement and structure to the other type. This change of arrangement of primary vascular tissues is usually referred to as vascular transition. The change occurs at successive levels of radicle—plumule axis. The place, where the change of orientation of vascular tissues occurs, is designated as transition region.

The exact region of vascular transition varies. The transition region may be at the top, centre or base of hypocotyl. Vascular transition may also occur at the top of radicle. The length of transition region is also variable. It may be short and the length ranges from one to three millimeter where the vascular transition occurs abruptly. In long transition region it may be several centimeters in length and the vascular transition occurs gradually. The transition region may be visible externally as a zone of depression or change in diameter. However in many plants the external morphology and transition region do not correspond exactly.

Eames and MacDaniels (1947) illustrated four types of root-stem transition in angiosperm among which three are from dicotyledons and one form monocotyledon. They described the dicotyledonous types in Type A, Type B and Type C, and the monocotyledonous type in Type D. In the diagrams (Fig. 16.1) the lower circles represent the transverse section of roots while the upper mosts are the transverse sections of stems. The three middle circles are successive transverse sections through transition regions. In the diagram three small circles grouped together represent xylem. In root the smallest circle is directed toward periphery to denote exarch protoxylem. In stem the smallest circle is toward centre to represent endarch protoxylem. The single oval figure present in the diagrams represents phloem.

**Type A** (Fig. 16.1.1A-1E): This type is represented by *Dipsacus, Mirabilis* and *Fumaria.* In the root there are four strands of phloem and four strands of xylem with exarch protoxylem. They are arranged radially. The stem shows four conjoint collateral vascular bundles with endarch protoxylem. The vascular bundles of stem are located in direct line above the phloem of root. Phloem during its upward course from root to stem has not changed its position. Each xylem strand of root, as it passes upward, forks radially into two halves. Among the two halves of xylem one swings to the right and the other to the left as they pass upward. Before joining with the phloem of stem inside the two halves of forked xylem rotate through 180°. Thus, the exarch xylem of root turns to endarch xylem of stem. So each vascular bundle of stem consists of an entire phloem, half xylem of one xylem strand and half xylem of another xylem strand of root.

**Type B** (Fig. 16.1.2A-2E): This type is exhibited by *Acer, Cucurbita, Phaseolus* and *Tropaeolum*. In the root there are two strands of phloem and two strands of xylem with protoxylem exarch. They are arranged radially. The stem shows four vascular bundles with conjoint collateral arrangement of vascular tissues with protoxylem endarch. The two strands of phloem of root, as they pass upward, divide and thus form four phloem strands. The entire phloem strand of root during its upward course elongates. Constriction appears in the middle and thus the phloem strand becomes dumb-bell shaped. The constriction becomes complete and as a result the number of phloem strands is doubled. As the phloem strands pass upward, each strand occupies the characteristic position in stem. The xylem strand of root also divides. During division, the xylem forks radially into two halves. Among the two halves of xylem one swings to the right and the other swings to the left as they pass upward. Before joining with the phloem of stem inside the two halves of forked xylem rotate through 180°. Thus the exarch xylem of root turns to endarch xylem in stem. So each vascular bundle of stem consists of half xylem strand and half phloem strand of root.

**Type C** (Fig. 16.1.3A-3E): *Medicago, Lathyrus* and *Phoenix* exhibit this type of vascular transition. In the root there are two strands of phloem and two strands of xylem with protoxylem exarch. They are oriented radially. The stem shows two conjoint collateral vascular bundles with endarch protoxylem. The collateral vascular bundles are located in direct line above the xylem strands of root. Xylem strands of root as they pass upward rotate through 180° and thus they become endarch in stem. Xylem strands of root do not divide and continue their course into stem directly. They rotate only to become endarch. Phloem strands of root as they pass upward elongate. Constriction appears in the middle and thus the phloem strands become dumb-bell shaped. The constrictions become complete and thus four strands of phloem are formed. Two strands of phloem, each derived from separate strands, unite above the xylem strand of stem. Each vascular bundle of stem is made up of one xylem strand and two half phloem strands of root. The number of vascular bundles of stem is equal to the number of phloem strands of root.

**Type D** (Fig. 16.1.4A-4E): *Anemarrhena*, a monocotyledon, exhibits this type of vascular transition. In the root there are four strands of phloem and four strands of xylem with exarch protoxylem. They are arranged radially. The stem shows two vascular bundles with conjoint collateral orientation of vascular tissues with protoxylem endarch. During vascular transition one diagonal pair of xylem strands (black-filled in figure) of root does not divide and continues its upward course into the stem directly. These two xylem strands rotate through 180° during their upward course to stem and thus they become endarch. Two vascular bundles of stem are located in direct line above the xylem strands (that do not split) of root. Each of the two other xylem strands of root, as passes upward, forks radially into two halves. Among the two halves of xylem one swings to the right and other to left as they pass upward. The two halves of xylem rotate through 180° and join with the entire xylem strand on both sides. Two of four phloem strands of root, as they pass upward, become close to each other and join above the entire and forked xylem strands in stem. Five vascular strands of root make a single vascular bundle of stem. Two phloem strands, one entire xylem strand and two forked xylem strands of root unite to form a single vascular bundle of stem. As a result the number of vascular bundle in stem is reduced to half the number of phloem strands present in the root.

**Significance of root-stem transition:** During vascular transition there is a reorientation of primary xylem and primary phloem. The reorientation of primary vascular tissues commonly takes place in the hypocotyl of the embryo in provascular strands.

Transition regions have their own specific orientation of vascular tissues. This orientation can not be compared with stem, root and leaf. So it becomes

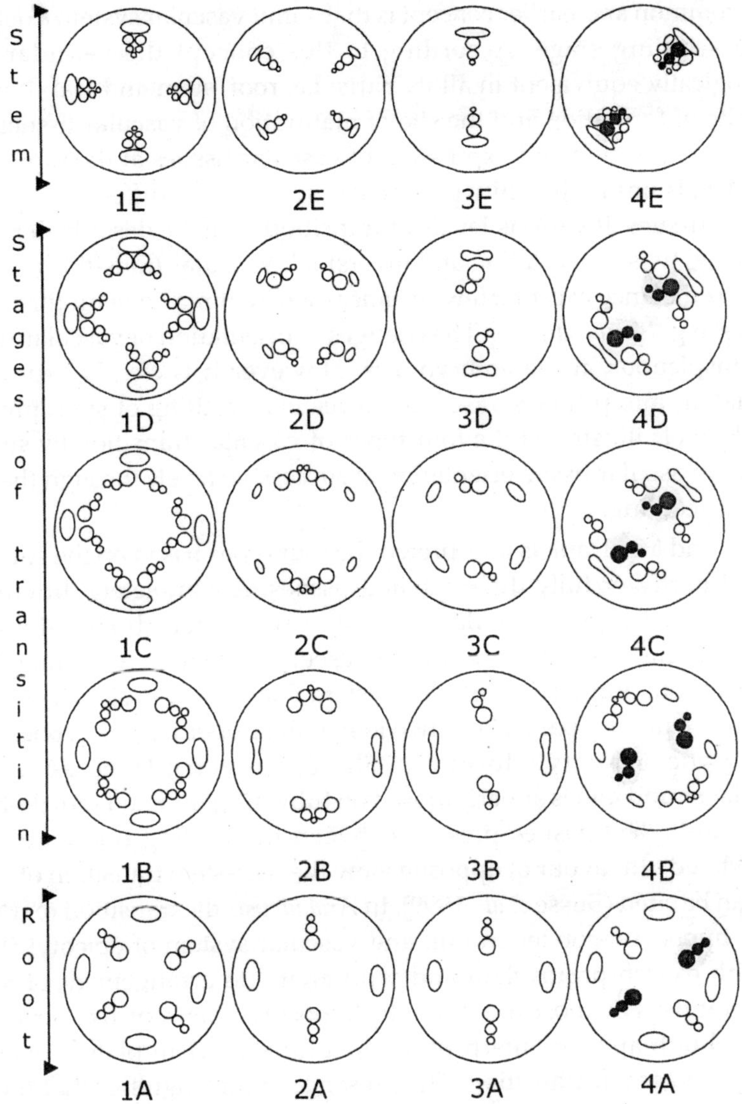


Figure 16.1

Diagrammatic representation of root-stem transition.

[ , & = vascular bundles of stem; , , & = xylem of root; = phloem; = split xylem; or = phloem about to split or unite]

Adapted after Emes and MacDaniels, 1947.

difficult to refer transition region in relation to embryo axis. Difficulty also arises to trace the relationship between transition region and the axes present above and below it. So several interpretations have been proposed regarding the structural and evolutionary significance of this region.

One common and earlier concept is that a unit vascular system exists in higher plants at seedling stage. According to this concept the vascular system is morphologically equivalent in all its parts, i.e. root, stem and leaf. It is regarded that the site of formation and the site of maturation of vascular tissues are same. Formation, maturation and expression of vascular tissues at the same site and at different levels imply the unity of vascular system. The difference in orientation of vascular tissues at various levels of transition 'may be described, figuratively, as branching, twisting, rotation and inversion'. Vascular transition occurs mostly over a short distance except a few instances where it may extend through several internodes, e.g. *Pisum sativum*. The changes of orientation of vascular tissues may not be conspicuous in the embryo axis. However it is clearly expressed upon differentiation into primary vascular tissues in a seedling of sporophyte. Eames and MacDaniels illustrated the four types of vascular transition by studying the diagrams of vascular tissue orientation at successive levels through the transition region of a seedling.

Eames and MacDaniels interpreted the four types based on the study of serial sections of partly or fully differentiated tissues. A completely different picture can be obtained if the vascular transition is studied during development. According to this opposite concept the vascular system in a seedling is initially not a continuous unit. The vascular system consists of radicular-hypocotylary unit and cotyledonary part. These two join in the upper part of hypocotyl to form radicle-hypocotyl-cotyledon unit. The epicotylary trace develops from provascular strand and is superimposed on fully differentiated radicle-hypocotyl-cotyledon unit. The tissues present between the uniting traces are 'mutually accommodated'. In favour of opposite view the root-stem transition of *Arabidopsis thaliana* can be cited (Busse *et al.*, 1999). In *Arabidopsis* the transition region consists of radicle-hypocotyl-cotyledon unit and vascular system of epicotyl. The root is diarch with exarch protoxylem and with alternate arrangement of xylem and phloem strands. For most of its length hypocotyl has root like structure. At a higher level pith appears interrupting the primary xylem plate. This marks the beginning of vascular transition. The presence of pith signifies that the region is unoccupied by vascular elements. Slightly above this level the two phloem strands divide into four strands. The split phloem strands move upward through the cotyledonary node. As they pass upwards two phloem strands become spatially associated with xylem. In the lower portion of cotyledon and in the region of midvein xylem strands complete the transition from exarch to endarch position. Each pair of phloem strands assumes a position above the endarch xylem. Thus the vascular strands become collateral completing transition. In radicle-hypocotyl-cotyledon unit, by sixth day, the procambial cells form cambium by periclinal division. Cambium is differentiated between primary xylem and primary phloem strands in the region of upper hypocotyl. This region also corresponds to the portion of cotyledonary node. Secondary vascular tissues are formed as a result of cambial activity and thus the primary vascular structure is disrupted. Then the vascular system of epicotylary trace begins to develop. After differentiation and maturation, the epicotylary trace is superimposed on radicle-hypocotyl-cotyledon unit. This happens after secondary thickening is initiated in the region of radicle-hypocotyl-cotyledon. Busse *et al.* examined by light and electron microscopy the transition region of *Arabidopsis* during vascular development and maturation.

Proponents of opposite view cited the anatomy of transition region of *Beta vulgaris* and *Daucus* earlier. They also exhibit radicle-hypocotyl unit and the commencement of epicotylary trace after secondary thickening in the region of hypocotyl-radicle (Esau, 1965 and Fahn, 1997).

The seedling of *Beta vulgaris* (Fig. 16.2) possesses two cotyledons and in between them there is an epicotylary shoot. On the lower side there exist the hypocotyl and the root. Anatomically the root shows diarch xylem and two phloem strands present alternating with xylem. The traces of each cotyledon consist of two vascular bundles that are partially fused along the protoxylem. In the root the protoxylem is exarch and this peripheral position is maintained up to hypocotyl. But at higher-level two elements of metaxylem are differentiated toward periphery on the lateral side of protoxylem. This orientation gradually leads to the formation of endarch xylem. The diarch xylem of root is thus gradually became four stranded that after fusion along the protoxylem form the cotyledonary traces. There are two phloem strands in the root. As they pass towards hypocotyl they branch and thus four phloem strands are differentiated. Each phloem strand associates with one metaxylem plate outside. Thus collateral vascular bundles are produced which persists in the cotyledons.

The epicotylary trace develops after the root-hypocotyl-cotyledon vascular strand is partly differentiated. The epicotylary traces are collateral with endarch xylem and they are connected with similarly oriented tissues present in the root. The epicotylary traces seem to be superimposed over the root-hypocotyl-cotyledon vascular traces. So in *Beta vulgaris* there exists no vascular transition between root and stem, but rather a mutual accommodation of vascular tissues.

In *Daucus carota,* like *Beta vulgaris,* the vascular continuity exists between radicle and cotyledon. The epicotylary traces join with root-cotyledon traces (Esau, 1940).

*Daucus carota* is dicotyledonous plant and each cotyledon of the seedling possesses three vascular bundles. The median bundle consists of xylem with exarch protoxylem that is continuous with the protoxylem of root. Two phloem strands are present lateral to the median xylem strand in root. The other two lateral vascular bundles are collateral with endarch xylem. These two bundles owe their origin from the diarch xylem of root. So the primary vascular tissues of cotyledon and root are continuous without any inversion. The epicotylary traces develop later. They are collateral and ultimately join with radicle-hypocotyl-cotyledon traces.

The ontogenetic studies of vascular transition of *Beta, Daucus* and *Arabidopsis* reveal that there is no root-stem transition and the phenomenon like twisting, inversion etc. are not operative as was illustrated by Eames and MacDaniels in the earlier or common concept. The modern concept, which is based on

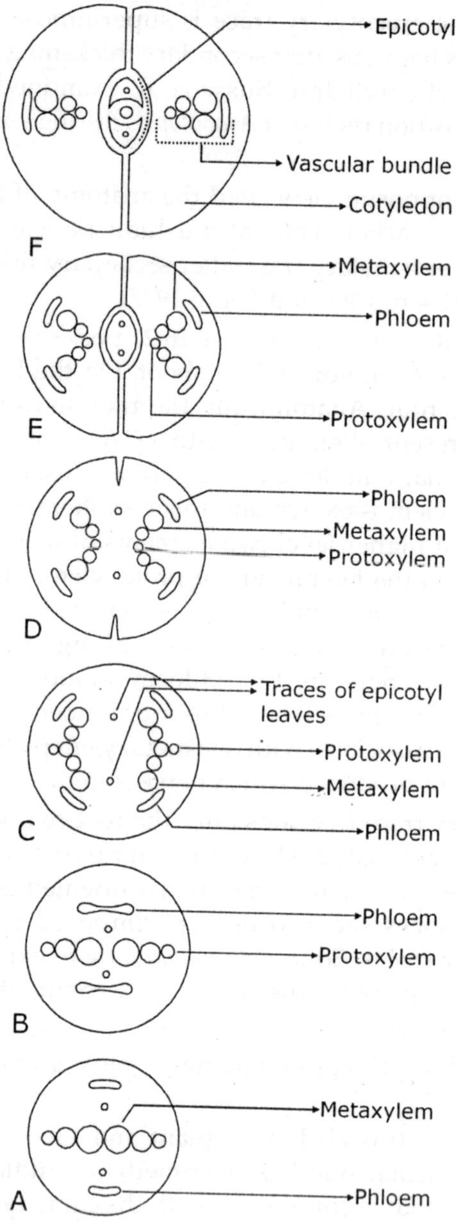


Figure 16.2

Diagram representing the transition region of *Beta vulgaris*. A. The diarch root. B-E. Arrangement of xylem & phloem in the transition region. F. Distribution of vascular bundle in cotyledons. Adapted after Esau, 1965.

developmental studies, believes in the double origin of vascular system. It reveals that during ontogeny vascular connections are established between radicle and cotyledon via hypocotyl. The epicotylary traces are developed later from procambium strand. The traces ultimately join with the fully differentiated radicle-hypocotyl-cotyledon unit and the tissues between the traces are mutually accommodated (Esau, 1965).

There is a physiologic interpretation of transition region (Esau. 1965). A seedling axis has two ends—root and shoot. Root end has its own root apical meristem. The derivative cells of this meristem form the structure of root including vascular tissues. The shoot end consists of cotyledons, epicotyl and foliar primordia. Epicotyl (if it is precocious), cotyledons and first foliar primordia influence in the formation of upper parts of the seedling axis. The derivative cells of root apical meristem form an impression at the upper part of seedling axis. The cells present at the base of upper part of seedling axis and the derivate cells of root apical meristem merge with each other thus forming a continuous vascular strand.

From evolutionary point of view it is postulated that the different arrangement of vascular tissues as exhibited by different plant parts are not equivalent. The alternate arrangement of xylem and phloem strands in root represents its primitive nature. In shoot the superimposed or collateral arrangement of vascular tissues is thought to be advanced. The different internal structures at the successive levels of transition region exhibit the evolutionary types. In this respect advanced evolutionary stages are exhibited by collateral arrangement of vascular tissues in shoot. In this region the primitive stage, i.e. alternate arrangement of vascular tissue in root is completely omitted.

An understanding of the structure of transition region is important due to followings.

- To interpret the homologies between root and shoot.
- To trace evolutionary stages of vascular tissues.
- To study the structure of radicle-hypocotyl-cotyledon unit.
- To study the relationship between epicotylary trace and radicle-hypocotyl-cotyledon unit.
- To make a comparative study of epidermis and cortex of root and shoot.
- To study the significance of pericycle.
- To study the differences between protoxylem and metaxylem, and protophloem and metaphloem.
- To study the developmental relationship between primary and secondary tissues.

CHAPTER **17**

# The Anatomy of Stems at Node

In stems there exist some differences in the arrangement of vascular tissues at the nodes and internodes. The difference is difficult to detect in monocotyledons because the vascular bundles are scattered in the ground tissue. The following discussion and terminology are based on the studies of nodal anatomy of dicot only.

In dicotyledon the vascular bundles are usually more or less in a ring and show different arrangements at the nodes and internodes. The vascular cylinders are generally continuous at the internode and their continuity is interrupted at the nodal region due to the emergence of bundles that terminate either at the leaf bases, axillary buds or stipules etc. At the node three types of bundles are recognized: (i) Leaf trace bundle: The single vascular bundle that connects the leaf base with the main vascular cylinder of stem is designated as leaf trace bundle. In a leaf there may be several leaf trace bundles that collectively are termed as leaf traces; (ii) Cauline bundle: The vascular bundles that entirely form the vascular system of stems are known as cauline bundles. Sometimes these bundles anastomose with each other and extend from stem to leaf as leaf traces; (iii) Common bundle: The vascular bundles, which run unbranched through a few successive nodes and internodes and ultimately terminate as leaf traces are called common bundles.

The arrangement of vascular tissues at the nodes is more complex than the internodes due to emergence of vascular traces to the leaves, buds, stipules etc. present at the node.

• Leaf trace and leaf gap (Fig. 17.1A, B & C): A leaf trace is defined as the cauline part of vascular tissue that departs from the stele of stem towards leaf base. Leaf trace may be a portion of cauline bundle as it occurs within the caulis, that is, stem. Cauline bundles entirely form the vascular system of stems. Sometimes cauline bundle departs from the stele towards leaf base thus forming leaf trace. Leaf trace is seen in the nodal region of stem. Leaf trace is an independent

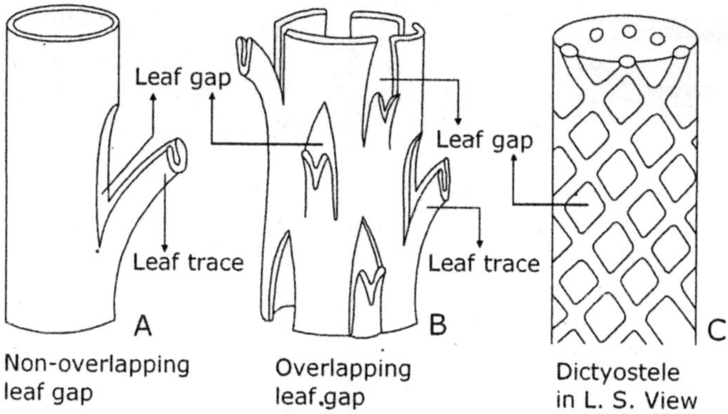


Figure 17.1

Diagrammatic illustration in three dimensional view of vascular cylinder at node with leaf gap (A & B) and dictyostele (C). A & B are redrawn after Sporne (1976, p. 15) and C from Chandurkar (1983, p. 129).

bundle that may occur through one or more nodes and internodes before bending away from the stem toward the leaf.

A leaf trace originates from the apical meristem of leaf primordium. After complete differentiation it joins with the vascular tissue of stem. Apart from leaf trace other vascular strands entirely occur in the stem. They constitute the cauline (= stem) bundle. They originate from apical meristem of shoot apex. Some cauline bundles may bend towards leaf thus forming leaf trace and these bundles are referred to as common bundles.

In the nodal region, a leaf trace bends away from the vascular cylinder of stem toward the petiole of leaf. From the base of petiole leaf trace extends into leaf blade where the trace forms vascular bundle of leaf. In the stem phloem occurs on the peripheral side of vascular cylinder. As the leaf trace is bent away from the vascular cylinder toward petiole phloem in the vascular bundle of leaf occurs on the abaxial side of leaf (Fig. 17.2A). As a result the vascular bundle of leaf has an inverted orientation of xylem and phloem in relation to vascular bundle of stem.

A cross-section through node reveals that there is a break in the vascular cylinder of stem, that is, the vascular cylinder is not in the form of a continuous ring. In this region the vascular tissue is interrupted and parenchyma cells fill the gap. Pith and cortex are continuous through this gap (Fig. 17.6). Such region is referred to as leaf gap and occurs opposite to leaf trace. A leaf gap is defined as the wide interfascicular region that is located opposite the upper part of leaf trace and is filled up with parenchyma cells through which pith and cortex become continuous.

There exist many variations in the number of leaf gaps and the number of leaf traces in different plants. The variation may be different in the same plant at

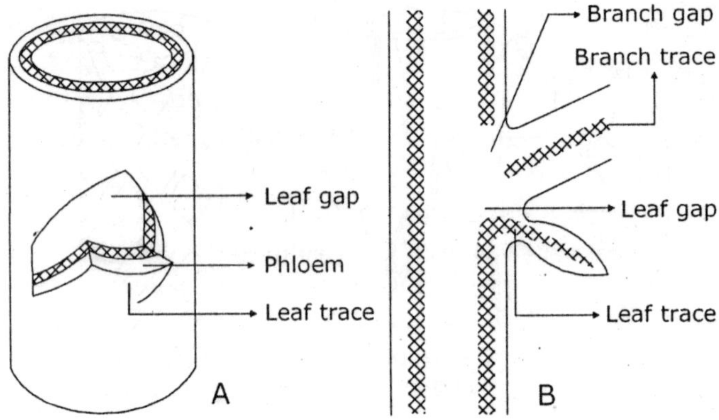


Figure 17.2

A. Diagram illustrating the position of phloem when a leaf trace enters to leaf base. B. Diagram showing the positions of leaf gap and branch gap in longitudinal sectional view of stem at node. Xylem crosshatched.

different levels, e.g. *Hypecoum procumbens*—the lower node of which has three traces per leaf whereas the upper nodes have single trace per leaf. The variations are subjects of comparative studies and therefore have taxonomic and phylogenetic significance. Atactostele is the characteristic of monocot stem and the arrangement of vascular strands at the node is highly complex. Comparatively the anatomy at the node of dicotyledonous stem is less complex. To describe the nodal anatomy of stem different terms are used. In the terminology the term lacuna replaces the term gap. Dicotyledonous nodes are very suitable to study the nodal anatomy and the different terminology used to describe the anatomy of nodes is chiefly based on dicotyledons. The following types of anatomy at nodes in dicots are recognized:

(a) Unilacunar single-trace node: This type of node exhibits one leaf trace to a leaf and the leaf trace is associated with one lacuna. Ex. *Spiraea*. Unilacunar nodes are exstipulate (Figs. 17.3 & 17.4A).

(b) Unilacunar two-trace node: This type of node has two leaf traces to a single leaf and one lacuna. The two leaf traces are associated with the single lacuna. Ex. *Clerodendron* (Fig. 17.3).

(c) Trilacunar node: This type of node has three leaf traces to a leaf and three lacunae. Each trace is associated with single lacuna (Fig. 17.4B). Ex. *Salix*. Among the three traces, the trace that occurs in the median position with reference to leaf is referred to as median trace. The others are called lateral traces. This type of node has stipules.

(d) Multilacunar node: This type of node has more than three traces to a leaf and more than three lacunas. Each trace is confronted with single lacuna. Ex. *Rumex*. Multilacunar node is exhibited in the plants that have sheathing leaf bases.

It is to note that the terms unilacunar-, trilacunar- and multilacunar node are applied with reference to a single leaf. In this sense in unilacunar node each leaf is associated with one lacuna. A node may consist more than one leaf. A cross-section of such node reveals the presence of more than one lacuna and more than one leaf trace. Such node is to be referred in relation to the number of lacuna(s) associated to a single leaf, that is, in trilacunar node one leaf is associated with three lacunas and in multilacunar node one leaf is associated with several lacunas. In other words, whatever may be the number of leaves present, a node is characterized with reference to the number of lacuna(s) present in a single leaf.

In unilacunar node a leaf has one trace, in trilacunar node three traces diverse to a single leaf and in multilacunar node several traces are present in a single leaf. Usually one gap is associated with one trace but in two-trace unilacunar condition two leaf traces confront to a single gap and in three-trace unilacunar condition three leaf traces are associated to a single gap (Figs. 17.3 & 17.4C).

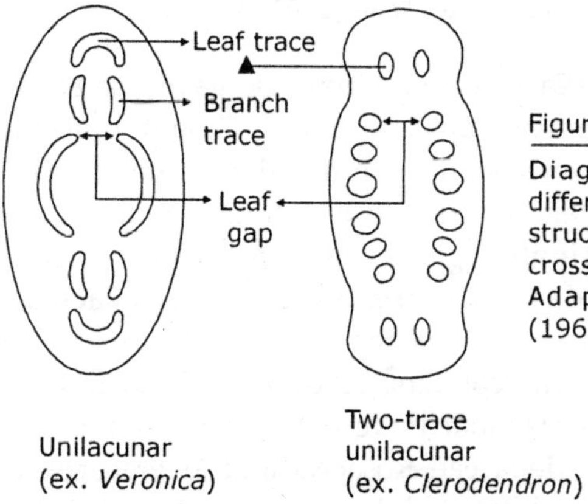


Figure 17.3

Diagram illustrating different types of nodal structures as seen in cross sections of stems. Adapted after Esau (1961, p. 212).

It is mentioned previously that a leaf gap is identifiable at the node. In internode also leaf gaps may be identified when a leaf trace has oblique course through a part of the internode. Studies on the cross-section of internode, if followed level by level, will reveal position(s) of leaf gap(s) and the continuity between ground tissue and pith.

A definite nodal type generally characterizes a taxon. Unilacunar node is found in Centrospermae and in certain members of Myrtaceae (*Eucalyptus*), Lauraceae (*Laurus*) etc. Trilacunar node occurs in Compositae (*Chrysanthemum*), Salicaceae (*Salix*), Brassicaceae (*Brassica*) etc. Multilacunar node is observed in Polygonaceae (*Rumex*) etc. The anatomy of the node is regarded an important aspect in the study of phylogeny in dicotyledons, since a definite nodal type characterizes a taxon. According to Sinnott (1914) trilacunar node is primitive among dicotyledons. During evolution it gave rise to unilacunar type by reduction in the number of gaps and traces. Reduction occurred either by the disappearance of the two lateral gaps with the associated two lateral traces or the two lateral traces became confluent with the median trace. In the latter case a single bundle is formed consisting of three traces. This bundle is associated with a single gap to

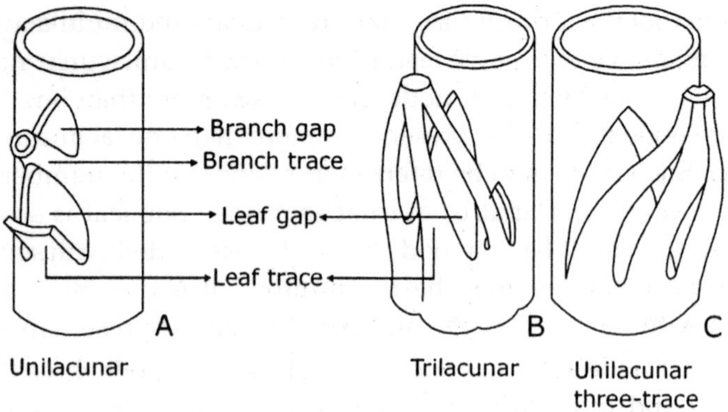


Figure 17.4

Diagram illustrating in three dimensional view of leaf trace, leaf gap, branch trace and branch gap. A & B are redrawn after Esau (1961, p. 209) and C is redrawn after Fahn (1997, p. 196).

form unilacunar node. Trilacunar condition also gave rise to multilacunar type by the formation of more new gaps and traces. Though unilacunar condition is considered as advanced, later studies on nodal anatomy by Bailey (1956), Fahn and Bailey (1957) and others reveal that unilacunar condition is primitive as this type is found in some primitive groups like pteridophyta, fossil gymnosperms like Bennettitales and Cordaitales, *Ginkgo* and *Ephedra*. Unilacunar two-trace node also seems to be primitive as it is represented by extinct Cordaitales and Bennettitales, and extant ferns, conifers, *Ephedra* and the primitive dicotyledonous genus *Austrobaileya* etc. Trilacunar node is widespread and is exhibited in the families like Winteraceae, Meliaceae, Rosaceae and Asteraceae etc. Degeneriaceae and Chenopodiaceae etc. exhibit multilacunar condition. *Lepidium latifolium* has unilacunar node with several leaf traces. It is regarded as more primitive than the unilacunar one-trace node.

Previously the nodal structures were described on the basis of number of leaf traces that are associated with each leaf, e.g. one-trace, two-trace, three-trace and multi-trace. Later the nodal anatomy is interpreted on the basis of leaf gaps that are associated with each leaf, e.g. unilacunar, trilacunar and multilacunar. Still later leaf gap and leaf trace—these two aspects were combined to describe a node, e.g. unilacunar-one trace, unilacunar-two trace etc.

Before the discovery of unilacunar two-trace node, trilacunar node was regarded as central type from which unilacunar - and multilacunar node arose. It is interpreted that unilacunar node is more advanced than trilacunar node. Later in the light of above facts it is now interpreted that the evolution of nodal structure.proceeded in the following two sequences: (i) Two trace unilacunar gave rise to trilacunar, which terminated in multilacunar condition; (ii) Two trace unilacunar, by the loss of one trace, gave rise to one trace unilacunar that formed trilacunar node by the addition of two new gaps associated with two traces. The multilacunar condition is derived from the trilacunar type by the addition of more new traces and gaps (Fig.17.5). This evolutionary sequence may be observed in a single family namely Chenopodiaceae. The trilacunar type may also give rise to one trace unilacunar condition.

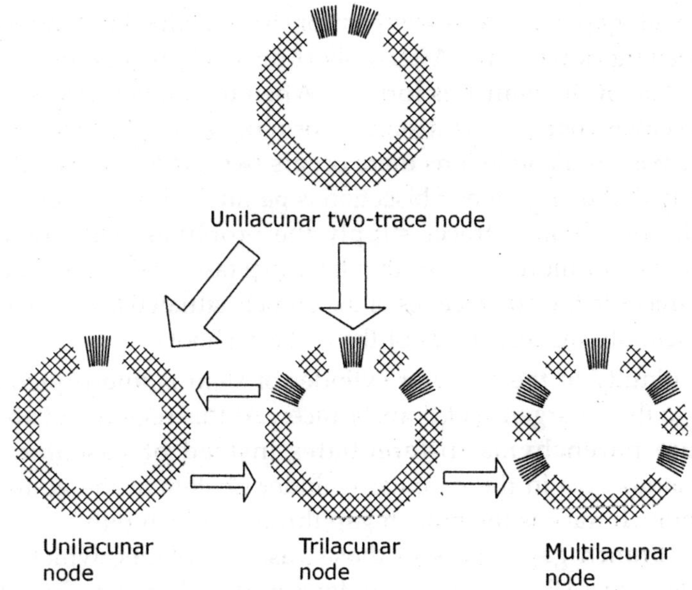


Figure 17.5

Diagrams illustrating the probable lines of development of nodal vascularization. Adapted after Fahn, 1997, p. 200. Xylem crosshatched. |||| = leaf trace.

- Branch trace and branch gap: Branch trace can be defined as the vascular trace that originates from the vascular cylinder of stem and enters to a branch. The position of the origin of a branch is from the axil of a leaf (Fig. 17.4A). In this position the vascular cylinder of stem is discontinuous. Parenchyma occupies this region as observed in a cross-section of stem at node. This interrupted region of vascular cylinder due to the presence of branch trace is designated as branch gap. Two branch traces are associated to a single leaf gap. Branch trace directly diverges to a branch from the main vascular cylinder of stem without running obliquely through the internode. The branch traces depart from the main vascular cylinder of stem toward right and left of the median leaf trace. After a short distance the traces coalesce and form a complete stele similar to main vascular cylinder. So the vascular bundle of branch has the same orientation of xylem and phloem in relation to the vascular bundle of stem.

It is previously mentioned that branch develops from axillary bud that originates on the stem at the axil of leaves. From the time of initiation the buds are connected by vascular traces to the vascular strands on the main axis. These vascular traces are referred to as bud trace or branch trace or ramular trace (Eames and MacDaniels, 1947). Axillary buds form axillary shoots whose first foliar structures are the prophylls. In dicotyledons, usually two branch traces emerge out from the vascular cylinder of stem. In exceptional cases the branch trace may be one (e.g. *Peperomia, Cayratia* etc.) or more than two. Sometimes medullary bundles may enter into the bud (Ex. *Dahlia*). In case of single branch supply the vascular cylinder appears as crescent or horseshoe shaped in cross sectional view with the opening downward. After a short distance the opening closes and the cylindrical stele of the branch is formed. When the branch traces are more than one they coalesce after a short distance, forming a complete stele similar to that of the main axis. In dicotyledons there occurs two prophylls, which are oriented in such a way that their plane of bisection is parallel with the plane of the axillant leaf. Initially two branch traces supply the prophylls. The branch traces are composed of one or more bundles that later increase in size due to the formation of vascular traces to the other leaves of the branch, situated above. Thus the branch traces, are actually the leaf traces of the axillary shoot.

The continuity of the vascular cylinder of stem is interrupted at the nodal region due to the emergence of branch traces. At this region and above the point of departure parenchyma differentiates instead of vascular tissues. The parenchymatous area in the vascular cylinder of stem at the node immediately above the branch trace is the branch gap through which pith and cortex become continuous. Branch gap occurs in those vascular plants, which have pith. In pteridophytes, where the vascular cylinder is protostele and devoid of pith, branch gap is absent. So the branch gap in association with leaf gap results in the formation dissected siphonostele.

There exists a definite correlation between leaf—and branch trace. As seen in cross-section of a stem at node a leaf trace occurs on the peripheral side. It is followed by branch trace towards the inner side (Fig. 17.3). A node that bears a single leaf exhibits the following sequences from periphery towards centre—leaf trace, branch trace and leaf gap on the side where the leaf is inserted. When there are two leaves in each node the above mentioned sequences of leaf—and branch trace and leaf gap are observed on the opposite sides in a cross-section of stem at node. In longitudinal section the positions of branch gap and leaf gap (Fig. 17.2B), and their relation can be observed individually.

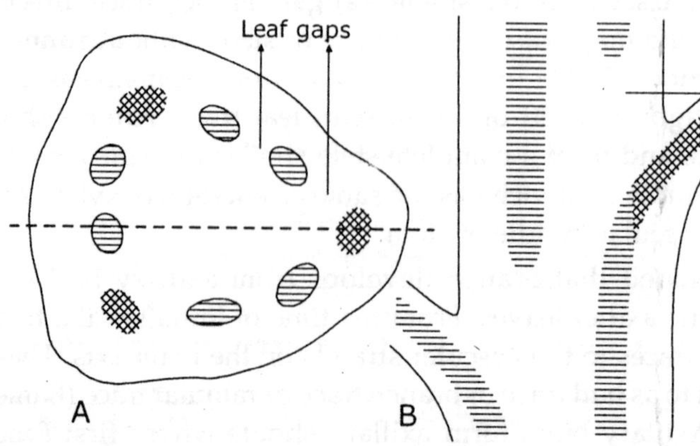


Figure 17.6

A. T. S. through node having dictyostele. B. L. S. of stem along the dashed line. Leaf traces = crosshatched. Primary vascular tissues = hatched.

Comparison between leaf trace and branch trace.

| Leaf trace | Branch trace |
|---|---|
| 1. Leaf trace is the cauline part of vascular tissue that departs from the stele of stem towards leaf base. | 1. Branch trace is the cauline part of vascular tissue that departs from the stele of stem towards branch. |
| 2. Leaf trace may run through one or more nodes and internodes before bending towards leaf base and it obliquely enters into leaf bases. | 2. Branch trace directly diverges to branch without running obliquely through the internodes. |
| 3. Leaf trace is associated with leaf gap. | 3. Branch trace is associated with branch gap. |
| 4. In a cross-section of stem at node a leaf trace occurs on the peripheral side of branch trace. | 4. A branch trace occurs towards inside of leaf trace as is revealed in cross-section of a stem at node. |

Comparison between unilacunar, trilacunar and multilacunar node.

| Unilacunar node | Trilacunar node | Multilacunar node |
|---|---|---|
| 1. Unilacunar node exhibits one leaf trace to a leaf and the trace is associated with one lacuna. | 1. Trilacunar node exhibits three leaf traces to a leaf and three lacunae. Each trace is associated with one lacuna. | 1. Multilacunar node exhibits more than three traces to a leaf and more than three lacunae. Each trace is associated with one lacuna. |
| 2. Unilacunar node is exstipulate. | 2. Trilacunar node has stipules. | 2. Multilacunar node has sheathing leaf base. |
| 3. Ex. Centrospermae. | 3. Ex. *Salix*. | 3. Ex. *Rumex*. |

Comparison between leaf gap and branch gap.

| Leaf gap | Branch gap |
|---|---|
| 1. It is formed when a vascular trace departs towards the petiole of leaf from main vascular cylinder. | 1. It is formed when branch traces depart to a branch from main vascular cylinder. |
| 2. It is formed above the leaf trace and each gap usually is associated with one trace. | 2. It is formed above the branch trace and each gap is associated with two branch traces. |
| 3. One, three or more numbers of leaf gaps may be associated to a single leaf and accordingly a node is said to be unilacunar, trilacunar or multilacunar. | 3. No such terminology exists to describe a node in relation to branch gap. |

CHAPTER **18**

# Primary Tissues in the Vegetative Body of Plants

Vascular plants exhibit a great diversity in the arrangement and structure of primary tissues in their stems, roots and leaves, a brief account of which is discussed below.

## Arrangement of primary tissues in dicotyledonous stem

In stem there exist three tissue systems – dermal, vascular and ground tissue, which respectively consist of epidermis, vascular bundle and all tissues except the former two. A great diversity is observed in the arrangement and structure of primary tissues, a brief account of them that commonly occur in mature stems of dicot is mentioned below beginning from the periphery of stem.

**Epidermis:** Externally, the stem is bounded by epidermis, the details of which are discussed in the chapter 12. To mention briefly it is the peripheral layer of stem and is uniseriate. The cells are more or less tabular or rectangular with thin primary walls. The cells are usually compactly set and contain living protoplast. Cutin is often deposited on the outer surface of cells to form cuticle. In stems, which carry on photosynthesis, the continuity of epidermis is interrupted by the presence of stomata. Glandular and nonglandular hairs are also observed on the epidermis.

**Cortex:** The cylindrical region that occurs between epidermis and vascular cylinder is the cortex. Various types of cell are observed in this region. It may consist of thin walled parenchyma cells only (ex. *Salicornia, Pelargonium* etc.). In many stems the peripheral layers of cortex, immediately below the epidermis, may consist of other types of cell. This layer is usually termed as hypodermis and its constituent cells differ in origin from the epidermis and structure from the cortical cells below. Usually the peripheral layer contains collenchyma that occurs either as a continuous band or discontinuously in groups or as isolated strips. Patches of collenchyma occur at the projecting ridges of stem (e.g. *Cucurbita,*

Fig. 32.2). The stems, which carry on photosynthesis, contain abundant chloroplastids in the peripheral collenchyma and cortical parenchyma, and these cells are known as chlorenchyma. Sclereids (ex. *Trochodendron*), oleoresin cells (ex. *Zingiber*), laticifers (ex. *Calotropis*) etc. may also occur here. The cortical parenchyma may contain starch, crystals etc.

**Endodermis:** It is the cylindrical layer of cells that delimit cortex from the stele. This layer is inconspicuous in aerial stems. It is prominent in underground stems and sometimes in aerial stem like *Piper* where casparian strips are also observed. It usually surrounds the entire vascular cylinder and in exceptional case it encircles individual vascular strand where the stem is polystele (ex. *Nymphaea*). The innermost layer of cortical cells of many dicotyledonous stems (ex. *Phaseolus*) contains abundant starch. This layer is called starch sheath. Starch sheath, in some stems, may be transformed into endodermis and develops casparian strips.

Endodermis consists of a layer of living cells that are compactly set and have characteristic wall structure. Lignin and suberin, in addition to cellulose, deposit on the transverse and radial walls of endodermal cells as strips or bands. This layer is called casparian strip or band. A mature endodermal cell contains suberin, cellulose and often lignin on the entire inner surface of walls.

**Function of endodermis:** (1) it accumulates fat, starch, protein granules, tannins and carbohydrates; (2) the suberized cell wall of endodermis acts as a barrier to the movement of iron, lead and copper either in or out of vascular tissues; (3) it excretes water into trachea as it possesses higher hydrostatic pressure than the neighbouring cortical cells; (4) it is impermeable to gaseous diffusion and so it acts as a barrier to gaseous exchange; (5) it controls the conduction between the central cylinder and cortex in rhizomes; (6) it forms a barrier of pathogens as it contains large amount of benzoquinones, naphthoquinones and anthroquinones that inhibit the growth of fungi and bacteria; (7) in *Cocculus, Paederia foetida* stem the cork cambium or phellogen originates from endodermis; (8) endodermis may behave as meristem termed proendodermis that divides to form a part of the cortex.

**Stele:** The tissues that lie internal to endodermis compose the stele. It includes the vascular and ground tissue systems. The ground tissue consists of pith or medulla. The pith is usually parenchymatous and sometimes may be lignified and pitted.

In dicotyledonous stems, the vascular tissues lie in between cortex and pith. It occurs in the form of a continuous or split cylinder. The phloem is situated on the peripheral or external side. In contrast to root, xylem and phloem of stem lie on the same radii. Each strand of the split vascular cylinder is designated as vascular bundle that is separated by interfascicular parenchyma. A vascular bundle with external phloem only is termed as collateral bundle. In some plants belonging to families Solanaceae, Cucurbitaceae, Apocynaceae etc. phloem is also present on the inner side of xylem. To distinguish it from external phloem, the inner phloem is termed as internal or intraxylary phloem. This phloem occurs as isolated strands either on the peripheral margin of pith (ex. *Calotropis, Strychnos*) or in close contact on the inner side of xylem of a vascular bundle. In latter case the vascular bundle with both internal and external phloem is termed as bicollateral vascular bundle (ex. *Cucurbita*). In dicotyledonous stem the collateral and bicollateral bundles are always open, i.e. cambium is present in between xylem and phloem. Usually the stele of dicotyledonous stem is monostele, i.e. the vascular bundles are arranged more or less in a ring and lie within a common endodermis. But variations occur when each vascular strand becomes completely enclosed by an endodermis. Ex. *Nymphaea* (Nymphaeaceae). In some genera medullary and cortical bundles occur.

## Arrangement of primary tissues in monocotyledonous stem

In the stem of monocot there exist three tissue systems—dermal, vascular and ground tissues, which consist of epidermis, vascular bundles and all tissues except the former two respectively. A brief account of the arrangement and structure is given below.

**Epidermis:** Epidermis is more or less like dicots. In contrast to dicot stem, the epidermis is usually devoid of glandular and nonglandular trichomes.

**Ground tissue:** In monocotyledonous stems all tissues except vascular bundles present internal to epidermis are described as ground tissue because there exists no real distinction between cortex and pith. In many genera of Liliaceae (ex. *Asparagus,* Fig. 31.12) and Amaryllidaceae there is differentiation of cortex and stele with distinct endodermis and pericycle. Peripheral layers of ground tissue, i.e. cells beneath the epidermis may be composed of sclerified parenchyma or sclerenchyma. Sclerenchyma, in the form of a continuous ring, composes the pericycle of *Asparagus.* The other portions of ground tissue are composed of parenchyma upon which the vascular bundles appear to be embedded.

Vascular bundles in monocot stem usually lie scattered throughout the ground tissue and this type of distribution is termed as atactostele. In many genera like *Coix, Triticum, Oryza* etc. they are arranged more or less in a ring. Each vascular bundle may be leptocentric or amphivasal (ex. *Dracaena, Aloe* etc.) or collateral and closed, i.e. cambium is absent (ex. *Zea, Asparagus* etc.). In collateral bundles the xylem are arranged in the form of capital 'Y' or 'V' or 'U', which appear in cross section. Large metaxylem vessel elements form the arms of the alphabets and the protoxylem composes the rest. In some genera protoxylem disintegrates and the cavity thus formed is termed as protoxylem cavity or lacuna or canal. Phloem groups are situated in the grooves of Y or V or U-shaped xylem, i.e. xylem encircles phloem on the three sides. A ring of sclerenchyma termed bundle sheath usually encircles each vascular bundle. In *Asparagus* the bundle sheath is absent.

## Arrangements of primary tissues in roots

A cross-section of root reveals the epidermis, cortex and vascular cylinder.

**Epidermis:** Epidermis of root is also referred to as epiblema or rhizodermis as it differs from that of shoot in origin, structure and function. Epidermal cells are compactly set, elongated, thin-walled with thin cuticle that covers the cell wall. Sometimes mucilage may occur superficially. Epidermal cells of the underground and exposed parts of root that persist for a long time may become thick-walled and contain lignin, cutin or suberin.

Except some hydrophytes, the most notable feature of epidermis is the presence of root hairs that are the tubular prolongations from the single cells of epidermis. They function in anchorage and in the absorption of water and salts. Usually all epidermal cells may give rise to root hairs. But in certain plants like *Phleum, Hydrocharis, Sinapis alba* etc. root hairs are formed from some special epidermal cells that can be distinguished from other epidermal cells by their smaller size and dense cytoplasm. These specialized cells are known trichoblast or piliferous cell. So, the root epidermis that bears root hairs is also termed as piliferous layer. Root hairs are usually short lived; long-lived root hairs are observed in some species of Asteraceae. They are unbranched and unicellular. Multicellular root hairs are observed in the aerial adventitious root of *Kalanchoë fedtschenkoi*.

Epidermis of root is typically single layered, i.e. uniseriate but exceptions are noted in the aerial roots of certain epiphytic orchids where the epidermis is several layered. This multilayered (Fig. 18.1) or multiple epidermis is specially termed as velamen that is derived from the initials of epidermis, i.e. dermatogen. Velamen consists of several layers of dead cells that are devoid of any contents and forms a circular sheath around the root. The cells are compactly arranged and the cell walls are strengthened by spiral or reticulate bands of lignin. The walls also contain many primary pit fields.

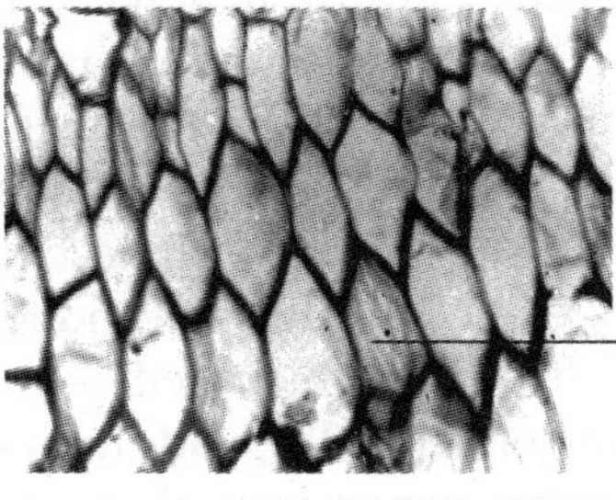


Figure 18.1

Photomicrograph of a portion of velamen of orchid root.

The inner limiting layer of velamen is known as exodermis. This layer is uniseriate and forms a partition between the thick-walled cells of velamen and thin-walled cells of cortex. Exodermis is considered as the outermost layer of cortex as it is derived from the periblem and not from the dermatogen that gives rise to velamen. Two types of cell are observed in the exodermis of orchid root—long and short, which are more or less alternately arranged. The small cells are thin walled and known as passage cells. The long cells are thickened on their tangential and radial walls only. Thus exodermis may be comparable with endodermis of root.

It is regarded that velamen is a protective tissue that prevents water loss from the inner cortical cells, thus the delicate thin-walled cortical cells are protected from desiccation. It was previously believed that velamen absorbs and conserves water from atmosphere, but later studies (Dycus and Knudson, 1957) revealed that velamen and exodermis are almost impermeable to water and solutes.

**Cortex:** Cortex of root is mainly composed of parenchyma cells. In many monocotyledons sclerenchyma develops in the cortex in addition to parenchyma. Schizogenous and lysigenous intercellular spaces are frequently formed in the root cortex of some members of Gramineae and Cyperaceae. In many hydrophytes the cells of root cortex are very regularly arranged to form aerenchyma.

Apart from epiphytic orchids, exodermis is also observed in *Iris, Smilax, Phoenix, Citrus* etc. In these genera this layer is either uniseriate or multiseriate and forms a kind of hypodermis, just beneath the epidermis. The outermost layer or layers of root cortex are differentiated into exodermis whose cell walls are thickened with suberin. In some cases lignin is also deposited on the cell walls and casparian strips are also formed. A mature cell of exodermis may contain living protoplast. It is regarded that exodermis protects the inner delicate cortical cells.

Collenchyma rarely occurs in roots, e.g. *Monstera* where it is present in the cortex. Root cortex is usually very wide and the cells often contain starch, crystals etc. Aerial roots of many epiphytes contain chloroplastids in the cortical cells. Several plants possess secretory cells, resin ducts and laticifers in cortex. Trichosclereids are found in the root cortex of *Monstera*.

**Endodermis:** The innermost layer of cortex is endodermis that delimits cortex from stele. Endodermis is uniseriate, compactly set and has casparian strips in young cells (Fig. 18.2). The band of suberin that runs around the radial and cross walls of endodermal cell is the casparian strip. The strip may be narrow, thread like or as wide as the wall itself. Suberinization occurs across the middle lamella of radial and cross walls, but not on the tangential walls. In roots where there is no secondary thickening suberin deposits as lamellae over the whole inner primary wall of endodermal cells including casparian strips. Later, cellulose is deposited centripetally on the inside of suberin lamella (ex. *Iris*). Thus the endodermal cells attain considerable thickness on radial walls. These thickened cell walls often become lignified on the inner radial and tangential walls in addition to cellulose and suberin. Later study with electron microscope confirms the thick-walled nature of endodermis and reveals the presence of thick wall in the region of casparian strips. The deposition of suberized material initially begins to those endodermal cells, which are adjacent to phloem strands. Later deposition spreads radially to other adjacent endodermal cells. The cells that are situated opposite the protoxylem are delayed in wall thickening, occasionally remain unthickened and have casparian strips only. These cells are known as passage cells, which may remain

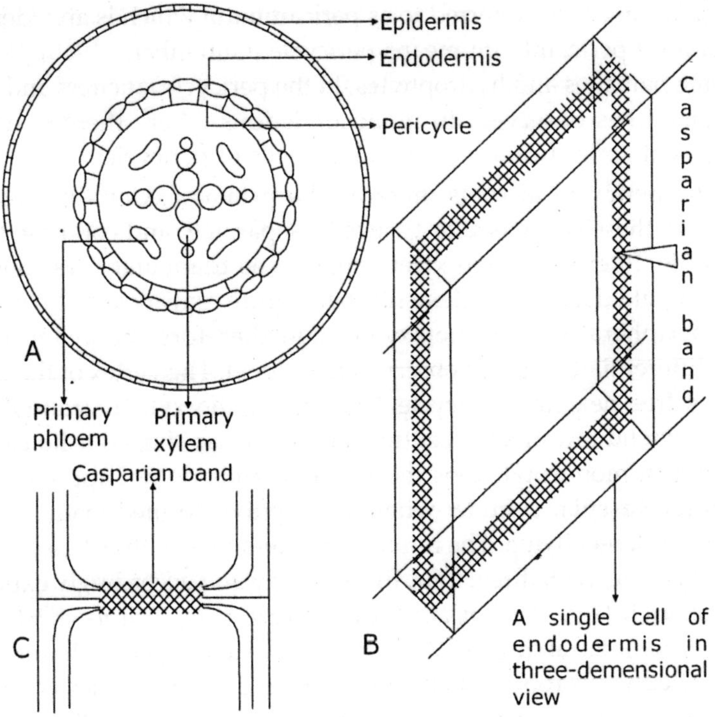


Figure 18.2

A. Schematic representation of root in transverse section showing the position of endodermis. B. Diagram showing casparian band in radial and tangential walls. C. Sectional view illustrating that the casparian band extends through the wall of contiguous cell. B. & C. adapted after Beck, 2005, p. 284.

thin walled throughout the entire life or later may become thickened like other endodermal cells. It is regarded that the casparian strip regulates the movement of materials in the root and the passage cells provide passages for the movement of substances between vascular cylinder and cortex. Plasmodesmata exist between cortical and endodermal cells and it is absent from the region of casparian strips. The main functions of endodermis are more or less the same as stems that were mentioned previously.

**Vascular cylinder:** The vascular cylinder consists of vascular tissues (i.e. xylem and phloem) and pith. Peripheral to the vascular tissue and lying immediately inside the endodermis, there is a cylindrical region of cells - termed pericycle. The pericycle is usually uniseriate and sometimes may be multiseriate (e.g. *Agave*, *Smilax*, *Salix* etc.). It consists of thin walled parenchyma and sometimes fibres. This layer is in direct contact with protoxylem and protophloem and can be distinguished prior to the differentiation of them. The pericycle retains its meristematic activity and gives rise to (i) the primordia of lateral roots in spermatophytes, (ii) in dicotyledons the portions of vascular cambium that give rise to secondary vascular tissues and (iii) the phellogen—the meristem of cork.

The pericycle is sometimes referred to as pericambium which is an older term. In German literature pericambium means pericycle (Guttenberg, 1960). Pericycle is absent in some parasites and hydrophytes. In the pericycle laticifers and secretory ducts are sometimes observed. In monocotyledons where there is usually no secondary growth, sclerification occurs in the cells of pericycle.

Below the pericycle in roots there lie the primary xylem and phloem. In contrast to stem these two vascular tissues lie separately and compose the radial vascular bundle. In cross section xylem appears as triangular rays consisting of thick walled, lignified tracheary elements that mature centripetally. So protoxylem lies on the peripheral side of metaxylem and therefore the xylem is exarch. Centripetal differentiation is also observed in phloem. The vertex of the triangular xylem rays is free near the pericycle. The rest of the xylem strands, in many dicotyledons, extend towards the centre and form a star-shaped solid central core of xylem. In most monocotyledons the xylem strands do not approach towards the centre of the vascular cylinder where large pith is formed made up of either parenchyma or sclerenchyma. The number of xylem rays with protoxylem groups varies. It may be one, two, three, four, five or many and accordingly expressed by the terms monarch (e.g. *Trapa natans*), diarch (e.g. *Beta, Daucus, Raphanus* etc.), triarch (e.g. *Pisum*), tetrarch (e.g. *Vicia, Ranunculus* etc.), pentarch and polyarch (e.g. monocotyledonous root). Usually the number of xylem arches is constant for a plant but exceptions are noted in *Libocedrus decurrens* where the root may be diarch, triarch, pentarch and hexarch. The aerial root of *Tinospora cordifolia* also has tri-, tetra-, or pentarch xylem. In polyarch roots the number of xylem arches may be as high as 100 or more (e.g. Palmae). In the root of *Triticum* there is a single large metaxylem vessel situated at the centre of vascular cylinder and the peripheral protoxylem are separated from the metaxylem by parenchyma. In *Iris, Zea* etc. a number of metaxylem vessels are arranged in a circle around the pith.

## Arrangements of tissues in leaves

Leaves are generally flat organs and to differentiate the two surfaces different terms are used. The surface that faces the stem upwardly is designated as upper surface or adaxial surface whereas the other surface is the lower or abaxial surface. Anatomically a leaf comprises the usual three main tissues systems—the dermal system, the ground tissue and the vascular system. The dermal system consists of upper and lower epidermises, the ground tissue comprises the mesophyll and the vascular system is represented by vascular bundles that form the veins in a leaf. A cross section of leaf reveals the followings (Fig. 18.3): (i) upper or adaxial epidermis; (ii) mesophyll where at the central region the vascular bundles appear to be embedded and (iii) lower or abaxial epidermis.

**Epidermis (abaxial and adaxial):** Usually the epidermis is uniseriate, but the leaves of *Ficus, Piper* and *Nerium* have multiseriate epidermis where the number of layers of cells varies from two to 16. It is regarded that the multiple epidermis protects the inner mesophyll from desiccation. In the leaves where both epidermises are multiseriate the adaxial epidermis consists of more number of

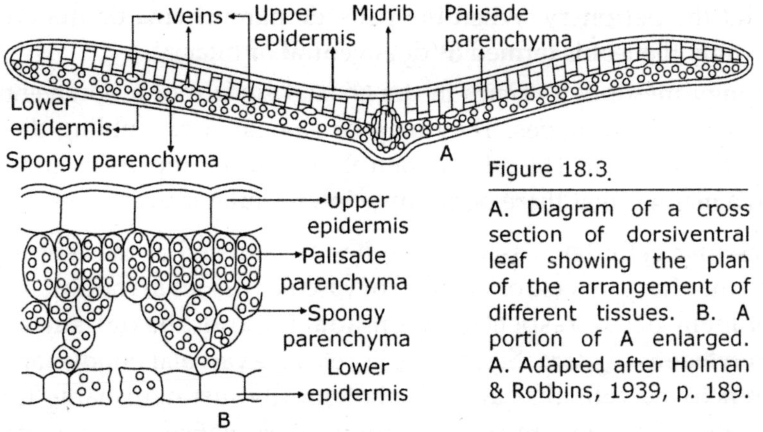


Figure 18.3. A. Diagram of a cross section of dorsiventral leaf showing the plan of the arrangement of different tissues. B. A portion of A enlarged. A. Adapted after Holman & Robbins, 1939, p. 189.

layer of cells than abaxial one. In multiple abaxial epidermis a sub-stomatal cavity is present (e.g. *Piper*), which is a space between the guard cells and mesophyll. In *Nerium* abaxial epidermis is multiseriate and the stomata are present in stomatal crypts only where the epidermis is uniseriate. Stomata may occur either on the adaxial (e.g. *Nymphaea*) or abaxial (e.g. *Ficus*) or both surfaces. Their positions in relation to other epidermal cells may be either at the same level or raised above or sunken (e.g. *Hakea*) or restricted to stomatal crypts (e.g. *Nerium*).

Usually epidermal cells are devoid of chloroplastids except the guard cells, but in aquatic plants like *Ranunculus*, *Phyllospadix* etc. chloroplastids occur in the epidermal cells. In the epidermis glandular or nonglandular hairs, crystals, cystoliths may occur. The epidermal cell walls usually contain cutin, occasionally lignin and wax.

**Mesophyll:** Ground tissue of a leaf is the mesophyll that lies internal to epidermises. Mesophyll consists of parenchyma cells. In mesophytic dicotyledons these parenchyma cells are of two types: palisade parenchyma and spongy parenchyma.

In transverse section of leaf the cells of palisade parenchyma are cylindrical, elongated or rod shaped. They are densely packed together and may be arranged in one or more rows. In longitudinal section of a leaf these cells appear as spherical and either separated or slightly attached to each other. Each palisade cell contains abundant chloroplasts. The palisade cells are usually present immediately below the adaxial uniseriate or multiple epidermis, but sometimes the epidermis and palisade may be separated by the presence of hypodermis. Though palisade occurs on the adaxial side only, the leaves of *Thymelaea* have abaxial palisade. Palisade parenchyma is present on both adaxial and abaxial sides in the leaf of *Dianthus*, *Atriplex*, and *Artemisia* etc. In these plants spongy parenchyma occurs in the form of strips at the centre of lamina. A leaf is said to be isobilateral or isolateral when palisade parenchyma occurs on both adaxial and abaxial surface. Usually monocotyledonous leaves belong to this category, and the abaxial palisade cells are less developed. The cylindrical leaves of *Hakea* have palisade tissue situated all round the periphery. When the palisade parenchyma occurs on the adaxial surface only the leaf is termed as dorsiventral or bifacial.

Spongy mesophyll occurs in the abaxial surface of a dorsiventral leaf and may be of various shapes. They may be elongated resembling the palisade or spherical or irregular. The cells contain abundant chloroplastids and are arranged in such a manner that there occur large intercellular spaces.

**Vascular cylinder:** Vascular bundles are situated at the central region of a leaf and they have inverse orientation of vascular tissues in contrast to stem. Xylem and phloem occur respectively on adaxial and abaxial surface. In some monocotyledons (*Cyperus, Saccharum, Zea* etc.) the vascular bundles are surrounded by parenchyma cells that are morphologically distinguishable from adjacent mesophyll cells. They are usually larger, thick walled and may contain fewer chloroplasts. This sheath of parenchyma around the vascular bundle of a leaf is termed as bundle sheath. These cells are concerned in conduction and storing of food materials. The bundle sheath is regarded either as endodermis due to occasional occurrence of casparian strips or as starch sheath, since these cells sometimes contain starch. Later study with improved techniques reveals that the $C_4$-dicarboxylic acid pathway, known as Hatch-Slack pathway occurs in the cells of bundle sheath.

Difference between dicotyledonous root and stem.

| Dicotyledonous root | Dicotyledonous stem |
|---|---|
| 1. Epidermis: also known as epiblema that is uniseriate, non-cuticularized, provided with unicellular root hairs and lacks stomata. | 1. Epidermis is cuticularised with epidermal appendages like hairs, scales etc.and stomata. |
| 2. Cortex is commonly parenchymatous and wider than stem. | 2. Cortex parenchymatous. The peripheral layers, i.e. the hypodermis may be collenchymatous and narrower than root. |
| 3. The stele is radial. Pericycle is uniseriate and consists of parenchyma only. | 3. The stele is siphonostele, dictyostele. Pericycle multiseriate and consists of sclerenchyma and parenchyma. |
| 4. The vascular bundle is radial. | 4. The vascular bundles are collateral open or bicollateral open and arranged more or less in a ring. |
| 5. Xylem is monarch to hexarch, protoxylem exarch and phloem occurs alternate to xylem strands. | 5. Protoxylem endarch and phloem occurs either on the periphery or both inner and outer side of xylem. |
| 6. Pith is either small or absent. | 6. Well-developed pith is present. |
| 7. Intrastelar secondary growth occurs with the help of wavy cambium ring that is secondary in origin. | 7. Intrastelar secondary growth occurs with the help of cambium ring that is partly primary and partly secondary in origin. |

Difference between dicotyledonous and monocotyledonous root.

| Dicotyledonous root | Monocotyledonous root |
|---|---|
| 1. Epidermis is single layered. | 1. Epiblema is uniseriate or multiseriate (ex. velamen of orchid root). |
| 2. Cortex is composed of parenchyma cells only. | 2. Cortex consists of parenchyma and in some cases sclerenchyma (e.g. stilt root of Maize) or collenchyma cells (e.g. *Monstera*) occur in addition to parenchyma. |
| 3. Pericycle gives rise to a portion of vascular cambium that gives rise to secondary vascular tissues and phellogen—the meristem of cork. | 3. Pericycle undergoes sclerification. |
| 4. Xylem is monarch to hexarch. | 4. Xylem is polyarch. |
| 5. Xylem strands in many dicots extend towards the centre and form a star shaped solid core of xylem. | 5. The xylem strands do not approach towards the centre of vascular cylinder. |
| 6. Pith is either small or absent. | 6. Large and well-developed pith occurs at the centre. |
| 7. Intra- and extrastelar secondary growth occur with the help of wavy vascular cambium and phellogen respectively. | 7. As there is no cambium there occurs no normal secondary growth. |
| 8. Fibres are present in primary phloem elements (ex. Malvaceae, Papilionaceae etc.). | 8. Fibres are usually lacking in phloem. |

Difference between dicotyledonous and monocotyledonous stem.

| Dicotyledonous stem | Monocotyledonous stem |
|---|---|
| 1. Internally it is differentiated into epidermis, cortex and stele. | 1. It is differentiated into epidermis and ground tissue where vascular bundles appear to be embedded. |
| 2. Epidermis: It bears hairs that are usually multicellular. | 2. Usually epidermis is smooth and sometimes interrupted by stoma. |
| 3. Cortex: The peripheral layers, i.e. hypodermis are generally collenchymatous. | 3. Ground tissue: The peripheral layers, i.e. the hypodermis, are composed of either sclerenchyma or sclerified parenchyma. |
| 4. The rest of cortex bears nonvascular tissues. | 4. In the rest of ground tissues vascular bundles are present. |
| 5. The innermost layer of cortex is endodermis. | 5. There exists no such endodermal layer. |
| 6. Below endodermis there occurs the pericycle. | 6. No such layer like pericycle exists. |
| 7. The stele is commonly siphonostele, dictyostele where the vascular bundles are arranged more or less in a ring around central pith. | 7. The stele is atactcstele, i.e. the vascular bundles are scattered in the ground tissue where there is no formation of pith. |

*Contd.*

| Dicotyledonous stem | Monocotyledonous stem |
|---|---|
| 8. The vascular bundles are more or less uniform in size. | 8. The vascular bundles are of different sizes, the smaller ones being restricted to the peripheral side. |
| 9. Each vascular bundle is commonly collateral open and occasionally bicollateral open, i.e. cambium is present in between xylem and phloem. | 9. Each vascular bundle is collateral and closed, i.e. cambium is absent. |
| 10. Usually a sclerenchymatous bundle cap occurs over each vascular bundle. | 10. Generally each vascular bundle is encircled by a sclerenchymatous bundle sheath. |
| 11. The proto- and metaxylem vessels are arranged in rows. | 11. The proto- and metaxylem are arranged in the form of capital 'Y'. |
| 12. There is no protoxylem lacuna. | 12. Usually a protoxylem lacuna is formed. |
| 13. In the phloem elements parenchyma is present. | 13. In phloem elements parenchyma is lacking. |
| 14. In between the vascular bundles there occurs the medullary rays. | 14. Ground tissue exists between the two vascular bundles. |

Difference between monocotyledonous root and stem.

| Monocotyledonous root | Monocotyledonous stem |
|---|---|
| 1. Internally it is differentiated into epiblema, cortex and stele. | 1. Internally it is differentiated into epidermis and ground tissue where vascular bundles appear to be embedded. |
| 2. Epiblema is uniseriate or multiseriate (e.g. velamen of orchid root), lacks cuticle and stoma, and provided with unicellular root hairs. | 2. Epidermis is uniseriate and provided with cuticle and stomata. |
| 3. Cortex is commonly parenchymatous, the peripheral layers of which may be sclerenchymatous (e.g. stilt root of Maize) or collenchymatous (e.g. *Monstera*). | 3. Ground tissue is parenchymatous, the peripheral layers of which, i.e. hypodermis may be composed of sclerenchyma or sclerified parenchyma. |
| 4. The stele is radial stele and remains encircled by endodermis. | 4. The stele is atactostele and there exists no endodermis. |
| 5. Pericycle is uniseriate and may undergo sclerification. | 5. Pericycle is lacking. |
| 6. The vascular bundle is radial and xylem polyarch. | 6. The vascular bundle is conjoint, collateral and closed, and leptocentric and many in number. |
| 7. Protoxylem is exarch. | 7. Protoxylem is endarch. |
| 8. Xylem and phloem strands are alternate to each other. | 8. Xylem and phloem occur together on the same radius. |
| 9. The protoxylem and metaxylem are arranged in rows. | 9. The proto- and metaxylem are arranged in the form of capital 'Y'. |
| 10. There is no protoxylem lacuna. | 10. Usually a protoxylem lacuna is formed. |
| 11. At the centre there occurs large and well-developed pith. | 11. Pith is lacking. |

Difference between dorsiventral, isobilateral and centric leaf.

| Dorsiventral leaf | Isobilateral leaf | Centric leaf |
|---|---|---|
| 1. Internally the leaf is differentiated into upper and lower epidermis and meso-phyll, which lies in between the two epidermises. | 1. Internally the leaf is differentiated into upper and lower epidermis, and meso-phyll, which occurs in between the two epidermises. | 1. Internally the leaf is differentiated into epidermis and mesophyll; a single epidermis is present all round the periphery and it encloses the mesopyll. |
| 2. The distribution of stomata is usually restricted to lower epidermis. | 2. The stomata are distributed on both the epidermises. | 2. The stomata are distributed all round the circumference of epidermis. |
| 3. Mesophyll is composed of palisade and spongy paren-chyma. The palisade cells are elongated and situated in transverse plane towards the upper epidermis. The spongy parenchyma cells are more or less isodiametric in shape and occur in between the palisade and lower epidermis with profuse intercellular spaces. | 3. The mesophyll cells are composed of one type of cells, i.e. the cells are either more or less isodiametric (simulate spongy parenchyma) or elongated (simulate palisade parenchyma); the cells lie in between the upper and lower epidermis with little inter-cellular spaces. | 3. The mesophyll may be composed of one type of cells or may be differentiated into palisade and spongy paren-chyma. In the latter case the palisade cells occur all round the periphery of the cylindrical leaf (ex. *Hakea*). |
| 4. The vascular bundles, in the transverse section of leaf, are of different sizes and not arranged in a transverse line. Ex. dicotyledonous leaf. | 4. In the transverse section of leaf the vascular bundles are more or less of same sizes and arranged in a transverse line. Ex. monocotyledonous leaf. | 4. In the transverse section of leaf the vascular bundles are present either in a line or in a ring. Ex. onion, *Hakea* etc. |
| 5. The vascular bundles are usually surrounded by bundle sheath consisting of a single layer of parenchyma. | 5. The vascular bundles are usually surrounded by bundle sheath, which may be one or two layered. In the latter case the outer sheath is composed of thin walled parenchyma cells. The cells of the inner sheath are smaller in cross-sectional diametre and thick walled in comparison to the outer sheath. The inner sheath is termed as mestome sheath and its cells are thickened with suberin. | 5. The vascular bundles are usually surrounded by bundle sheath, which is single layered. Sometimes sclerenchymatous fibres occur on both sides of vascular bundles. |

CHAPTER **19**

# Anomalous Structure in the Primary Body of Stem

A great diversity in the arrangement of vascular tissues is exhibited by the angiosperms in their stems. The arrangements, which exist in majority of the plants, are considered as normal. The variations are regarded as anomalous though the so-called anomalous structure is normal for the particular species. The anomalous structure, which occurs in the primary body of a plant, is defined as primary anomalous structure. These anomalies occur in both monocotyledonous and dicotyledonous stem, and may persist throughout the life of the plant. The followings are the common anomalous structures that occur in stems.

1. *Occurrence of scattered vascular bundles in dicotyledonous stem*: (Fig. 19.1C): Normally the vascular bundles, in the dicotyledonous stem, are arranged more or less in a ring. The followings are a few variations where the vascular bundles are more or less irregularly scattered in the parenchymatous ground tissue of the stem: Ex. *Thalictrum* (Ranunculaceae), *Podophyllum peltatum* (Podophyllaceae), *Papaver orientale* and *P. somniferum* (Papaveraceae), *Bougainvillea* (Nyctaginaceae), *Piper betle* and *Peperomia langsdorfii* (Piperaceae), *Nymphaea* (Nymphaeaceae) etc.
2. *Occurrence of vascular bundles arranged in a ring in monocotyledonous stem* (Fig. 19.1D): In monocot stems the stele is atactostele, i.e. the vascular bundles are scattered in the ground tissue. But the followings are the exceptions where the vascular bundles are arranged more or less in one or two rings. Ex. *Coix*, *Triticum*, *Oryza* (Poaceae), *Tamus communis* (Dioscoreaceae) etc.
3. *Occurrence of cortical bundles* (Fig. 19.1A): In dicotyledonous stem the vascular bundles occur in the stele. But there are some dicotyledonous stems where accessory vascular bundles are found in the cortex of stem in addition to stelar bundles. The vascular bundle, which runs through the cortex independent of the ring of bundles present in the stele, is referred

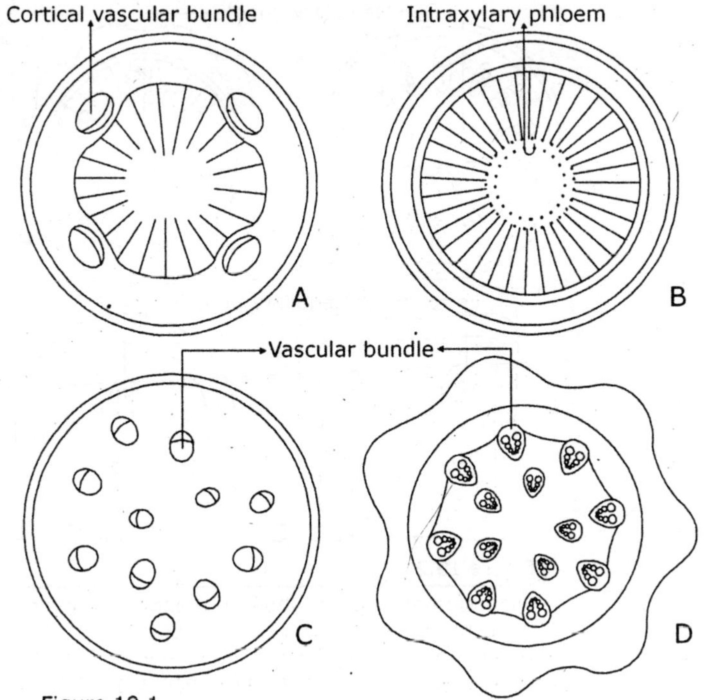


Figure 19.1

Diagram showing anomalous structure. A. Cortical vascular bundles in the stem of *Calycanthus occidentalis* (diagrammatic cross-section). B. Intraxylary phloem in the stem of *Asclepias curassavica* (diagrammatic cross-section). C. Cross-section of the stem of *Peperomia langsdorfii* (dicot) showing scattered vascular bundles (diagrammatic). D. Cross-section of the stem of *Tamus communis* (monocot) showing two rings of vascular bundle (diagrammatic). A. & B. adapted after Metcalfe and Chalk (1983). C. & D. are redrawn after Vasishta (1981).

to as cortical bundles. In most species cortical bundles are leaf trace bundles. Metcalfe and Chalk (1950) listed thirty-seven families where cortical bundle occurs. Mention may be made of *Bombax malabaricum* (Bombacaceae), *Calycanthus occidentalis, Chimonanthus* (Calycanthaceae) where four pericyclic bundles (leaf traces) occur with internal phloem and external xylem; *Campanula pyramidalis* (Campanulaceae); *Crepis, Senecio, Vernonia* (Compositae); *Cotyledon* (Crassulaceae); *Momordica* (Cucurbitaceae); *Armeria, Plumbago europaea* (Plumbaginaceae); *Peltiphyllum* (Saxifragaceae); petiole of *Eryngium campestre, Siler trilobum* (Umbelliferae); *Nyctanthes arbortristis* (Oleaceae) where four inversely oriented cortical vascular bundles occur with external xylem and internal phloem and these bundles are not directly connected with the main axial ring (Majumdar, 1941) and in some genera of Rutaceae and Polygonaceae etc.

4. *Occurrence of medullary bundle* : The supplementary bundle, which occurs in the pith of dicotyledonous stem in addition to the normal ring of bundles,

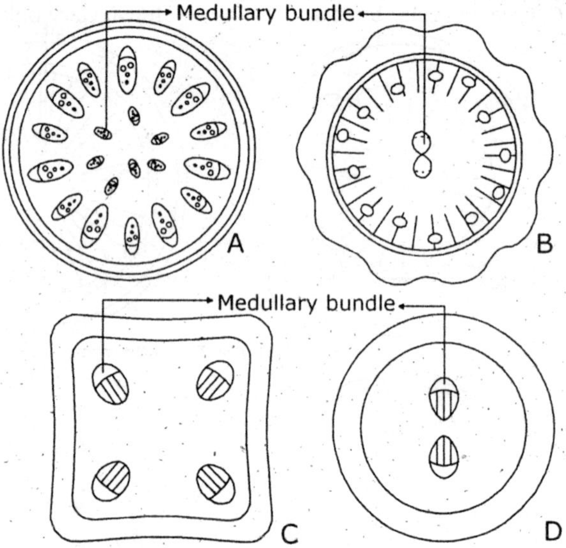


Figure 19.2

Diagrammatic figures showing medullary bundles: A. in the stem of *Aralia* (after Sing *et al.*, 1994), B. in the stem of *Achyranthes aspera*, C. in the inflorescence axis of *Achyranthes coynei* (after Vasishta, 1981) and D. in the young stem of *Bougainvillea* (adapted after Tayal, 1994).

is referred to as medullary bundles (Fig. 19.2). Metcalfe and Chalk (1950) reported thirty-seven families where medullary bundles occur and a few examples are mentioned below: *Acanthus* (Acanthaceae) where collateral and inversely oriented medullary bundles occur; *Achyranthes aspera* (Amaranthaceae) with two medullary bundles at internode and *Cyathula prostrata* with four medullary bundles; *Obione portulacoides, Atriplex hortensis* (Chenopodiaceae); *Crepis, Dahlia, Lactuca, Sonchus* (Compositae); *Greenovia, Echeveria* (Crassulaceae); *Raphanus sativus* (Cruciferae); *Anthocleista* (Potaliaceae); *Bougainvillea, Mirabilis longiflora, Neea, Pisonia* (Nyctaginaceae); *Apium, Eryngium, Oenanthe* (Umbelliferae) etc.

5. *Occurrence of internal bundles*: The supplementary bundles, which occur below the normal ring of bundles and above the medullary bundles, are referred to as internal bundle. These bundles are not so deeply situated in the pith to be classified as medullary bundles. Ex. *Rumex crispus, R. orientalis* etc.

6. *Occurrence of internal phloem or intraxylary (Perimedullary, Medullary) phloem*: The phloem that occurs at the margin of pith in the form of a continuous cylinder or strands is referred to as intraxylary phloem or internal phloem. It is also known as medullary phloem due to its occurrence at the margin of pith. Metcalfe and chalk (1950) reported the presence of internal phloem in twenty-seven families and a few examples are mentioned below: *Wrightia, Vinca, Thevetia, Landolphia* (Apocynaceae); *Calotropis* (Asclepiadaceae);

*Convolvulus, Ipomoea, Evolvulus* (Convolvulaceae); *Acanthus, Barleria* (Acanthaceae); *Solanum, Nicotiana* (Solanaceae) etc.

7. *Absence of vessels in angiosperm*: The xylem of angiosperms is characterized by the presence of vessels. Exceptions are noted in families Tetracentraceae (*Tetracentron*), Trochodendraceae (*Trochodendron*) and Winteraceae (*Drimys, Pseudowintera, Bubbia, Zygogynum*) where vessels are absent.

8. *Occurrence of polystele in dicotyledons*: Usually the stele of dicotyledonous stem is monostele, i.e. the vascular bundles are arranged more or less in a ring and lie within an endodermis. But variations occur when each vascular strand becomes completely enclosed by an endodermis. As a result the stele becomes polystele. Ex. *Nymphaea* (Nymphaeaceae); *Parnassia palustris* (Parnassiaceae) whose nodes are polystelic; *Dianthera americana* where seven steles occur and each stele is represented by a vascular strand, which is completely enclosed by an endodermis. One stele is situated at the centre and the remaining six steles are arranged towards the periphery of stem. Polystelic condition is also reported from *Justicia*. The stolons of Gunneraceae exhibit polystele. The fruit stalks of Malvaceae, Meliaceae, Moraceae, Sterculiaceae and Theaceae are also polystelic.

9. *Occurrence of separate xylem and phloem bundles*: Usually the two vascular tissues xylem and phloem occur together to form bundles. In abnormal cases bundles consisting of either xylem or phloem strands only occur in addition to the normal bundles. In *Cuscuta* fully developed phloem occurs interspersed between other normal collateral bundles. Incomplete bundles consisting of phloem only are observed in *Rumex crispus, Ricinus communis, Xanthium strumarium, Antigonon leptopus, Achyranthes aspera, Mirabilis jalapa* etc., Maheswari (1930). In *Boerhaavia diffusa* incomplete bundles are formed as a result of centripetal differentiation of phloem from procambium cells (Maheswari). In these procambium cells xylem is not formed on the inner side, instead the innermost undifferentiated cells divide tangentially to form cambium. In *Utricularia* and *Polypompholyx* the two vascular tissues are not associated in bundles, instead they lie in alternate strands attached to a ring of sclerenchyma, consisting of fibres only. The two vascular tissues of *Circaeaster agrestis* stem like that of the main root lie separately at alternate radii. In this species there is a diarch xylem plate and phloem is situated on both sides.

CHAPTER **20**

# Secondary Growth in Dicotyledonous Stem and Root

Secondary growth can be defined as a phenomenon where, after the completion of primary tissue formation, more secondary tissues are developed to supplement dermal, vascular and mechanical tissue system in certain plants.

The meristem of primary tissue divides. The daughter cells, after differentiation and maturation, form primary tissues of plant body. Primary tissues include epidermis, cortex, primary phloem, primary xylem and pith, which are observed in the cross section of young stems and roots.

In many plants, vegetative development is completed after the maturation of primary tissues. However, in many herbaceous and woody dicotyledons, formations of new tissues continue even after the maturation of primary tissues. The production of these new tissues is attributable to the lateral meristem, which includes cork and vascular cambium. These cambia produce new tissues for effective protection, conduction and mechanical strength—a phenomenon termed secondary growth.

Secondary tissues are formed by the cambium, which is normally present in dicotyledonous roots and stems. Though some monocot stems possess nonvascular cambia, which produce secondary tissues, it is regarded as anomalous. So normal secondary growth occurs in dicots only, and it causes increase in thickness both in intrastelar and extrastelar region of roots and stems.

## Intrastelar secondary growth in dicot stems

In dicotyledonous stem, the vascular bundles are arranged more or less in a ring. The bundles originate from procambium strand. The procambium cells divide; the inner derivatives are differentiated into primary xylem and the peripheral ones form primary phloem. A strip of procambium remains in between primary xylem and phloem and this undifferentiated procambium is known as cambium. The cambium, along with peripheral primary phloem and inner primary xylem, forms collateral open vascular bundle of dicot stem.

At the onset of secondary growth, the union of fascicular and interfascicular cambium forms a normal vascular cambium ring. The cambium, which is normally present in between primary xylem and primary phloem is the fascicular cambium, which originates from procambium and so primary in origin. The interfascicular cambium originates from primary permanent medullary ray cells, situated between the primary vascular bundles and so secondary in origin. The interfascicular cambium differentiates along the line of fascicular cambium and so their union produces a normal cambium ring (Fig. 20.1). The cell lineages during secondary growth in dicotyledonous stem are illustrated in Box 20.3.

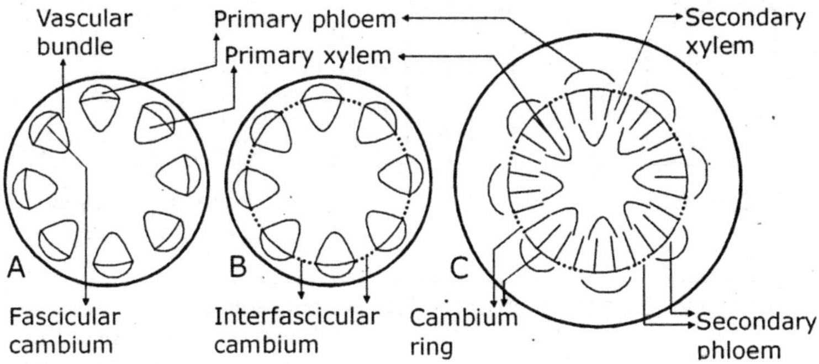


Figure 20.1

Diagrammatic representation showing different stages (A, B & C) of normal intrastelar secondary growth in dicotyledonous stem.

The cambium ring divides tangentially and produces daughter cells on the peripheral and inner side. The peripheral cells are differentiated into secondary phloem, while the inner derivatives form secondary xylem (Figs. 20.2 & 20.3). In between the two secondary tissues, i.e. xylem and phloem, the continuously dividing cambium is present. Secondary tissues are produced both at fascicular and interfascicular region. Normally equal proportion of secondary xylem are differentiated in all segments of the cambium ring on the inner side while the secondary phloem are formed towards periphery in equal amount in all the segments of the cambium ring but they (secondary xylem and phloem) are not of equal quantity. Generally, more amount of secondary xylem is differentiated in contrast to secondary phloem. The cells of cambium ring also divide radially to produce ray cells or medullary cells. The ray parenchyma cells are arranged radially in secondary xylem and phloem.

The cells of cambium possess all the properties of meristematic cells, which are: (i) compactly set with little intercellular spaces, (ii) cell wall is thin, (iii) rich cytoplasm, (iv) cytoplasm contains less reserve food, ergastic substances and crystals, (v) proplastid present, (vi) vacuoles present and (vii) nucleus large.

During secondary growth a large amount of secondary tissues are produced within the stele by the activity of cambium ring. As more and more secondary tissues are produced, the primary xylem gradually approaches to the centre and appears as small patches below the secondary xylem. The primary phloem occupies the peripheral position above the secondary phloem (Fig. 20.4).

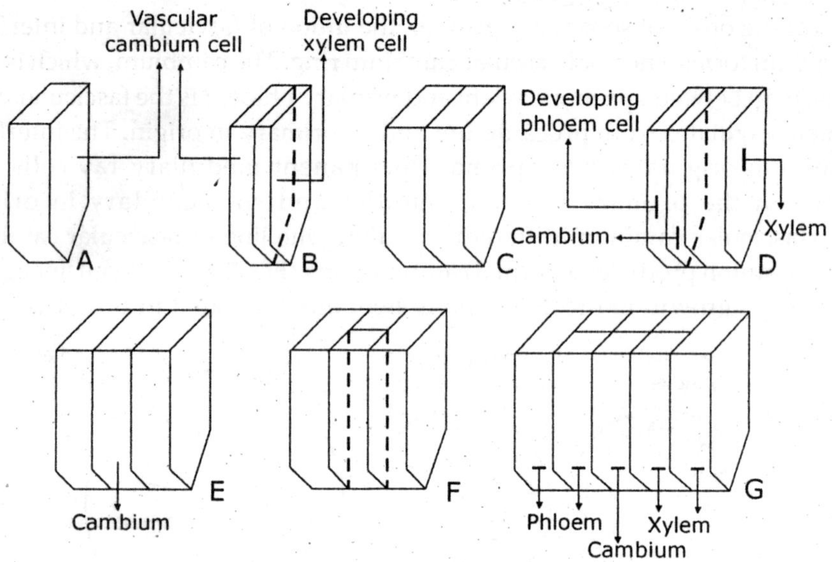


Figure 20.2

Diagrams illustrating formation of secondary xylem and phloem from a vascular cambium cell. The lower half of a fusiform initial is drawn only to simplify the diagram. A. Vascular cambium before division. B. As a result of periclinal division two cells are formed; the peripheral cell remains as cambium and the inner cell develops into secondary xylem. C. Cambium and xylem cell enlarge. D. Division by a periclinal wall in cambium results in the formation of peripheral phloem cell. E. Peripheral phloem, cambium and xylem cell enlarge. F. Division by an anticlinal wall in fusiform initial results in two fusiform initials. G. Formation of secondary xylem and secondary phloem by cambium. Adapted after Mauseth, 2009, p. 165.

## Extrastelar secondary growth in dicot stem or periderm formation

In intrastelar secondary growth a considerable amount of secondary vascular tissues are produced. As a result a pressure is developed within the stele and it is transmitted to extrastelar region when the endodermis is ruptured, cells of cortex are crushed; the epidermis is stretched and tends to be ruptured. So to withstand this pressure a special protective tissue is formed usually replacing the peripheral tissues of stems. These protective tissues are known as periderm, which is formed as a result of extrastelar secondary growth.

Periderm originates from the permanent cells present in the epidermis, cortex or phloem and so secondary in origin. In these permanent tissues certain cells become meristematic, which is termed as cork cambium or phellogen. The cells of phellogen are uniseriate, polygonal or rectangular in shape, compactly set with little intercellular spaces. They contain large nuclei, vacuoles and sometimes chloroplastid and tannin. In contrast to vascular cambium with ray and fusiform initials, phellogen consists of one type of cells. The cork cambium divides mainly tangentially or less commonly radially. The entire cork cambium ring may divide or the division may be restricted to certain cells only. The daughter cells, thus produced, lie at radial rows.

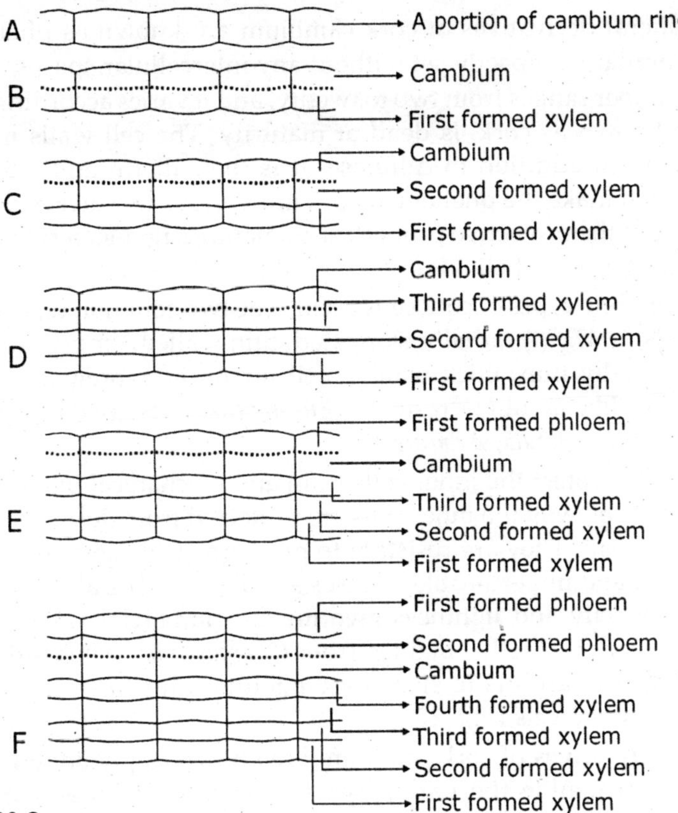


Figure 20.3

Diagrammatic representation of the stages (A-F) of formation of secondary phloem and secondary xylem from a portion of cambium ring. The recently formed secondary xylem and phloem are in contact with the dividing cambium. The first formed xylem and phloem are towards the centre and periphery respectively.

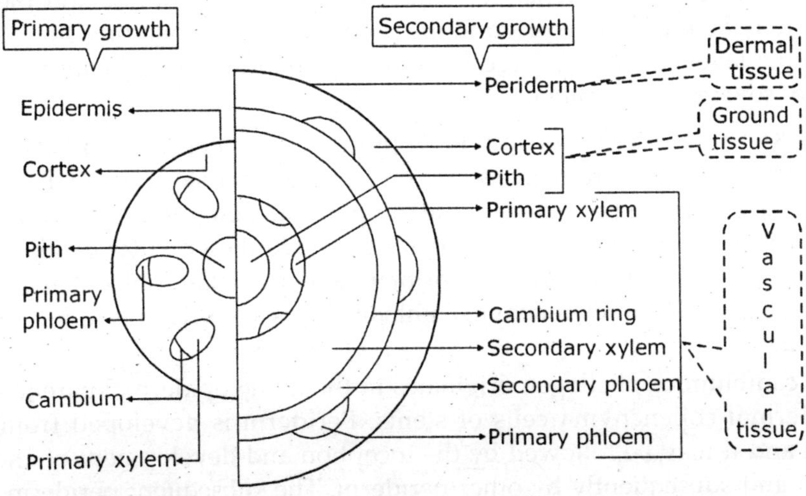


Figure 20.4

Diagrammatic representation of cross section of stems (in part) showing primary and secondary tissues.

The peripheral derivatives of cork cambium are known as phellem. These cells are rectangular, compactly set without any intercellular spaces, lie at radial rows whose number ranges from two to twenty, and it varies according to species. Phellem, also known as cork, is dead at maturity. The cell walls may contain suberin or lignin in addition to cellulose. It is the suberin and wax, the fatty substances, which make the phellem or cork impervious to water and gases, thus conferring upon it the protective properties. Sometimes non-suberized cells occur in the phellem—termed phelloids.

In plants different types of cork cells occur. The two common types are (Eames and MacDaniels, 1947): (i) radially elongated, thin walled, empty and very light cork and (ii) radially thin, thick walled, cell lumen is impregnated with dark staining materials like tannin or resin, ex. *Eucalyptus*. These two types may occur in the same plant, ex. *Betula, Arbutus*.

The cork cells protect the inner cells from any mechanical injury and owing to their suberized cell wall the inner cells are protected from desiccation. Phellem or cork cells are dead. They are resistant to pressure, acids and other chemicals. They are resilient and impermeable to gases and liquids. The above features and the strength, elasticity and lightness give much commercial value to the cork. The commercial cork is chiefly obtained from *Quercus suber*. The bottle corks that are used as stoppers are made sizes by tangential unit to stop leakage from vertically oriented lenticels (Fig. 32.3).

The inner derivatives of cork cambium are known as phelloderm. They are living cells and resemble the parenchyma of cortex, but they can be easily differentiated from the cortical parenchyma by their radial arrangement. The cell wall of phelloderm is thin and composed of cellulose only. In some plants phelloderm possesses chloroplastids, can photosynthesize and store starch.

Phellem, phellogen and phelloderm are collectively known as periderm, which forms a protective covering on the peripheral side and is impervious to air and water (Fig. 20.5). Intrastelar secondary growth may continue and the sole periderm may be unable to withstand the pressure of the expanding stele. Then, in the deeper layers of cortex, successive layers of periderm may be formed with the help of several cork cambia.

• BARK: The cork or suberized tissue present at the periphery of stems and roots, which originates from the secondary lateral meristem or phellogen, is termed as bark. It is the protective tissue of roots and stems and now a days this term is used to designate all tissues present outside the vascular cambium. In this sense it includes periderm, cortex, primary and secondary phloem. In mature barks a large quantity of sclerenchyma, phloem fibres and sclereids (e.g. *Alstonia*) are present.

Cork cambium or phellogen originates in the living epidermal parenchyma or hypodermal collenchyma cells of stems. Periderm is developed from this meristem and it may be followed by the inception and development of another phellogen and subsequently by other periderm. The subsequent periderm may be in the form of complete cylinders around the stem parallel to first formed periderm. The barks, thus developed, are known as ring bark (Ex. *Vitis, Clematis*

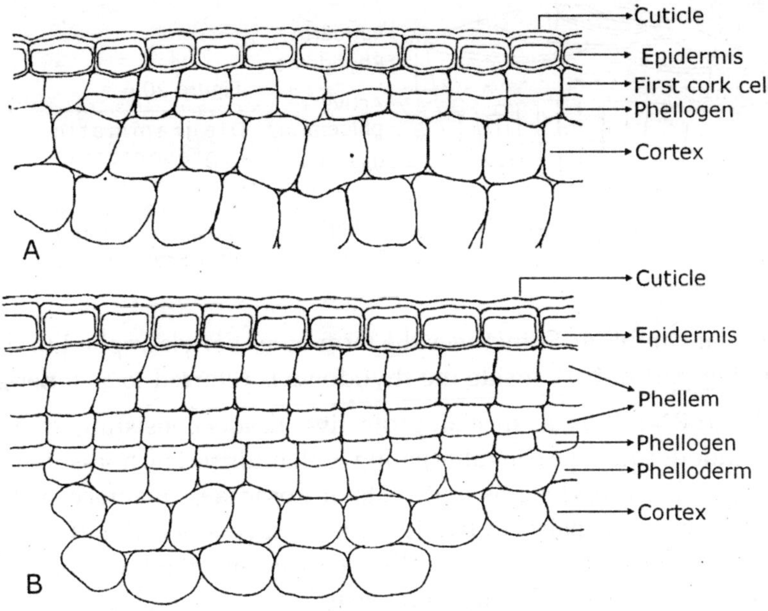


Figure 20.5

Cork cambium or phellogen in the cross section of a stem (A) and phellem, phellogen and phelloderm in B. (After Holman and Robbins, 1939, p. 112).

etc.). In *Ficus*, *Pyrus* etc. additional phellogen develops in the form of separate overlapping arcs and so the periderm assumes the form of scales or shells. These barks are known as scale bark.

Barks are composed of dead cells. When the subsequent inner periderm develops, the outer layers are sloughed off. In ring bark the peripheral dead cells are sloughed in the form of hollow cylinders where as in scale bark they are peeled off as scales.

The barks protect the inner tissues from desiccation, mechanical injury, entry of pathogen and the strain set up due to the formation of secondary tissues at the stele.

• RHYTIDOME (Fig. 20.6): It is a special type of protective tissue composed of successive bands of periderm with enclosed cells, which are either cortical parenchyma or secondary phloem. At the time of extrastelar secondary growth a periderm is formed at the peripheral region of stems. In *Robinia* this may be followed by successive inception and development of phellogen in more and more deep region of stem which gives rise to bands of periderm. The cells of phellem eventually become suberized. The successive bands of periderm enclose either cortical tissues or secondary phloem. Due to the formation of inner periderm, the peripheral periderm bands and the enclosed cells die. The peripheral successive bands of periderm together with the enclosed cortex or secondary phloem and all the tissues present external to innermost phellogen are collectively referred to as rhytidome. The term bark is applied to all tissues present external

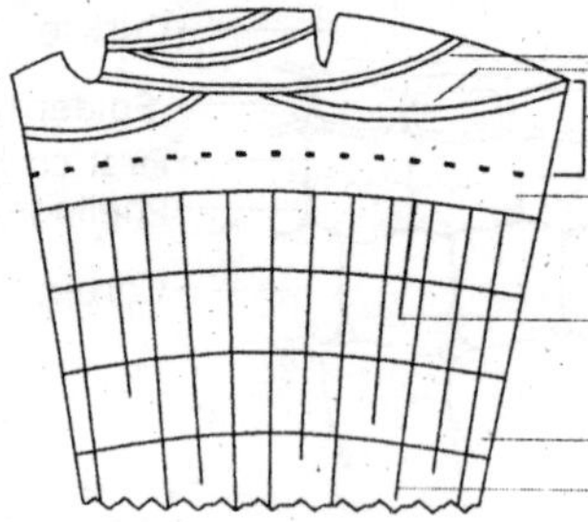


Figure 20.6

Diagrammatic representation of rhytidome in transverse section (adapted after Esau, 1965).

to the vascular cambium. Sometimes the term outer bark is applied to rhytidome and the living part of bark inside the rhytidome is referred to as inner bark.

• POLYDERM: It is a special protective tissue consisting of uniseriate suberized endodermis-like cells alternating with multiseriate non-suberized cells. It occurs in underground stems and roots of Rosaceae, Onagraceae, Myrtaceae and Hypericaceae. It consists of twenty or more layers of cell that are suberized and non-suberized. The peripheral cells are dead and the inner cells including the suberized cells contain living protoplasts.

At the time of formation, a special phellogen is differentiated at the pericycle. The phellogen forms tissues centrifugally by tangential divisions. The tissues, thus formed, consist of thin walled non-suberized parenchyma cells, which alternate with uniseriate endodermoid cells. The latter becomes cork cells. During the formation of cork casparian strips appear on the wall that later undergoes more extensive suberization. Polyderm is formed at the pericycle and it is exposed to surface after the death of the cortical tissues. It protects the inner tissues. The inner tissues including the suberized cells remain alive. The non-suberized cells are concerned with food storage.

• LENTICEL (Fig. 20.7): It is the raised opening or pore on the epidermis or bark of stems and roots through which atmospheric gases diffuse into and out of the living cells. They occur on young branches and on the surface of fruits (ex. apple and pear). In some plants like *Anabasis*, *Haloxylon*, *Vitis* etc., though periderm is formed lenticels are not developed. They may appear as dots, little flecks, rough dark patches, lens-shaped raised or dome shaped corky spots or fissure on the fruit, root and stem surfaces. The orientation of fissure recognizes the transverse and longitudinal arrangement of lenticels.

In cross sectional view lenticels appear as loose aggregate of cells which are protruded above by rupturing the epidermis of roots and stems. These cells may be suberized or non-suberized and are termed as complementary cells or filling cell (Wutz, 1955), which are loosely arranged, more or less spherical, thin walled and larger in size than the neighbouring cells.

Below the complementary cells phellogen is present and it is continuous with the rest of phellogen, which is situated between phelloderm and phellem of periderm. Beneath the phellogen, phelloderm is situated and it is also continuous with the rest of periderm.

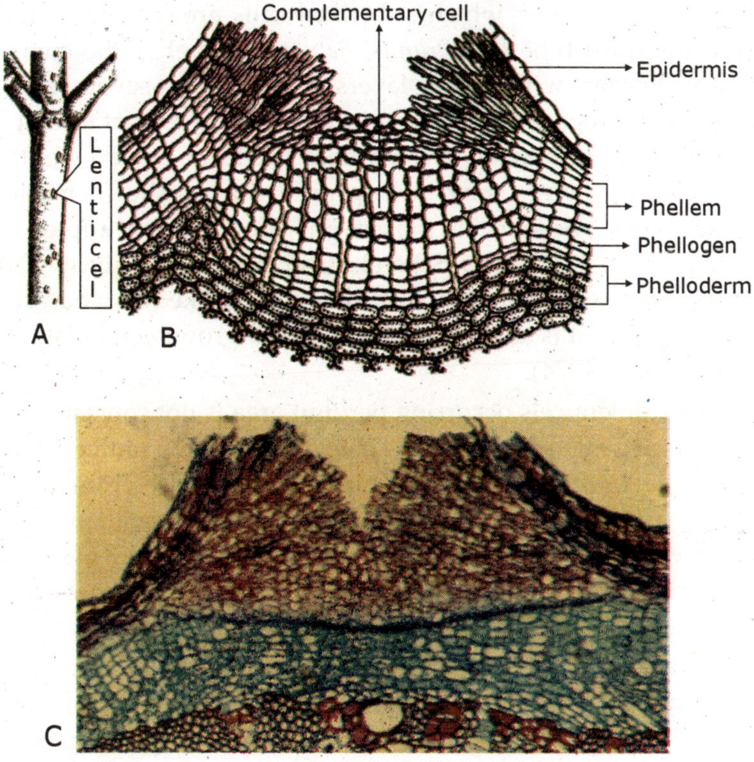


Figure 20.7

A. A portion of stem showing lenticels. B. Cross-section of *Sambucus* stem showing lenticel. C. Microphotograph of lenticel of *Aristolochia* stem in transverse sectional view.

In some plants, within the complementary cells some cells are produced which are compact and suberized. These cells are termed as closing layer. It is believed that the loose aggregate of complementary cells is held in position by the closing layer. The closing layer may be broken due to the formation of new complementary cells. This layer may be formed again at the end of growing season.

At the region of lenticels, complementary cells with profuse intercellular spaces interrupt the continuity of phellem. The closing layer, though compact contains intercellular spaces. The phellogen also possesses similar gaps. There is continuity of the intercellular spaces between complementary cells, closing layers, phellogen, phelloderm and inner tissues. So it is believed that gaseous exchange between inner tissues and atmosphere occurs through lenticels like stomata.

Three types of lenticels are recognized (Esau, 1965) in dicotyledonous stems on the basis of structure of complementary cells and the arrangement of closing layers: (i) in the first type (ex. *Pyrus, Populus, Magnolia* etc.) the complementary cells are suberized with air spaces. These tissues exhibit annual variation of loose aggregate of thin walled cells and compact thick walled tissue, (ii) the second type (ex. *Tilia, Quercus* etc.) exhibits a loose aggregate of non-suberized complementary cells below which suberized cells are formed to form closing layer; (iii) in the third type (ex. *Prunus, Betula, Fagus* etc.), the complementary cells regularly alternate with closing layers. The former is several cells wide and consists of loose aggregate of non-suberized cells. The cells of closing layers are suberized, compact and narrow, which hold the loose complementary cells in places. The closing layers may be ruptured due to the formation of new complementary cells. The lenticels, present on fruit surfaces of apples, are atypical consisting of areas of corky cells. Scan electron microscopic study of the lenticel of potato tubers reveals the presence of some thread like waxy outgrowths from the sides of the pores; it is suggested that these outgrowths regulate the gaseous exchange (Hayward, 1974).

The number of lenticels occurring per unit area is not constant. Initially and usually lenticels arise beneath stomata. They may develop under each or a group of stomata or irrespective of them. Lenticels may arise from deeply seated phellogen and from successive phellogen, which has no relation to the position of stomata. Generally lenticels occur irregularly scattered over the surface; sometimes transverse and longitudinal arrangement are also noted.

Usually lenticels develop from the phellogen. The phellogen originates at the sub-stomatal region. The sub-stomatal cells divide in all directions and lose chlorophyll. The derivatives towards the cortex form the phellogen of lenticel, which gradually becomes continuous with the rest of the phellogen of periderm. The phellogen of lenticel divides and the cells produced towards the exterior, as well as those formed during the divisions of sub-stomatal cells, are termed as filling cells or complementary cells. As more and more complementary cells are formed, they exert a pressure on the epidermis and as a result the epidermis ruptures. The complementary cells protrude above through the ruptured epidermis. The exposed cells gradually dry up, die and the phellogen of lenticel forms the closing layers centrifugally in addition to complementary cells. The phellogen of lenticel, by repeated divisions, forms phelloderm towards the interior. This phelloderm becomes continuous with that formed by the phellogen of periderm.

When lenticel originates from the deep-seated successive phellogen, certain localized region functions as phellogen of lenticel irrespective of the position of stomata. In some cases lenticel may be formed from the phellogen, which produced phellem for a while. In this case the phellogen, after forming cork for sometime forms complementary cells.

## Secondary growth in dicot root

Radial vascular strands with exarch protoxylem are the characteristic of roots, i.e. primary xylem and phloem lie separate and alternate to each other. In dicot roots secondary tissues are formed both at intrastelar and extrastelar region.

Intrastelar secondary growth occurs by the secondary cambium, as the primary cambium is absent in roots. Secondary cambium originates from the permanent tissues present in the stele. Strips of cambia differentiate below each primary phloem. Cambia are also formed above the protoxylem, near or at the pericycle. These strips of cambia extend laterally and join with each other. As a result a wavy cambium ring with ridges and furrows appear. The ridges lie overarching the protoxylem whereas the furrows (Fig. 20.8) are present below the primary phloem. So according to the number of xylem and phloem present in the stele the number of ridges and furrow appear.

The wavy cambium ring divides in all segments and produces secondary tissues at the peripheral and inner side. The peripheral derivatives are

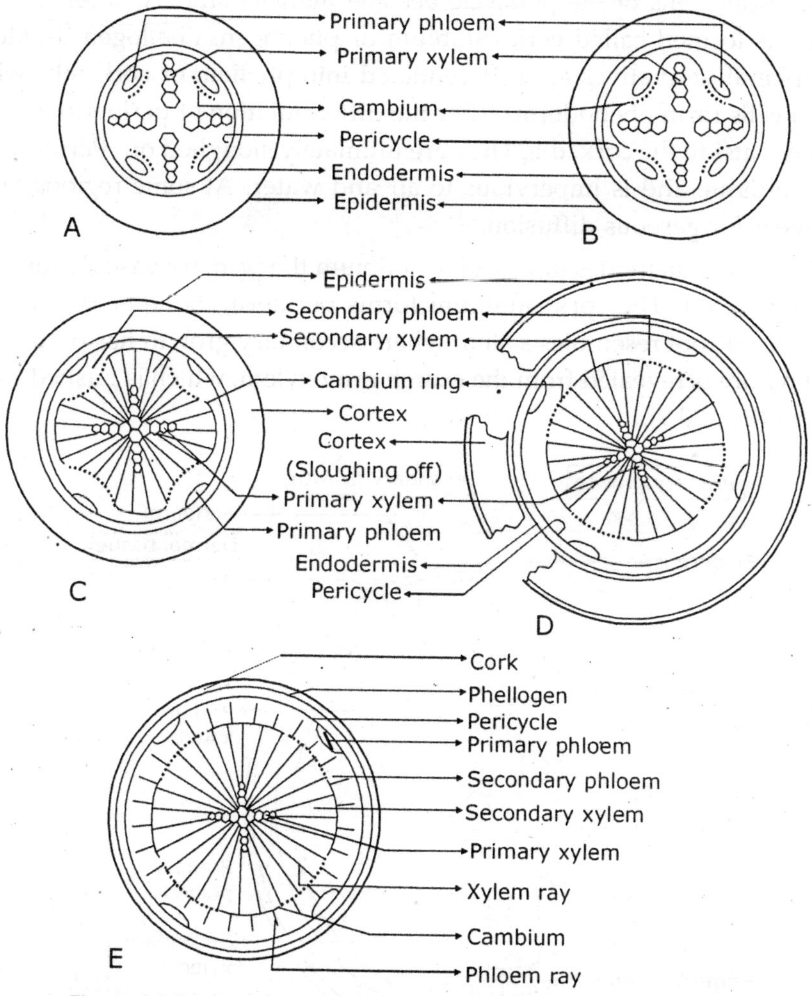


Figure 20.8

Diagrammatic figures (A, B, C, D & E) showing the different stages of secondary growth in dicotyledonous root. A. The primary structure of root and the appearance of cambium segment below the primary phloem. B. The wavy cambium ring. C. The formation of secondary xylem and secondary phloem. D. Sloughing off the cortex and E. The formation of cork.

differentiated into secondary phloem while the secondary xylem is formed from the inner cells. Normally cambial cells at the furrowed region divide more in contrast to ridged region. As a result the wavy cambium ring becomes more or less circular and it lies in between the secondary xylem and phloem. Usually more secondary xylem is differentiated in comparison to phloem. All elements of xylem (tracheids, trachea, xylem fibre and xylem parenchyma or xylem ray) and phloem (sieve tube, companion cells, phloem fibre and phloem parenchyma or phloem ray) are differentiated in the secondary vascular tissues.

Periderm is formed in roots. Soon after the formation and division of vascular cambium some cells of the pericycle become meristematic. As a result another cambium is formed called cork cambium or phellogen. Phellogen divides and the peripheral derivatives are differentiated into phellem or cork cells whereas the inner cells form phelloderm. After the differentiation of periderm all tissues, which lay outside the cork, die. They are ultimately sloughed off. Periderm is the protective tissue and is impervious to air and water. At some regions lenticels may appear for gaseous diffusion.

As a result of activities of vascular cambium the primary vascular tissues are widely separated. The cork cambium forms periderm. So after the secondary growth an old root resembles a stem where secondary growth has occurred. But the root nature is revealed from the exarch protoxylem, which is pushed towards the centre (Fig. 20.9).

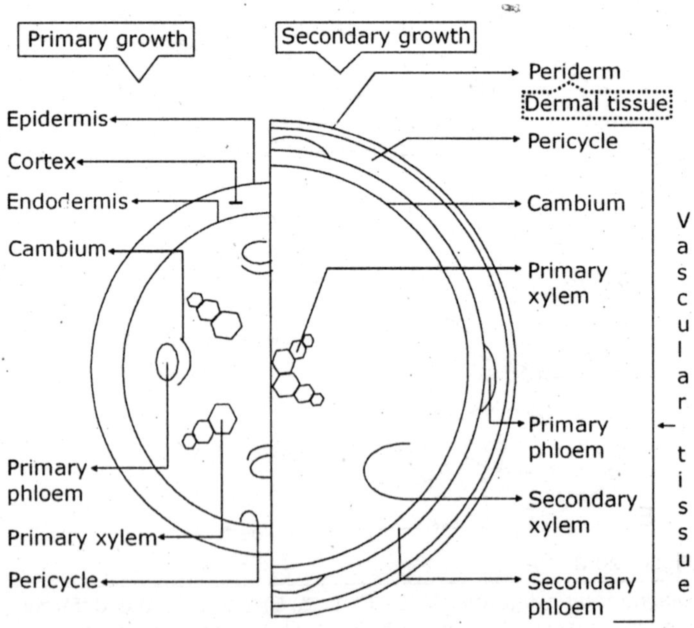


Figure 20.9

Diagrammatic representation of cross-section of root (in part) showing primary and secondary tissues.

## Secondary xylem

Secondary xylem is a complex tissue, known also as wood. The study of wood by preparing sections for microscopic observations is defined as xylotomy. Secondary xylem is derived from the vascular cambium. It develops in stems and roots of gymnosperm and angiosperm-dicotyledonous plants as a consequence of secondary growth. The cells composing wood are mainly thick walled. Cell cytoplasm deposits the thickening materials during differentiation. Later cytoplasm dies and leaves the cells of wood devoid of any living contents.

Woods are divided into two main groups—porous and non-porous based on the presence or absence of vessels or pores. Non-porous wood predominate in gymnosperm where predominantly tracheids with a small amount of parenchyma compose the wood. Example: pines, spruces and firs where the texture of wood is uniform and a carpenter works with ease. So non-porous wood or gymnospermous wood is referred to as softwood. In contrast angiosperm dicotyledonous wood exhibits a variety of cell types including thick-walled fibres. It is sometimes difficult to work with such type of wood. So porous wood or angiosperm-dicotyledonous wood is referred to as hardwood. These two terms hardwood and softwood are traditionally used in timber trade. But the terms, however, are misleading. Woods with soft and hard texture can be found in both groups of plants. The terms have no relation to the relative hardness or softness of timber. These terms—hardwood and softwood are applicable only to porous and non-porous wood respectively. As for example the wood of *Ochroma, Bombax ceiba* and *Pterocymbium tinctorium* are soft to very soft. These plants belong to angiosperm and their wood is referred to as hardwood by definition as the wood is porous. On the other hand the woods of *Cedrus deodara* and *Pinus roxburghii* are very hard. These plants belong to gymnosperm and their wood is referred to as softwood by definition as the wood is non-porous.

Wood contains all the types of cells that are observed in primary xylem but no new ones. The only difference between primary- and secondary xylem is the origin and arrangement of elements of xylem. Secondary xylem originates from vascular cambium whereas procambium gives rise to primary xylem. The arrangement of wood cell types is either 'along-the-axis'—called axial system or 'across-the-axis'—termed radial system. The arrangement reflects to that of the fusiform- and ray initials of cambium that form wood cell types. Fusiform initial gives rise to elements of axial system and ray initial forms the cells present on the radial system (Box 20.1).

The cell types of wood can be easily studied in transverse section (t.s. = cut across the axial plane) and in longitudinal section (l.s. = cut along the axial plane), Figs. 20.10 & 20.19. The images observed in the sections of wood through an optical microscope reveal the shape of the cells present in wood. The knowledge of shapes in t.s. and l.s. provides one to envision the three dimensional shape of a cell (Fig. 20.10A).

The cambium that forms wood is composed of spindle shaped fusiform initials and the isodiametric ray initials. The ray initials form the ray parenchyma cells,

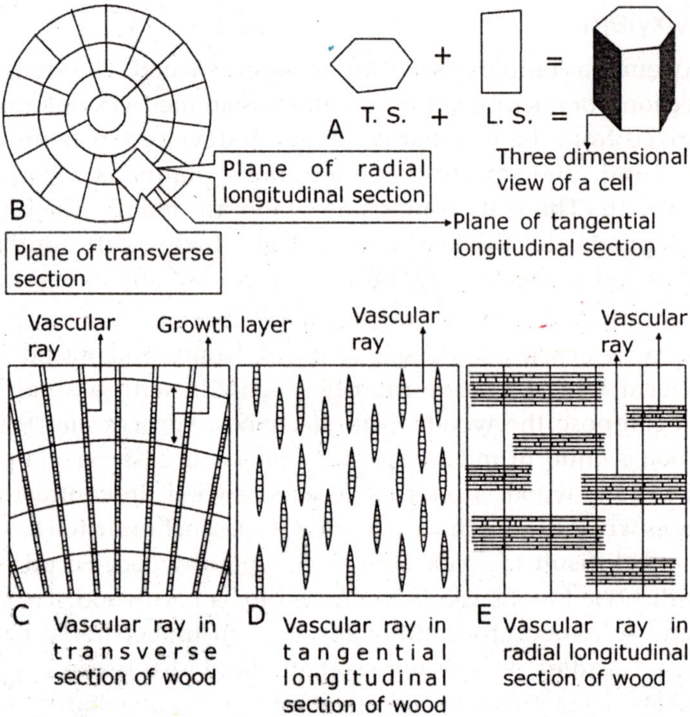


Figure 20.10

A. With the images of l. s. & t. s. the three dimensional view of a cell can be obtained. B. Diagram pointing the planes of section of wood. C., D. & E. The appearance of ray parenchyma in different planes.

which compose the horizontal or radial tissue system. The fusiform initials lie parallel to the long axis and give rise to tracheary elements, fibres and axial parenchyma that form the vertical or axial tissue system. The living cells of axial and radial systems are interconnected to form a continuous system.

Box 20.1

Types of wood cells present in dicot and gymnosperm wood

| | Dicotyledonous wood | Gymnospermous wood |
|---|---|---|
| Axial system | | |
| Tracheids | Present | Present |
| Vessels | Present | Absent (except *Ephedra*, *Welwitschia* & *Gnetum*) |
| Fibres | Present | Rare |
| Parenchyma | Present | Rare |
| Radial system | | |
| Ray parenchyma | Present | Present |
| Ray tracheids | Present | Present |

## Ray parenchyma

Ray parenchyma cells are of various shapes and the two most common forms are where the longest axis of the cell is either vertical or radial. The former one, where the longest axis is oriented vertically, is called erect or upright ray cells. The latter one is termed as procumbent ray cells where the longest axis is oriented radially. Ray cells may also be square or isodiametric. Square cells are considered as morphologically equivalent to erect cells. In a radial longitudinal section the parenchyma cells appear as fine horizontal lines. So a transverse or radial longitudinal section reveals the longitudinal axis of ray cells. The transverse section of ray cells is obtained from tangential longitudinal section (TLS). In TLS ray parenchyma appears variously, for instance: (i) Uniseriate: the ray cells are entirely one cell in width, Ex. *Salix, Pinus* etc. (ii) Biseriate: any portion of the ray masses is two cells wide and (iii) Multiseriate: any portion of the ray masses is more than two cells in width, e.g. *Quercus.* The TLS reveals that the biseriate and multiseriate ray gradually become uniseriate at both its upper and lower edges.

Uniseriate and multiseriate rays both may be either homocellular or heterocellular (Fig. 20.11). In dicotyledons the homocellular rays consists of (1) erect cells only, (2) procumbent cells only, (3) square cells and (4) erect and square cells only. In heterocellular type procumbent and square or procumbent and erect cells occur. The entire ray system may consist of either homocellular or heterocellular types, or of combinations of two.

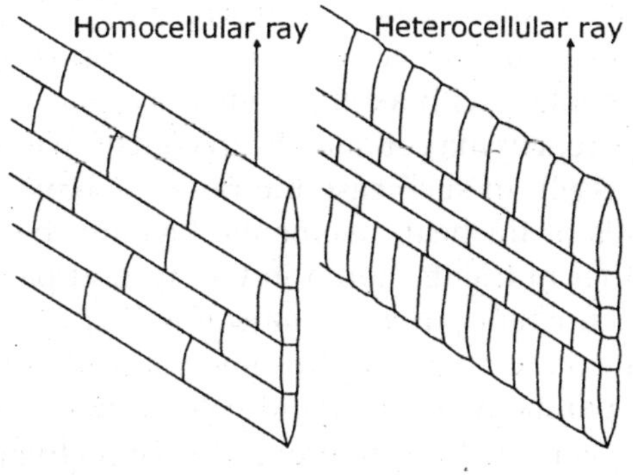


Figure 20.11
Diagrammatic figures of ray parenchyma cells in tangential and radial longitudinal sectional view. Redrawn from Cutler, 1978, p. 20 & 21.

In gymnosperm xylem rays are almost exclusively uniseriate. The ray cells may be (Fahn, 1987) (i) homocellular: rays consist of parenchyma cells only, ex. *Thuja* and (ii) heterocellular: where the ray cells comprise parenchyma cells and ray tracheids, ex. *Pinus.* The ray tracheids (Fig. 20.12B) have bordered pits and are devoid of protoplasts in contrast to ray parenchyma. They are multiseriate only if they contain resin canals. Ray tracheids are horizontal, rectangular cells that look somewhat like parenchyma cells. They have lignified walls and in *Pinus* the lignifications may be in form of teeth or band that project in the cell lumen.

In angiosperm ray cells may be homogeneous (ex. *Populus*) where all the cells are procumbent, i.e. the longest axis is oriented in radial direction or heterogeneous (ex. *Olea*) where the ray cells are procumbent and square or oriented

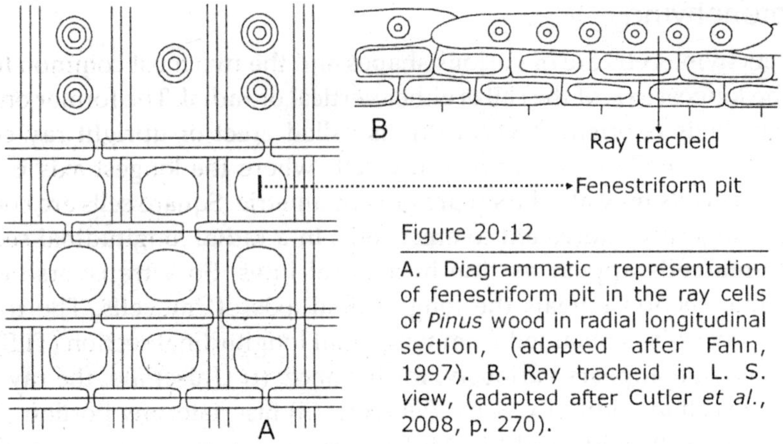


Figure 20.12

A. Diagrammatic representation of fenestriform pit in the ray cells of *Pinus* wood in radial longitudinal section, (adapted after Fahn, 1997). B. Ray tracheid in L. S. view, (adapted after Cutler *et al.*, 2008, p. 270).

vertically. These two terms—homogeneous and heterogeneous are more or less similar to homocellular and heterocellular ray cells respectively of gymnospermous wood.

The ray cells of *Pinus* have very large pits with very narrow border that appears as more or less circular or rectangular areas throughout the width of the cells. These are known as fenestriform pits (Fig. 20.12A) whose surface view is obtained in radial longitudinal section.

Initially ray parenchyma cells are living and serve in storage and aeration. They store carbohydrates and other nutrients. These substances are transported radially within the wood over short distances. The upright cells have direct connection with axial cells. The ray/axial interface exhibit many forms. When the upright ray parenchyma cell is in contact with axial parenchyma plasmodesmata exist between them. The upright parenchyma cell may be in contact with the axial tracheary elements (tracheid or vessel). In such case the ray cells have very thin walls while pits occur on the walls of tracheary elements. Pits are the routes of transport of carbohydrates and other nutrients stored in the upright parenchyma cells. The stored starch in the procumbent parenchyma cell is first digested into sugar and then transferred to axial conducting cells. In some plants there exists no connection between procumbent parenchyma cells and cells of axial system. In such cases the transfer of nutrients is routed through upright parenchyma cells.

## Axial parenchyma

The distribution of axial parenchyma is very characteristic in dicotyledonous wood. Parenchyma may lie independent of vessels or they are distinctly associated with them. These two forms are known as apotracheal and paratracheal respectively. The common apotracheal forms are (Fig. 10.2) (i) diffuse parenchyma: parenchyma cells appear as single cell or as small uniseriate band throughout the growth ring (ex. *Quercus*); (ii) banded or metatracheal parenchyma: parenchyma cells appear concentric bands (ex. *Hicoria*) and (iii) boundary parenchyma: parenchyma may appear at the beginning or end of the growth ring and accordingly termed as initial parenchyma (ex. *Ceratonia, Zygophyllum*) and terminal parenchyma (ex. *Magnolia, Salix* etc.)

Paratracheal parenchyma may be (Fig. 10.2): (i) Scanty (ex. *Acer*): The parenchyma cells do not form a continuous sheath surrounding the vessel; (ii) Vasicentric parenchyma (ex. *Tamarix*): The parenchyma cells form a continuous sheath around the vessel of different width; (iii) Aliform parenchyma (ex. *Acacia*): Vasicentric parenchyma extends laterally as wings.

## Annual ring

Most trees and shrubs of temperate origin show characteristic growth layers, often called growth rings or annual rings of secondary xylem. Growth rings appear as concentric or eccentric rings seen in transverse section of stems; each ring represents a year's growth of secondary xylem. The concentric rings are formed at the straight parts of stems under uniform conditions whereas the eccentric rings appear as a result of adverse natural calamities. The activity of cambium is a seasonal phenomenon. The periodic activity and quiescence of cambium lead to the formation of annual or growth rings. These rings are distinguishable with an unaided eye because of the differences in structure and colours between the secondary xylem formed at the early and late parts of a growth season. As each ring represents a year's growth, the approximate age of a plant can be ascertained by counting the number of growth-rings (Fig. 20.13).

Figure 20.13

A. Microphotograph of cross-section of *Pinus* stem showing annual ring. B. Cross-section of a portion of stem showing annual ring, sap wood and heart wood.

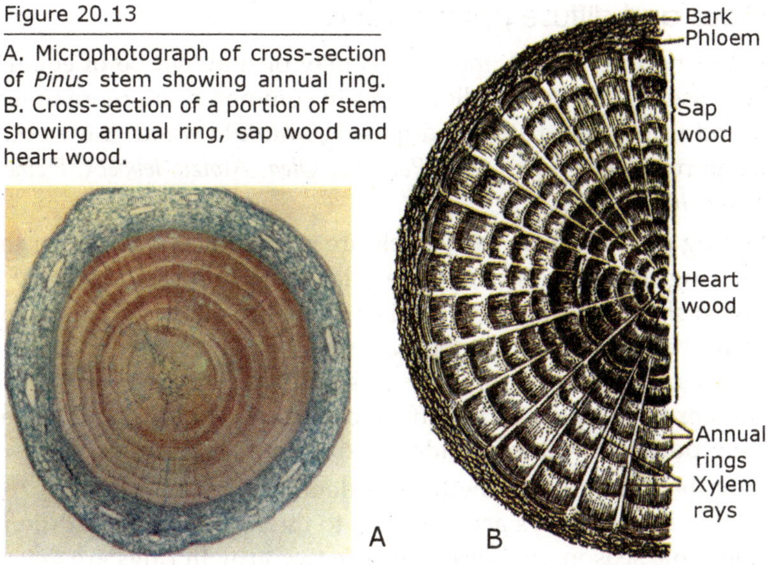


In temperate perennial plants, the activity of cambium begins in the spring. The wood produced during this period is the early wood, often called spring wood. The spring wood is less dense, consists of thin-walled elements and possesses vessels with large lumen. The cells of wood, produced at the late parts of growth season, possess thicker walls and vessels with smaller diameter. These are late wood, also known as summer or autumn wood. The early wood and late wood, formed in one growth season, together constitute the growth ring or annual ring. The early wood and late wood of the same season gradually merge with each other and with adjacent growth rings also. But a sharp line of demarcation exists between the late wood of one growth season and early wood of next season. In *Tilia, Ceratonia, Zygophyllum* etc. a few layers of wood parenchyma is present between the growth rings and these parenchyma form the division line between the rings. It was possible to determine accurately the age of *Pinus longifolia, Tectona grandis, Terminalia tomentosa, Acacia catechu* and *Bombax malabaricum* (Chowdhury, 1939). The annual rings were carefully counted and the false rings, which occasionally develop in the above species, were properly located and omitted from counting. There was a common belief that growth rings are formed only in deciduous trees and the evergreen trees are without them. However, Chowdhury (1939) reported the formation of growth rings in both evergreen (e.g. *Michelia champaca*) and deciduous trees (e.g. *Cedrela toona, Albizia lebbeck* and *Dalbergia sissoo*). Annual rings are also formed in tropical genera namely, *Bursera* (Burseraceae), *Citharexylum* (Verbenaceae), *Rapanea* (Myrsinaceae) and *Swietenia* (Meliaceae).

In some plants woods are formed without growth-rings, e.g. *Baccharis* (Asteraceae), *Laguncularia* (Combretaceae), *Rhizophora* (Rhizophoraceae), *Manilkara* (Sapotaceae) and *Pisonia* (Nyctaginceae) etc.

## Ring porous and diffuse porous wood

In a growth ring (Fig. 20.14), the vessels may or may not be of equal diameter. When the vessels, sometimes called pores in commerce, are more or less of same diameter and uniformly distributed in early and late wood, the wood is said to be diffuse porous wood; ex. *Acer, Populus, Olea, Albizia lebbeck, Dalbergia sissoo* and *Michelia champaka* etc.

In the ring porous wood, the vessels are not of equal in diameter throughout the growth ring. The vessels with larger diameter are present on early wood and the late wood shows vessels with smaller diameter. These variations in diameter can be easily being observed in the transverse sections of stem of *Cedrela toona, Tectona grandis, Quercus, Fraxinus* etc. It is observed that ring porous woods are few in number in comparison to diffuse porous wood. From phylogenetic point of view ring porous wood is considered to be more advanced.

Usually the number of growth rings denotes the approximate age of a tree. However, this is not always accurate due to the formation of additional growth rings in the same season. In *Tamarix aphylla* two growth rings are developed in a single season (Fahn, 1958). More than one growth rings formed per annum is also reported in *Avicennia*. Additional growth rings may be developed due to adverse natural calamites like draught, frost, defoliation, flood, and mechanical injury or due to infection, and biotic factors that disturb the plants when the activity of cambium is either checked or stimulated; when stimulated additional rings are formed, termed false annual ring. When two or more rings are formed in one season they are often called as double or multiple annual rings respectively (Eames and MacDaniels, 1947). False rings may be formed both in temperate and

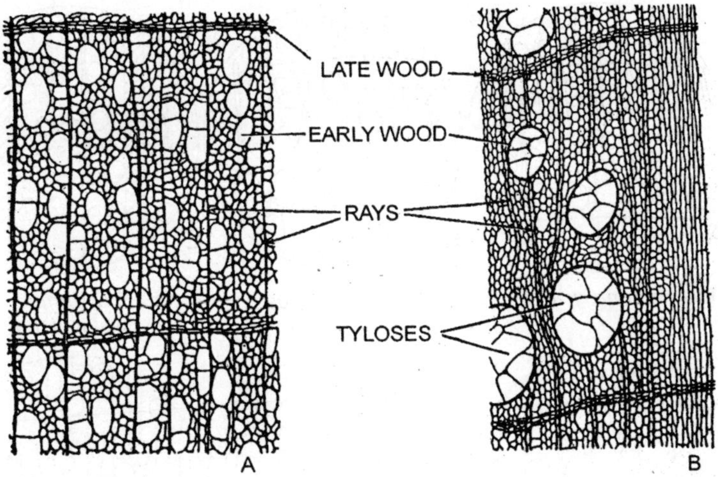


Figure 20.14

Transverse section of wood showing (A) diffuse porous and (B) ring porous wood.

tropical species. Researches reveal that day length, temperature, hormones etc. influence the activity of the cambium. It is not very difficult to distinguish the false growth marks from the true ones. In *Cedrela toona* the growth marks, either true or false, are always associated with concentric parenchyma bands. Vessels occur on either side of the growth marks of the parenchyma bands. In true marks, the vessels occurring on different sides of parenchyma bands are markedly of different diameter while the vessels on either side of parenchyma bands of false marks are more or less of same diameter.

In most trees of tropical origin except a few genera there is no distinguishable growth ring. The plants like *Acacia tortilis, Thymelaea hirsuta,* where there are no growth rings, produce same type of secondary wood throughout the growth season. In the South Californian pine *Pinus radiata* the cambium is not fully dormant even in winter. The Indian evergreen and deciduous trees do not always show distinguishable growth rings. The concentric rings in wood, as seen in *Shorea robusta* and *Eugenia jambolana,* have no relation with annual growth activity and so not indicative regarding the age of the tree (Chowdhury, 1939).

The arrangement and pattern of distribution of vessels differ and these are easily observed in transverse sections of wood. For instance, *Eucalyptus* shows single vessel; vessel clusters are present in *Pistacia*. The vessels whose walls are appressed with each other are often called multiples. They are present in *Salix*. There are more types that are illustrated in Fig. 20.15.

The size, shape, arrangements and the relative amount of constituent elements of secondary xylem of most trees are genetically labeled. Thus secondary xylem provides characters to identify an unknown timber. With the aid of these characters a key can be formulated by which timbers can be identified. A key can be made by using punched-card system and the computer (Fahn, 1997, p. 369).

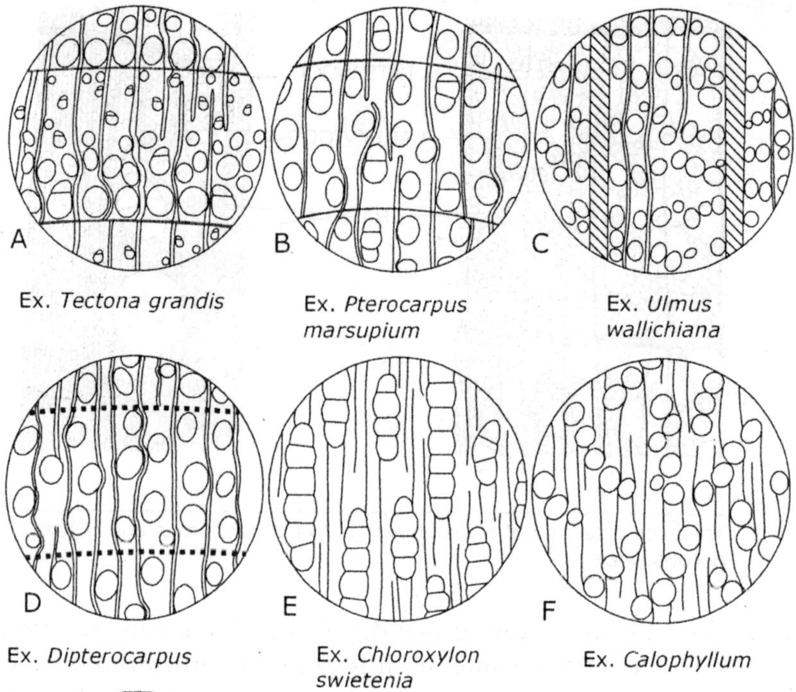


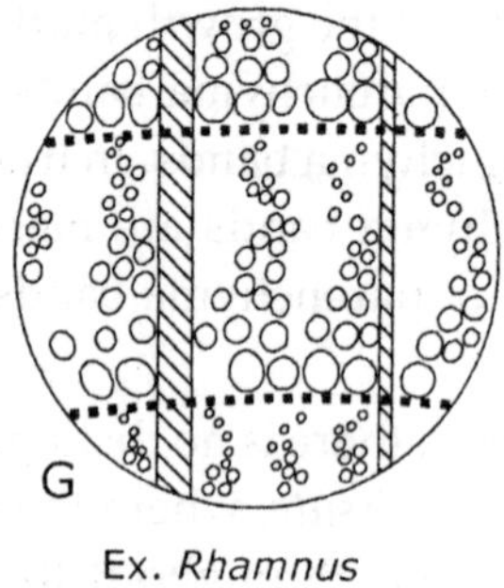


Figure 20.15

Diagram illustrating the pore arrangement in wood. A. Pores solitary and in radial multiples; ring porous. B. Pores solitary and in radial multiples; diffuse porous. C. Pores in tangential clusters. D. Pores exclusively solitary. E. Pores in long radial multiples. F. Pores in oblique groups. G. Pores flame like; ring porous. Adapted after Rao & Juneja, 1971, p. 14.

Usually a dichotomous key is formulated using the same principle as is followed in plant taxonomy (vide Lawrence, 1967 p. 226 for details). Dichotomous key is very popular and is used all over the world. But it has limited scope and is applicable only to a small group of timber. Fahn (p. 369-370) illustrated a dichotomous key to identify woods. Rao and Juneja (1971) proposed a dichotomous key to identify fifty important timbers of India. They schematically represented a dichotomous key for eight timbers only and commented that 'even a novice may not have difficulty in following the principle and working of a dichotomous key' — p. 31. The dichotomous key is illustrated in Fig. 20.16. It is customary to present a dichotomous key in a running form and it is illustrated in Box 20.2. For details consult Fahn and Rao and Juneja.

*Importance of studying growth rings*

Analysis of growth rings, also referred to as dendrochronology, has many uses and a few are mentioned below:

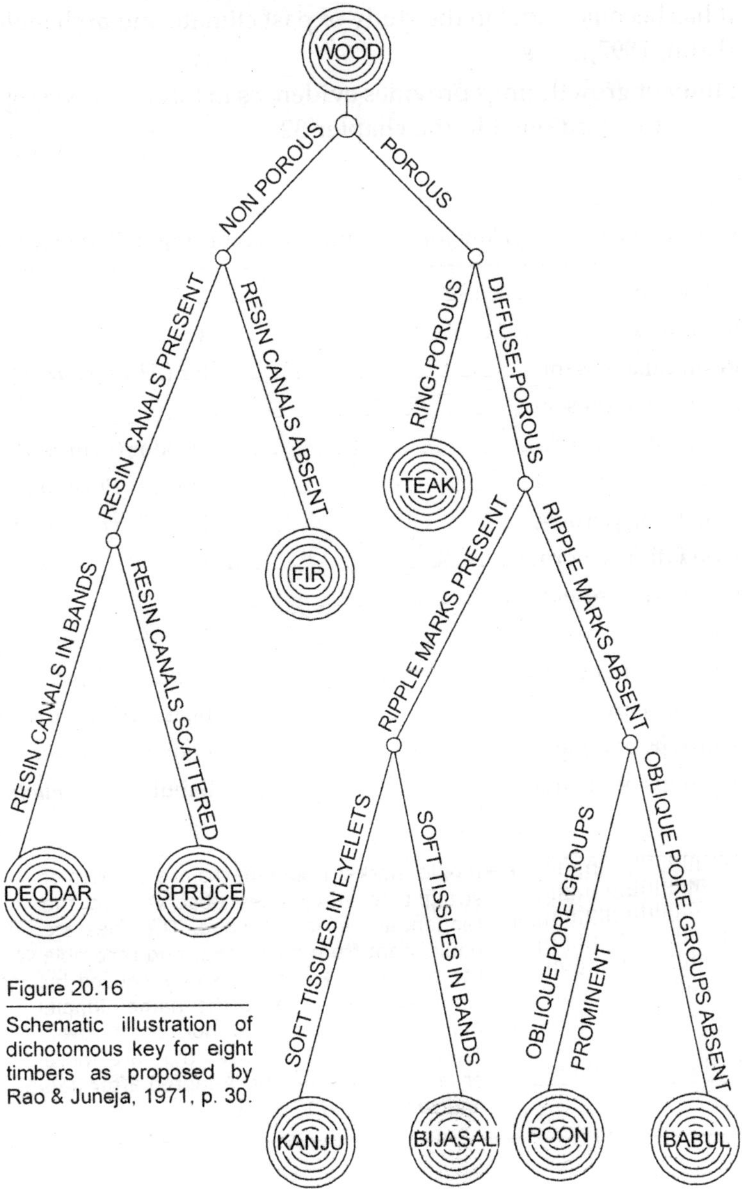


Figure 20.16
Schematic illustration of dichotomous key for eight timbers as proposed by Rao & Juneja, 1971, p. 30.

- Foresters often estimate the age of a tree by counting the numbers of growth rings.
- From the growth-ring-patterns the quality of timber can be ascertained.
- Study of growth-rings is used as a means of dating of wood, e.g. oak. Oak wood is used in house construction and in the backing of old pictures. A chronology of at least one thousand years has been established in oak wood.
- Growth-ring analysis is also used as a check of radio-carbon dating.

- It has become a tool in the study of past climate and archaeological dating (Fahn, 1997).
- Study of growth rings provides evidences in forensic investigation—a few cases are mentioned in the chapter 32.

Box 20.2

Illustrating dichotomous key for eight timbers according to Rao & Juneja, 1971.

| Character | Lead |
|---|---|
| 1. Wood non porous | 2 |
| 1. Wood porous | 4 |
| 2. Resin canal absent | Fir (*Abies pindrow*) |
| 2. Resin canal present | 3 |
| 3. Resin canals in bands | Deodar (*Cedrus deodara*) |
| 3. Resin canals scattered | Spruce (*Picea smithiana*) |
| 4. Wood ring porous | Teak (*Tectona grandis*) |
| 4. Wood diffuse porous | 5 |
| 5. Ripple marks present | 6 |
| 5. Ripple marks absent | 7 |
| 6. Soft tissue in eyelets | Kanju (*Holoptelea integrifolia*) |
| 6. Soft tissue in bands | Bijasal (*Pterocarpus marsupium*) |
| 7. Pore groups oblique | Poon (*Calophyllum*) |
| 7. Pore groups not oblique | Babul (*Acacia nilotica*) |

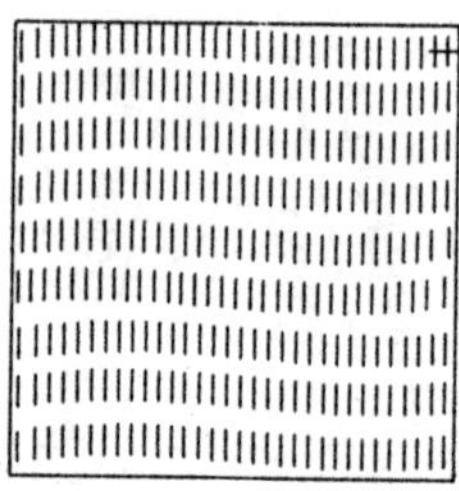

Ripple marks: ripple marks are a series of straight to wavy lines observed on the tangential surface of wood. The lines are equidistant from each other and resemble the ripple marks on the sands caused by wind or water - hence the name. Ripple marks are arranged in horizontal rows. They are due to the rays that are more or less of same height. Adapted after Rao and Juneja, 1971, p. 21.

## Tylosis

In many plants several axial or ray parenchyma surround the tracheary elements. Some of them protrude into tracheary elements through the pit. These ingrowths are called tyloses (singular—tylosis or tylose). Several tyloses may be formed in the vessel formed by the surrounding parenchyma. The tyloses increase, come in contact with one another, completely fill and block the vessel. As a result the tracheary elements become inactive. The nucleus and some amount of cytoplasm flow to the tyloses. The walls may remain thin or lignin may deposit. The lumens of tracheary element with numerous tyloses (Fig. 20.17) appear as network (ex. *Quercus*).

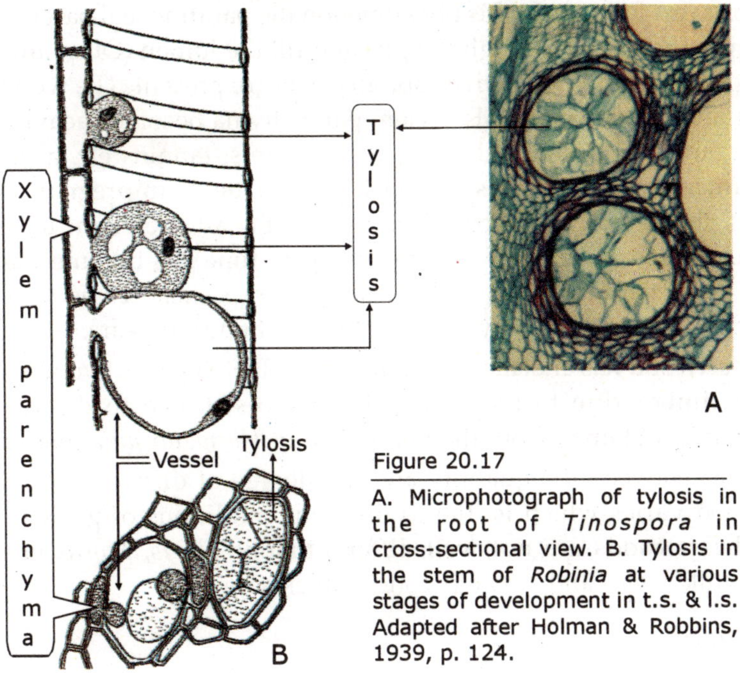


Figure 20.17

A. Microphotograph of tylosis in the root of *Tinospora* in cross-sectional view. B. Tylosis in the stem of *Robinia* at various stages of development in t.s. & l.s. Adapted after Holman & Robbins, 1939, p. 124.

## Tylosoid

In gymnosperm, the epithelial cells, i.e. resin producing parenchyma surrounding the resin ducts, sometimes enlarge as tylosis-like intrusions and block the duct. These ingrowths are termed as tylosoids (ex. *Pinus*). In angiosperm (ex. *Vitis, Bombax* etc.) parenchyma proliferates into the neighboring sieve tubes in a tylose-like manner. These proliferations are also known as tylosoids. Tylosoids of gymnosperm and angiosperm differ from tyloses in not protruding through pits.

## Sapwood and heartwood

The wood of some old trees where considerable amount of secondary xylem has developed, becomes differentiated into two regions – the peripheral region, called sapwood and the inner zone, termed heartwood.

The sapwood also called alburnum, in contrast to heartwood, is of lighter colour and consists of the active outer secondary xylem which is mainly concerned with translocation and storage of food. These are the youngest formed wood and are softer in texture than the heartwood. The sapwood contains living parenchyma and water-filled tracheary elements; it is full of 'xylem sap' – from where the name is derived.

The inner heartwood also termed as duramen, in contrast to sapwood, is dark colored, hard in texture and contains inactive elements of primary and secondary xylem. Tyloses develop in the tracheary elements of some wood and block the lumen. The lumen may also be blocked by another method referred to as gummosis (ex. *Prunus*). In this phenomenon the paratracheal parenchyma cells produce gum that flows through the pits and fill the lumen (Olien and Bukovac, 1982). In gymnosperm, non-active aspirated pits are present. The wood loses cell sap, water and reserve materials. Xylem parenchyma becomes heavily lignified. Various organic compounds like oils, resins, gums, tannins etc. accumulate in the cell lumen, which imparts the dark color. Due to impregnation of these substances, the heartwood becomes more durable and resistant to decay. The living xylem elements of sapwood in course of time die become inactive and gradually converted to heartwood. However, in *Sequoia sempervirens* the ray parenchyma of heartwood may remain alive for hundred years.

The heartwood has enormous commercial value. The heartwood yields good quality of timber due to its durability and resistance to decay. The dye haematoxylin is obtained from the heartwood of *Haematoxylum campechianum*.

The ratio between the quantity and the degree of difference of heartwood and sapwood varies which is due to different conditions of growth. In *Abies*, *Picea* the heartwoods are not well differentiated. *Taxus*, *Morus* possess thin sapwood while *Fagus* and *Acer* etc. consist thick sapwood.

### Reaction wood

Reaction wood is developed in stems and branches of a tree that are under stress. Gravity causes lateral stress. In response to such stress dicots and gymnosperm plants produce reaction wood and thus prevent the branches from drooping and becoming pendant. Reaction wood appears in three different forms according to the nature of plants. They are referred to as tension wood, compression wood and contrasting wood.

• *Tension wood*: Tension wood, as the name implies, develops under tension. It is formed in dicotyledonous deciduous trees on the upper side of leaning hardwood tree branches. It is also formed on the curved side of the stem. A cross-section of a branch having tension wood reveals that the wood is lighter in colour than normal wood. The growth rings are eccentric and much wider on the upper side of a branch than normal wood. Gelatinous fibres compose tension wood. The cell wall of these fibres have little or no lignin and high cellulose content (Fig. 20.18) that fills most of the volume of lumen of cells. The content of hemicellulose is also different from that of normal wood. The tension wood has fuzzy appearance. This is due to the fact that when tension wood is sawed the fibres tend to pull out giving a fuzzy or villous surface.

• *Compression wood*: The reaction wood that is formed in conifers is referred to as compression wood. Compression wood is formed on the underside of softwood tree branches in response to compression stress. A cross-section of a branch having compression wood reveals that the wood is darker in colour than normal wood. The growth rings are wider than normal. In this wood the tracheids are short. In cross-sectional view the tracheids are round and spaces occur between the corners of them. The structure of the secondary wall of tracheid is noteworthy. The secondary wall is composed of $S_1$ and $S_2$ (=outer & middle layer respectively)

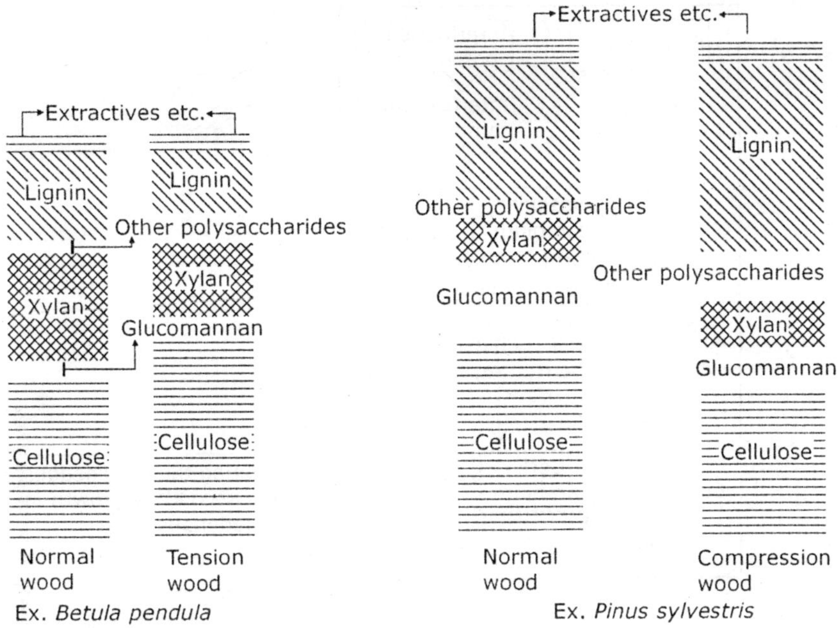


Figure 20.18

Diagrammatic comparison of the main structural constituents in normal wood, compression wood and tension wood. Glucomannan, xylan and other polysaccharides together represent the relative mass percentage of hemicelluloses. Adapted after Kettunen, 2006, p. 108 & 110.

layers only. $S_3$ (the innermost layer) layer is absent. The compression wood has less cellulose (Fig. 20.18). It is enriched in lignin; as a result it has higher ductility than the normal soft wood.

- *Contrasting wood*: The wood that develops on the opposite side of reaction wood is referred to as contrasting wood. In softwood conifer trees it develops on the outer side and in hardwood deciduous trees it develops on the under side of the leaning branch. The contrasting wood exhibits thin growth rings, long tracheids with square or rectangular cross-section-walls and thick $S_3$ (inner layer) layer in the structure of the secondary wall.

## Secondary phloem

The vascular cambium ring produces secondary phloem on the peripheral side. The elements of secondary phloem are arranged in vertical and horizontal manner and thus compose the axial and radial system of plants. In some dicotyledonous plant, growth rings composed of phloem, are observed but the demarcation lines are not prominent.

Sieve elements, phloem fibre and phloem parenchyma represent the vertical system. They are developed from the fusiform initials of cambium. The ray initials in the cambium divide and produce xylem rays inwardly and phloem ray on the peripheral side. The phloem rays compose the radial system and are continuous with the xylem rays.

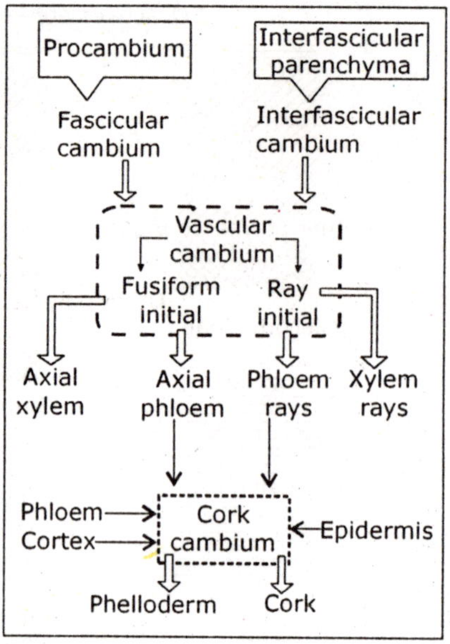


Box 20.3

Illustrating the cell lineages during secondary growth in dicotyledonous stem.

In conifers sieve cells, albuminous cells, parenchyma cells and fibres represent the vertical system. Uniseriate phloem rays, which consist of parenchyma cells only, comprise the radial system and in some cases albuminous cells are found in association with parenchyma cells. The ray cells may be in contact with sieve cells lengthwise. The sieve cells overlap one another at their ends and in these regions, there are more sieve areas. The parenchyma cells comprising the vertical system may store starch, resins, fats, tannins and sometimes crystals. Calcium oxalate crystals are found in *Larix.* Fibres are absent from secondary phloem of *Pinus.* Resin ducts are frequently observed in secondary phloem of conifers. The resin, obtained from *Abies balsamea,* is popularly known as Canada balsam and is commonly used as mounting medium in microscopical preparation because it possesses the same refractive index of a glass.

In dicotyledons, the vertical system of secondary phloem consists of sieve tubes, companion cells, phloem fibres, phloem parenchyma and the horizontal system comprises uniseriate or multiseriate ray parenchyma cells only. In both systems sclereids, laticifers and crystals are observed. The secondary phloem of

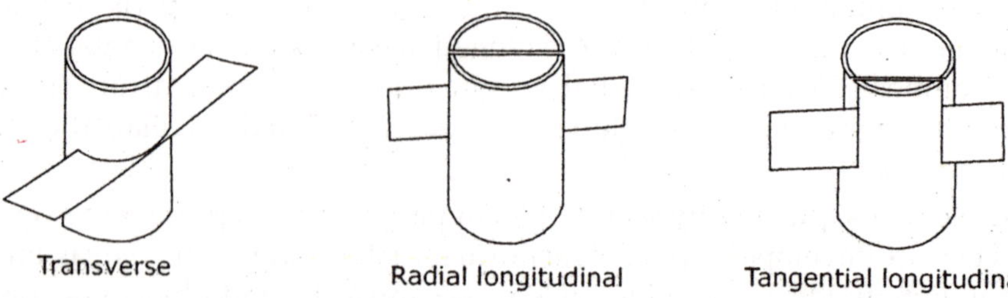


Figure 20.19

Diagram showing the planes in which sections are cut.

*Pyrus malus* contains fibres that do not develop from the fusiform initials of cambium; instead they arise from non-functioning phloem. These fibres are termed as fibre-sclereids (Esau *et al.*, 1953).

Sclereids are present in functioning and nonfunctioning phloem. They may be the sole sclerified element of the secondary phloem (e.g. *Fagus*), or they may be associated with fibre (e.g. *Prunus*).

In the sieve tubes the sieve areas are well developed in the secondary phloem. *Vitis, Populus,* in their sieve tubes, possess oblique end walls with compound sieve plates. Simple sieve plates with slightly oblique end walls are found in *Acer*.

Comparison between spring wood and autumn wood.

| Spring wood | Autumn wood |
|---|---|
| 1. This wood is produced in early period of growth, also called early wood. | 1. This wood is produced in the late part of growth season, also called late wood or summer wood. |
| 2. It is less dense and consists of thin walled elements. | 2. It is denser and consists of thick walled elements. |
| 3. In the wood, the vessels possess large lumen. | 3. The vessels in autumn wood are with smaller diameter. |

Comparison between ring porous wood and diffuse porous wood.

| Ring porous wood | Diffuse porous wood |
|---|---|
| 1. In it the vessels are of different diameter. | 1. In it the vessels are more or less equal in diameter. |
| 2. The vessels are not uniformly distributed throughout the wood. | 2. The vessels are uniformly distributed throughout the wood. |
| 3. Vessels with wide and smaller diameter are formed in the early and the later part of the growth season respectively. | 3. Vessels with more or less equal diameter are formed throughout the growth ring. |
| 4. Vessels with wide diameter of early wood and vessels of smaller diameter of late, summer or autumn wood are distinguishable. | 4. Vessels of early wood and late wood are indistinguishable. |
| 5. The development of vessel is sudden and rapid. | 5. The development of vessel is slow. |
| 6. The vessels are longer in length than those of diffuse porous wood. | 6. The vessels are shorter in length than those of ring porous wood. |
| 7. The rate of transport of water in plant with ring porous wood is ten times faster than those with diffuse porous wood. | 7. The rate of transport of water in plants with diffuse porous wood is slower than those with ring porous wood. |

Comparison between apotracheal and paratracheal parenchyma.

| Apotracheal parenchyma | Paratracheal parenchyma |
|---|---|
| 1. The axial parenchyma of secondary xylem is situated independent of vessel. | 1. The axial parenchyma of secondary xylem is distinctly associated with the vessel. |
| 2. The apotracheal parenchyma may be : (a) Diffuse: parenchyma cells appear as single cell or as small uniseriate band throughout growth ring (ex. *Quercus*). (b) Banded: parenchyma cells appear as concentric rings (ex. *Hicoria*). (c) Boundary: parenchyma appears at the beginning or end of the growth ring (ex. *Magnolia*, *Ceratonia* etc.). | 2. The paratracheal parenchyma may be: (a) Scanty: the parenchyma cells do not form a continuous sheath surrounding the vessel, ex. *Acer*. (b) Vasicentric: the parenchyma cells form a continuous sheath around the vessel of different width, e.g. *Tamarix*. (c) Aliform: vasicentric parenchyma extends laterally as wings (ex. *Acacia*). |

Comparison between primary and secondary phloem.

| Primary phloem | Secondary phloem |
|---|---|
| 1. It is present either on the peripheral side or internal to primary xylem termed internal or intraxylary phloem when it is primary in origin. | 1. It is formed either on the peripheral side of secondary xylem but internal to primary phloem or embedded in secondary xylem termed interxylary phloem or internal to secondary xylem called intraxylary phloem when it is secondary in origin. |
| 2. It originates from procambium of the apical meristem. | 2. It is formed from the cambium—the lateral meristem (fascicular and interfascicular) and accessory cambium (e.g. intraxylary phloem of *Tecoma*). |
| 3. According to mode of differentiation protophloem and metaphloem are distinguished in it. | 3. There exists no such differentiation into protophloem and metaphloem. |
| 4. In comparison to secondary phloem the sieve tubes are larger and have narrow lumen. | 4. The sieve tubes are shorter and wider than that of primary phloem. |
| 5. The sieve elements of primary phloem are arranged in parallel to the long axis of the organ. | 5. The elements of secondary phloem comprise both vertical and horizontal system. |
| 6. Storied and nonstoried arrangement of the element cannot be distinguished. | 6. Storied and nonstoried arrangement of the element can be distinguished as they are developed from storied and nonstoried fusiform initial of cambium. |
| 7. No growth rings are formed. | 7. In some dicotyledonous trees growth rings are formed. |

Comparison between sapwood and heartwood.

| Sapwood | Heartwood |
|---|---|
| 1. It is also known as alburnum and is situated on the peripheral side of secondary xylem. | 1. It is also known as duramen and is situated on the inner side of the secondary xylem. |
| 2. It is of lighter colour in contrast to heartwood and consists of active secondary xylem. | 2. It is of darker colour in contrast to sapwood and consists of inactive primary and secondary xylem. |
| 3. It is softer in texture and later formed wood than heartwood. | 3. It is harder and early formed wood than sapwood. |
| 4. Tylosis is not usually present in it. | 4. Tylosis is frequently formed in the vessels and blocks the lumen. |
| 5. Translocation and storage of food occur through vessels and xylem parenchyma respectively. | 5. Vessels are nonfunctional due to accumulation of gums, resins, tannins etc. into the lumen and xylem parenchyma becomes heavily lignified. |
| 6. The sapwood is gradually transformed to heartwood. | 6. Heartwood is more durable and resistant to decay than sapwood. |

Difference between periderm and polyderm.

| Periderm | Polyderm |
|---|---|
| 1. It is usually formed on the aerial stems and roots. | 1. It is usually formed on the underground stems and roots. |
| 2. It is composed of three layers of cell namely phellem, phellogen and phelloderm collectively termed as periderm. | 2. It is composed of alternating layers of cells composed of uniseriate endodermoid cells and multiseriate parenchyma, collectively termed as polyderm. |
| 3. Suberin and sometimes lignin and wax are deposited. | 3. Suberin is deposited on endodermis like cells. |
| 4. Casparian strips are not formed. | 4. Casparian strips are formed before suberization in endodermis-like cells. |
| 5. The phellogen originates either at epidermis, hypodermis, cortex or phloem. | 5. The phellogen originates at the pericycle. |
| 6. The phellogen forms centrifugally phellem and centripetally phelloderm cells. | 6. The phellogen centrifugally forms polyderm. |
| 7. Phellem cells become cork. | 7. Endodermis-like cells become cork. |
| 8. Phellem and phelloderm are respectively concerned with protection and storage. | 8. The outermost dead cells and nonsuberized cells are respectively concerned with protection and storage. |
| 9. Ex. *Alstonia*, *Vitis*, *Clematis* etc. | 9. Ex. Rosaceae, Onagraceae etc. |

Difference between primary xylem and secondary xylem.

| Primary xylem | Secondary xylem |
|---|---|
| 1. It is present in the primary plant body. | 1. It is present superimposed over the primary plant body. |
| 2. It originates from the procambium of the apical meristem. | 2. It originates from the cambium (fascicular and interfascicular) of the secondary lateral meristem. |
| 3. It is of two types: (i) protoxylem: it is the first formed xylem and (ii) metaxylem: it develops later in the primary body. | 3. There exist no such types in it. |
| 4. According to position of protoxylem it may be exarch, mesarch and endarch. | 4. There exist no such types in it. |
| 5. The tracheids and trachea of it are narrower and longer than secondary xylem. | 5. The tracheids and trachea are wider and shorter than those of primary xylem. |
| 6. The xylem parenchyma is of one kind only. | 6. The parenchyma cells are of two types – axial and ray parenchyma. |
| 7. Generally no tylosis is formed in vessels. | 7. Tylosis is formed in vessels of some dicotyledons. |
| 8. Heartwood and sapwood are not formed. | 8. Heartwood and sapwood are generally formed. |
| 9. No growth rings are formed. | 9. Annual rings are formed in trees. |

CHAPTER **21**

# Secondary Growth in Monocotyledonous Stem

Most monocotyledons consist entirely of primary tissues. The usual vascular cambium is absent from this group and so there is no normal secondary growth. However, in some monocots, the thickening and elongation of stem occurs through primary thickening meristem (Ball, 1941; Clowes, 1961), diffuse secondary thickening (Tomlinson, 1961) and secondary thickening meristem (Philipson *et al.*, 1971).

**Primary thickening meristem:** This meristem is observed in palms, in the rhizomes of *Musa* and in the bulbs of *Allium cepa* etc. In these plants, the shoot apex is not large and produces only a small part of the primary body. A considerable thickening occurs below the shoot apical meristem (Fig. 7.5). This is due to the intensive cell division of primary thickening meristem. This meristem lies below the young leaf bases and originates by periclinal division of the cells situated below the region of attachment of young leaf primordia. The meristem appears as a flat zone in longitudinal section of the developing embryo. The zone gradually assumes a concave form in the mature young plant. The meristem consists of several layer of cells which are rectangular, elongated and oriented parallel to the surface of shoot apex. The derivatives of meristem are the ground parenchyma cells. Sometimes localized mitotic activity within this meristem forms procambial strands, which run horizontal to the surface. These procambial strands gradually mature into vascular bundles. The term meristematic cap has been coined by Zimmermann and Tomlinson (1976) to designate the zone of procambium formation. Primary thickening meristem contributes mainly to the increase in width of stem and later it causes the elongation of young stem. In palms, *Musa* and a number of other monocotyledons with a similar growth habit the procambium originates from two sources – the shoot apical meristem and meristematic cap.

**Diffuse secondary thickening:** In palm stem the ground parenchyma cells, close to and distant from the shoot apex (ex. *Roystonea*, *Actinophloeus* etc.), expand along with proportional increase of the intercellular spaces, thus causing diffuse secondary growth. Sometimes these ground parenchyma and the procambium cells, destined to produce the outer fibres of bundle sheath, divide, expand and thus contribute to the diameter of stem. The diffuse secondary thickening occurs after the completion of expansion and elongation caused by primary thickening meristem.

**Secondary thickening meristem:** Secondary thickening with this meristem occurs in a number of monocotyledonous species such as, *Xanthorrhoea*, *Dracaena*, (Figs. 21.1 & 31.23) *Cordyline*, *Aloe*, *Yucca*, *Kingia*, *Dioscorea* etc. This meristem is a type of vascular cambium (Cutter, 1971), which originates in the parenchyma cells present on the peripheral sides of the entire mass of primary vascular bundles. This region may be the inner layer of cortex or pericycle. Though this lateral meristem is distant from the stem apex, in seedling stages it develops in continuity to the primary thickening meristem. In mature plants these two meristems are found to be discontinuous (Stevenson, 1980).

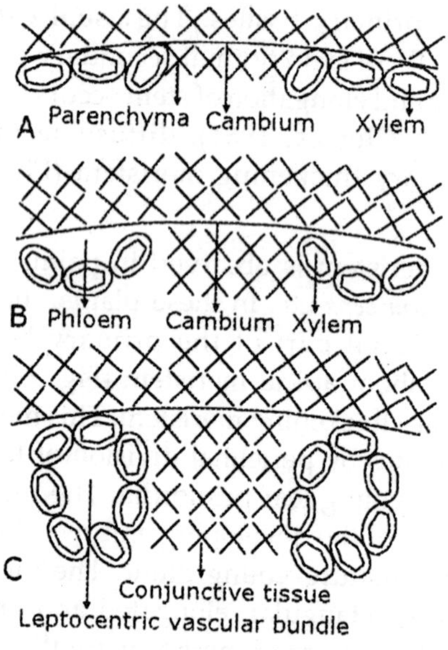


Figure 21.1

Diagrams illustrating the formation of conjunctive tissue and leptocentric vascular bundle in *Dracaena*. A. The cambium forms parenchyma only on the peripheral side and parenchyma and xylem on inner side alternately. B. The cambium situated above the xylem forms phloem inwardly. C. After a period of activity the cambium situated above phloem in B forms xylem. Thus a leptocentric vascular bundle is formed. The parenchyma formed by the cambium on the inner side is the conjunctive tissue. Adapted after Tayal, 1994.

The secondary thickening meristem consists of several tiers of cells that may be of different shapes, such as fusiform, rectangular, polygonal or only with one end tapered. This cambium divides tangentially, some of the peripheral derivatives divide repeatedly and the others form secondary cortex. These parenchyma cells usually remain thin walled and sometimes may contain crystals. In *Xanthorrhoea* these cells secrete resin, which form a sheath around the stem. A large number of tissues are formed to the inner side. The inner tissues form the ground parenchyma and vascular bundles. The ground parenchymatous tissues are termed as conjunctive tissues, which may sometimes become lignified. The secondary vascular bundles remain embedded in the conjunctive tissue. The bundles are either collateral (e.g. *Kingia*) or amphivasal (e.g. *Dracaena*, *Aloe* etc.). In *Kingia*, the phloem is surrounded by xylem only on three sides and so in cross section the vascular bundles appear as U-shaped. The xylem consists of small amount xylem parenchyma that may be lignified and tracheids only. Some localized region of the tiers of cambium is involved in the formation of a single bundle (Cheadle, 1937). The conjunctive tissues with embedded secondary vascular bundles show radial arrangement in contrast to primary strands, which are irregularly arranged.

Protective tissues are developed in the perennial monocotyledons by periderm and storied cork formation.

Periderm is formed with the origin of cork cambium or phellogen and the subsequent formation of phellem and phelloderm from it like dicotyledons. The periderm formation of *Dracaena, Aloe* and *Yucca* has no difference than those of dicotyledons.

**Storied cork:** It is a special type of protective tissue composed of cells, which are radially arranged in files and have suberized walls. The storied corks are observed in *Curcuma, Cordyline* and in many palms. These corks are formed by the activity of storied meristem. This meristem originates at the outer cortex. In contrast to phellogen, the storied meristem does not form a regular uninterrupted cylinder. The cortical parenchyma cells, which are destined to be the initials of this meristem, become three to eight layered by periclinal division. The derivatives of this meristem become suberized and are termed as storied cork. The cork cells are arranged in irregular tangential bands and radial files. Sometimes these bands may coalesce tangentially and radially when they enclose some undividing cortical cells, which also become suberized. Thus the storied cork along with enclosed suberized cells form a hard protective tissue (Fig. 21.2).

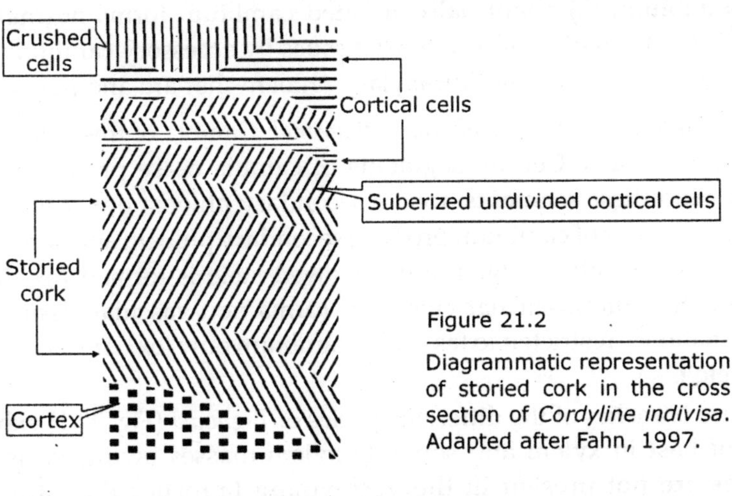


Figure 21.2

Diagrammatic representation of storied cork in the cross section of *Cordyline indivisa*. Adapted after Fahn, 1997.

CHAPTER **22**

# Anomalous Secondary Growth

Anomalous secondary growth can be defined as a deviation of normal secondary growth and production of secondary vascular tissues and secondary non-vascular tissues by normal cambium with abnormal activity, or by accessory (additional) cambia or by abnormally situated cambia with normal activity.

There are two criteria of normal/usual type of secondary growth. (i) Fascicular and interfascicular cambia together form a normal cambium ring. (ii) All the segments of cambial ring produce secondary phloem outside and secondary xylem inside, secondary xylem being in excess of secondary phloem. Deviation from the above two criteria will result in the formation of abnormal arrangement of tissues. Anomalous secondary growth occurs due to (1) the abnormal activity of normal cambium, (2) abnormally situated cambium forms normal secondary tissues, (3) the formation of secondary tissues by accessory cambium and (4) the formation of interxylary and intraxylary phloem that are anomalous structure.

• *The activity of normal cambium is abnormal*: Cambia of this category function mostly in two ways. Certain segments of cambia cease producing secondary xylem; instead these segments donate secondary phloem only towards exterior. The other segments of cambium produce secondary phloem and secondary xylem normally. As a result a ridged and furrowed stele is formed (ex. *Bignonia*). In other cases the interfascicular cambium forms non-vascular tissues. Vascular-tissue formation is restricted to fascicular cambium only (ex. *Aristolochia, Tinospora, Clematis* etc.).

Mature branches of the different species of *Bignonia* (Fig. 22.1) exhibit (in t. s.) wedges of bast in xylem and separate xylem-masses owing to fission. These anomalies are not present in the very young branches that show a ring of secondary xylem where the vessels are with small lumina.

Mature branches of *Bignonia* have four wedges of phloem that form a definite pattern. The wedges of bast form an orthogonal cross if it is imagined that wedges of phloem continue as far as pith and intersect with one another at right angles at the pith.

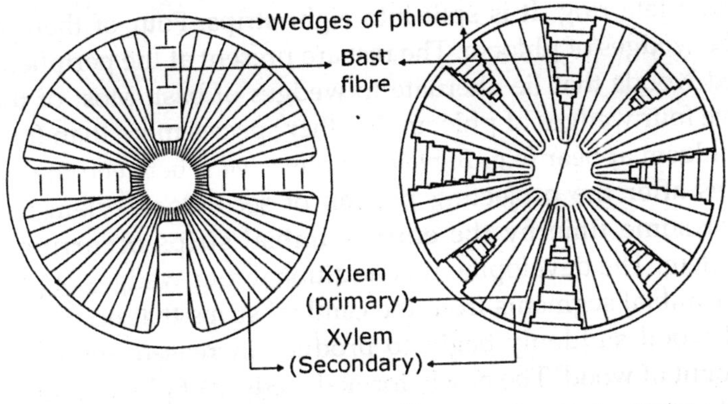


A. *Bignonia capreolata* B. *Bignonia aequinoctialis*

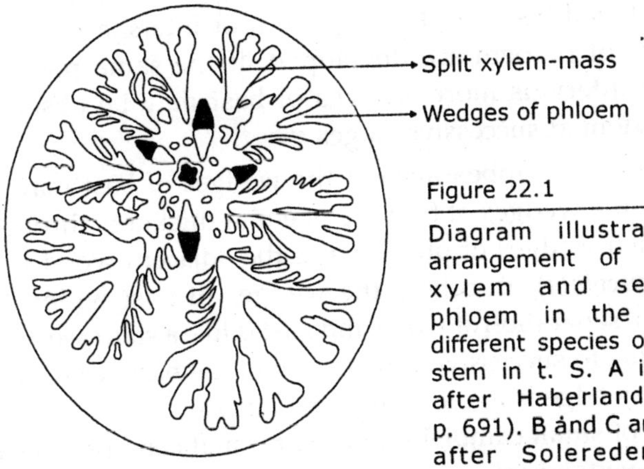


C. *Bignonia catharinensis*

Figure 22.1

Diagram illustrating the arrangement of secondary xylem and secondary phloem in the stele of different species of *Bignonia* stem in t. S. A is adapted after Haberlandt (1914, p. 691). B and C are adapted after Solereder (1908, p. 606). Vide text for details.

The primary vascular bundles of *Bignonia capreolata* are conjoint, collateral and open. The stems have normal ring of primary vascular bundles and the bundles are of different sizes. Fascicular cambium exists in the primary vascular bundles. During intrastelar secondary growth the interfascicular parenchyma, by dedifferentiation, forms interfascicular cambium. Fascicular- and interfascicular cambium together form a normal cambium ring. Initially the stem shows normal cambial ring and normal ring of vascular bundles. All the segments of cambium ring produce secondary phloem outside and secondary xylem inside, the latter being in excess of secondary phloem. The xylem has vessels with small lumen. As soon as the cambial ring starts producing vessels with large lumen the formation of wedges of phloem commences. Four opposite small segments of cambial ring cease to form secondary xylem. Instead, these segments donate secondary phloem outside only. The four alternate segments of cambial ring continue to produce normal secondary xylem and phloem. As a result four wedges of phloem are formed in the cylinder of secondary xylem. The wedges of phloem are symmetrically arranged and corresponding in position to the larger primary vascular bundles. As the stem grows in thickness the wedges of secondary phloem become deeper in xylem. At this stage the vascular cambium is no longer in the form of a complete ring. It is split into eight strips. Four of them occur at the bottom of the wedges of phloem. The rest are present at the peripheral margin of the projected woods that lie alternate to wedges of bast. Most of the species of *Bignonia* have four wedges of phloem. In other species more than four wedges of secondary phloem appear (e.g. *Bignonia aequinoctialis*) between the original four. This happens after the projecting portion of secondary xylem has grown in thickness for some time by the entire cambial strips occurring external to secondary xylem. The development of additional four wedges of phloem occurs when small cambial segments from the cambial strips that occur at the margins of projected wood suddenly begin to produce increased amount of bast and reduced amount of wood. The newly formed wedges of phloem do not penetrate so deeply into the secondary xylem as the original wedges. The recently formed wedges form angles of 45° with original wedges (Fig. 29.16A). Thus four more wedges may be developed bisecting the original four interspaces. As a result a total of eight wedges of phloem are developed. If the same phenomenon is repeated at frequent intervals more wedges of bast may be observed in the transverse section of stem at successive stages.

In transverse section the appearance of the wedges of bast may be of two kinds. In simpler case the wedges of bast are of equal thickness throughout. The boundary between such wedges of phloem and the adjacent wood is a straight line (e.g. *Bignonia capreolata*). In others the wedges of bast widen from base upwards and the line of separation of wood and bast is like a staircase, e.g. *Bignonia aequinoctialis*. This is due to successive broadening of the bast-wedges from base to periphery. The bast-wedge is in the form of an inverted arrow. The sequential change in the activity of small cambial region of normal cambium results in the development of such wedge-bast.

In another anomaly the mass of wood exhibits fission, e.g. *Bignonia catharinensis* stem of sufficient age. The xylem-mass has wedges of bast from the beginning. At later stages fission occurs by cell-division and dilatation in the parenchyma of the wood and pith.

Anomaly in the stele of the different species of *Bignonia* stem is due to presence of followings that form the characteristics of the genus.

- Four or more wedges of phloem that are traversed by bast fibres.
- The peripheral boundary of wood is ridged and furrowed.
- The vascular cambium is not in the form of a continuous ring.
- The cambia are in the form of strips and the number of strips corresponds to the sum total of projecting woods and wedges of bast.
- The cambium ring does not function uniformly around the stem.
- Occurrence of wedge-bast at regular intervals.
- Cambium is confined to the inner surface of wedge-bast and to the peripheral surface of the projecting portions of wood.
- Cambium is absent from the radial surfaces, i.e. the lateral boundary between wood and wedge-bast.

- As cambium is absent from the lateral boundary between wood and wedge-bast the wedge of phloem appears to be sliding along the lateral surfaces of the furrow during growth in thickness.
- In fresh materials of *Bignonia* stem it is observed that fissures appear on the lateral surfaces of wedge-bast that subsequently close again. This is due to ease of sliding of wedge-bast in the furrow.

The stems of *Bignonia, Doxantha* etc. of sufficient age, in addition to original wedges of bast, exhibit split or fissured xylem. The cells of pith and xylem parenchyma divide and expand to split the xylem mass. The wood of *Bignonia* is dichotomously split.

In some other members of Bignoniaceae such as *Paragonia* (Fig. 22.2), *Pyrostegia, Petastoma, Clytostoma, Amphilophium* etc. a normal cambium ring is formed and it functions normally until a thin ring of secondary phloem and secondary xylem is formed at the peripheral and center side respectively. After a brief period of activity four small segments of cambium, located opposite to each other, become unidirectional, i.e. produce secondary phloem on the peripheral side only. These segments of cambia do not produce secondary xylem at all and as a result remain stationary within the stem and push out narrow wedges of secondary phloem on the peripheral side. The other four cambium segments, those situated alternate to unidirectional cambia, continue as bi-directional meristem, i.e. produce secondary xylem and phloem normally. These four bi-directional cambia gradually move outward as more and more secondary xylem are produced. After a period of activity, it is noticed that the four unidirectional cambia are deep within the secondary xylem masses, which are produced by the bi-directional cambia. The narrow wedges of secondary phloem, formed by the unidirectional cambia, slide past the secondary xylem on their sides. In this way, the shoot grows and secondary growth continues. After the shoot grows for a period, small segments within the bi-directional cambia are converted to unidirectional one, which also produces narrow wings of secondary phloem only on the peripheral side. This process is repeated several times. In mature and large stems, where this process has occurred many times, it is noticed that there are several strips of unidirectional cambia situated at different sites of varying depths in the secondary xylem. These cambia have produced and pushed out narrow wings of secondary phloem.

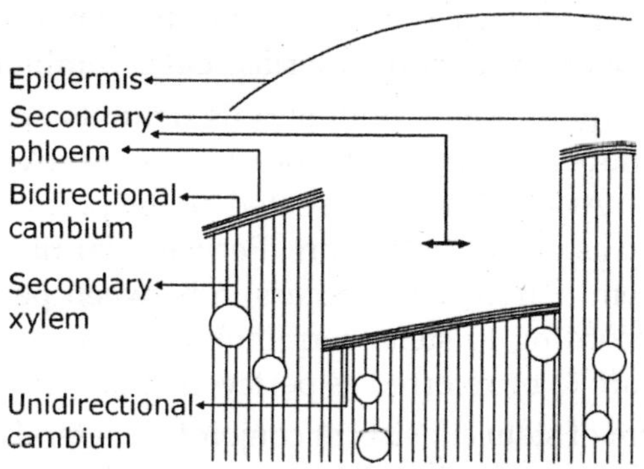


Figure 22.2

Diagrammatic representation of a portion of *Paragonia* stem showing unidirectional and bidirectional cambium.

Usually, in *Bignonia* and *Doxantha*, the numbers of wedges of phloem correspond to the numbers of orthostichies present in the stem and the furrows alternate with the orthostichies. An experiment was performed on *Doxantha unguis-cati* (Fig. 22.3) with four orthostichies to test the relation between the orthostichies and the wedges of phloem with very interesting results by Dobbins, 1969 (Cutter, 1971). Removal of leaves from two or four orthostichies was used as experimental material to study the activity of the vascular cambium. In the control plant, where no leaves were removed, the four wedges of secondary phloem were produced. In the experimental plant from where leaves from all four orthostichies were removed, the wedges of phloem were not formed and little amount of secondary tissues was developed. In the other plant, from which leaves were removed from two successive orthostichies only, the wedges of bast were not produced at the two large primary vascular bundles corresponding to the treated orthostichies. These observations clearly indicate the effect of leaves on the activity of vascular cambium.

In *Aristolochia* stem anomaly in the stele is due to the production of secondary ray parenchyma at the interfascicular region by interfascicular cambium.

The stele of *Aristolochia* is siphonostele. Endodermis delimits stele on the peripheral side. Just below the endodermis there occurs a cylinder of sclerenchyma that is composed of fibrous cells. As the fibres are located on the periphery of the vascular cylinder it is referred to as perivascular fibre. Perivascular fibres are not related ontogenetically to phloem. Parenchyma occurs in between perivascular fibre and vascular strands. According to stelar terminology perivascular fibre and the subjacent parenchyma are to be referred as pericycle (Esau, 1965, p. 409).

The primary vascular system of stem consists of conjoint, collateral and open strands. In transverse section the vascular bundles are arranged in an oval ring surrounding parenchymatous pith. The individual bundles are of different sizes, wedge-shaped and remain separated by wide interfascicular regions. The smallest bundles are the trace bundles.

Each vascular bundle has fascicular cambium on the periphery of which there occurs the primary phloem. Primary xylem is situated inside the fascicular cambium. At the onset of secondary growth the parenchyma cells present at the primary medullary ray become meristematic. The meristematic cells originate at the level of fascicular cambium by dedifferentiation. This secondary meristem is the interfascicular cambium. Fascicular- and interfascicular cambium join with each other thus forming a continuous cambium ring.

Production of secondary vascular tissue is restricted to fascicular cambium only. The fascicular cambium divides tangentially and the inner derivative cells form the secondary xylem mother cell. The peripheral derivative cells are the secondary phloem mother cell. Secondary xylem mother cell, secondary phloem mother cell and fascicular cambium form a cambial zone. The secondary xylem mother cell and secondary phloem mother cell later differentiate into secondary xylem and secondary phloem respectively. As a result of formation of secondary vascular tissues, the primary phloem is pushed towards the peripheral side and the primary xylem is pushed inside.

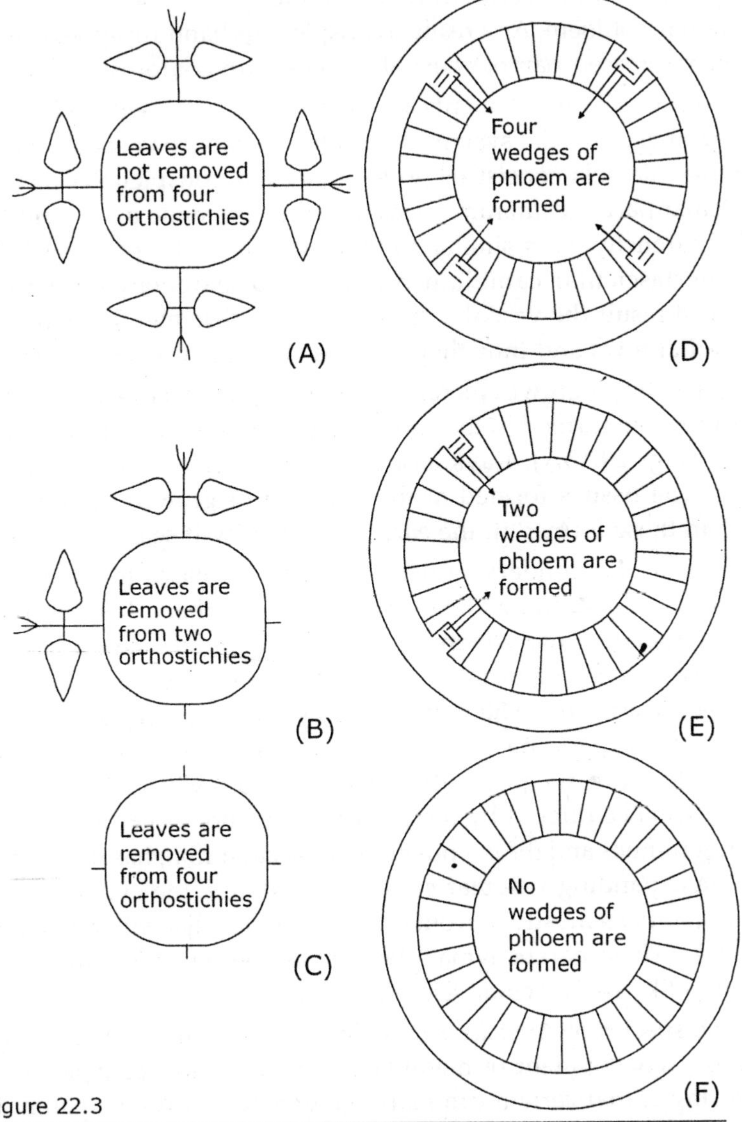


Figure 22.3

Experiment with *Doxantha unguis-cati* (Vide text for details). (A)-(C): Diagrammatic representation of treated shoots. (D), (E) & (F): T. S. of shoots at 28th internode. (D): Four wedges of phloem are formed at the intact shoot with four orthostichies of leaves. (E): Two wedges of phloem are formed at the shoot with leaves removed from two orthostichies. (F): No wedges of phloem are formed at the shoot with all leaves removed. Adapted after Cutter, 1971.

The primary phloem and secondary phloem consist of sieve tubes, companion cells and phloem parenchyma. Phloem parenchyma is abundant and is arranged in radial system. Sieve tube and companion cell compose the axial system. The most conspicuous feature is that phloem contains no fibres. In secondary phloem sieve tubes and phloem parenchyma occur as bands. Bands containing sieve tubes and associated parenchyma alternate with tangential bands of parenchyma. At later stages sieve tubes cease to function and get crushed by the formation of more secondary phloem. As a result a conspicuous banding appears in the phloem where uncompressed parenchyma alternates with compressed cells.

In the xylem all four elements are present where vessel, tracheids and xylem fibre are arranged in axial system. The radial system consists of abundant xylem parenchyma. Vessels consist of protoxylem and conspicuous metaxylem. The interfascicular part of cambium forms rays that are made up of parenchyma cells only. The parenchyma is similar to that present in the primary interfascicular region. Interfascicular cambium produces ray parenchyma both inside and outside. As a result the vascular strands remain discrete. Ray parenchyma cells occur in radial rows and thus they compose the radial system of stems.

The markings of growth increment are conspicuous in secondary xylem. The elements of xylem formed in the early and late part of growth season exhibit size difference (Fig. 29.15A). Rays associated with xylem also exhibit such size difference. The tissues formed at the late part of growth season are relatively smaller than those formed in the early part of growth season.

As a result of secondary growth the stem increases in circumference. The individual vascular strand widens towards the periphery. It is observed that new rays are interpolated into widening of vascular wedges. Continuous secondary growth causes squeezing of pith. At later stage pith cells and the associated ray cells are partly crushed. The continuous cylinder of perivascular fibres offers resistance to the expanding vascular system. As pith and medullary rays form a continuous structure, the pith cells and associated ray cells are crushed probably owing to resistance offered by the perivascular fibre. Secondary vascular tissue formation continues and the cylinder of perivascular fibre has not enough strength to resist the expanding vascular strands. Eventually the cylinder ruptures (Fig. 29.15A). In happens mostly in front of rays and the adjacent parenchyma cells fill up the breaks. These parenchyma cells in some species differentiate into sclereids that function like perivascular sclerenchyma.

In some species of *Aristolochia*, in old stems, xylem groups become fissured owing to the development of medullary rays. The most conspicuous feature of cambium ring is that certain cambial segments form ray-like parenchyma only. As the stem increases in circumference due to production of secondary vascular and non-vascular tissues new cambial segments develop and these cambial segments also donate rays of parenchyma. Thus a fluted vascular cylinder originates. In *Aristolochia triangularis* (Fig. 22.4A) the vascular bundles have forked structure towards periphery owing to development of rays of parenchyma by certain cambial segments. So the vascular cylinder has fan-shaped appearance.

In *Tinospora* stem the anomaly in the stele is due to formation of secondary ray parenchyma at the interfascicular region by interfascicular cambium.

The stele of *Tinospora* stem is siphonostele (Fig. 22.4B). Endodermis delimits stele on the peripheral side. Just below the endodermis there occurs a cylinder of sclerenchyma. This is pericycle and is composed of many layers of cells. In some regions the sclerenchymatous pericycle arches over the vascular bundles.

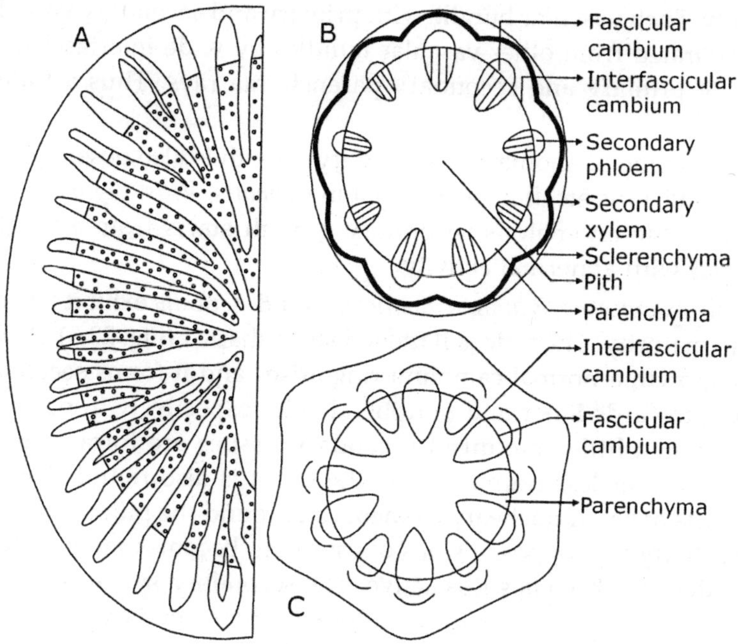


Figure 22.4

Diagrammatic cross-section of A. *Aristolochia triangularis* showing forked vascular bundles; B. *Tinospora* stem and C. *Clematis* stem showing dissected stele. A. & C. adapted after Eames & MacDaniels, 1947.

In the stele the primary vascular bundles are arranged more or less in a ring and remain surrounding central parenchymatous pith. Each vascular bundle is conjoint, collateral and open. The vascular bundles remain separated by wide interfascicular region.

The intra-fascicular cambium has primary phloem on the outside and primary xylem inside. During secondary growth certain cells of the interfascicular region at the level of intra-fascicular cambium become meristematic by dedifferentiation. These secondary meristematic cells are the interfascicular cambium. The interfascicular cambia join with intra-fascicular cambia and thus a continuous cambium ring is formed.

All segments of cambium ring do not function normally. The intra-fascicular part of cambium divides tangentially. The inner derivative cells are xylem mother cells. The peripheral derivative cells are phloem mother cells. Xylem and phloem mother cells together with intra-fascicular cambium form a cambial zone within vascular bundles. Later phloem mother cell differentiates into secondary phloem and xylem mother cell differentiates into secondary xylem. As a result of production of secondary vascular tissues the primary xylem is gradually pushed towards the centre and the primary phloem is gradually pushed towards the periphery. The interfascicular part of cambium ring forms rays of parenchyma only. The interfascicular cambium donates parenchyma both inside and outside. Production of secondary vascular tissues is restricted to individual vascular bundles only. Each vascular bundle with primary and secondary vascular tissues remains separated from other vascular bundles by wide interfascicular regions composed of primary and secondary parenchyma rays. Thus a fluted stele is formed.

The above type of anomalous secondary growth is also observed in *Clematis* (Fig. 22.4C) and *Vitis* etc. It is to note that in *Clematis, Vitis, Bignonia, Aristolochia* and *Tinospora* the anomalous secondary growth starts from a single normal cambial layer with abnormal activity.

*Prestonia macrocarpa, Bauhinia divaricata* and *B. sericella* exhibit flat and ribbon like stem at maturity. The stele is flattened strap shaped (Fig. 22.5). At the time of secondary growth a normal cambium ring arises and it forms secondary xylem and phloem towards inner and peripheral side respectively in normal way. At early stage when little amount of secondary tissues are formed, the vascular cylinder is more or less round and so the shoot is also round. The secondary xylem consists of small tracheary elements all around the pith uniformly. After a period of activity two opposite sides of cambium becomes more active, and the alternate sides of it becomes less active. The secondary xylem produced by the alternating segments of cambium is different. The secondary xylem, present at the less active region of cambium, shows small tracheary element. The xylem at the more active region of cambium exhibits vessels with large lumen. As growth continues the stele gradually becomes flattened strap or band shaped and the shoot becomes flat and ribbon like. This type of stem is usually observed in those species of *Bauhinia,* which are vines. By becoming flat, they maintain the ability to flex and increase conductivity by forming vessels with large lumen.

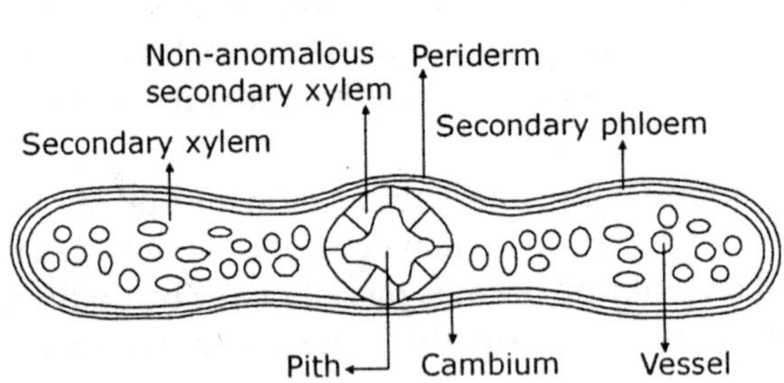


Figure 22.5

Transverse section of *Bauhinia* stem (diagrammatic). Adapted after Mauseth, 1988.

• *Abnormally situated cambium forms normal secondary vascular tissues*: Cambia of this category produce secondary vascular tissues in normal fashion, but their positions are anomalous. These cambia form discrete vascular cylinders and their arrangements differ according to species. Most of the climbing species of *Serjania* exhibit a principal vascular cylinder surrounded by small peripheral vascular cylinders. The vascular cylinders and the pericycle compose the stele. Endodermis delimits the stele on the peripheral side.

In *Serjania* stem several cambial layers are present from the first. In the cross-section of stem the cambia are arranged in various ways. Different species of *Serjania* (Fig. 22.6) exhibit the various arrangements of cambia. In the stele of *Serjania caracasana* (Fig. 22.6D) stem a central principal cambial cylinder occurs.

Several minor cambial cylinders surround the principal cambial cylinder. In *Serjania corrugata* (Fig. 22.6B) there is no central cambium. In the stele five to seven peripheral cambial cylinders occur in a circle and the cambia are approximately equal. In all species individual cambial layer behaves in normal fashion, i.e. donates secondary xylem on the inside and secondary phloem on the peripheral side. As a result several cylinders of secondary vascular tissues are formed. Each vascular cylinder consists of a cambial cylinder, secondary phloem on the peripheral side of cambium and secondary xylem on the inside of cambium that encloses parenchymatous pith. The vascular strands remain separated by parenchyma.

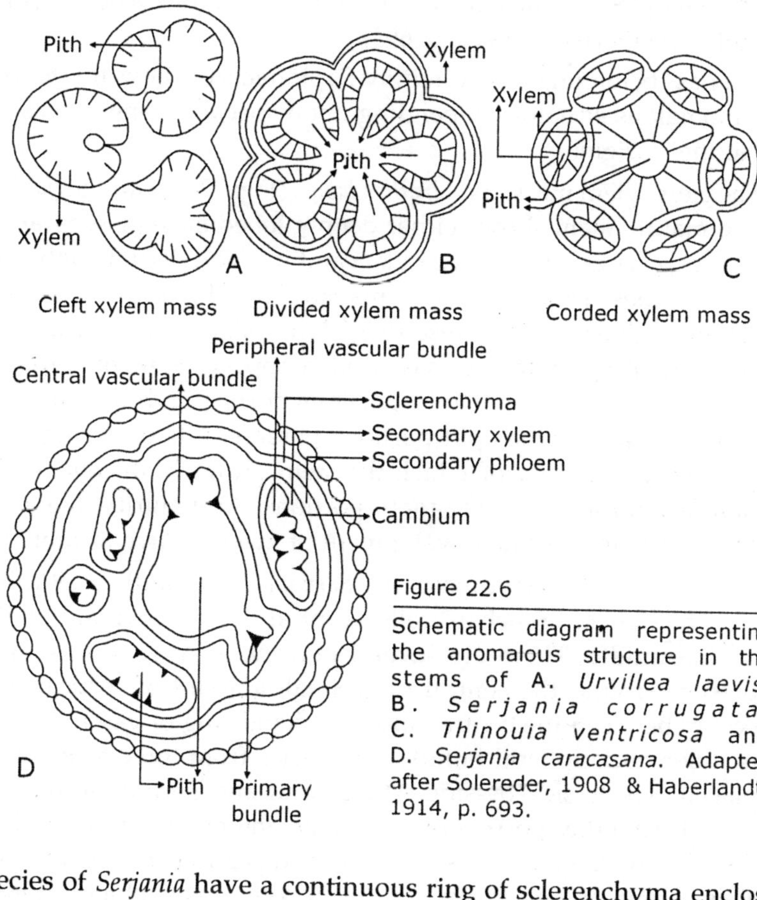


Figure 22.6

Schematic diagram representing the anomalous structure in the stems of A. *Urvillea laevis*, B. *Serjania corrugata*, C. *Thinouia ventricosa* and D. *Serjania caracasana*. Adapted after Solereder, 1908 & Haberlandt, 1914, p. 693.

All species of *Serjania* have a continuous ring of sclerenchyma enclosing all vascular strands. The ring is composed of fibrous cells and is wavy in climbing species. The ring occurs at the outer limit of bast. This wavy fibrous cylinder is regarded as pericycle.

Procambial strand differentiates into primary vascular bundles. Normally primary bundles differentiate in a circle. In case of *Serjania* it is observed that primary leaf-trace bundles at certain points of circle are indented. These bundles tend to be abstricted from the circle of vascular bundles. Thus individual groups of primary leaf-trace bundles are formed. During secondary growth a continuous cambial cylinder develops linking the absricted groups of primary leaf-traces. As a result several separate cambial cylinders develop in the developing stele. This happens at a very early stage of development.

The individual cambial layer behaves in normal fashion and ultimately gives rise many vascular cylindrical strands in the stele. The principal cambial cylinder forms the central vascular strand of the stem. The peripheral cambial cylinders form vascular cylinders that surround the principal vascular strand. It is to note that like *Bignonia* the cleavage of wood in *Serjania* is not a secondary process. The formation of several cylinders of wood is predetermined. Apical meristem forms procambial strands that differentiate into primary vascular bundles where cambial cylinders develop. So the stele of *Serjania* consists separate woody cylinders from the beginning (Haberlandt, 1914, p. 693).

The lianas, belonging to the family Sapindaceae, show the following four types of anomalous structure depending upon special arrangements of vascular bundles at their origin:

i. The cleft xylem mass: Ex. *Urvillea laevis* (Fig. 22.6A) and *Serjania piscatoria*. The mature stem shows either deeply lobed stele or separated vascular strands, each possesses a vascular cambium and discrete pith. The primary body of these stems shows normal structure but superficially grooved at certain regions due the presence of depressions in the cortex. At the onset of secondary growth, strips of cambia appear below the grooves and the axis subsequently split into portions corresponding to the number of grooves. Each of the split portions possesses its own ring of cambium, through the activity of which the stem grows in thickness. Usually the stem is split into three or more portions. At maturity, the axis becomes ribbed due to unequal development of xylem at five or more points.

ii. The compound xylem mass: Ex. *Serjania fuscifolia* (Fig. 29.15B), *Paullinia* etc. The stem or branch exhibits a central ring of vascular bundles present at the middle. There are several peripheral rings of vascular bundles present around the central ring. Small amount of cortical parenchyma occurs between the central and peripheral ring of vascular bundles. The number of peripheral bundle ring varies from three to ten and they are placed closely side-by-side. The central and peripheral bundle rings of vascular bundles possess pith and have their own cambium ring, through the activity of which the stem or branch grows permanently in thickness. This cambium ring forms secondary xylem and phloem towards inner and peripheral side in normal way. The vascular bundles of central and peripheral rings join each other at the nodes. The leaf traces at first run through the central ring and subsequently pass into one of the peripheral rings. At maturity, the stem or branch appears to be a cable like structure. This is an adaptive type of anomaly where the plant is benefited when exposed to stretching or torsion. In an estimate, it is recorded that this type of anomaly occurs in 91 out of 172 species of *Serjania* and 16 out of 122 species of *Paullinia*.

iii. The divided xylem mass: Ex. *Serjania corrugata* (Fig. 22.6B). The young stem possesses five to seven bundle rings and they lie side-by-side in a circle. Each bundle ring is in the form of incomplete circle and encloses pith at the centre. In young stems, the peripheral pith is continuous with central pith. The bundle rings possess permanent growth in thickness by a ring of cambium present in them. The divided xylem mass, at maturity, encloses pith. In contrast to compound, the divided xylem mass exhibits no central ring vascular cylinder.

iv. The corded xylem mass: Ex. *Thinouia* (Fig. 22.6C & 29.15D). The stem exhibits a central and around it several peripheral rings of vascular bundles. Each vascular cylinder possesses ring of cambium with bast and wood that encloses pith. At young stage, the stems grow in thickness with central vascular cylinder only, i.e. by the appearance of a normal cambium ring with normal activity. This normal growth in thickness occurs up to the fifth or sixth year after which new strands of accessory cambia originate at the cortex external to the original vascular ring. The central vascular cylinder still continues to grow. The new strands of accessory cambia, though abnormal in position, produce normal phloem and xylem towards exterior and interior respectively. The ring of vascular bundles encloses pith at the centre. The newly formed peripheral ring of vascular bundles is connected with one another but remains separated from the original vascular ring. This anomaly also occurs as a secondary complication in older stems of a few species of *Serjania* and *Paullinia,* which exhibit compound or divided xylem mass. The newly formed secondary vascular cylinder may be cylindrical (e.g. *Thinouia*) or flat transversely (e.g. *Paullinia*). At maturity, the stems appear as cords.

- *Formation of secondary tissues by accessory cambium*: *Baugainvillea, Amaranthus, Boerhaavia, Achyranthes, Celosia* etc. exhibit accessory cambia that form vascular and non-vascular tissues. This type of anomaly is illustrated below taking the example of *Boerhaavia diffusa* (Maheswari, 1930).

In *Boerhaavia diffusa* (Fig. 22.7) endodermis delimits stele of stem. All segments of endodermis are not always distinct. Just below the endodermis there occurs the pericycle. It is narrow and consists of one or two cell layers. In older stems just below the endodermis a few scattered fibres are present. The position of fibres is of special interest because they locate the endodermis when the latter is indistinct. The other structures of stele include parenchymatous ground tissue, conjunctive tissues and primary vascular bundles.

A cross-section of young stem shows that the primary vascular bundles are of different sizes. The bundles are arranged in three rings. All vascular bundles are derived from procambial strands.

A young stem of *Boerhaavia diffusa* exhibits (in t. s.) two large central vascular bundles in the innermost ring. These are the largest bundles in the stem. The bundles have a tangential diameter about twice as long as the radial. Each vascular bundle is conjoint, collateral and has intra-fascicular cambium that behaves in

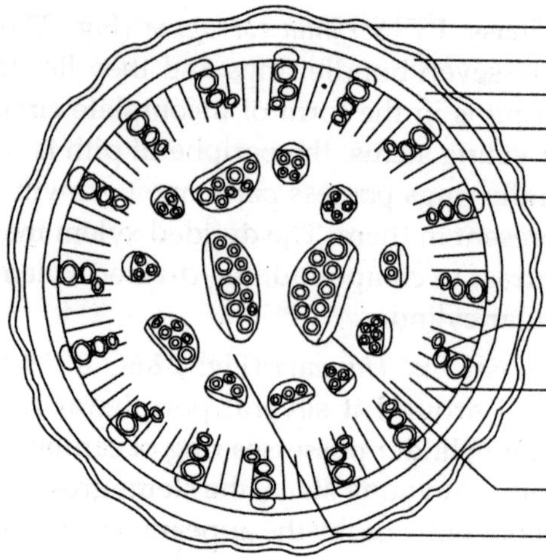


Figure 22.7

Diagrammatic transverse section of *Boerhaavia* stem.

normal fashion. The cambium divides tangentially and the inner derivative cells differentiate into secondary xylem. The peripheral derivative cells form secondary phloem. The secondary vascular elements are arranged in axial and radial rows. Production of secondary vascular tissues is restricted to individual vascular bundles only. Each vascular bundle has limited amount of secondary vascular tissue formation. As secondary growth continues the primary phloem cells become crushed. The dead remnants of crushed phloem appear as cup over the later formed secondary phloem. No interfascicular cambium develops in the inner ring of vascular bundles.

The middle ring consists of six to fourteen loosely arranged primary vascular bundles. The bundles are conjoint, collateral and open. Little amount of secondary vascular tissues are produced in them by intra-fascicular cambium. Interfascicular cambium is not formed in the middle ring of vascular bundles.

The outermost ring consists of 15-20 or more primary vascular bundles. The bundles are small, even minute and occur at the periphery of stele. Each bundle is conjoint, collateral and open. These bundles are also formed from procambial strands but are belated in development in comparison to vascular bundles of other two rings. Each vascular bundle has intra-fascicular cambium. Interfascicular cambium develops at interfascicular region by dedifferentiation. Interfascicular cambium joins with intra-fascicular cambium on both of its lateral sides. This results in the formation of a complete cambial cylinder. It is this cambial cylinder that is responsible for all subsequent anomalous secondary growth.

The cambial cylinder is very active. It divides tangentially. The intra-fascicular part of cambial cylinder forms secondary vascular tissues only. The peripheral derivatives form secondary phloem. The inner derivatives of intra-fascicular cambium differentiate into secondary xylem. The interfascicular part of cambial cylinder also divides tangentially. The peripheral derivatives form parenchyma cells. The inner derivatives differentiate into conjunctive tissue and storage parenchyma. The conjunctive tissues are always formed internally. It consists of elongated living cells that later are transformed to sclerenchyma by lignin deposition on their walls. These sclerenchyma cells are similar to fibres and serve for food storage (Maheswari). With further secondary growth the cambial cylinder donates a broad zone of anomalous wood inwardly. The anomalous wood consists of secondary xylem formed by intra-fascicular cambium, lignified conjunctive tissue and lignified adjacent cells of pith. In mature stele lignified conjunctive tissues and secondary xylem become confluent forming a zone of wood.

The activity of original vascular cambial cylinder after a period declines. Another cambial cylinder arises at the peripheral parenchyma cells formed by the original cambial cylinder. This new cambium donates conjunctive tissues inside and parenchyma outside. Certain segments of new cambium form secondary xylem inside and secondary phloem outside. These secondary vascular tissues are always formed opposite to each other, as a result collateral vascular bundles are formed.

After a period of activity the function of new cambium ceases. Then another new additional cambium cylinder arises from the peripheral parenchyma cells produced by its predecessor. The function of new additional cambium is similar to that of the previous cambial cylinders. In this way four to five additional cambial cylinders originate and accordingly growth rings are produced. All the supernumerary cambia have the same function. As a result the anomalous wood broadens.

It is to note that in the stele of *Boerhaavia diffusa* the original cambial cylinder does not develop *de novo* in the pericycle during secondary growth. Instead the cambial cylinder is formed by the union between intra-fascicular cambium of peripheral small bundles and interfascicular cambium formed at interfascicular region of peripheral bundles. Later additional/accessory cambia originate from parenchyma produced by original cambial cylinder or from the parenchyma of their products.

In *Chenopodium*, isolated phloem patches termed phloem islands are found embedded in the conjunctive tissues, usually above the clusters of vessels. These phloem patches are formed centrifugally by the accessory cambium. Later there arise arcs of new cambium that forms conjunctive tissues to the inside and buries the phloem.

• *Formation of interxylary phloem* (Fig. 22.9): The secondary phloem strands or layers that remain surrounded by secondary xylem are referred to as interxylary phloem. Interxylary phloem is also termed as included phloem/interxylary soft bast as it remains embedded in wood. Eames and MacDaniels (1947, p. 311) illustrated the following two methods by means of which interxylary phloem becomes embedded in secondary xylem.

In *Combretum, Leptadenia* (Figs. 22.8B & 22.9A-D) and *Entada*, during secondary growth a normal cambium ring is formed by the union of intra-fascicular and

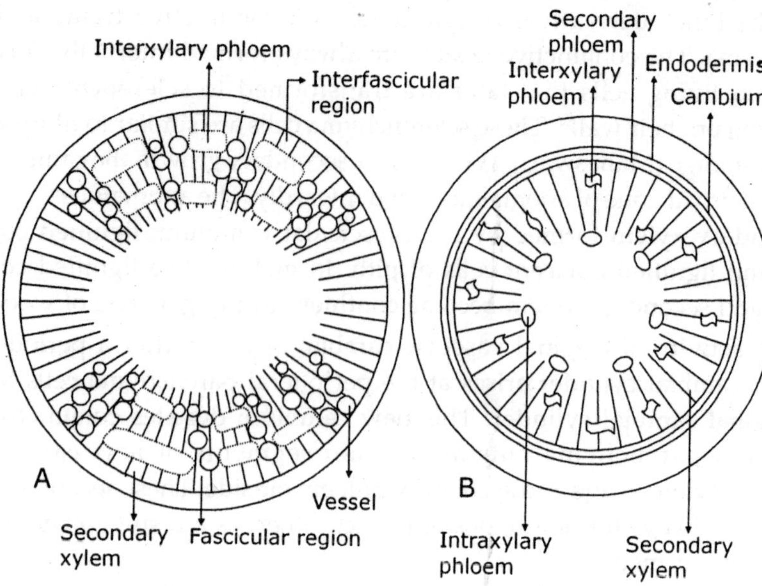


Figure 22.8

A. *Thunbergia coccinea* stem in transverse section (diagrammatic) showing interxylary phloem. Adapted after Solereder, 1908. B. *Leptadenia* stem in t.s. (diagrammatic) showing intraxylary and interxylary phloem. Adapted after Tayal, 1994.

interfascicular cambium. The cambial ring functions normally, i.e. produces secondary xylem inside and secondary phloem outside. Subsequently certain small segments of cambial ring donate secondary phloem toward the inside for a brief period. Normally these segments produce secondary xylem. During the production of interxylary soft bast these segments form secondary phloem in place of secondary xylem. The production secondary phloem continues and after a brief period of such activity these abnormal segments of cambial region regain their normal activity, i.e. donate secondary xylem inside. As a result the inwardly formed phloem gets buried in the secondary xylem. It is to note that the islands of soft bast are developed from the inner derivative cells of cambium.

The other form, e.g. *Strychnos* (family: Strychnaceae) has siphonostele in the stem. The primary vascular bundles are conjoint, open and bicollateral. The inner phloem of the vascular bundle is referred to as intra-xylary phloem or intra-xylary soft bast. To avoid confusion with interxylary phloem the term internal phloem replaces intra-xylary phloem (Committee on Nomenclature, 1957). Internal phloem is normally primary in origin and develops from procambial strands along with primary outer phloem and primary xylem. The internal phloem of *Strychnos* is in the form of isolated strands that form a ring completely surrounding the pith. In certain species the isolated bast occurs opposite all or some of the vascular bundles of stem.

During secondary growth a normal cambial ring is formed by the union of intra-fascicular and interfascicular cambium. The cambial ring has normal function and forms secondary xylem toward inside. Secondary phloem is produced toward outside of stem. The formation of secondary tissues continues for some time. The interxylary phloem appears at a later stage of the growth of wood. During secondary growth certain small segments of cambial ring form the peripheral phloem strands toward the outside as a part of their normal function. Later the phloem strands become embedded in the secondary xylem in the following way (Fig. 22.9E-H). Such segments of cambium cease to be active. The rest of the cambial segments continue to produce secondary vascular tissues. The derivative cells of once active cambial segments differentiate into secondary vascular tissues. Thus the original cambial ring becomes interrupted. Later it is repaired by the production of new complementary strips of cambia on the outer side. Strips of cambia arise opposite the cambial segments that cease to function. The new cambial strips develop as secondary meristem. They originate either at the pericycle or in the phloem that is few rows away from the original cambium. Strips of cambia and the active segments of original cambia unite with their edges thus forming a complete cambial ring. This cambial ring has normal activity. The secondary xylem produced by this cambial ring encloses the secondary phloem formed by the once active original cambial segment. This process is repeated several times and as a result numerous scattered strands of interxylary phloem are formed embedded in secondary xylem.

It is to note that in *Strychnos* both internal phloem and interxylary phloem are formed. The former is primary in origin and is regarded as primary anomalous structure. Interxylary phloem is formed as a result of anomalous secondary growth and the islands of soft bast in the wood are formed by the cambium on its outer side. So the interxylary phloem is the normal secondary soft bast.

Interxylary phloem is also formed in several species of *Thunbergia* stem. The islands of soft bast occur in the form of bands in the wood (Fig. 22.8A). The characteristic of stem is that the bands of soft bast only arise in the interfascicular region. The bands remain surrounded by secondary xylem. The characteristic of wood is that vessels only occur at intra-fascicular region. The wood present at interfascicular region above and below of interxylary phloem consists of fibres and tracheids only.

Interxylary phloem of *Thunbergia* is formed either from the inner derivative cells of cambium (like *Combretum*) or from the outer derivative cells of cambium (like *Strychnos*). In *Thunbergia coccinea, T. grandiflora, T. parva, T. sinuata, T. mysorensis* etc. the islands of soft bast are developed from the inner side of the cambium. In *T. capensis, T. cyanea, T. hispida, T. hirta* etc. wedges of bast are formed and from these some of the islands arise. 'These wedges are partly formed by tissue produced from the inner side of the cambium and partly by tissue given off externally by the cambium'—Solereder (1908, p. 622). It is to note that in the latter case the interxylary phloem is the normal secondary soft bast.

• *Formation of intra-xylary phloem* (= internal phloem): The cylinder of phloem strands or isolated phloem strands that occur below xylem strands towards pith are termed as intra-xylary phloem. Committee on Nomenclature (1957) suggests

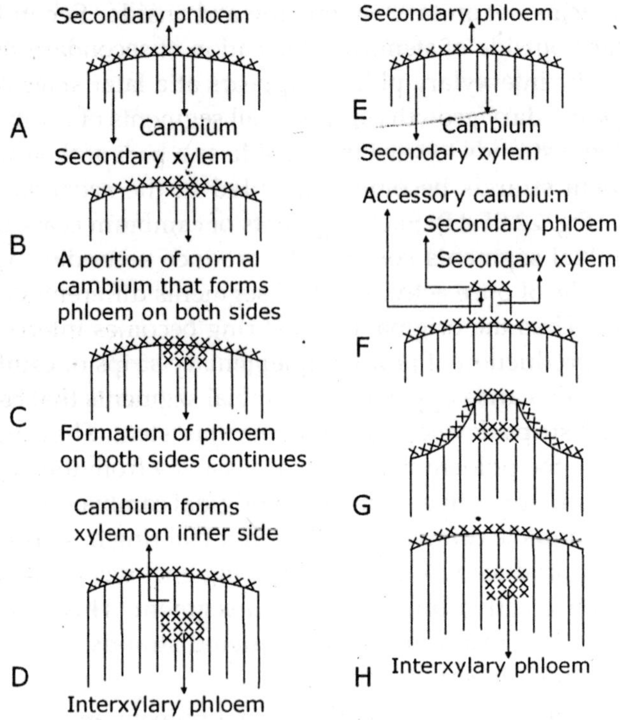


Figure 22.9

Schematic representation of interxylary phloem formation. A-D. *Leptadenia* stem. E-H. *Strychnos* stem. The accessory cambium joins with normal cambium on two sides.

the term internal phloem in place of intra-xylary phloem. Isolated phloem strands occur in *Calotropis, Strychnos* etc. Cylindrical phloem strand is present in *Asclepias curassavica*.

Normally internal phloem is primary in origin, i.e. it develops from provascular strands. Exceptions are observed in *Campsis radicans* (=*Tecoma radicans*) and *Campsis grandiflora* (=*T. andrepens*) where internal phloem is secondary in origin. In the above mentioned species internal phloem is developed as a result of anomalous secondary growth.

The stele of *Tecoma radicans* is siphonostele. The primary vascular bundles are arranged more or less in a ring and each vascular bundle is collateral, open with protoxylem endarch. During secondary growth intra-fascicular cambium and interfascicular cambium unite to form complete cambium ring. The cambium ring has normal function, i.e. produces secondary phloem on the peripheral side and secondary xylem towards inner side, the secondary xylem being in excess of secondary phloem. As a result primary vascular bundles are pushed towards the centre. After a period of activity two strips of additional cambia originate below secondary xylem on two opposite sides of pith. Each strip of cambium functions abnormally. Each cambial strip donates secondary phloem towards the centre or pith side and secondary xylem towards the peripheral side. As growth continues, two arcs of secondary vascular bundles are formed at the margin of pith. The bundles show inverse orientation of wood and bast. The bast that develops towards pith side is referred to as intra-xylary phloem/internal phloem. Subsequently the two patches of intra-xylary phloem crush the pith cells to a narrow band. In developing stems the primary vascular bundles are conspicuous towards the inner margin of secondary xylem formed by the normal cambial ring. A few layers of parenchyma cells are observed between the wood formed by the normal cambial ring and additional cambium/accessory cambium.

It is to note that anomaly in the stele of *Tecoma* is due the development of two arcs of medullary bundles with inverse orientation of wood and bast (Figs. 31.25, 31.25A) in contrast to normal vascular bundles. Anomaly in the stele is also owing to the formation of internal phloem/intra-xylary phloem by additional/accessory cambium during anomalous secondary growth.

*Anomalous secondary growth in Dracaena stem*: *Dracaena* is arborescent in habit, and belongs to the monocotyledonous family Agavaceae. The vascular bundles of monocotyledons are closed, i.e. intra-fascicular cambium is absent. So monocotyledons lack normal secondary growth from a vascular cambium. In *Dracaena* (Figs. 31.23 & 21.1) the stele is atactostele. The primary vascular bundles are distributed over the ground tissue without having any definite arrangement. The primary vascular bundles are not compactly arranged on the ground tissue and the interfascicular region is moderately wide. The ground tissue is composed of parenchyma cells and the parenchyma shows no radial seriation of cells. Each primary vascular bundle is leptocentric/amphivasal, i.e. the xylem completely surrounds phloem and there exists no intra-fascicular cambium. Each vascular bundle is circular or oval in transverse section. The xylem consists of tracheids only. Protoxylem with annular and spiral thickening is present.

The cross-section of mature stem of *Dracaena*, where certain amount of secondary growth has occurred, exhibits secondary vascular bundles and ground tissue commonly termed as conjunctive tissue. Conjunctive tissues are parenchymatous, the walls of which are thin and may sometimes become thickened or even lignified. The conjunctive tissue exhibits radial arrangement of cells and thus aids in differentiation from the primary interfascicular ground tissue. The secondary vascular bundles are to some extent arranged in radial rows. The vascular bundles are more or less compactly arranged and anastomose in some regions in contrast to primary vascular bundles. Each vascular bundle is oval in t. s. and leptocentirc/amphivasal like primary vascular bundles. The secondary phloem is small in amount in comparison to primary phloem. The secondary phloem elements are sieve tubes, companion cells and phloem parenchyma. The sieve tubes are short with transverse end walls and the sieve plate is simple. The tracheary elements are composed of tracheids and xylem parenchyma. The tracheids are long and the associated xylem parenchyma may be lignified. The tracheids exhibit scalariform thickening and lack annular and spiral protoxylem. The secondary bundles appear to lie embedded in the conjunctive tissues. The secondary tissues (secondary vascular bundles and conjunctive tissues) surround the primary vascular bundles and ground tissue.

Internally the stele of *Dracaena* stem before secondary growth remains surrounded by many layered parenchymatous cortex. The peripheral layer of stele is the pericycle that cannot be easily distinguished. The secondary growth of *Dracaena* is brought about by a special type of vascular cambium, also termed as secondary thickening meristem. This cambium originates in the older regions of stem that consists of parenchyma cells situated outside the vascular bundles. Cambium originates either from cortical parenchyma or pericycle. The cambium contributes to the growth of stem in diameter. The cambium is active in the part of axis that has ceased to elongate. The cambium does not function like the vascular cambium of dicotyledons.

Each cambial cell, as seen in l. s. may be fusiform with both ends tapered or rectangular. Sometimes one end of a cambial cell may be truncated and the other end is tapering. The cambial cells divide tangentially. Initially the cambial cells donate cell towards inside only. Later a small amount of tissue is produced towards outside. Tangential divisions in individual cambial cell continue resulting in the formation of radial series of derivatives. The peripheral derivative cells may continue tangential divisions. Some of the derivatives may cease to divide. These cells then transform to secondary cortex. The inner derivate of cambial cells forms a radial series of cells. Some of the derivative cells differentiate into thin walled parenchyma cells that form ground tissue. The ground tissue is termed as conjunctive tissue. Later the thin walls of ground tissue may be thickened or lignified. The other derivatives of cambial cells differentiate into vascular strands. The cambial initials form longitudinal files of single cell where vascular bundles develop. Anticlinal, periclinal and irregular longitudinal divisions in the files of cells result in the formation of leptocentric vascular bundles. The vascular bundles interrupt the radial seriation of conjunctive tissues that are interfascicular secondary parenchyma. Each cell of parenchyma has moderately thick and pitted wall. So each cell of secondary parenchyma may be regarded as conducting parenchyma.

It is to note that in *Dracaena* anomalous secondary thickening is brought about by a special cambium – termed secondary thickening meristem. Due to the activity of this meristem conjunctive tissue and secondary vascular bundles originate. The continuous activity of the meristem results in the formation of indefinite amount of secondary tissues. Sometimes weakly developed growth rings are observed. But the relation between rings to annual growth is yet to be established. Continuous formation of secondary tissues results in the increase of diameter in *Dracaena* stem. It is reported that in *Dracaena draco*, the stem attained a girth of 45 feet. The height of the tree was 70 feet and the age was estimated to be of six thousand years old (Haberlandt, p. 688; Willis, 1973, p. 388).

Haberlandt grouped all the anomalous forms of secondary growth into two categories, namely adaptive and non-adaptive. The adaptive anomaly is an adaptation to definite external conditions. In contrast the non-adaptive anomaly is not an adaptation to definite external condition (ex. *Amaranthus, Mirabilis, Boerhaavia, Chenopodium* etc.). It merely represents variations of design. The adaptive anomalies are the characteristic of plants that are lianas [ex. *Aristolochia, Tinospora, Bignonia, Serjania* and *Bauhinia* (Fig. 29.16B & C) etc.], plants that have fleshy roots (ex. beet root, Fig. 29.17) which principally serve for storage and plant with submerged stem where cambium devotes most of its derivatives to produce tissues that add buoyancy (ex. *Aeschynomene aspera, A. indica* and *Herminiera* etc.).

Inextensibility, inflexibility and incompressibility are the essential mechanical requirements of a liana. A climbing plant is often thrown into a number of folds or frequently becomes extensively twisted over a living supporting organ. It hangs freely and so is exposed to pulling strain. As a result a twiner becomes inextensible. Sometimes the hanging stems are shaken by violent winds. So a liana becomes inflexible. The twiners are often exposed to radial compression. This happens when supporting organ grows in thickness. All the above mentioned mechanical requirements of liane-stems are achieved by adaptive anomalous secondary growth.

The stelar organization of liane-stems is constructed like a twisted rope or cable. This cable-like structure possesses high degree of strength combined with pliancy. In a liana pliancy is often accomplished through splitting of the lignified xylem into isolated strands. Pliancy is also achieved by the interpolation of softer tissues like secondary phloem into lignified xylem. The anatomical features like splitting of xylem and interpolation of softer tissues into xylem is common to all liane-stems, but they are brought about in different ways.

In the stele of *Bignonia* four wedges of phloem split the xylem cylinder. The resulting lobation of secondary xylem increases the pliancy of stem. There are eight lateral sides where wedges of phloem and secondary xylem are in contact with each other. These sides are the planes of weakness of stele but they are not detrimentally weak. Because parallel islands of bast fibres are differentiated in the wedges of phloem and these mechanical cells add strength to the planes of weakness.

In *Aristolochia* stem a fluted vascular cylinder is formed. In the young stem pliancy is brought about by the development of broad medullary rays at interfascicular region. In mature stems the vascular strands are split by the development of rays of parenchyma.

In *Serjania* several vascular strands are formed from the beginning and they are held together by parenchymatous ground tissue. The presence of parenchyma between the vascular strands increases the pliancy of stem.

In *Strychnos, Thunbergia,* pliancy is brought about by the formation of interxylary phloem. It is to be noted that the development of inter-xylary-leptome-strands is not confined to liane-stems only. They do occur in a number of woody plants that are not lianas. So Schenck (Haberlandt) concluded that the formation of interxylary phloem is not an adaptive anomaly. It merely represents a variety of design.

*Gnetum*, a gymnospermous liane, also shows adaptive anomalous secondary growth (Fig. 22.10A). In this case, in addition to primary cambium ring, some accessory cambium develops in succession encircling the primary cambium. These cambium-rings function normally, i.e. produce xylem at inner face and phloem at outer face. Thus certain amount of parenchyma and secondary phloem become embedded between the xylem, formed by two successive cambium rings. Due to the presence of softer tissues *Gnetum* becomes flexible.

• INTERXYLARY PERIDERM (Fig. 22.10B): The periderm, which is formed in the wood and medullary tissues, is described as interxylary periderm. The formation of periderm at the periphery of stems and roots, and its protective function are generally recognized. But the periderm in the wood and medullary tissues has received little attention because it is less prevalent in plants in contrast to outer periderm. Interxylary periderm is formed in the roots and rhizome of the following plants: *Aconitum* (Ranunculaceae), *Sedum* (Crassulaceae), *Epilobium*, *Oenothera* (Onagraceae), *Mertensia* (Boraginaceae), *Salvia* (Labiatae), *Crepis*, *Artemisia* (Compositae), *Geranium* (Geraniaceae), *Polemonium* (Polemoniaceae) etc.

The interxylary periderm is formed in addition to normal peripheral periderm. It develops in between the two concentric rings of secondary wood

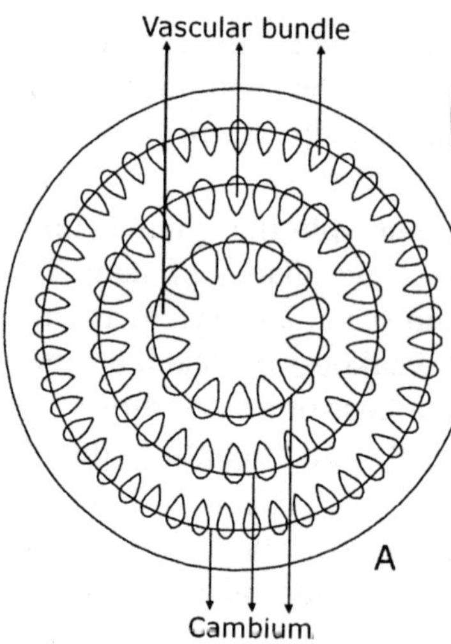


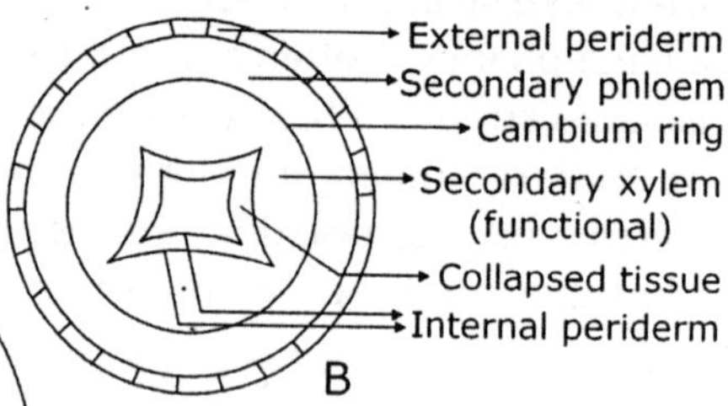


Figure 22.10

A. Diagrammatic representation of *Gnetum* stem showing concentric rings of cambium and vascular bundles. adapted after Chandurkar, 1983.

B. Diagrammatic representation of transverse section of a root showing interxylary periderm and enclosed collapsed tissue.

formed at different seasons. The interxylary periderm is formed by phellogen that is differentiated from the outer parenchymatous part of the wood of the previous year. It may arise from the deeper portions also. The phellogen forms phellem and phelloid on the peripheral side. The suberized cork and non-suberized phelloid layers alternate each other. Normally two or three layers of cork and a similar number of phelloid layers are present in each interxylary periderm. There is little or no differentiation of phelloderm.

The phelloid cells present on the inner layers collapse soon. As growth continues, rings of secondary xylem are formed by vascular cambium ring.

Accordingly continuous and concentric rings of interxylary periderm are formed. So the mode of formation of interxylary periderm is centrifugal in contrast to peripheral periderm, which is centripetal in development. After the formation of interxylary periderm the tissues present on the inner side die. On the peripheral side on inner periderm new vascular tissues are developed by the vascular cambium. These tissues are protected by the external periderm situated on the peripheral side and inner or interxylary periderm present on the inner side. In the old roots the number of rings of interxylary periderm commonly ranges from one to ten and the number may be as high as twenty (e.g. *Epilobium angustifolium*). Usually interxylary periderm is formed in spring and it was observed that a single periderm is formed each year during the life of the organ except the first one. Therefore, by counting the number of rings of interxylary periderm the approximate age of the organ can be determined. Moss (1940) regards that the interxylary periderm protects the organ from desiccation and plant pathogen, and thus provides efficient functioning of perennating structures.

Comparison between periderm and interxylary periderm.

| Periderm | Interxylary periderm |
|---|---|
| 1. Present on the peripheral parts. | 1. Present in the wood and medulla. |
| 2. It is formed in roots and stems. | 2. It is formed in stems, roots and rhizomes. |
| 3. It is composed of phellem, phellogen and phelloderm. | 3. It is composed of phellem, phelloid and phellogen. |
| 4. Phellogen arises either at the cortex or secondary phloem. | 4. Phellogen arises in the parenchymatous zone of the outer portion of wood. |
| 5. The mode of formation is centripetal. | 5. The mode of formation is centrifugal. |
| 6. It may be formed in continuous or discontinuous rings. | 6. It is commonly formed in continuous and concentric rings. |
| 7. More than one ring may be formed in a year. | 7. One ring is formed in each year. |
| 8. The rings of periderm do not reveal the age of a plant. | 8. The concentric rings reveal the age of the organ. |
| 9. It protects inner tissues. | 9. It protects and helps to perennate. |

CHAPTER **23**

# Cambium: A General Account

Cambium is a group of actively dividing cells that divide predominantly periclinally to form permanent tissues both in stems and roots of most dicotyledons and gymnosperms at intra- and extrastelar region. Cambium is an example of lateral meristem and during secondary growth it comes to form a thin cylinder or strip of cells. It may be vascular- , accessory- and cork cambium.

**Vascular cambium:** Vascular cambium occurs in the stele. The cells of cambium remain surrounding the primary xylem in both stems and roots. The vascular cambium is one cell thick permanent secondary meristem. It is often described as cylindrical and develops between primary xylem and – phloem. It is present in stems and roots and in their branches. It extends from tips to the bases of stems and roots. It may even extend into leaf petioles in some woody plants.

In the stele of stem, the cylinder of cambium consists of fascicular- and interfascicular cambium. In a vascular bundle of dicotyledonous stem the fascicular cambium occurs between primary xylem and – phloem. The interfascicular cambium originates from the parenchyma tissue that occurs between the two vascular bundles. Fascicular and interfascicular cambium together compose a cylinder of cambial cells. The cylinder of cambial cells, once complete, divides chiefly by periclinal division. The inner derivatives differentiate into a cylinder of secondary xylem and the peripheral derivative cells form a cylinder of secondary phloem.

Cambial initials have primary walls. The walls are thin, non-lignified and have primary pit fields only. Plasmodesmata pass through the pit fields. Plasmodesmata connect the protoplast of adjacent cambial cells or cambial derivative cells. It is regarded that the symplastic connections or plasmodesmata are the path of hormone transmission to the sites of developing xylem and phloem cells. The submicroscopic structure of the cells of cambium, in general, resembles those of other meristematic cells. Each cell contains nucleus, ribosomes and dictyosomes, Golgi bodies, endoplasmic reticulum, mitochondria, spherosomes etc. Variations in cell organelles are observed and this can be seasonal or taxonomic. As for example in the early phases of dormancy some species have plastids with abundant starch grains while species of other taxa lack them. Cyclosis, i.e. streaming of cytoplasm occurring during cytokinesis, is more conspicuous in fusiform initials than ray cell initials. In arborescent species, the plastids of cambial cells do not form grana although grana are formed in the plastids of herbaceous species. In *Fagus sylvatica,* the cells of resting cambium contain lipid droplets and protein bodies. Electron microscopic study reveals that unlike most other meristematic cells, the cells of vascular cambium are highly vacuolated. It is thought that vacuoles may be formed by the dissolution of stored lipids and proteins when the cambium resumes its activity after resting period (Kidwai *et al.,* 1969).

The cambial cells occur in two forms—fusiform initial and ray initial.

**Fusiform initial:** The term fusiform initial refers to a single cell. Fusiform initials owe their name to the spindle shape and it is observed in tangential longitudinal section (TLS). It is several times longer than wide. The middle part of a fusiform initial is approximately prismatic and the ends are wedge shaped. The wedge is pointed and it is revealed in TLS. In radial longitudinal section (RLS) the wedge appears as truncated. A fusiform initial may be very long and there is report that it may attain a length of 8.7mm in *Sequoia sempervirens* (Fahn, 1997). Though the fusiform initials are uninucleate, Ghouse *et al.* (1977) report polynucleate fusiform cambial cells in *Psidium guajava.*

The fusiform initials compose the axial system of cambium. So they are considerably elongated in the longitudinal plane of the axis. The tangential wall of fusiform initial is conspicuously thin at the time of cell division and becomes thick during resting period. The radial wall also becomes thick due to formation of primary pit fields. The thick radial wall has distinctly beaded appearance and the thin areas are the sites of plasmodesma. The fusiform cells are arranged in two ways—on the basis of which two types of cambium are distinguished.

- Storied- or stratified cambium (Fig. 23.1B): In TLS the initials are arranged in horizontal regular file having a stratified structure. The initials are usually short and all of about the same length. The ends of the initials occur approximately at the same level. The length of initials ranges from 140 μ to 520 μ. Ex. *Wisteria, Tamarix, Robinia* etc.

- Non-storied or non-stratified cambium (Fig. 23.1A): In longitudinal tangential view the tapered ends of fusiform initials overlap each other in a random arrangement. Ex. *Rhus.* Non-storied cambium is more common and longer than the storied type. The length of the initials ranges between 320 μ to 2300 μ and it may be as long as 6200 μ, e.g. vesselless dicotyledons. Initials of 1000 μ to 8700 μ are reported from gymnosperm (Fahn). It is regarded that these initials, during phylogenetic evolution gave rise to storied initials by reduction.

**Ray initial:** The term ray initial refers to a group of initial cells. In contrast fusiform initial refers to a single cell. A single cell of ray initial is referred to as ray cell initial (Beck, 2005, p. 166).

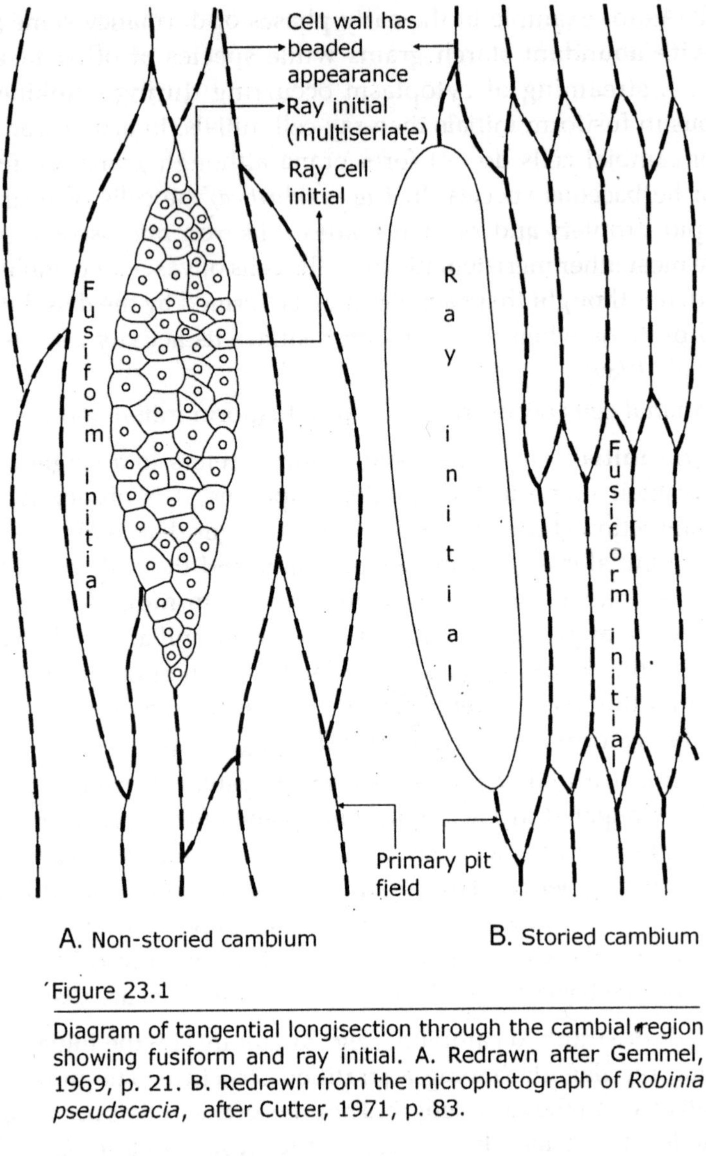


Figure 23.1

Diagram of tangential longisection through the cambial region showing fusiform and ray initial. A. Redrawn after Gemmel, 1969, p. 21. B. Redrawn from the microphotograph of *Robinia pseudacacia*, after Cutter, 1971, p. 83.

The ray initials compose the radial system of cambium. A ray cell initial is roughly cuboidal or isodiametric in shape and much smaller than fusiform initial. At the time of cell division the radial dimension becomes greater than the longitudinal dimension. The radial dimension also becomes greater than the width of fusiform initials. Ray cell initial together with its derivative rays may be uniseriate, biseriate or multiseriate depending upon one, two or more longitudinal rows of cells composing them. The height, shape and the number of ray cell initial vary. Ray initial is best studied in TLS and RLS. In TLS multiseriate ray appears to be fusiform.

**Origin of vascular cambium:** In stems the vascular cambium and the primary vascular tissues differentiate from procambium. Procambium develops from the derivative cells of apical meristem. Transverse sections of a growing vegetative shoot apex reveal the presence of a cylinder of cells that are highly cytoplasmic and more densely staining. This ring of cells is regarded as a residuum of the meristematic tissue of apical meristem and so termed as residual meristem. Within the residual meristem more densely staining regions are present and these regions have a topographic relationship with leaf primordia. This region constitutes procambium that develops as leaf trace. The remainder of residual meristem forms the interfascicular parenchyma. In longitudinal section of vegetative apical shoot of angiosperm and gymnosperm it is observed that the procambial ring or strand is continuous and develops acropetally (Figs. 23.2 & 23.3). Procambial strands exhibit two waves of differentiation, that is, differentiation of protophloem on the peripheral side and differentiation of protoxylem towards the inner edges in normal angiosperm. In monocotyledons the two waves of differentiation meet

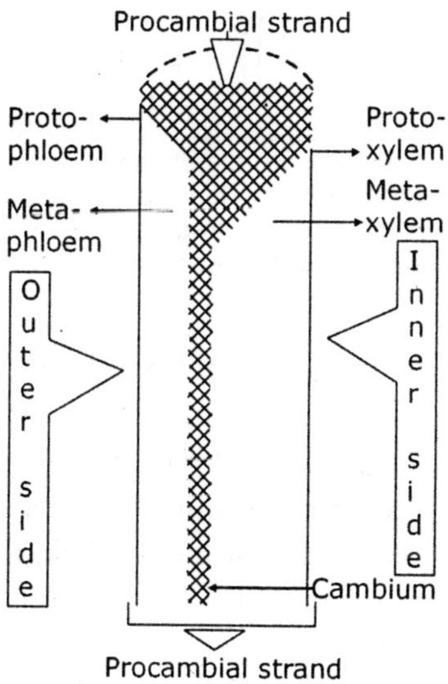


Figure 23.2

Schematic diagram of longisection of procambial strand from the apex to illustrate the mode of origin and development of xylem and phloem. Proto-phloem originates earlier than proto-xylem. The edges of procambium strand differentiates before than centre. The undifferentated region of procambium becomes cambium (Adapted after Gemmel, 1969).

and the whole procambial strand form the primary vascular tissue. In most dicotyledons these waves do not meet. A zone of procambial meristematic cells remains. This zone that occurs between primary xylem and – phloem is the vascular cambium. If it is not in the form of a continuous ring, a continuous ring of cambium is formed by dedifferentiation of interfascicular parenchyma into interfascicular cambium and their subsequent lateral union with fascicular cambium. Procambium gives rise to permanent tissues and it is customary to designate the tissues as primary. The derivative cells of cambium are designated as secondary. In normal dicotyledonous stem the ring of vascular cambium is composed of fascicular and interfascicular cambium. In the normal dicotyledonous root the vascular cambium is wavy. There is no distinction between fascicular and interfascicular cambium. The ridges of the wavy cambium occur overarching the protoxylem while the furrows lie below the primary phloem.

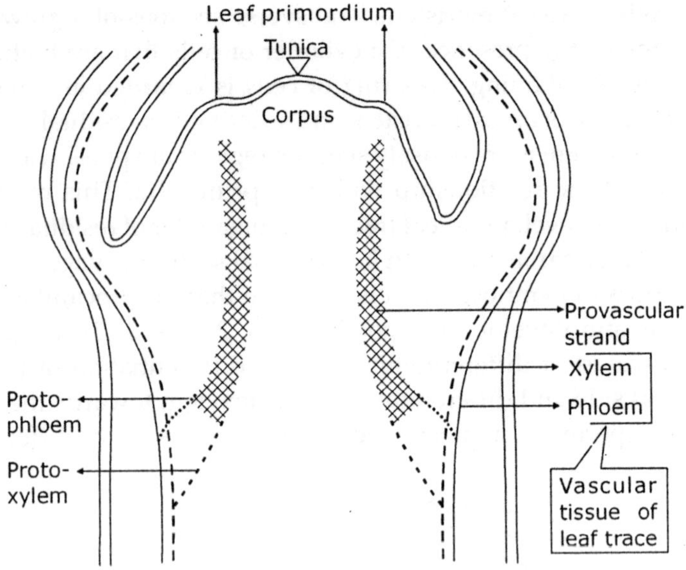


Figure 23.3

Diagram of longisection of angiosperm shoot apex illustrating the topographical relationship between leaf primordium and provascular strand, and representing the linkage of differentiating provascular strand and existing leaf trace. (Redrawn after Gemmel, 1969, p. 14).

**Activity of vascular cambium:** The vascular cambium is one cell thick and the cells of cambium are compactly set without having any intercellular spaces. Cambium and its immature derivatives form a cambial zone where it is difficult to differentiate the cambial initial. Two types of mitotic divisions characterize an active cambium—periclinal and anticlinal. As a result of periclinal divisions new cells of secondary xylem and—phloem are produced. Anticlinal divisions give rise to new cambial initials.

At the time of secondary growth the fusiform initial divides periclinally. One of the two cells thus formed remains as fusiform cambial initial whereas the other is an immature cambial derivative that is added to the cambial zone (Fig. 23.4). The fusiform cambial initial continually cuts of new cambial derivatives to the exterior and to the interior. The derivative cells may differentiate directly into new secondary phloem to the exterior and new secondary xylem to the interior. In this case the cambial zone is narrow. But more commonly the cambial initial and its derivative cells divide further. As a result a wide cambial zone is formed. Continued periclinal divisions in fusiform cambial initials and in their derivative cells result in radially oriented files of cells of similar shape that mature to secondary vascular tissues. Such radial oriented files of cells are conspicuous in the secondary xylem of conifers where the secondary xylem is largely composed of tracheids. In angiosperms (Fig. 23.5) such radial files of cells are less conspicuous due the formation of vessels. Vessels often increase greatly in diameter and as a result distortion of files occurs. As the fusiform initials compose the axial system

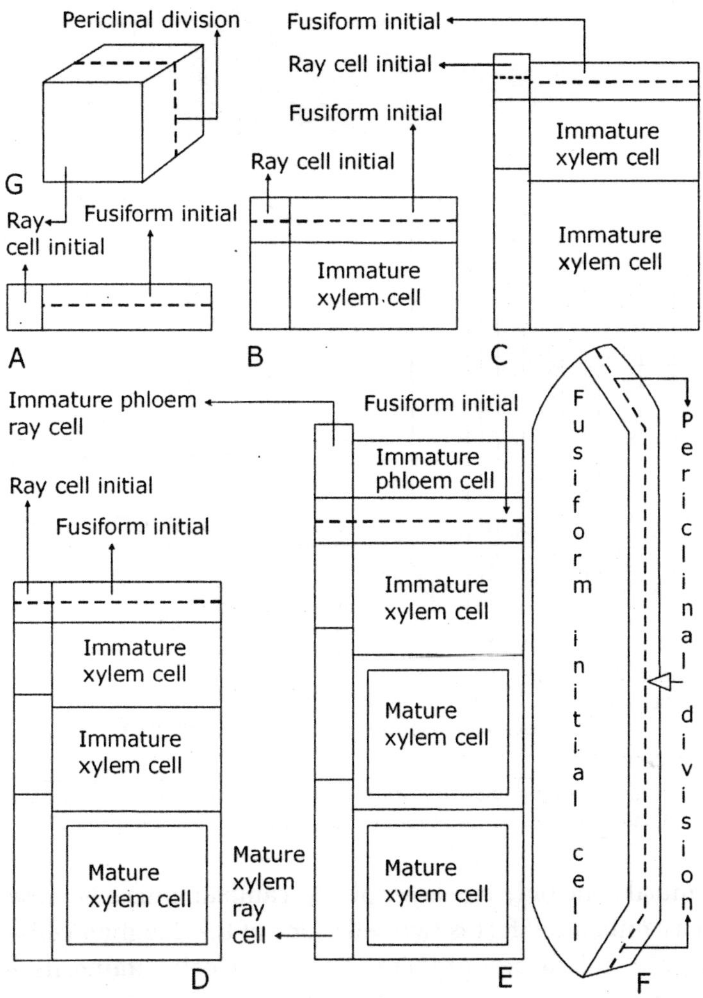


Figure 23.4

Schematic drawings of the production of secondary xylem and secondary phloem by cambial initial cells (A-E). Vide text for details. Broken lines indicate periclinal division. (Adapted after Beck, 2005). F. & G. are fusiform initial and ray cell initial in three dimensional view respectively (redrawn after Esau, 1961).

of vascular cambium, their derivatives mature into the elements of vascular tissues that compose the axial system. The derivative cells mature into tracheids, trachea and xylem fibre of xylem and sieve tube, companion cell and phloem fibre of phloem as they compose the axial system of plants.

The ray cell initial divides periclinally to form two cells. One of the daughter cells remains as cambial ray initial. The other cell differentiates into either xylem parenchyma if it is the inner cell or phloem parenchyma if it is the outer cell. Ray initials produce mostly xylem and phloem parenchyma. In gymnosperm ray initials form albuminous cell. The ray initial and its derivatives compose the radial system of plants.

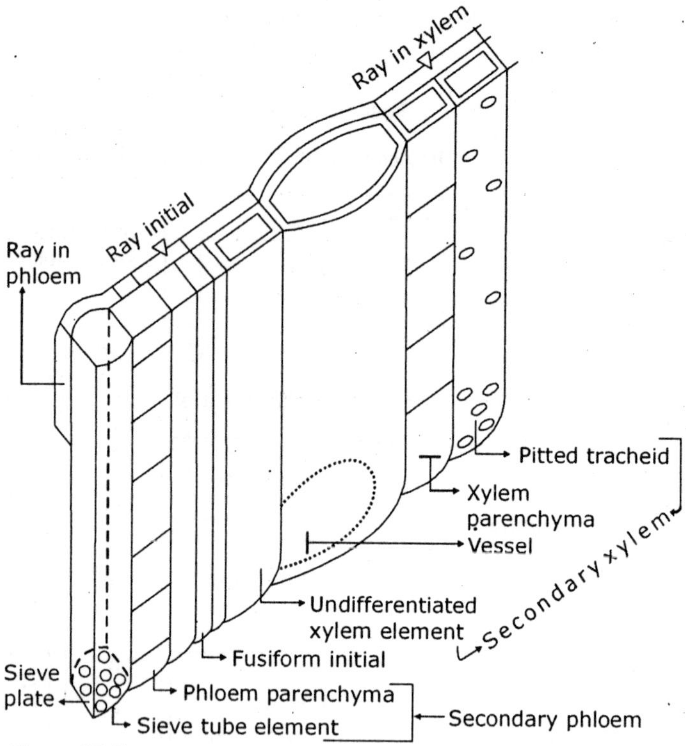


Figure 23.5

A diagrammatic interpretation of the production of secondary vascular tissues by cambium (Adapted after Ray, 1963, p. 137).

In a cambial zone only one layer of true cambial initials is present. Periclinal division in a cambial cell forms two cells. One of the daughter cells differentiates into secondary xylem or – phloem. The other cell retains its meristematic properties. This process continues throughout the life of a plant and therefore the division of cambium is limitless. This virtual immortality of cambial cells ideally displays the 'continuing meristematic residue' of Newman (1965).

The derivatives of cambial initials are incipient vascular tissues. They differentiate into secondary xylem cells and secondary phloem cells. Normally the amount of secondary xylem formed is in excess in comparison to secondary phloem. The ratio of differentiation between secondary xylem and secondary phloem is 3 to 1 in conifers. In dicotyledons the ratio is variable and may be as great as 10 to 1. In an experiment *Eucalyptus camaldulensis* was exposed to labeled $^{14}CO_2$ that was incorporated in secondary vascular tissues. The incorporation of $^{14}CO_2$ indicates that the ratio of layers of secondary xylem and – phloem produced by the cambium is 4 to 1 (Cutter, 1978).

The cambial activity is related to rainfall and temperature in tropical and temperate zone respectively. In tropical zones the vascular cambium of some species is continually active throughout the entire life. In temperate zones the vascular cambium remains dormant in winter. It activates in spring and produces secondary vascular tissue. Activation and cessation of cambial activity occur at an earlier age in ring porous trees (e.g. *Quercus*) than in diffuse porous trees (e.g. *Fagus*). There is experimental evidence that day-length affects the duration of cambial activity. In short-day condition the cambium remains dormant in *Robinia pseudacacia*. Activation/reactivation of cambial activity followed by a period of dormancy occur over the entire life of a plant. It leads to the formation of growth rings that reveal approximate age of plant. Thus the age of *Pinus aristata* and *Sequoiadendron* was estimated to be more than three thousand years (Cutter, 1971, 1978).

The cambial activity causes the increase in girth of axis. The circumference of vascular cambium also increases to cope up with the increase in girth of axis. This is accomplished by the formation of new fusiform- and ray initials. Formation of the fusiform cambial initial occurs in the following ways.

- Fusiform initials in a storied cambium divide by anticlinal divisions and thus new initials are added to cambium.
- Short fusiform initials divide by radial anticlinal divisions where the partition walls occur parallel to long axes (Fig. 23.6A-C).
- The production of new initials from long non-storied fusiform initials occurs by oblique anticlinal divisions with walls of various degrees of inclination (pseudo-transverse). The daughter cells thus formed elongates by apical intrusive growth of overlapping ends.
- The elongation and intrusive growth of ray cell initials may form new fusiform initials. The elongation of ray cell initials occur parallel to the long axis of plants (Fig. 23.6D-E).

Fusiform initials give rise to ray initials in the following ways.

- An entire or part of short fusiform initial divides by a series of transverse anticlinal divisions thus forming a tier of ray cell initials (Fig. 23.6F).
- Fusiform initial cuts off a single cell on its lateral side by an arcuate wall. In the small lateral daughter cell series of transverse anticlinal division occur and as a result a new ray initial is formed (Fig. 23.6G).
- Fusiform initial cuts off a single cell at its end (Fig. 23.6H).
- A fusiform initial is reduced to a ray cell initial (Fig. 23.6I).

In the above cases a uniseriate ray initial is formed. The ray initials may divide by a series of longitudinal anticlinal divisions to form biseriate or multiseriate rays.

During secondary growth as the cambium increases in circumference a balance between the number and distribution of fusiform- and ray initial is always maintained, because rays are the passageways for the transport of nutrients to cambium and its immediate incipient derivatives.

In case of dicotyledons the wavy cambium of root donates secondary xylem towards interior and secondary phloem towards exterior. The portion of the wavy

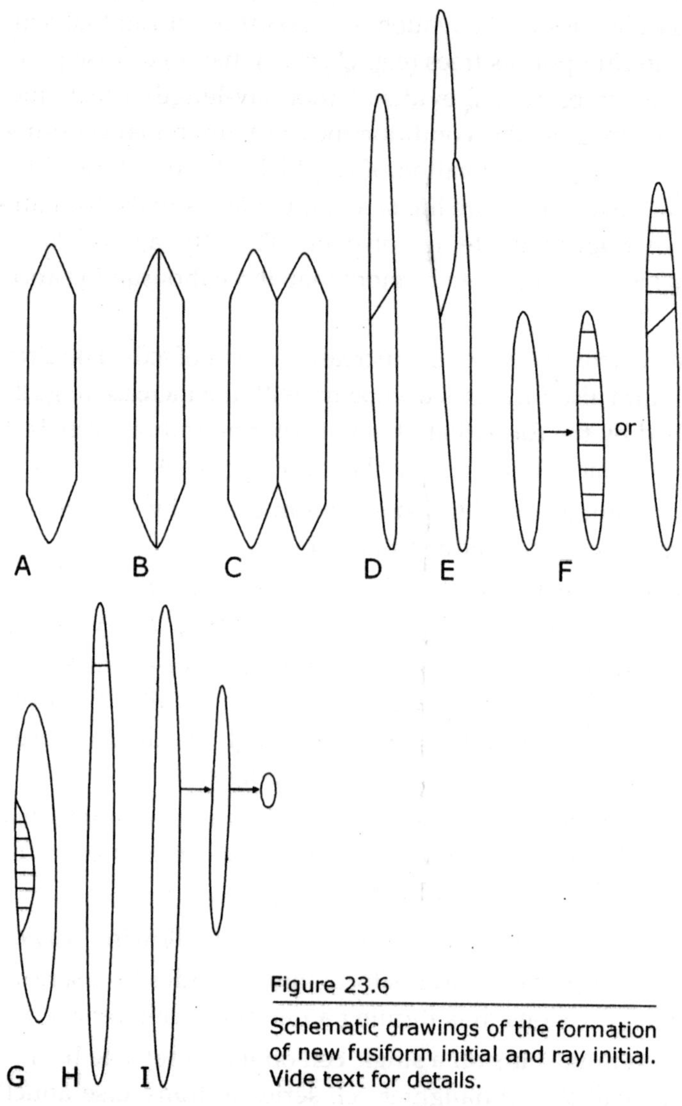


Figure 23.6

Schematic drawings of the formation of new fusiform initial and ray initial. Vide text for details.

cambium ring that occurs at the furrow forms more secondary xylem. As a result at later stage of development the wavy ring becomes more or less circular.

In dicotyledonous stem the fascicular cambium becomes active before the differentiation of interfascicular cambium (Fig. 23.7A-D). In many plants at the interfascicular region there originate provascular strands. The strands arise close to the initial vascular bundles (Fig. 23.7B). At later stage differentiating provascular strands and initial vascular bundles may contact to each other laterally thus forming vascular cambium continuous across the vascular bundles (Fig. 23.7C & D).

In many woody herbs (Fig. 23.8A & B) interfascicular cambium originates from ground meristem when it differentiates early. In later stages mature interfascicular parenchyma by dedifferentiation forms interfascicular cambium. The fascicular cambium becomes active initially in the vascular bundles. When the vascular

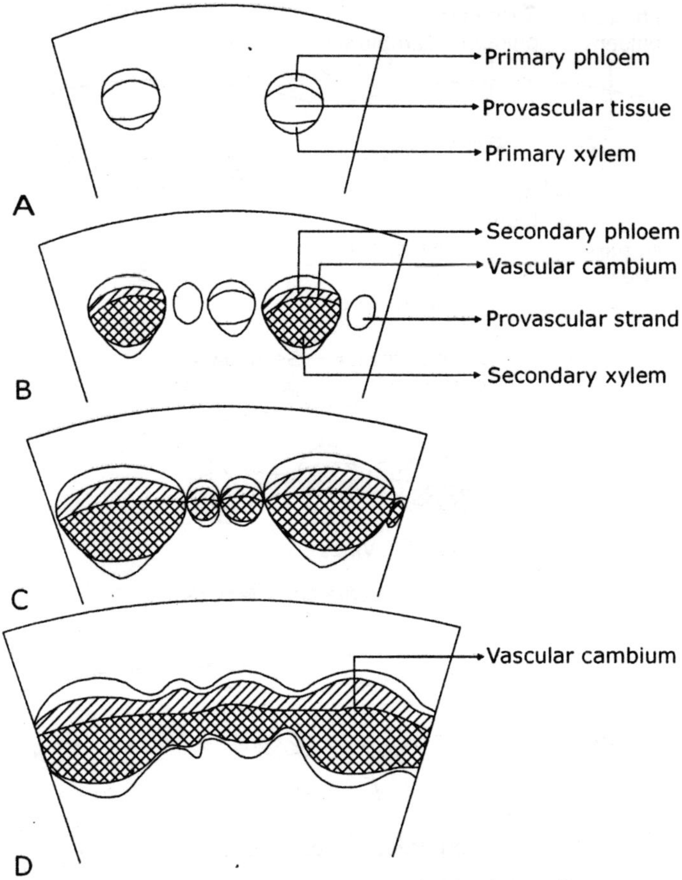


Figure 23.7

Diagram illustrating stages of formation of provascular strand, vascular cambium and secondary vascular tissues. Secondary xylem = crosshatched. Secondary phloem = obliquely hatched. Adapted after Beck, 2005, p. 155 (Vide text for details).

cambium becomes continuous and complete, the derivative cells of cambium cause the increments of secondary vascular tissues in normal secondary growth.

In herbaceous dicotyledons, e.g. *Geum, Agrimonia* etc. (Fig. 23.8D) the fascicular cambium is active only. In *Ranunculus, Impatiens* etc. (Fig. 23.8E) cambium does not develop usually or if it develops it remains inactive. In annuals the vascular cambium remains active only during the growth of the plant and ceases its activity before the plant dies. The vascular cambium functions throughout the life of woody perennials. It is to note that vascular cambium is absent from most monocotyledons, pteridophytes and some herbaceous dicotyledons.

In woody lianas, e.g. *Aristolochia, Clematis* etc. (Fig. 23.8C) the derivatives of fascicular cambium differentiates into characteristic secondary xylem and—phloem. But the derivative cells of interfascicular cambium differentiate into

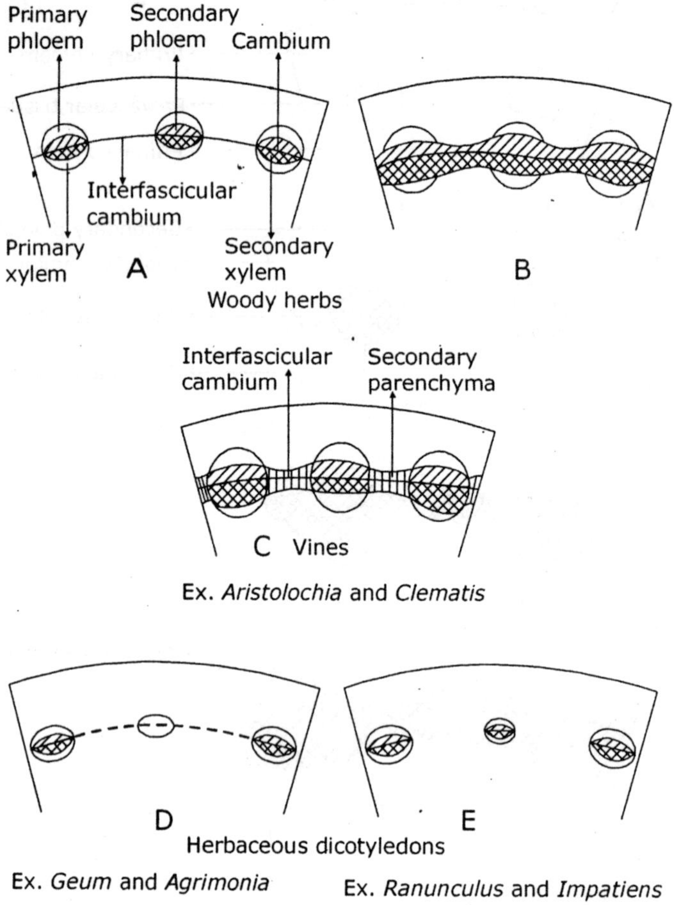


Figure 23.8

Diagram showing fascicular cambium, formation of interfascicular cambium, formation of secondary xylem (= crosshatched) and secondary phloem (= obliquely hatched). Adapted after Beck, 2005, p. 156.

parenchyma cells only. This type of cambial activity is regarded as abnormal. Anomaly in the activity of cambium is also observed in *Serjania, Paullinia. Thinouia* etc. where in stems the cambium ring is split into several isolated segments that form separate vascular strands, often in a regular pattern. In *Bauhinia* stem different sectors of cambium donate unequal amount of secondary xylem and as a result the stem becomes lobed. In *Bignonia* stem symmetrically located segments of cambium cease to form xylem inwards and form phloem in excess outwards. As a result xylem becomes separated by radial slits.

**Accessory cambium:** Accessory cambium develops both in stems and root. In the stem of *Amaranthus, Boerhaavia, Mirabilis* etc. this secondary cambium originates on the peripheral side surrounding vascular bundles. Accessory cambium divides and the derivative cells differentiate into conjunctive tissue and secondary vascular bundles. In *Tecoma* stem two arcs of accessory cambium develop at the peripheral margins of pith. The cambium produces vascular tissues that have inverse orientation, i.e. secondary xylem on the peripheral side and secondary phloem (intraxylary phloem) on the inside. Accessory cambia also develop in monocotyledon (ex. *Dracaena*). In *Dracaena* accessory cambium forms thin-walled conjunctive tissue and leptocentric vascular bundles towards the inner side.

In the root of *Beta vulgaris* concentric rings of accessory cambia originate in succession outwards. Most of the derivative cells store nutrients and as a result *Beta* root swells.

Cheadle (1937) reported the existence of vascular cambium in monocotyledon. The cambium occurs at the peripheral region of the stem. The peripheral parenchyma divides to form the vascular cambium. The cambial cells are narrow and elongated, ex. *Agave*. The cells may be fusiform, polygonal or rectangular in shape. The cambium divides and the inner derivatives differentiate into vascular bundles, consisting of both xylem and phloem. The peripheral derivatives form parenchyma only. The vascular bundles, thus formed, remain embedded in the ground parenchyma.

**Cork Cambium:** Cork cambium, also known as phellogen, is a secondary lateral meristem. It originates from the permanent cells of epidermis, hypodermis, cortex and phloem by dedifferentiation. The cells of phellogen are compactly set without any intercellular spaces and rectangular or radially flattened in cross-sectional view. In longitudinal section they are rectangular or polygonal and sometimes irregular in outline. Each cell of phellogen is highly vacuolate and may contain chloroplast and tannins.

The phellogen divides periclinally and the derivative cells differentiate into peripheral phellem (also called cork) and inner phelloderm. Cork cells have suberized cell walls that are impervious to air and water, and have protective properties. Phellem, phellogen and phelloderm are collectively called periderm. Periderm forms a protective layer in stems and roots.

The following factors influence the cambial activity.

- Hormone: It was first shown by Jost, 1893 that the activity of stem-cambium (Cutter, 1971) is greatest just below the growing leaves. In later years after the discovery and identification of indole-3-acetic acid (IAA) and after various experiments it was definitely concluded that the stimulus that activates the cambium is hormonal in nature. Hormone is produced in young buds and leaves and it is translocated downwards thus stimulating the cambium to divide. It was previously mentioned that the activity of cambium is highest in spring in the trees of temperate regions. It gradually declines as summer advances. In spring new buds develop where hormones are produced. Hormones move basipetally and then the growth activity of cambium is initiated and promoted. Later researches reveal that the interactions among IAA, gibberellic acid (GA) and 6-furfurylaminopurine (Kinetin) stimulate the division of cambium. In an experiment (Wareing *et al.*, 1964) the shoots of *Acer pseudoplatanus* were completely disbudded. It was treated with IAA only and the formation of xylem was observed.

It was noted that narrow zone of xylem was formed where the vessels were lignified only. The other disbudded shoots were treated with GA alone. In this case xylem was produced and they were non-lignified and without vessels. But a normal and wide zone of xylem was formed when the disbudded shoots were treated both with IAA and GA. This reveals that there exists a correlation between hormones in initiating the cambial activity.

• Gravity: In the branches of gymnosperm and angiosperm a special type of wood is formed termed reaction wood. In gymnosperm branches the reaction wood is formed on the lower side and is specially called compression wood. In the branches of angiosperm the reaction wood is formed on the upper side and is specially termed as tension wood. The reaction wood is the collective term of tension wood and compression wood. Various experiments reveal that these unequal proportions of wood are formed with respect to gravity (Cutter, 1971). Gravity is the stimulus that causes the formation of reaction wood and there are ample evidences that auxin is present behind the process. In an experiment plants were grown in klinostat and it was noted that no reaction wood is formed. This proves that reaction wood is formed in response to gravity.

• Day length: In short days late wood is formed only, e.g. *Robinia* (Wareing *et al.*, 1956). In this wood no vessels are formed. The vessels, if formed, are few in number and with small diameter. In long days early wood is formed. In this wood many vessels are developed and they have large diameter. In long days tracheids with large diameter are formed in conifers in contrast to short days where tracheids with narrow diameter are formed (Larson, 1962). In long days the needles of conifers elongate and there is no doubt that this is associated with auxin production. The production of auxin promotes to form tracheids with large diameter. In an experiment (Larson, 1964) short day (i.e. long night) plants were given a low intensity of light for a short time during the long night period. It was observed that tracheids of wide diameter were formed. It indicates that the cambial activity is influenced by day length and it is a true photoperiodic phenomenon (Cutter, 1971). Temperature influences the activity of cambium. It was observed that in *Robinia* seedling the activity of cambium was continuous for a long period when the plants were kept in a high temperature and short days (Waisel and Fahn, 1965). There exists a correlation between high summer temperature and cambial activity in the tropical tree *Polyalthia longifolia* (Paliwal *et al.*, 1975).

• Pressure: The vascular cambium always remains compressed by the tissues present on internal and peripheral side of it. It is needed for the normal functioning of cambium. In an experiment with *Pinus strobus* and *Populus trichocarpa* (Brown, 1964) the pressure was released on the peripheral side. Releasing a strip of bark from the tree by incisions while the apical end is still attached with the tree did this. Aseptic measures were taken and the operated zone was wrapped in polythene to prevent infection and reduce desiccation. After a few days it was noticed that an unorganized callus was formed on the inner side of the free strip of the bark and the function of cambium was not normal. In another experiment the pressure was released according to the previous experiment and then the pressure was applied by binding the strip of bark with the tree. Aseptic measures were taken as usual and after a few days it was noticed that no callus tissue was formed and the cambium functioned normally. This implies that pressure influences the normal cambial activity.

The dominant features of vascular cambium are in its origin at a definite position in stems and roots, and production of xylem and phloem in relation to poles. The xylem is formed towards the inner side and the phloem is produced towards the peripheral side. This inherent polarity was demonstrated by Siebers (1971) in the following ingenious experiment with the hypocotyls of *Ricinus communis* (Fig. 23.9). A rectangular block of tissue from the interfascicular region of hypocotyl, where the interfascicular cambium was supposed to originate after several days, was excised out. This tissue was grafted to the original position but

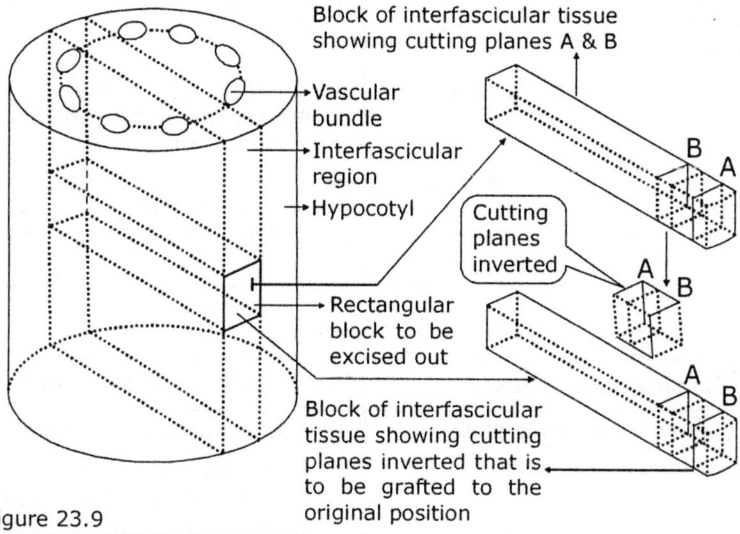


Figure 23.9

Diagram illustrating the experiment of Siebers, 1971 on the hypocotyl of *Ricinus communis.* See text for explanation. Adapted after Cutter, 1978.

with inverted cutting planes. Rotating the block of tissue with cutting planes through 180° did this. The successful grafts showed that xylem was produced on the peripheral side and phloem was formed towards the inner side. This proves that the production of vascular tissues by the cambium is predetermined and the same tissues are formed in the same position even when the polarity of cambium is changed.

Comparison between procambium and vascular cambium.

| Procambium | Vascular cambium |
|---|---|
| 1. Procambium is the meristem from where primary xylem and primary phloem develop. | 1. Vascular cambium is the meristem from where secondary vascular tissues, i.e. secondary xylem and secondary phloem develop. |
| 2. Promeristem gives rise to procambium. | 2. Procambium at later stages of differentiation forms vascular cambium. |
| 3. Protoplast of procambium stains deeply. | 3. Protoplasts of vascular cambium do not stain strongly. |
| 4. Procambium is not differentiated into long and short cells. | 4. Vascular cambium is differentiated into long fusiform initial and short ray cell initial. |
| 5. The procambial cells have gabled ends that they show in radial longitudinal section (RLS). | 5. The vascular cambial cells have flat ends that they show in RLS. |
| 6. Procambium occurs in all vascular plants. | 6. Vascular cambium is present in most dicotyledons, gymnosperms and in a few monocotyledons. |
| 7. Procambium originates from meristematic tissues only. | 7. Vascular cambium may originate from permanent primary parenchyma tissue by dedifferentiation, e.g. interfascicular cambium. |

Comparison between fascicular and interfascicular cambium.

| Fascicular cambium | Interfascicular cambium |
|---|---|
| 1. It occurs within vascular bundles. | 1. It occurs between two vascular bundles. |
| 2. It develops from procambium. | 2. It develops from primary parenchyma present between two vascular bundles by dedifferentiation. |
| 3. It remains sandwitched between primary phloem and primary xylem towards exterior and interior respectively. | 3. It remains embedded in interfascicular parenchyma. |
| 4. It forms secondary vascular tissues only. | 4. It forms secondary vascular tissues. It may also form secondary parenchyma that is considered as anomalous (e.g. *Aristolochia*). |

Comparison between fusiform initial and ray initial.

| Fusiform initial | Ray initial |
|---|---|
| 1. It refers to a single cell. | 1. It refers to a group of cell. Each cell is referred to as ray cell initial. |
| 2. Each cell has spindle shape. | 2. Each cell has cuboidal shape. |
| 3. Each cell is several times longer than breath. | 3. Each cell is short and more or less isodiametric. |

*contd.*

| Fusiform initial | Ray initial |
|---|---|
| 4. Fusiform initial and its derivative cells may have a storied or nonstoried appearance. | 4. Ray initial and its derivative cells may be uniseriate, biseriate and multiseriate. Multiseriate ray appears to be fusiform. |
| 5. The derivative cells of fusiform initial compose axial system of plant. | 5. The derivative cells of ray initial form radial system of plant. |
| 6. Tracheids, trachea, xylem fibre, sieve tube, companion cell and phloem fibre develop from fusiform initial. | 6. Xylem ray parenchyma and phloem ray parenchyma develop from ray initials. |
| 7. Fusiform initial develops from procambium. | 7. Ray cell initial develops from fusiform initial. |

Comparison between vascular cambium and cork cambium.

| Vascular cambium | Cork cambium |
|---|---|
| 1. It is also referred to as cambium. | 1. Anatomically it is termed as phellogen. |
| 2. Procambium gives rise to cambium. | 2. The permanent tissues of epidermis, cortex or phloem give rise to phellogen by dedifferentiation. |
| 3. Cambium originates from meristemetic tissue and so it is customary to regard it as primary meristem. | 3. Phellogen originates from permanent tissues by dedifferentiation and so it is customary to regard it as secondary meristem. |
| 4. Each cell of vascular cambium may be spindle shaped or cuboidal. | 4. The shape of phellogen may be rectangular or polygonal. |
| 5. The cells comprising cambium may be long— termed fusiform initial or short— termed ray cell initial. | 5. There exists no such size difference. |
| 6. Cambium and its derivative cells are arranged in stratified or non-stratified manner. | 6. Phellogen and its derivative cells are always arranged in stratified manner. |
| 7. The derivative cells of cambium are vascular tissues that occur in the stele only. | 7. The derivative cells of phellogen are non-vascular tissues and occur mostly in extrastelar region. |
| 8. The peripheral derivatives of cambium are secondary phloem. | 8. The peripheral derivatives of phellogen are phellem. |
| 9. The inner derivatives of cambium are secondary xylem. | 9. The inner derivatives of phellogen are phelloderm. |
| 10. Cambium with peripheral phloem and inner xylem together compose open vascular bundle of stem. | 10. Phellogen with peripheral phellem and inner phelloderm together compose periderm of root and stem. |
| 11. Cambium, secondary phloem and secondary xylem compose the vascular tissue system of plant. | 11. Periderm forms the protective tissue system of plants. |

CHAPTER **24**

# Axillary Bud

The lateral branch of seed plants, in their nascent state, is referred to as axillary bud that is commonly present on the stem at the axil of leaves.

The lateral bud primordia usually arise in close association with leaf primordia. Opinion varies regarding the place of origin of bud primordia. Foliar origin is suggested by Garrison (1955) in dicotyledons. In monocot (e.g. *Musa, Tradescantia*) the bud develops below leaf primordia and is axial in origin. In most cases bud primordia arise on stem at the axil of leaf primordia and sometimes may be displaced towards the leaf base during growth (Esau, 1965). Whatever may be the place of origin, i.e. foliar or axial, the buds arise exogenously in relation to superficial tissues. Champagnat (1961) reported epidermal origin of axillary buds.

Bud primordia usually arise later than the subtending leaves, which are commonly the second or third plastochron. [The term plastochron was adopted to designate the time interval between the formation of leaf primordium and the next on a shoot]. At this level the outer cells of apical meristem divide anticlinally and the cells, thus formed, delimit the meristem of axillary bud from apical meristem. These cells possess all the characteristics of meristematic cells and divide in various planes, predominantly anticlinal and periclinal. As a result, the apical meristem of lateral branch is organized resembling that of the parent shoot and elevated above the surface. Initially the bud meristem is formed in continuity with shoot apical meristem. Later the elongation of internodes separates them from each other. Vacuoles appear in certain cells situated above the axillay meristem and these cells separate the bud meristem from apical meristem. As growth continues the bud meristem begins to give rise to leaf primordia. Normal growth and development do not occur immediately. It happens when the axillary bud attains a position some distance away from shoot apical meristem due to normal growth of shoot.

In some species lateral bud is formed in the axil of older leaves. Such bud arises either from differentiated cell like parenchyma or from meristematic cells present at the internode. In the former case the cells owe their origin from the apical meristem.

In the derivatives of apical meristem, vacuoles appear and thus partly or fully differentiated parenchyma tissues are formed. At the time of bud formation, these parenchyma cells regain the power of cell division and form the meristem of axillary bud. This phenomenon can be interpreted as dedifferentiation when the meristematic activity is resumed in differentiated cells. In the latter case (i.e. origin from meristematic cells) the cells are continuous with the shoot apical meristem. In some cells vacuoles appear and these differentiated cells separate a group of meristematic cells from the shoot apical meristem. This isolated group of meristematic cells is designated as detached meristem. This meristem remains dormant or inhibited until stimulated to form meristem of axillay bud.

In unusual cases axillary bud may arise from the subtending leaf of the first plastochron (ex. *Hydrocharis*). In the axil of all leaves of a shoot, bud may not be formed, but in some species they arise sporadically or in a regular sequence. In *Alternanthera philoxeroides* two axillary buds arise at a node from the opposite leaves. These buds are of unequal size and have different potentialities of growth. In *Ulex* the lateral buds develop as thorns.

In potato plant it is observed that, during the initiation of axillary bud, the divisions are quite regular. This orderly division occurs at base and on lateral side of the incipient bud. As a result a meristematic zone composed of a series of curved layers are formed. These layers are more or less parallel to each other and are designated as shell-zone because of its shell-like configuration (Fig. 24.1A). Regarding the function it is suggested that the cells of shell-zone divide and give an active upward thrust to the developing bud to protrude above the surface. The upward growth of the bud (Fig. 24.1B) is followed by the appearance of the leaf primordia beginning with the prophyll.

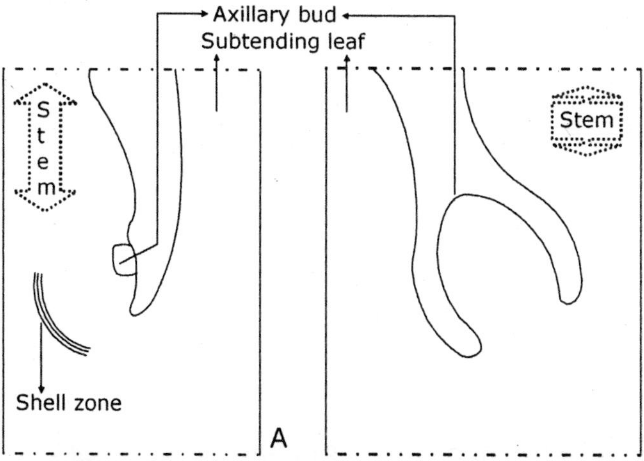


Figure 24.1

Diagrammatic representation of the origin of axillary bud in longitudinal section. A. The initiating cell of axillary bud. B. Axillary bud at later stage. The line drawing is based on micrographs. After Esau, 1961.

Axillary buds are connected by vascular traces to the vascular strands of main axis. These vascular traces are referred to as bud trace or branch trace. If the bud primordia originate close to apex, the bud traces differentiate earlier than the vacuolation of peripheral meristem. At this stage the bud trace appears as meristematic strand lying between the vascular tissues of main axis and protruding bud. Careful observation reveals that bud trace first originates in the bud primordium and then gradually extends towards the vascular strand of main axis, i.e. the differentiation of bud trace is basipetal.

In some species axillary bud arises in the axil of older leaves. In this case bud traces differentiate through partly differentiated or vacuolated ground tissues. The differentiation of bud traces in this case is also basipetal, i.e. traces originate in the bud and gradually extend towards the vascular strand of main axis. Sometimes the dormant axillary buds are removed further as a result of secondary growth of the main shoot. Here also the basipetal differentiation of bud traces occurs by the addition of vascular elements formed by vascular cambium.

Factors determining the initiation and organization of cells into a bud meristem are mainly the hormone and the interaction between buds and subtending leaves. Complex hormonal regulations are definitely involved in the process. So long as the apical bud causes apical growth, the axillary buds do not develop—this phenomenon is known as apical dominance. It is suggested that apical buds dominate in auxin content than the lateral buds, but how auxin causes the inhibition of development of lateral buds is still controversial. At initial stage the development of axillary bud is stimulated by the leaf primordia and further growth occurs due to interaction between bud and its subtending leaf. Competition among the axillary buds for nutrients is also an important factor.

# Chapter 25

# Plastochron

Plastochron [also spelt as plastochrone, Cutter (1971)] is defined as the time interval between the formation of one leaf primordium and the initiation of next leaf primordium.

Plastochron is a developmental measure and is expressed as 'plastochron index'. The index indicates the developmental status of each leaf and relates the observation to time. The developmental index is the plastochron age of a shoot. Plastochron Index (*P. I.*) can also be defined as a measure of plant growth and is used to determine the rate of growth based upon the initiation of successive leaf primordia.

Plastochron Index provides a morphological time scale rather than chronological age in studies relating to morphological and physiological development of a whole plant or plant organ. Similarly Leaf Plastochron Index (*L. P. I.*) provides a morphological time scale rather than chronological in studies relating to morphological and physiological development of a leaf.

In developmental studies plastochron is used as the unit of developmental scale. It is observed that morphologically similar leaves may be of quite different chronological age, while leaves with similar chronological age may also have quite different stages of development. Leaves of same plastochron age reveal similar morphology and development. In developmental studies it is to relate an aspect of developing leaf directly to time. When the aspect is leaf development the length and breadth are measured in relation to time and the leaf remains intact. The aspect may be the dry weight, development of mesophyll tissue etc. and in this case the leaf is to be sacrificed to obtain data. In this type of study one has to deal with many parallel samples when each is sacrificed at a given time for the required data. These parallel samples must be of same plastochron because *P. I.* eliminates the variation due to growth rate and creates a linear scale to measure development based on morphology.

For plastochron study the plants must be genetically uniform, grown under exactly controlled conditions and in vegetative growth. The *Xanthium* leaf is one of the most thoroughly investigated leaves among dicotyledons using *P. I.* & *L. P. I.* The estimation of *P. I.* fails when flowering is induced in *Xanthium* (Erickson *et al.*, 1957).

Plastochron Index applies to shoot. So it is very useful in developmental studies in which interest centres on characteristic of the shoot. The characteristic may be the rate of metabolic process, content of a biochemical constituent, total fresh weight or dry weight of shoot etc. and these can be determined at various stages of development. The development of shoot apex in *Zea* was studied using *P. I.* Abbe *et al.* (1951) demonstrated the progressive increase in mass of the apical meristem of *Zea* during growing season. In *Hypericum* Zimmermann in 1928 demonstrated cyclical changes in the area of apical meristem and these changes are correlated with the initiation of leaf using *P. I.* During one plastochron the shoot apex undergoes changes in its morphology and this change is referred to as plastochronic change (Fig. 25.1). The shoot apex is in the form of a small rounded mound before the initiation of a new leaf primordium. As growth continues the shoot apex widens. Later leaf buttresses are initiated on the sides of apical meristem. A leaf buttress constitutes the base of a leaf and it forms a lateral prominence on the sides of apical meristem (Esau, 1965). Leaf primordium grows upward from leaf buttress. After the formation of new leaf primordium the shoot apex again becomes in the form of a small rounded mound. During plastochronic change the volume and surface area of the shoot apices also change. The changes are expressed as minimal-area phases and maximal-area phases. Later these phases are respectively referred to as minimal phases and maximal phases (Esau, 1965) or minimal area and maximal area (Fahn, 1997). The following plants are suitable for the study of plastochronic changes in the shoot apices of dicotyledons with opposite leaves - *Coleus, Lonicera, Syringa* etc. (Fahn).

*L. P. I.* is used to study the development of leaf. *L. P. I.* is also used to determine the oxygen uptake, chlorophyll content, and fresh weight of *Xanthium* leaves at various stages of development (Michelini, 1955). Maksymowych (1956) studied the histogenesis of *Xanthium* leaf applying *L. P. I.* The study revealed the relationship between marginal meristem activity and mesophyll development.

The plastochron index is estimated in the following way:

$$P.\ I. = n + \frac{\log L_n - \log 10}{\log L_n - \log L_{n+1}} \quad \ldots \quad \text{Equation} \quad \ldots \quad \ldots (1)$$

*P. I.* = Plastochron Index; $n$ = it is the serial number of that leaf, which is 10 mm or just exceeds 10 mm length (counted from the base of the shoot); $n + 1$ = it is the succeeding leaf with a length that is less than 10 mm; $L_n$ = length of $n$ leaf; log 10 = it is the predetermined reference value of leaf length.

Specifically a leaf is $n$ plastochrons old when the leaf is 10 mm in length. The $n + 1$ leaf is $n + 1$ plastochrons old when it is 10 mm long and so on. But it is seldom observed that a leaf is exactly as long as 10 mm . For instance leaf 5, $n$, is 18 mm in length and leaf 6, i.e. $n + 1$, is 5 mm long at a given time. This indicates that the plant is more than 5 plastochrons but less than 6 plastochrons old. In this

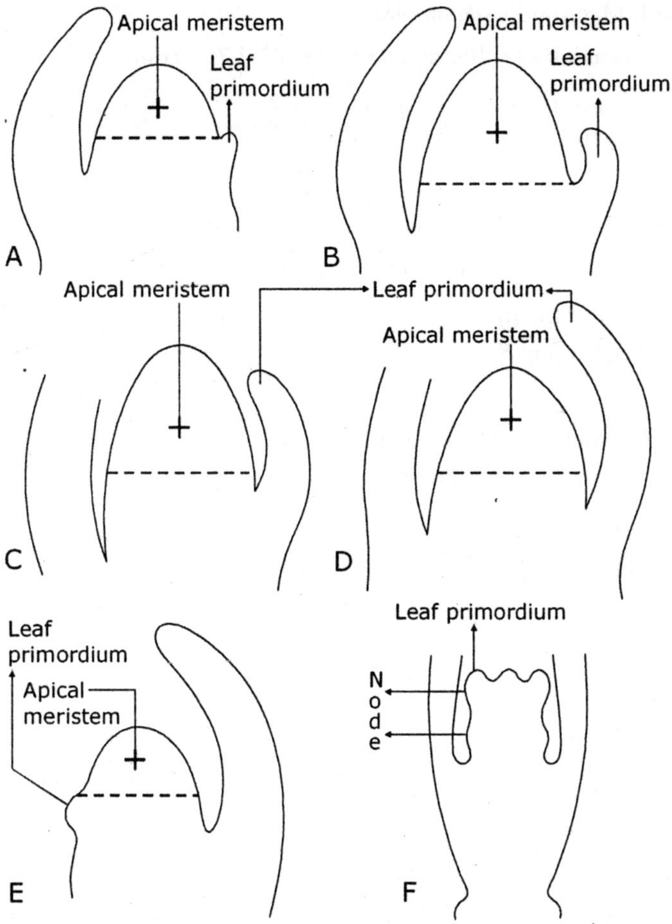


Figure 25.1

Diagrammatic illustration of the variation in size of apical meristem during a plastochron in l. s. view. The increase in length of apical meristem is illustrated in A-C. The decrease in length of apical meristem before the production of next leaf primordium is shown in figure D & E. F. Diagram showing the position of leaf primordium and nodes in a shoot apex in longitudinal view. Redrawn after Beck, 2005, p. 89 & Esau, 1965.

case the equation (1) is used to estimate the plastochron index of a plant in relation to leaf 5 at a given time. Examples 1, 2 and 3 illustrate the procedure to estimate *P. I.* of a shoot putting the length value of two successive leaves in equation (1).

The derivation of plastochron index for shoot development in *Xanthium* is described below (Erickson *et al.*).

The experimental plant was *Xanthium italicum*. Seeds were collected and allowed to germinate in pots. *Xanthium* is a short-day plant, i.e. the plant needs 8 hours of light and 16 hours of dark to flower. The short-day plants can be kept in vegetative growth by treating them as long-day/short-night plants only. *Xanthium* plants were kept in lighted green house. The long-day plants need 16hrs of

**Example 1.** Material *Xanthium* plant.

$n = 6$; $Ln$ = Leaf 6 (L6) = 10mm; $Ln+1$ = Leaf 7 (L7) = 5mm

$$P.I. = 6 + \frac{\log 10 - \log 10}{\log 10 - \log 5} = 6 + \frac{1-1}{1-0.699} = 6 + 0 = 6$$

**Example 2.** Measurement taken from example 1 later

$n = 6$; $Ln$ = Leaf 6 (L6) = 12mm; $Ln+1$ = Leaf 7 (L7) = 7mm

$$P.I. = 6 + \frac{\log 12 - \log 10}{\log 12 - \log 7} = 6 + \frac{1.079-1}{1.079-0.845} = 6 + \frac{0.079}{0.234} = 6 + 0.338 = 6.338$$

**Example 3.** Material *Xanthium* plant.

$n = 5$; $Ln$ = Leaf 5 (L5) = 18 mm; $Ln+1$ = Leaf 6 (L6) = 5mm

$$P.I. = 5 + \frac{\log 18 - \log 10}{\log 18 - \log 5} = 5 + \frac{1.255-1}{1.255-0.699} = 5 + \frac{0.255}{0.556} = 5 + 0.458 = 5.458$$

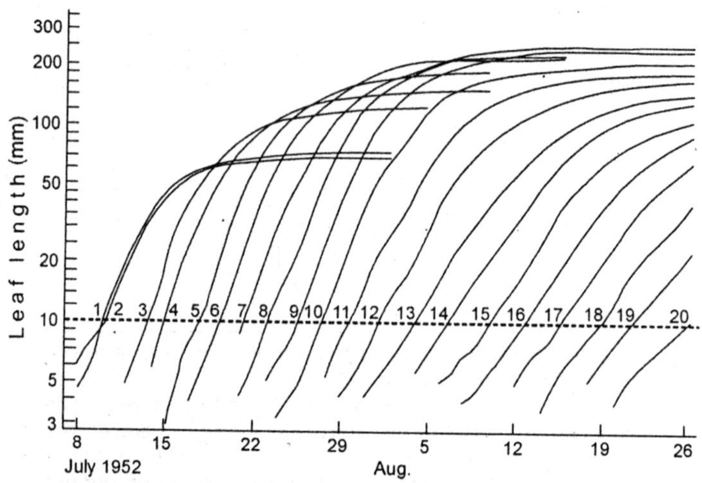


Figure 25.2

Diagram representing the length of successive leaves of *Xanthium* plant plotted logarithmically against time. The leaves are numbered serially in order of their appearance. Each growth curve is plotted from the measurements of a single leaf. Redrawn as per illustration of Erickson *et al.*, 1957, p. 299.

illumination per day. Light was provided to *Xanthium* before dawn and dusk in such a way that the plants got 16 hrs of illumination per day. Thus the plants were kept strictly in vegetative growth. About 20-30 plants were selected and each leaf of plants was measured daily from July 8 to August 26, 1952. The length of a leaf with petiole was measured only. Measurement was taken daily at a fixed time. The length of leaf was plotted logarithmically against time. The graph shows sigmoid growth curves of the leaves. This signifies that the leaf growth was exponential in the early portion. From the curve it also appears that the relationship between logarithm of leaf length and time for one plant was linear up to approximately 70 mm of leaf length (Fig. 25.2). From the above data Erickson *et al.* developed the equation (1) using *Xanthium*. Consult Erickson *et al.* (1957, p. 298) and Maksymowych (1973, p. 5) for details of derivation of the equation 1. In the equation (1) the reference value of leaf length is 10 mm, i.e. Log 10 at a given time. Ten millimeter length is considered for: (1) the 10 mm length of a leaf is large enough to be measured accurately; (2) the leaf can be measured without injuring the shoot apex and (3) the leaf of *Xanthium* of this order grows exponentially. In Fig. 25.2 the reference value of leaf length, i.e. 10 mm is designated with dashed horizontal line. In the reference value of leaf length logarithm of any base may be used. So the equation (1) may be rewritten as:

$$P.I. = n + \frac{\log L_n - \log L_\lambda}{\log L_n - \log L_{n+1}} \quad \ldots \text{ Equation} \quad \ldots \ldots (2)$$

*P. I.* = Plastochron Index; $n$ = serial number of leaf that is just the predetermined value of leaf length or exceeds the value; $\log L_n$ = natural logarithm of length of $n$ leaf; $\log L_{n+1}$ = natural logarithm of the succeeding leaf with a length that is less than the predetermined value of leaf length and $\log L_\lambda$ = the predetermined reference value of leaf length.

*L. P. I.* is estimated according to the following formula: *L. P. I.* = *P. I.* – *i* where *i* = serial number of *i*-th leaf. Example 4 illustrates the procedure to estimate the *L. P. I.* from the data obtained from examples 1 to 3. In the examples the plastochron age of the entire plant is respectively 6, 6.338 and 5.458. In the example 2 the age of leaf 5 in plastochron unit can be estimated in the following way: plastochron index of the entire plant minus the leaf number 5, i.e. $LPI_5 = PI - 5 = 6.338 - 5 = 1.338$.

| **Example 1:** | **Example 2.** | **Example 3.** |
|---|---|---|
| *L.P.I.* for L7 = 6 – 7 = – 1 | *L.P.I.* for L7 = 6.338 – 7 = – 0.662 | *L.P.I.* for L6 = 5.458 – 6 = – 0.542 |
| *L.P.I.* for L6 = 6 – 6 = 0 | *L.P.I.* for L6 = 6.338 – 6 = + 0.338 | *L.P.I.* for L5 = 5.458 – 5 = +0.458 |
| *L.P.I.* for L5 = 6 – 5 = + 1 | *L.P.I.* for L5 = 6.338 – 5 = 1.338 | *L.P.I.* for L4 = 5.458 – 4 = 1.458 |

**Example 4.**

Estimation of *L. P. I.* from the data obtained from examples 1, 2 & 3. In examples 1, 2 & 3 L4, L5, L6 & L7 are the leaf numbers of 4, 5, 6 & 7 respectively. Data partly obtained from httpp://biomath.biology.usu.edu/Fipse/Labs/Plantstruc/plastochron.htm

*L. P. I.* has negative value in leaf primordia that originate next to the reference leaf. In other words when a leaf primordium is shorter than the reference length *L. P. I.* has negative value and when the leaf primordium is longer than the reference leaf *L. P. I.* has positive value. In example 1 the reference value of leaf length (L6) was 10 mm. So the *L. P. I.* is zero. In example 2 L7 is shorter than L6 and L5 is longer than L6. Accordingly the *L. P. I.* for L7 has negative value and L5 has positive value. A few data recorded by Erickson *et al.* to formulate *P. I.* and *L. P. I.* is represented in Table 25.1.

| Date | *Leaf number (n)* 3 | 4 | 5 | 6 | 7 | 8 | 9 | 10 | *P. I.* |
|---|---|---|---|---|---|---|---|---|---|
| July | | | *Leaf length* (mm) | | | | | | |
| 21 | 52 | 54 | 40.4 | **18.6** | **8.5** | 4.1 | | | 6.79 |
| 22 | 55 | 60 | 42 | 30 | **12.7** | **5.8** | | | 7.30 |
| 23 | 57 | 64 | 54 | 45.2 | **19.9** | **9.0** | 4.9 | | 7.87 |
| 24 | 57 | 67 | 62 | 47 | 29.4 | **12.7** | **6.1** | 3.2 | 8.33 |
| 25 | 58 | 70 | 67 | 58 | 40.3 | **17.6** | **7.5** | 4.0 | 8.66 |
| 26 | 59 | 70 | 72 | 42 | 42 | 24.7 | **11.1** | **5.1** | 9.13 |
| | | | *Leaf Plastochron Index* (*L. P. I.*) | | | | | | |
| 21 | 3.79 | 2.79 | 1.79 | 0.79 | -0.21 | -1.11 | | | |
| 22 | 4.30 | 3.30 | 2.30 | 1.30 | 0.30 | -0.70 | | | |
| 23 | 4.87 | 3.87 | 2.87 | 1.87 | 0.87 | -0.13 | -1.13 | | |
| 24 | 5.33 | 4.33 | 3.33 | 2.33 | 1.33 | 0.33 | -0.67 | -1.67 | |
| 25 | 5.66 | 4.66 | 3.66 | 2.66 | 1.66 | 0.66 | -0.34 | -1.34 | |
| 26 | 6.13 | 5.13 | 4.13 | 3.13 | 2.13 | 1.13 | 0.13 | -0.87 | |

Table 25.1

Table showing leaf lengths of a *Xanthium* plant measured on successive days by Erickson *et al.*, (1957, p. 301) to estimate Plastochron Index (*P. I.*) and *L. P. I.* assuming the predetermined reference length ($\lambda$) as 10 mm. One can put the data in equation (1) and find the answer using Microsoft Excel / *fx* calculator for the value of logarithm of any base.

Conventionally a plastochron is defined as the period between initiations of two successive leaves. In a broad sense it might be the period between the successive stages of initiation, development, maturity or intermediate stage of development of two leaves as the stage of reference.

The main purpose of studying plastochron is to obtain many leaves having same morphology and development. When one is sacrificed to obtain data, the other leaf is ready to provide data if required. This is because leaves of same plastochron age grow exponentially.

CHAPTER **26**

# Development of Leaf

In the shoot system of plants the leaf is the principal lateral appendage of stem. A leaf has determinate growth. The foliage leaves are of two types – microphyllous- and macrophyllous leaf (Smith, 1955) and the leaves are also called microphyll and megaphyll respectively. The microphyllous leaf with a few exceptions has a single vein running from base to apex. The vein is not branched. In contrast the macrophyllous leaf has veins that are branched. The venation may be open and closed.

As regards phylogenetic origin there are two theories – Telome- and Enation theory. The former states that the leaves originated from telome trusses. A telome truss consists of telome and mesome. The ultimate dichotomy of fossil plants like *Rhynia, Hornea* etc. is called telome. The stalk that holds the telome or internode below the telome is called mesome. Telome and mesome together constitute telome truss.

Zimmermann in 1930 hypothesized the telome theory and it is well illustrated in the literature of Stewart (1983). They were of opinion that the primitive land plants were much like *Rhynia,* composed of leafless dichotomously branched stem. Later the hypothetical plant branched and the branching was in three dimensions. All branches were more or less equal in length. During evolution some of the branches grew more rapidly than the other branches – a phenomenon called 'overtopping'. The other slowly growing branches orient themselves in one plane – a phenomenon called 'planation'. Later photosynthetic tissues developed between telome and mesome – a phenomenon called 'webbing' (Fig. 26.1). Finally the hypothetical plant consisted of one or several main shoots, each supporting photosynthetic side branches. This type of plant is exhibited by the fossils *Pseudosporochnus nodosis* and *Cladoxylon scoparium* (Stewart).

According to telome theory macrophyllous leaf originated from telome trusses. Telome trusses oriented themselves in one plane and fused with similar lateral members. Thus a leaf with a petiole and blade is formed. The blade shows open dichotomous venation. The lycopodian microphyllous leaf with a single vein, according to telome theory, originated from telome truss. The telome truss

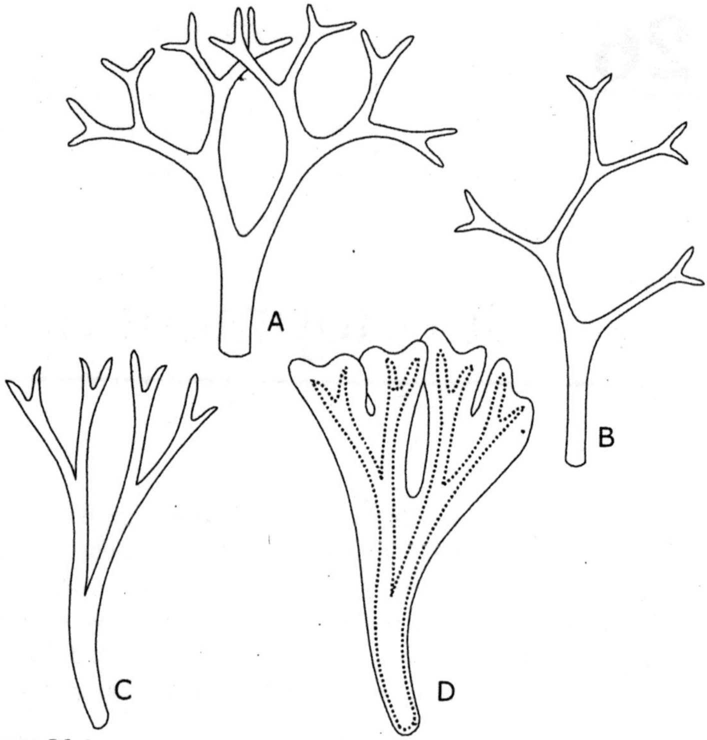


Figure 26.1

Diagram illustrating the probable line of evolution of macrophyllous leaf. A. The hypothetical dichotomously branched leafless original plant. B. Overtopping, i.e. some branch overtopped other branch. C. Planation, i.e. orientation of branches in one plane. D. Webbing, i.e. flattening and fusion of branches and subsequent development of extensive photosynthetic surface between the branches. Adapted after Mauseth, 1988.

reduced to a single telome. One arm of dichotomy had determinate growth that formed the microphyllous leaf. The other arm of dichotomy overtopped. The leaf has a single unbranched vein (Fig. 26.2).

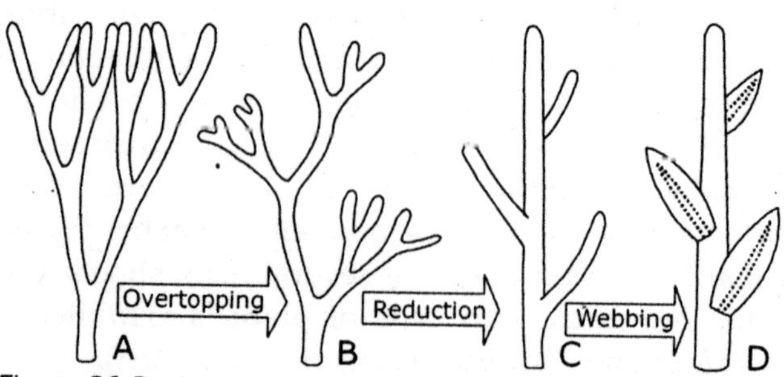


Figure 26.2

Diagrams showing the phylogenetic origin of microphyllous leaf according to telome theory. A. A portion of dichotomously branched leafless shoot. B. A portion of shoot showing overtopping of some branches. C. A telome truss reduced to single telome. D. Webbing to form leaf. Adapted after Smith, 1955.

According to enation theory the microphyllous leaf originated as lateral simple outgrowth (enation) of stem. The fossil plants *Sawdonia* and *Asteroxylon* exhibit the enations. The enations were just minor projections. During evolution the enations increased in surface area for photosynthesis and developed stomata. At first these outgrowths had no vascular connection. Later there appeared vascular strand that ran to the base of leaf. Still later there appeared vascular connection that ran to the tip of leaf (Figs. 26.3 & 26.3A). The microphyllous leaf is about 5 to 10mm in length and it is their maximum length. Stewart published that only Lycopodiales and the fossil group Drepanophycales have them.

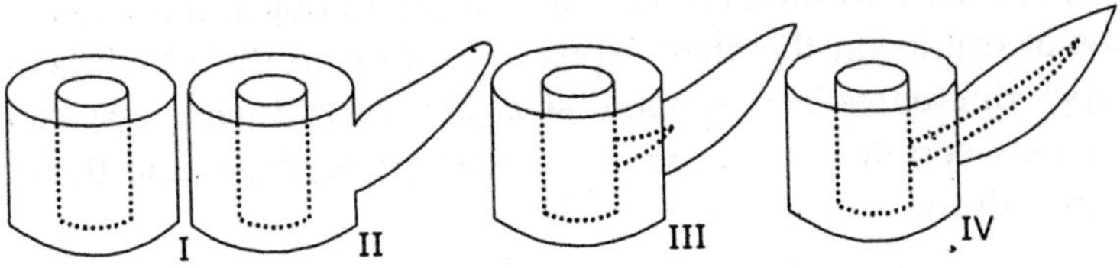


Figure 26.3

Diagram illustrating the enation theory of phylogenetic origin of microphyllous leaf. I. A portion of smooth surface of leafless shoot. II. A portion of shoot with enation without vascular connection, ex. *Psilophyton princeps*. III. A portion of shoot showing vascular connection that ran to the base of leaf, ex. *Asteroxylon*. IV. A portion of shoot showing vascular connection that ran to the leaf tip, ex. *Arthostigma*. (Adopted after Smith, 1955).

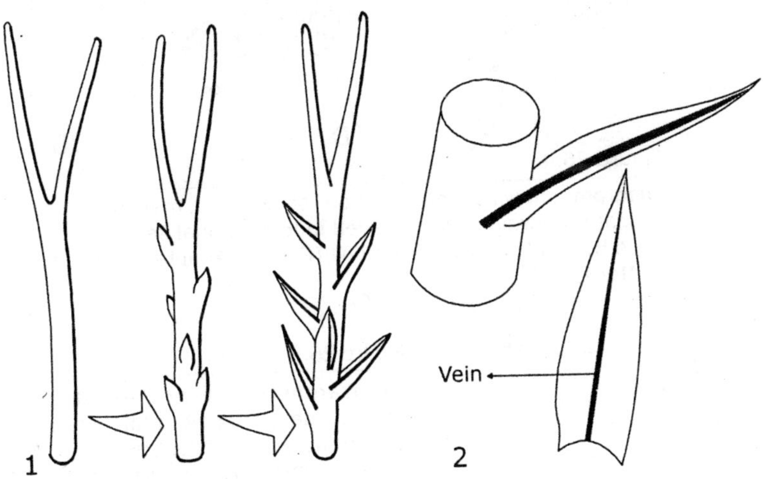


Figure 26.3A

1. A diagrammatic summary of enation theory on the evolution of one-veined leaves from leafless branches. 2. Leaf of *Lycopodim* with a single vein. Adapted after Meyer *et al.*, 1965, p. 270.

Whatever may be the interpretations of phylogenetic origin, a leaf like stem and root shows the dermal, ground and vascular tissue systems. In centric leaves the dermal system consists of epidermis. In dorsiventral and isobilateral leaves the dermal system consists of upper- and lower epidermis. The ground tissue comprises mesophyll tissue. In dorsiventral leaf mesophyll is differentiated into palisade tissue and spongy parenchyma. The former occurs below upper epidermis and the later lies between the lower epidermis and palisade tissue. Vascular tissue system comprises vascular bundles that are embedded in the mesophyll tissue and form the veins. Each of the tissue systems has separate meristem. The meristems according to position are called apical-, adaxial-, marginal-, plate- and intercalary meristem. The above meristems occur in a leaf primordium and can be observed in longitudinal and transverse sections (Figs. 26.4 & 26.5).

Leaf primordia are formed on the flanks of shoot apical meristem. They have definite sites of origin on the shoot apex according to the phyllotaxis of plant. The shoot apex fluctuates in shape and size (Fig. 25.1) during a plastochron- the time interval between the formation of one leaf primordium and the initiation of next leaf primordium.

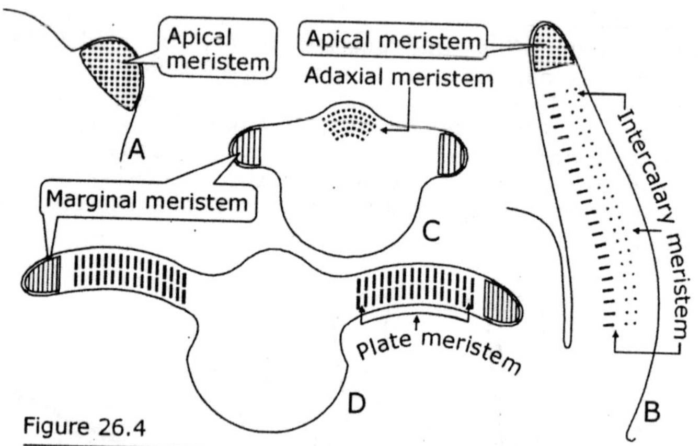


Figure 26.4

Diagram showing the position of various meristem of leaf primordium, as revealed in longitudinal sections (A & B) and transverse sections (C & D). Adapted after Cutter (1971).

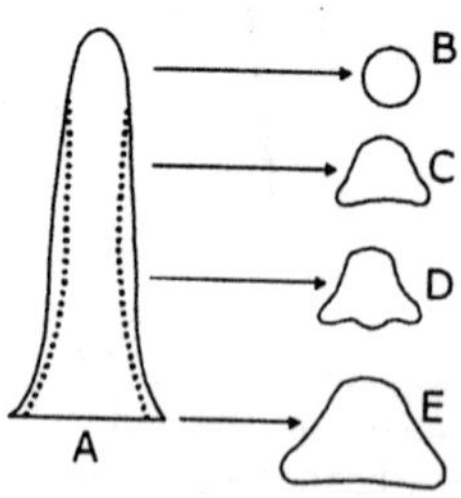


Figure 26.5

Diagrammatic representation of the outline of longitudinal section (A) and transverse section (B, C, D & E) of leaf primordium in successive stages of development. The dotted lines indicate development of lamina. Adapted after Cutter, 1971.

Schmidt 1924 postulated that the shoot apical meristems of gymnosperm and angiosperm consist of two distinct zones—tunica and corpus. Tunica is the surface layer and it encloses the corpus. The tunica may consist of one layer (monostratose) or two to four (multistratose), the former being most common. The corpus always consists of many layers. These two zones—tunica and corpus are distinguished on the basis of orientation of cell division. The chief mode of cell division of tunica is anticlinal irrespective of monostratose or multistratose. In anticlinal division the cell walls are laid down perpendicular to the surface of tunica. Anticlinal division causes the tunica to grow as a sheet but not in thickness. In contrast, corpus in addition to anticlinal division, has division with a periclinal wall, i.e. the cell wall is laid down parallel to the surface of tunica. So the cells of corpus can divide in any direction; as a result it can grow in three dimensions in contrast to tunica that can grow in two dimensions only.

The initiation of leaf commences with a lateral protrusion on the apical meristem. This is leaf buttress. Later leaf primordium develops on leaf buttress. The initial lateral protrusion results from periclinal division. The two growth zones—tunica and corpus variously participate in the development of leaf primordium. In monostratose tunica the first divisions to form leaf occur in corpus (e.g. *Scrophularia nodosa*). In multistratose tunica (e.g. *Vinca minor*) the initial divisions occur in the innermost layer of three-layered tunica. Tunica and corpus both are involved in the initial divisions in the development of *Acacia* phyllode though tunica is multistratose. Generally in most dicotyledons the first periclinal divisions occur in the subsurface layers, i.e. cells of one or more layers below. The surface layer in dicotyledons does not form the inner tissue of a leaf. It divides by its characteristic anticlinal divisions along with the growth of leaf primordium, which originates as a result of periclinal divisions of inner tissues. Protoderm of a young leaf differentiates from the surface layer.

In gramineae and many other monocotyledons the tunica is monostratose and some members have two-layered tunica. Whatever may be the number of layers the tunica divides periclinally. As a result the main portion of leaf primordium develops from cells situated at surface and subsurface layers of the shoot apex.

Variations are also noted in the initiation of leaf primordia in gymnosperm. In *Zamia* and in many other conifers periclinal divisions occur in the surface and subsurface layers of the apex. In *Taxodium distichum,* during the initiation of leaf primordium, periclinal divisions occur in the subsurface layer of the apex. The surface layer divides with anticlinal walls.

Cunninghame *et al.* (1986) studied the relationship between the distribution of periclinal divisions in the shoot apex and leaf initiation in *Pisum* and *Silene* using light and electron microscope. They concluded that periclinal divisions in the tunica and corpus, mainly in the latter cause the formation of leaf primordia. With the aid of periclinal cytochimeras (the term periclinal cytochimera has already been introduced in the discussion of tunica-corpus theory in the chapter 7), Dermen (1947, 1951) was able to detect in the apical meristem three layers of cells in *Vaccinium* (cranberry) and *Pyrus malus* (apple). Biseriate tunica and the peripheral layer of corpus composed the three layers. All the three layers take part in the formation of leaf. The surface layer of tunica forms the epidermis of leaf entirely by anticlinal divisions. Second layer of tunica and the peripheral layer of corpus contribute the rest of tissues of leaf. These tissues later differentiate into photosynthetic tissues (mesophyll) and vascular tissues (vascular bundle). So the cells of tip and margin of leaf are the derivatives of surface layer of tunica. The tissues, present at the central part of leaf, i.e. between the two epidermises, are the derivatives of the inner layer of tunica and all the layers of corpus.

The periclinal divisions of superficial tunica in monocotyledons and the periclinal divisions of the subsurface layers of dicotyledons form the leaf primordium. Further divisions of adjacent and underlying cells usually follow the initial periclinal divisions. As a result the leaf primordium is elevated above the surface of the apical meristem. Then several meristems, mentioned previously, originate in the leaf primordium and function either simultaneously or sequentially, and contribute tissues in the growth of a leaf primordium to form leaf. The growth includes apical growth, intercalary growth, marginal growth, adaxial growth and lateral growth.

Apical growth occurs by the activity of apical meristem. In vascular cryptogams an apical meristem is formed at the tip of leaf primordium and it has prolonged activity. In ferns like *Lygodium* the apical meristem can function for an extended period (Muellar, 1982). In seed plants, though there occurs the apical meristem, the apical growth ceases relatively early. Exceptions are noted in pinnate leaves of *Gaurea* (Meliaceae) where prongled apical growth occurs like *Lygodium* (Steingraeber *et al.*, 1986). The development of young primordium and the subsequent extension of leaf occur by cell division throughout the young primordium. Growth of leaf ceases first at the tip and last at the base. The majority of leaf growth in length is brought about by intercalary growth. Intercalary meristem occurs at the base of leaf primordium of most monocotyledons. The intercalary meristem present in the developing leaf primordium of *Nuphar lutea* causes the elongation of petiole and thus the lamina is raised above the water surface (Cutter, 1978). The leaf of *Narcissus* elongates due to intercalary meristematic activity of leaf primordium. In this leaf apparently there is no apical growth.

Leaf primordium manifests apical growth during second plastochron; ex. *Oryza sativa* (Kaufman, 1959). At this stage the leaf primordium appears as short cone or just a small protrusion projecting from the apical meristem. Lyndon (1968) estimated the increase in length of leaf primordium of *Pisum* and it was about 60 μm in one plastochron. During the elongation of leaf primordium it is frequently observed that the young primordium bends over the apical meristem. This is due to greater vertical growth on the abaxial side of leaf primordium, i.e. side away from the shoot apex.

Greater growth also occurs on the two lateral sides of simple, young and dorsiventral leaf primordium. As a result small bulges or outgrowths appear on the two lateral sides. The outgrowths form wing-like strips on the margins of leaf primordium and this growth pattern is usually referred to as marginal growth. Due to marginal growth the lamina of a leaf is formed. The marginal growth does not occur in the basal portion of leaf primordium from where the petiole is destined to develop. In *Oryza* (Kaufman, 1959) marginal growth is not distinct in leaf primordium and in *Narcissus* (Denne, 1960) there is no distinct marginal growth in the leaves.

Marginal growth occurs by the activity of marginal meristem. Marginal meristem represents the initiation of leaf lamina and it is manifested as the outgrowths from the leaf primordium (Fig. 26.5). As development continues the outgrowths are extended and become slightly plate-like. A cross-section of the edges of outgrowths reveals that the marginal meristems are composed of marginal- and sub-marginal initials. Foster (1936) illustrated the future destiny of the derivatives of the initials (Fig. 26.6).

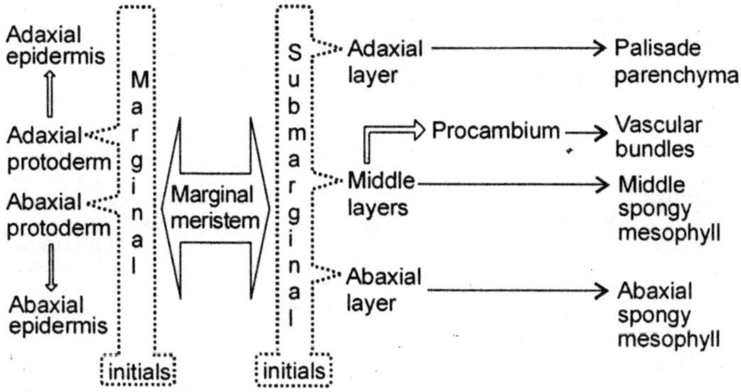


Figure 26.6

Diagram illustrating the cell lineages involved in the development of leaf of *Nicotiana tabacum*. Adapted after Fahn, 1997.

In young lamina the outermost cell layer of margin is the marginal initial. The marginal initials form the protoderm. Abaxial and adaxial protoderm layers both are the derivatives of marginal initials. Generally in angiosperms the marginal initials divide anticlinally, i.e. cell walls are laid down perpendicular to surface of initials. So the protoderm layers grow as a sheet, but not in thickness. Foster (1937) and Sharman (1942) reported the occurrence of periclinal divisions in marginal initials of bud scales of *Rhododendron* and in certain monocotyledons. In *Prunus, Pelargonium, Dracaena* and *Veronica* the leaves are variegated. In these leaves the marginal initials form protoderm. Protoderm divides periclinally and forms the white margins of leaves (Renner *et al.*, 1942). The marginal initials produce the whole lamina in *Eubrachion* of the family Viscaceae (Bhandari, 1969) and *Daphne* of the family Thymelaeaceae (Hara, 1957).

The cells situated just below the marginal initials are the sub-marginal initials. These initials divide in various planes and donate cells to the inner layers of developing lamina. The sub-marginal initials, after divisions, establish the number of layers of cells in the lamina. The future destiny of the derivatives of sub-marginal initials is much variable and a few are illustrated in Fig. 26.6-26.9. In *Xanthium* the marginal growth is initiated when the leaf primordium is at about 0.22 mm long. The activity of marginal meristem is continued for about 23 days. As a result six layers of cells are established in the leaf (Maksymowych, 1959). The young leaf primordium when observed in its cross-sectional view reveals

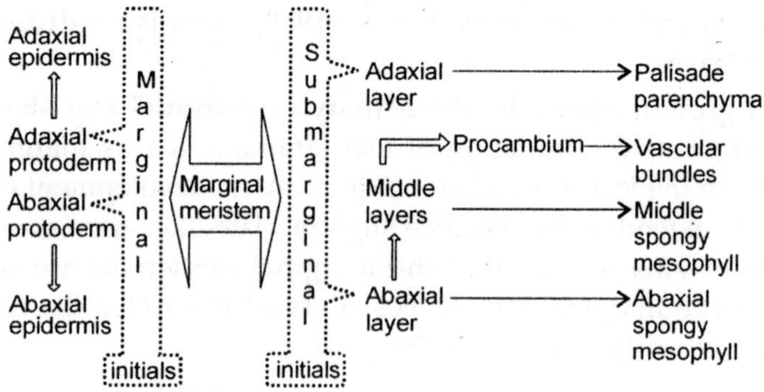


Figure 26.7

Diagram illustrating the leaf histogenesis of *Carya*. Adapted after Fahn, 1997.

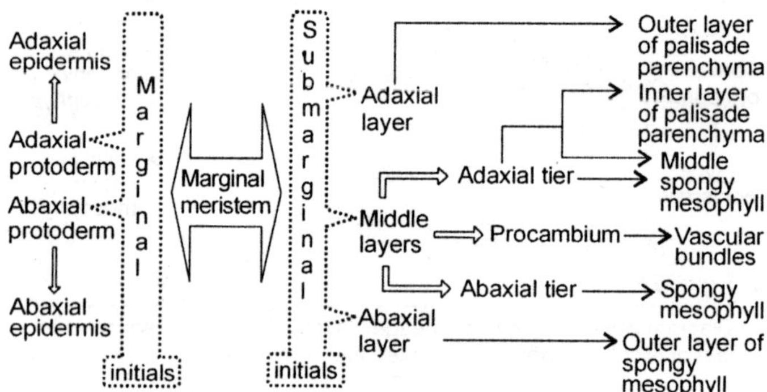


Figure 26.8

Diagram illustrating the leaf histogenesis of *Pelargonium*. Adapted after Fahn, 1997.

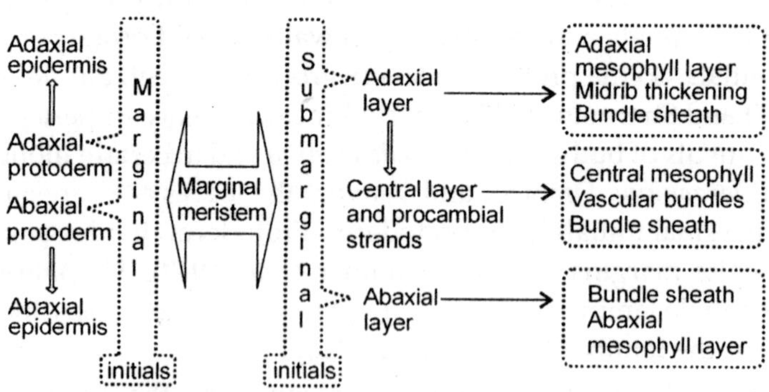


Figure 26.9

Diagram illustrating the leaf histogenesis of *Oryza*. Adapted after Fahn, 1997.

the number of cell layers and the occurrence of marginal meristem at the edges of the flanges. The flanges consist of the tissues that project from the future midrib.

The leaf marginal meristems with marginal and sub-marginal initials resemble shoot meristems with tunica and corpus. The marginal initials are equivalent to tunica whereas the sub-marginal initials represent the corpus. This analogy is based on the patterns of cell division. The marginal initials and tunica generally divide anticlinally. The sub-marginal initials and corpus divide in various planes.

The leaf primordium of many dicotyledons has adaxial growth. This growth is due to the activity of adaxial meristem (Foster, 1936), Fig. 26.4C. Adaxial meristem originates as a strip of cells below the adaxial epidermis. The chief mode of division of adaxial meristem is periclinal. The cells, thus formed, contribute to the thickness of leaf especially to growth in thickness of the midrib and petiole. The phyllode of *Acacia* and the leaf of *Acorus* are flattened laterally instead of dorsiventrally. The adaxial meristem in these plants is extremely active. As a result the laterally flattened lamina-like structure is formed. Moreover in both genera marginal growth is suppressed. It favours the activity of adaxial meristem and so the foliar organ extends in median plane.

In the early stages of development the activity of marginal meristems establish a certain number of layers of cells in the leaf lamina. These cells divide anticlinally, i.e. at right angles to the surface thus forming 'plate' of cells. These cells are referred to as plate meristem (Fig. 26.4D), the activity of which causes the expansion of lamina laterally. The whole lamina functions as a plate meristem and the derivative cells contribute to lateral growth. The leaf marginal meristems do not function indefinitely. So the leaves have a limited and determinate width. The marginal meristem forms definite number of layers of cells. These derivative cells are the plate meristems and consist basically the entire lamina. As the cells of the lamina divide anticlinally the layer of cells grow in surface area but not in thickness. Though the cell division in plate meristem is predominantly anticlinal, exception is noted in *Xanthium* where some periclinal divisions do occur thus forming additional cell layers and causing growth in thickness of the leaf.

In the developing lamina the marginal meristem forms a certain number of layers of mesophyll cells. Usually the adaxial layers of mesophyll give rise to palisade tissues. The abaxial layers of mesophyll form the spongy cells. The middle layer of cells situated between adaxial and abaxial layers form the procambium of veins. In some cases spongy cells may develop from parts of middle layer. It is to note that the layer, which is destined to form procambium varies according to plants.

CHAPTER **27**

# Abscission

Abscission is a process that initiates the removal of a part of plant following the formation of a protective layer or scar tissue that prevents the entry of pathogen and protects the inner tissues from desiccation.

Plants are shedding organisms. Trees shed inactive xylem internally as heartwood. They shed organs externally. Example: Root, branch, twig, bark, fruit, flower, bud scales and trichomes. Leaf abscission is a very common phenomenon. In annual herbaceous plants abscission is confined to bracts and floral parts like calyx, corolla and androecium. Perennials particularly the woody ones shed leaves, fruits, floral parts, branches etc. Some plants like *Araucaria, Salix babylonica* etc. abscise shoots, a process called cladoptosis (from the Greek clados = branch, ptosis = fall). Leaf abscission is the best example of shedding in trees.

Evergreen trees shed leaves in any season. Deciduous trees shed leaves in autumn, i.e. at the approach of winter. Gymnosperm and many dicotyledons show this periodic defoliation. Some changes occur at the base of shedding parts, which ultimately cause the separation. The exposed part is protected by the formation of protective layer. The protective layer, and the region of separation, i.e. abscission layer are collectively termed as abscission zone. This zone initiates the process of abscission (Fig. 27.1). A single abscission zone occurs in dicot leaves. A fruit and monocot leaf may have more than one abscission zone. In palms the lamina is abscised by the first abscission zone. The second abscission zone abscises the leaf base. Abscission zone is designed and pre-positioned to facilitate shedding.

In leaves, abscission zone is formed at the base of simple leaves and compound leaves. In compound leaves this zone may also be formed at the base of petiolule of leaflet. In sessile leaf, this zone may be the only remnant of petiole. The abscission zones in the petioles may have different colour and groove or constriction by means of which they may be differentiated from the rest. The cells in this zone are of same type as found in the other parts of a plant. Sclerenchyma cells are few or weakly developed. Collenchyma cells are absent. This zone can be between 15 to 40 cells wide. The separation layer is quite small. Parenchyma cells are thin walled, small and compactly set with no intercellular spaces. A few cells may have remained in a cell division phase. These cells divide and the derivatives also prepare themselves for the abscission process. The abscission zone may not be active as soon as it is formed. It remains quiescent. It becomes active, when needed.

The cells in the abscission zone contain abundant starch grains. The cells are dense with cytoplasm and organelles. Each parenchyma cell is living and produces the materials needed for abscission. Dictyosomes play a major role in the development of abscission zone. Gilliland *et al.* (1976) observed that dictyosomes transport digestive enzymes for the destruction of cell walls of pedicel. It is to be mentioned that dictyosomes transport materials for the construction of cells of pedicel. Dictyosomes, that transport wall materials at the time of construction, transport digestive enzymes for the destruction of same cell walls of pedicel at the time of abscission.

The abscission layer or separation layer is formed either earlier in the season or as the leaf approaches the time of falling from the plant. In the former case the abscission layer remains inactive until it is needed. It normally happens in those plants that grow in more variable climatic conditions. In the growing season if bad conditions arise, the trees shed their leaves. The abscission layer consists of vascular bundle and parenchyma. The xylem, in most cases, consists of tracheids only. Fibres are few. When vessels are present, they are either narrow or discontinuous, e.g. *Populus deltoides*. The parenchyma is thin walled and is oriented transversely across the petiole. Some changes occur during the formation of abscission zone. Prior to abscission, compounds like proteins, fats etc. are exported from the leaf. Chlorophyll production is stopped. The remaining chlorophyll breaks down and the nitrogen thus released is transported into the stem. Sugars, amino acids, minerals, nucleic acids etc. are transported from leaf to stem. The vessels gradually become nonfunctional due to deposition of gums and formation of tylosis.

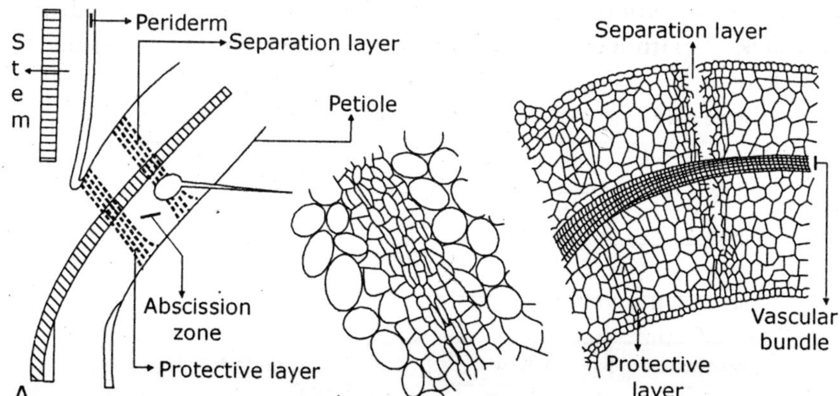


Figure 27.1

Abscission of leaf. A. Petiole showing abscission-zone in l. s. view (diagrammatic). B. Magnified view of separation layer. C. A portion of petiole magnified to show separation and protective layer.

The growth regulator auxin is produced in the leaf and is transported to the leaf base. As long as auxin is produced in the leaf and is transported across the abscission zone, cells of the abscission layer remain inactive. As the vessels gradually become inactive, leaves do not get adequate nutrients to produce auxin. As a result abscission layer begins to develop. Moreover decline in auxin production and transport accelerates the activity of ethylene—the only gaseous plant hormone. Ethylene stimulates the development of abscission layer (Moore, 1994). Abscisic acid promotes abscission. Indole-3-acetic acid inhibits abscission. When the levels of auxin including Indole-3-acetic acid decrease in a leaf the principal accelerators of abscission, abscisic acid and ethylene become active.

The abscission process begins with the decline of auxin production and formation of abscisic acid and ethylene. In the cells of abscission layer the wall degrading enzymes like pectinase and cellulase are produced. These enzymes cause the degradation of the middle lamella, the intercementing material between the cells, and cause hydrolysis of cellulose in the walls. The calcium bridges across the cell wall materials are removed. The cells in the abscission layer disconnect from each other. Starch, present in the cells, is converted to osmotically active substances and as a result cells absorb water. At turgid condition the cells expand. As cells expand a fracture line begins to develop in the separation layer. The fracture line gradually grows by the pushing and pulling of expanding cells, gravity and wind tugging. The fracture line follows the path of middle lamella between cells (Fig. 27.2). Normally, a single fracture line occurs in the abscission layer. The xylem elements are not degraded. With the formation of fracture line the separation layer disappears. The xylem elements only remain to hold the leaf. The leaf ultimately falls to the ground when the xylem elements break. This happens when the leaf sways by wind and the weight of the leaf exceeds the strength of vascular element. Animal actions also can break the xylem elements to allow leaf fall.

After the fall, tissues on the tree side are exposed. The vascular tissues are clearly visible and the wound is more or less smooth. They are subjected to moisture loss and attack of pathogenic organisms. The protective layer, formed in the

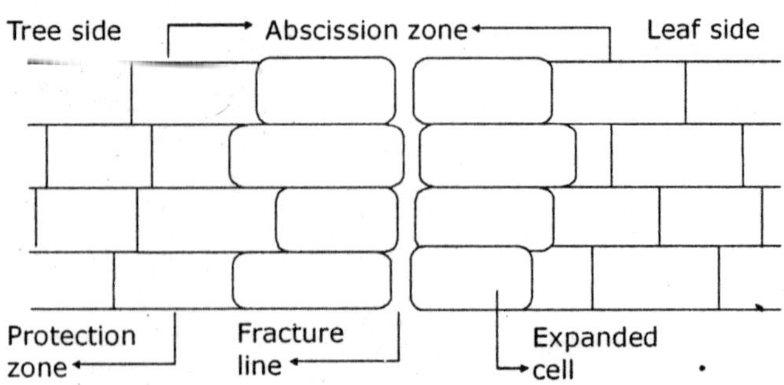


Figure 27.2

Diagram showing cells in the abscission zone of a leaf base and the fracture line. The expanded cells disrupt the cell to cell connections.

abscission zone, protects the tissues from desiccation and the entry of fungal spores, bacteria, pests and other pathogenic organisms. The protective layer is formed in the abscission zone next to separation layer on the tree side. This layer consists of a group of cells some of which are able to undergo cell division. These cells divide and form a sealing layer. After the completion of wall formation, the walls become filled with gum, suberin, lignin etc. In this region tylosis is formed in the vascular elements. Thus this layer forms an impenetrable corky barrier and protects the inner tissues on the tree side. The cells, situated on leaf side and exterior to protective layer, die.

The photoperiod and temperature are among the factors that cause the formation of abscission layer. Long night, frost, below freezing air temperature activate the formation of abscission layer. Leaf senescence is dependent on abscission, which is triggered by plant hormone. Abscission is the last step of leaf senescence. Senescence, also called aging, refers to the process that naturally terminates in the death of organ. In this process the quite inefficient leaves that cannot perform photosynthesis because of shading by the upper portion of shoot are shed. In the senescent leaf catabolism exceeds anabolism. In these leaves there is massive export of soluble metabolites to the stem. Frosts and heavy freezes at night may damage the living tissues of leaves. In this case the plants use the senescence sequence to remove the valuable resources from leaves to stems before they die and shed.

In autumn, when the duration of sunlight decreases, the dark period increases and cold weather sets in, the deciduous trees start shedding their leaves en masse. This phenomenon is so striking that it gives autumn another name – 'autumn fall' or simply 'fall'. Before fall, leaves change their colour. This is due to absence of chlorophyll. The colour of carotenoids is expressed as the green chlorophyll previously masked it. A wide spectrum of leaf colour ranging from yellow to red occurs. This is due to the presence of yellow xanthophylls and orange carotene. Microscopical examination of autumnal leaves shows the pigments. Chemical analysis detects many toxic chemicals like heavy metals in the shedding leaves.

Opinion varies regarding the fall of leaves of deciduous trees in fall. The common belief is that by this phenomenon plants conserve moisture in winter. In this season the ground water freezes, the plants do not get adequate water; so water conservation becomes necessary. By the fall the broad leaves of deciduous trees will not be able to give off large amount of moisture through transpiration. Opponents of this view cite the example of those plants, which do not shed their leaves at all. Moreover, the leaves of evergreen trees, which are anatomically adapted to survive extremes of cold temperature, do shed their leaves at the different seasons. Another popular view is that the leaves are shed because of competition for nutrients. But studies on abscission in the leaves of tropical rain forest reveal that they are regularly shed from those plants that are in the richest soil. The other view says that the leaves are shed to prevent shading of those of lower down and when they are damaged. But in palms the leaves are shed as it grows.

Naturally question arises why do leaves fall in fall? The answer is now emerging—a plant sheds its leaves to relieve the accumulated wastes. Excretion is one of the essential properties of all living organism. Like animals plants do not possess any excretory mechanisms. The fall is the excretory mechanism of a plant. Moreover by fall the plant conserves the valuable resources. It was previously mentioned that many compounds are transported to the stem before leaf fall. Leaf abscission is highly ordered series of events, which prepare a plant for a resting period in winter. It helps trees to survive.

It is now revealed that apart from photosynthesis and transpiration, leaves have another function, i.e. excretion (Ford, 1986). This property is now utilized to purify contaminated soil. A new line of study has developed called phytoremediation (also known as bioremediation or phytobioremediation). It is a process to purify the contaminated soil by plants. In this line of study, the deciduous plants that are metal tolerant are planted in the contaminated soil. The plants grow and absorb the toxic metals. Thus the soil gets purified. The toxic metals are excreted from the plant through the process of abscission.

To sum up, the process of abscission is now considered as a strategy for survival to the trees. It was previously mentioned that during the process the trees conserve the valuable resources needed to them. In abscission, the leaves are abscised passing through the senescence process. So they are designed to be disposable. During the 'fall' leaves change colour. The great diversity of colours makes the trees most popular for enjoying autumn colour. The fall of leaves, to people, is the beauty of 'fall'.

CHAPTER **28**

# Healing of Wounds

Plant organs often are wounded by adverse calamities of nature, grazing animals and human interference etc. As a result the living cells are exposed and they are subjected to desiccation and attack by bacteria, fungi and other pathogenic microorganisms etc. Higher plants have their own arrangement of healing of wounds.

There are different methods of healing of wounds. When the wound is superficial the exposed cells dry up and eventually die. Suberin deposits in the cells situated below the wound and thus a protective layer is formed. Some plants have laticifers and secrete latex. Latex covers the superficial damaged surface where it coagulates and thus forms a protective cover over the wound.

In plants having deep wounds callus or wound tissue is formed. Callus is composed of masses of soft parenchyma cells. It develops over and below the damaged surface of stem and root. It is formed by the division of parenchyma cells that occur either in the phloem, in the cortex or in the vascular rays. Most frequently cambium donates callus. During the formation of callus the cambium cells proliferate and produce masses of parenchyma. The peripheral cells of callus become suberized thus sealing the exposed surface of damaged tissue. Periderm (also called wound periderm) develops in the callus. Wound periderm forms a protective bark that inhibits the entry of pathogen. Below the periderm a recognized cambium occurs and it forms new vascular tissues normally. Suberinization seems to create an internal condition that is favourable for periderm formation.

Experiments with potato tuber and sweet potato root reveal that suberinization is preceded by an accumulation of phenolic substances, particularly of chlorogenic acid (Esau, 1965). It also reveals that lignification is associated with wound healing. In potato tuber wounding stimulates mitosis and the wound healing is accompanied by suberinization and wound cork formation. Increased temperature and high humidity are favourable for wound healing. Type of soil determines the number of layers formed in the wound cork of potato. It is observed in potato tuber that during wound healing suberin or cutin deposits over the damaged surface. Thus the surface is sealed and the formation of phellogen follows. Experiment with potato tuber reveals that covering of cut surface is necessary for phellogen formation. In an experiment the cut surface was covered with paraffin wax. It was observed that phellogen formed without prior suberinization. Later researches reveal that in wounded potato tubers rapid increase in endogenous gibberellins like substances occur (Cutter, 1978).

In trees where wound is deep, the uninjured living cells below the wound are stimulated to divide. As a result callus is formed. It overgrows the wound. Phellogen appears on the peripheral side of callus and divides to form wound cork. Callus with wound cork appears as swelling over the stem surface. Such structure is usually referred to as knot.

In trees where the wound is due to pruning of branches callus is formed on the circumference of wounds (Fig. 28.1A, B). This happens in the early growing season. Production of secondary xylem continues in the uninjured surrounding tissues by the cambium. Annual ring is also formed in undamaged area around the scar. The base of branch dies, shrivels up and loses the capacity of growth. As more and more growth rings are formed the base of branch gets buried more and more deeply. The cambium layer becomes continuous over the scar and thus the wound is completely covered by the secondary xylem (Fig. 28.1C). In a timber the base of branch appears as knot.

In case of deep wound where the cambium is damaged, normal callus is formed in the damaged area provided the uninjured tissues are protected from

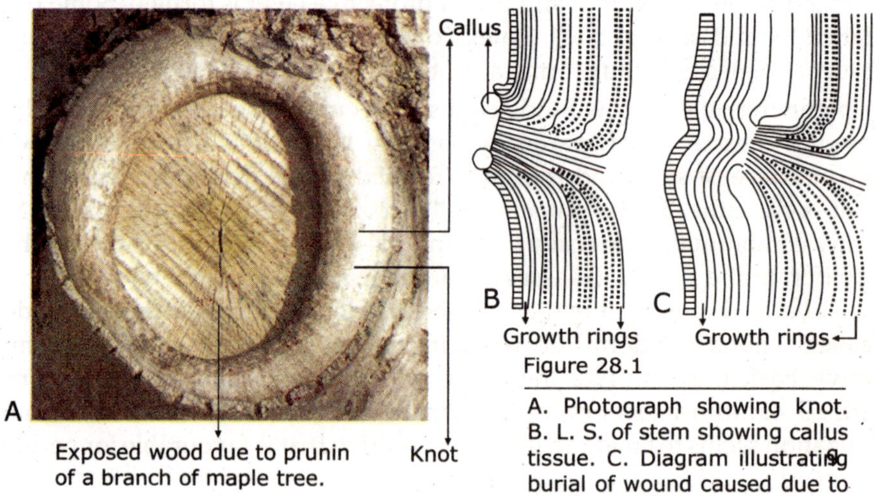


Figure 28.1

A. Photograph showing knot. B. L. S. of stem showing callus tissue. C. Diagram illustrating burial of wound caused due to pruning. Barks are hatched.

drying out immediately. In the callus a new cambium develops and it unites with the uninjured cambium. Thus the cambium becomes continuous and the production of secondary tissues continues.

The differentiation of cambium in callus and its subsequent fusion with neighboring cambium form the basis of grafting. In grafting the junction of stock and scion is filled up with callus. Within callus cambium differentiates and it fuses with the cambia of stock and scion. Thus a continuous cambium is formed and it donates normal vascular tissues.

CHAPTER **29**

# Ecological Plant Anatomy—Adaptive Features

There exists a close relationship between plant structure and the physical environment in which the plant grows. The scientific study of relationships that plants have with each other and with the environment, in which they live, is known as plant ecology. Ecologists study ecosystem. An ecosystem is natural unit and dynamic whole composed of plants that live in it and a natural environment. Here a continuous interaction occurs between plants and between plants and the non-living physical factors of environment upon which they all are interdependent.

In an ecosystem there exists a definite habitat. A habitat is characterized by temperature, humidity, illumination, soil conditions and water supply. A habitat, by tradition, regarded as normal and favourable for plant growth where there exist good water and nutrient supply and a medium temperature range. Mesophytes occur in this habitat. Plants also live, compete and survive under conditions, which are extreme in one respect or another in contrast to normal habitat. These plants thrive because of a combination of several morphological, physiological and anatomical adaptations. Thus they differ from mesophytes with regards to morphology, physiology and anatomy. The adaptations are so closely interrelated that the anatomy alone cannot be dealt with without considering morphology and physiology. The adaptations are thought to be of ecological benefit. The adaptations in many plants are genetically labeled, often dominant and form the diagnostic character of the family, genus and species.

## Xerophytes

Xeric habitat characterizes xerophytes (xero = dry, phytes = plants). Xerophytes evolved to survive in an ecosystem where there is deficiency in available water. This includes the areas that are subjected to drought like deserts where low rainfall is the norm. The areas may also include physiologically dry soil (not physically dry) where uptake of water becomes difficult due to salinity in soil water. Therefore xerophytes have evolved a wide variety of adaptations. These adaptations primarily aim to limit water loss, conserve water and obtain water as much as possible from the environment.

Xerophytes have developed the following three main adaptive strategies to survive—drought avoidance, drought tolerance and succulence. Each of the adaptations is different but effective against extremely low water availability, desiccating winds, and high and fluctuating temperatures. Different types of plant have these adaptive strategies and among them there exist little morphological, physiological and taxonomical relationship.

The drought avoiding xerophytes are ephemeral-annuals. They live only for a short time period. Germination, growth, flowering, fruiting and dying occur after a heavy rain. These plants tide over drought by the dormant seeds. The seeds germinate only after a heavy rain. Growth occurs and leaves are formed. The leaves remain alive so long water is available. They wither away after the soil dries out. So these plants remain leafless most of the time. These plants have large and brightly coloured flowers to attract pollinating insects. After fruiting the seeds enter into deep dormancy and thus these plants avoid drought. The drought avoiding plants do not need any xeromorphic adaptations because they have adequate water during their short period of activity. Mention may be made of the following ephemeral-annuals: *Dithyrea californica*, *Atriplex pseudocampanulata* (completes its life cycle within 70 days), *Wahlenbergia communis* (completes its life cycle within 36 days), *Abronia villosa*, *Castilleja chromosa* (Schrophulariaceae), *C. exserta* etc. The ephemeral-annuals, more accurately, are to be referred to as simply 'ephemerals' because most of them complete the entire life cycle within few months, some in just weeks.

The drought tolerant xerophytes are perennials. They have structural modifications that help them to adapt to the environment.

The roots are very well developed and spread out over a large area to collect maximum water. They may be laterally extensive and shallow. They may be radially elongated. The roots may also be long, vertically deep, e.g. *Acacia*. Most of the shallow and radial roots are within three feet of the surfaces, e.g. Cactus. These roots allow quick acquisition of large quantities of superficial water when it rains. Cacti store water both in root and stem. These plants have adaptations by means of which they can survive years of drought on the water collected from a single rainfall. *Prosopis*, a drought tolerant plant, develops long vertical deep root systems. The vertical roots may be as long as 80 feet. These roots draw water from deep underground reservoirs. Some cacti develop special roots after rain to absorb more water. Many xerophytes have both shallow and deep taproots and respectively allow collecting surface water and absorbing water from water table or directly above or below it.

Many xerophytes have root hairs extended to the tip to increase water uptake. The root hairs of many xeric grasses, e.g. *Andropogon foveolatus*, *Panicum turgidum* etc. have rhizosheaths. Rhizosheaths (Fahn, 1997) are also known as sand grain root sheaths. The root hairs secrete mucilage where the moving sand grains become attached. The attached sand grains form a sheath around the root hairs. With the increase in number of root hairs the sheath covers the whole root. The following functions are attributed to rhizosheath: (1) it is equivalent to cork layer that is formed in dicotyledonous root and prevents the loss of water from inner tissues, (2) it accelerates water absorption as it contains mucilage that has high absorptive power for water and (3) it is associated with nitrogen fixation (Wullstein *et al.*, 1979). The cortex is usually thin (Fig. 29.1) and therefore there exists small distance between xylem and soil water. The endodermis has wider casparian bands. So endodermis is more efficient in its function of radial diffusion of water from stele to the cortex. The xylem is well developed and as a result rapid transport of water occurs following absorption. Xerophytes have evolved methods by means of which they maintain a high osmotic pressure in root tissues. This enables them to increase water absorption. Many cacti accumulate hydrophilic colloids in their root cortex. This reduces the water potential of the root-tissues and thus accelerates the absorption of water by osmosis.

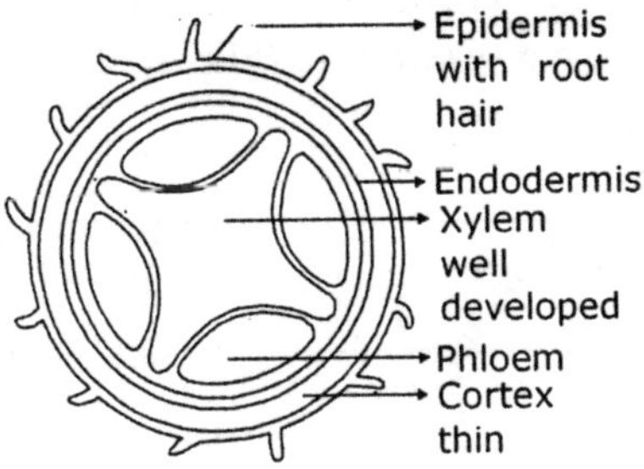


Figure 29.1

Diagrammatic illustration of transverse section of a root of xerophyte.

Some members of Juncaceae and many of the Restionaceae show hydromorphic features in the roots and xeromorphic features like abundant sclerenchyma in stems. The roots show air cavities in the cortex like hydrophytes. This remarkable combination of dual adaptations of hydromorphic roots and xeromorphic stems is probably beneficial to the plant because the stems are sometimes exposed to strong drying winds when the roots become too cold to deliver enough water to meet evaporation losses (Cutler, 1978).

One of the adaptations among the xerophytes is surface reduction. The leaves are very much reduced and absent in many xerophytes. The functions like transpiration and photosynthesis of leaf are surrendered to stem. Such stems grow by marginal growth like leaves and acquire the structure of leaves (ex. *Rhipsalis, Ruscus* etc.). Such stems are called phylloclades.

The bulbous habit is often associated with xerophytes. In many plants like *Narcissus, Scilla, Tulipa, Haemanthus* etc. leaves and flowers develop for a limited period in each year (Cutler, 1978). The plants survive through underground stems. Swollen underground stems occur in many species of Asclepiadaceae. Rhizomes occur in the species of *Iris*. Corms occur in *Crocus, Watsonia* etc. The above plants grow actively when water is available. The aerial organ like leaf remains alive so long water is available and withers away when the water is used up. So the aerial organs show little adaptations to xeric condition.

Internal adaptations in xerophytes include—mechanisms to provide for the storage of water and to develop mechanical tissues to resist collapse and tearing on drying. The plants that are adapted to store water are described as being succulent (ex. many species of Crassulaceae, *Aloe* etc.). The plants that develop mechanical cells to resist tearing and disruption of tissues as a result of excessive desiccation are usually described as sclerotic (ex. *Hakea*, *Leptocarpus*, *Ulex* etc.).

Stems of woody xerophytes are efficiently insulated by periderm to limit water loss. The tissues of periderm may be lignified and often impregnated with resins. Spines may be present on the surfaces of stem. Many xerophytes have hypodermis, the cells of which are composed of chlorenchyma cells. These cells, when enclosed in rigid and lined channels, add structural rigidity, e.g. *Leptocarpus* stem. Hypodermis may also be composed of thick walled cells like fibres and sclereids. These cells may be present as a continuous sheet or in patches. Hypodermal fibres may develop on the peripheral side of thin-walled chlorenchyma, e.g. *Ecdeiocolea* stem. Hypodermal sclerenchyma protects the inner tissues from high intensity of light and thus limits the loss of water. It also provides mechanical support. Leafless xerophytes and xerophytes with reduced leaves have palisade like cells in the outer cortex. These cells are compactly arranged and compose the photosynthetic tissues, e.g. *Casuarina* (Fig. 29.2).

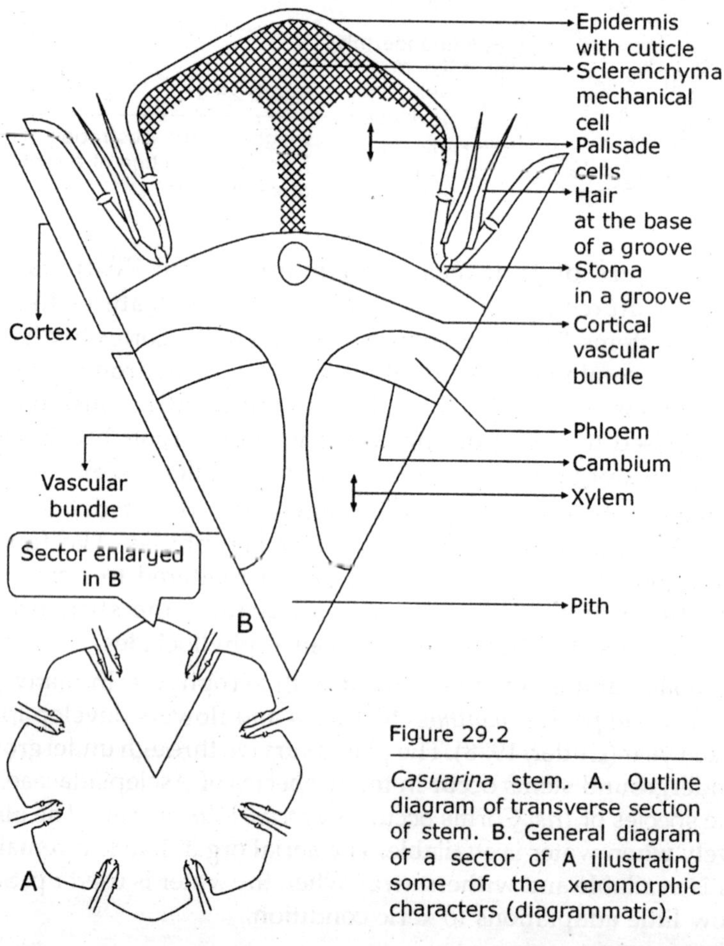


Figure 29.2

*Casuarina* stem. A. Outline diagram of transverse section of stem. B. General diagram of a sector of A illustrating some of the xeromorphic characters (diagrammatic).

Xerophytes have well-developed vascular tissues with long xylem vessels. The tracheids have relatively thicker walls than those present in mesophytes. Annual rings are well developed. Wood of many xerophytes is ring porous. The conduction through this wood is ten times more than diffuse porous wood. In *Pinus*, oils and resins are produced. Latex develops in *Euphorbia*.

Leaf polymorphism is observed in many xerophytes. After heavy rain broad leaves are formed and narrow leaves follow when the soil dries out. Xerophytes have evolved mechanisms to limit water losses. Loss of water is due to transpiration from exposed aerial parts. The size of the transpiring surface determines the amount of water loss. The smaller is the surface usually the lower is the transpiration. The reduction of transpiring surface can be accomplished in the following ways. The twigs, branches and mature leaves may be shed during drought. The plant-core remains alive. Immature leaves are more resistant to drought than mature leaves. So the former is seldom shed. The leaves may be long, slender, dissected or greatly reduced to spines. The scale or needle-like leaves are common in Coniferae and Ericaceae. Leaves with reduced surface are noted in *Calluna*, *Thuja* and *Asparagus*. Such leaves are called microphylly. Leaves are entirely absent in some Cactaceae, Asclepiadaceae and Euphorbiaceae. Most Restionaceae have non-functional leaves (Cutler, 1978). During midday sun small and narrow leaves heat up less rapidly than larger ones. As a result transpiration decreases. The long, slender leaves are usually vertically oriented to reduce the amount of heat absorbed. Plants reduce the exposed surface to sun or drying winds by rolling or folding of leaves. Plants also achieve this by rotating and orienting leaves away from maximum exposure to wind or sun. The leaves of xeromorphic species of *Stipa* and *Ammophila* remain folded constantly thus hiding the stomata. *Ammophila arenaria* (Fig. 29.3) leaf entirely rolls up towards the upper surface where stomata are located thus reducing the surface of moist tissue that is exposed to air and enclosing the stomata when dry condition prevails. In many xeromorphic species dead hairs or hair-like projections cover the leaf surface and they form an insulating layer, e.g. *Artemisia*, *Kleinia*, Elaeagnaceae etc. Such structures are called trichophyllous (Vasishta, 1968). Such leaves are gray or white owing to the hair covers. The white colour aids in deflecting heat from leaves. Moreover leaf hairs provide shade to leaves. It is interpreted that hairs create pockets on the leaf surface where water vapour accumulates. As a result the diffusion of water from the leaf is reduced. Most xerophytes have hairs that have thickened walls and some also have thick cuticle. These hairs, e.g. *Gahnia*, *Ammophila* and *Erica* reduce water losses in contrast to thin-walled hairs that increase water loss under some conditions. Moreover hairs are good deterrents against insect feeding and egg laying (Levin, 1973). Hairs also give protection against predator when plants become the only source of moisture for animals during drought.

Transpiration may be cuticular or stomatal. The magnitude of cuticular transpiration is governed by the formation of cuticle and cuticular layers. Many xerophytes such as Cactaceae have a heavy cuticle and thick cuticular layers.

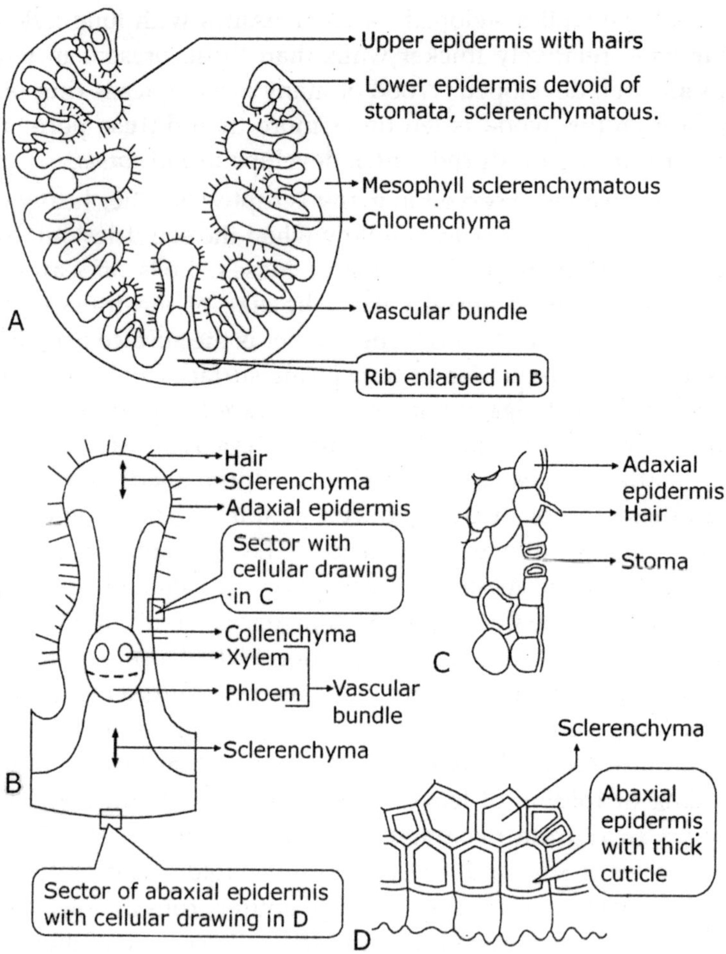


Figure 29.3

*Ammophila arenaria*. A. A cross section of leaf illustrating the rolling towards upper surface (diagrammatic). B. A portion of A showing details (diagrammatic). C. A portion of B showing stoma. D. A portion of B showing lower epidermis & sclerenchyma. Adapted after Cutler *et al.*, 2008, p. 141.

These cuticular layers almost entirely arrest the cuticular transpiration. Leaves are also covered with lipids to reduce moisture evaporation (ex. *Ricinus, Calotropis* etc.). Due to the presence of cutin and lipids the two epidermises of a leaf become impervious to water loss. The thickness of cuticle is related to xeric conditions. In a study it is observed that *Prosopis velutina*, when grown in natural xeric condition, have cuticle ten times thicker in comparison to that grown in indoors (Bleckmann *et al.*, 1980).

Transpiration is a vital process in all plants. When the water potential inside a leaf is higher than the environment, the water vapour will diffuse out of the leaf. If a plant loses too much water, wilting will result. In case of permanent wilting plants will die. So the xeromorphic leaves have adaptations that decrease the water potential in order to reduce water loss. The functioning and position of stomata govern the magnitude of stomatal transpiration. In many xeromorphs stomata open for a brief period only. In others stomata remain closed in the day. They open at night (ex. *Camellia thea*) when the relative humidity is high and temperature is low. As a result there is less transpiration. Usually xeromorphs have well developed and often numerous stomata. It is interpreted that carbon dioxide enters rapidly through these stomata during rare wet periods. Xeromorphic leaves may be epi- or hypostomatic. The stomata are often protected to restrict water loss. An individual stoma may be sunken (ex. *Aloe*). In *Nerium oleander* groups of stoma occur in a groove or depression on the abaxial surface of leaf (Fig. 29.4). The surface of groove is lined with hairs. In the groove the relative

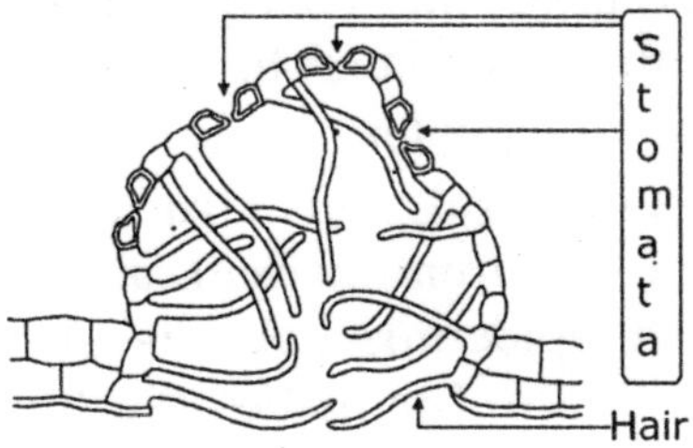


Figure 29.4

*Nerium oleander*. A portion of abaxial surface of leaf showing stomata in cross sectional view present in a groove. The groove is lined by hairs to limit water loss.

humidity always remains high thus reducing the diffusion gradient within the chamber. This reduces evaporation of moisture. In *Ficus, Nerium* etc. the stomata are restricted to well-protected crypts to reduce water loss. In *Agave* (Fig. 29.5), *Dasylirion* etc. cuticle forms ridges inside the pore of a stoma. Thus the canals

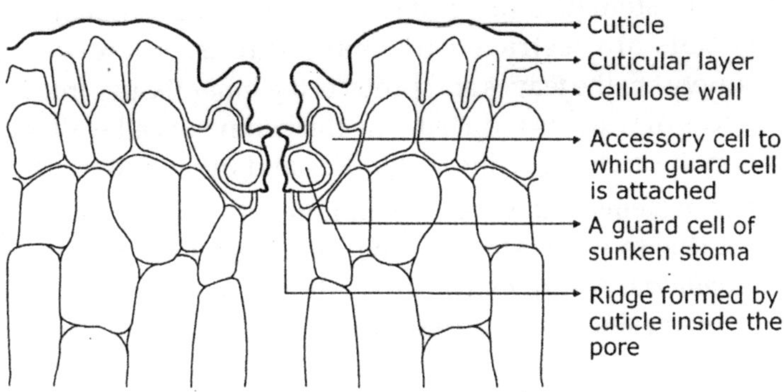


Figure 29.5

*Agave* leaf. A portion of transverse section of leaf showing epidermis with strongly thickened walls and sunken stoma. Adapted after Burström *et al.*, 1964, p. 108.

that communicate between open air and intercellular space become narrower. As a result water loss is reduced. Multiseriate epidermis occurs in *Ficus elastica* and *Nerium*. In mature leaves sometimes it is difficult to distinguish a multiple epidermis and a hypodermis. This structure reduces evaporation of water through epidermis and diminishes the intensity of light that reaches the photosynthetic tissue.

Xeromorphic leaves, like many mesophytes, have mesophyll differentiated into adaxial palisade and abaxial spongy tissue. But the palisade mesophyll is well developed and it is often correlated with high light intensity. Though the palisade mesophyll is typically adaxial, they may occur on abaxial side also, e.g. *Nerium, Ficus, Atriplex portulacoides, Artemisia, Myoporum, Sonneratia alba* etc. (Mauseth, 1988). In these leaves spongy tissue occurs in between the adaxial and abaxial palisade tissue. In many xeromorphs the mesophyll palisade replaces spongy tissue where the leaves have palisade mesophyll only, namely *Greggia camportum, Sphaeralcea incana* etc. (Vasishta, 1968). Due to the loss of spongy mesophyll, the palisade parenchyma becomes smaller and packed together. As a result the volume and surface area of leaf apoplast diminish. So each cell loses less water to the apoplast. The mesophyll cells of *Pinus* have peg-like ingrowths (plicate) to increase the surface of photosynthesis. Xeromorphic leaves have larger bundles of vascular tissues as compared to mesophytes.

Drought-resistant leaves are hard and rigid. This structure is called sclerophylly. Sclerophylly prevents leaf tissues from mechanical deformation during shrinkage. Thick cuticle occurs on the epidermises. As a result the leaves become leathery with a hard and glossy surface. The tissues of scleromorphic leaves are small-celled and dense. The palisade mesophyll may be multiseriate. The spongy tissues have little intercellular spaces. Development of sclerenchyma is very common. The hypodermis of *Pinus* needle is composed of sclerenchyma. In many leaves lignification occurs in mesophyll. In *Stipa*, the major part of the leaf tissue is composed of sclerenchyma. The abaxial epidermis of *Ammophila arenaria* is without stomata and lignified as the major portion of mesophyll. The collenchyma cells are restricted to small strands. Sclerenchyma provides mechanical support to the leaves and protects the inner tissues from high intensity of light. Thus loss of water is reduced. The thickening of cell wall is caused by the conversion of polysaccharides into celluloses and other materials.

Succulence in stems and leaves is common within xerophytes. The succulent organs thicken due to water storage. The succulent stem may attain a spherical shape, e.g. *Mamillaria*. A sphere has the smallest surface area in relation to tissue volume. So by increasing the thickness of an organ the relative surface is reduced. The surface reduction is favourable for the water balance.

Succulent stems have little differentiation of ground tissue. Parenchymatous spherical cells compose the ground tissue. Stem succulence occurs as a result of proliferation of xylem parenchyma and by primary thickening growth, e.g. Cactaceae. The mechanical tissue and vascular tissue are poorly developed. The cell walls are often mucilaginous. The succulent leaves, also referred to as malacophyllous xerophytes, are similar on all sides. The epidermis of such leaves is multilayered, large-celled and occupies the major portion to the leaf volume, e.g. *Begonia, Zebrina* (Fig. 29.6), and other members of Commelinaceae, Piperaceae etc. It is interpreted that the water present in these layers protects the central mesophyll against water losses during dry periods. The succulent leaves of *Aloe* and *Haworthia* have epidermis with thick outer walls. The epidermis has cuticle

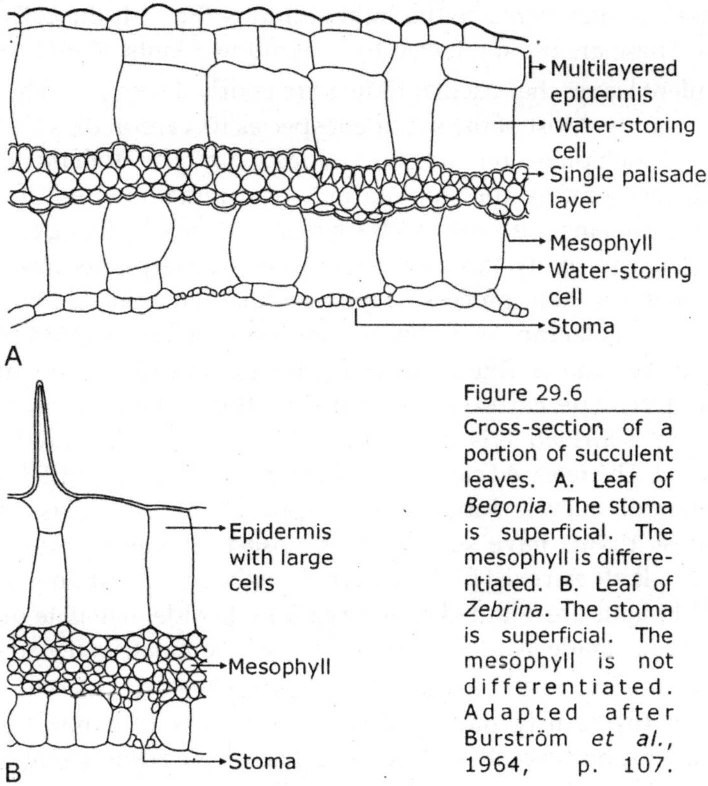


Figure 29.6

Cross-section of a portion of succulent leaves. A. Leaf of *Begonia*. The stoma is superficial. The mesophyll is differentiated. B. Leaf of *Zebrina*. The stoma is superficial. The mesophyll is not differentiated. Adapted after Burström *et al.*, 1964, p. 107.

and cuticular waxes. The stomata are sunken. In *Aloe,* the stoma (Fig. 26.7) is variously protected thus regulating and minimizing water loss during dry periods. Above each stoma there is raised rim that forms a suprastomatal cavity. The cavity has a constricted opening to the atmosphere. It is interpreted that during fovourable conditions of growth the cavity has a role in enhancing evaporation. Due to the presence of narrow opening in the cavity the structure has a venturi effect that lowers pressure above the stoma and assists transpiration (Cutler, 1978).

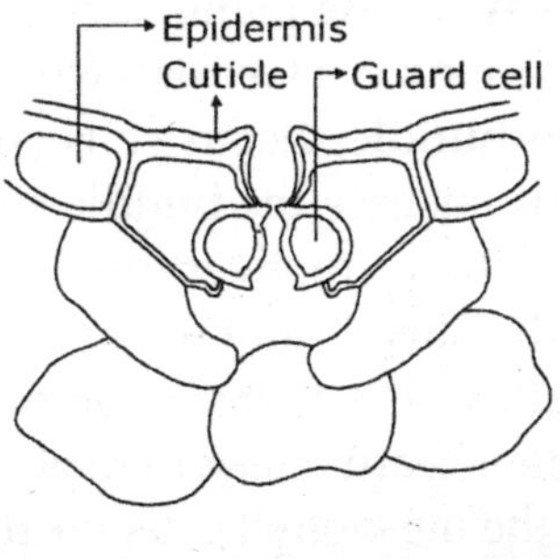


Figure 29.7

A part of cross section of the succulent leaf of *Aloe somaliensis* illustrating the adaxial sunken stoma, thick cuticle and thick outer wall of epidermal cell. Adapted after Cutler *et al.*, 2008, p.140.

Some leaves, e.g. *Haworthia* and *Lithops* have characteristic features. The leaves have chlorenchyma and water storing mesophyll. The leaf tips are translucent. The leaves remain underground, the tips being above the ground level only. The photosynthetic tissues perceive the light stimulus through the cells present on the leaf tips. These are often referred to as 'window plants' (Cutler, 1978).

In succulent leaves the vascular tissues are poorly developed. The mechanical cells are also scanty. Most of the succulent species fix carbon dioxide in the dark. The stomata of such plants remain closed during day. The stomata open at night. During high temperature the stomata may also remain closed at night when carbon dioxide exchange becomes nil. Such plants survive by fixing carbon dioxide made available internally through respiration. These plants have adapted a specialized photosynthetic process, called Crassulacean Acid Metabolism (CAM). Plant having Crassulacean Acid Metabolism is referred to as CAM plant. In this plant carbon dioxide is fixed through phosphoenolpyruvate in the dark. Phosphoenolpyruvate carboxylase catalyzes the carboxylation. As a result oxaloacetic acid is formed. It is then reduced to malate. During daytime malate is decarboxylated. The released carbon dioxide is fixed in ribulose-1,5-bisphosphate and enters the Calvin cycle. Ribulose-1,5-bisphosphate carboxylase catalyzes the reaction. These plants have sufficient ribulose-1,5-bisphosphate carboxylase activity. CAM plants can exist for long periods without any carbon dioxide uptake in light. CAM plants are adapted to store carbon dioxide in malate during night. Carbon dioxide is made available during day when malate is decarboxylated.

Plants belonging to the family Crassulaceae have CAM. The term CAM derives from the family name. Certain members of other families like Bromeliaceae, Euphorbiaceae, Aizoaceae, Liliaceae etc. also exhibit CAM.

Many xerophytes have an alternative pathway of photosynthetic carbon fixation in contrast to Calvin cycle that operates in mesophyte. This alternate pathway is known as Hatch-Slack pathway (Hatch *et al.*, 1966) or $C_4$-dicarboxylic acid pathway or simply $C_4$-pathway. Plants having this pathway are referred to as $C_4$-plants. This pathway is characterized by the formation of dicarboxylic acids of 4-carbon compounds as primary products of photosynthesis in contrast to 3-carbon compounds of the Calvin cycle.

The chloridoid—eragrostoid and panicoid taxonomic divisions of the Gramineae, Centrospermae, Compositae, Euphorbiaceae etc. carry out photosynthesis by using Hatch-Slack pathway. These plants have specific type of anatomy, referred to as 'Kranz' anatomy. Kranz is a German word—meaning garland or wreath. In this type specialized bundle-sheath cells surround the vascular bundles of leaf. Chlorenchyma cells encircle the bundle-sheath. These radially aligned chlorophyllous mesophyll cells were given the German name 'Kranz' (Haberlandt, 1914). The cells of bundle-sheath contain large chloroplasts and starch. The essence of Hatch-Slack pathway is that $C_4$-compounds carry carbon dioxide from mesophyll cells to bundle-sheath cells where photosynthesis occurs. The $C_4$-pathway starts (Fig. 29.8) in the mesophyll cells for the transport of carbon dioxide. Carboxylation occurs in the mesophyll cells where carbon dioxide and its receptor phosphoenolpyruvate condense to transient 4-carbon compound—oxaloacetate. The enzyme phosphoenolpyruvate carboxylase catalyzes the condensation. Oxaloacetate is rapidly reduced to malate or aminated

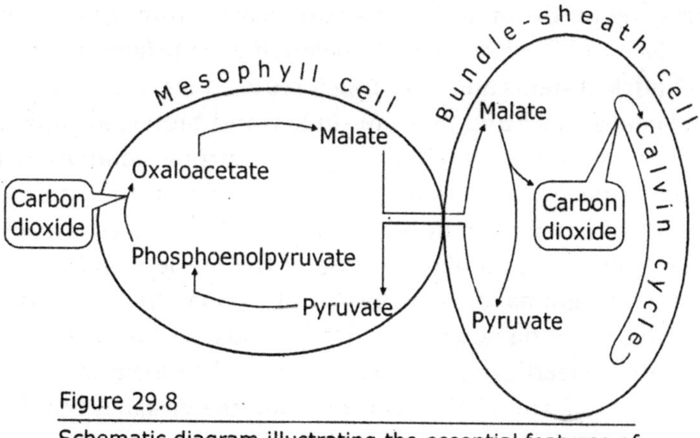


Figure 29.8

Schematic diagram illustrating the essential features of Hatch-Slack pathway and the concentration of carbon dioxide in bundle-sheath cell.

to aspartate depending upon species. $C_4$-compounds, formed in the mesophyll, enter into bundle-sheath cells where they are decarboxylated. The released carbon dioxide enters the Calvin cycle. Pyruvate formed during decarboxylation returns to the mesophyll cell for another round of carboxylation. It is to note that the decarboxylation of the $C_4$-compounds in the bundle-sheath cells maintains a high concentration of carbon dioxide at the site of photosynthesis.

In Calvin cycle carboxylation occurs by the condensation of carbon dioxide and ribulose-1,5-bisphosphate to form a transient 6-carbon compound, which rapidly hydrolyzes to two molecules of 3-phosphoglycerate. Ribulose-1,5-bisphosphate carboxylase catalyzes the condensation. Ribulose-1,5-bisphosphate carboxylase is also an oxygenase when it catalyzes the addition of oxygen to ribulose-1,5-bisphosphate to form phosphoglycolate and 3-phosphoglycerate. The oxygenase and carboxylase reactions occur at the same site and compete with each other. Under normal atmospheric conditions at 25°C the rate of carboxylase activity is four times greater than oxygenase. In higher temperature and when there is high oxygen and low carbon dioxide the Ribulose-1,5-bisphosphate carboxylase switches to oxygenase activity. This enzyme catalyzes the oxygenation of Ribulose-1,5-bisphosphate thus forming phosphoglycolate and 3-phosphoglycerate. The recycling of phosphoglycolate leads to the release of carbon dioxide and consumption of oxygen in a process called photorespiration. It is interpreted that photorespiration is seemingly a wasteful process because the organic carbon is converted to carbon dioxide without the production of adenosine triphosphate (ATP).

Xerophytes have evolved method to minimize the wasteful reaction of photorespiration. They have adapted $C_4$-pathway. This pathway maintains a high concentration of carbon dioxide in the bundle-sheath cells at the site of Calvin cycle. This accelerates the carboxylase reaction of Ribulose-1,5-bisphosphate carboxylase in relation to oxygenase reaction. Thus the $C_4$-plants take advantage of high temperature and minimize the oxygenation of Ribulose-1,5-bisphosphate.

Plants growing on the mountainous parts have xeromorphic adaptation, e.g. *Pycnophyllum molle* and *P. micronatum* etc. belonging to the family Caryophyllaceae. They have cylindrical stems and very much reduced leaves. Stomata are present only on the adaxial surface and they are sunken and protected amongst papillae. But the species like *Oxalis exidua* show little xeromorphic adaptation. In *O. exidua* the stomata are superficial; hairs and papillae occur on the epidermal surface (Cutler, 1978). The chlorenchymatous mesophyll cells are not compact. The leaves have all the characteristics similar to those of the mesic members of the genus. But the anatomy of stem have characteristic of a liana. The vascular bundles are separate. During secondary growth the interfascicular cambium produces parenchyma only instead of secondary xylem and phloem. By this device the stem can twist and deform without compressing the vascular bundles when the stem penetrates the cracks between rocks. *Azorella compacta*, another plant growing on mountainous environment, has very shiny leaves. The shiny leaves reflect ultraviolet light. The plant possesses contractile roots, which help the plant to be firmly anchored in the frost heave. Many plants growing in mountainous environment have sap that is mucilaginous nature. This acts as a kind of antifreeze (Cutler, 1978).

Halophyte often shows xeromorphic adaptations. They grow on locations with a high content of salts that make the soils entirely sterile. Many halophytes exhibit succulence that is normally associated with drought. The leaf area is strongly reduced, e.g. *Salsola*, *Glaux*, *Mesembryanthemum* etc. In some plants leaves are absent, e.g. *Salicornia*. The succulent leaves have salt glands through which excretion of salt occurs. Salt gland regulates the salt content of plants and prevents salts accumulating in the protoplasm. Moreover salt glands protect the leaves against strong sunlight and insect feeding.

Succulence and hard-leaf characters are combined in the tree and shrubby species of halophytes. The leaves have thick cuticle and the palisade tissue is many layered. The mesophyll tissue is weakly differentiated.

## Hydrophyte

Hydrophyte embraces all the different and varying plant forms that adapted to live under direct influence of free water or in waterlogged soil, e.g. a small number of ferns and angiosperm. Gymnosperms have not migrated into aquatic habitats. Hydrophytes, also called hydrophytic plants, are classified according to forms: Submergent, Floating and Emergent. Submergent hydrophytes grow and remain entirely submerged below the surface of water throughout life. Floating hydrophytes have leaves or shoots that float on top of water surface. Emergent hydrophytes have roots anchored to hydric soil, but a portion of shoot lies above water surface.

The different form of hydrophyte exhibits the following characteristics that help them to survive in water and waterlogged soil.

- Increase in leaf surface that helps plants for floatation.

- The presence of air chambers that assist in buoyancy of the plant.
- Stomata, if present, nonfunctional because the plants need not regulate transpiration due to abundance of water in their environment.
- Reduction of xylem because the hydrophytes need not conserve water.
- Reduction in mechanical and protective tissue as water pressure supports them.
- Hydrophytes have thin cuticle or lack it. Because the functions of cuticle like protection from strong sunlight, conservation of water and deterrence of insects are not required by hydrophytes. They are protected from excess sunlight by surrounding water; no question arises for water conservation as they have abundant free water in their environment; the hydrophytic angiosperms suffer little predation by aquatic animals. Due to the thinness of cuticle the leaves absorb nutrients from the water directly. But the major bulk of minerals are absorbed by roots and transported to the leaves. As there is no transpiration pull like land plants the transportation of minerals occur through root pressure. Hydathodes are very common in the leaf tips of many submerged hydrophytes. Transportation of water occurs as a bleeding stream ending in hydathodes. The opening of hydathodes may be large.
- Stomata when present may be many in number and occur on upper surface of floating leaves (epistomatous).
- Roots are small and reduced in all respects as water can diffuse directly into leaves.
- Formation of special roots and these are negatively geotropic and designed to function in gas exchange. Such roots are termed as pneumatophore and present in halophytes.
- Hydric soils have diluted oxygen and minerals and the hydrophytes adapted to absorb nutrients directly from water.
- The piliferous layer scarcely has (or no) root hairs. Root hairs are entirely absent from *Ceratophyllum*.
- The piliferous layer is not cutinized/suberized and so absorptive in function. Moreover the turgid piliferous zone provides mechanical strength, as mechanical cells are absent in hydrophytic roots.
- The cortex of root is wide and provided with air cavities.
- Xylem of root and stem is poorly developed and has thin walls. Vessels are absent in many hydrophytes. In the root of *Potamogeton* (Fig. 29.9A) the vascular tract in the centre is occupied by a large xylem cavity. The reduction in xylem is due to the fact that in hydrophytes no elaborate system of water absorption is required as all the epidermises of plant body are capable of water and mineral absorption.
- The stem of submerged hydrophytes is very weak and flexible. It moves according to the direction of flow of water currents.
- The epidermis in the stem of submerged hydrophytes is uniseriate. It is not cuticularized and in some cases the cells have cellulose walls only. As

a result the epidermal cells can absorb mineral salts and gases dissolved in water directly.

- The cortex is wide and has large air chambers. The air chambers may be arranged symmetrically e.g. *Myriophyllum* (Fig. 29.9B) or scattered irregularly all over the cortex, e.g. *Elodea, Hydrilla, Hippuris* etc.
- The cells of cortex have low osmotic pressure. The cells contain chloroplastids that help in photosynthesis.
- The vascular tissues in stem of submerged hydrophytes are not well developed. Endodermis and pericycle are indistinct. The vascular tissues are mostly composed of phloem.

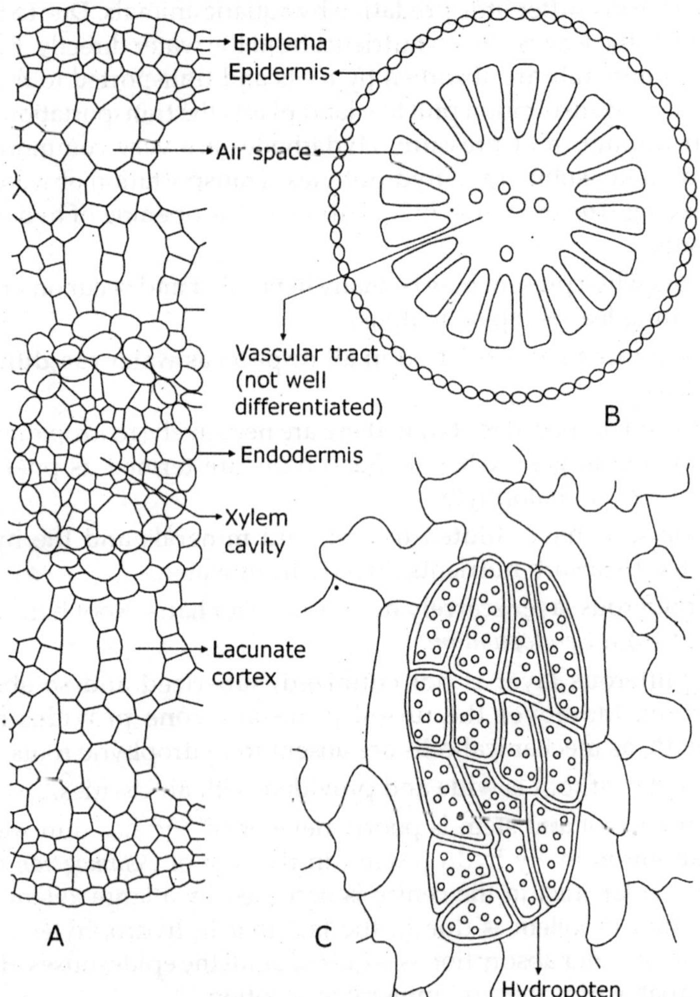


Figure 29.9

Diagram illustrating A. A portion of cross-section of *Potamogeton* root, B. Cross-section of *Myriophyllum* stem (diagrammatic) and C. A portion of leaf surface of *Sagittaria guayanensis* showing hydropoten (redrawn after Bhattacharyya, 1982; p. 71).

- Though the endodermis is not distinct many hydrophytic angiosperms have the innermost layer of cortex that functions as endodermis. This endodermoid layer is present in submerged roots, stems and leaves. It is assumed that this layer is involved in channelizing the flow of water through reduced xylem. In hydrophytes water is pushed up by root pressure produced as the endodermal cells pump ions into the xylem. This is due to the absence of transpiration pull like land plants. If there had been no endodermis like layer, the pressurized water could fill up the air chamber thus affecting buoyancy of the plant. So with the endodermoid layer the flow of water is effectively channelized to the leaves.
- Plants growing in running or tidal water have very linear leaf, e.g. *Posidonia, Zostera* etc. The leaves of *Ceratophyllum* are segmented. The leaves of *Ranunculus, Myriophyllum* are finely dissected. The submerged leaves of monocotyledons are often band-like and undivided, e.g. *Potamogeton*, Najadaceae etc. Such shapes of leaves offer little resistance to water that move around them.
- The leaf epidermises of submerged hydrophytes lack cuticle and contain abundant chloroplasts. As a result the epidermal cells have absorptive function and help in photosynthesis. The epidermises lack stoma and differ in no way from the ground tissue.
- The mesophyll cells of submerged leaves are not differentiated into palisade and spongy tissue. They contain abundant chloroplasts and well-developed air spaces. The submerged species of *Potamogeton* (Fig. 29.10B & C) has a single layer of mesophyll cells between the two epidermises. Mesophyll cell is absent in *Elodea* (Fig. 29.10A).

  [The term 'epidermis' is singular noun. The plural is *epidermi*. The plural *epidermi* was rarely used by anatomists and so became obsolete. In this *Title* the plural 'epidermises' is used following the suggestion of Mauseth (1988, p. 167).]
- Hydrophytes have strongly developed air lacunae in their roots, stems and leaves. They are referred to as air chambers. Though they are typically called air chambers, but they are not filled with air. The gases, generated by the plant, fill the chambers. The air chambers assist in buoyancy of the plant. Beside this, the air chamber and intercellular spaces function as an aeration system. This system allows the diffusion of oxygen to the non-chlorophyllous part of the plant. Oxygen is generated in the upper chlorenchyma and then it is diffused down to the roots. The roots always remain in contact with water and so they need internal supply of oxygen. The soils, where the roots are anchored, must periodically dry slightly to maintain soil oxygen content. It never occurs, so the root requires internal supply of oxygen. The air chambers also facilitates in regulating photosynthetic gases—carbon dioxide and oxygen. Moreover the air chambers along with the turgid tissues compensate the deficiencies of mechanical elements, which are very much reduced in hydrophytes.

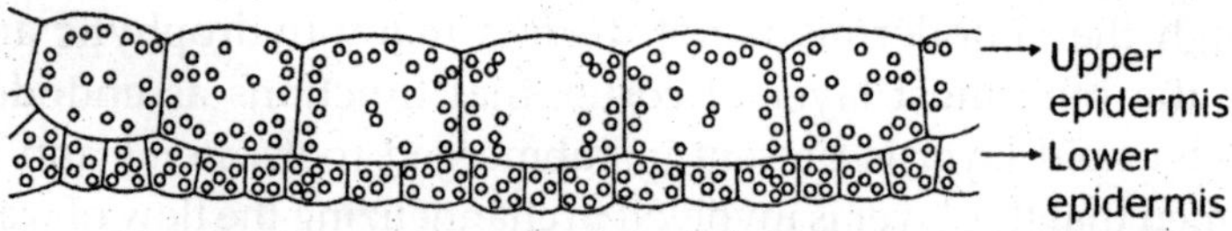


A. Portion of *Elodea* leaf in cross-section (after Singh *et al.*, 1994).

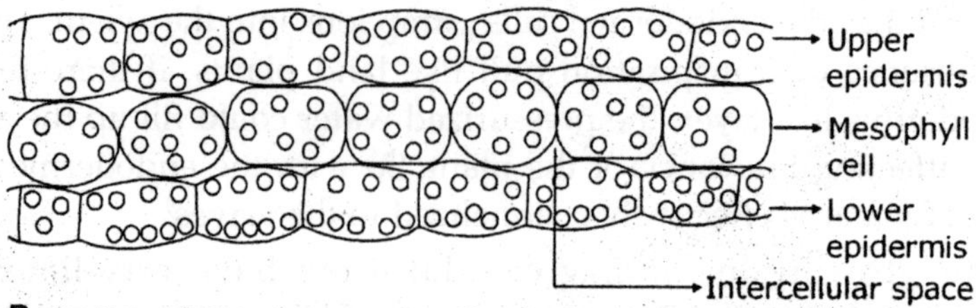


B. Portion of *Potamogeton pusillus* leaf in cross-section.

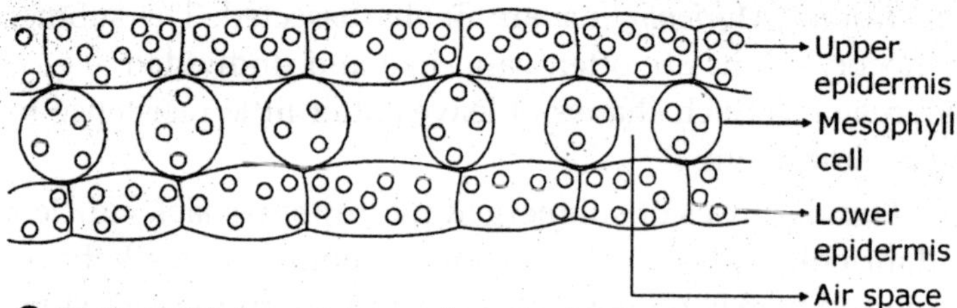


C. Portion of *Potamogeton nodosus* leaf in cross-section.

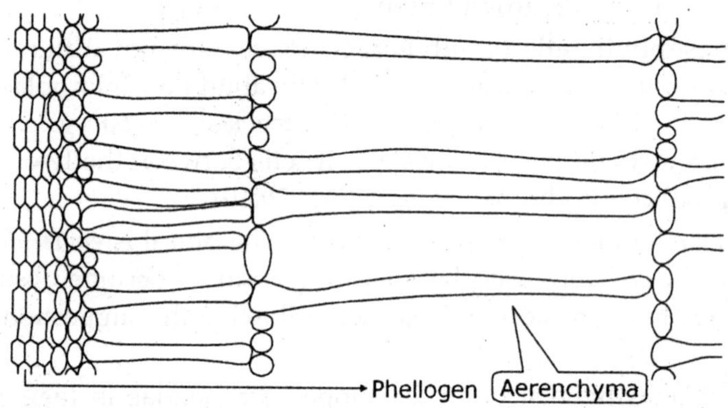


D. Portion of cross-section of pneumatophore of *Jussiaea peruviana* (Redrawn after Fahn, 1997, p. 292).

Figure 29.10

Diagram illustrating absence (A) and presence (B & C, redrawn after Vasistha, 1968) of mesophyll cells in hydrophytic leaves. D. Diagram showing phellogen and peripheral aerenchyma.

- Fibres are mechanical element and provide elastic flexibility. Hydrophytes exposed to wave action or tidal fluctuations have fibre bands on leaf margins, which are liable to tear in wave actions.
- In submerged hydrophytes there are localized groups of cells on the epidermises that are concerned with absorption of water and minerals. This multicellular absorbing structure is termed as hydropoten. Hydropoten is observed in *Sagittaria* (Fig. 29.9C), *Echinodorus*, *Ceratophyllum*, *Myriophyllum* etc.

- Laticifers are present in many hydrophytes in their rhizomes, petioles, pedicel and peduncle etc. Ex. *Sagittaria* (Fig. 29.11A), *Limnophyton* etc.
- Periderm is formed in many submerged dicotyledons, e.g. *Lycopus*, *Jussiaea* (Fig. 29.10D), Onagraceae, Lythraceae etc. Here phellogen, the cork cambium, instead of producing dense phellem on the peripheral side, gives rise to aerenchyma with abundant intercellular spaces. The aerenchyma consists of thin radial elongated cells. The radial lamellae of cells separate the wide air ducts. Periderm with dense phellem in land plants and periderm with peripheral aerenchyma in hydrophytes both have same function, i.e. to insulate the inner tissues from external environmental influence.

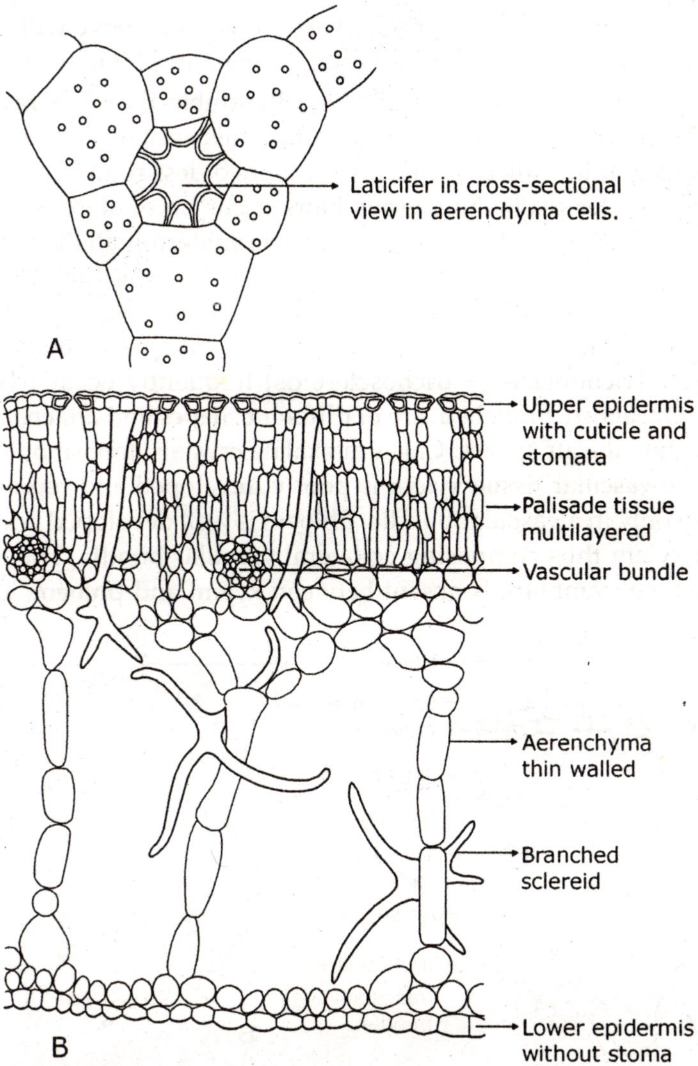


Figure 29.11

A. *Sagittaria guayanensis* (Redrawn after Bhattacharyya, 1982).
B. Cross-section of a portion of floating leaf of *Nymphaea alba* (adapted after Burström *et al.*, 1964; p. 116).

- The leaves of floating hydrophytes, e.g. Nymphaeaceae (Fig. 29.11B), *Hydrocharis* etc. exhibit a remarkable combination of xeromorphic and hydromorphic features. The upper surface of leaf is exposed to strong sunlight. So the upper portions of leaf have xeromorphic adaptations like (1) upper epidermis with cuticle, (2) sunken stomata, (3) dense and many layered upper palisade tissue and (4) presence of idioblast-sclereids. In contrast the lower portions of leaf exhibit hydromorphic adaptations like (1) lower epidermis with thin cuticle, (2) absence of stoma from lower epidermis, (3) large intercellular spaces with (4) thin walled parenchyma. Species of *Ranunculus, Polygonum amphibium* etc. are amphibious plants. These plants are able to live both in air and under water. The hydromorphic and xeromorphic adaptations that are mentioned above with reference to floating leaves are also encountered in these amphibious plants.
- The petiole of *Nymphaea* (Fig. 29.12A & B) exhibits the following anatomical characteristics: (1) The uniseriate epidermis contains chloroplastid and consists of thin walled compactly set more or less round cells with cuticle on their outer walls. Numerous hairs occur on epidermis and they are multicellular and not branched. (2) Beside epidermis all the tissues present in the petiole compose the ground tissue. The peripheral two or more layers of ground tissue are composed of collenchyma. Rest of cells of the ground tissue are parenchymatous, thin walled and enclose large air spaces. Trichoblasts (= trichosclereids) frequently occur in the spaces. Trichoblasts are branched, are of different shapes and attached to the cells bordering the air spaces. Calcium oxalate crystals deposit on trichoblasts. (3) The vascular tissue xylem is poorly developed and represented by a cavity only in a vascular bundle. Phloem is well developed and associated with xylem thus composing collateral vascular bundle. The bundles are closed, i.e. cambium is absent between xylem and phloem. The bundles

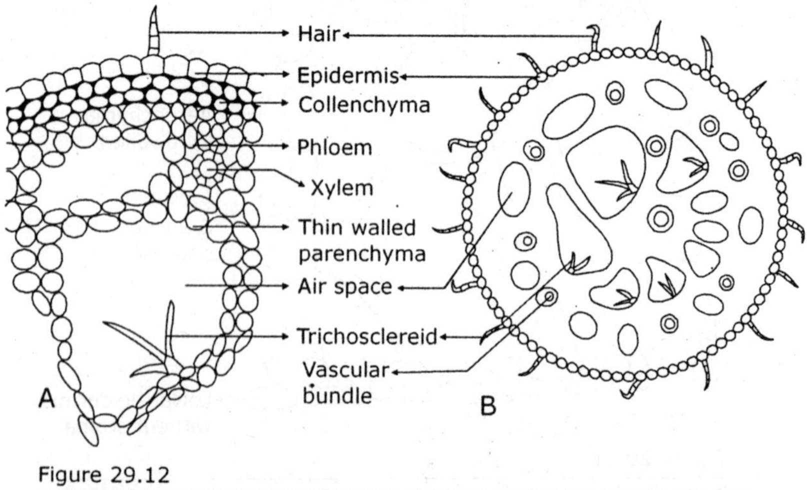


Figure 29.12

*Nymphaea*. A. A portion of cross-section of petiole. B. Diagrammatic cross-section of entire petiole.

are of different sizes. The smaller bundles have a single patch of phloem whereas two patches of phloem occur in large vascular bundles. The bundles are scattered over the ground tissue. They occur on the partition wall between air spaces. (4) The petiole of Nymphaeaceae exhibits hydromorphic adaptations due to presence of (a) numerous large air spaces that assist the petiole to remain erect, (b) reduced development of xylem, (c) well development of phloem and (d) thin cuticle on the outer wall of epidermal cells. Collenchyma in hypodermal layer, turgid aerenchyma and large air spaces supplement the mechanical cells. The occurrence of scattered vascular bundle in the petiole is regarded as anomaly, because it is a dicot where circular arrangement of vascular bundle is the norm in stem and petiole.

Plants inhabiting in acid bogs develop anatomical features that help them to survive because in water nitrogenous salts are absent and concentration of minerals is low. Such adaptations are meant for trapping animals and are observed in insectivorous plants, e.g. Lentibulariaceae and Droseraceae. In *Dionaea* (Droseraceae) the upper portion of leaf blade has quadrangular shape. Specialize hinge or motor cells occur along the midrib. The leaf margin projects as long teeth close together. The edge of blade is green and the rest of the surface is covered with reddish dots. The dots are digestive glandular hairs that are revealed under microscope. The two halves of the blade are bent upwards and in cross-sectional view appear as V-form. On each half of blade there are three long hairs—also called trigger hairs. The hairs are very sensitive and hinged at the base. The hairs require two or three tactile stimuli to cause immediate folding of leaf. When a prey stimulates the trigger hairs the leaf folds vigorously. The marginal teeth cross one another and so the prey cannot escape. The two halves squeeze together tightly. Reddish glandular hairs secrete digestive enzymes that act upon the proteids of the prey. Absorption follows and when the process is complete the leaf opens again.

## Epiphyte

Epiphyte forms an ecologic group of plants that grow in substrate where there is no soil. They grow and remain attached to another plant. Epiphytes use the other plant merely for support. They do not grow parasitically. Trunks and branches of trees serve as supports of epiphyte but cannot retain water for their provision. The habitats of epiphyte are always dry except during rainfall. The climate, whether humid or arid, has no special effect on habitat. So epiphytes are regarded as special kind of xerophytes (Burström *et al.*, 1964).

There exists no relation between an epiphytic seed plant and substrate. As for example *Tillandsia usneoides* (so named because the plant resembles the lichen *Usnea*) belonging to the family Bromeliaceae grows on telephone wires, depending entirely upon air-borne nutrients. The whole plant is thickly covered with scaly hairs with thin cuticle, which collect and absorb rainwater. The plant is not adapted to store water.

Epiphytes are common among monocotyledons especially among the families Orchidaceae, Bromeliaceae and Araceae. These seed plants are usually succulent. The plants have three kinds of root—clinging root, absorbing root and freely suspended aerial roots (Pandey, 1979, p. 377). An epiphyte is attached to the supporting plant with the help of clinging root. Moreover the clinging roots form a network where humus accumulates in meshes. The absorbing roots originate from clinging root and draw nutrition from humus. The hanging aerial roots are covered by a velamen. Velamen (Figs. 29.13, 18.1) is a multilayered tissue formed by periclinal divisions of protoderm. So velamen is regarded as multiple or

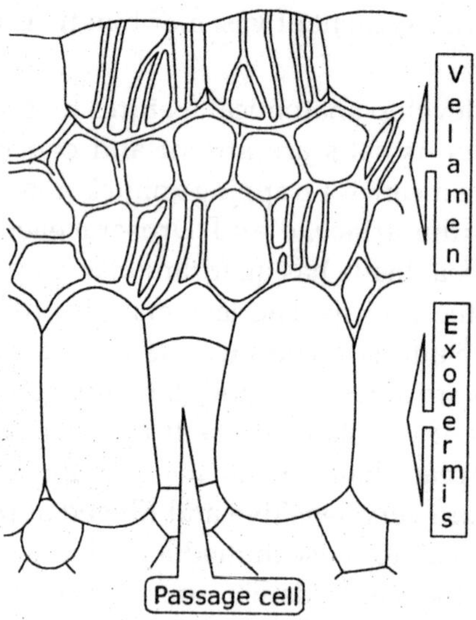


Figure 29.13

*Vanda* root. A portion of velamen with exodermis in cross-sectional view. The cells of velamen have thickening on the walls. Adapted after Cutter, 1971, p. 24.

multiseriate epidermis. Each cell of velamen is thick walled, devoid of any contents, nonliving, horizontally elongated or polygonal in shape. The cells are compactly set and air-filled. Some cells have band like or reticulate secondary wall thickenings. The cells have fissures and pores on their walls. Velamen becomes filled with water during precipitation. Exodermis occurs just below velamen. The continuity of exodermis is interrupted by the presence of thin walled passage cells. It is assumed that water from velamen is transported through passage cells to the inner tissues that are living, but this has not been verified experimentally. Though velamen is interpreted as absorptive tissue, but some physiological studies on orchid velamen indicate that velamen is concerned with mechanical protection of inner tissues and prevention of excess loss of water from cortex (Dycus *et al.*, 1957).

Special structure termed pneumathode occurs in the velamen. Pneumathode has some loosely packed cells with dense spiral wall thickenings. These cells form an outlet of the ventilating system of epiphyte. Exchange of gases is facilitated through pneumathodes when velamen is saturated with moisture.

Aerial roots regularly have chloroplasts in the cortex. In *Taeniophyllum*, the main photosynthetic activity is restricted to cortex of aerial roots. Apart from normal functions, leaves are also concerned in water absorption and it is of special importance among epiphytes. Epiphytic Bromeliaceae develops scale-like hairs with thin cuticle that facilitate water uptake.

Some special features of the anatomy of suspended aerial epiphytic roots are mentioned below taking *Vanda* (Fig. 29.14) as an example: (1) Presence of protective tissue – velamen, (2) chlorophyllous cortex, (3) aerenchyma in cortex, (4) radial and polyarch xylem with exarch protoxylem, (5) centralized mechanical cells like sclerenchymatous conjunctive tissue and (6) sclerotic pith in mature root. The anatomy of *Vanda* root is discussed in chapter 31 (Fig. 31.37).

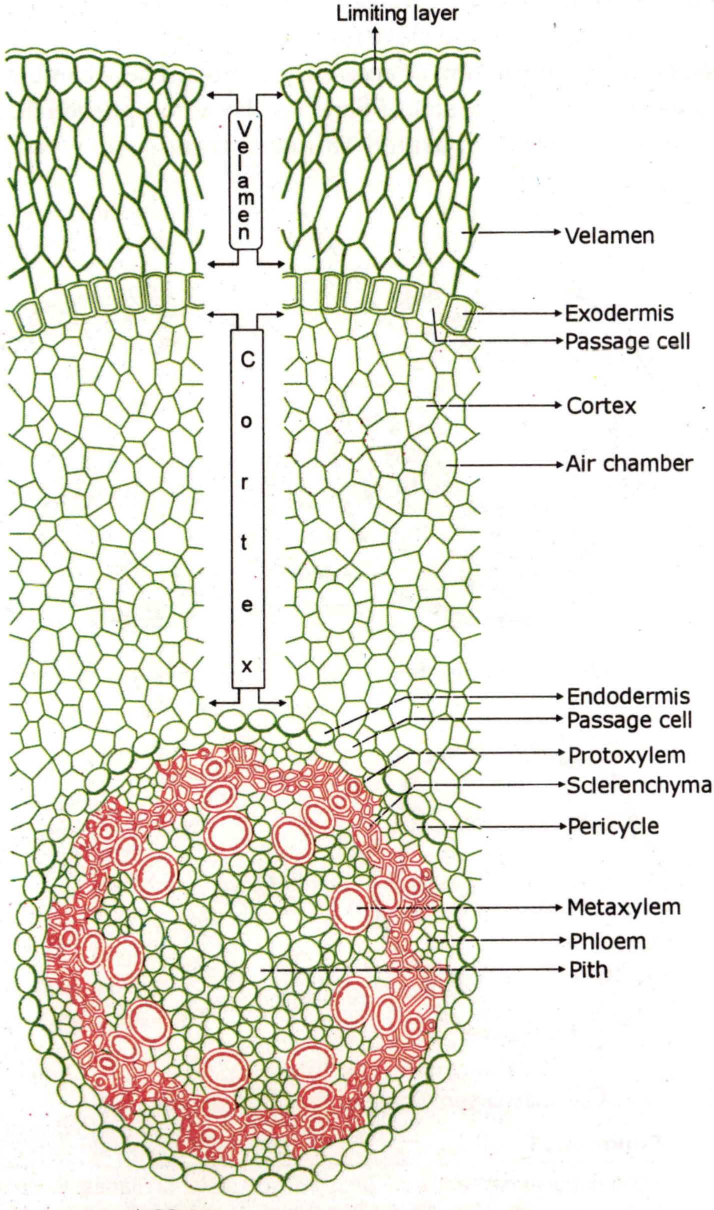


Figure 29.14

*Vanda* root. A portion of cross-section of root.

## Lianas

Liana (=liane) represents an ecologically specialized plant form. The plants are climbing and the specialized lianes have winding stems. It is regarded that winding is governed by light requirement. Lianes have stem that is much narrower than the length.

The winding stems are strong and pliant. The vascular system of stem is well developed and has wide vessels and sieve tubes. With the increase in length of the stem the resistance to water transport also increases. This is compensated by the development of vessels and sieve tubes with wide lumina.The pliancy is often accomplished by the formation of accessory cambia. These cambia form isolated vascular strands that split the stele. Pliancy is also accomplished through splitting of lignified xylem into separated thinner strands. In woody lianas, e.g.

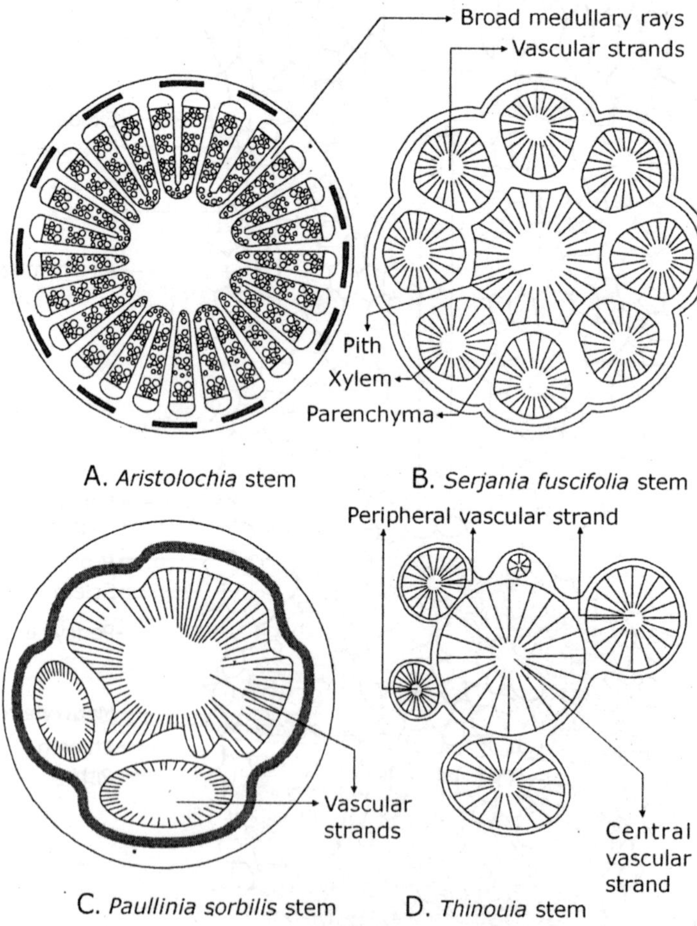


Figure 29.15

Schematic illustrations of anatomy of stele of lianas in cross sectional view. Vide text for details. Adapted after Soleredor, 1908, p. 235; Burström, *et al.* 1964, p. 120 and Mauseth, 1988, p. 372).

Aristolochiaceae (Fig. 29.15A) and Menispermaceae pliancy is brought about by the formation of broad medullary rays by interfascicular cambium. In Sapindaceae, e.g. *Serjania* (Fig. 29.15B), *Paullinia* (Fig. 29.15C), *Thinouia* (Fig. 29.15D) etc. the cambium ring is split into several isolated segments. The isolated cambium segments form separate vascular strands, often on a regular pattern. In between the vascular strands parenchymatous ground tissue occurs. At later stage the stem resembles a twined rope. The stem of *Bauhinia* (Fig. 29.16B, C) becomes lobed owing to unequal rate of secondary growth in different sectors of cambium. The lobation of xylem increases the pliancy of the stem. In *Bignonia* (Fig. 29.16A) stem symmetrically located segments of cambium cease to form xylem inwards and donate phloem only on the peripheral side at an increased rate. As a result the xylem is split and becomes separated by radial slits. Thus the pliancy is brought about during the continued secondary growth.

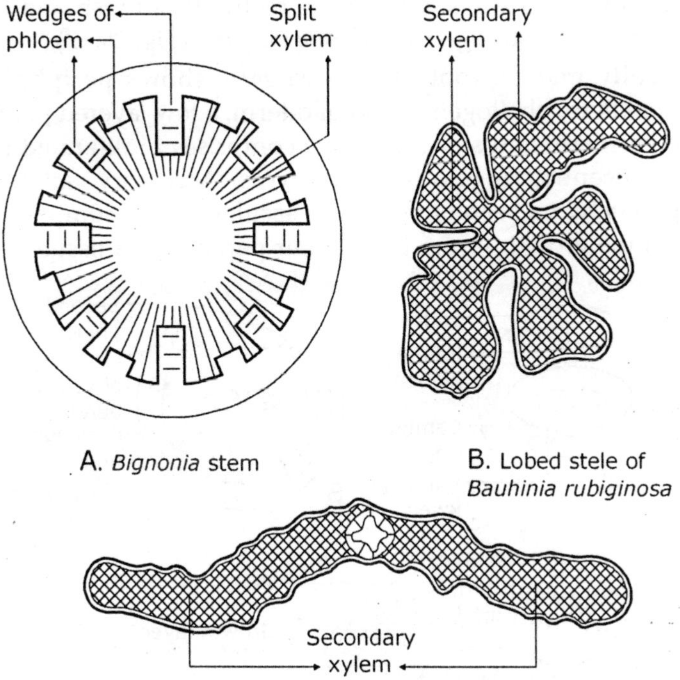


A. *Bignonia* stem

B. Lobed stele of *Bauhinia rubiginosa*

C. Band-shaped stele of *Bauhinia blumenaviana.* Such stem clasps and flexes with ease over the supporting object during twining.

Figure 29.16

Schematic illustrations of the anatomy of stele of lianas in cross-sectional view. Vide text for details. Adapted after Chandurkar, 1983, p. 207.

## Storage organs

Biennials, besides seeds, have fleshy and swollen root or stem that store nutrients. The biennial herbs store nutrients in the first year. This is for the development of the flowering shoot in the following year. Many important food plants have storage organs.

The anatomy of all storage organs has the following common features: (1) normal or accessory cambia cause the thickening of the storage organs. It is to note that formation of accessory cambia is a taxonomic character that is inherited by a species. Formation of accessory cambia is not an adaptation to a specific function. The dilatation of cambial rays and the proliferation of other parenchyma may fortify the thickening. (2) Tracheary elements are poorly developed. Comparatively phloem elements are well developed. (3) Parenchyma cells are the storage tissues in storage organs. They are thin-walled and formed profusely in various ways.

Ample development of parenchyma cells that store nutrients, ample development of sieve elements through which large transport of organic compounds occur and rapid development of the storage organ can be accomplished in various ways. One of the methods is illustrated below on the basis of anatomy of the storage root of *Beta vulgaris* (Fig. 29.17).

Anatomically mature root of *Beta vulgaris* shows peripheral periderm consisting of phellem, phellogen and phelloderm, a few layered parenchymatous cortex below periderm and stele. The stele comprises proliferated pericycle and five to eight concentric rings of growth layers. Accessory cambium, vascular tissues and parenchyma present between vascular tissues compose each growth layer. Between the growth-layers there occur parenchyma tissues.

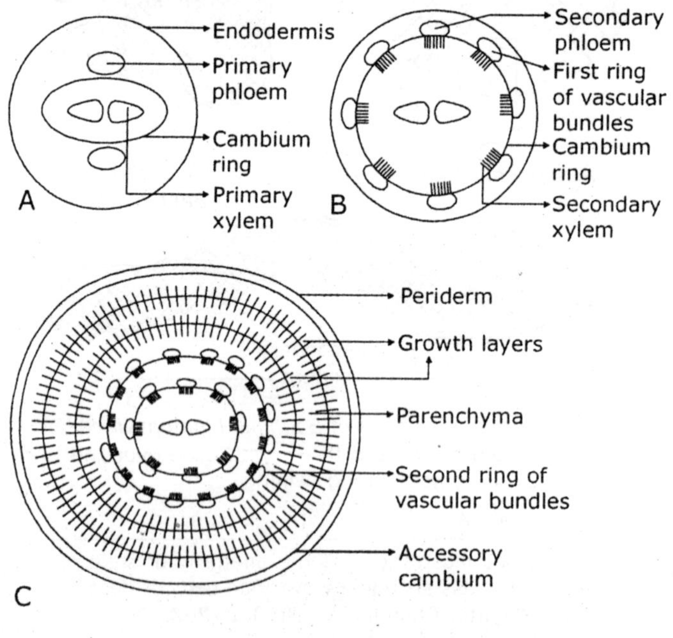


Figure 29.17

Diagrammatic illustrations (A-C) of storage tissue formation in *Beta vulgaris*. Vide text for details. Adapted after Tayal, 1994, p. 227.

Primary stele of the root is radial with diarch xylem. At the time of secondary growth a regular cambium is formed. Phellogen originates on the peripheral side and forms periderm. Periderm covers the surface of root and forms a protective layer. Pericycle proliferates. Parenchyma originates between periderm and stele. Later five to eight accessory cambia originate in succession outwards. These cambia divide almost simultaneously. The derivative cells differentiate into narrow strands of vascular tissues and parenchyma. Broad cambial rays separate the narrow vascular strands. Tracheary elements are weakly differentiated. The vascular strands are mainly composed of parenchyma. The origin of successive rings of accessory cambium, their divisions and formation of vascular strands and parenchyma, and proliferation of parenchyma are very rapid. Thus the *Beta* root grows in thickness rapidly. All the parenchyma cells store nutrients that mainly contain sucrose.

## Hygrophyte

Hygrophytes are land plants living in plentiful supply of water and high humidity in air. These are shade plants and form the herbal ground vegetation in forests on humid habitats, e.g. hydromorphic forms of *Oxalis* (Fig. 29.18), *Circaea* and some ferns.

Hygrophytes have the tendency to develop large leaf surface. These plants grow in shades and so large leaf surface is favourable for photosynthesis at low light intensity. The only ecologically limiting factor that operates on hygrophyte is light. As water and humidity are abundant the hygromorphs are not adapted to limit transpiration.

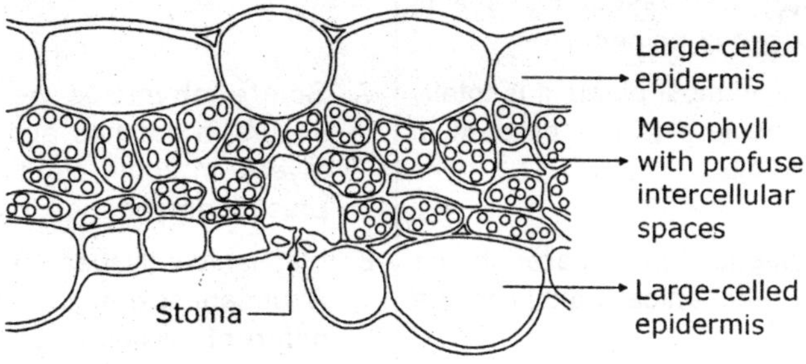


Figure 29.18

A portion of cross section of hygrophytic leaf of *Oxalis*. Mesophyll undifferentiated. Adapted after Burström *et al.*, 1964, p. 112.

Anatomically the leaves have weakly differentiated mesophyll with profuse intercellular spaces. Palisade tissue may be absent. The xylem is poorly developed, as water is abundant. Mechanical tissues may not develop. The epidermises have thin cuticle. Living hairs may be present to increase the transpiring surface. Stomata are present and they may be flush or raised above the surface. Poor development of tracheary elements and mechanical cells signify the savings of cell wall building materials.

Comparison of anatomical adaptations of hydrophytes and xerophytes.

| Hydrophyte | Xerophyte |
|---|---|
| 1. Epidermis: epidermis has thin cuticle and may contain chloroplastids. | 1. Epidermis has thick cuticle, cuticular wax and may contain hairs. Sometimes it may contain lignin. |
| 2. Stoma: stomata are absent on submerged plants. Floating leaves are epistomatic. Hydathodes are present on leaf tips. Gaseous diffusion occurs throughout the whole epidermis. | 2. Most plants are hypostomatic and have sunken stoma. Stoma may remain in grooves and may be guarded by hairs to minimize water loss. *Ammophila* leaf is epistomatic and it rolls upward to block stoma. |
| 3. Mesophyll: mesophyll cells are not differentiated. They are loosely arranged and contain air spaces. | 3. Mesophyll tissues are differentiated into upper palisade and lower spongy cells. The cells are compactly arranged and lack air spaces. |
| 4. Air chamber: aerenchyma is present in leaves and stems, and forms large air spaces. Air chambers give buoyancy to the plants. Gaseous diffusion occurs through air chambers. | 4. Air chambers are not formed in xerophytes. |
| 5. Storage tissue: most cells of leaves and stems always remain turgid as they always remain surrounded by water. | 5. Succulent xerophytes store water in parenchyma cells and have mucilage. |
| 6. Vascular tissue: xylem is poorly developed and has thin walls. Vessels are absent in many hydrophytes. Vascular tissue is mainly composed of phloem. | 6. Xylem and phloem are well developed. |
| 7. Sclerenchyma: in most plants it is totally absent. Trichosclereids are present in leaves and petiole. | 7. Sclerenchyma is well developed. It occurs in patches and sheets of many layers. In some plant sclerenchyma is present among mesophyll cells. |
| 8. Mechanical tissue: turgid parenchyma cells provide mechanical rigidity to the plant. | 8. Mechanical tissues are well developed. Abundant sclerenchyma cells present in different positions of different organs provide mechanical rigidity to the plant. |

CHAPTER **30**

# Development of Flower

A flower is regarded as a compressed shoot appearing periodically on a plant with longer or shorter intervals and carrying successive basal sterile lateral organs (sepals and petals), upper fertile lateral organs (stamen and carpel) intended for the propagation of plants through seeds and the sterile staminodes and carpellodes, where present.

Arber (1946) opines that all the sterile and fertile lateral organs are simply the different forms of the lateral member termed leaf. Most Botanists agree that a flower is morphologically equivalent to a shortened branch system bearing appendages. It is also agreed that floral parts are homologous to leaves.

A period of vegetative activity precedes the formation of flower. The vegetative shoot apex, terminal and lateral, causes the vegetative growth of a plant. At the time of flowering the vegetative shoot apex switches to reproductive apex. The reproductive apex is destined to form either an inflorescence or a flower. During transition to reproductive apex, the vegetative shoot apex undergoes various morphological, anatomical and physiological changes. Transition to reproductive apex occurs gradually.

Indeterminate growth is the characteristic feature of shoot apex. In contrast the reproductive apex is relatively short and determinate. According to the position of shoot apex, which is destined to form reproductive apex, the flowers may be borne on the apex of main shoot, or on the lateral branches, or on both. The activity of vegetative shoot apex promotes longitudinal growth of the axis. In contrast the reproductive apex forms a large surface area of meristematic tissue that gives rise the floral parts or flower primordia.

## Transition to flowering

Reproductive apex exhibits plastochron size fluctuations during the development of floral parts/primordia. The apex may widen and become slightly flattened (Pophan and Chan, 1952). In *Liriodendron* the apex becomes conical and elongated (Cutter, 1971). Smith (1966) reported the horizontal and vertical enlargement of the apex in *Carex*. The apex of *Chrysanthemum*, during the formation of capitulum, increases in area nearly 400-fold within a few hours (Schwabe, 1959).

During transition to flowering apex the above-mentioned changes in shape is linked with the rise of mitotic index in the shoot apex. The increased mitotic activity is observed in the cells present between the rib meristem and central mother cell zone. Gradually mitoses spread over the zone of central mother cell. As a result cells are added to tunica. In the study of transition to flowering, plants are induced to flower. In an experiment with *Xanthium*, which is one of the most thoroughly investigated angiosperm leaves and flowers, it is observed that a significant increase in mitotic index occurs within 24 hours after the beginning of inductive treatment (Cutter, 1971). In one occasion the mitotic index increased within 16 hours after the beginning of the inductive treatment (Thomas, 1963). Bhar *et al.* (1969) observed the increase of mitotic index in *Pharbitis nil* on the second day after inductive treatment. Several species of short day plant exhibit increased rate of mitosis following induction. The cells, situated below the central zone and distal to rib meristem, show increased mitotic activity (Wetmore *et al.*, 1959). Apart from mitosis the amount and distribution of endoplasmic reticulum change. This was observed in *Chenopodium*, 3 hours after the end of the first inductive long night (Cutter, 1971). The enzyme phosphatase and the number of dictyosomes per cell increase.

[Mitotic index is the percentage of cells where nuclei are in mitotically dividing stage in the total number of cells present at the time of fixing. Fixing is a process of immediate killing of tissues and to preserve the contents of cells as closely as possible to their living state. Mention may be made of the following chemicals that are frequently used as cytological fixatives: Ethyl alcohol, Nawaschin-A, Nawaschin-B, Liwitsky's fixative, Carnoy's fixative, Bouin's fixative etc.]

The typical tunica-corpus organization of vegetative shoot apex, advanced by Schmidt in 1924, becomes modified during the transition to flowering. In the floral apex Gregoire (see Gemmell, 1969) recognized two zones. The outer zone consisted of darkly staining cells and this was named as mantle. The inner zone was composed of large parenchyma cells and that was named as core. These two zones cannot be analogized with the usual tunica-corpus organization. So, according to Gregoire, the outer and inner zones were respectively named as meristematic mantle and parenchymatous core (Fig. 30.1). The large cells of core become highly vacuolated during transition. The association between mantle and vacuolated large cells of core are termed as mantle-core type of organization. Mantle-core type of organization gradually becomes established during transition to flowering. Thus a mature floral apex shows outer mantle and inner parenchymatous core. The mantle is composed of dense cells and consists of tunica and a large part of the corpus. The parenchymatous core is largely derived from the rib meristem situated below the apex.

Before transition to flowering the apex of *Calendula officinalis* is dome-shaped with two-layered tunica. The apex is gradually transformed to a broadly based cone by horizontal and vertical growth. The rib meristem is conspicuous. The

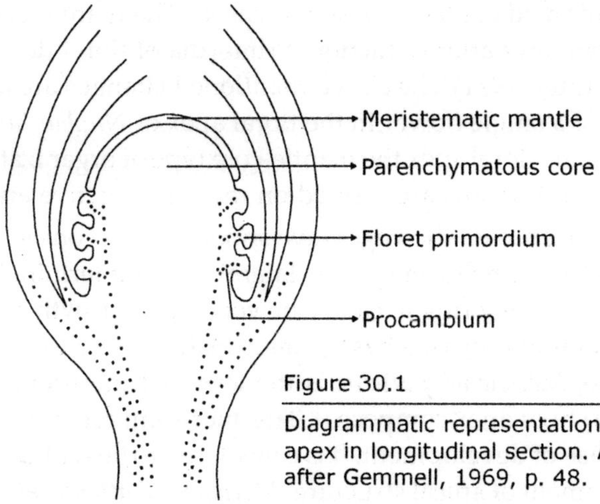


Figure 30.1

Diagrammatic representation of floral apex in longitudinal section. Adapted after Gemmell, 1969, p. 48.

apex loses the organization of vegetative apex. At this stage the mantle-core organization is yet to be established. *Calendula* has capitulum inflorescence. Before flowering begins the inflorescence apex assumes the characteristic shape by the increase of central parenchymatous core. Primordia of involucral bracts appear first on the inflorescence apex. After further growth primordia of floret appear on the periphery of inflorescence apex in acropetal sequence (Cutter, 1971). Gradually floral primordia develop over the whole surface of inflorescence apex (Fig. 30.2).

The shoot apex of *Xanthium* has single-layered tunica. The cells present on the outermost layer of corpus are large. The rib meristem is conspicuous. *Xanthium* plants were given inductive photoperiod to study the transition of vegetative apex to reproductive apex. *Xanthium* is short day/long night plant. The vegetative apices of *Xanthium* plant were treated with one inductive long night. The apices showed a change in form two days after induction. Increased mitotic index was observed in the cells present between the rib meristem and central zone. In these

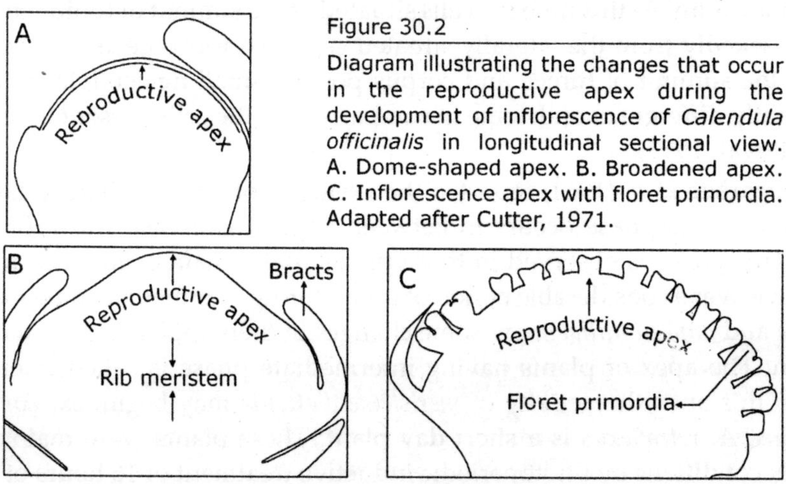


Figure 30.2

Diagram illustrating the changes that occur in the reproductive apex during the development of inflorescence of *Calendula officinalis* in longitudinal sectional view. A. Dome-shaped apex. B. Broadened apex. C. Inflorescence apex with floret primordia. Adapted after Cutter, 1971.

cells the ribonucleic acid content is also increased. The mantle-core organization is established seven days after induction. Primordia of floret developed ten days after induction (Cutter, 1971).The above-mentioned changes are also observed in the development of a single flower in the floral apex of *Nuphar lutea* of the family Nymphaeaceae. In the floral apex the mantle-core type of organization is gradually established. The floral organs are formed on the reproductive apex.

Buvat (1955) interpreted the vegetative shoot apex to consist of *anneau initial* and *méristème d'attente*. The former is the lateral and sub terminal meristem, and considered as the initiating ring to cause the vegetative growth of the shoot apex. The *anneau initial* or initiating ring has the main histogenic role in the development of apical structure. *Méristème d'attente* is the terminal meristem and the central cells of vegetative shoot apex compose it. This meristem remains dormant during the vegetative phase of development. It is considered to be without any histogenic role in the development of apical structure. *Méristème d'attente*, also called waiting meristem, supposedly becomes active in the formation and development of inflorescence apex and terminal flower at the time of flowering.

The above interpretation regarding the apical zonation is based on the study of mitotic index within the apical region and on the observations of cytological characteristics of the apical cells. The vegetative meristems have the following characteristics: (1) produce foliage leaves continuously; (2) possess vacuolated cells; (3) the tunica cells contain little ribonucleic acid in the cytoplasm and (4) the corpus cells have small nucleolar volume. The following features are present in the reproductive apices: (1) the tunica cells have increased nucleolar volume; (2) the central cells have greater activity and (3) the central axial cells have more ribonucleic acid (Cutter, 1971). In a study with *Lupinus*, Buvat (1952) reported that the cells in tunica became more meristematic in appearance at the onset of flowering. At this time the cells present at the summit of corpus also become more meristematic in appearance. During this period the cells of tunica and corpus contain homogenous condriome and numerous small vacuoles. The apex of *Aster sinensis* shows cytological differences after the formation of third leaf primordium. At this time the cells situated at the summit of tunica and corpus differ markedly from the laterally situated cells that form the *anneau initial*. The cells at the summit of tunica and corpus possess small nucleoli, large vacuoles and slightly different chondriome in comparison to the cells present in the *anneau initial* (Buvat, 1952).

Nougarède *et al.* (1964) reported the existence of 'intermediate' phase in the shoot apex. This phase occurs in between vegetative phase and reproductive phase. This phase is observed in those plants that remain or induced to remain vegetative over a considerable period of time. During intermediate phase *méristème d'attente* and *anneau initial* show some changes in their appearance, zonation and function. The apex of plants having intermediate phase is called intermediate apex. In this apex the activity of *méristème d'attente* may begin, ex. *Amaranthus retroflexus*. *A. retroflexus* is a short day plant. These plants were maintained in long day conditions by photoperiodic inductive treatment of 16 hours of light for

60 days. The apices of such plant show the combined characteristics of vegetative meristems and reproductive apices that are mentioned in the previous paragraph. Cutter (1971) opines that the intermediate apex represents a compromise between vegetative and reproductive apices. Heslop-Harrison *et al.* (1969) considers the intermediate stage of *Cannabis sativa* as a very slow transition to flowering. The characteristic of intermediate apex is attributable to gibberellin that increases in apical tissues (Bernier, 1966). It is observed that plants, treated with gibberellin, have growth only on the central zone in the apex. It is to note that it is the environmental conditions, to which the plants have previously been subjected, determine the transition to flowering.

The above discussed structural, anatomical and cytological changes in the transitional apex are no doubt accompanied and preceded by certain biochemical and physiological changes. A few changes are mentioned below.

Physiologists have shown that synthesis of ribonucleic acid (RNA) or deoxyribonucleic acid (DNA) is the necessary preliminary to flowering. The cytoplasmic RNA content is usually restricted to the flanks of vegetative shoot apices. It increases and uniformly spreads throughout the apices when the apices are induced to flower. Example: *Chenopodium album* (Gifford *et al.*, 1962). *C. album* is a short day/long night plant. In this plant the concentration of RNA increases after two long inductive nights and again after four long inductive nights. This increase of RNA content is also observed in *Pharbitis* and *Xanthium* within one day after one long inductive night. The increase of RNA is a photoperiodic phenomenon. It is revealed by the fact that when the long inductive night of *Xanthium* is interrupted with a few minutes of light the RNA content is not increased.

[The presence of RNA is detected by pyronin. Pyronin is a dye that stains RNA. Sections of shoot apex were stained with pyronin. Pyronin combines with RNA and the latter stains pink. Extracting RNA from sections with the enzyme ribonuclease can test the specificity of the dye pyronin. These sections show no stain of RNA. The layer where RNA is present can be detected with the dye pyronin. Another dye pyronin – methyl green was evolved for localization of RNA and DNA in a tissue (Sharma, 1985). The more highly polymerized DNA combines with methyl green and so the DNA stains green. The cytoplasm and nucleolus stains pink. The less polymerized RNA also stains pink. To locate RNA in these sections, RNA is extracted form some parallel sections and the sections were stained with pyronin. A comparison between the test section and parallel section will locate the presence of RNA.]

Following floral induction Barbar *et al.* (1968) demonstrated changes in the content of soluble protein in the vegetative apex. It was observed that after four inductive short days the concentration of total protein was much greater than the vegetative apices. Following five inductive short days there was decrease in staining for histones.

Corson (1969) demonstrated that in *Datura stramonium* the vegetative apex enters the transition phase when 7 or 8 leaf primordia are formed. The apex shows four zones on the basis of staining property. The summit one, two peripheral flank zones and one central zone compose the four zones of the apex as is revealed in longitudinal sections of vegetative and transition shoot apices. Corson analyzed the mitotic index of the different zones. The mitotic index of the summit zone of vegetative apex was significantly lower than the other three zones. The mitotic index in the summit and central zones of transitional apices increased significantly. The increase of mitotic index was accompanied with the increase in RNA, total protein and phospholipid.

## Formation of floral organs

The formation of flower or an inflorescence occurs during and after the transition of shoot apex to flowering phase. The introduction of scan electron microscope has greatly advanced the study of development of floral organs in a flower and floral primordia in an inflorescence. In an inflorescence apex a series of small, leaf-like primordia of involucral bracts appear first. Subsequently the floral primordia are formed in acropetal sequence. Floral primordia gradually develop over the whole surface of the inflorescence apex. In the apex of an individual flower the floral organs, in most cases, appear in distinct acropetal sequence (Fig. 30.3). The recently formed organ is closest to the reproductive apex.

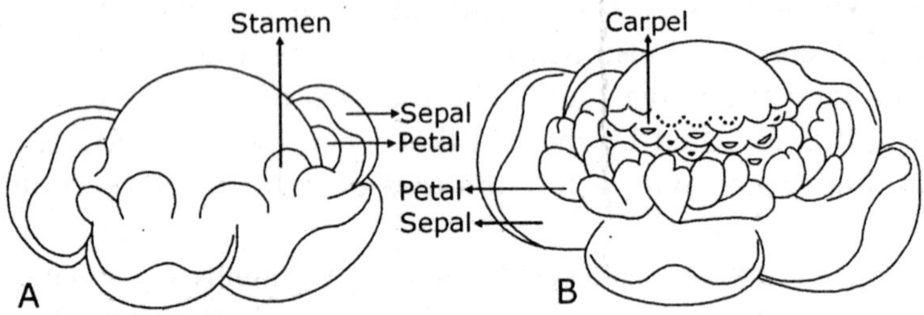


Figure 30.3

Diagrammatic illustration of the two stages – A and B in the development of the entire flower of *Ranunculus trilobus*. Adapted after Fahn, 1997.

Exceptions are noted in certain genera and families where basipetal or centrifugal development of the primordia of stamen is noted in the zone of stamens. Ex. *Paeonia*, Bixaceae, Dilleniaceae, Tiliaceae etc. There are many other exceptions that are mentioned in the literature of Fahn (1997).

The apical meristem of an individual flower and inflorescence consists of a central zone of cells covered by an outer zone of cells. The cells of the central zone have large vacuoles. The cells of outer zone are smaller than the cells of central zone. The cells of outer zone are rich in protoplasm and they divide anticlinally. It is noted that the number of layers of cell in the outer zone of reproductive apical meristem is larger than that is present in apical meristem of vegetative apex.

The various lateral members of a flower like sepal, petal, stamen and carpel originate from the second and third layers of apical meristem, present in the reproductive apex of an individual flower. Ex. *Nuphar*, *Aquilegia* etc. In *Portulaca* the second layer of apical meristem gives rise to the lateral organs of a flower by periclinal divisions. Leaf, the other lateral member of vegetative shoot apex, also shows similar type of development. This led Cutter (1971) to interpret that all the lateral members of the shoot apex are 'different manifestations of a leaf-like structure'—p. 223.

In angiosperm there exist a number of variations in the development of flower. No single example can be cited as a typical for the whole group. Boke in 1948, illustrated the development of flower of *Vinca rosea* (Fig. 30.4). In the reproductive apex the sepals are formed in spiral sequence. After the evocation of sepals the apex enlarges. Petal primordia (five in number) are formed on the enlarged apex. The petals unite during development. The stamen primordia originate from the basal region of corolla tube. After the formation of stamen the floral apex reduces considerably. The second layer of tunica and the peripheral layers of corpus of

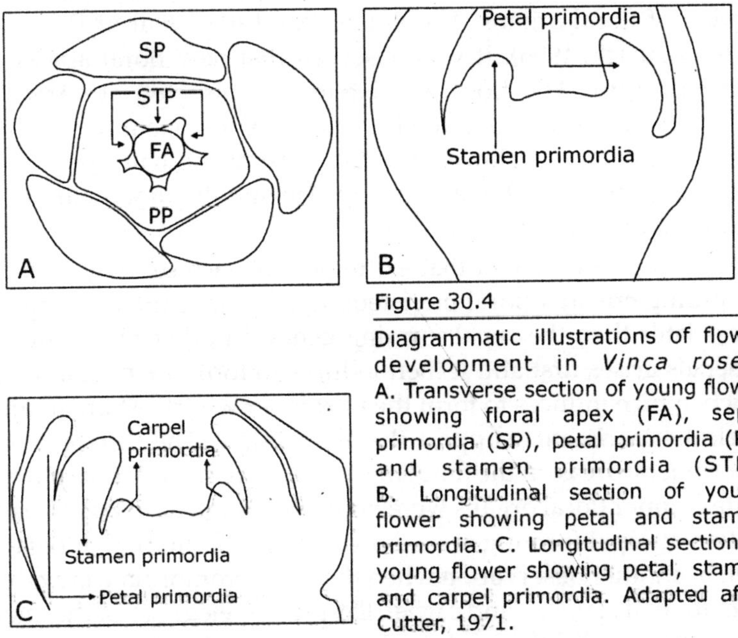


Figure 30.4

Diagrammatic illustrations of flower development in *Vinca rosea*. A. Transverse section of young flower showing floral apex (FA), sepal primordia (SP), petal primordia (PP) and stamen primordia (STP). B. Longitudinal section of young flower showing petal and stamen primordia. C. Longitudinal section of young flower showing petal, stamen and carpel primordia. Adapted after Cutter, 1971.

floral meristem divide periclinally and form the primordia of sepals, petals and carpels. The filament of stamen is formed as a result of intercalary growth at the base of stamen primordium. The diameter of floral meristem increases again at the end of stamen formation. The cells at the periphery of the meristem form a meristematic ring. On the two opposite sides of meristemaic ring carpel primordia are formed.

The floral members and vegetative leaves exhibit similar growth pattern. Sepals and petals undergo apical, marginal and intercalary growth. The apical growth is of shorter duration. Increased activity is observed in plate meristem. Marginal growth occurs in stamen and carpel. Thus, the leaf and the lateral members of a flower exhibit similar mode of growth and origin.

The shoot apical meristem forms diverse type of lateral organs. Normally it forms leaves and after switching over to floral meristem it forms all the floral organs. Naturally it is assumed that the shoot apical meristem undergoes various changes during ontogeny. Wardlaw (1961) suggested that in the apical meristem during development there is a 'sequential evocation of genes'. These genes, when expressed, form the different lateral organs. The expression of each gene is dependent on the preceding one. Heslop-Harrison (1964) holds more or less same opinion that each set of gene expresses to develop each phase of flower development. The expression of each set is dependent on the preceding act. Heslop-Harrison considers that each set of floral gene becomes depressed during ontogeny thus allowing the other set of genes to express.

The sequential evocation of genes causes the floral meristem to pass through different physiological states. The successive different physiological states regulate the formation of each kind of floral organ. Microsurgical experiments were designed to test the formation of various physiological states and consequently the formation of floral organs. The young floral meristem of *Primula bulleyana* was bisected (Cusick, 1956). It was observed that new floral apices developed from each half apex. The floral meristem was bisected at various stage of development. It was observed that when the floral meristem was bisected before sepal formation, complete flower developed from the half apices. In the later stages of development the flower develops lacking the organ adjacent to which incision was made.

It is previously mentioned that expression of each set of gene is dependent on the preceding one in a floral meristem. In this sense the petal forming genes will express later than the sepal forming genes. In other words the peripheral whorl of sepals arises first and it then induces to form the next inner whorl, i.e. petal, which in turn induces to form the next inner whorl—stamens and so forth. Microsurgical experiments support the above view. The different floral organ forming peripheral areas on the floral meristem of *Nicotiana tabacum* were resected and the fate of inner floral organs were noted (McHughen, 1980). It was observed that resection at sepal forming areas prevented the formation of petals, stamens and carpel. In another observation when the petal promordia were resected there are no formation of stamens and carpel. Despite the elegance of the microsurgical experiments, the results obtained on the basis of molecular genetics reveal that the formation of petal or carpel is not dependent on the preceding one. Studies on ABC model of flowering (discussed later) reveal that any type floral organ can be induced to develop in any whorl at will. Though the conclusion derived from microsurgical experiments is inconsistent with the finding based on molecular genetics, it is consistent with the fact that the development of floral organs in the inner whorl is dependent on the formation of organs in preceding whorl regardless the type of organ produced in that whorl (Howell, 1998).

Two major factors of the environment control the induction of reproductive growth in plants. These are temperature and photoperiod. They are described in a tremendous bulk of Internet Websites and literatures and will be mentioned here in brief.

Some amount of vegetative development precedes the formation of flowering. A plant must attain certain age before it is 'ripe to flower'. Klebs introduced the term 'ripeness-to-flower' in 1913 (Devlin, 1969, p. 507). Later this term is used in the study of vernalization and to denote the time when a plant becomes sensitive to photoperiodic induction. The age varies markedly with species. In the seedling stage plants like *Chenopodium rubrum*, *Pharbitis nil* etc. can be induced to flower. *Xanthium strumarium* and *Hyoscyamus niger* need a few weeks of age when they can be induced to flower. The forest trees need tens of years of vegetative growth before the formation of any flower. Environmental stimulus causes the transition to flowering of the vegetative shoot apex. When a plant is 'ripe to flower' a period of exposure to low temperature is required for flowering.

The stimulus of transition to flowering of vegetative apex by exposure to low temperature is called vernalization. Vernalization induces or accelerates flowering by low temperature. Normally annuals are vernalized at seedling stage and biennials perceive the cold temperature stimulus after the first season of growth (ex. *Digitalis, Apium graveolens* etc.). The spring-flowering species like *Viola* and *Primula* perceive the low temperature stimulus during the preceding winter.

Vernalization is well studied in cereal crops like wheat, rye (*Secale cereale*) etc. These crops have several varieties. There exists marked difference in reproductive behavior between the varieties. Vernalization may be illustrated by the following example. The seeds of winter variety of wheat when sown in autumn, they flower in early summer. The seeds germinate and the seedlings attain some growth before winter. At this stage the seedlings consist of large number of leaves that remain close to the ground in the form of a mat. This stage of wheat seedling is referred to as rosette stage. The growth is minimum at this stage in winter. Normal growth is resumed with the advent of spring and the crop is produced following flowering in summer. However if the seeds of winter variety of wheat are sown in late spring, they will grow up to the rosette stage. These plants fail to form flower. In contrast, the spring variety of wheat shows a different reproductive behavior. The seeds of spring variety of wheat when planted in spring, they grow and produce crop following flowering in early summer. The seeds of winter variety can be converted to spring variety by vernalization treatment in the following way. The seeds of the winter variety were moistened with water and allowed to germinate. The germinated seeds were then stored for a few weeks at 0° to 5°C. Such treated seeds when sown in the spring, they grow and produce crop following flowering in the early summer. The rosette stage is omitted and the seedling responded like spring variety. The winter variety of wheat needs low temperature (0°-5°C) for vernalization but the plants like millet, sorghums, maize, soybeans etc. need a high temperature of 20°-30°C for 5-10 days (Datta, 1980).

The growing tip consisting of the meristematic tissues is the site of vernalization. Certain plant like *Arabidopsis thaliana* can be induced to flower at the stage of seed. Researches reveal that the embryo is the seat of response of vernalization in a seed. The effect of vernalization is transmitted within the plant.

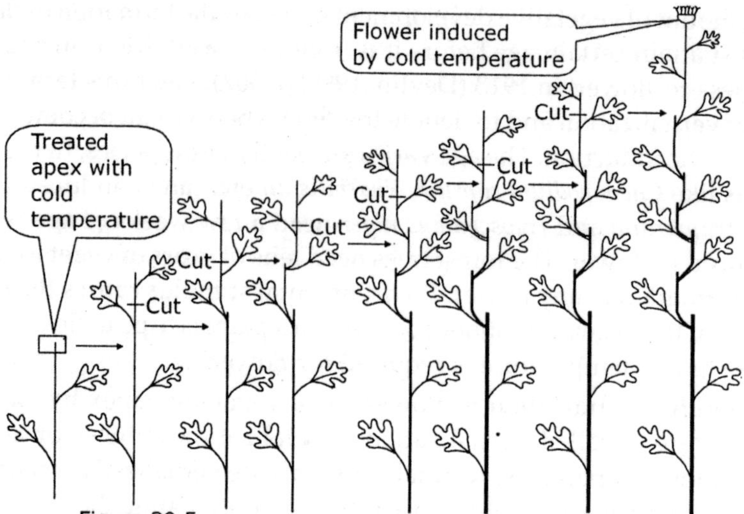


Figure 30.5

Diagrammatic illustration of induction of flowering in *Chrysanthemum*. The terminal apex of the far left figure is treated with cold temperature. The effect of treatment propagates. Vide text for details. After Burgess, 1989.

The following experiment with *Chrysanthemum* reveals it (Fig. 30.5). A plant was grown and the apex was treated with chilled temperature. The treated apex was allowed to grow until it forms two more leaves. The apex and the terminal leaf were removed. The axillary bud, which is present in the second leaf from tip, was allowed to grow until it forms two more new leaves. The apex and the terminal leaf were removed and the axillary bud present in the next leaf was allowed to grow to form two more new leaves. The apex and the terminal leaf were again removed and this process was repeated until seven pairs of leaves were produced. It is to note that all the seven pairs of leaf are developed from the original apex that is treated with cold temperature. The terminal bud and leaves of the 7$^{th}$ pair were not removed. The terminal bud was allowed to grow. It was observed that it produced flower (Burgess, 1989, p. 211).

Once an apical meristem is vernalized it seems to be permanent. All growth forms from it are also vernalized. The effect of vernalization in *Chrysanthemum* is confined to the treated apex only. Flowering is not induced in non-chilled apices of the same *Chrysanthemum* plant. This argues against the presence of a transmissible stimulus as a result of induction of vernalization. In favour of the presence of transmissible stimulus the following grafting experiment is cited (Burgess). *Hyoscyamus niger* was vernalized and a leaf from it was grafted on to unvernalized *Hyoscyamus*. It was observed that in unvernalized *Hyoscyamus* flowering was induced. The effect of vernalization was transmitted to the unvernalized *Hyoscyamus* that led to the induction of flowering. The transmissible stimulus is common to more than one species. It was revealed from the experiment that the grafting of vernalized leaf of *Hyoscyamus* was successful on *Petunia* and *Nicotiana*.

Later researches suggest that the effect of vernalization is mediated through a hormone. Melchers (1936) postulated the existence of a flower-forming hormone called vernalin on the basis of the transmission of vernalization stimulus across a graft union between vernalized and unvernalized *Hyoscyamus* plant. Vernalin is hypothetical hormone that has not been extracted as yet even in crude form. Florigen is hypothetical flowering hormone. Cajlachjan (1936) coined the word 'florigen' for the hypothetical unisolated flowering hormone. Vernalin is assumed to be the precursor of florigen (Fig. 30.6) or catalyser of its formation (Lang, 1952).

Lang (1957) first reported that exogenous gibberellic acid (GA) could cause flowering in vernalized-requiring plants under non-inductive environmental condition and in many species of long day plants. Chailakhyan (1970) emphasized the role of GA on the flowering of many plants. Precursors of GA are produced and accumulated during vernalization (Chailakhyan, 1966). These precursors change into gibberellins in long days. So it is regarded that the precursors of gibberellins are hormone, which is assumed to be the vernalin. Vernalin is

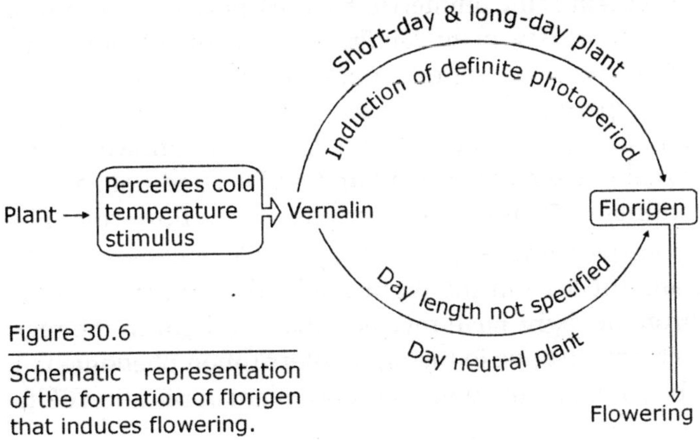


Figure 30.6

Schematic representation of the formation of florigen that induces flowering.

produced on the meristematic shoot apex and seems to move only with dividing cells (Chailakhyan, 1968).

After vernalization if a plant gets definite photoperiod flowering occurs. Garner and Allard (1920) termed the response of plants to the length of day as photoperiodism. Garner and Allard classified angiosperms into three main groups according to day-length that affects the ways of flowering. These groups are referred to as short-day plant (SDP), long-day plant (LDP) and day-neutral plant (DNP). SDP requires day length that is reduced to a certain critical value. In LDP flowering occurs only if the day-length exceeds a certain critical value. In DNP the day-length does not exert any effect on the time of flower initiation. Though the SDP, LDP and DNP are classified on the basis of critical day-length, it is regarded that photoperiodism is a response to the duration and timing of light and dark periods. Photoperiodism occurs in natural 24-hours light-dark cycles and so the duration and timing of light and dark periods are referred to 'critical day-length' and 'critical night-length'. In LDP critical day-length is the minimum day-length required for flowering. In SDP, the critical day-length is the maximum day-length on which flowering can occur. In contrast the critical night-length in SDP is the minimum period of darkness required for flowering, whereas in LDP the critical night-length is the maximum period of darkness required for flowering. So in LDP, for flowering, the day length must be longer than critical day-length and the dark period must be shorter than the critical night-length. On the contrary in SDP, for flowering, the day length must be shorter than the critical day-length and the dark period must be longer than the critical night-length. So it is the day lengths—shorter or longer than the critical day-length will determine whether flowering will occur. The distinction between SDP and LDP in not based on the absolute values of the critical day-length. The following illustration will explain the above discussion. *Xanthium strumarium* is a short-day plant. The critical day-length of this plant for flowering is about 15.5 hours. *Hyoscyamus niger* is a long-day plant. The critical day-length of this plant for flowering is about 11 hours. When 12 to 14-hour of photoperiods in 24-hour light-dark cycles was given to *Xanthium strumarium* and *Hyoscyamus niger*, flowering occurred in both species.

Later researches reveal that the single most important controlling factor for flowering is the length of the uninterrupted dark period. For example, to stimulate flowering in the short-day plant *Xanthium strumarium* 8-hour day and 16-hour night are required. Such appropriate photoperiod is referred to as inductive day or night. In an experiment with *Xanthium* a brief flash of light interrupts the continuous 16-hour inductive dark period. This inhibits flowering. On the contrary it is observed that when a LDP is kept under SDP condition, a brief flash of light during the continuous 16-hour dark period induces flowering (Fig. 30.7).

Using action spectrum technique it is possible to determine the wavelengths of light that are effective in the night interruption phenomenon. Experiments with monochromatic light have revealed that red light in the region of 660 nm wavelength is most effective in the night interruption phenomenon. The effect of red light can be counteracted with the subsequent application of 'far-red' light in

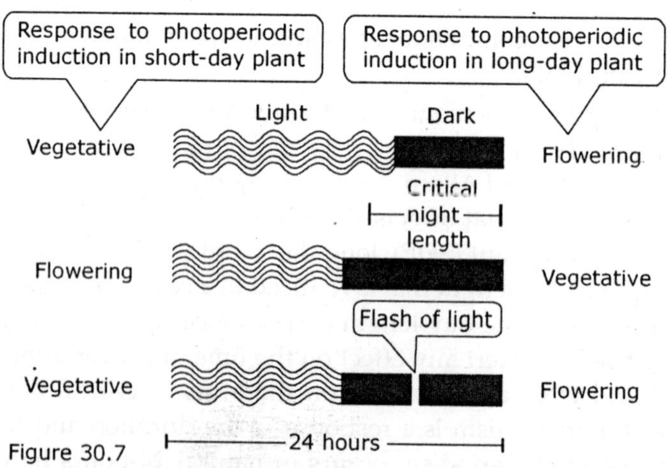


Figure 30.7

Schematic representation on the effect of night interruption by a flash of light in long-day and short-day plant. Adapted after Galston, 1963, p. 100.

the region of 730 nm wavelengths. The effect of red and far-red light is depicted in Fig. 30.8. In SDP when the long inductive night was interrupted by red light the flowering was inhibited, while by the subsequent application of far-red after red promoted flowering. The experimental plant may be exposed several times to red followed by far-red and accordingly the inhibition and promotion of flowering is observed. The effect of light on flowering is always dependent on the last quality of light exposure, i.e. when it is far-red flowering occurs, but when it is red, flowering is inhibited.

The above behavior was later interpreted on the basis of the presence of phytochrome. Phytochrome is a pigment and it is involved in floral initiation. It is a water-soluble chromoprotein and bluish in color. It is a dimeric chromoprotein

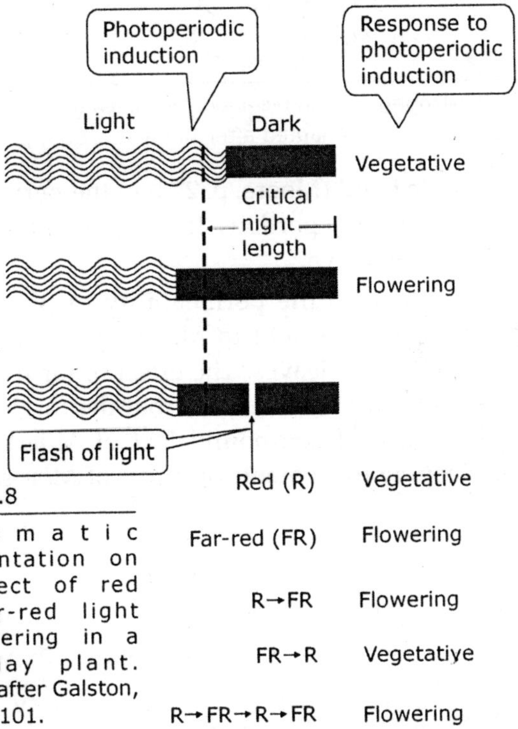


Figure 30.8 Schematic representation on the effect of red and far-red light on flowering in a short-day plant. Adapted after Galston, 1963, p. 101.

and consists of two chromophores, each of which consists of linear tetrapyrrole compound covalently linked to a cysteine residue (Moore, 1994). Phytochrome exists in two states that are referred to as $P_r$ and $P_{fr}$. Each form is photointerconvertible, i.e. each form is capable of being converted to the other form by absorbing light of appropriate wavelength. $P_r$ form absorbs light maximally in the red region of 660 nm wavelength. $P_{fr}$ absorbs light maximally in the far-red region of 730 nm wavelength and it serves as the photoreceptor. The 'red, far-red reversible photoreactions' of flowering occur through these inter-convertible forms of phytochrome. It is to note that both $P_r$ and $P_{fr}$ also show absorption in the blue region at about 370 nm and 400 nm respectively but their action spectra for the photo-transformations are restricted to 660 nm and 730 nm respectively.

$P_r$ is metabolically inert form. The active form of phytochrome is $P_{fr}$. Phytochrome is synthesized in $P_r$ form. It is red-light absorbing form and red light converts $P_r$ to $P_{fr}$. $P_{fr}$ may then exert its action by promoting flowering. $P_{fr}$ is the far-red absorbing form. $P_{fr}$ is reverted to $P_r$ by far-red. $P_{fr}$ may also revert to $P_r$ in the dark. $P_{fr}$ may be destroyed without exerting its action (Fig. 30.9).

Evidence suggests that during photo-transformation of $P_r$ to $P_{fr}$ in one chromophore, a proton migrates from pyrrole ring to its ethylidene group. This

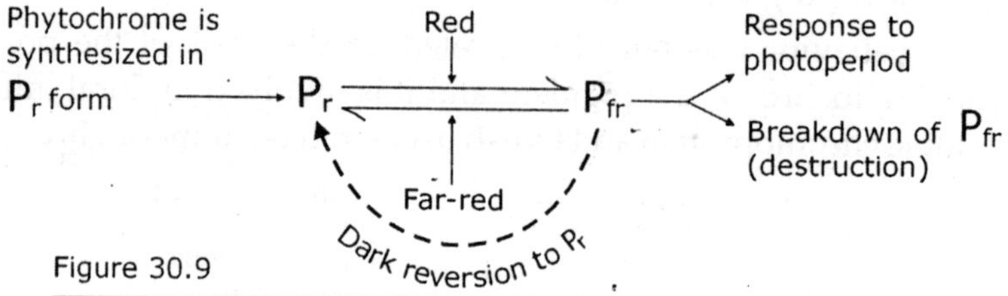


Figure 30.9

Diagram illustrating the interrelationships between $P_r$ and $P_{fr}$ form of phytochrome. Adapted after Burgess, 1989, p. 221.

causes a shifting in double bond (Moore, p. 272) in the pyrrole ring 1 (Fig. 30.10). Changes also occur in the protein portions. It is regarded that protein accounts $P_r$ being inactive and $P_{fr}$ being active.

Phytochrome has effect on the permeability of plasma membrane. For example, the exposure of far-red light to *Mimosa pudica* at the end of the day inhibits the sleep movements of leaves. The effect of far-red can be nullified by the application of red light. It is now known that the turgidity and flaccidity of pulvinus—the swollen base of compound leaf of *Mimosa* is responsible for seismonasty, i.e. the sleeping movements of leaves of *Mimosa*. Evidence suggests

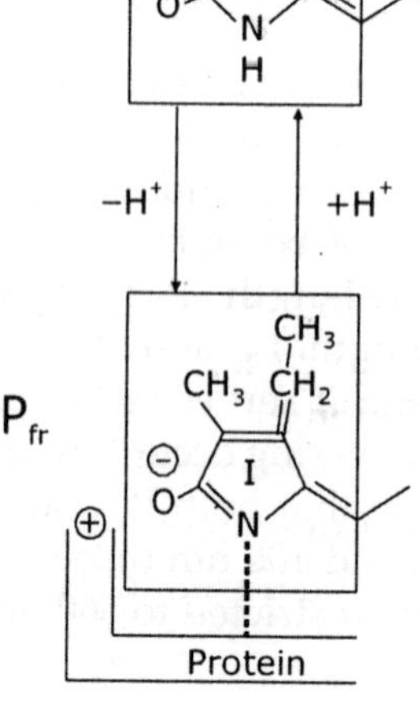


Figure 30.10

Diagram illustrating the proposed structure of chromophore (phytochromobilin) of phytochrome, its phototransformation mechanism and linkage to protein. Redrawn after Moore, 1994, p. 273.

that the movement of potassium ions controls the turgor of cells present in the pulvini. Hence it is regarded that phytochrome regulates the permeability of plasma membrane. So far as transition to flowering is concerned, it is suggested that phytochrome changes the permeability of plasma membrane of cells present in the shoot apex. As a result the entry of growth hormones like gibberellins, cytokinin etc. is regulated. The level of growth regulators determines the response to flowering.

The leaf perceives the light stimulus. A single leaf of *Xanthium*, more precisely one eighth of one leaf is enough to cause the plant to flower. Evidence indicates that the stimulus developed as a result of inductive photoperiod is transmissible. The site of perception and the site of responses are distantly placed. This led to postulate the hypothetical flower forming hormone—'florigen'. The concept of florigen complex is now based on the following four groups of hormones and physiologically active compounds—indole acetic acid, phytokinins, gibberellins and anthesins. Gibberellins and anthesins form an integrated system of metabolism that precedes flowering. Gibberellins are necessary for the formation and elongation of stem and anthesins are necessary for flower formation. The above mentioned four groups of hormone influence flowering (Chailakhyan). Lang (1965) after extensive grafting experiments concluded that the florigen complex is same in LDP, SDP and DNP.

## Phylogenetic origin of flower

Most Taxonomists agree that angiosperms form a natural group. They are monophyletic—that is, their origin is from one ancestor only. But the ancestor is to be identified definitely (Stewart, 1983). The ancestral organs from where the extant floral parts evolved are not known. Wilson (1982) regards that angiosperms evolved from gymnospermous ancestors during the Lower Cretaceous Period at about 135 million years ago. The oldest fossils of flower are reported from Albian Epoch. The fossils are about 120 million years old (Dilcher, 1979). The fossils are casts of conduplicate carpel. The conduplicate carpel resembles a folded leaf lamina with appressed margin. Among the living angiosperms the conduplicate carpel is found in *Drimys*, Winteraceae and *Degeneria* (Degeneriaceae). It is now known that Winteraceae and Degeneriaceae have many primitive characters. So it is assumed that the ancestor of carpel is similar to that of *Drimys* and *Degeneria*.

The conduplicate carpel is an elongated structure without any distinction into stigma, style and ovary. Innumerable ovules occur in longitudinal rows along the ovary wall. Numerous papillae are present along the margins of the folded lamina. The papillose surfaces of the margins close the opening of the carpel and thus form a stigmatic crest. Pollen grains are received at the longitudinal crest. The stigmatic crest in *Drimys membranea* extends longitudinally from the tip along one side up to the base of ovary. Pollen tube grows through the papilla that acts as stigmatic surface. During evolution the lower and middle portions of conduplicate carpel stopped producing papillae and the middle region elongated. As a result stigma and style evolved.

The fossil flower, reported by Dilcher, has carpel only; the other floral organs like sepals, petals and stamens are absent. The fossil flowers had scars at the region of other floral parts because they had fallen away. The living gymnosperm has microsporophyll that is leaf like. So it is assumed that the ancestral flower had broad stamen. This type of stamen is found in primitive angiosperms, e.g. *Drimys*, *Degeneria*, *Austrobaileya* and *Himantandra*. In these genera the stamens are flat and slightly thickened. They lack the characteristic filament and anther; instead four long sporogenous tissues are present (Fig. 30.11). In *Degeneria* the sporangia were deeply sunken (Majumder, 1963). During evolution the sporangia became superficial.

Lastly it is to mention that due to lack of fossil evidences, the evolution of flower is hypothesized on the basis of comparative morphology between gymnosperm and extant angiosperms that have primitive features.

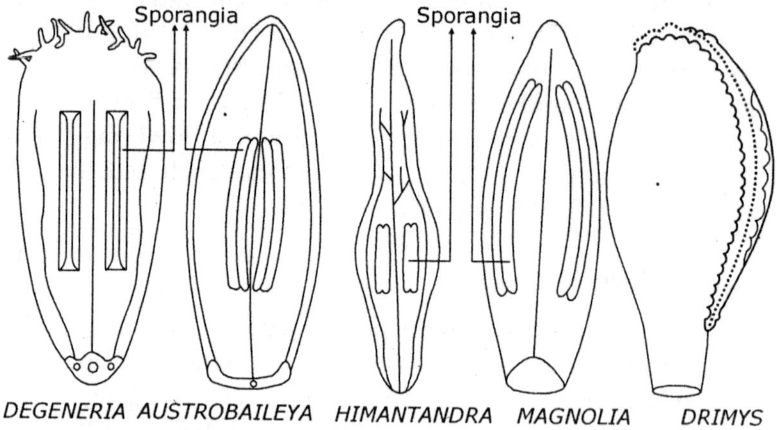


Figure 30.11

Diagram showing stamens and a carpel (*Drimys*) of extant angiosperms that are regarded to have primitive features.

## ABC model of flower development

The ABC model of flower development in angiosperm demonstrates the presence of three classes of genes that regulate the development of floral organs. The genes are referred to as class A genes, class B genes and class C gene. These genes and the interaction between them induce the development of floral organs. Many literatures on molecular genetics and Internet Websites provide articles on ABC model. In the following essay the basic concept of ABC model will be discussed in brief. The analysis of ABC model is based on the use of molecular genetics and formulated on the observation that mutants induce right floral organs to develop in wrong whorls.

In the flower of angiosperms there are usually four concentric whorls of organs, i.e. sepal, petal, stamen and carpel that are formed in whorl 1, whorl 2, whorl 3 and whorl 4 respectively, the whorl 1 being on the peripheral side. In the whorl 1 class A genes when expressed induce the development of sepals. The interaction between class A and class B genes induce the development of petals in the whorl 2. Stamens are formed in the whorl 3 as a result of interaction between class B and class C genes. In the whorl 4 class C gene induces the formation of carpel. So the summary of ABC model is: class A genes together and class C gene alone are responsible for the development of sepals and carpel respectively. The class B genes and class A genes function cooperatively to determine the development of petals. The class B genes and class C gene act together to induce the development of stamens (Fig. 30.12).

Coen *et al.* (1991) formulated the ABC model. While analyzing the mutations affecting flower structure Coen *et al.* identified the class ABC genes that direct

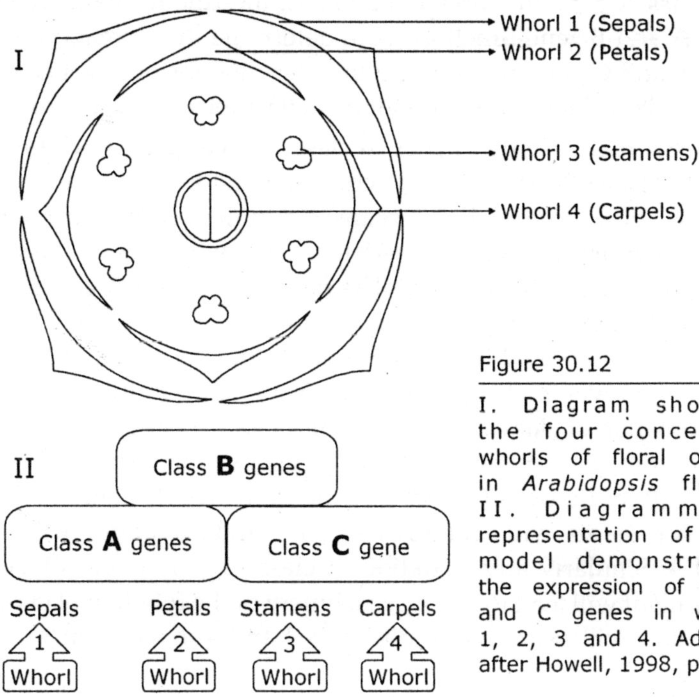


Figure 30.12

I. Diagram showing the four concentric whorls of floral organs in *Arabidopsis* flower. II. Diagrammatic representation of ABC model demonstrating the expression of A, B and C genes in whorls 1, 2, 3 and 4. Adapted after Howell, 1998, p. 196.

flower development. They also formulated the molecular models of how floral meristem and organ identity may be specified. They have shown that the distantly related angiosperm plants use homologous mechanisms in pattern formation of floral organs. Ex. *Arabidopsis thaliana* and *Antirrhinum majus.*

The following two have led to formulate ABC model: (1) the discovery of homeotic mutants (homeotic genes identify specific floral organs and help the organ to develop in respective whorl. The homeotic mutant has inappropriate expression—that is, it induces right organ to develop in wrong whorl. As for example—petals emerge in the whorl where normally stamens develop) and (2) the observation that each of the genes that induce the formation of an organ in a flower has an effect on two groups of floral organs, i.e. sepal and petals or petals and stamens.

Class A, B and C genes are homeotic genes. They determine the identity of different floral organs and induce the organs to develop in their respective whorls.

The homeotic mutants have defects in floral organ development and induce the right organs to develop in wrong whorls / place, i.e. one floral organ develops in the whorl, which is the normal position of another floral organs. Petals, for example, develop in the whorl where stamens are normally to be formed.

In each whorl of a flower there is one or more homeotic genes and their cooperative functions determine the organ to be formed in that whorl. For example, the activity of class A genes is restricted to whorls 1 and 2. The class B genes have function in whorls 2 and 3. The class C gene functions in whorls 3 and 4. Another way of describing the function of class A, B and C genes is that—in whorl 1, the class A gene-function alone determines the formation of sepals; in whorl 2, class A and B gene-functions both determine the formation of petals; in whorl 3, class B and C gene-functions both determine the emergence of stamens and in whorl 4, class C gene-function alone determines the carpel formation.

In *Arabidopsis* there are two genes in class A, two genes in class B and one gene in class C (Table 30.1). The most characteristic feature of these homeotic genes is in the identification of floral organs and in the determinacy of position / whorl of their emergence in a floral meristem. The two genes of class A and the two genes of class B act cooperatively. The function of class A genes is confined to whorls 1 and 2. Similarly the function of class C gene is restricted in whorls 3 and 4. This can be interpreted in another way. In the whorls 1 and 2 the function of class A genes prevents class C gene from functioning in the same whorls. Similarly the function of class C gene prevents class A genes from functioning in the whorls 3 and 4. Any mutation in class A genes with defects in floral organ development will invite class C gene to express in whorls 1 and 2. The class C gene, in class A mutants, will express in whorls 1 and 2 in addition to the normal whorls 3 and 4. Similarly any mutation in class C gene with defects in floral organ development will lead to the encroachment of the function of class A genes. The class A genes will express in the whorls 3 and 4 in addition to the normal whorls 1 and 2.

| Class | *Arabidopsis* |
|---|---|
| Class A | *APETALA 1 (AP1), APETALA 2 (AP2)* |
| Class B | *APETALA 3 (AP3), PISTILLATA (P1)* |
| Class C | *AGAMOUS (AG)* |

Table 30.1

Illustrating Homeotic genes in *Arabidopsis*. Adapted after Howell, 1998, p. 197.

The following three examples of homeotic mutant genes will illustrate the above discussion (Fig. 30.13). (1) The flower of *Arabidopsis* with class A mutants, such as *apetala 1(ap 1)* shows the following pattern of floral organs (Fig. 3.13.II): whorl 1 shows bract-like structure with carpelloid characteristics; whorl 2 shows stamens; whorl 3 shows stamens and whorl 4 shows carpel. The pattern of floral organ formation in whorls 1 and 2 is changed. In *ap 1* mutants the activity of two genes of class A is lost. So the class C gene expressed in whorls 1 and 2 in addition to whorls 3 and 4. As a result carpelloid organ developed in whorl 1 and stamens formed in whorl 2. In the whorls 3 and 4 stamens and carpel respectively are formed similar to wild type (Fig. 30.13I). (2) Example: Flower of *Arabidopsis* with class B mutant, such as *apetala 3* (*ap 3*): The flower shows sepals only both in whorls 1 and 2, while the whorls 3 and 4 show carpel only (Fig. 30.13III). Class B mutant contains loss-of-function genes and as a result class A genes express in whorls 1 and 2; and class C gene alone expresses in whorls 3 and 4. In *ap 3* mutants

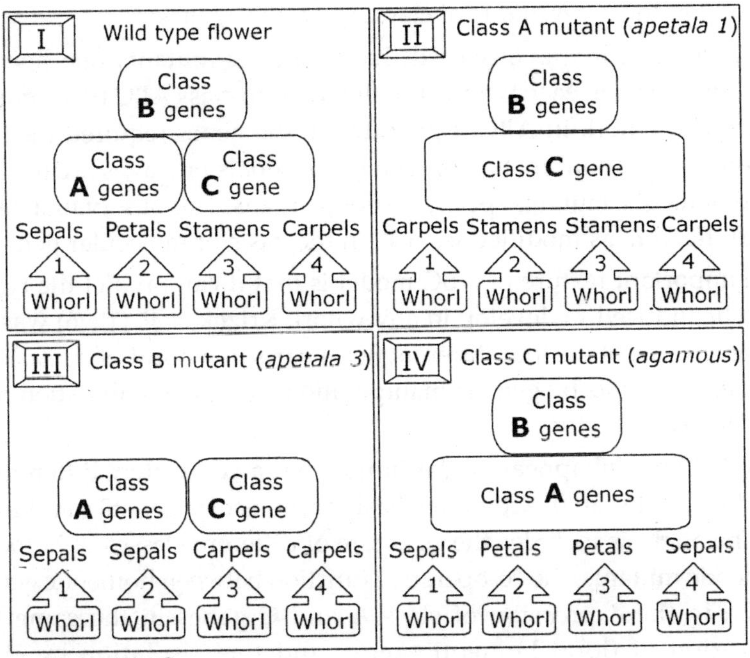


Figure 30.13

Diagrammatic illustrations of the floral organs formed in wild type and homeotic mutants of *Arabidopsis* flower. I. Wild type flower. The individual and cooperative expressions of class A, B and C genes are in the formation of sepals, petals, stamens and carpels in whorls of 1, 2, 3 and 4 respectively. II. Class A mutant, *apetala 1*. This homeotic mutant contains loss-of-function A genes. Class C gene expresses alone in whorl 1 in addition to whorl 4. Class C gene and class B genes together express in whorls 2 and 3. As a result stamens are formed in both whorls 2 and 3. III. Class B mutant, *apetala 3*. This homeotic mutant contains loss-of-function B genes. Class A genes express in whorls 1 and 2. As a result sepals are formed in both whorls 1 and 2. Class C gene expresses in whorls 3 and 4. So carpels are formed in both whorls 3 and 4. IV. Class C mutant, *agamous*. This homeotic mutant contains loss-of-function C gene. Class A genes express in whorls 3 and 4 in addition to whorls 1 and 2. As a result petals and sepals are formed in whorls 3 and 4 respectively. In whorl 3 class A and class B genes together form petals. In whorls 1 and 4 the sole group of class A genes expresses and so sepals are formed. [Howell (1998, p. 199) provides the scan electron micrograph of the phenotypes of the above illustrated mutants.]

in whorl 2, sepals are formed instead of petals and in whorl 3, carpel is formed instead of stamens. (3) In *Arabidopsis* the class C gene contains the sole gene *agamous* (*ag*). *Arabidopsis* flower with *agamous* (*ag*) mutant consists of many sepals and petals. The reproductive organs – stamens and carpel are not formed in the whorls 3 and 4. Class C gene with *ag* mutant contains loss-of-function gene. As a result class A genes express in whorls 3 and 4 in addition to 1 and 2. In *ag* mutant sepals and petals are formed in whorls 3 and 4 instead of stamens and carpel. The literature of Howell provides the scan electron micrograph of flower phenotypes of the floral homeotic mutants of class A, B and C genes.

In *Arabidopsis* it was observed that in all the mutants one homeotic gene remains functional in each whorl. The flower with class ABC triple mutant shows sepals in each whorl. In ABC triple mutant, the genes required for floral organ formation become nonfunctional. As a result sepals or leaves are formed in each whorl, as homeotic mutants specify no floral organs. This observation led Botanists to regard 'flowers as modified leaves' on the basis of molecular genetics.

The important feature of ABC model is that it can predict the type of floral organ to be induced to develop in any whorl. Krizek *et al.* (1996) was successful to induce any one of the four different floral parts in whorl 1 of *Arabidopsis* flower. This became possible by genetic manipulations of right combination of homeotic selector genes.

The ABC model appears to be simple, but a completely different picture is obtained when it is analyzed on the basis of molecular genetics and in molecular terms. The analysis includes the structure of different classes of homeotic genes, the homeotic mutants, the co-operative function between homeotic genes, mutual exclusion in the expression of class A and C genes in the same whorl, the identification of floral homeotic genes and their isolation by cloning, the production of MADS box protein by homeotic mutants, the study of genes that mediate the interaction between floral meristem and floral organ development, presence or absence of different classes of transcription factors etc., the details of which can be obtained in the literatures on molecular genetics.

*Arabidopsis thaliana* belongs to the family Brassicaceae and has become the model organism for understanding the genetics and molecular biology of flowering plants like mice and *Drosophila* in animal researches due to following reasons: (i) it has five chromosomes (n = 5) and so this small-size-genome is advantageous in gene mapping and sequencing; (ii) the size of plant is small and so can be cultivated in a small space and requires modest indoor facilities; (iii) it has rapid life cycle and takes about six weeks from germination to mature seeds; (iv) an individual plant produces several thousand seeds; (v) 'The *Arabidopsis* genome is among the smallest in higher plants, with a haploid size of about 100 megabases (mb) of DNA. With a small genome size it was expected that there would be fewer problems with gene duplication'—Howell (p. 19) and (vi) it is easily transformable with T-DNA mediated transformation.

In 2004 ABCE model has been formulated. The characterization of *sepallata 1, 2, 3* triple mutant in *Arabidopsis* has led to the above formulation. It is regarded that the class E genes have important role in the development of floral organs (Wikipedia, 2007).

Comparison between shoot apical meristem and floral meristem.

| Shoot apical meristem | Floral meristem |
|---|---|
| 1. It forms vegetative organs of a plant. | 1. It forms reproductive organs of a plant. |
| 2. It has indefinite growth. | 2. It has definite growth. |
| 3. It promotes longitudinal growth of the axis. | 3. It forms a large surface area of meristematic tissue where floral primordia develop. |
| 4. It has unlimited life. It produces vegetative organs continuously. | 4. It has limited life. It produces inflorescence and floral primordia and then dies. |
| 5. It is not compressed. The vegetative organs like leaves are widely spaced up on the stem. | 5. It is very compressed and tends to be flat so that all floral parts appear to be borne approximately at the same level. |
| 6. The growth of shoot meristem is not dependent on the duration of light. | 6. The growth of floral meristem is dependent on definite critical photoperiod. |
| 7. The apex of plumule is the shoot meristem. | 7. Shoot meristem is transformed to floral meristem. |
| 8. It consists of outer tunica and inner corpus layers. | 8. It consists of outer meristematic mantle and inner parenchymatous core which are respectively equivalent to tunica and corpus. |
| 9. It does not exhibit much size fluctuations during development. | 9. It exhibits much size fluctuations during development. The apex may be conical, elongated and flattened etc. |
| 10. The mitotic activity remains more or less constant. | 10. It exhibits increased mitotic activity as is revealed by mitotic index. |
| 11. The tunica cells contain little RNA. | 11. The central axial cells have more RNA. |
| 12. The corpus cells have small nucleolar volume. | 12. The tunica cells that form the surface layer of outer meristematic mantle have increased nucleolar volume. |
| 13. The cytoplasmic RNA content is usually restricted to the flanks of vegetative shoot apices. | 13. It contains more RNA than shoot meristem and the RNA uniformly spreads through out the apices. |

CHAPTER **31**

# Anatomy of Stems, Roots and Leaves of Some Plants

There exist wide diversities between stems regarding anatomical structures. This is also observed in roots and leaves. Differences in internal organization are due to different stages of development of the organs and formation of certain tissues in response to the environment where the plants normally grow. An outline of the internal structure of some plants are discussed below with comments citing examples from primary body, secondary growth, normal anatomical structure, anomalous forms and tissues formed to cope up with environment. Plants that can be obtained with ease are only selected for the study.

**Stem of *Helianthus annuus*** (Sunflower. Family Asteraceae: internal structure before secondary growth).

The transverse section through the internode of stem is more or less circular in outline and exhibits the following arrangement and structure of tissues from periphery towards centre (Figs. 31.1, 31.1A).

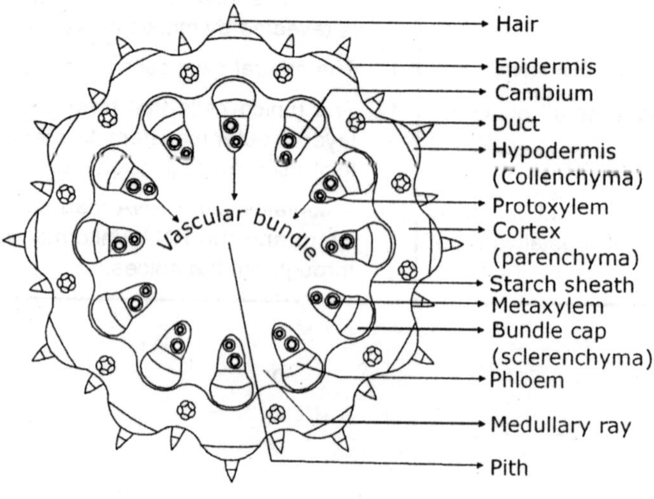


Figure 31.1

Diagrammatic cross-section of *Helianthus annuus* stem.

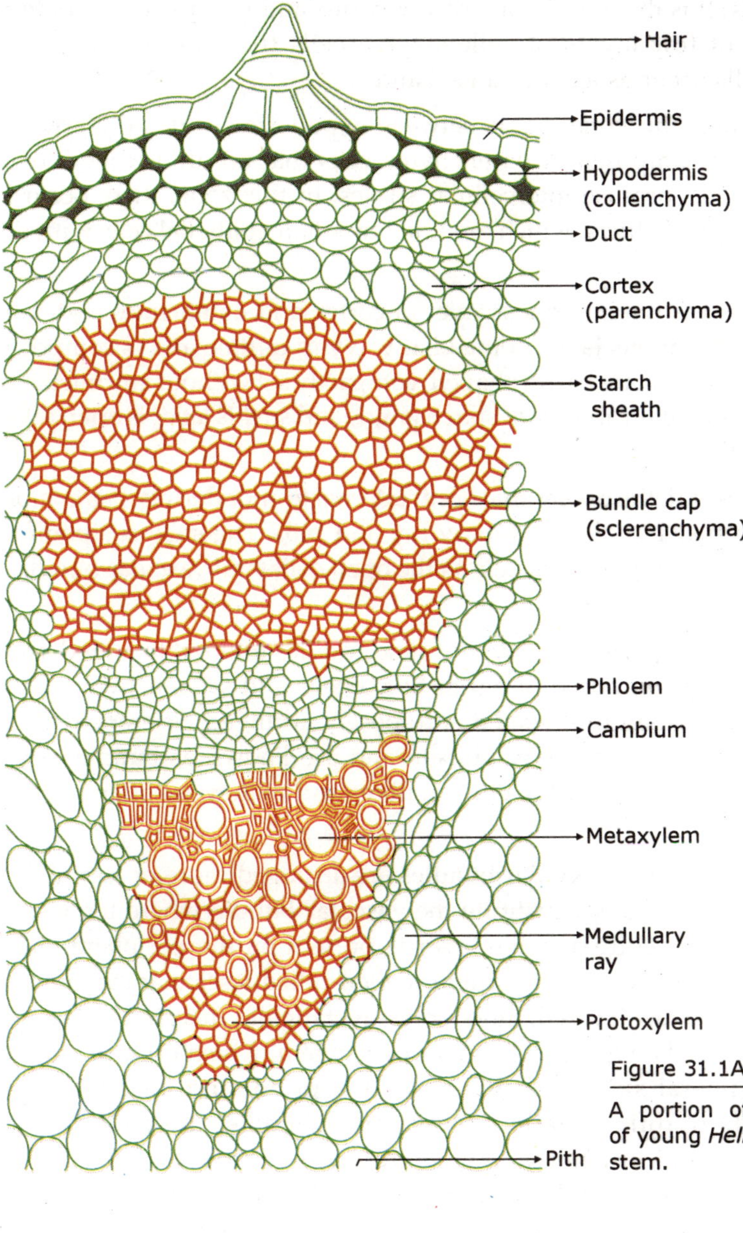


Figure 31.1A

A portion of cross-section of young *Helianthus annuus* stem.

**Epidermis:** The outermost layer is the epidermis. It is uniseriate and composed of living parenchyma cells. The cells are more or less tabular in shape, contain vacuolated protoplast and remain attached with the adjacent cells by their radial walls. There are no intercellular spaces and the continuity of the epidermis is sometimes interrupted by the presence of stomata. The outer walls are cuticularized and many multicellular hairs occur on them.

**Cortex:** The zone between the stele and epidermis is referred to as cortex. It is differentiated into hypodermis, general cortex and endodermis.

**Hypodermis:** It is the peripheral layer of cortex and is adjacent to epidermis. It is composed of few layers of collenchyma cells that are of angular type. Collenchyma cells occur as a continuous band.

**General cortex:** This zone occurs between hypodermis and endodermis. It is many layered and composed of parenchyma cells that are thin walled, oval or spherical, large with profuse intercellular spaces. In this zone, there occur here and there the secretory ducts, which are lined by small epithelial cells with dense protoplasm.

**Endodermis:** It is the innermost layer of cortex and demarcates cortex from the stele. It is a continuous layer, somewhat wavy in outline and is composed of parenchyma cells, which are barrel-shaped, compactly set without any intercellular spaces. These cells contain abundant starch grains and so this layer is often referred to as starch sheath.

**Stele:** It consists all the tissues that occur internal to endodermis. It is composed of pericycle, vascular bundle, medullary ray and pith. The stele is siphonostele due to the presence of pith and eustele as the open collateral vascular bundles are arranged more or less in a ring.

**Pericycle:** It is the peripheral layer of stele and lies internal to endodermis. It is many layered and composed of parenchyma cells and sclerenchyma patches. The parenchyma cells are thin walled and loosely arranged. Sclerenchyma cells occur as patches over the primary phloem and are known as bundle cap. The sclerenchyma cells are actually the phloem fibres and so they are often referred to as hard bast.

**Vascular bundle:** The vascular bundles are arranged more or less in a ring. Xylem and phloem together constitute the vascular bundle where the former is situated towards the centre and the latter is present on the peripheral side. In between the two, there occurs the cambium often called fascicular cambium. The cells of cambium are few layered, more or less rectangular in appearance and are storied. Xylem consists of protoxylem and metaxylem, which occur towards the centre and peripheral side respectively. As the proto- and metaxylem show centrifugal mode of differentiation, the vascular bundles are endarch. In the protoxylem elements, tracheids and trachea are of smaller diameter with characteristic spiral and annular thickenings. In the metaxylem, tracheids and trachea are of larger diameter with pitted and other type of thickenings.

**Medullary rays:** The cells of medullary rays are parenchymatous of various shapes, and occupy regions between the two vascular bundles. These narrow wings of parenchyma at the interfascicular region are in continuation with pith. In this region the secondary meristem interfascicular cambium originates at the time of secondary growth.

**Pith:** It is also known as medulla that is the innermost central region of stele. It is large and composed of parenchyma of various shapes with abundant intercellular spaces.

**Comment:** Cuticularized epidermis protects the inner tissues and checks the loss of water. Collenchyma cells present at hypodermis add mechanical rigidity to the stem. Xylem and sclerenchymatous bundle caps are mechanical cells. The sclerenchymatous bundle caps form the flanges of I-girder and they form a composite I-girder that provides mechanical rigidity against inflexibility. Secondary increase in thickness of the stem occurs through cambium. Dicotyledonous nature of stem is revealed due to the presence of collateral and open vascular bundle without bundle sheath and their ring like arrangement, and large parenchymatous pith.

## Stem of *Cucurbita* (Family: Cucurbitaceae)

In transverse section, the stem is wavy in outline with distinct ridges and furrows. The following internal tissue organization is exhibited from periphery towards the centre (Fig. 31.2).

**Epidermis:** It is uniseriate and composed of living parenchyma cells. The cells are more or less tabular in shape, contain vacuolated protoplast and remain attached with the adjacent cells by their radial walls. There are no intercellular spaces and the continuity of epidermis is sometimes interrupted by the presence of stomata, which are situated at the apices of small projections and raised above the general level of epidermis. The outer walls of epidermal cells are cuticularized and many multicellular hairs occur on them.

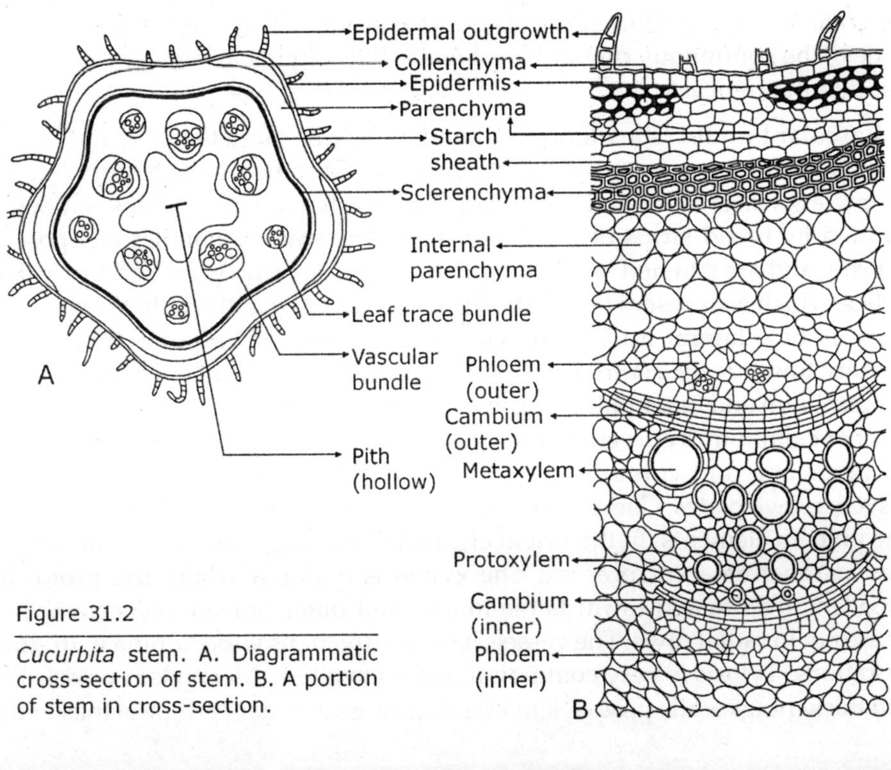


Figure 31.2

*Cucurbita* stem. A. Diagrammatic cross-section of stem. B. A portion of stem in cross-section.

**Cortex:** The zone between the stele and epidermis is referred to as cortex. It is differentiated into hypodermis, general cortex and endodermis.

**Hypodermis:** It is the peripheral layer of cortex and lie adjacent to epidermis. It consists of collenchyma and chlorenchyma cells. The collenchyma cells are not in a continuous band and interrupted by patches of assimilatory tissues like chlorenchyma extending to the epidermis. The collenchyma, below the ridges, is several layered, which become few layered internal to furrow. Chloroplasts may occur in some collenchyma.

**General cortex:** This zone occurs between hypodermis and endodermis. It is few layered and composed of parenchyma cells that are thin walled, larger in size and contains lesser number of chloroplasts than chlorenchyma cells.

**Endodermis:** It is the innermost layer of cortex and demarcates cortex from the stele. It is a continuous layer, somewhat wavy in outline and is composed of parenchyma cells, which are barrel-shaped, compactly set without any intercellular spaces. These cells contain abundant starch grains and so this layer is often referred to as starch sheath.

**Stele:** It consists of all the tissues that occur internal to endodermis. It is composed of pericycle, vascular bundle, medullary rays and pith. The stele is siphonostele due to the presence of pith.

**Pericycle:** It is the peripheral layer of stele and lies internal to endodermis. It is many layered and consists of sclerenchyma and parenchyma cells. The sclerenchyma is present internal to endodermis, few layered and occurs as a continuous band. Internal to this layer a few layers of parenchyma are present. These cells form a continuous ring and extend up to the outer ring of vascular bundles. The continuous ring of peripheral sclerenchyma and inner parenchyma constitute the pericycle.

**Vascular bundle:** The number of bundles may be more than ten and they are arranged more or less in two distinct circles accompanied by accessory bundles. Each circle consists of about five bundles. The bundles are of two sizes and widely separated by broad strips of medullary rays. The smaller bundles compose the outer ring and are situated below the ridges. The inner ring is formed by the large bundles, which are present below the furrows—meeting at the pith. Each vascular bundle is bicollateral and open. The arrangement of vascular tissues from periphery towards the centre is as follows: outer phloem, outer cambium, xylem, inner cambium and inner phloem. The outer cambium is many layered and storied. The inner cambium is curved and few layered. The outer phloem is more massive than inner one and contains sieve tubes, which are wide with conspicuous transverse sieve plates. Companion cells and phloem parenchyma are also present in the phloem elements. In the xylem elements tracheids, trachea, xylem fibre and xylem parenchyma are present. The xylem is endarch where the proto- and metaxylem are situated towards the inner—and outer phloem respectively, and they are not in radial rows. The vessels have conspicuous wide lumina with simple perforations. Tylosis is very common in the metaxylem. Metaxylem vessels have pitted thickenings while protoxylem vessels possess annular and spiral thickenings.

**Medullary ray:** The cells of medullary ray are parenchymatous of various shapes and occupy regions between the two vascular bundles. They are broad and continuous with the pith. In this region interfascicular cambium originates at the onset of secondary growth.

**Pith:** The pith is hollow, i.e. there is no tissue at the centre.

**Comment:** Epidermis with cuticle is protective tissue. Stoma is meant for gaseous diffusion. At hypodermis the chlorophyllous parenchyma is the photosynthetic tissue and the collenchyma cells are mechanical cells. The dicot nature of the stem is revealed due to the presence of pith in the stele, arrangement of vascular bundles in rings, bicollateral and open vascular bundles. Anomaly in the stele is due to the formation of secondary xylem and phloem in individual vascular bundles only in mature stems. The outer cambium forms the secondary vascular tissues only. At interfascicular region parenchyma cells are only formed. The plant is a climber and it is revealed by the presence of large metaxylem vessel, wavy pericycle and, presence and formation of parenchyma between vascular bundles. The xylem is the mechanical cell and provides mechanical rigidity against inflexibility. Presence of parenchyma at the interfascicular region adds flexibility to the stem during climbing.

## Stem of *Leonurus* (Family: Lamiaceae)

The transverse section of stem is more or less square in outline and shows the following internal tissue organization from periphery towards the centre (Figs. 31.3, 31.3A).

**Epidermis:** It is single layered, composed of horizontally flattened cells, compactly set with each other without any intercellular spaces and cuticularized on the outer walls. The epidermis is interrupted by stomata and numerous multicellular unbranched hairs occur on it here and there.

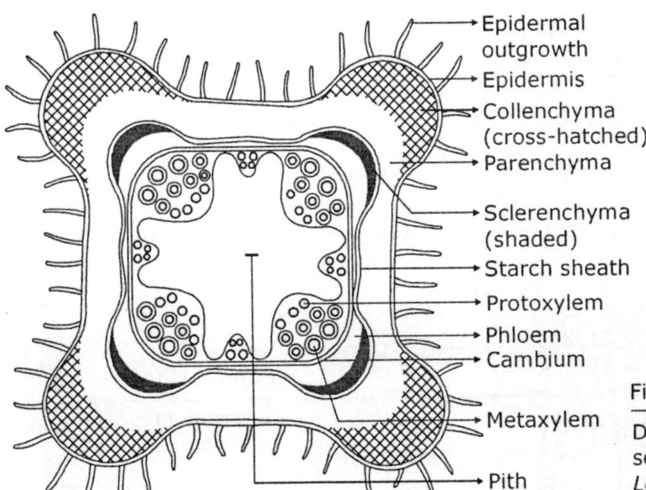


Figure 31.3

Diagrammatic transverse section of young stem of *Leonurus*.

**Cortex:** The peripheral layer of cortex is hypodermis. It is composed of collenchyma cells. The four ridges of stem are densely packed with collenchyma cells. Collenchyma cells do not form a continuous cylinder. Chlorenchyma cells interrupt it. These cells usually occur beneath the stomata. Just below the hypodermis there lie parenchyma cells. They are loosely arranged with intercellular spaces. The innermost layer of cortex is starch sheath or endodermis, the cells of which are barrel shaped, compactly set and contains abundant starch. The endodermis is conspicuous in the stems of those plants, which are at flowering stage.

**Stele:** It is limited externally by endodermis. Below endodermis pericycle or perivascular cylinder is situated. It is composed of sclerenchyma, which is in the form of a continuous cylinder. It is few layered but the number of layers is increased below the four corners. The vascular bundles are arranged in a ring and they are situated more towards the periphery. The bundles situated below the ridges, are large and they alternate with smaller bundles. They are collateral and open, phloem being situated on the peripheral side. Xylem occurs towards the centre and is endarch, i.e. metaxylem is situated towards the periphery and protoxylem lies towards the centre. In between xylem and phloem there lies the cambium. It is composed of fusiform initials that are more or less rectangular in shape. The vascular tissues contain the usual elements like vessels, xylem parenchyma, tracheids and fibres in xylem and sieve tube, companion cell and phloem parenchyma in phloem. In young stems a continuous cylinder of cambium is observed. The union of fascicular and interfascicular cambium forms this. All segments of cambium show normal secondary growth, i.e. the cambium produces secondary xylem inwardly and secondary phloem on the peripheral side. Pith is large and composed of parenchyma cells with intercellular spaces.

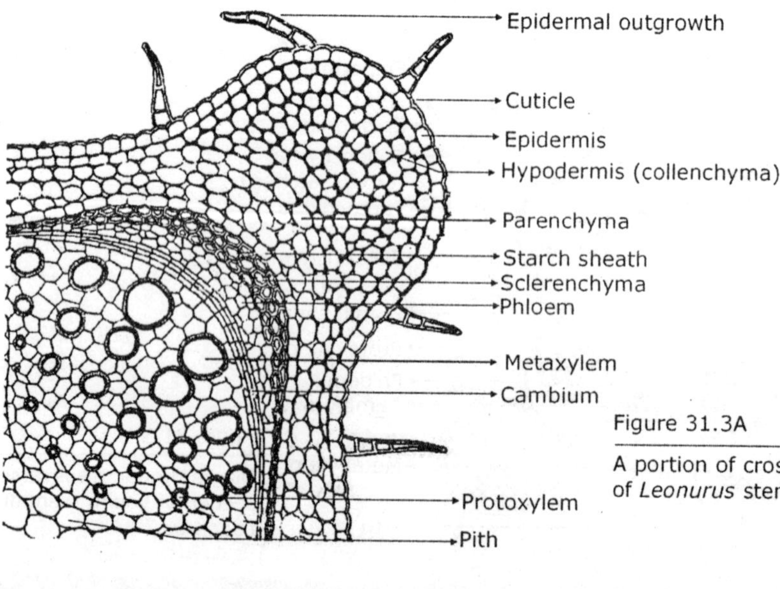


Figure 31.3A

A portion of cross-section of *Leonurus* stem.

**Comment:** Epidermis with cuticle is protective in function. Stoma is meant for gaseous diffusion. Chlorenchyma at hypodermis is photosynthetic tissue. Collenchyma at hypodermis is the mechanical cells. Aggregates of collenchyma at the four corners form the diagonally opposite I-girders that resist the stem suffering from lateral bending strain. Aggregates of collenchyma form the flanges of I-girder. Sclerenchyma at the pericycle also adds mechanical strength. The endarch xylem, open vascular bundle and normal secondary tissue formation indicate its dicotyledonous stem nature.

## Stem of *Leucas* (Family: Lamiaceae)

The transverse section of stem shows that the internal organization of tissues is similar to those present in *Leonurus*. In *Leucas* (Fig. 31.4) the aggregates of collenchyma at the four corners of stem consist of less number of layers in comparison to *Leonurus*.

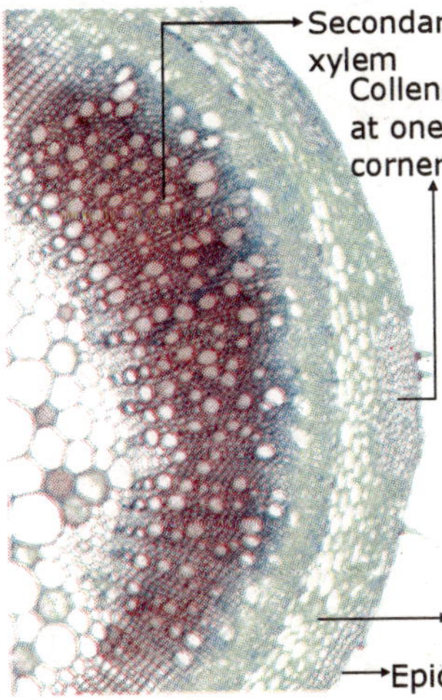


Figure 31.4

Microphotograph of a portion of *Leucas* stem.

## Stem of *Nyctanthes* (Family: Oleaceae)

The transverse section of stem is more or less square in outline and reveals the following tissue organization from periphery towards the centre (Figs. 31.5, 31.5A).

**Epidermis:** It is uniseriate and composed of tabular, compactly set parenchyma cells, the peripheral layers of which are covered with thick cuticle. The epidermis is frequently interrupted by stoma and numerous multicellular hairs occur here and there.

In older stems periderm is formed, consisting of phellogen, peripheral phellem and inner phelloderm. At some regions, lenticels are present.

**Cortex:** It is many layered and composed of parenchyma cells with conspicuous intercellular spaces. The peripheral layer, i.e. hypodermis, may be chlorenchymatous. The innermost layer is endodermis. It is conspicuous and composed of barrel shaped cells. In the cortex there occur four cortical vascular bundles at the four corners of stem. The bundles are collateral and open, and inversely oriented, i.e. xylem is situated on the peripheral side and phloem occurs towards the centre side.

**Stele:** It is siphonostele. Internal to endodermis there occur a few layered parenchymatous pericycle. The vascular bundles are collateral, open with endarch xylem. The xylem is in the form of a closed cylinder traversed by narrow rays. The phloem is situated internal to pericycle and in the secondary phloem fibres and stone cells are present. In between xylem and phloem, there occurs cambium. The cells of cambium are storied in arrangement.

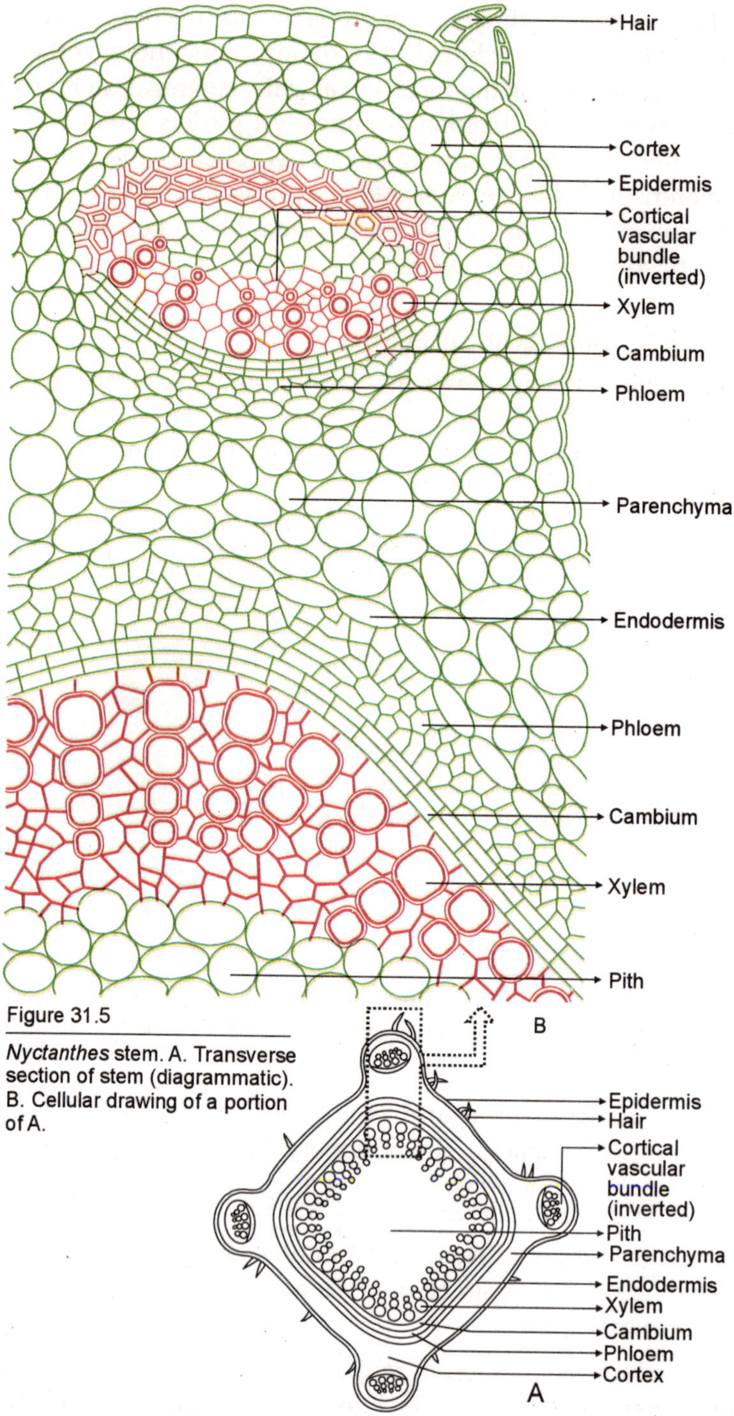


Figure 31.5

*Nyctanthes* stem. A. Transverse section of stem (diagrammatic). B. Cellular drawing of a portion of A.

**Pith:** It is composed of many layered parenchyma cells with intercellular spaces.

**Comment:** Epidermis and periderm are protective tissue. Lenticel is meant for gaseous diffusion. The dicot nature of stem is revealed due to the presence of collateral and open vascular bundle with endarch protoxylem and large pith. The chlorenchyma cells present at hypodermis are photosynthetic tissue. Anomaly in the cortex is due to the presence of cortical bundles present at the four corners of stem. The bundles show inverse orientation, i.e. xylem is situated towards the

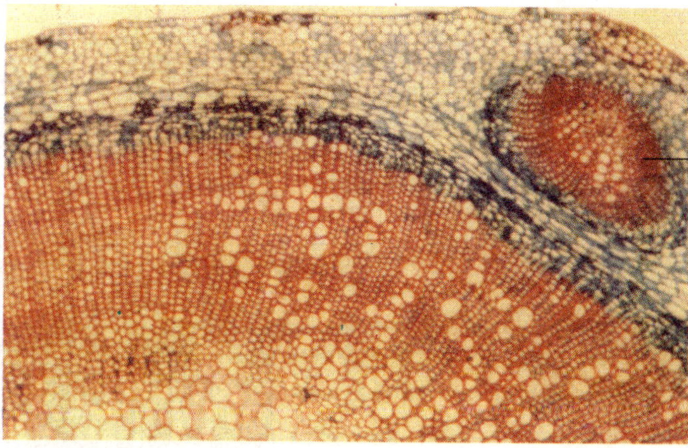


Figure 31.5A Photomicrograph of a portion of *Nyctanthes* stem.

periphery and phloem occurs towards the centre. The cortical bundles form diagonally opposite I-girder and provide mechanical strength against inflexibility. The closed cylinder of xylem also adds mechanical strength. Cortical bundle with inversely oriented vascular tissue is the characteristic of stem.

## Stem of *Calotropis procera* (Family: Asclepiadaceae)

Transverse section through the internodes of a young stem reveals the following internal tissue organization from periphery towards the centre (Fig. 31.6).

**Epidermis :** It is single layered. The cells are thin-walled, horizontally flattened with cuticularized outer walls. Waxy depositions are present over the cuticle. The continuity of epidermis is interrupted by the presence of stoma.

**Cortex:** The peripheral layer of cortex is hypodermis. It is two to three layered and composed of collenchyma cells. The collenchyma is of lacunate type. The inner cortex is parenchymatous. The cells are large, loosely arranged with conspicuous intercellular spaces. It is many layered and contains chloroplastids. The innermost layer of cortex is endodermis. It is made up of barrel shaped cells. The cells are compactly arranged. Starch grains are present in these cells and so this layer is also termed as starch sheath.

**Stele:** Below endodermis there lies the pericycle. It is many layered. It is composed of parenchyma cells. The cells are polygonal in shape and compactly arranged. Circular or oval patches of fibres composed of cellulose are present as wide zone in the pericycle here and there. Below the pericycle there occur numerous discrete vascular bundles. The bundles are compactly arranged. In between the bundles ray cells are present. Each bundle is collateral and open with endarch protoxylem. The primary phloem is situated on the peripheral side. In the primary xylem, metaxylem and protoxylem are very distinct. In between primary xylem and primary phloem cambium is present. The cambium appears to be in the form of a continuous ring, because the vascular bundles are closely arranged. Moreover it may be in a ring when the interfascicular cambium unites with the fascicular cambium during secondary growth. Below the primary xylem there occurs distinct patches of internal or intraxylary phloem here and there.

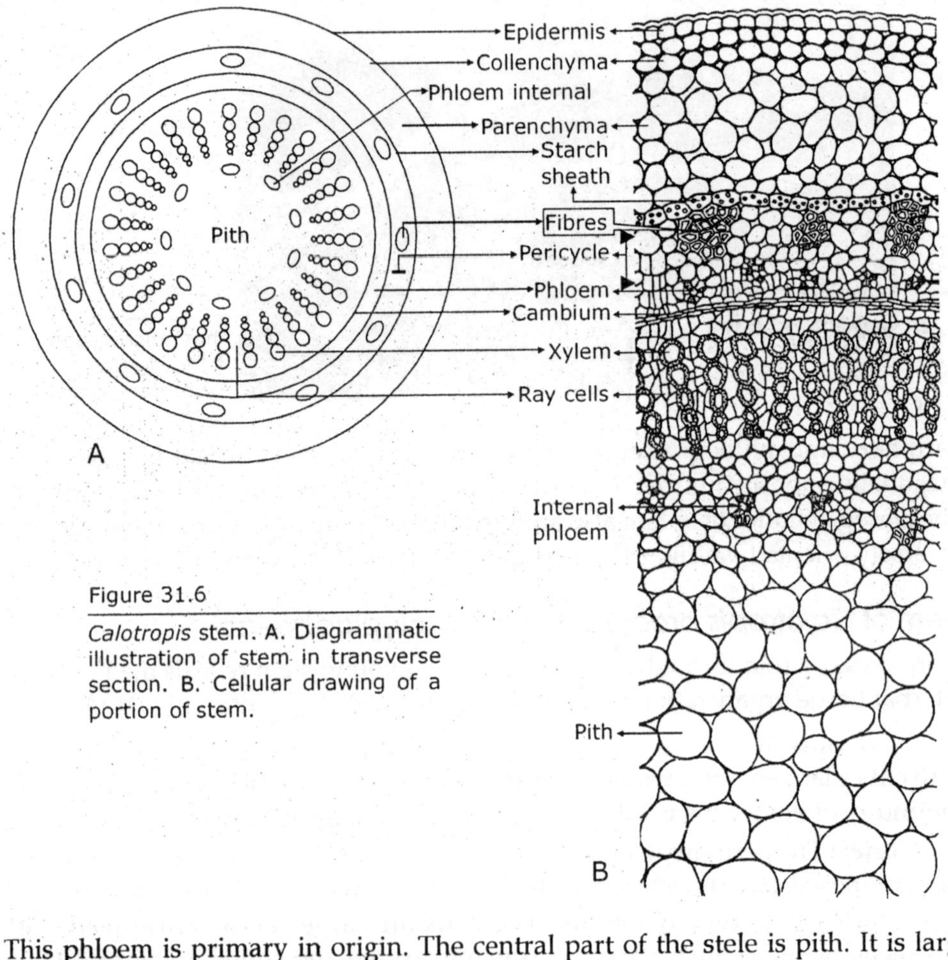


Figure 31.6

*Calotropis* stem. A. Diagrammatic illustration of stem in transverse section. B. Cellular drawing of a portion of stem.

This phloem is primary in origin. The central part of the stele is pith. It is large and composed of parenchyma with intercellular spaces. Latex vessels are observed in their cross sectional view.

**Comment:** Epidermis with cuticle and waxy depositions is protective in function and it is the characteristic of a xerophyte. Xerophytic nature is also revealed by the presence of latex vessels. Cellulosic fibres at the pericycle and xylem are mechanical cells. They provide mechanical strength against inflexibility. Anomaly in the stele is due to the presence of intraxylary phloem. Conjoint, collateral and open vascular bundle with endarch protoxylem reveal its dicotyledonous nature.

## Stem of *Enhydra fluctuans* (Family: Asteraceae)

The transverse section through the internode of stem is more or less circular with wavy outline. It shows the following internal tissue organization from periphery towards the centre (Fig. 31.7).

**Epidermis:** It is single layered and composed of thin walled, small parenchyma cells. The cells are compactly arranged with a very thin layer of cuticle on outer walls.

**Cortex:** The peripheral layer is the hypodermis. It is two to three layered and composed of parenchyma cells. The cells are thick walled with little intercellular spaces. This layer lies uninterrupted below the epidermis. Next to this layer there lies a zone of parenchyma. The cells are thin walled and contain chloroplastids. The cells are arranged in such a way that they enclose numerous large and conspicuous air spaces. The air chambers are of various shape and size, and occur here and there. The innermost layer of cortex is the endodermis. It is wavy in outline. The cells are barrel shaped and compactly set. The cells contain starch grains and so this layer is also called starch sheath.

**Stele:** It is siphonostele. The vascular bundles are arranged in a ring. Below the ridges of endodermis there lie the bundles. Between the endodermis and a bundle there occurs a patch of sclerenchyma. It forms a small bundle cap. Bundle caps and the parenchyma present between and below the furrowed region of endodermis compose the pericycle. Each vascular bundle is collateral and open. The phloem is towards the peripheral side and the xylem occurs towards the centre side. The latter is composed of protoxylem, metaxylem and xylem parenchyma. The xylem shows endarch arrangement. In between xylem and phloem cambium is present. Parenchyma cells are present between the vascular bundles and at the interfascicular region. These cells are thin walled with intercellular spaces. The central portion of stele is hollow due to disintegration of cells and so the pith is hollow.

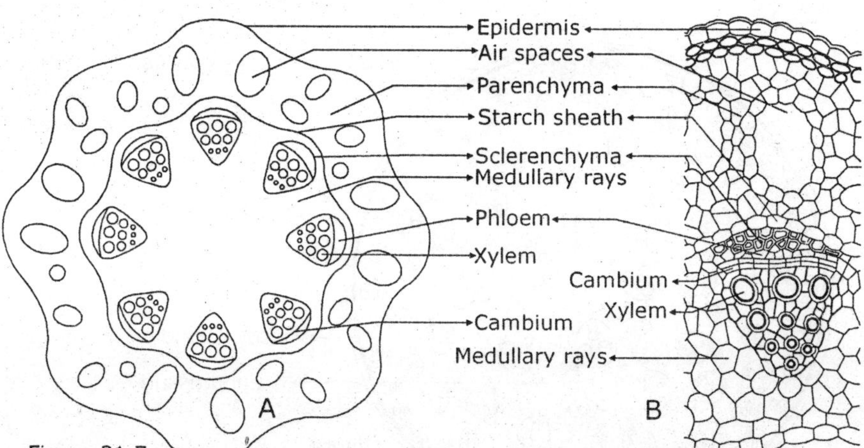


Figure 31.7

*Enhydra* stem. A. Diagrammatic representation of stem in transverse section. B. Cellular drawing of a portion of stem.

**Comment:** It is a dicot stem due to the presence of collateral and open vascular bundles that are arranged in a ring. The stem shows hydrophytic characters like air chambers and reduced mechanical tissue. The air chambers are meant for aeration and add buoyancy to the plant. The small patches of sclerenchymatous bundle cap are the mechanical cells and form the flanges of composite I-girders. The other mechanical cells are xylem and the thick walled parenchyma present at sub-epidermal region. Presence of collenchyma at hypodermis is the characteristic of dicot stem. In *Enhydra* it is replaced by thick walled parenchyma present in that region. The mechanical tissues provide strength against inflexibility.

## Stem of *Zea mays* (Maize. Family: Poaceae)

In transverse section, the stem is more or less circular in outline. The following internal tissue organizations are exhibited from periphery towards the centre (Figs. 31.8, 31.8A).

**Epidermis:** The outermost layer is the epidermis the continuity of which is sometimes interrupted by stomata. It is single layered and composed of parenchyma cells, which are tabular in shape and compactly set without any intercellular spaces. The outer wall is cuticularized and devoid of any epidermal appendages.

**Ground tissue:** In this stem all tissues except the vascular bundles present internal to epidermis are referred to as ground tissue because there exists no real distinction between cortex and pith. It is composed of hypodermis, parenchyma and stele that consists the vascular bundles.

**Hypodermis:** It is the peripheral layer that lies internal to epidermis. It is few layered and composed of sclerified parenchyma and thin walled parenchyma. The sclerified parenchyma is not in a continuous band and is interrupted by the thin walled parenchyma.

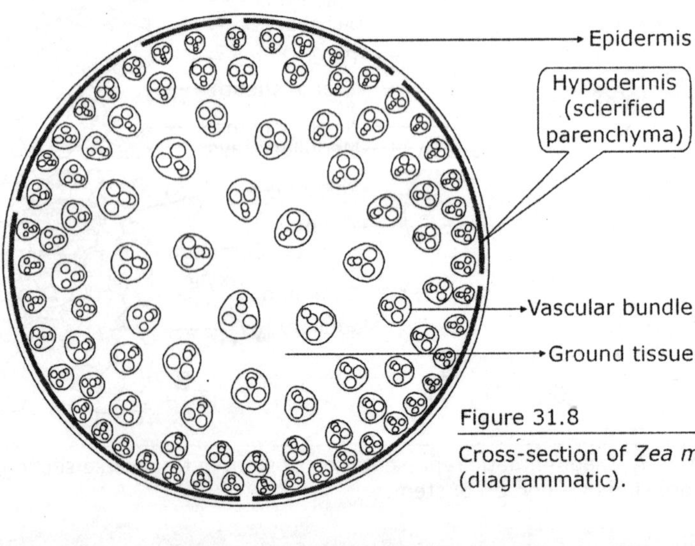


Figure 31.8

Cross-section of *Zea mays* stem (diagrammatic).

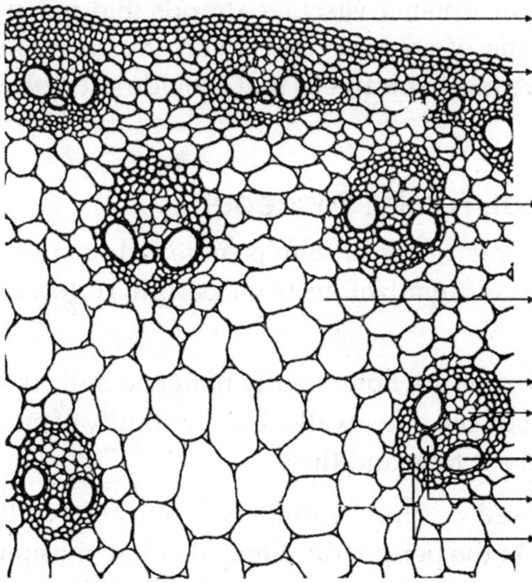


Figure 31.8A

A portion of *Zea mays* stem in cross-sectional view.

**Parenchyma:** Internal to hypodermis there occur the parenchyma and vascular bundles. The parenchyma is present up to the centre. The peripheral parenchyma, i.e. situated internal to hypodermis are smaller, polygonal in shape with little intercellular spaces in contrast to the inner parenchyma which are large, spherical with conspicuous intercellular spaces.

**Stele:** It is atactostele where numerous vascular bundles are distributed irregularly in the parenchymatous ground tissue. The peripheral bundles are smaller, more numerous, closely arranged to each other and some of them are in contact with hypodermis. The other bundles are large, fewer in number and widely spaced.

**Vascular bundle:** The bundles are collateral and closed where the cambium is absent. Each bundle remains surrounded by a sheath of sclerenchyma known as bundle sheath. The bundles are more or less oval in outline. The xylem consists of meta- and protoxylem that are arranged in the form of letter capital 'Y'. The metaxylem vessels occur at the two arms of 'Y' whereas the protoxylem vessels are present at the base. The metaxylem vessels have wider lumina with pitted thickening in contrast to protoxylem vessels that have narrow lumen and annular and spiral thickening. The innermost protoxylem vessel disintegrates and thus a lacuna is formed known as protoxylem cavity. A few tracheids are present near the protoxylem. Parenchyma cells occur in the xylem elements. Phloem occurs above the metaxylem vessels and in between the two arms of 'Y'. The phloem elements are composed of sieve tubes and companion cells only. Phloem parenchyma is lacking.

**Comment:** Cuticularized epidermis protects the inner tissues and checks the loss of water. The stem is monocot due to the presence of atactostele, collateral and closed vascular bundle with sclerenchymatous bundle sheath. There is no demarcation between stele and cortex. Sclerified parenchyma at hypodermis, is the mechanical cell. The presence of isolated vascular strands and parenchyma between them are the characteristic of an inflexible organ. Xylem vessel and sclerenchymatous bundle sheath are mechanical cells that provide strength against inflexibility.

## Stem of *Triticum vulgare* (Wheat. Family: Poaceae)

The transverse section through the internodes of stem is more or less circular and reveals the following internal tissue organization from periphery towards the centre (Figs. 31.9, 31.9A).

**Epidermis:** It is uniseriate. The cells are horizontally flattened and compactly arranged. Distinct cuticle is present on the outer walls. The continuity of epidermis is interrupted by stoma that is present here and there.

**Ground tissue:** It is composed of thin walled parenchyma cells with conspicuous intercellular spaces. On the peripheral side and below the epidermis there lie the patches of sclerenchyma. It is free from the epidermis and is few layered. It alternates with parenchyma cells containing chloroplastids. The chlorophyllous cells are present below the stoma present on the epidermis. The centre of ground tissue is hollow.

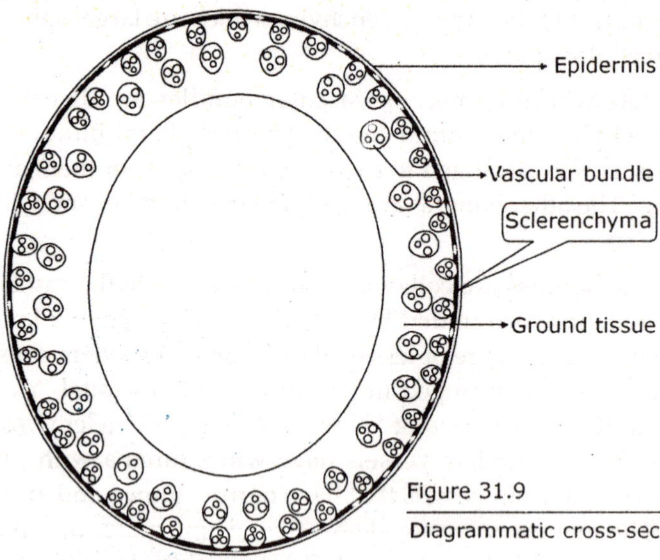


Figure 31.9

Diagrammatic cross-section of *Triticum* stem.

**Vascular bundle:** It is arranged in two rings. The bundles of the outer ring remain embedded in the peripheral sclerenchymatous patches. The inner ring of bundles remains embedded in the parenchymatous ground tissue. The inner bundles are larger than peripheral bundles. Each bundle of both the rings is collateral and closed. It remains surrounded by bundle sheath made up of sclerenchyma. The peripheral bundles remain attached with epidermis by sclerenchymatous bundle sheath extension. Each vascular bundle is oval in shape and is similar to those present in *Zea mays* (Maize).

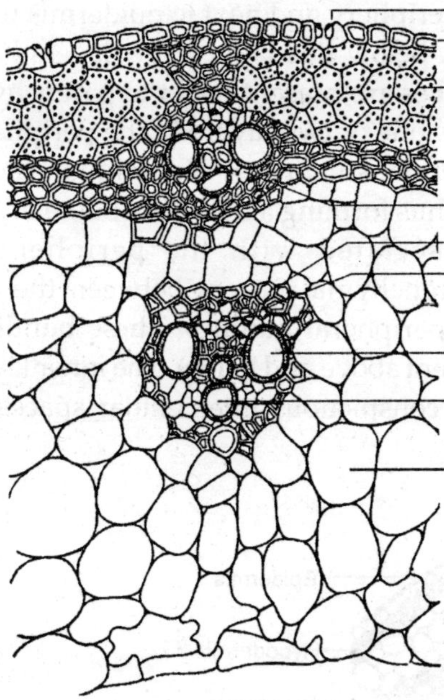


Figure 31.9A

A portion of *Triticum* stem in transverse section.

**Comment:** Epidermis with cuticle is protective in function. Stoma is meant for gaseous diffusion. Patches of chlorophyllous cells are the photosynthetic tissue. Each vascular bundle is collateral and closed, and it is the characteristic of monocot stem. Patches of sclerenchyma, sclerenchymatous bundle sheath and xylem are the mechanical cells that provide strength against inflexibility.

## Stem of *Leptochloa chinensis* (Aquatic grass. Family: Poaceae)

The transverse section of stem is more or less circular in outline and shows the following internal tissue organization from periphery towards the centre (Figs. 31.10, 31.10A).

**Epidermis:** It is single layered. The cells are more or less round, compactly arranged with cuticle on outer walls.

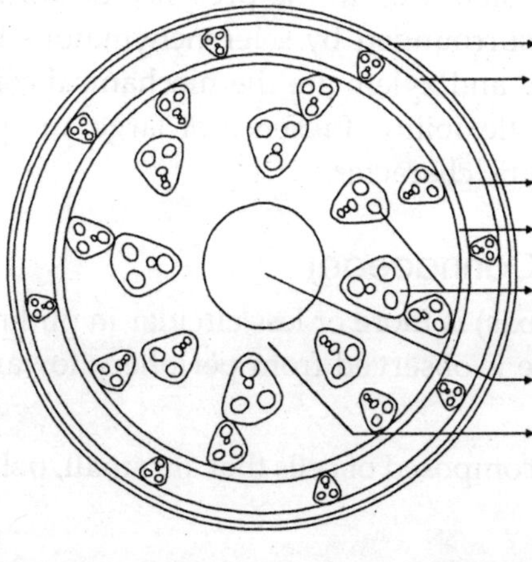


Figure 31.10

Transverse section of aquatic grass *Leptochloa chinensis* (diagrammatic).

**Ground tissue:** Towards the periphery and next to epidermis there lie two or three layers of sclerenchyma cells. Below this layer a few layers of parenchyma with conspicuous intercellular spaces are present. Next to parenchyma there lies another layer of sclerenchyma. It is one or two layered. The peripheral and inner sclerenchyma layers form a continuous cylinder. Numerous vascular bundles remain scattered in ground tissue thus forming atactostele. Some of the peripheral and inner bundles remain associated with the peripheral and inner sclerenchymatous layers. The parenchyma present between the two layers of sclerenchyma also embeds some peripheral bundles. These bundles may be in contact with the sclerenchyma present above and below. The inner layers of ground tissues are parenchymatous with conspicuous intercellular spaces. The ground tissue at the centre is hollow.

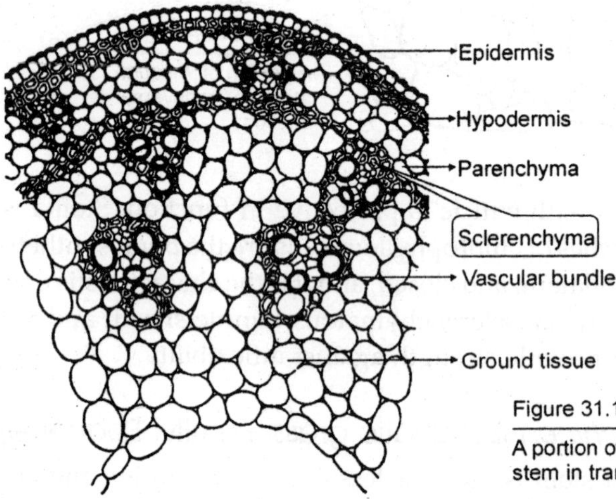


Figure 31.10A

A portion of *Leptochloa chinensis* stem in transverse section.

**Vascular bundle:** Each bundle is collateral, closed and surrounded by sclerenchymatous bundle sheath. It is similar to those present in other monocots like *Zea mays*.

**Comment:** It is a typical monocot stem due to the presence of atactostele, collateral and closed vascular bundle surrounded by sclerenchymatous bundle sheath. The two layers of sclerenchyma and xylem are the mechanical cells that provide mechanical strength against inflexibility. Presence of large air space at the centre of ground tissue is hydrophytic character.

## Scape of *Canna indica* (Family: Cannaceae)

The transverse section of scape (floral axis) is more or less circular in outline. The following plan of arrangement of tissue is observed from periphery towards the centre (Figs. 31.11, 31.11A).

**Epidermis:** It is single layered and composed of cells that are small, polygonal with cuticle on their outer walls.

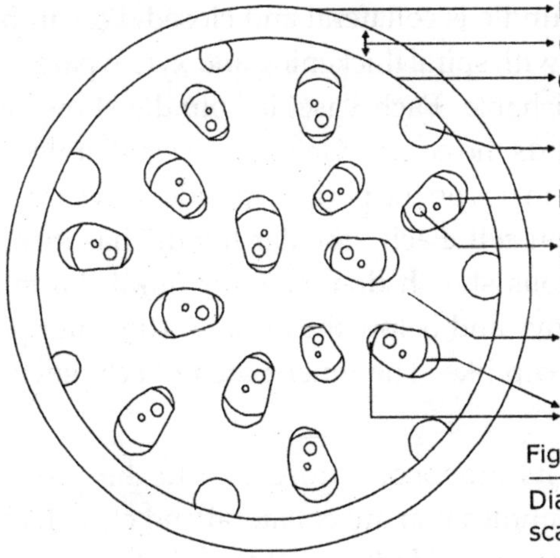


Figure 31.11

Diagrammatic representation of the scape of *Canna* in transverse section.

**Ground tissue:** The peripheral layers form the cortical region. The cells are parenchymatous, two to four layered and situated just below the epidermis, thus forming a small cortex. The cells are large and polygonal. The innermost layer of cortex contains chlorophyll. It forms a continuous band and is referred to as chlorophyllous tissue. It is uniseriate. Alternate patches of sclerenchyma and parenchyma remain attached to chlorophyllous layer. The remaining portions of ground tissues are parenchymatous.

**Vascular bundle:** The bundles lie embedded in the ground tissue in an irregular manner. They are scattered in arrangement. The stele is atactostele. The

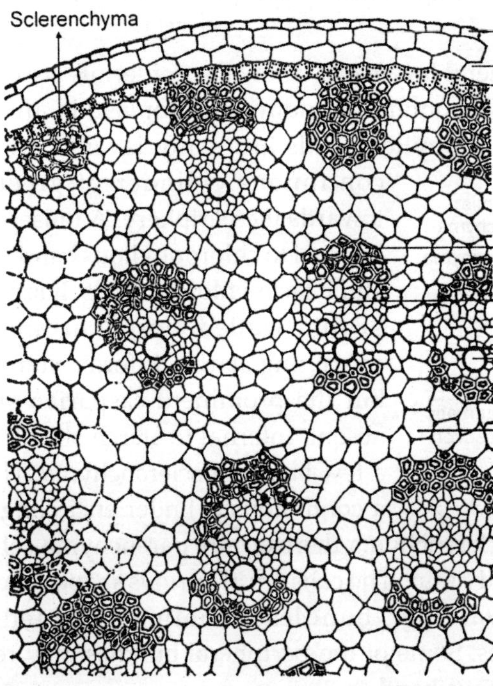


Figure 31.11A

A portion of *Canna* scape in transverse section.

bundles are of various sizes. Each bundle is collateral and closed, i.e. cambium is absent. The xylem consists of vessels with spiral thickening and xylem parenchyma; tracheids and fibres are indistinguishable. Each vascular bundle shows a large and prominent vessel situated towards the centre. One or two smaller vessels are present towards the periphery. Phloem lies on the peripheral side of xylem. Phloem consists of sieve tube and companion cell. Each vascular bundle has incomplete bundle sheath. The sclerenchymatous sheath does not surround the bundles. Sclerenchymatous patches occur above and below the bundle. The outer patch of sclerenchyma is large, distinct and cap like. The inner patch of sclerenchyma is small and not well developed.

**Comment:** The scape reveals its monocot nature due to the presence of scattered arrangement of vascular bundles that are collateral and closed, massive ground tissue where the bundles are embedded. The chlorophyllous tissues are photosynthetic tissue. Sclerenchyma patches are the mechanical cells that provide strength against inflexibility. Each vascular bundle has a single large vessel and incomplete bundle sheath.

## Stem of *Asparagus* sp. (Family: Liliaceae)

The transverse section of stem is more or less circular in outline and reveals the following internal tissue organization from periphery towards the centre (Figs. 31.12, 31.12A).

**Epidermis:** It is uniseriate. The cells are more or less round, thin walled, compactly arranged with cuticle on outer walls. Stomata are present on the epidermis.

**Ground tissue:** The peripheral layers just below the epidermis are composed of parenchyma cells. The cells contain abundant chloroplastids and enclose intercellular spaces. The chlorophyllous cells are many layered and form a continuous cylinder of green tissue towards the periphery. A uniseriate layer internal to chlorophyllous band is present and it lacks chloroplastids. It is composed of thin walled closely arranged cells containing starch. It forms starch sheath but the prominent casparian strips are absent. Internal to starch sheath there lie a few layers of sclerenchyma. They form a continuous cylinder encircling the rest of the ground tissue. Next to sclerenchyma there occurs parenchyma cells. It fills up the central position of ground tissue. The cells are of various shapes and enclose intercellular spaces. The vascular bundles remain embedded in the parenchymatous ground tissues. Some of the peripheral bundles may remain in contact with the sclerenchymatous band.

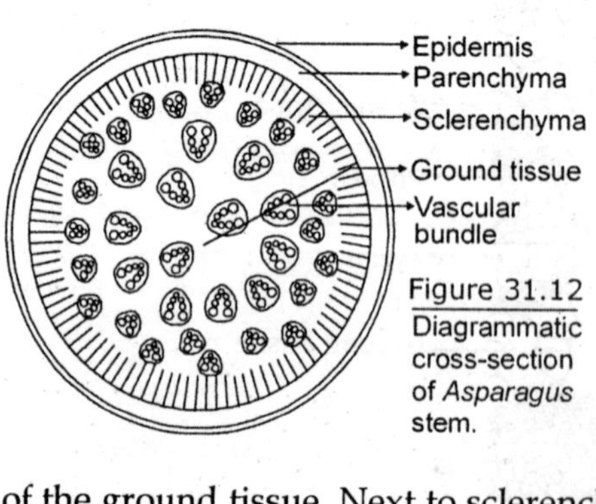


Figure 31.12 Diagrammatic cross-section of *Asparagus* stem.

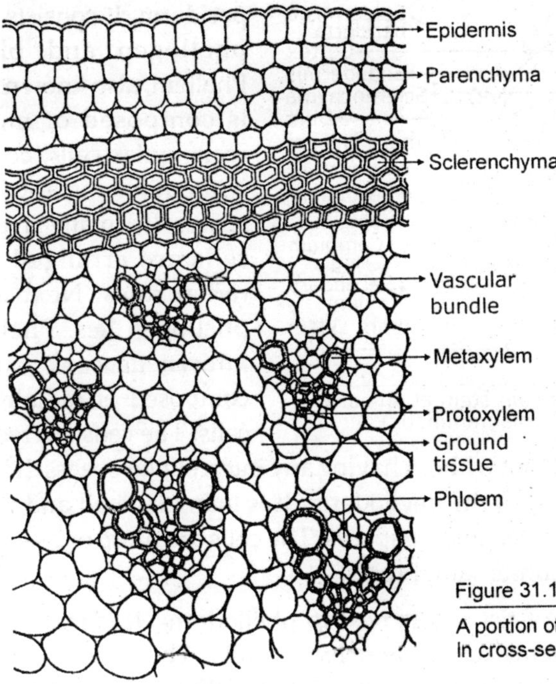


Figure 31.12A

A portion of *Asparagus* stem in cross-sectional view.

**Vascular bundles:** Numerous discrete vascular bundles are present scattered in the central ground tissue thus forming atactostele. The peripheral bundles are smaller than those present at the centre. The vascular bundles are devoid of bundle sheath. Each vascular bundle is collateral and closed. The protoxylem and metaxylem are well developed and they are arranged in the form of 'V'. The metaxylem vessels form the arms of 'V' whereas protoxylem form the base of the angle. Phloem occurs in between the two upper arms of 'V'.

**Comment:** Peripheral chlorophyllous band is the photosynthetic tissue. The monocot nature of stem is revealed by the presence of atactostele and collateral and closed vascular bundles. The vascular bundles are devoid of bundle sheath which is usually present in other monocots. The cylinder of sclerenchyma and xylem are the mechanical cells that provide strength against inflexibility. The presence of wide band of chlorenchyma and sclerenchyma in the form of cylinder reveal the xerophytic nature of stem.

## Stem of *Xanthium strumarium* (Family: Asteraceae)

Stem at an early stage of secondary growth was selected. The transverse section of stem is more or less circular in outline and exhibits the following internal organization of tissues from periphery towards the centre (Figs. 31.13, 31.13A).

**Epidermis and Periderm:** The outermost layer, i.e. the epidermis is uniseriate and composed of tabular cells with cuticularized outer walls. The cells are compactly arranged without having any intercellular spaces. Numerous multicellular hairs are present here and there. Next to it there lies the zone of

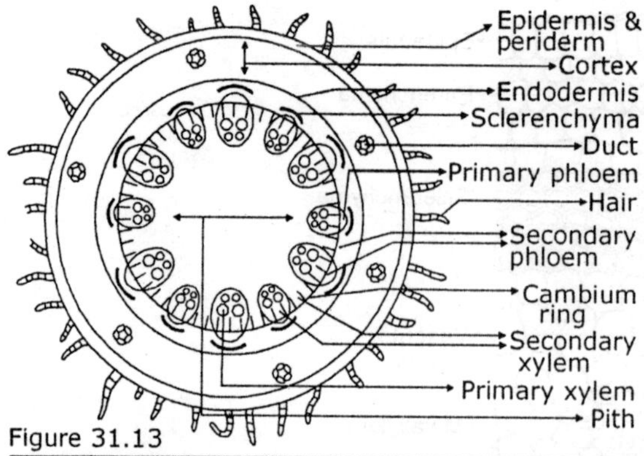


Figure 31.13

Transverse section of *Xanthium* stem at an early stage of secondary growth (diagrammatic).

periderm. It consists of phellem, phellogen and phelloderm. Phellem, also known as cork cell, is composed of more or less rectangular cells, compactly set without having any intercellular spaces. It is many layered and the cells are arranged in distinct radial rows. Next to phellem, phellogen lies. Phellogen, also known as cork cambium, is composed of radially elongated cells. The cells are rectangular in shape and are compactly set without having any intercellular spaces. It is uniseriate. Next to phellogen there lies the phelloderm. It is many layered and composed of parenchyma cells with intercellular spaces. The cells are more or less isodiametric and are arranged in definite radial rows.

**Cortex:** This zone of tissue lies next to phelloderm. It is many layered and composed of parenchyma and collenchyma cells. The cells are more or less isodiametric or oval in shape and in contrast to phelloderm not arranged in radial

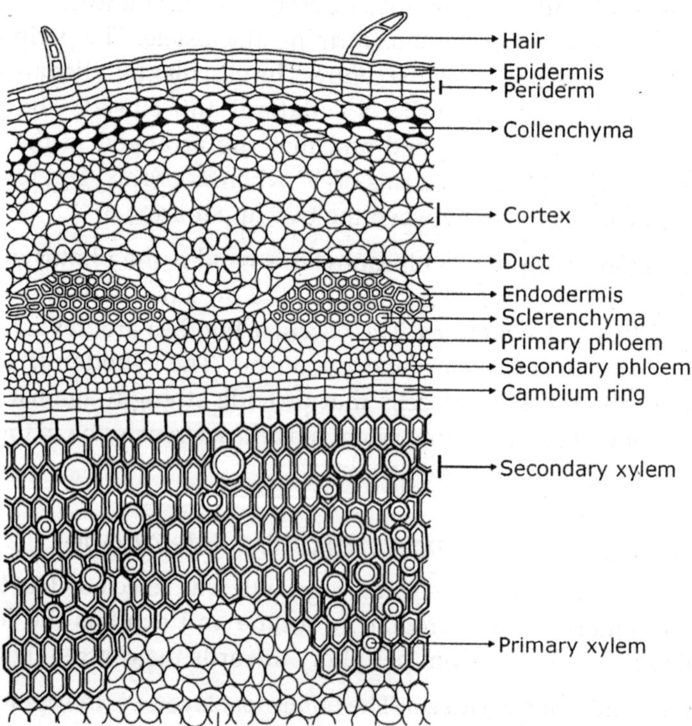


Figure 31.13A

Cellular diagram of a portion of cross-section of *Xanthium* stem at an early stage of secondary growth.

rows. Hollow cavities are present here and there in the cortex and they remain surrounded by epithelial cell. The innermost layer of cortex is the endodermis. It is composed of parenchyma cells that are barrel shaped and compactly set.

**Stele:** It consists of all the tissues that occur internal to endodermis. These are primary and secondary vascular tissues and the tissues present between them, and pith. The vascular bundles are arranged in a ring. Above each vascular bundle and next to endodermis there lies a patch of sclerenchyma. This is known as bundle cap. The bundle cap and the intervening parenchyma cells form the peripheral layers of stele or pericycle. Each vascular bundle is collateral and open with endarch protoxylem. In between xylem and phloem there occurs the fascicular cambium. Medullary rays separate the vascular bundles. It is composed of thin walled parenchyma cells and many layered. In the medullary ray there occurs the interfascicular cambium. The fascicular and interfascicular cambium are joined to each other forming a complete cambium ring. All the segments of cambium ring have produced secondary xylem towards the inner side and secondary phloem towards the peripheral side. Secondary xylem shows distinct vessels and fibres. The secondary xylem is present in the form of a continuous circular broad band. Each vascular bundle shows the following tissues from periphery towards the centre: primary phloem, secondary phloem, cambium (fascicular), secondary xylem and primary xylem. The remaining central part of stele is the pith or medulla. It is composed of more or less isodiametric parenchyma cells with intercellular spaces. This region is confluent with the medullary rays.

**Comment:** Periderm is impervious to air and water and so protective in function. The dicot stem nature is revealed due to the presence of collateral and open vascular bundle and their ring arrangement. The stele shows normal secondary growth. A normal cambium is present and it produced secondary xylem towards the centre and secondary phloem towards the periphery. The continuous band of xylem is the mechanical tissue that provides strength against inflexibility.

## Stem of *Bignonia* sp. (Family: Bignoniaceae)

The transverse section of stem is ridged and furrowed in outline and reveals the following tissues from periphery towards the centre (Figs. 31.14, 31.14A and 31.14B).

**Epidermis:** It is uniseriate, parenchymatous, thin walled, columnar and compactly set without any intercellular spaces. The peripheral walls are cuticularized. The epidermis is wavy and in the furrows, stomata may be present.

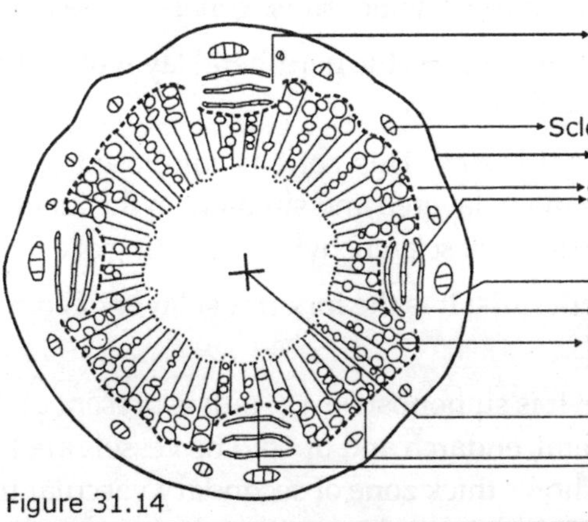


Figure 31.14

Transverse section of *Bignonia* stem (diagrammatic). X 20

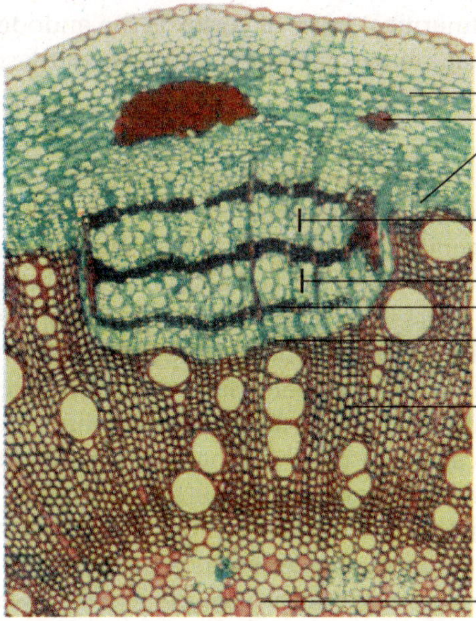


Figure 31.14A

Photomicrograph of a portion of the cross-section of *Bignonia* stem.

In older stems the epidermis is replaced by periderm, which is composed of cork cambium or phellogen, phellem and phelloderm. The latter two are situated on the peripheral and inner side of phellogen respectively. Phellogen consists of a few layers of cells that are thin-walled, rectangular and arranged in a storied manner. The cells of phellem are thick-walled and compactly set whereas the cells of phelloderm are parenchymatous, thin-walled with intercellular spaces. The continuity of the periderm is sometimes interrupted by the presence of lenticels, which are composed of ruptured epidermis, loose complementary cells, phellogen and phelloderm.

**Cortex:** The zone between the stele and epidermis is referred to as cortex. It is differentiated into hypodermis, general cortex and endodermis.

**Hypodermis:** It is the peripheral layer of cortex and composed of two to three layers of collenchyma.

**General cortex:** This zone is composed of parenchyma and chlorenchyma, which are many layered and situated on the inner and peripheral side respectively. Isolated patches of sclerenchyma consisting of a few cells are present here and there.

**Endodermis:** It is the innermost layer of cortex and not conspicuous in older stems.

**Stele:** It is siphonostele, i.e. pith is present at the centre. The vascular bundles are collateral, endarch and open. The vessels are large in diameter. In older stems, the stele shows thick zone of secondary vascular tissues, i.e. secondary xylem and phloem. The peripheral margin of the secondary xylem is ridged and furrowed, which are four in number and situated alternately. In the furrow, narrow wings

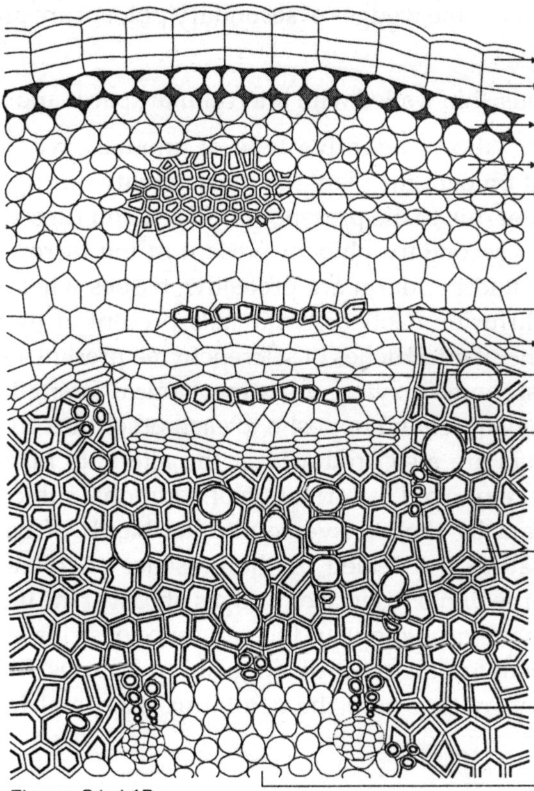


Figure 31.14B

Cellular drawing of a portion of transverse section of *Bignonia* stem. The cells are not drawn in proportionate to diagrammatic and photomicrograph for clarity.

of secondary phloem are present. The stele shows four wedges of secondary phloem, located opposite to each other. In the wedges of phloem, bast fibres are present and they are sclerenchymatous, one or two layered, and occur as transverse parallel bands across the secondary phloem. The secondary xylem is composed of tracheids, fibres, xylem parenchyma and vessels with large lumen. The cambium is not in the form of a continuous ring. Strips of cambia are present at the base of wedges of phloem and on the peripheral sides of ridged region of secondary xylem. Cambium is absent from the lateral sides of the wedges of phloem, which slide past the secondary xylem. Secondary phloem is also present on the peripheral sides of ridged region of secondary xylem.

The stele shows anomalous secondary growth where the cambium is normal in position but abnormal in activity. The union of fascicular and interfascicular cambium forms a normal cambium ring, and it functions normally until a thin ring of secondary vascular tissues is formed. After a brief period of activity four small segments of cambia, located opposite to each other become unidirectional and form four wedges of secondary phloem. The alternate four segments of the unidirectional cambia continued as bi-directional meristem, i.e. produced secondary xylem and secondary phloem normally. As a result, a ridged and furrowed stele appear where the wedges of secondary phloem are present in equal thickness throughout.

**Pith:** Parenchymatous, thin walled with intercellular spaces and many layered.

**Comment:** The stele shows adaptive type of anomaly where by forming opposite four wedges of phloem the plant attains flexibility during climbing. The climbing nature is evident due to the presence of ridged and furrowed secondary xylem, narrow wings of secondary phloem present in the furrow, vessels with large lumen. Epidermis and periderm are protective tissues whereas the lenticel is meant for gaseous diffusion. Collenchyma cells present at hypodermis add mechanical rigidity to the plant. Patches of sclerenchyma present at the cortex are mechanical cells. The secondary xylem adds mechanical strength to the plant. The ridged regions of xylem remain opposite to each other thus forming I-girder across the stem. I-girders are also produced at the alternate furrowed zone of xylem and this region is reinforced by formation of additional sub-cortical I-girder at the wedges of phloem by the transverse parallel bands of bast fibres. The dicotyledonous stem nature is revealed by the presence of collateral and open vascular bundles and large pith. Presence of wedges of bast at the furrowed region of xylem in equal thickness throughout and the presence of strips of cambia at the base of wedges of phloem are the most distinguishing features. The stele shows anomalous secondary growth where the cambium is normal in position but abnormal in activity.

## Stem of *Aristolochia indica* (Family: Aristolochiaceae)

The transverse section of a young stem is wavy in outline and shows the following internal tissue organization from periphery towards the centre (Figs. 31.15, 31.15A and 31.15B).

**Epidermis:** It is single layered; parenchymatous and has cuticle. The cells are tabular and compactly arranged.

**Cortex:** The peripheral layers of cortex are the hypodermis. It is composed of two to three layers of collenchyma cells. The inner layers of cortex are parenchymatous with intercellular spaces and consist of five to six layers of cells. Many of the cortical cells are chlorenchymatous. Innermost layer of cortex is the endodermis. It is conspicuous and wavy in outline.

**Stele:** The stele is surrounded by pericycle. It consists of sclerenchyma cells, two to three or more layered and wavy in outline. Below it vascular bundles occur.

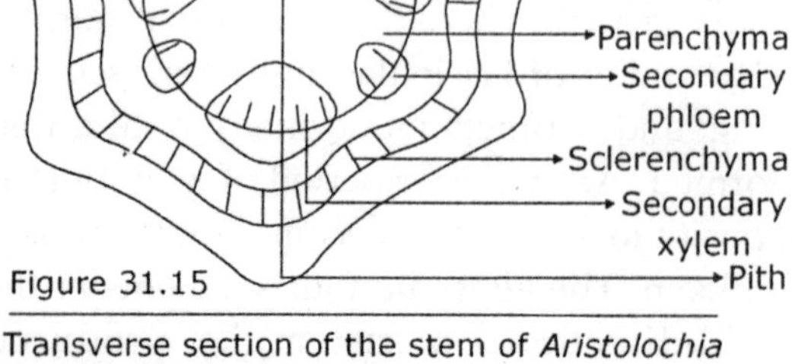


Figure 31.15

Transverse section of the stem of *Aristolochia indica* (diagrammatic).

The bundles are of different shapes and sizes and remain arranged more or less in a ring. Each bundle is collateral and open, xylem being situated on the centre side and phloem occurs on the peripheral side. A strip of cambium lies in between xylem and phloem. The protoxylem is endarch and xylem contains large vessels. At the centre pith occurs. It consists of many layers of parenchyma cells with intercellular spaces. It is large.

The stele shows secondary tissues. A complete cambium ring is present. The ring consists of fascicular and interfascicular cambium. The cambium that is present in between xylem and phloem is the fascicular cambium. The interfascicular cambium occurs between the vascular bundles. The production of secondary tissues like secondary xylem and – phloem is restricted to fascicular cambium of individual bundle only. Thus each vascular bundle contains both primary and secondary vascular tissues. Each vascular bundle shows the following tissues from periphery towards the centre: primary phloem, secondary phloem, fascicular cambium, secondary xylem and primary xylem. Secondary vascular tissues like secondary xylem and secondary phloem are not formed at the interfascicular region. The interfascicular cambium has produced parenchyma cells only.

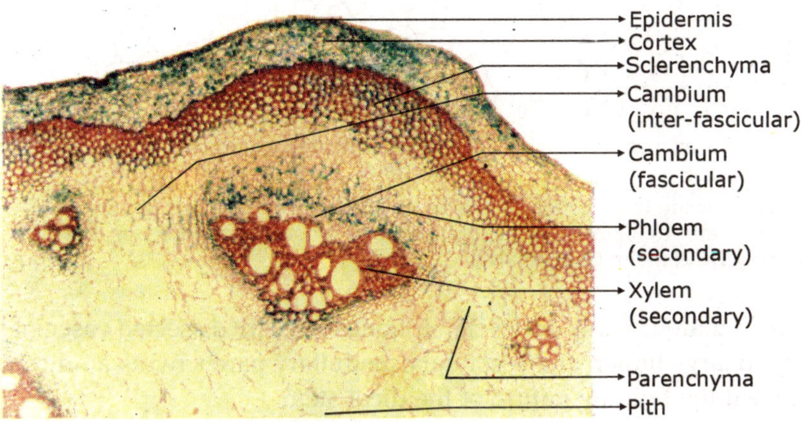


Figure 31.15A

• Photomicrograph of the cross-section of *Aristolochia indica* stem.

**Comment:** Cuticularized epidermis is protective in function. Collenchyma at hypodermis, is mechanical cell. The cortical cells with chloroplastids are the photosynthetic tissues. The sclerenchymatous pericycle adds mechanical rigidity to the plant. The stele shows adaptive type of anomalous secondary growth where the cambium ring is normal in position but the function of interfascicular cambium is abnormal. The interfascicular cambium has produced medullary parenchyma only. It is softer tissue and is produced alternating with the secondary vascular tissues. The primary and secondary xylem in a vascular bundle are the mechanical tissues. These dissected strands of mechanical cells, during twining, may overlap each other by compressing intervening secondary parenchyma. Thus the secondary medullary parenchyma adds flexibility to the stem during twining. The presence

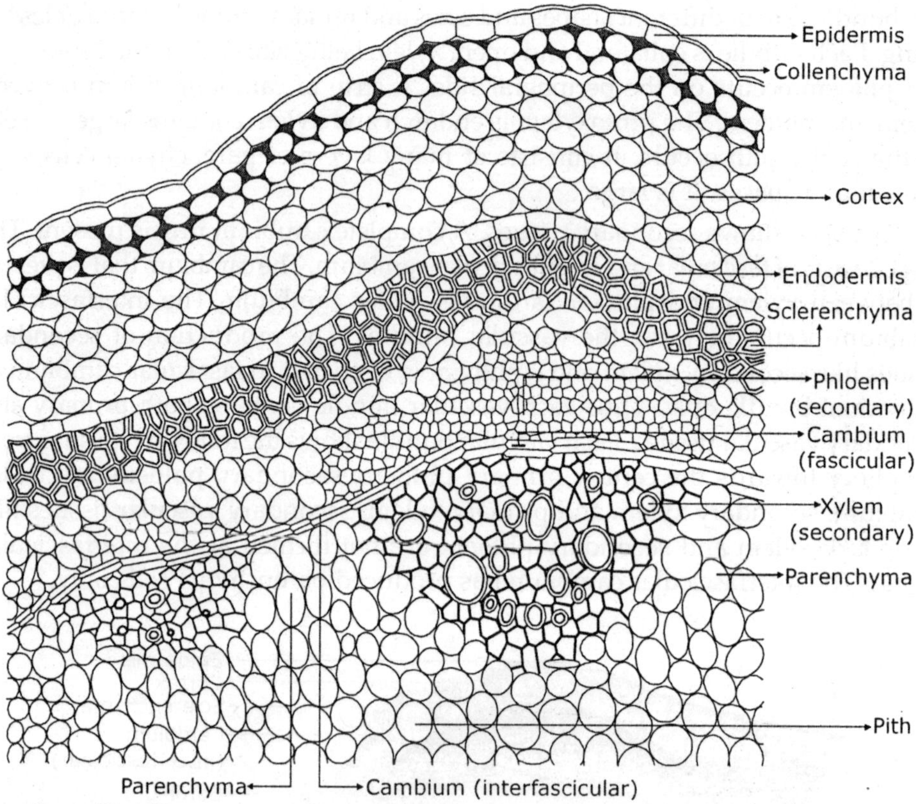


Figure 31.15B

Cellular drawing of a portion of *Aristolochia* stem. The cells are not drawn in proportionate to diagrammatic drawing and photomicrograph for clarity. Courtesy Dr. Nandini Chakrabarti.

of sclerenchymatous wavy pericycle, siphonostele with dissected vascular strands, vessels with large lumen and secondary medullary parenchyma at interfascicular region reveal the twining nature of the dicot stem.

## Stem of *Serjania* sp. (Family: Sapindaceae)

The transverse section of stem is wavy in outline and shows the following anatomical characters (Figs. 31.16, 31.16A).

**Epidermis:** It is uniseriate and composed of parenchyma cells. The cells are compactly arranged and have cuticle on outer walls.

**Cortex:** The peripheral layers of cortex are the hypodermis. It is two to three layered and consists of collenchyma cells. The inner layers of cortex are composed of parenchyma cells with intercellular spaces. They are five to six layered. The innermost layer of cortex is the endodermis. It is uniseriate and composed of barrel shaped parenchyma cells that are compactly set.

**Stele:** It is siphonostele with endarch protoxylem. The stele appears to be polystelic. But actually it is monostele as all the so-called steles are the vascular

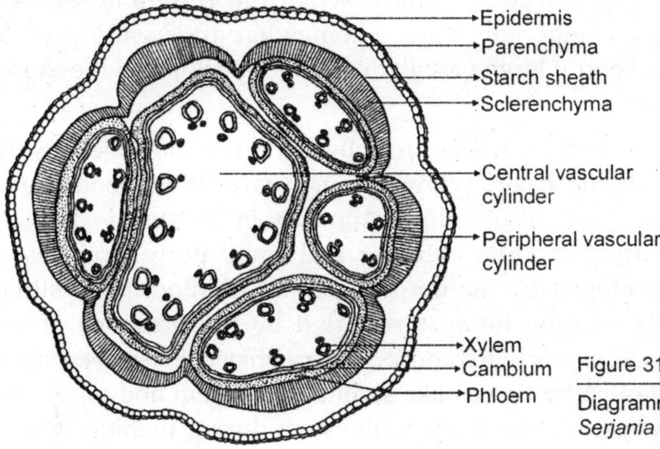


Figure 31.16

Diagrammatic cross-section of *Serjania* stem.

cylinders and lie within a common endodermis. Each vascular cylinder is composed of peripheral phloem, cambium and xylem towards the centre. Each cylinder encloses certain amount of parenchyma cells. There is a large central vascular cylinder surrounded by small peripheral cylinders of vascular tissues. A pericycle is present surrounding the entire peripheral and central vascular cylinders. It is present next to endodermis. It is composed of two to three or more layers of sclerenchyma cells. The pericycle is continuous and wavy in outline.

The stele shows abnormal secondary growth where the position of cambium is abnormal but its function is normal. There are several cambium rings and each is present in a vascular cylinder. It signifies that the cambium ring is not normal but segmented to form several cambium rings. This happened at a very early stage when the activity of cambium has not started. The several cambium-rings

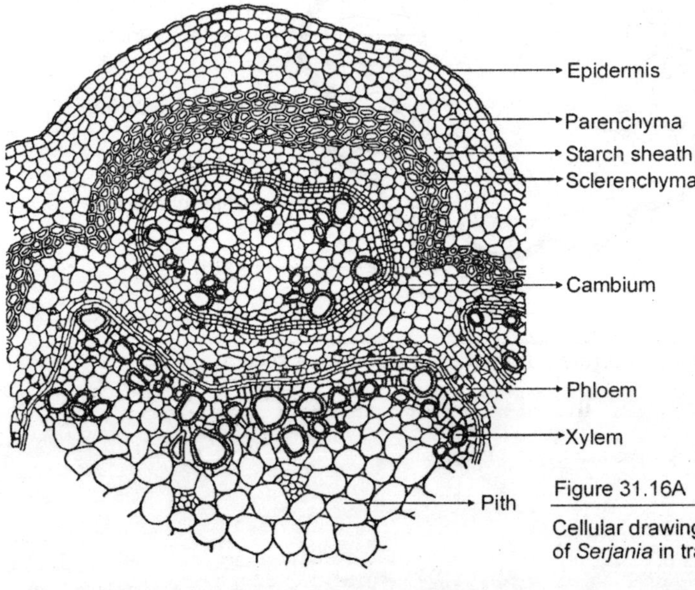


Figure 31.16A

Cellular drawing of a portion of stem of *Serjania* in transverse section.

have normal activity, i.e. produce secondary xylem inwardly and secondary phloem towards the peripheral side. Thus a somewhat apparently polystelic condition develops. The central large vascular cylinder develops by the union of several segments of cambia.

**Comment:** Epidermis with cuticle is protective in function. Collenchyma at hypodermis and sclerenchyma at pericycle are mechanical cells. Siphonostele with endarch protoxylem reveals its dicot nature. Anomaly in the stele is due to the presence of a large central vascular cylinder and small peripheral vascular cylinders that have developed by the normal activity of abnormally situated cambium segments. Its twining habit is revealed by the presence of wavy sclerenchymatous pericycle, parenchyma and secondary phloem in between vascular cylinders. The presence of softer tissues like secondary phloem and parenchyma between vascular cylinders add flexibility to the stem during twining. The stele shows apparently polystelic condition and adaptive type of anomalous secondary growth.

## Stem of *Tinospora* sp. (Family: Menispermaceae)

The description and comment of *Tinospora* stem (Figs. 31.17, 31.17A) are more or less similar to *Aristolochia* stem, discussed previously (Fig. 31.15).

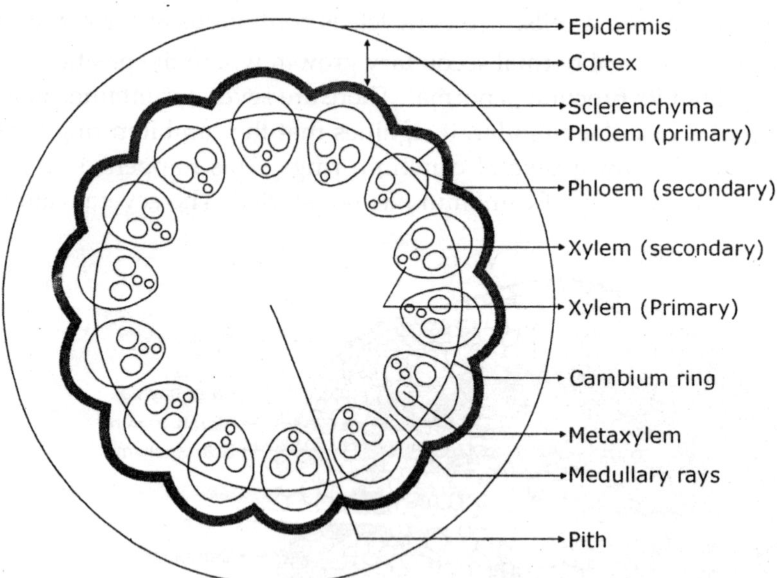


Figure 31.17

Stem of *Tinospora* (diagrammatic) in transverse section.

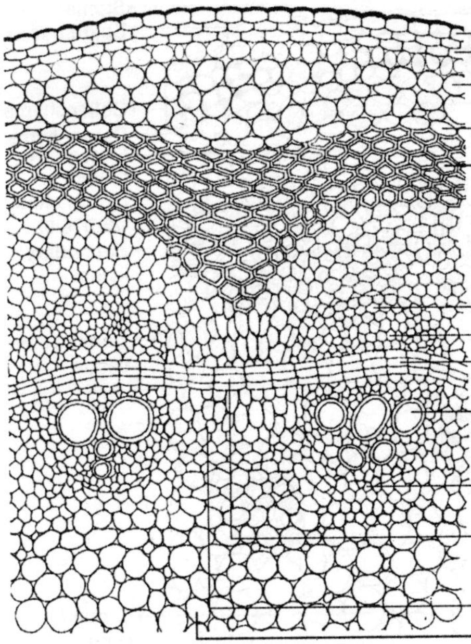


Figure 31.17A

Diagram showing transverse section of a portion of *Tinospora* stem at an early stage of secondary growth. Adapted after Sahu, 1967.

## Stem of *Boerhaavia* sp. (Family: Nyctaginaceae)

The transverse section of stem is more or less wavy circular in outline and reveals the following internal tissue organization from periphery towards the centre (Figs. 31.18 and 22.7).

**Epidermis:** It is single layered, parenchymatous, compactly set and the peripheral walls are cuticularized. The continuity of epidermis is interrupted by the presence of stomata at the grooves. In older stem, periderm is present. It is composed of phellogen with peripheral phellem and inner phelloderm.

**Cortex:** It is differentiated into hypodermis, general cortex and endodermis.

**Hypodermis:** It is few layered, collenchymatous, that is not in a continuous band and interrupted by patches of parenchyma.

**General cortex:** It is few layered; the peripheral layers are chlorenchymatous and the inner layers are parenchymatous. The cells are thin walled with distinct intercellular spaces.

**Endodermis:** It is single layered, clearly defined and composed of thick walled tabular cells.

**Stele:** It includes the vascular bundles, conjunctive tissue and pericycle, and ground tissue. The ground tissues are parenchymatous and the vascular bundles are embedded in them. The pericycle is few layered and parenchymatous. Isolated patches of sclerenchyma are present on the peripheral layer of pericycle. In the young stem, the vascular bundles are arranged more or less in three rings. The

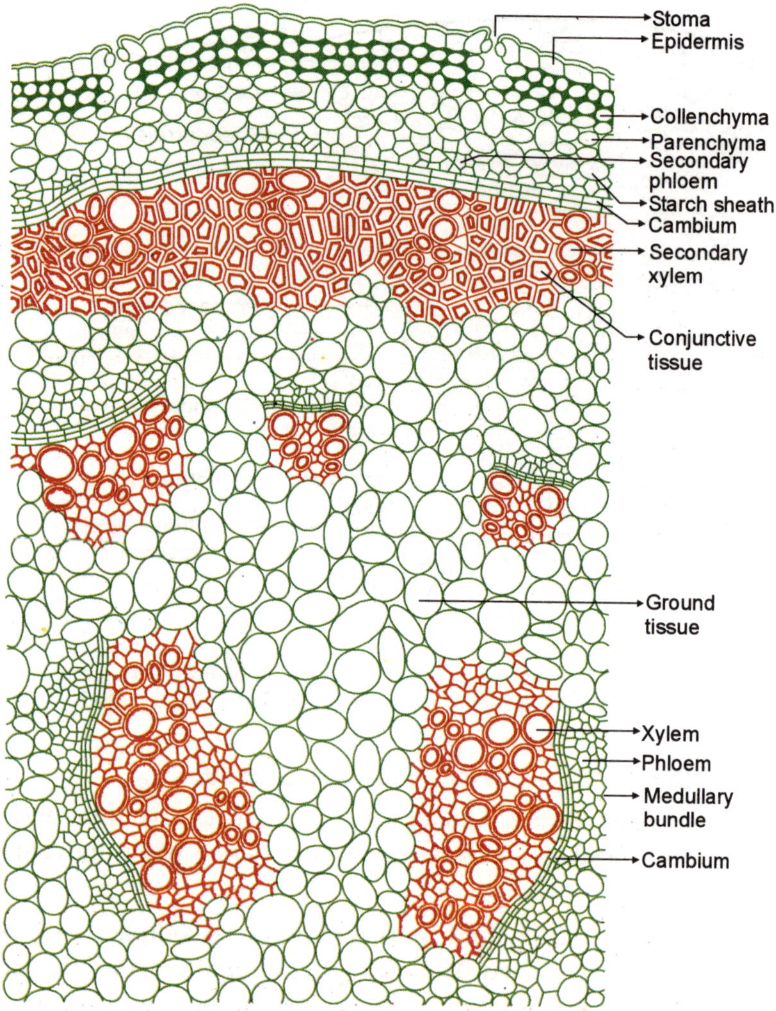


Figure 31.18

A portion of transverse section of *Boerhaavia* stem.

two large medullary bundles represent the inner ring. The middle ring consists of six to 14 smaller vascular bundles, and in the outer ring there occur 15 to 20 or more small bundles. All bundles are collateral, open with endarch protoxylem. The two medullary bundles of the first ring are more or less oval in shape and show little or no secondary vascular tissue formation. In the inner or middle ring the production of secondary vascular tissues are restricted to the fascicular cambium only up to a limited extent. The peripheral bundles are widely separated and connected to each other by a zone of sclerenchymatous conjunctive tissue. Initially these bundles are very small, separate and have their own fascicular cambium. Interfascicular cambium differentiates and forms a complete cambium ring by uniting with the fascicular cambium. The fascicular cambium forms secondary vascular tissues. The interfascicular cambium at some region produces secondary xylem and – phloem where new vascular bundles are formed; the remaining portions of the interfascicular cambium give rise to parenchyma cells, that are termed as conjunctive tissue. These tissues, at maturity, become lignified, compactly set and become confluent with the secondary xylem. So it appears that the secondary vascular bundles are embedded in the sclerenchymatous conjunctive tissue.

**Comment:** The vascular bundles are arranged in three rings; each bundle is collateral, open and devoid of bundle sheath. It indicates the dicotyledonous nature of stem. The stele shows non-adaptive type of anomalous secondary growth. Production of secondary vascular tissues is restricted to medullary bundles only. In the peripheral bundles a complete cambium ring is present by the union of fascicular and interfascicular cambium. But the interfascicular cambium produces conjunctive tissue in addition to new secondary vascular bundles. Collenchyma at hypodermis and sclerenchymatous conjunctive tissues are mechanical cells. The continuous band of sclerenchyma with secondary xylem acts as a cylinder and the vascular bundles form I-girders across the stem, and these protect the stem from lateral bending strain.

## Stem of *Mirabilis* sp. (Family: Nyctaginaceae)

The transverse section of stem is more or less rectangular in outline with rounded corners and reveals the following internal tissues from periphery towards the centre (Figs. 31.19, 31.19A).

**Epidermis:** It is uniseriate; the cells are compactly set and covered with cuticle on their outer walls. It is discontinuous at the regions where stomata are present. It possesses epidermal outgrowths that are multicellular and present here and there.

**Cortex:** It consists of collenchyma, chlorenchyma and parenchyma cells. The peripheral layer of the cortex is the hypodermis and consists of collenchyma cells.

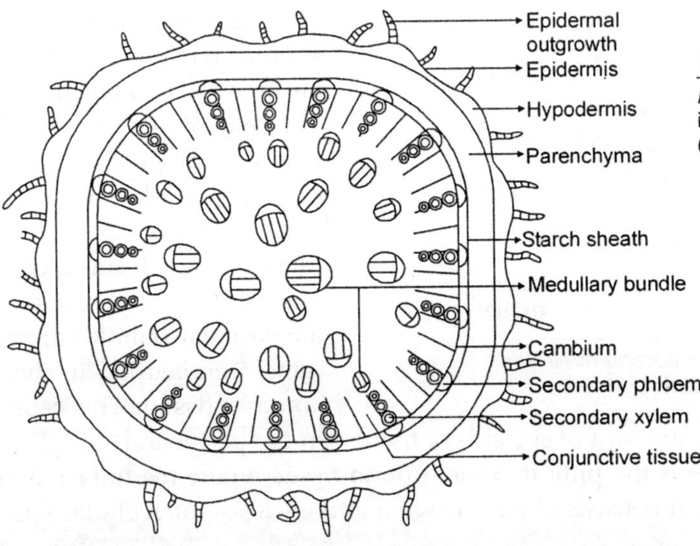


Figure 31.19

*Mirabilis jalapa* stem in transverse section (diagrammatic).

It is two to four layered. The cells are thickened at the corners. Below collenchyma chlorenchyma occurs. It is few layered. The inner cortex is parenchymatous. The cells are oval or spherical in shape, loosely arranged and have intercellular spaces. The innermost layer is the endodermis. This uniseriate layer is colourless and can be identified at ease.

**Stele:** Below endodermis there lies the pericycle. It is composed of one or two layers of cells. The cells are thin walled. There are numerous vascular bundles. The outermost or peripheral bundles are arranged more or less in a ring. The other bundles are scattered and are called medullary bundles. The bundles occurring towards the periphery are smaller in size and more crowded in contrast to bundles that occur towards the centre. The central bundles are larger and more spaced out. Each vascular bundle is collateral and open with endarch protoxylem. In the medullary bundles the formation of secondary vascular tissues is restricted to individual vascular bundles only. Little amount of secondary xylem and secondary phloem is formed in them. In the outermost ring of vascular bundle a distinct cambium ring is formed. It originates by the union of fascicular and inter-fascicular cambium. The fascicular cambium produces secondary xylem towards the inner side and secondary phloem towards the peripheral side. The interfascicular cambium produces tissues mainly inwardly. These tissues at the time formation are thin walled and elongated. Later they become lignified and thus become sclerenchymatous. These tissues are known as conjunctive tissue. The cells of this tissue are rectangular, compact and without any intercellular spaces. Conjunctive tissue along with secondary xylem forms a continuous band of sclerenchyma and the vascular bundles appear to remain embedded in the conjunctive tissue. The cambium ring produces little amount of secondary tissues on the peripheral side. The central portion of stele is the pith. It is the ground tissue where medullary bundles remain embedded. It consists of parenchyma and encloses intercellular spaces.

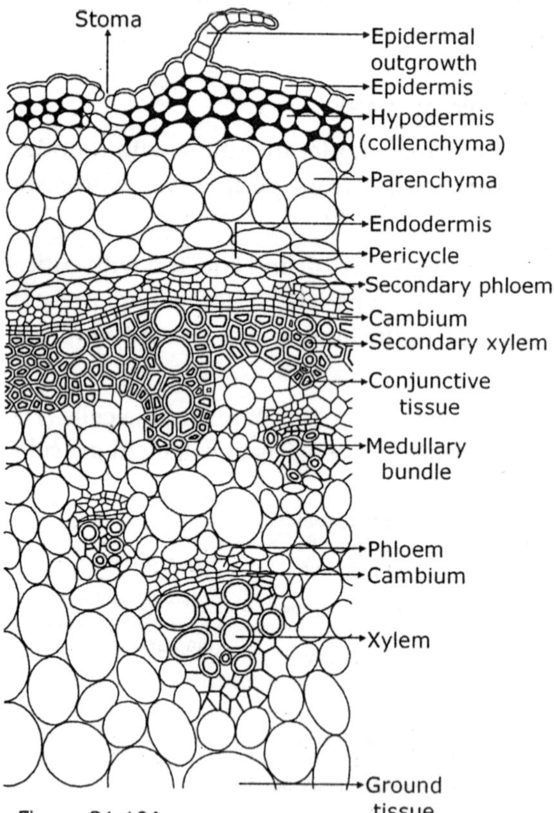


Figure 31.19A

Cellular drawing of a portion of *Mirabilis jalapa* stem in transverse section.

• **Comment:** The cuticularized epidermis is protective in function. The stoma at epidermis is meant for gaseous diffusion. Hypodermal collenchyma cells are mechanical cells. The continuous band of sclerenchymatous conjunctive tissue adds mechanical rigidity against inflexibility. The stele shows non-adaptive type of anomalous secondary growth. In medullary bundles little amount of secondary vascular tissues are produced. The major bulk of secondary tissue is the conjunctive tissue formed at the peripheral side by the cambium ring. The dicotyledonous nature of the stem is revealed by the presence of vascular bundles that are conjoint, collateral and open with endarch protoxylem.

## Stem of *Chenopodium* sp. (Family: Chenopodiaceae)

The transverse section of the young stem is ridged and grooved in outline and shows the following internal structure from periphery towards the centre (Figs. 31.20, 31.20A).

**Epidermis:** It is uniseriate and composed of compactly arranged parenchyma cells with cuticle on outer walls.

**Cortex:** The peripheral layers of cortex are the hypodermis. It consists of collenchyma and chlorenchyma cells. Collenchyma occurs below the ridges. Chlorenchyma is present beneath the furrows. The inner layers of the cortex are five to six layered, parenchymatous and may contain chloroplastids. The innermost layer is endodermis, which is uniseriate and the cells are barrel shaped.

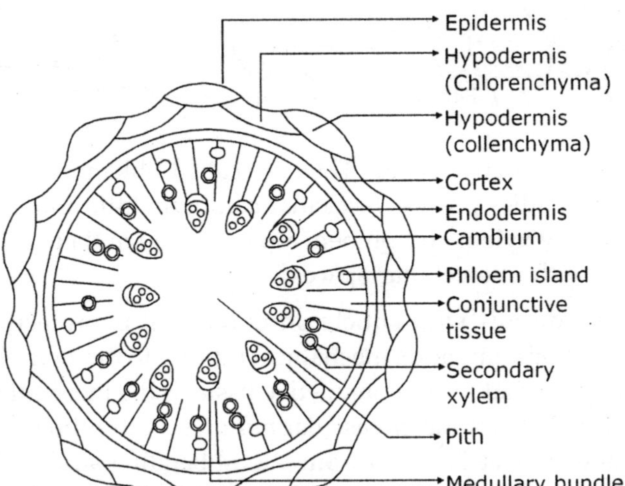


Figure 31.20

Transverse section of *Chenopodium* stem (diagrammatic).

**Stele:** It is siphonostele and the primary bundles are more or less arranged in a ring. Each vascular bundle is collateral and open, and devoid of bundle sheath. The protoxylem is endarch. Below the endodermis there occurs a cambium termed accessory cambium. It is in the form of a continuous ring. Below it, conjunctive tissue is present. It originates from accessory cambium. It is many layered, sclerenchymatous, compactly set and lignified. It is in the form of a band of continuous ring. The accessory cambium also produces secondary vascular bundles that remain embedded in the conjunctive tissue. Sometimes it becomes difficult to differentiate secondary xylem without vessels from conjunctive tissue. Groups of phloem occur as isolated strands. They are thin walled and so quite

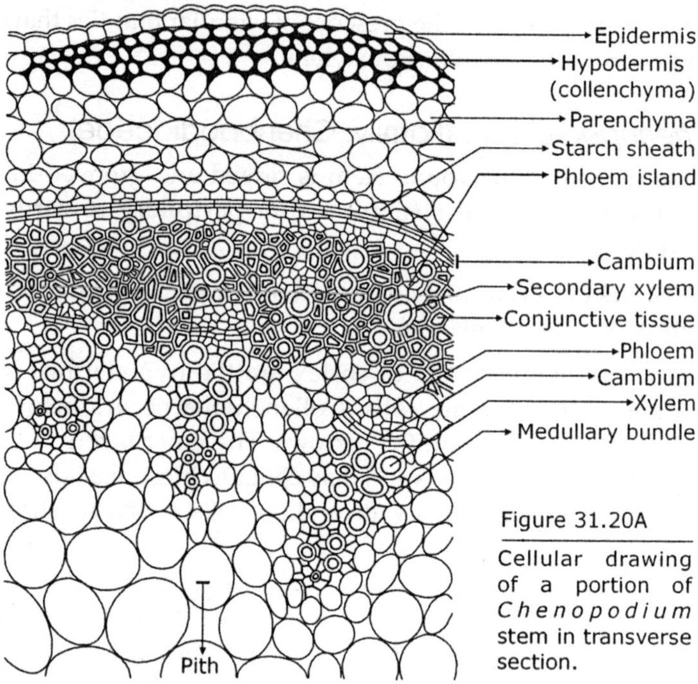


Figure 31.20A

Cellular drawing of a portion of *Chenopodium* stem in transverse section.

prominent in the conjunctive tissue. They are called phloem islands. Though the primary vascular bundles possess fascicular cambium little amount of secondary tissues, i.e. secondary xylem and phloem are produced in them. The pith is large, parenchymatous with intercellular spaces.

**Comment:** Epidermis with cuticle on outer wall is protective in function. Collenchyma at hypodermis adds mechanical rigidity to the stem. The continuous band of sclerenchyma that forms conjunctive tissue is the mechanical cell and acts like a cylinder. Isolated strands of vascular bundles, both primary and secondary form I-girders and protect the stem from lateral bending strain. The stele shows anomalous secondary growth. It is non-adaptive type of anomaly. The secondary tissues like conjunctive tissue and secondary vascular bundles are produced from the accessory cambium that originates from the pericycle. As more and more conjunctive tissues are formed the primary vascular bundles are pushed towards the centre. So the primary bundles appear as medullary bundles. Little amount of secondary vascular tissues is formed in these bundles. The dicot nature of stem is revealed by the presence of vascular bundles that are collateral, open, devoid of bundle sheath and arranged more or less in a ring.

## Stem of *Amaranthus* sp. (Family: Amaranthaceae)

The transverse section of young stem is more or less circular in outline and reveals the following anatomical structure from periphery towards the centre (Figs. 31.21, 31.21A).

**Epidermis:** It is uniseriate and composed of parenchyma cells. The cells are compactly arranged with cuticle on outer walls. The outline of epidermis may be wavy. The continuity of it may be interrupted by stomata. It occurs in grooves.

**Cortex:** The peripheral layers of it are the hypodermis. It is two to three layered and consists of collenchyma cells. Some of the cells may contain chloroplastids. The inner cells of the cortex are parenchymatous and consist of five to six layers of cells with intercellular spaces. Some cells may contain chloroplastids. The innermost layer of the cortex is the endodermis. It is uniseriate and consists of barrel shaped cells that are compactly arranged.

**Stele:** It is siphonostele. There are many vascular bundles and they are not arranged in a ring. They remain scattered in the pith. The vascular bundles are of different shape and size. The central bundles are larger. Each vascular bundle is collateral and open, phloem being on the peripheral side and xylem towards the inner side. In between xylem and phloem there occurs the cambium. Though the bundles possess cambium little amount of secondary vascular tissues are formed in them.

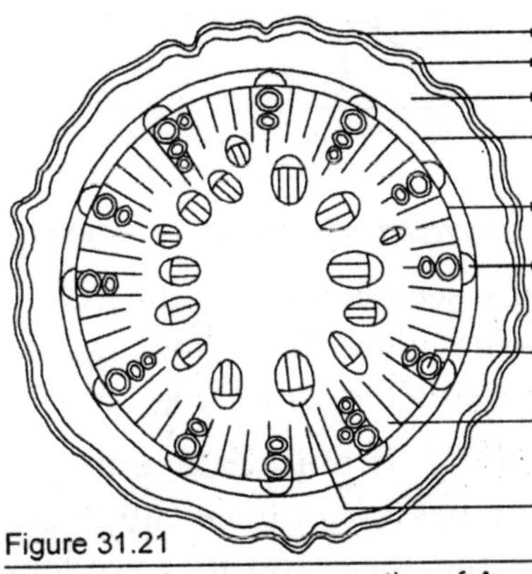


Figure 31.21
Diagrammatic transverse section of *Amaranthus* stem.

The stele shows anomalous secondary growth. Below the endodermis at the pericyclic region a circular zone of tissue is present. This is conjunctive tissue. It is thin walled (thick walled and sclerenchymatous in mature stems), compactly arranged with little intercellular spaces. Secondary vascular bundles are embedded in the conjunctive tissue. The conjunctive tissues are wavy in outline on the inner side. The secondary vascular bundles and conjunctive tissues are produced by the activity of an accessory cambium produced in the pericycle. The accessory cambium is in the form of a continuous ring and so the conjunctive tissue with embedded secondary vascular bundles is formed in the form of a continuous cylinder. The pith is large and consists of parenchyma cells with intercellular spaces.

**Comment:** The dicotyledonous nature of stem is revealed by the presence of vascular bundles, which are collateral and open, and devoid of bundle sheath. There is no definite arrangement of primary or medullary vascular bundles. The stele shows non-adaptive type of anomalous secondary growth. Little amount of

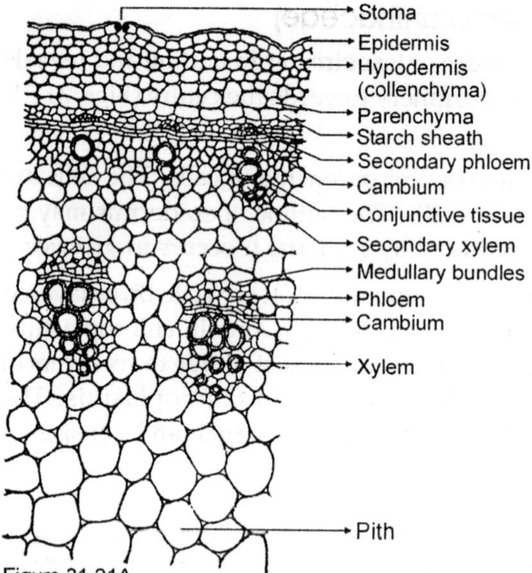


Figure 31.21A

A portion of *Amaranthus* stem in cross-sectional view.

secondary vascular tissues is produced in primary vascular bundles. Major bulks of secondary tissues are produced at the pericyclic region and they are conjunctive tissues with embedded secondary vascular bundles. The collenchyma at hypodermis and thick walled conjunctive tissue of mature stems are mechanical cells. The primary (medullary bundle) and secondary vascular bundles form I-girders across the stem and thus add mechanical rigidity to the stem. The chlorophyllous cells of cortex are the photosynthetic cells and the stoma at epidermis helps in gaseous diffusion.

## Stem of *Achyranthes aspera* (Family: Amaranthaceae)

The transverse section through the internode of stem shows a number of ridges and furrows in outline. The following internal organization of tissue is seen from periphery towards the centre (Figs. 31.22, 31.22A).

**Epidermis:** It is uniseriate and composed of tabular cells. The cells are compactly arranged with thick cuticle on their outer walls. Epidermal hairs are present here and there and they are long and multicellular. The continuity of epidermis, at some regions, is interrupted by stoma.

**Cortex:** The peripheral layers are composed of collenchyma and chlorenchyma cells. The collenchyma cells occur below the ridges and are many layered. Chlorenchyma, i.e. parenchyma containing chloroplasts is present below the furrow and few layered. The rest of the cortex is parenchymatous. The cells are circular or oval and enclose little intercellular spaces. The innermost layer of cortex is the endodermis. It is distinct and composed of uniseriate, radially elongate and compactly set cells. It is wavy in outline.

**Stele:** The peripheral layer of it, i.e. the pericycle is composed of parenchyma and sclerenchyma. They occur in patches and remain alternating each other just below the endodermis. Below pericycle there lie the vascular bundles and they are arranged in a ring. Each bundle is small, collateral and open. Xylem lies towards the centre side and phloem is towards the peripheral side. In between xylem and phloem a strip of fascicular cambium occurs. Wide medullary rays separate the vascular bundles at the primary stage. During secondary growth interfascicular cambium differentiates and it forms a complete cambium ring by uniting with the fascicular cambium. The fascicular cambium produces normal secondary

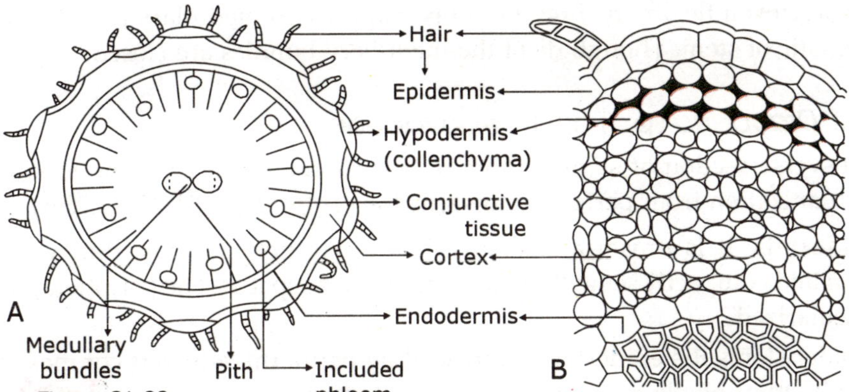


Figure 31.22

A. Transverse section of *Achyranthes* stem (diagrammatic). B. Cellular drawing of a portion of epidermis, cortex and endodermis.

tissues like secondary xylem towards the centre and secondary phloem towards the periphery. The interfascicular cambium produces conjunctive tissue inwardly. The tissues are sclerenchymatous; the cells are more or less rectangular, compactly set without any intercellular spaces. Conjunctive tissue and xylem form a continuous band of sclerenchyma where the vascular bundles appear to be embedded. Little amount of tissues are produced on the peripheral side of the cambium ring. The central part of the stele is pith. It is composed of parenchyma cells with conspicuous intercellular spaces. At the centre of it there lie two vascular bundles. These are medullary bundles. Each vascular bundle is collateral, closed and endarch. Xylem occurs towards the centre and phloem lies towards the periphery. The two bundles are in contact by their xylem and appear to form a single bundle. A parenchymatous sheath surrounds the two medullary bundles.

**Comment:** Cuticularized epidermis is protective in function. The stele shows anomalous secondary growth by the formation of conjunctive tissue. This anomaly is non-adaptive type. Hypodermal patches of collenchyma are mechanical cells. The continuous band of sclerenchymatous conjunctive tissue provides mechanical

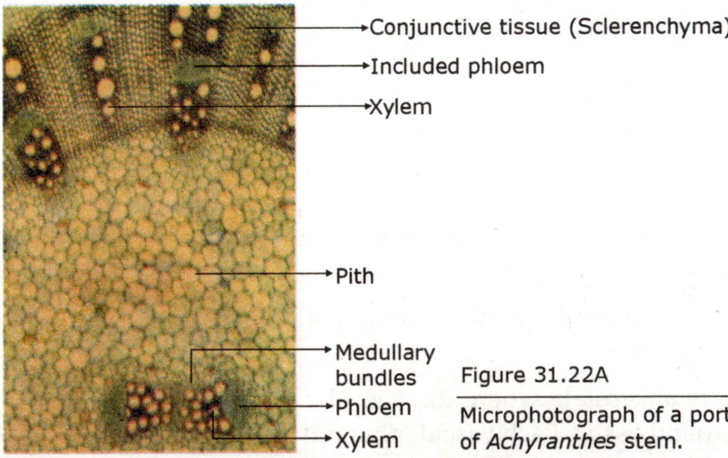


Figure 31.22A

Microphotograph of a portion of *Achyranthes* stem.

strength against inflexibility. Presence of two appressed medullary bundles is the characteristic of stem. Though dicot the medullary bundles are closed.

## Stem of *Dracaena* sp. (Family: Agavaceae)

The transverse section of stem is more or less round and reveals the following tissue organization from periphery towards the centre (Figs. 31.23, 31.23A).

**Epidermis:** It is uniseriate. The cells are parenchymatous, thin walled, columnar and compactly set without any intercellular spaces. The peripheral walls are cuticularized.

In older stems the epidermis is replaced by periderm, which is composed of cork cambium or phellogen, phellem and phelloderm. The latter two are situated on the peripheral and inner side of phellogen respectively. Phellogen consists of a few layers of cells that are thin walled, rectangular and arranged in a storied manner. Phellogen originates either in the epidermis or in the cortex and divides tangentially and donates peripheral phellem and inner phelloderm. The cells of phellem are thick walled and compactly set whereas the cells of phelloderm

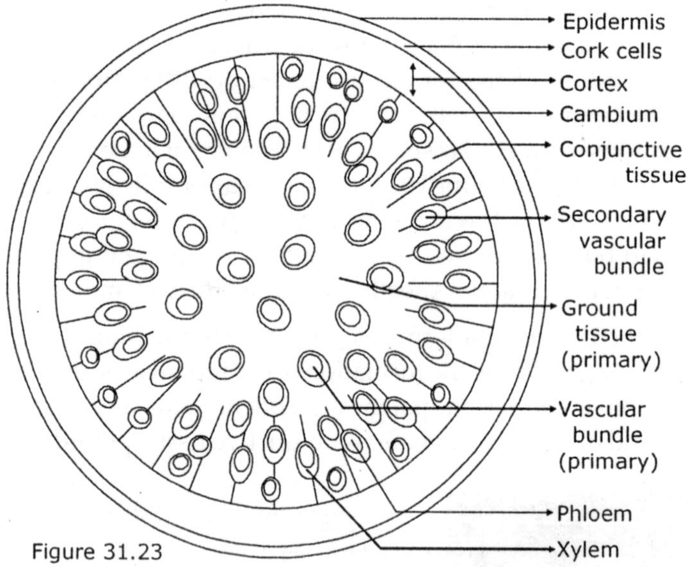


Figure 31.23

Transverse section of stem of *Dracaena* (Diagrammatic).

are parenchymatous, thin walled with intercellular spaces. The continuity of periderm is sometimes interrupted by the presence of lenticels, which are composed of ruptured epidermis, loose complementary cells, phellogen and phelloderm.

**Cortex:** It is many layered, parenchymatous, thin walled with profuse intercellular spaces.

**Stele:** It is atactostele where the vascular bundles are irregularly scattered within the ground tissue. Each vascular bundle is circular or nearly oval in outline and leptocentric, i.e. phloem is surrounded by xylem and the fascicular cambium is absent. The primary vascular bundles are loosely arranged. The stele shows the presence of conjunctive tissue in between the cortex and ground tissue. Cambium is present internal to cortex. This cambium produced the conjunctive tissues inwardly. These tissues are parenchymatous, thin walled with profuse intercellular spaces and are arranged in longitudinal radial rows, and so they are easily distinguishable from the internal ground tissue. A few cells are formed by the cambium towards the periphery. The peripheral cells remain parenchymatous whereas secondary vascular bundles are developed in the inner conjunctive tissue. The secondary vascular bundles are compactly set, oval in outline and also leptocentric or amphivasal like the primary one. The xylem consists of parenchyma and tracheids only. Vessels are absent.

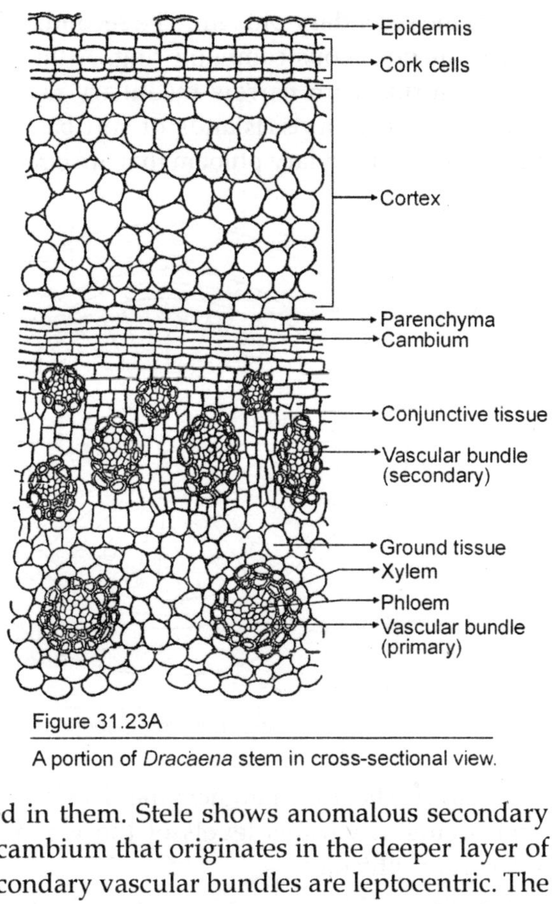


Figure 31.23A

A portion of *Dracaena* stem in cross-sectional view.

**Comment:** Epidermis and periderm are protective tissue. Lenticel is meant for gaseous diffusion. The monocot nature of stem is revealed by the presence of atactostele and leptocentric vascular bundles. Anomaly in the stele is due to the presence of conjunctive tissue formed by the cambium. The tissues are parenchymatous and the secondary vascular bundles remain embedded in them. Stele shows anomalous secondary growth where growth occurs by a cambium that originates in the deeper layer of cortex or pericycle. Primary and secondary vascular bundles are leptocentric. The bundles form I-girders and provide mechanical strength against inflexibility. Atactostele, leptocentric primary and secondary vascular bundles and parenchymatous conjunctive tissue are the characteristic of the stem.

## Stem of *Strychnos* sp. (Family: Strychnaceae)

The transverse section of stem is more or less round or oval in outline. The outline may be wavy also. The followings are the internal tissue organization of stem from periphery towards the centre (Figs. 31.24, 31.24A).

**Epidermis:** It is single layered. The cells are parenchymatous, compactly set and the peripheral walls are cuticularized. In older stems, periderm is observed. It is composed of phellogen with peripheral phellem and inner phelloderm thus constituting the periderm. The cork cells are with wide lumina and thick walls, the thickening being on one side only.

**Cortex:** It is composed of many-layered parenchyma cells that are thin walled with conspicuous intercellular spaces. The innermost layer is the endodermis that is composed of tabular cells and not conspicuous in mature stems.

**Stele:** It is siphonostele. The peripheral layer of it, i.e. pericycle is composed of a few layered loose continuous rings of stone cells. Internal to this layer there occurs the secondary phloem. The secondary phloem is present on the peripheral side of a continuous cambium ring. A thick zone of wood is present below the cambium ring with narrow medullary rays. The vessels are medium sized and may occur in groups. Islands of soft bast are present in the wood. These are included or interxylary phloem that are circular to oval in outline and surrounded

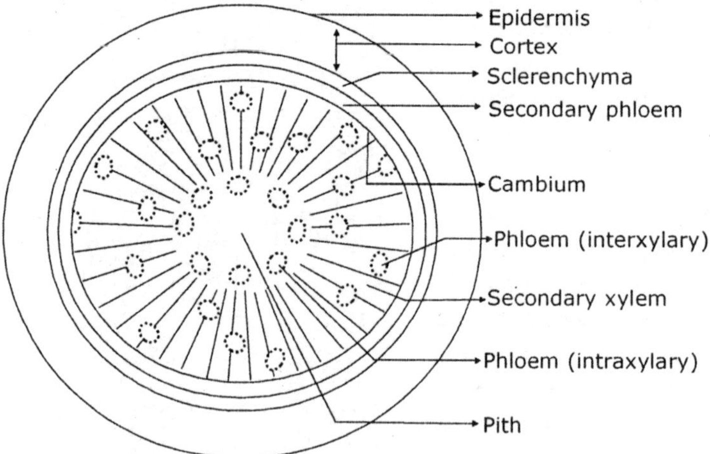


Figure 31.24

Diagrammatic transverse section of stem of *Strychnos*.

by a narrow sheath of parenchyma. The interxylary phloem is secondary in origin and occurs at various levels of the wood. Vide chapter 22 for the origin of interxylary phloem.

**Pith:** It is composed of parenchyma cells with conspicuous intercellular spaces and islands of soft bast, which occur on the peripheral side. This intraxylary soft bast, also called intraxylary phloem, is primary in origin and situated opposite to vascular bundles as isolated bundles. These bundles of soft bast are many in number and occur in the form of a ring.

**Comment:** The dicot nature of stem is revealed due to the presence of siphonostele, collateral and open vascular bundle with endarch protoxylem and thick zone of secondary wood. Epidermis and periderm are protective tissue. Stele shows anomalous secondary growth by the production of phloem islands (interxylary phloem) within the secondary xylem. Stele also shows internal phloem or intraxylary phloem that is an anomalous structure. This internal phloem lies

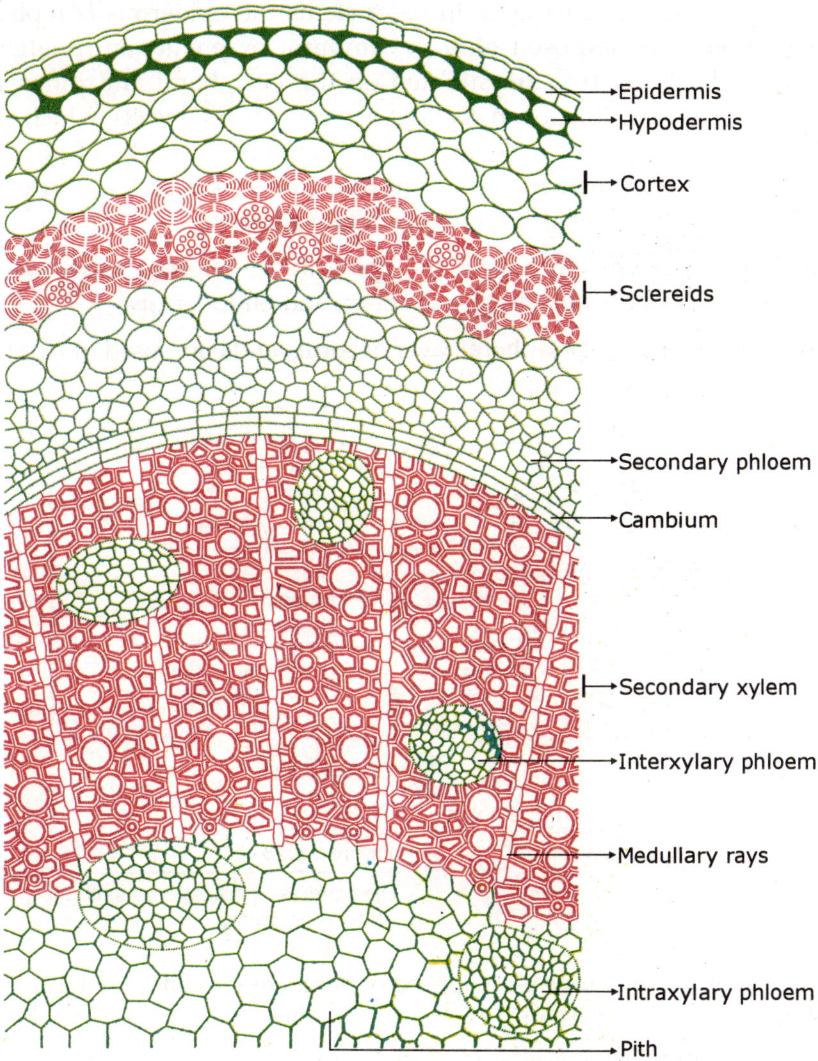


Figure 31.24A

A portion of transverse section of *Strychnos* stem.

below the secondary xylem and at the peripheral margin of pith. The internal phloem is primary in origin. The continuous band of sclerenchyma and thick zone of secondary wood are mechanical cells. They form I-girder and provide strength against inflexibility. Presence of both interxylary—and intraxylary phloem is the characteristic of the stem.

## Stem of *Tecoma* sp. (Family: Bignoniaceae)

The transverse section of stem is more or less circular in outline and reveals the following tissues from periphery towards the centre (Figs. 31.25, 31.25A).

**Epidermis:** It is uniseriate, parenchymatous, and the cells are thin-walled. The cells are tabular and compactly set without any intercellular spaces. The peripheral walls are cuticularized. In older stems the epidermis is replaced by periderm, which is composed of cork cambium or phellogen, phellem and phelloderm. The latter two are situated on the peripheral and inner side of phellogen respectively. Phellogen consists of a few layers of cells, which are thin walled, rectangular and arranged in a storied manner. The cells of phellem are thick walled and compactly set whereas the cells of phelloderm are parenchymatous, thin walled with intercellular spaces.

**Cortex:** The zone between stele and epidermis is referred to as cortex. It is differentiated into hypodermis, general cortex and endodermis.

**Hypodermis:** It is the peripheral layer of cortex and composed of two to three layers of collenchyma.

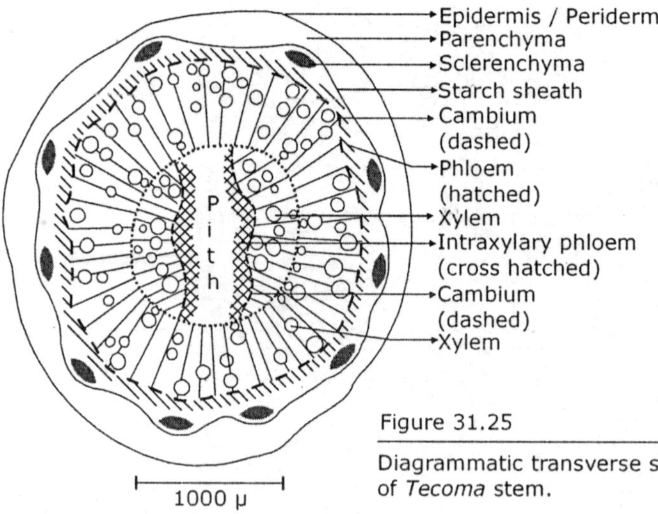


Figure 31.25

Diagrammatic transverse section of *Tecoma* stem.

**General cortex:** This zone is composed of parenchyma cells and many layered. Intercellular spaces are present.

**Endodermis:** It is the innermost layer of cortex and conspicuous in older stems. A few patches of sclerenchyma occur below endodermis.

**Stele:** It is siphonostele. There is a continuous thick zone of secondary xylem, secondary phloem on the peripheral side of it and cambium in between secondary xylem and secondary phloem. At the margin of pith, there occur two arcs of vascular bundle with inverse orientation of wood and bast as compared with the normal ring of bundles. At the margin of pith, two arcs of intraxylary phloem are present and they crush the pith to a narrow band. Strip of cambium is present between intraxylary phloem and xylem, which is situated towards the peripheral side.

The stele shows both normal and anomalous secondary growth. Normal secondary growth occurs by the normal cambium ring formed by union of fascicular and interfascicular cambium. This normal cambium ring functions normally, i.e. produces secondary xylem and secondary phloem towards centre and peripheral side respectively. Anomaly in the stele is due to the formation of intraxylary phloem. Strips of cambia appear in two arcs at the margin of pith and below the secondary xylem formed by normal cambium. These cambia, though abnormal in position, but also function abnormally, i.e. produce secondary xylem towards the peripheral side and secondary phloem towards the pith. These

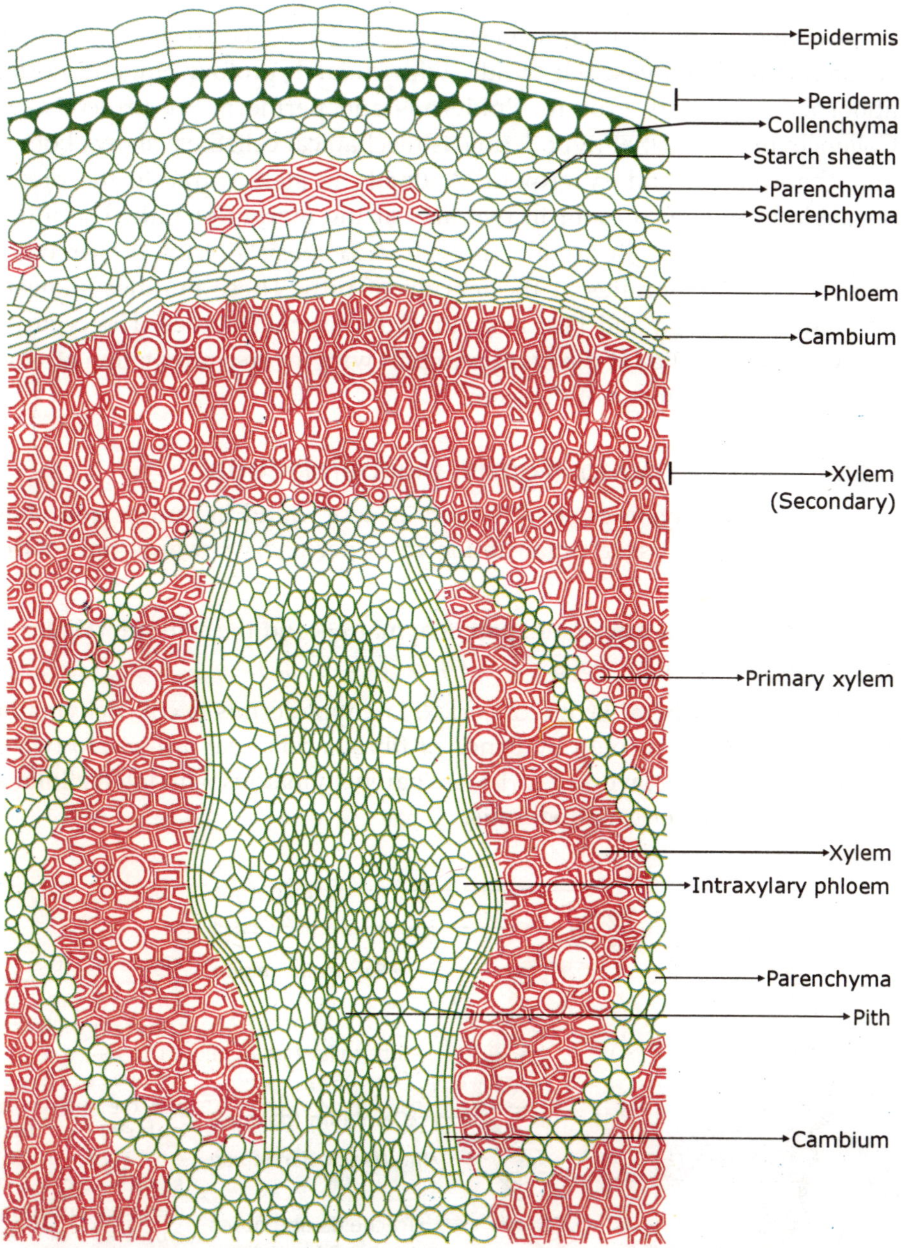


Figure 31.25A

Cellular drawing of a portion of *Tecoma* stem in cross-sectional view. Courtesy Dr. Nandini Chakrabarti.

secondary phloem patches produced by two arcs of cambia are intraxylary phloem that is secondary in origin. The secondary xylem, formed by normal and strips of cambia, merges with each other. In younger stems, few-layers of parenchyma may be present between the xylem produced by the two cambia.

**Pith:** Parenchymatous and it is crushed to a narrow band by two arcs of intraxylary phloem.

**Comment:** The dicot nature of stem is revealed by the presence of siphonostele, collateral and open vascular bundle with endarch protoxylem, and normal and abnormal secondary growth. Epidermis and periderm are protective tissues. Collenchyma present at hypodermis is mechanical cell. Stele shows both normal and abnormal secondary growth. Anomaly in stele is due to the presence of intraxylary phloem present in two arcs at the margin of pith. The intraxylary phloem is secondary in origin. Anomaly also consists of the occurrence of two arcs of vascular bundles at the margin of pith. These bundles show inverse orientation of wood and bast in contrast to normal bundles. The occurrence of two vascular bundles at the margin of pith with inverse orientation of wood and bast is the characteristic of the stem.

## Stem of *Thunbergia* sp. (Family: Thunbergiaceae)

The transverse section of stem is more or less round in outline and shows the following internal tissue organization from periphery towards the centre (Figs. 31.26, 31.26A).

**Epidermis:** It is uniseriate, parenchymatous and the cells are tabular. Cuticle is present on the outer walls.

**Cortex:** Peripheral layers of cortex, i.e. hypodermis is collenchymatous, 2-3 layered and is interrupted by bands of fibrous cells that are one to two cells in thickness. The inner layers of cortex are parenchymatous with intercellular spaces. The innermost layer is endodermis.

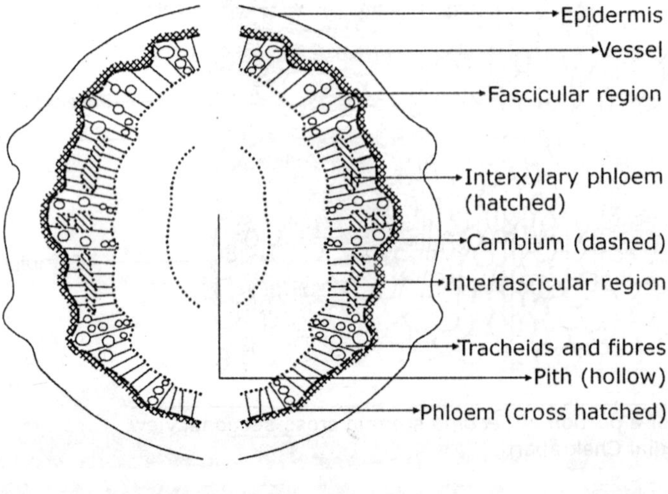


Figure 31.26

*Thunbergia* stem. Diagram illustrating cross-section (diagrammatic). Interxylary phloem is formed on two opposite sides only. Gap denotes unspecified number of layers where interxylary phloem is not formed.

**Stele:** Siphonostele, surrounded by pericycle where isolated bast fibres consisting of few cells are present. The wood shows different types of tissue distribution at fascicular and interfascicular region. The fascicular region consists of vessels, tracheids and fibres. The vessels are large in diameter and are endarch in arrangement. Vessels are completely absent from the interfascicular region.

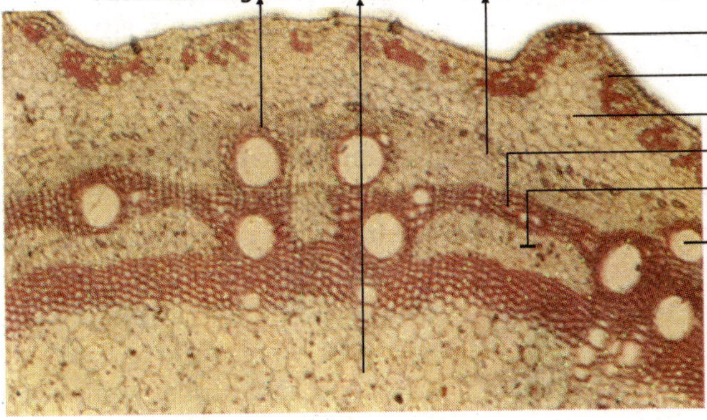


Figure 31.26A

Photomicrograph of a portion of *Thunbergia* stem.

The interfascicular region shows the presence of interxylary phloem that is in the form of bands. They alternate with corresponding bands of wood on the same radii. The transverse bands of interxylary phloem remain surrounded by tracheids and fibres on its upper and lower sides. On the two lateral sides vessels are present in addition to tracheids and fibres. The interxylary phloem is present on the two opposite sides only of the stem. The other two alternate sides show secondary xylem inwardly and secondary phloem on the peripheral sides, but no interxylary phloem. Pith is large, parenchymatous and hollow at the centre.

**Comment:** It is a dicot stem due to the presence of siphonostele, collateral and open vascular bundle. The stele shows adaptive type of anomalous secondary growth. Anomaly in stele is due to the presence of transverse bands of interxylary phloem at the interfascicular region only. The stem is a liane as is revealed by the presence of soft bast, i.e. interxylary phloem and vessels with large lumina. The soft bast adds flexibility to the stem during twining. Hollow pith, presence of soft bast in the form of bands at the interfascicular region and vessels with large lumen are the characteristics of *Thunbergia*.

## Stem of *Hydrilla* sp. (Family: Hydrocharitaceae)

The transverse section of stem is more or less circular and shows the following tissue organization from periphery towards centre (Fig. 31.27).

**Epidermis:** It is composed of tabular cells. The cells are parenchymatous, compactly arranged with a very thin cuticle on outer walls. It is uniseriate.

**Cortex:** The peripheral two or three layers are parenchymatous. The cells are thin-walled and loosely arranged with conspicuous intercellular spaces. Below this layer there occurs a large zone composed of aerenchyma. Large air spaces are

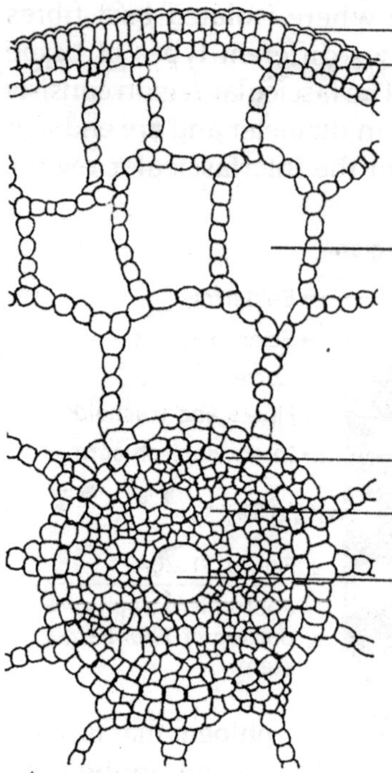


Figure 31.27

Cellular drawing of a portion of *Hydrilla* stem. Adapted after Vasishta, 1981.

present here and there and a lacunate cortex is formed. The partition walls between air spaces are composed of uniseriate parenchyma cells. The last layer of cortex is parenchymatous with intercellular spaces. The endodermis is indistinct.

**Stele:** The pericycle is indistinct. The vascular bundle is composed of xylem and phloem. The xylem consists of a single cavity that is encircled by xylem parenchyma. Xylem vessel is absent and is represented by the cavity only that is present at the centre. Phloem is situated on the peripheral side and is composed of well-developed sieve tubes with companion cells. Pith is absent and in this region there lies the xylem cavity. So the stele is protostele.

**Comment:** The stem shows hydrophytic characters due to the presence of large air chambers forming lacunate cortex, very thin cuticle on the epidermal layer, absence of mechanical cells and xylem vessel. The vessel is represented by a central cavity only.

## Stem of *Casuarina equisetifolia* (Family: Casuarinaceae)

The transverse section of stem shows the outline is thrown into furrows and ridges. The furrows are narrow and deep. The ridges are prominent. The following internal tissue organization is present from periphery towards the centre (Figs. 31.28, 31.28A).

**Epidermis:** It is uniseriate and composed of horizontally flattened cells that are compactly arranged. The cells have thick cuticle on outer walls. The epidermis is interrupted by the presence of stoma. It is present only in the furrowed regions of epidermis. At the base of furrow a number of long hairs are present.

**Cortex:** The peripheral layers form the hypodermis. It is composed of parenchyma cells. It is few layered and present all round the periphery. Below this layer patches of sclerenchyma are present below each ridge. It is absent from the region of furrow. The inner portion of cortex is composed of parenchyma cells. The cells are much elongated and palisade like. The cells contain abundant chloroplasts. Below each ridge there lies a single vascular bundle. This is leaf trace bundle occurring in the cortex and also called cortical bundle. The endodermis is indistinct.

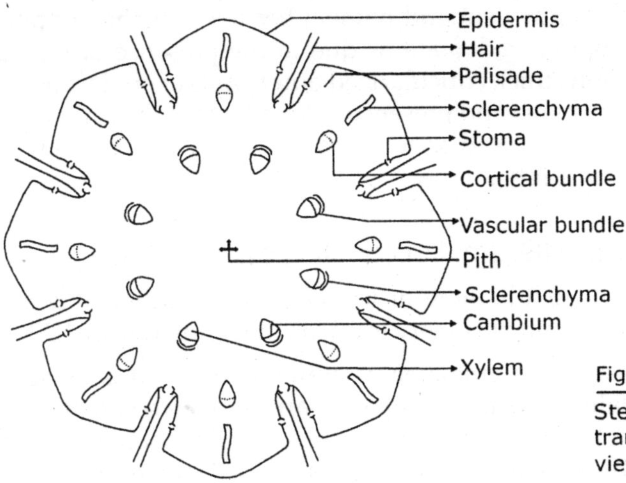


Figure 31.28

Stem of *Casuarina* in transverse sectional view (diagrammatic).

**Stele:** The vascular bundles are arranged more or less in a ring. Each bundle is collateral, open with endarch xylem. The phloem is towards the periphery and the xylem lies towards the centre. Meta- and protoxylem can be differentiated. Protoxylem lies towards the centre and the metaxylem occurs towards the phloem. A strip of cambium is present in between xylem and phloem. This is fascicular cambium. Above each bundle there lies a patch of sclerenchyma. It is composed of phloem fibres. Each cortical bundle is similar to stelar bundle except that the former is closed, i.e. cambium is absent in between xylem and phloem. Pith is present at the centre. It is conspicuous and many layered, and composed of parenchyma cells. The cells are more or less round and oval with prominent intercellular spaces.

**Comment:** The stem is a phylloclade as it shows the combination of leaf and stem character. The leaf character is revealed by the presence of stoma and palisade like chlorophyllous cells. The stem nature is due to the presence of cortical and stelar vascular bundles that are arranged in a ring. Each bundle is conjoint,

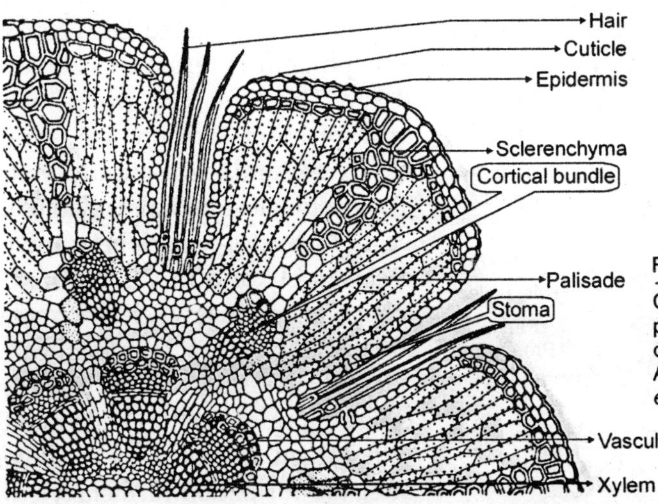


Figure 31.28A

Cellular drawing of a portion of cross-section of *Casuarina* stem. Adapted after Santra *et al.*, 2005.

collateral with endarch protoxylem and phloem lies towards the periphery. The phylloclade shows xerophytic adaptations due to the presence of stoma in the side of narrow furrow only, thick cuticularized outer walls of epidermis, presence of sub-epidermal sclerenchyma and presence of long hairs in the furrow to reduce transpiration. Sclerenchyma and xylem are mechanical cells that provide mechanical rigidity against inflexibility.

## Root of *Cicer arietinum* (Gram. Subfamily: Papilionaceae)

The transverse section of root is more or less circular in outline and shows the following internal tissues from periphery towards the centre (Figs. 31.29, 31.29A).

**Epidermis:** It is also called epiblema or piliferous layer. It is uniseriate and composed of horizontally flattened, compactly arranged parenchyma cells. The cells are devoid of cuticle. Transverse section through the root hair region shows the presence of root hairs here and there. Root hairs are the tube like prolongation of some epidermal cells.

**Cortex:** It is multiseriate and composed of parenchyma cells. The cells are of various shapes and enclose profuse intercellular spaces. The cells contain abundant leucoplasts. The cortex is large and the innermost layer of it is the endodermis. It is uniseriate and encloses all tissues present within the stele. The cells of endodermis are barrel shaped, compactly set and have casparian thickening on the radial walls.

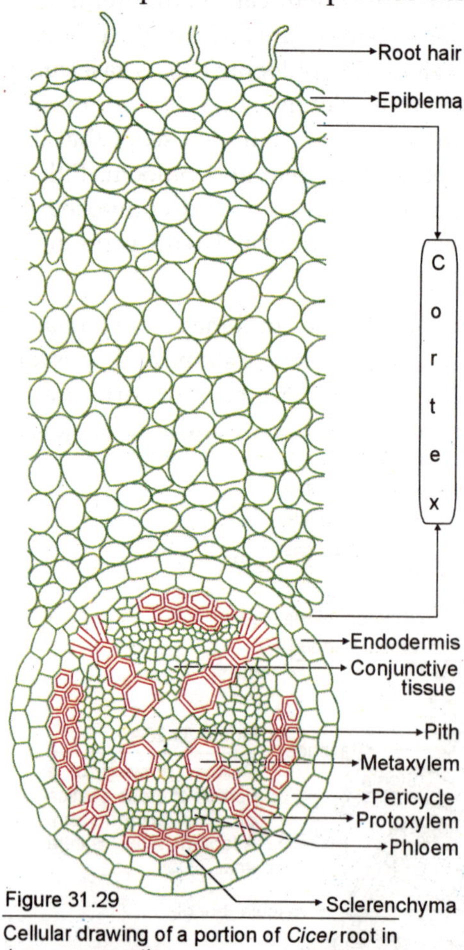


Figure 31.29

Cellular drawing of a portion of *Cicer* root in transverse section.

**Stele:** It includes pericycle, vascular tissues and ground tissue. The pericycle is uniseriate and composed of thin walled parenchyma. The vascular bundle is radial; xylem and phloem are separate and occur at alternate radii. The parenchyma situated between xylem and phloem is termed as conjunctive tissue. There are four groups of xylem and phloem, which are arranged alternately and so the bundle is tetrarch. The protoxylem and metaxylem vessels occur towards the periphery and centre respectively and this is referred to as exarch xylem. A few sclerenchyma cells are present over each phloem patch. At a very early stage a few parenchyma cells are present at the centre and this is pith, which at later stage is obliterated after the formation of metaxylem. The four-metaxylem vessels are grouped at the centre and so in mature roots pith is completely absent.

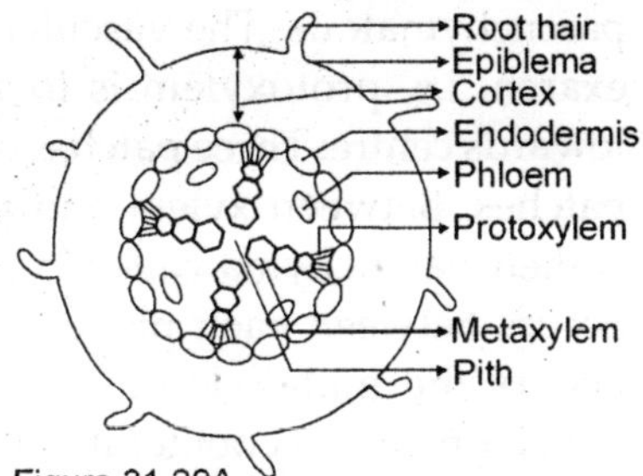


Figure 31.29A
Diagrammatic transverse section of the root of *Cicer*.

**Comment:** It is a root due to the presence of radial vascular bundle. It is a dicot root because the xylem is tetrarch and pith is absent. Epiblema with root hair is absorptive and protective in function. Cortical parenchyma helps in storage. In the stele the sclerenchyma patch and tetrach xylem are the mechanical cells only. This centralized mechanical cell provides mechanical strength against inextensibility.

## Root of *Pisum sativum* (Pea. Subfamily: Papilionaceae)

A transverse section of root is more or less round and reveals the following anatomical structure from periphery towards the centre (Figs. 31.30 & 31.30A).

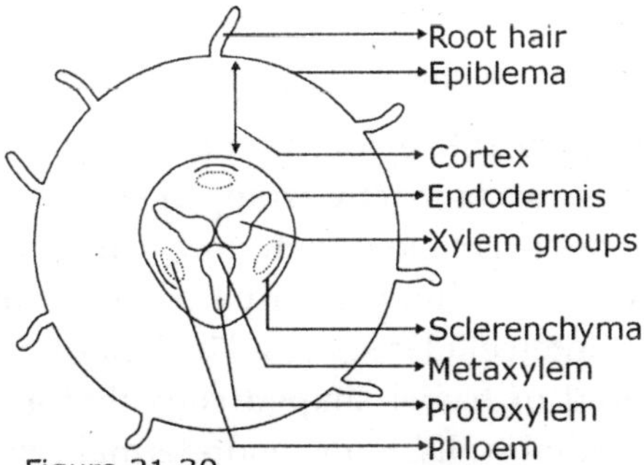


Figure 31.30
Transverse section of root of *Pisum* (diagrammatic).

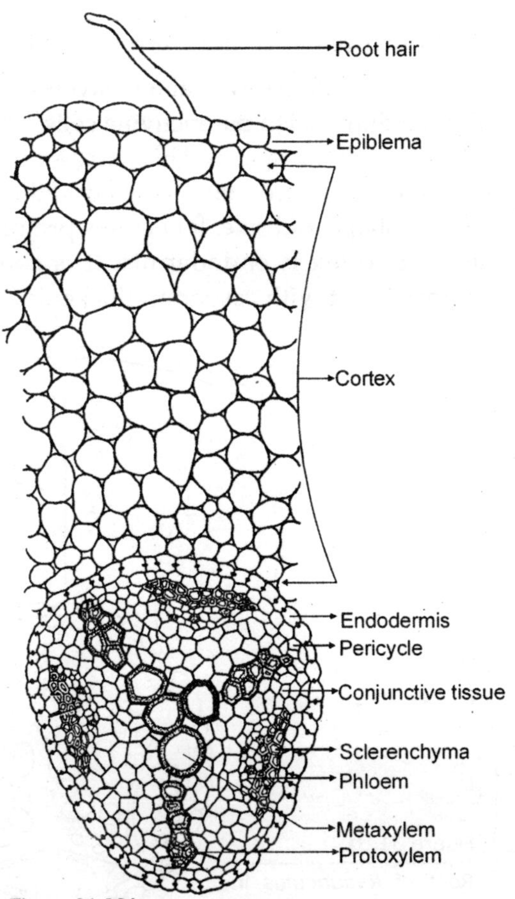


Figure 31.30A
Cellular drawing of a portion of cross-section of *Pisum* root.

**Epiblema:** It is uniseriate. The cells are more or less horizontally flattened and compactly set. Root hairs are present here and there. It is unicellular and prolongation of epidermal cells.

**Cortex:** It is large and composed of parenchyma cells of various sizes and shapes with intercellular spaces. The innermost layer is the endodermis. It is uniseriate. The cells are barrel shaped, compactly set and have distinct casparian strips on the radial walls.

**Stele:** Next to endodermis there lies the pericycle. It is uniseriate and parenchymatous. The vascular bundle is radial with triarch xylem. The xylem is exarch, i.e. protoxylem is towards the peripheral side and metaxylem occurs towards centre. Three patches of phloem are alternately arranged with three xylem patches. Between xylem and phloem patches there lies a small amount of thin walled parenchyma called conjunctive tissue. A small amount of sclerenchyma occurs between each phloem patch and the pericycle. Pith is absent and there occurs the metaxylem. At an early stage small pith is present, but during development it is obliterated and the metaxylem vessels occupy this region.

**Comment:** It is a root due to the presence of radial vascular bundle. It is a dicot root because the xylem is triarch and pith is absent. Epidermis with root hair helps in absorption. Cortex is storage tissue. Sclerenchyma patches above phloem and xylem are the only mechanical cells that provide mechanical strength against inextensibility.

## Root of *Ranunculus* sp. (Buttercup. Family: Ranunculaceae)

The transverse section of root is more or less circular in outline and shows the following internal tissues from periphery towards the centre (Figs. 31.31, 31.31A).

**Epidermis:** It is single layered and composed of horizontally flattened cells.

**Cortex:** The peripheral few layers that occur next to epidermis are composed of small thin walled parenchyma cells. The cells are compactly arranged and have little intercellular spaces. This narrow zone of tissue is called exodermis. The inner layers of cortex are many layered and composed of thin walled parenchyma of various shape and size. It encloses prominent intercellular spaces. The innermost layer of cortex is endodermis. It is composed of barrel shaped cells that are compactly set with casparian thickening on radial walls. Some endodermal cells

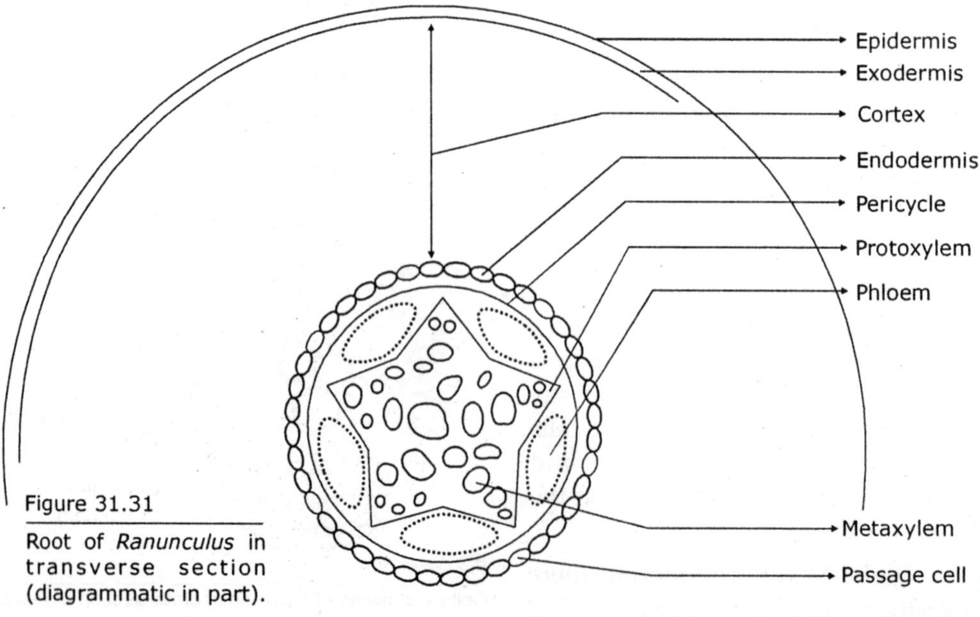


Figure 31.31

Root of *Ranunculus* in transverse section (diagrammatic in part).

may undergo secondary thickening. The remaining unthickened cells are known as passage cells and they occur against the protoxylem groups.

**Stele:** Next to endodermis there lies the pericycle. It is single layered and made up of parenchyma cells. The vascular bundles are radial. The xylem strand is tetrarch or pentarch. It is alternately arranged with four or five small patches of phloem. Protoxylem is exarch, i.e. it is towards the peripheral side and metaxylem is situated towards the centre. In between xylem and phloem there lies the conjunctive tissue made up of parenchyma cells with scanty intercellular spaces. The metaxylem vessels occupy the central position and so there is no pith.

**Comment:** It is a dicotyledonous root due to the presence of radial vascular bundle with tetra- or pentarch xylem where pith is absent. The cortex is storage tissue. Xylem is the only mechanical cell that provides mechanical strength against inextensibility.

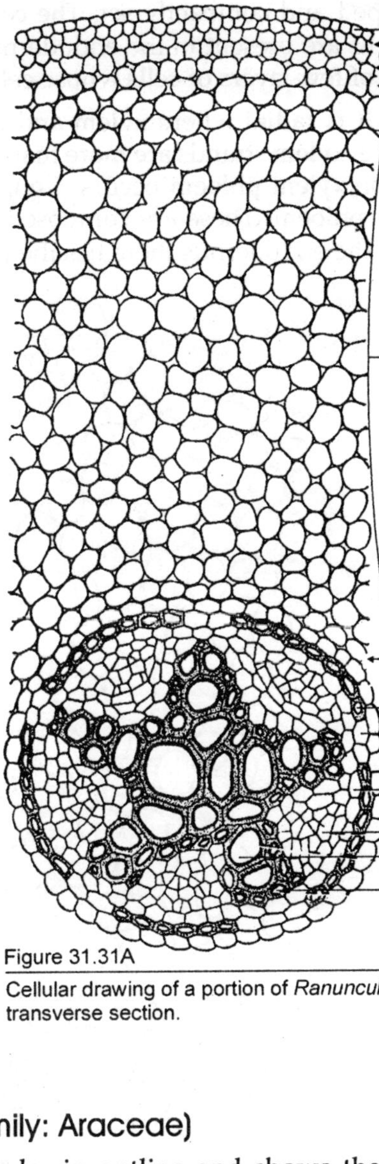


Figure 31.31A

Cellular drawing of a portion of *Ranunculus* root in transverse section.

## Root of *Colocasia* sp. (Arum. Family: Araceae)

The transverse section of root is circular in outline and shows the following arrangement of tissues from periphery towards the centre (Figs. 31.32, 31.32A).

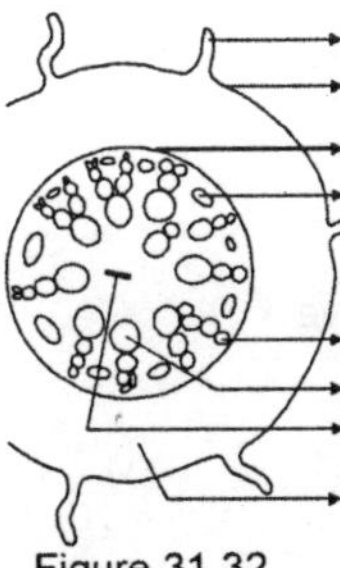


Figure 31.32

Diagrammatic transverse section of *Colocasia root*.

**Epiblema:** It is single layered. The cells are horizontally flattened and compactly arranged without any intercellular spaces. Unicellular root hairs are present here and there.

**Cortex:** It is composed of parenchyma cells, many layered and large. Numerous intercellular spaces are present in it and they are formed schizogenously. The innermost layer is endodermis, the cells of which are barrel shaped and compactly set. The cells have conspicuous casparian strips. Passage cells are present in the endodermis. Mature root shows exodermis. It is composed of few layers of cells whose cell walls have undergone suberization.

**Stele:** It is radial where xylem and phloem lie at different radii. Xylem is polyarch, i.e. xylem strands are more than six in number. Protoxylem is exarch. In between the xylem phloem occurs. Conjunctive tissues are present in between xylem and phloem. These are composed of small patches of parenchyma cells. Just below the endodermis there lies the pericycle. It is uniseriate and composed

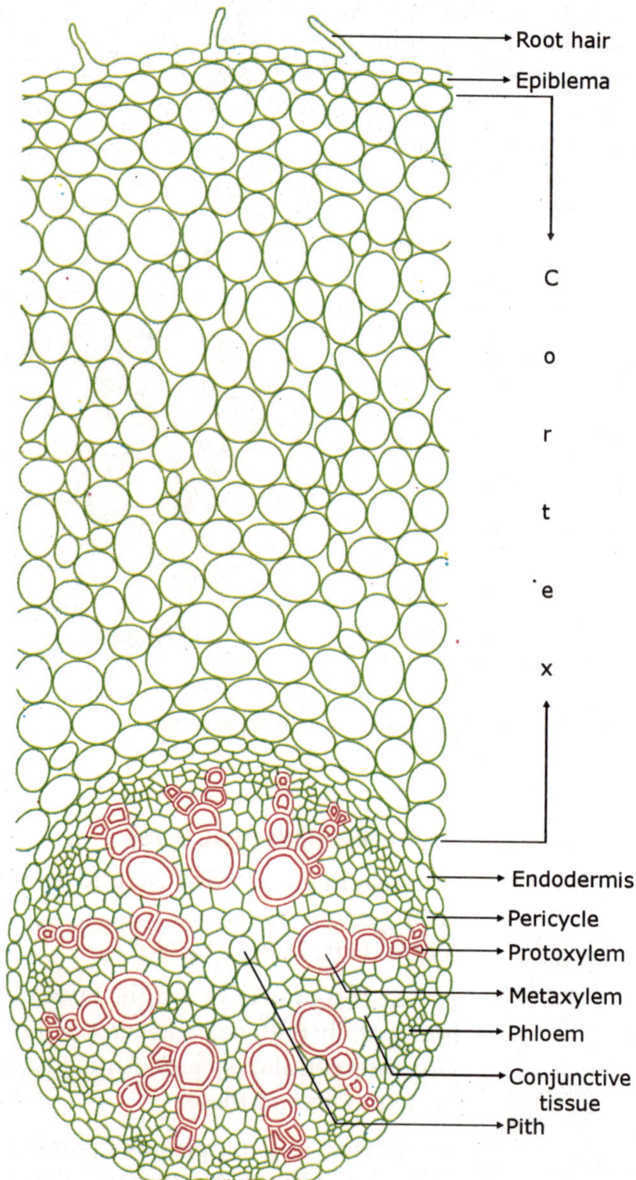


Figure 31.32A

Cellular drawing of a portion of *Colocasia* root in transverse section.

of parenchyma cells. At the centre of stele pith is present. It is large and composed of parenchyma cells.

**Comment:** It is a monocot root due to presence of radial vascular bundle, polyarch and exarch xylem, and large pith. Epiblema is protective and absorptive in function. The exodermis protects the inner tissues when epiblema is decayed. Xylem is the mechanical cell. Centralized mechanical cells provide strength against inextensibility.

## Root of *Zea mays* (Maize. Family: Poaceae)

The transverse section of root is circular in outline and shows the following internal organization of tissues from periphery towards centre (Fig. 31.33).

**Epiblema:** It is uniseriate; the cells are thin walled and horizontally flattened; the cells are compactly set and devoid of cuticle. Unicellular root hairs are present here and there.

**Cortex:** It is composed of parenchyma cells, many layered and large, and encloses air spaces. Numerous large air spaces are present and they are formed

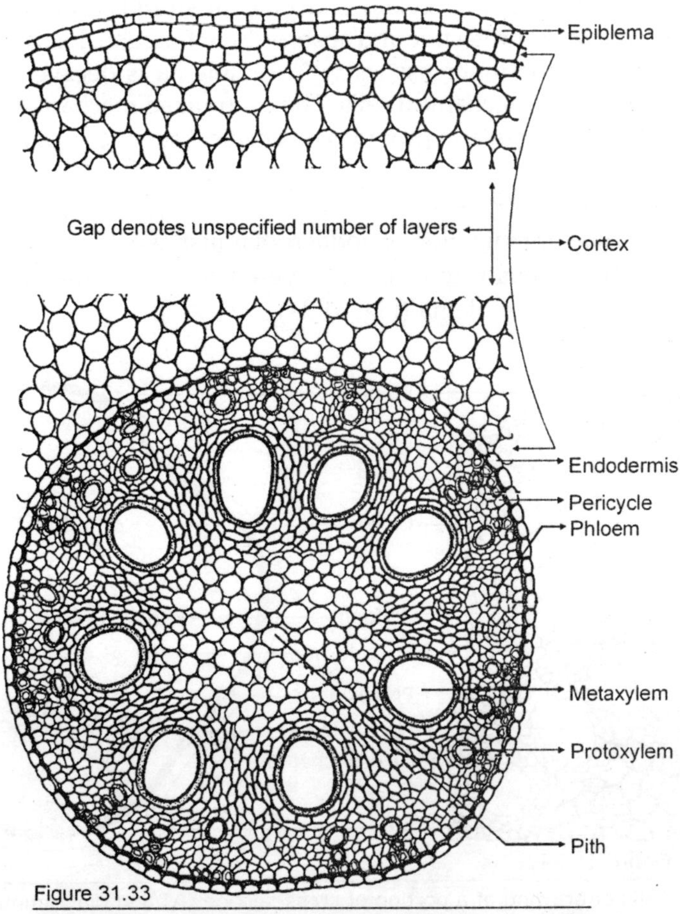


Figure 31.33

Cellular drawing of a portion of *Zea* root in transverse section.

schizogenously. Mature roots show the exodermis. It is the peripheral layer of cortex and composed of cells whose walls have undergone suberization. They differ from cork cells, whose cell walls are also suberized, in possessing living cell contents. The innermost layer of cortex is endodermis. It is composed of compactly set barrel shaped cells that have conspicuous casparian strips. Passage cells are present here and there in the endodermis. In mature roots the cells of endodermis may have additional layers on their walls when the casparian strips become inconspicuous.

**Stele:** Next to endodermis there occurs the pericycle. It is uniseriate. It is composed of parenchyma and sclerenchyma cells. The vascular bundle is radial with protoxylem exarch. The bundles are polyarch with numerous xylem and phloem strands. Parenchyma cells those are in contact with xylem may undergo sclerosis. Protoxylem and phloem occur towards the pericycle whereas the metaxylem lies towards the centre. The central portion of stele is the pith. It is large and composed of parenchyma cells with intercellular spaces. On the periphery of pith there lies a ring of metaxylem vessels. The pith cells contain abundant starch grains.

**Comment:** The root shows the characteristics of monocot due to the presence of radial stele with polyarch xylem and phloem strands, exarch protoxylem and large pith. Epiblema is protective and absorptive in function. Exodermis is protective layer and protects the inner tissues when epiblema is decayed. The xylem and sclerenchyma are the mechanical cells. The centralized mechanical cell reveals that it is an inextensible organ and provides mechanical strength against inextensibility.

## Stilt root of *Zea mays*

The internal tissue organization of nodal root is just like the root of *Zea mays* except that it has a continuous ring of many-layered sclerenchyma at the periphery of cortex just below the epiblema (Fig. 31.34).

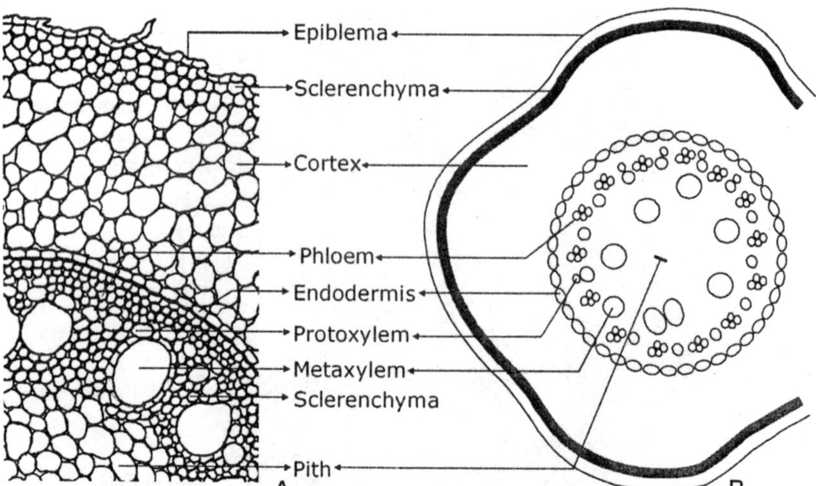


Figure 31.34

Cellular drawing of a portion of cross-section (A) and diagrammatic representation (B) of the stilt root of *Zea mays*.

**Comment:** It is a root due to the presence of radial stele with exarch protoxylem. The monocot nature is revealed due to the presence of polyarch xylem and phloem strands, and large pith. Sclerenchyma at the periphery is the mechanical cell and it reveals that it is an inflexible organ. The presence of centralized mechanical cells like xylem and sclerenchyma surrounding them reveal that the monocot root is an inextensible and incompressible organ. Radial stele, exarch protoxylem and polyarch vascular strands with peripheral and centralized mechanical cells (sclerenchyma) are the characteristic of stilt root of maize where inflexibility, incompressibility and inextensibility are operative.

## Root of *Canna* sp. (Family: Cannaceae)

The transverse section of root is more or less circular in outline and shows the following internal arrangements of tissue from periphery towards the centre (Figs. 31.35, 31.35A).

**Epidermis:** It is uniseriate. The cells are tabular, thin walled, compactly set and parenchymatous without any intercellular spaces. Cuticle is absent from the peripheral walls. Transverse section through the root hair region shows the presence of root hairs, which are the tube like prolongation of the epidermal cells. The epidermis with root hairs is also termed as epiblema or piliferous layer.

**Cortex:** This zone occurs in between epidermis and stele. It consists of many layered, thin walled parenchyma cells with conspicuous intercellular spaces. This zone stores starch grains. In mature root, the peripheral cortex consists of compactly set parenchyma cells. The inner cortex is composed of parenchyma with conspicuous inter-cellular spaces. The middle cortex shows many air spaces. The spaces are linearly elongated. They occur radially in all directions. The innermost layer of inner cortex is endodermis, which is uniseriate and composed of barrel shaped, compactly set parenchyma cells. The endodermal cells possess casparian thickening and its inner walls are thickened.

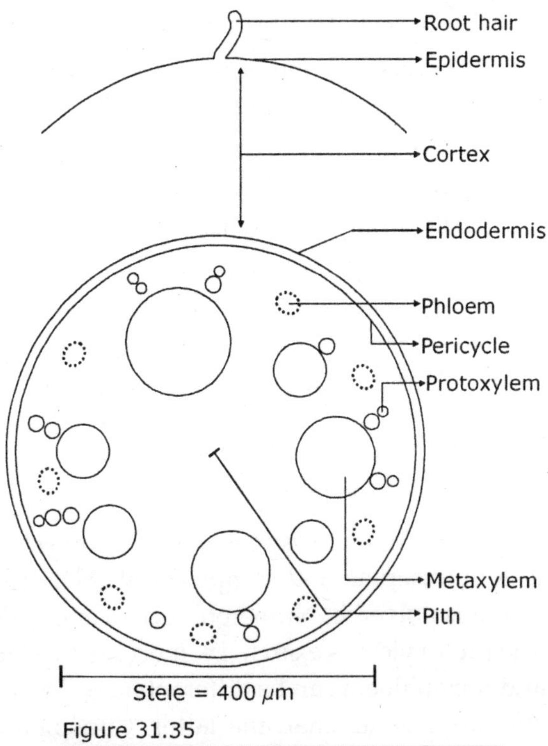


Figure 31.35

A portion of root of *Canna* (diagrammatic).

**Stele:** It includes pericycle, vascular tissue and ground tissue. The pericycle is uniseriate and parenchymatous. The vascular bundle is radial, polyarch with

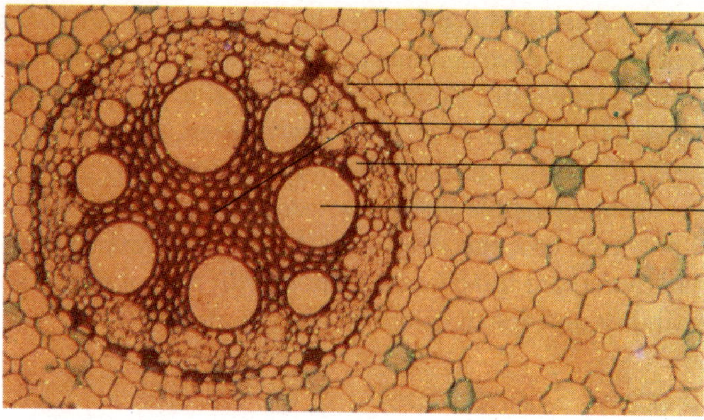


Figure 31.35A

Microphotograph of a portion of cross-section of *Canna* root.

exarch protoxylem. There are about 12 protoxylem and approximately seven metaxylem vessels of different diameter. Conjunctive tissues are present in between xylem and phloem. Pith is large and conspicuous. It is composed of sclerenchyma cells.

**Comment:** It is a monocot root due to the presence of radial vascular bundle, polyarch xylem with exarch protoxylem and large pith. Linearly elongated air space at the middle cortex and sclerenchymatous pith is the characteristic of the root. Xylem and sclerenchymatous pith are the mechanical cells that provide strength against inextensibility.

## Root of *Smilax* sp. (Family: Smilacaceae)

The transverse section of root is more or less circular in outline and reveals the following internal tissue organization from periphery towards the centre (Fig. 31.36).

**Epidermis:** It is uniseriate. It is composed of parenchyma cells that are compactly arranged. The outer wall of the cells is round and cuticularized.

**Cortex:** The outermost layer of the cortex is exodermis. This layer lies just internal to epidermis. It is uniseriate and the cells are very much thick walled. The inner cortex is composed of parenchyma cells. The cells are more or less round and oval, and enclose intercellular spaces. The cells contain starch grains. The innermost layer of cortex is the endodermis. It is single layered and the cells are thickened on their radial and inner walls. It is discontinuous at the region of passage cells that are present here and there.

**Stele:** The outermost layer is the pericycle. It occurs just internal to endodermis. It is multiseriate and composed of sclerenchyma cells. The stele is radial. The xylem is polyarch. Numerous xylem and phloem remain alternate to each other. The protoxylem is exarch, i.e. it occurs towards the peripheral side. Protophloem and metaphloem can be differentiated. The former that occurs towards the outer side are smaller than the latter. Conspicuous pith is present. It is large and composed of parenchyma cells that are full of starch grains.

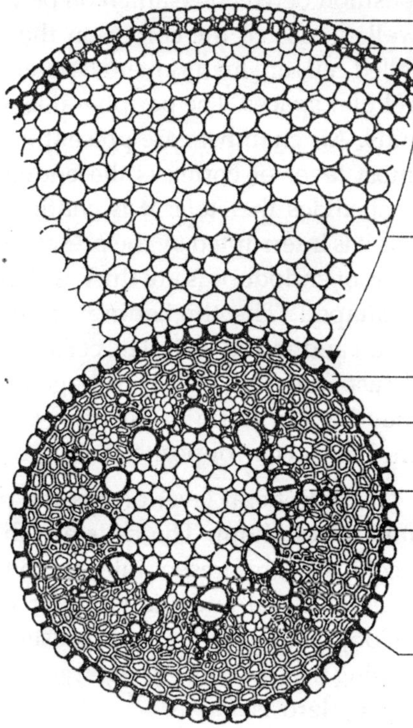


Figure 31.36

Cellular drawing of a portion of cross-section of the root of *Smilax*.

**Comment:** It is a root due to the presence of radial stele. Polyarch xylem and large pith reveal that it is monocot root. In contrast to other roots it has multiseriate sclerenchymatous thick walled pericycle and the cells of pith are full of starch grains. Centralized mechanical cells like xylem and pericycle reveal that it is an inextensible organ and provide mechanical strength against inextensibility.

## Root of *Vanda* sp. (Orchid root. Family: Orchidaceae)

The transverse section of root is circular in outline and shows the following tissue arrangements from periphery towards the centre (Figs. 31.37 and 29.13, 29.14).

**Velamen:** It consists of several layers of cells and forms a sheath around the cortex. It is multiple or multiseriate epidermis the outermost layer of which is the limiting layer. The cells of velamen are non-living and compactly set. The cells are devoid of any contents. In some cells strip like thickening (Fig. 29.13) can be seen on their walls. The cells have perforated walls that act like a sponge. It soaks up water that wets the limiting layer. It usually occurs during rainy season. In dry weather air fills the cells. The cells are large, thick walled and horizontally flattened. Some cells are polygonal.

**Cortex:** The peripheral and outermost layer of it is exodermis. It consists of long and short cells. At some regions the long and short cells alternate. The short cells are thin walled and called passage cell. The long cells are thick walled. The thickening occurs on outer tangential and radial surface. The exodermis occupies

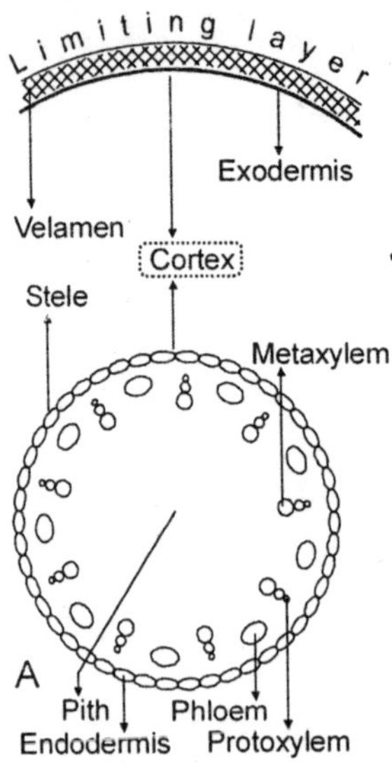


Figure 31.37
Diagrammatic illustration of a portion of *Vanda* root.

a limiting position between velamen on peripheral side and thin walled cells of the cortex on the inner side. The thin walled cells are many layered, parenchymatous with intercellular spaces. The peripheral layers contain chloroplastids. Air chambers are present in the cortex. The innermost layer of cortex is the endodermis that completely encircles the stele. It is continuous and at some regions it is interrupted by the presence of passage cells. The cells of endodermis are thick walled, compact and barrel shaped. The thickening occurs on inner tangential wall and radial wall. The cells contain starch grains and casparian strips. The passage cells are thin walled. They have a definite position in relation to protoxylem. They occur opposite to the protoxylem.

**Stele:** The peripheral layer of it is the pericycle. It is continuous throughout. The cells are thick walled except the cells that are present internal to the passage cell. The stele is radial and the vascular bundles are polyarch with protoxylem exarch. Conjunctive tissues are present. They are sclerenchymatous and remain surrounding the phloem. Pith is large, many layered and parenchymatous that may undergo sclerosis.

**Comment:** Velamen forms a sheath of cortex and thus protects the inner tissues. It gives mechanical strength by the presence of strip like thickening on the walls of some cells. It has also absorptive function; the water thus absorbed is transported to cortex through passage cells of exodermis. The other cells of exodermis are thick walled and are mechanical cells. Parenchymatous cortex is storage in function. Chlorophyllous parenchyma in the cortex is the photosynthetic tissue. The passage cells of endodermis allow radial diffusion of water. Radial stele with polyarch vascular bundle and exarch protoxylem is the characteristic of monocot root. The presence of spongy and absorptive velamen is the characteristic of epiphytic root. The centralized mechanical cells like xylem and sclerenchymatous conjunctive tissue provide mechanical strength against inextensibility as the root hangs freely and bears its own weight. Presence of multiple epidermis (=velamen), exodermis with passage cells, radial vascular bundle with polyarch xylem strands, exarch protoxylem and sclerenchymatous conjunctive tissue are the characteristic of epiphytic orchid root.

## Pneumatophore of *Rhizophora* sp. (Family: Rhizophoraceae)

The transverse section of breathing root or pneumatophore is more or less circular in outline and reveals the following internal organization of tissues from periphery towards the centre (Figs. 31.38, 31.38A).

**Periderm:** It consists of phellem, phellogen and phelloderm cells. Phellem cells are cork cells and are few layered. It is outermost layer and may consist the uniseriate epidermal layer at an early stage. The epidermis and cork cells are interrupted by the presence of lenticel. It is composed of ruptured epidermis and cork cells. In this region there lie the complementary cells that are parenchymatous and very loosely arranged. Phellogen is cork cambium that forms cork cells and loose complementary cells on the peripheral side. Phelloderm cells are the inner derivatives of phellogen and more or less like cortical cells.

**Cortex:** It is few layered and composed of parenchyma cells. The cells are more or less round. They are arranged in such a way that several cells enclose a large air space. Abundant large air spaces are present in this region and they compose a well-developed intercellular space system. The innermost layer of cortex is endodermis. It is uniseriate and composed of barrel shaped cells that are small and compactly arranged.

**Stele:** Next to endodermis there lies the pericycle. It is multiseriate. The uniseriate peripheral layer of it is parenchymatous and compactly arranged. The inner few layers are sclerenchymatous. The vascular bundles are collateral and open, i.e. cambium is present in between xylem and phloem. The protoxylem is endarch where it is towards the centre. The vascular tissues form a continuous cylindrical band below the pericycle. In the band phloem occurs towards the peripheral side, xylem lies towards the centre and cambium is situated in between xylem and phloem. Pith is large and composed of parenchyma cells with intercellular space. It fills up the entire centre portion of root.

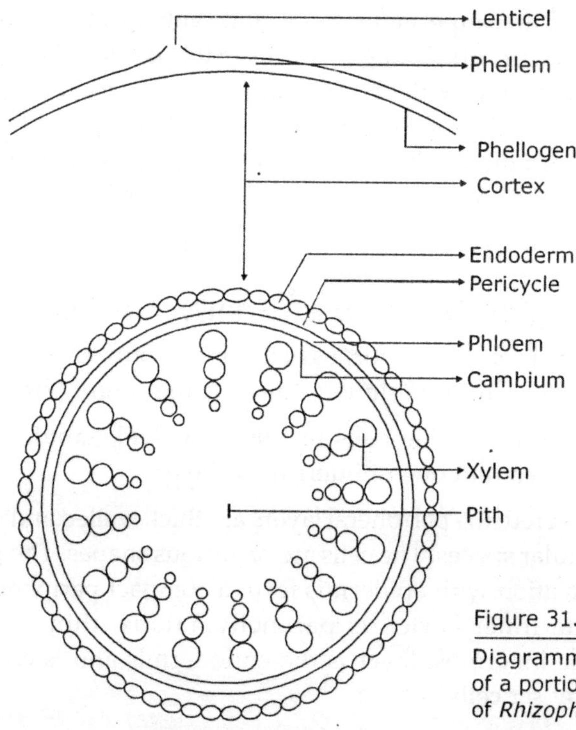


Figure 31.38

Diagrammatic cross-section of a portion of pneumatophore of *Rhizophora*.

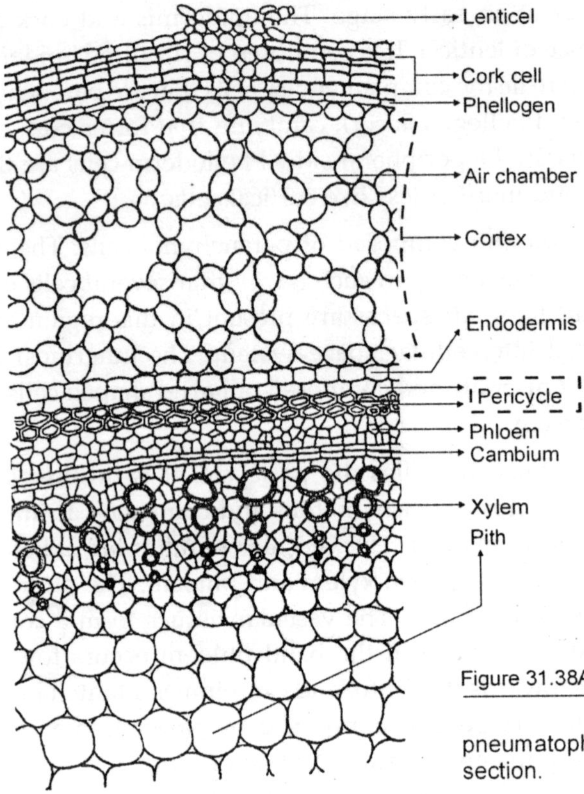


Figure 31.38A

pneumatophore in transverse section.

**Comment:** Cork cells are impervious to air and water and so protective in function. Gaseous diffusion occurs through lenticels. Stele is like a dicotyledonous stem where collateral and open vascular bundles are present. Cork cells interrupted by lenticels, large air spaces at the cortex and the stele like stem are the characteristic features of negatively geotropic and halophytic root, i.e. pneumatophore. Sclerenchyma at pericycle and xylem are the mechanical cells that provide strength against inflexibility.

## Root (aerial and young) of *Tinospora* sp. (Family: Menispermaceae)

The transverse section of root is more or less circular in outline and shows the following tissue organization from periphery towards the centre (Fig. 31.39).

**Epidermis:** It is uniseriate. The cells are thick walled, tabular in shape, compactly set and form a continuous uninterrupted layer.

**Cortex:** It is many layered; the peripheral layers are thick walled and compactly set without any intercellular spaces. The cells are of various shapes. The peripheral layers of cortex in association with epidermis form a compact peripheral zone of thick walled cells. The inner cortex is parenchymatous, thin walled with intercellular spaces. The innermost layer is the endodermis and is composed of barrel-shaped compactly set cells.

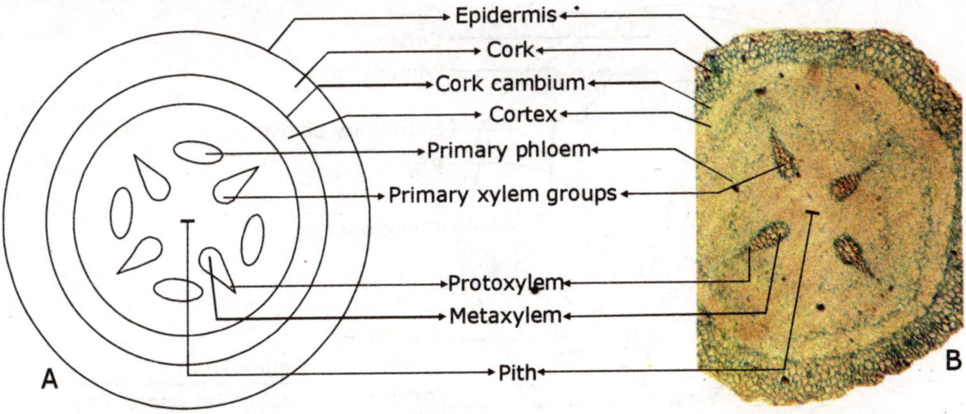


Figure 31.39
Root of *Tinospora* in transverse sectional view. A. Diagrammatic. B. Photomicrograph of a portion.

**Stele:** Internal to endodermis there occurs a few layered, thin walled parenchymatous pericycle. The vascular bundle is radial and the xylem is tri-, tetra- or pentarch. Protoxylem is exarch. In between the xylem, phloem occurs and it corresponds to the number of xylem present in the stele. Conjunctive tissues occur in between xylem and phloem. Pith is small and parenchymatous.

**Comment:** It is a dicotyledonous root due to the presence of radial vascular bundle with exarch protoxylem and the number of xylem is less than six. The presence of peripheral thick-walled cell reveals that it is an inflexible organ and aerial in nature. Absence of root hair is the characteristic of aerial root. The thick-walled cells provide strength against inflexibility. The centralized xylem is mechanical cell and provides strength against inextensibility. This dicotyledonous aerial root shows the remarkable combination of inextensibility and inflexibility.

## Root (aerial and mature) of *Tinospora* sp.

The transverse section of root is polyhedral circular in outline and shows the following tissue organization from periphery towards the centre (Figs. 31.40, 31.40A).

**Periderm:** It is composed of cork cambium phellogen, phellem and phelloderm. It forms a continuous layer. The continuity at some region is interrupted by the presence of lenticels. It consists of ruptured phellem, loose complementary cell, phellogen and phelloderm. Phellem is situated on the peripheral side of phellogen. The cells of phellem are more or less rectangular or polyhedral, thick-walled and many layered with little intercellular spaces. The cells of phellogen are thin-walled; two to three layered and are arranged in storied manner. Phelloderm lies below the phellogen. The cells of phelloderm are parenchymatous with conspicuous intercellular spaces.

**Cortex:** It is composed of few layered, thin-walled, compactly set parenchyma cells. The innermost layer is endodermis, which is conspicuous.

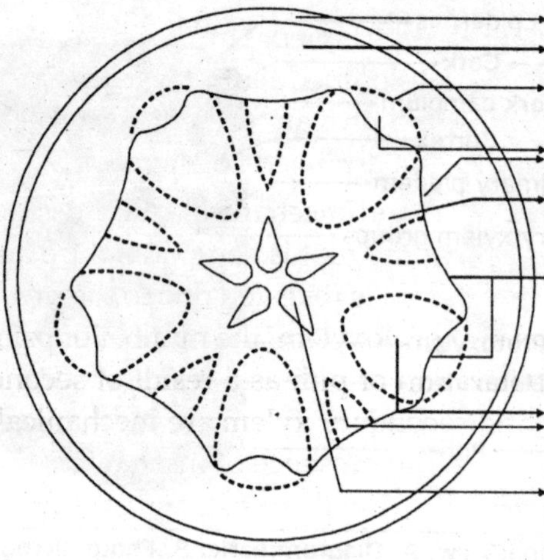


Figure 31.40
Root of *Tinospora* (mature) in transverse sectional view (diagrammatic).

**Stele:** The stele shows primary and secondary vascular tissues. The primary vascular bundle is radial with tri-, tetra- or pentarch xylem. They are grouped together at the centre obliterating the pith. Internal to endodermis there occurs an inconspicuous pericycle. A continuous somewhat wavy cambium ring is present and on the peripheral side of it, there occurs the secondary phloem. Primary phloem is present as crushed patches over the secondary phloem. A thick zone of secondary xylem is present at the inner side of cambium. The secondary xylem is

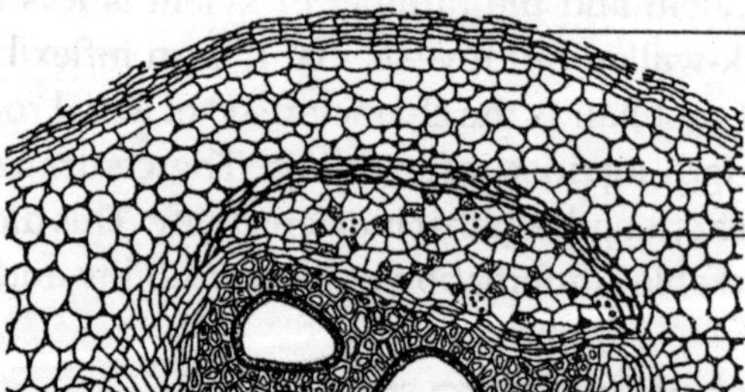


Figure 31.40A
Cellular drawing of a portion of transverse section of *Tinospora* root showing secondary tissues.

composed of tracheids, fibres and tracheae with large lumen. The most characteristic of the root is the occurrence of broad vascular rays or medullary rays among xylem and phloem passing through the cambium. These rays are present corresponding to protoxylem. The secondary vascular tissues are arranged collaterally.

**Comment:** Thick walled cork is protective and mechanical cell. Presence of thick walled cork at the periphery signifies that the root is aerial in nature and so inflexible organ. The dicotyledonous nature of root is revealed due to the presence of radial primary vascular bundle with exarch protoxylem, the number of primary xylem strand is less than six and the obliteration of pith as a result of secondary vascular tissue formation. The primary and secondary xylem are mechanical cell that provides strength against inextensibility. The presence of mechanical cells at the periphery and centre side reveals that the aerial root is an inflexible and inextensible organ that exhibit the remarkable combination where both inflexibility and inextensibility are operative.

## Root of *Beta vulgaris* (Family: Chenopodiaceae)

The transverse section of root is more or less circular in outline and reveals the following internal organization of tissues from periphery towards the centre (Fig. 31.41).

**Periderm:** It consists of phellem, phellogen and phelloderm. Phellem is the outermost layer and consists of a few layers of cells. Phellogen is the cork cambium that divides. The peripheral derivatives are the phellem. The inner derivatives are the phelloderm. Phelloderm is the innermost layer of periderm and the cells are more or less like cortical cells.

**Cortex:** It is parenchymatous; few layered and is confluent with stelar tissues.

**Stele:** It shows pericycle and numerous concentric rings of growth layers. Each growth layer consists of vascular bundles and parenchyma. The vascular bundles are collateral, i.e. xylem and phloem occur on the same radius. Phloem is situated on the peripheral side. The xylem contains very scanty lignified elements. The vascular bundles are mainly composed of parenchyma. Bands of parenchyma are present

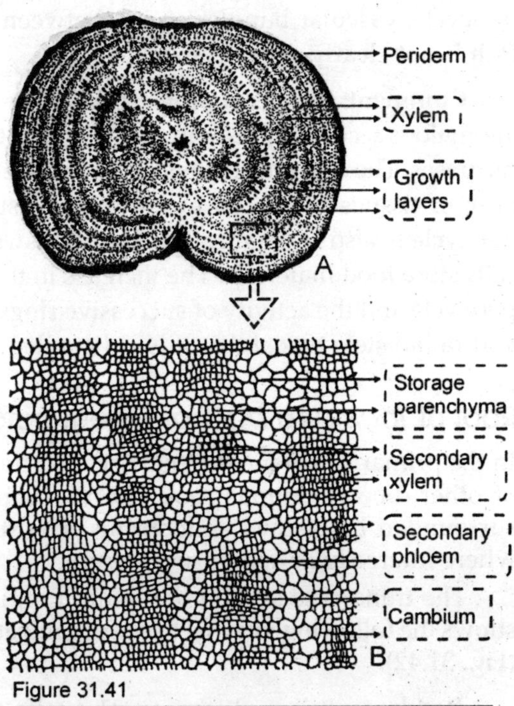


Figure 31.41

*Beta* root. A. Diagrammatic transverse section of root. B. A portion of the root in cross-sectional view.

between the vascular bundles. In between the growth rings parenchyma cells are also present. The stele exhibits anomalous secondary growth. The primary xylem is diarch (Fig. 29.17). The first cambium ring originates from the parenchyma present in between xylem and phloem and pericycle. The cambium though in a continuous ring produces separate vascular bundles separated by bands of radial parenchyma. Thus a ring of vascular bundle is produced. After a period of activity the first ring of cambium ceases its function. A second cambium ring originates from the phloem parenchyma produced by the first ring. Like the first ring the cambium also produces a ring of vascular bundle that are separated by bands of radial parenchyma. After a period of activity its function also ceases. Then a third cambium ring originates from the pericycle. It also behaves like the first and second cambium ring. When its activity ceases another cambium arises from the pericycle. This also has similar function. Thus concentric rings of cambia arise in succession and thus several rings of vascular bundles are produced. Narrow bands of radial parenchyma separate the vascular bundles produced by the first cambium ring whereas the vascular bundles produced by the later successive cambium ring are separated by wide bands of radial parenchyma. Concentric rings of cambia that originate from pericycle enclose cells of pericycle that undergo repeated divisions thus forming more pericyclic layers of parenchyma. This is proliferated pericycle. It alternates with the rings of vascular bundles and parenchyma present between the vascular bundles. In unstained cross section the ring of proliferated pericycle appears to be dark red and thus can be distinguished from the rings of vascular bundles, which are of lighter colour. The parenchyma cells present in proliferated pericycle, vascular bundles and in between vascular bundles are storage tissues. Pith is not clearly demarcated.

**Comment:** Periderm protects the inner tissues from desiccation. Stele shows anomalous secondary growth. It is adaptive type of anomaly and occurs for storage purposes. The concentric rings of cambia are abnormal in origin and activity. Major bulk of tissues produced by cambia are storage parenchyma. The proliferated pericycle is also storage tissue. The root swells when all the storage parenchyma cells store food materials. The increase in thickness is also due the proliferation of pericycle and the activity of successive rings of cambia. The diarch primary xylem and radial stele reveal that it is a dicotyledonous root.

## Root of *Ipomoea batatas* (Sweet potato. Family: Convolvulaceae)

In the primary state the internal organization of root is more or less similar to that of other dicotyledonous root. The vascular bundle is pentarch or hexarch and surrounded by distinct endodermis. The stele remains encircled by wide cortex where intercellular spaces are present.

The transverse section of mature root is more or less circular in outline and shows the following internal tissue organization from periphery towards the centre (Fig. 31.42).

**Periderm:** It is multiseriate. The cells are more or less rectangular in shape and compactly arranged. It is composed of phellem, phellogen and phelloderm.

The outermost few layers are phellem that develops from the phellogen or cork cambium. The phellogen originates from the pericycle and it forms phelloderm towards the inner side. The cells of phelloderm are parenchymatous. Below the periderm there lies a wide zone of parenchyma with conspicuous intercellular spaces.

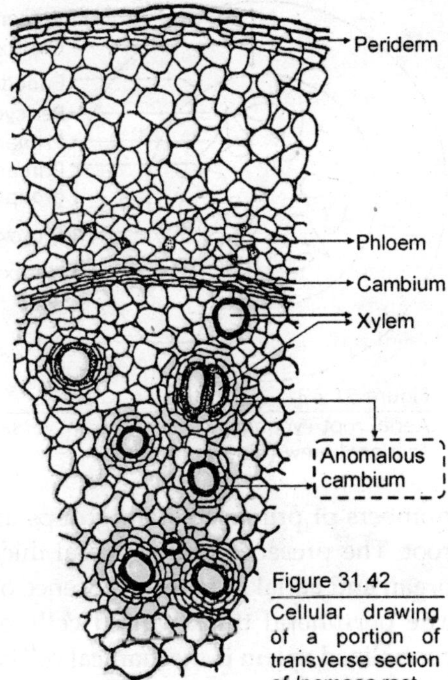


Figure 31.42 Cellular drawing of a portion of transverse section of *Ipomoea* root.

**Vascular tissue:** A cambium ring is present in between xylem and phloem. The secondary vascular tissues like secondary xylem and phloem are normally formed by the cambium. At a later stage, an anomalous cambium originates in the form of a ring surrounding a vessel or a group of vessels. These rings of cambia divide and the inner derivatives are differentiated into secondary xylem elements and the peripheral derivatives form the sieve tubes and laticiferous elements. The anomalous cambia originate from the ground parenchyma present surrounding the vessel or a group of vessels. Tyloses are often present in the vessels with large lumen. The normal cambium and anomalous cambia produce considerable amounts of parenchyma on the peripheral and inner side. These are storage parenchyma.

**Comment:** Periderm is protective tissue and it is pericyclic in origin. Anomaly in the stele is due to the formation of anomalous cambia. The normal cambium and anomalous cambia form large amount storage parenchyma that causes the root to swell.

## Root (aerial and young) of *Ficus* sp. (Family: Moraceae)

The transverse section of root is more or less circular in outline and shows the following tissue organization from periphery towards the centre (Fig. 31.43).

**Epidermis:** It is uniseriate. The cells are thick walled, tabular in shape, compactly set and form a continuous uninterrupted layer.

**Cortex:** It is many layered; the peripheral layers are thick walled and compactly set without any intercellular spaces. The cells are of various shapes. The peripheral layers of cortex in association with epidermis form a compact peripheral zone of thick walled cells. The inner cortex is parenchymatous, thin walled with intercellular spaces. The innermost layer is endodermis and composed of barrel-shaped compactly set cells.

**Stele:** Internal to endodermis there occurs a few layered, thin walled parenchymatous pericycle. The vascular bundle is radial and the xylem is polyarch

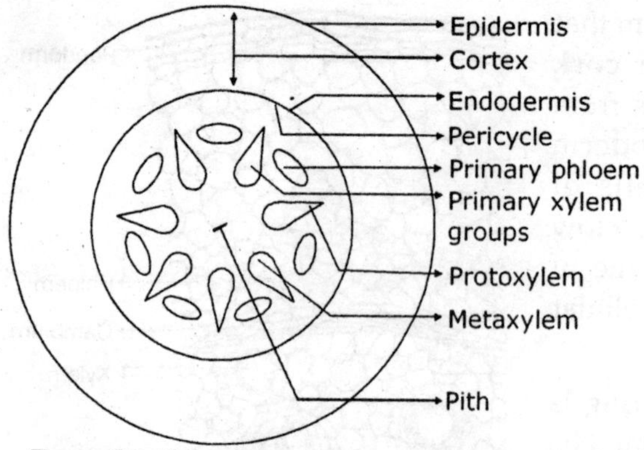


Figure 31.43
Aerial root (young) of *Ficus* in transverse sectional view (diagrammatic).

and the number of primary xylem groups is more than six. Protoxylem is exarch. In between the xylem, phloem occurs and it corresponds to the number of xylem present in the stele. Conjunctive tissues occur in between xylem and phloem. Pith is large and parenchymatous.

**Comment:** It is a root due to the presence of radial vascular bundle with exarch protoxylem. Anomaly in the stele is due to the presence of more than six numbers of primary xylem groups and large pith though it is a dicotyledonous root. The presence of peripheral thick walled cell reveals that it is an inflexible organ and aerial in nature. Absence of root hair is the characteristic of aerial root. The peripheral thick-walled cells provide strength against inflexibility. The centralized xylem is mechanical cell and provides strength against inextensibility. This dicotyledonous aerial root shows the remarkable combination of inextensibility and inflexibility.

## Root (aerial and mature) of *Ficus* sp.

The transverse section is more or less circular in outline and shows the following internal tissue organization from periphery towards the centre (Figs. 31.44, 31.44A).

**Epidermis and Periderm:** The outermost layer is epidermis. It is uniseriate. The cells are horizontally flattened, compactly arranged with thick cuticle on their outer wall. Next to epidermis there lies the periderm. It consists of phellem, phellogen and phelloderm. The peripheral layer is the phellem or cork cell. It is many layered and consists of thick walled somewhat rounded cells. The cells enclose little intercellular spaces and are radially arranged. Phellogen, also called cork cambium, lies below the cork cells. The cells are thin walled and compactly arranged. It produces phellem on the peripheral side and phelloderm on the inner side. Phelloderm is few layered and composed of more or less isodiametric parenchyma cells. The cells enclose intercellular spaces and are radially arranged.

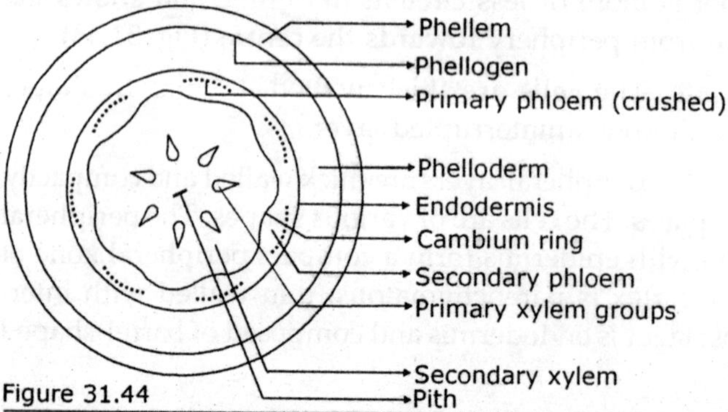


Figure 31.44
Aerial root (mature) of *Ficus* in transverse sectional view (diagrammatic).

**Cortex:** It is few layered and composed of parenchyma cells with intercellular spaces. It differs from phelloderm, as the cells are not radially arranged. The innermost layer is endodermis. It is uniseriate, compactly arranged and composed of barrel shaped cells.

**Stele:** Next to endodermis there lies the pericycle. It is few layered and parenchymatous. The vascular bundle is radial in arrangement with protoxylem exarch. A complete cambium ring is present in the stele. Below the ring there lies a complete cylinder of secondary xylem produced by the cambium. It is many layered and composed of vessels, fibres and tracheids. On the peripheral side

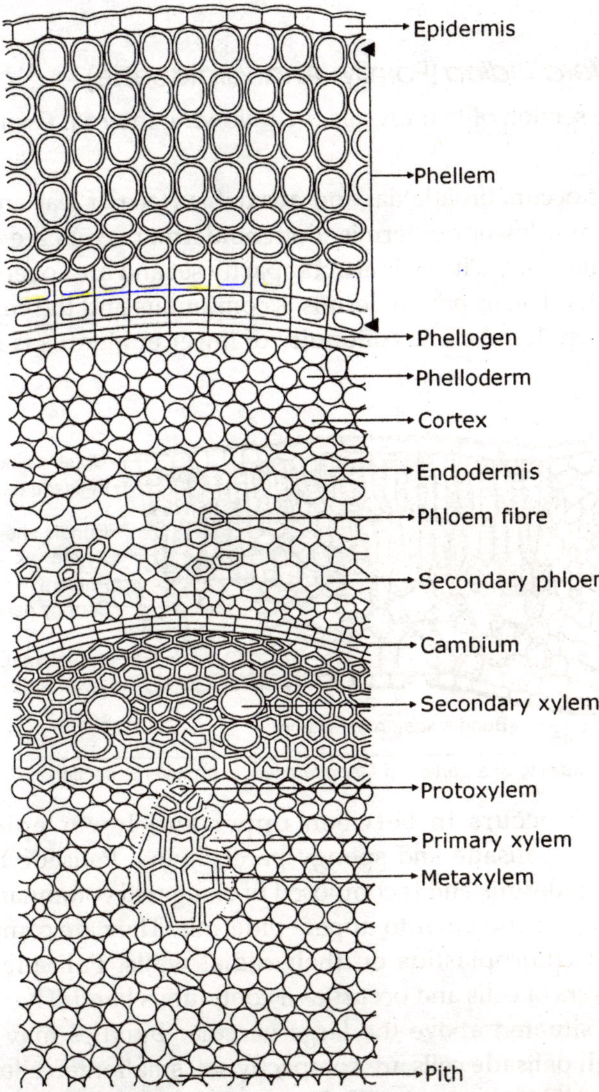


Figure 31.44A

Diagram showing transverse section of aerial root of *Ficus*. Adapted after Pandey, 1978.

cambium produced secondary phloem. It is few layered and consists of sieve tube, companion cell, and phloem fibre and phloem parenchyma. In between secondary phloem and pericycle the patches of primary phloem remain appressed. Below the secondary xylem there occurs the primary xylem. The centre is occupied by pith. It is small, parenchymatous with few intercellular spaces.

**Comment:** Thick walled periderm is protective in function. The cylinder of secondary xylem and primary xylem are mechanical cells. It is a root due to the presence of radial vascular bundle. Its aerial nature is revealed due to the presence of peripheral cork cells and centralized mechanical cells. The latter provides mechanical strength against inextensibility as the root hangs freely and bears its own weight.

## Leaf of *Mangifera indica* (Family: Anacardiaceae)

A thin transverse section of leaf reveals the following internal tissue organization (Fig. 31.45).

**Epidermis:** It occurs on adaxial and abaxial surface of leaf and accordingly termed as upper and lower epidermis. Both epidermal layers are uniseriate and composed of tabular cells. The cells are compactly set and are covered with cuticle on their outer walls. The upper epidermis is continuous and the cuticle is thick in contrast to lower epidermis. The continuity of lower epidermis is interrupted by stomata.

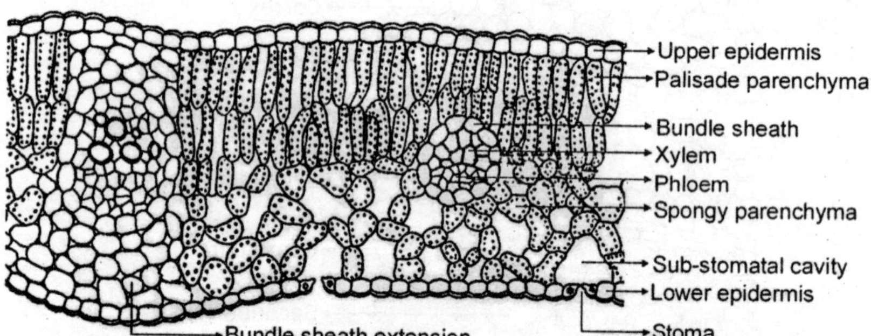


Figure 31.45

Cellular drawing of a portion of *Mangifera* leaf in transverse section.

**Mesophyll:** It occurs in between upper and lower epidermis. It is distinguishable into palisade and spongy parenchyma. Palisade tissue lies just internal to upper epidermis and is composed of elongated columnar parenchyma, which is arranged perpendicular to upper epidermis. They are compactly set and contain abundant chloroplastids on their radial walls. Palisade parenchyma consists of two layers of cells and occurs as a continuous band. The cells of bundle sheath extension situated above the large vascular bundles may interrupt the continuity. Though palisade cells are compactly set, small intercellular spaces are present and each cell is in contact with the spaces. These spaces are in continuity with the intercellular spaces formed by the spongy parenchyma. Spongy parenchyma is present towards the lower epidermis. They are many layered and occur as a continuous layer except at the bundle sheath extension region. The cells are more or less isodiametric, loosely arranged with conspicuous intercellular spaces and enclose small air spaces. They contain lesser number of chloroplast in contrast to upper palisade.

**Vascular bundle:** The bundles are collateral, closed with mesarch xylem. Phloem occurs towards the lower epidermis and xylem lies towards the upper epidermis. The bundles are of various sizes. A layer of bundle sheath made up of parenchyma surrounds the smaller bundles. In the large bundles, in addition to bundle sheath, there occur thin or thick walled parenchyma cells above and below, and they connect the bundle with upper and lower epidermis. This is bundle sheath extension where the mesophyll tissue is discontinuous. In the xylem element trachea with annular and spiral thickening, and xylem parenchyma are present. In the phloem, sieve tubes with narrow lumen, companion cell and phloem parenchyma occur.

**Comment:** Epidermis with cuticle protects the inner mesophyll tissues. Stoma present on the lower epidermis is meant for gaseous diffusion. Mesophyll cells are photosynthetic tissue. It is differentiated into palisade and spongy parenchyma, and it is the characteristic of dicotyledonous dorsiventral leaf. Inverted vascular bundle is the characteristic of leaf. Stoma at lower epidermis, differentiated mesophyll cells and inverted vascular bundles are the characteristic of typical dorsiventral mesophytic leaf.

## Leaf of *Ficus* sp. (Family: Moraceae)

A vertical section through the leaf shows the following anatomical characters (Fig. 31.46).

**Epidermis:** There are two epidermal layers—upper and lower. The upper epidermis is multiseriate and the cells are compactly set. The outermost layer consists of elongated cells with cuticle on the outer walls. This layer is continuous

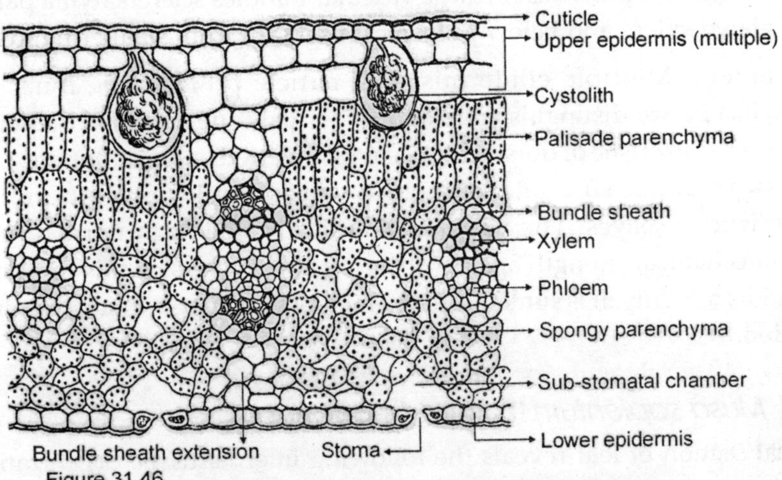


Figure 31.46
Cellular drawing of a portion of leaf of *Ficus* in transverse section.

and devoid of stomata. The lower layers of upper multiple epidermis are interrupted by the lithocysts. Lithocyst is a large cell that contains cystolith. It is composed of a cluster of crystals of calcium carbonate and remains attached to the roof of lithocyst by means of a stalk. The cells of lower layers of upper epidermis are larger than the outermost layer. All cells lie above the upper palisade cells. The lower epidermis is also multiseriate and consists of cells that are compactly arranged. The cells are more or less of same size. The outermost layer of lower epidermis has cuticle on the outer walls but the degree of cutinization is less than the upper epidermis. The lower epidermis is interrupted by stomata. The stomata are sunk and below each stoma there exists the sub-stomatal chamber. In some species cystoliths also occur on the lower side. They are smaller in size than those occurring on the upper side. The lithocysts are larger than the adjacent cells and some of its portion may intrude the spongy parenchyma cells.

**Mesophyll:** It lies between upper and lower layers of multiple epidermis. The cells situated towards the upper epidermis are coloumnar, two or more layered and contain abundant chloroplast. These are called palisade cells that compose the chief photosynthetic region of the leaf. The cells that occur towards the lower epidermis are more or less isodiametric and may be irregular in shape. The cells are loosely arranged and have profuse intercellular spaces. The cells are called spongy parenchyma and contain lesser number of chloroplasts in comparison to palisade parenchyma. The mesophyll cells may be interrupted by bundle sheath extensions.

**Vascular bundle:** The vascular bundle remains embedded in the mesophyll tissue and the size of it varies. The bundles are collateral and closed. The bundles are surrounded by bundle sheath (border parenchyma), which is parenchymatous and colourless. The large bundles have bundle sheath extension. Each vascular bundle has xylem that occurs towards the upper epidermis and phloem that lies towards the lower epidermis. In large vascular bundles sclerenchyma patches are present above xylem and below phloem. It is absent from smaller bundles.

**Comment:** Multiple epidermis with cuticle protects the inner tissues. Mesophyll cells are distinguishable into upper palisade and lower spongy cells. This is the characteristic of dorsiventral leaves. Inverted vascular bundle, i.e. xylem towards the upper side and phloem towards the lower epidermis is the characteristic of leaves. The vascular bundles beside from normal functions, provide mechanical strength against shearing stress. The stoma is present on the lower epidermis only. It is sunken and characteristic of leaves that have xerophytic adaptation.

## Leaf of *Musa sapientum* (Family: Musaceae)

A vertical section of leaf reveals the following internal tissue organization (Fig. 31.47).

**Epidermis:** It consists of upper and lower epidermis. The upper epidermis is uniseriate and consists of cells that are thin walled, more or less round and compactly arranged with prominent and well-developed cuticle on outer walls. The lower epidermis is also uniseriate and the cells are like the upper epidermis. But the cuticle is not prominent and well developed like upper epidermis. The continuity of lower epidermis is interrupted by the presence of stomata. Presence of stoma is restricted to lower epidermis only and it is slightly sunken. Below each stoma a prominent sub-stomatal chamber is present. Below the upper and lower epidermis there occur one or more layers of cells. This layer is termed as sub-epidermal layer. The cells of this layer are thin walled, parenchymatous and devoid of chlorophyll. These cells have water storage function. The cells of sub-epidermal layer below upper epidermis are much larger than those present below the lower epidermis.

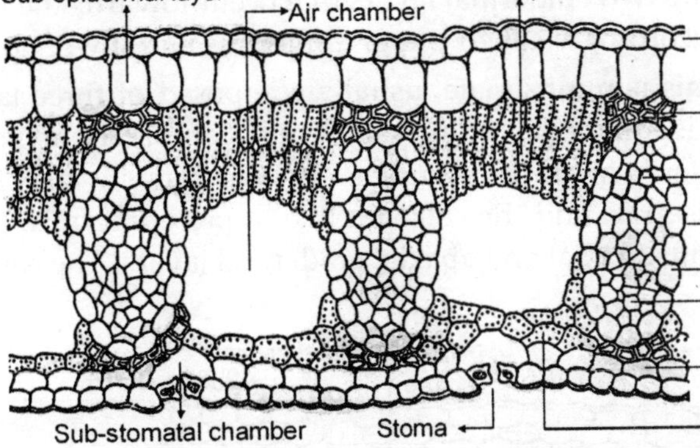


Figure 31.47
Cellular drawing of a portion of transverse section of *Musa* leaf.

**Mesophyll:** It is differentiated into palisade tissue and spongy parenchyma. The palisade tissues are two to three layered and occur below the upper epidermis. The cells are columnar and compactly arranged with little intercellular spaces. The cells contain chloroplastids on their radial walls. The palisade layer is discontinuous above the vascular bundles where patches of sclerenchyma cells are present. The spongy parenchyma is situated below the palisade zone. The cells are smaller than the palisade cells. The shape of the cell is more or less isodiametric. The cells that are present below the lower epidermis are almost round. These cells contain chloroplastids. The zone of spongy parenchyma is interrupted by the presence of large air spaces. The spaces develop lysigenously. These spaces alternate with the vascular bundles. The spongy layer is also discontinuous below the vascular bundle where sclerenchyma is present.

**Vascular bundle:** Each vascular bundle is oval in shape. The bundles are collateral and closed. Xylem occurs towards the upper epidermis and is poorly developed. The phloem occurs towards the lower epidermis. Each vascular bundle is surrounded by a bundle sheath made up of parenchyma. Sclerenchymatous patches occur above and below of each vascular bundle.

**Comment:** The leaf nature is revealed by the presence of collateral and inverted vascular bundle where xylem occurs towards the upper epidermis and phloem lies towards the lower epidermis. The cambium is absent. The leaf is dorsiventral due to the differentiation of mesophyll into palisade and spongy parenchyma. Large air spaces are meant for gaseous diffusion. *Musa,* though monocot, possesses differentiated mesophyll tissues, which are the characteristic of dicotyledonous leaf. Each vascular bundle with sclerenchymatous caps forms I-girder where the patches of sclerenchyma form the flanges and the vascular bundles the web. These I-girders provide mechanical strength against shearing stress.

## Leaf of *Nerium odorum* (Family: Apocynaceae)

A vertical section through the lamina of leaf reveals the following internal organization of tissues (Fig. 31.48).

**Epidermis:** There are two epidermal layers that occur towards the upper and lower side. They are accordingly called upper epidermis and lower epidermis.

The upper epidermis is multiseriate, usually composed of three layers. The cells are more or less isodiametric, thin walled and compactly arranged. The peripheral layer is covered with thick cuticle on their outer walls. The lower epidermis is also multiseriate and the cells are like upper epidermis. Epidermal outgrowths occur on both adaxial and abaxial epidermal layers. The continuity of

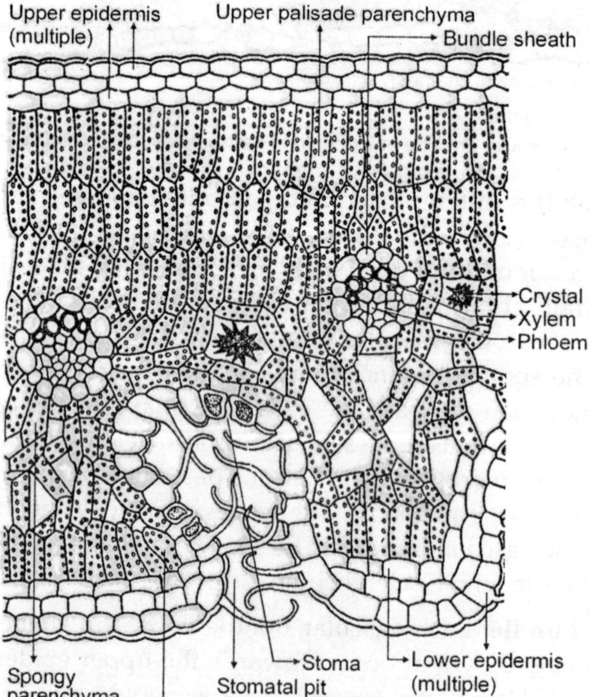


Figure 31.48

A portion of *Nerium* leaf in transverse section.

lower epidermis is interrupted by the presence of stoma that is present on lower epidermis only in cavities. The cavities are called stomatal pit. It is formed by the incurvation of epidermis. The cavities are lined by epidermal cells, which are discontinuous at the region of stomata. The stoma is slightly elevated above the epidermal layer that borders the cavities. Numerous unicellular hairs are present in the cavity and they are the prolongations of cells that border the cavity.

**Mesophyll:** It is composed of palisade parenchyma and spongy parenchyma. The palisade cells occur below the upper and lower epidermis. On the upper side the palisade cells are three or more layered. The cells are columnar, compactly arranged with few intercellular spaces. The cells contain abundant chloroplastids that usually occur along the radial walls of the cells. In contrast to the upper palisade layer the inner palisade is single layered and the cells are smaller. The spongy cells occur in between upper and lower palisade layers. It is many layered. The cells are more or less isodiametric and thin walled. The cells are very loosely arranged in such a way that they enclose large air spaces. In the mesophyll there occurs the calcium oxalate crystals here and there.

**Vascular bundle:** The bundles are collateral and closed, i.e. cambium is absent and xylem and phloem lie radially. The xylem occurs towards the upper epidermis and phloem is towards the lower epidermis. Each bundle is surrounded by bundle sheath. It is composed of parenchyma cells.

**Comment:** The leaf is dicot and dorsiventral due to the presence of differentiated mesophyll cells into palisade and spongy parenchyma. Unlike other dicots palisade cells occur below both upper and lower epidermal layers. Xylem is the only mechanical cell. The leaf shows xerophytic adaptations due to the presence of (i) abaxial and adaxial multiple epidermal layers, (ii) thick cuticularization on the outer wall of cells composing both the epidermal layers, (iii) occurrence of multilayered palisade cells, (iv) occurrence of stoma in stomatal pit and (v) presence of hairs in the pit to reduce the rate of transpiration.

## Leaf of *Polianthes tuberosa* (Tube rose; Family: Agavaceae)

A vertical section through the leaf reveals the following anatomical features (Fig. 31.49).

**Epidermis:** It occurs on adaxial and abaxial surfaces of a leaf and accordingly termed as upper epidermis and lower epidermis. The epidermal layers are single layered. The cells are rectangular in shape with rounded outer walls. The outer walls are cuticularized. The cells are compactly set. The continuity of both epidermal layers is interrupted by stomata. A sub-stomatal chamber is present below each stoma. Stomata are flush with the epidermis.

**Mesophyll:** It lies in between the two epidermal layers and is many layered. It consists of similar type of cells. The cells are more or less isodiametric. The cells contain abundant chloroplastids. The cells cannot be differentiated into palisade parenchyma and spongy parenchyma, but the cells are of spongy type. The cells enclose intercellular spaces that are continuous to the external environment through sub-stomatal chamber and stomata.

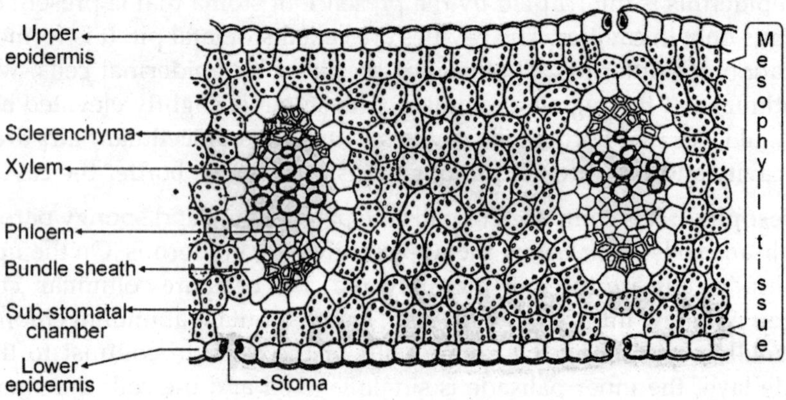


Figure 31.49

A portion of *Polianthes* leaf in transverse section.

**Vascular bundle:** The bundles are more or less of same size and shape. They are situated at regular intervals. Each vascular bundle is collateral and closed, i.e. cambium is absent. The xylem occurs towards the upper epidermis and the phloem lies towards the lower epidermis. Above xylem and below phloem there occurs a patch of sclerenchyma. Each vascular bundle is surrounded by a bundle sheath. It is composed of parenchyma cells. The cells are thin walled and chloroplastids are absent from these cells.

**Comment:** It is an isobilateral leaf due to the presence of stomata on both upper epidermis and lower epidermis, and the mesophyll cells are not differentiated into palisade parenchyma and spongy parenchyma. The mesophyll cells with chloroplastids are photosynthetic tissue. Stomata are meant for gaseous diffusion. Sclerenchyma patches, present above xylem and below phloem, are mechanical cells. They form I-girders and add mechanical strength against shearing stress.

## Leaf of *Zea mays* (Family: Poaceae)

A vertical section through the lamina of leaf reveals the following internal organization of tissues (Fig. 31.50).

**Epidermis:** Epidermis occurs on upper and lower surface of leaf. The upper epidermis is uniseriate. The cells are more or less oval with cuticularized outer walls. It is discontinuous where stoma is present. Below each stoma sub-stomatal chamber is present. Upper epidermis possesses the characteristic bulliform cells. These cells occur in groups and can be easily differentiated from the adjoining cells by their larger size. The cells are hyaline and empty. The lower epidermis is just like the upper epidermis but it is devoid of bulliform cells. It is also interrupted by stoma.

**Mesophyll:** It is composed of similar type of cells. It is not differentiated into palisade and spongy parenchyma. The cells are more or less isodiametric and enclose small intercellular spaces. The intercellular spaces are connected with each

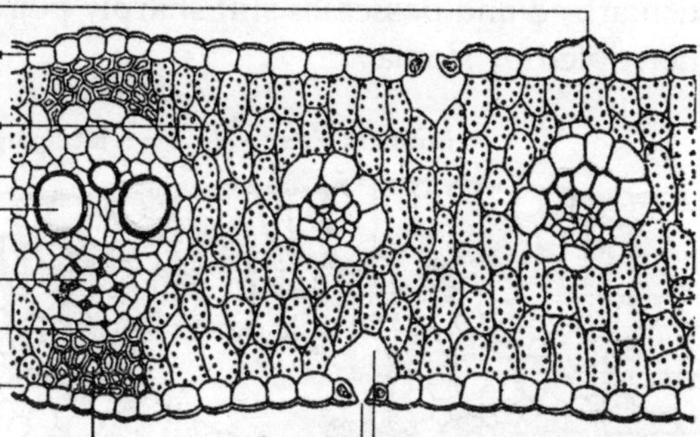


Figure 31.50

Cellular drawing of a portion of *Zea* leaf in transverse section.

other and form a continuous system. Mesophyll cells are connected to the external environment through the spaces, sub-stomatal chamber and stoma. Chloroplastids are present in the cells.

**Vascular bundle:** The bundles may be large and small. They are arranged in parallel and alternate to each other. Each bundle is collateral and closed, i.e. cambium is absent. Xylem occurs towards the upper epidermis and phloem is situated towards the lower epidermis. The large bundles show distinct xylem and phloem. There are two metaxylem and one protoxylem vessel. Metaxylem is more or less oval in shape. The lysigenous protoxylem cavity is towards the upper epidermis. Each vascular bundle remains surrounded by a bundle sheath. The sheath is parenchymatous, the cells of which contain starch and plastids. The cells are thin walled. In large vascular bundle the bundle sheath is connected with upper epidermis and lower epidermis by patches of sclerenchyma. The cells may be of several layered, compactly arranged and appear as a cap over and below the vascular bundle. These cells are called bundle sheath extension.

**Comment:** The leaf is monocot because the mesophyll is composed of similar type of cells. Presence of stoma on both epidermal layers and the undifferentiated mesophyll are the characteristic of isobilateral leaf. The numerous vascular bundles remain in parallel. Presence of bulliform cell in the upper epidermis is also the characteristic of a monocot leaf. The stoma is meant for gaseous diffusion. Presence of starch in the bundle sheath signifies that it is a storage tissue. The large vascular bundles with their sclerenchymatous bundle sheath extensions form I-girder. It gives mechanical strength against shearing stress.

## Leaf of *Bambusa* sp. (Family: Poaceae)

A transverse section through leaf shows the following anatomical characters (Fig. 31.51).

**Epidermis:** It occurs on adaxial and abaxial surfaces of leaf and accordingly termed as upper epidermis and lower epidermis. The upper epidermis is cuticularized. Bulliform cells occur here and there on the upper epidermis. The cells are present between the vascular bundles. These cells may not be at the usual level of other epidermal cells. Silica containing cells are also present. The lower epidermis is also cuticularized and possesses stiff sharply pointed hairs. It is not continuous and is interrupted by stoma.

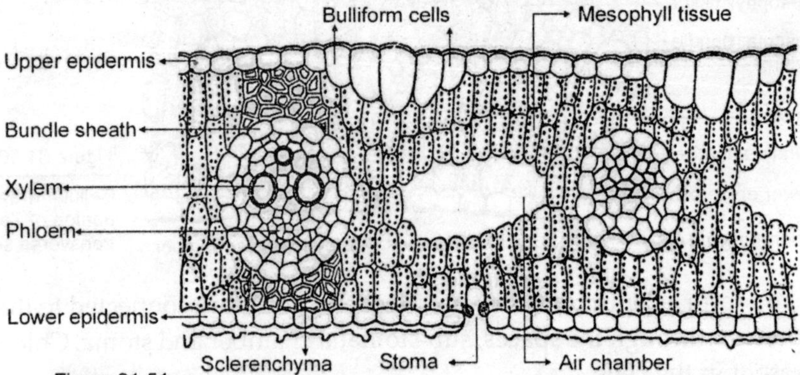


Figure 31.51

Cellular drawing of a portion of leaf of *Bambusa* in transverse section.

**Mesophyll:** It is composed of more or less elongated cells without having any distinction into spongy and palisade tissue. The cells are parenchymatous, compactly arranged and contains chloroplasts. A central row of cavities is present in the mesophyll.

**Vascular bundle:** The size of bundle varies. Bundles are arranged in parallel. They are closed and collateral. The bundles are inverted, i.e. xylem occurs on adaxial side and phloem is situated on the abaxial side. Each bundle is surrounded by bundle sheath. The larger bundles have patches of sclerenchyma on adaxial and abaxial side. The sclerenchyma patches may be in contact with upper and lower epidermis.

**Comment:** The epidermis with cuticle protects the mesophyll tissues. Bulliform cells help in water storage. The mesophyll cells are not distinguishable into palisade and spongy tissue. This is the characteristic of an isobilateral leaf. The inverted vascular bundle is the characteristic of leaves. The bundles are arranged in parallel and it is the characteristic of monocot leaf. The bundles provide mechanical strength against shearing stress. Presence of air chamber in the mesophyll helps in gaseous diffusion.

## Leaf of *Phoenix sylvestris* (Datepalm; Family: Palmae)

Cross section of leaf reveals the following internal organization of tissues (Fig. 31.52).

**Epidermis:** It occurs on adaxial and abaxial surface and accordingly termed as upper and lower epidermis. Both epidermal layers are uniseriate and composed of tabular cells. The cells are thin walled, compactly arranged and covered with thick cuticle on their outer walls. The continuity of both epidermises is interrupted by the presence of stomata. A uniseriate, thin walled parenchyma with scanty chlorophyll occurs as sub-epidermal layer just internal to both the epidermises.

These sub-epidermal layers are referred to as upper and lower sub-epidermal layer according to their position. They are distinct from mesophyll cells and can be regarded as special hypodermis.

**Mesophyll:** It occurs between the two sub-epidermal layers and is not differentiated into palisade and spongy tissue. The cells are polygonal and more or less isodiametric, and enclose small intercellular spaces. Chloroplastids are present in these cells and so they are sometimes referred to as chlorenchyma. Sclerenchyma cells are present in groups internal to both sub-epidermal layers. They occur more or less in parallel series.

**Vascular bundle:** Vascular bundles are located in the mesophyll tissue and arranged in parallel. They are variable in size. Each bundle is collateral with xylem present towards upper epidermis and phloem being situated towards lower epidermis. The bundles are closed, i.e. cambium is absent. A uniseriate thin walled hyaline parenchyma cell surrounds the smaller bundles. This band of parenchyma is known as bundle sheath and it is distinct from mesophyll cells. The larger bundles are also encircled by bundle sheaths, which are made up of sclerenchyma and they may be continuous with the sclerenchyma patches present internal to both epidermises.

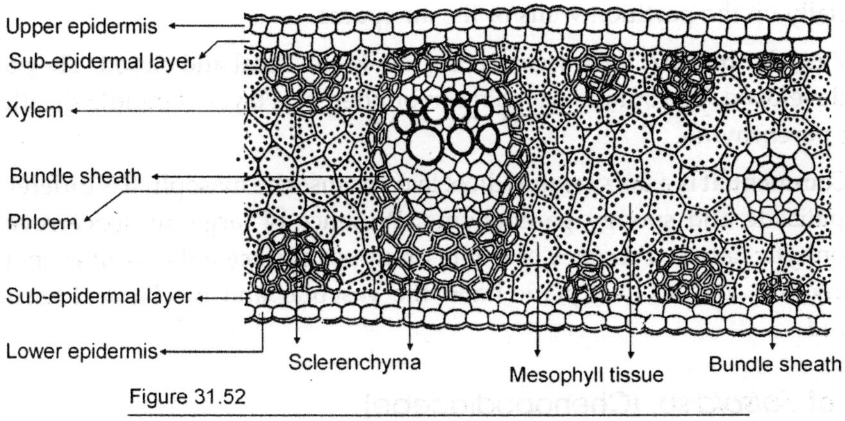


Figure 31.52

Cellular drawing of a portion of leaf of *Phoenix* in transverse section.

**Comment:** Epidermis with cuticle protects the inner mesophyll tissues. Stoma is meant for gaseous diffusion. Presence of stoma at both epidermal layers is the characteristic of isobilateral leaf. Undifferentiated homogeneous mesophyll is also the characteristic of an isobilateral leaf. The patches of sclerenchyma situated at the periphery of mesophyll form I-girders and provide mechanical strength against shearing stress. Inverted vascular bundle is the characteristic of a leaf. The parallel arrangement of bundles reveals its monocotyledonous nature. Sclerenchymatous bundle sheath and sclerenchyma at the two edges of large vascular bundles are mechanical cells. The leaf exhibits the characteristic of monocot isobilateral xeric leaf due to the presence of stoma on both epidermal surfaces, parallel arrangement of inverted vascular bundles with sclerenchymatous bundle sheath, presence of patches of sclerenchyma at the periphery of mesophyll and homogeneous undifferentiated mesophyll.

## Leaf of *Nymphaea stellata* (Family: Nymphaeaceae)

A vertical section through lamina of leaf reveals the following internal tissue organization (Fig. 29.11B).

**Epidermis:** It consists of upper and lower epidermis. Both epidermal layers are uniseriate. The cells are more or less round and compactly set. The cells of upper epidermis are cuticularized on their outer walls. Waxy depositions are present over the cuticle. The continuity of upper epidermis is interrupted by the presence of stoma. Below each stoma sub-stomatal chamber is present. The lower epidermis is devoid of stoma and contains slime glands.

**Mesophyll:** It consists of upper palisade and lower spongy cells. The palisade cells are three or more layered. The cells are columnar, enclose intercellular spaces and contain chloroplastids. Below palisade layer there lies the spongy parenchyma. It is many layered and many large air spaces occur in this region. The cells of spongy parenchyma are more or less isodiametric. Some of the cells may form the partition wall between the air spaces. Trichosclereids are present here and there especially on the partition walls of the air spaces.

**Vascular bundle:** Each vascular bundle is collateral and closed. They are not well developed. The bundles remain surrounded by a layer of bundle sheath made up of parenchyma.

**Comment:** It is a dorsiventral dicot leaf because the mesophyll is differentiated into palisade and spongy parenchyma. Presence of large air space reveals its hydrophytic nature. Presence of stoma on upper surface only signifies that it is a floating leaf. Due to its aquatic habit the vascular and mechanical tissues are reduced.

## Leaf of *Salsola* sp. (Chenopodiaceae)

A vertical section through leaf reveals the following internal tissue organization (Fig. 31.53).

**Epidermis:** It is single layered and composed of horizontally flattened thin walled cells. The cells are compactly arranged and have thin cuticle on their outer walls.

**Mesophyll:** It is differentiated into palisade and parenchyma cell. Palisade cells are elongated and contain abundant chloroplastids. The palisade layer is uniseriate and the cells are compactly arranged. It forms a band of chlorophyllous tissue all round the periphery next to epidermis. The chlorophyllous band is interrupted at more than one places by parenchyma cells. Below the palisade layer there occur the parenchyma cells. The cells are large, thin walled, centrally located and enclose conspicuous intercellular spaces.

Figure 31.53

A. Outline of the cross-section of *Salsola* leaf. B. A portion of the leaf in transverse section.

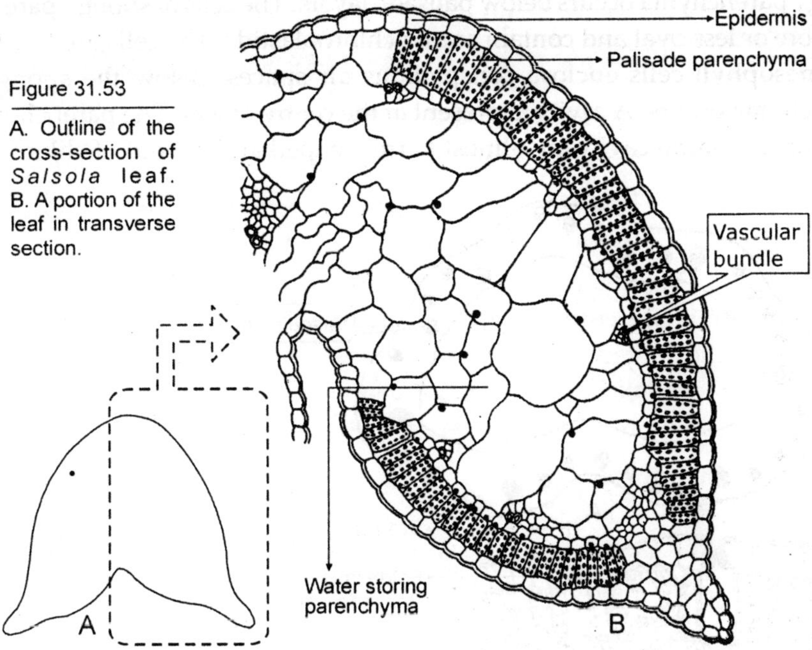


**Vascular bundle:** Next to palisade layer there occur the vascular bundles. The vascular tissues are poorly developed. The mechanical cells are scanty. At the centre a large vascular bundle is present. In it xylem and phloem can be distinguished. Phloem occurs towards the lower side.

**Comment:** *Salsola* is halophyte and the leaf is succulent with xerophytic adaptation. The peripheral band of palisade is photosynthetic tissue. The central parenchyma cells store water. Vascular tissues are reduced with scanty mechanical cells. The vascular bundles are inverted, i.e. phloem is towards the lower side and it is the characteristic of a leaf.

## Leaf of *Allium cepa* (Family: Alliaceae)

Transverse section through the leaf appears more of less circular in outline and reveals the following internal tissue organization from periphery towards the centre (Fig. 31.54).

**Epidermis:** It is continuous all round the circular peripheral layer. It is uniseriate and composed of parenchyma cells. The cells are more or less round, compactly arranged with well-developed cuticle on outer wall. The continuity of epidermis is interrupted by the presence of stoma that is present all round the epidermal surface. The stoma is slightly sunken.

**Mesophyll:** It is composed of palisade tissue and spongy parenchyma. The palisade cells occur towards the peripheral side. It is two layered and the cells are columnar. Abundant chloroplastids are present in them. Below the epidermis palisade layers form a continuous chlorophyllous band all round the surface. The spongy parenchyma occurs below palisade layers. The cells of spongy parenchyma are more or less oval and contain scanty chloroplastids. The cells are few layered. The mesophyll cells enclose conspicuous air spaces. Below the spongy layer parenchyma occurs. A cavity is present at the centre of leaf. The parenchyma cells that border central cavity are almost barrel-shaped and devoid of chloroplastids.

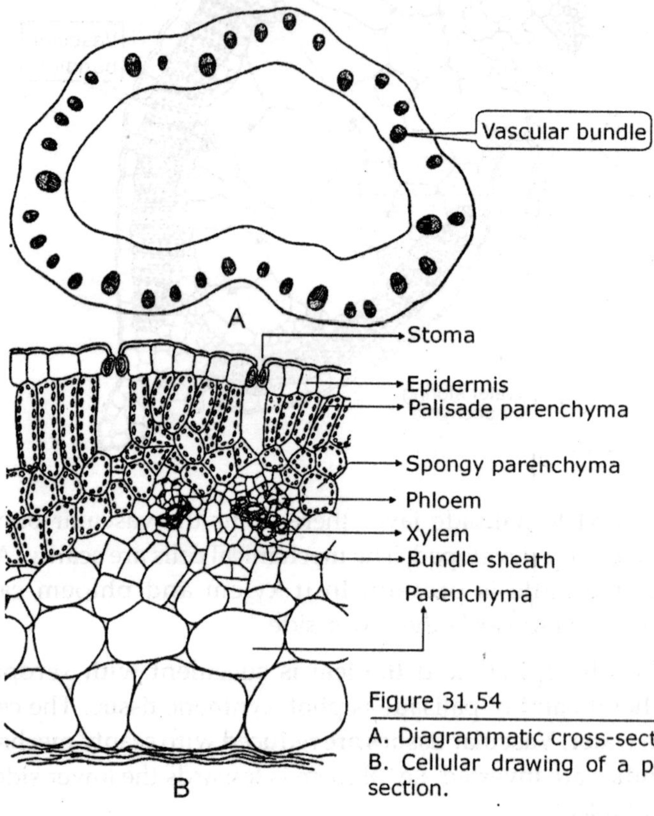


Figure 31.54

A. Diagrammatic cross-section of leaf of *Allium*.
B. Cellular drawing of a portion of transverse section.

**Vascular bundle:** There are numerous vascular bundles that occur below palisade but among spongy parenchyma. The bundles are scattered in arrangement. Each bundle is collateral and closed. Xylem occurs towards centre and phloem is directed towards the periphery. Each bundle is surrounded by bundle sheath. It is single layered and parenchymatous.

**Comment:** The leaf is centric because epidermis is continuous all round the surface and is not differentiated into epidermal layers. Stoma is distributed all round the epidermal surface and meant for gaseous diffusion. Though monocot the mesophyll is differentiated into palisade and spongy cells like dicots. The arrangement of xylem and phloem is like a stem, i.e. the phloem is directed towards the periphery. The vascular bundles are scattered and this arrangement reveals its monocot nature. Xylem is the only mechanical cell and forms the flanges of an I-girder. It gives mechanical strength against inflexibility.

## Leaf of *Vallisneria* sp. (Family: Hydrocharitaceae)

A vertical section through leaf shows the following internal tissue organization (Fig. 31.55).

**Epidermis:** It consists of lower and upper epidermis. They are composed of horizontally flattened cells that are compactly arranged with thin cuticle on their outer walls. The stomata are absent from the epidermal layers.

**Mesophyll :** The mesophyll is not differentiated. It is composed of parenchyma cells. The cells are more or less isodiametric and of various sizes. They are arranged in such a way that they enclose large air spaces. Undifferentiated mesophyll with air spaces is called lacunate mesophyll.

**Vascular bundle:** The bundles may be large and small. The vascular tissues are not clearly differentiated into xylem and phloem.

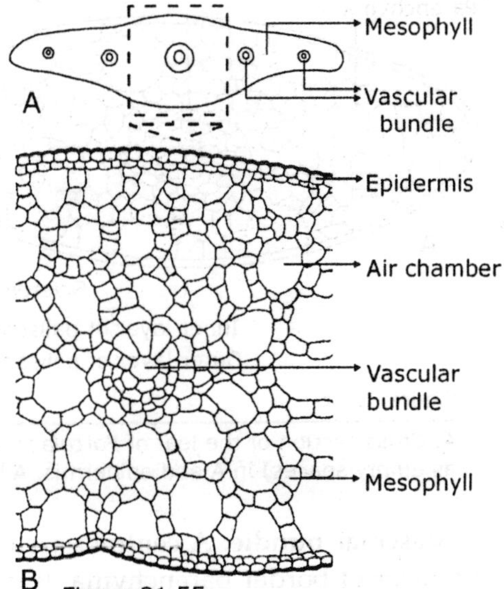


Figure 31.55

A. The outline of the vertical section of *Vallisneria* leaf.
B. A portion of the vertical section of *Vallisneria* leaf.
Adapted after Vasishta, 1981.

**Comment:** *Vallisneria* is hydrophyte. The leaf shows hydrophytic characters due to the presence of lacunate mesophyll, reduced vascular tissue and absence of stoma. The air spaces are meant for aeration. They add buoyancy to the plant.

## Leaf of *Portulaca* sp. (Family: Portulacaceae)

Vertical section of leaf is more or less oval in outline and shows the following internal tissue organization from periphery towards the centre (Fig. 31.56).

**Epidermis:** It is uniseriate and composed of compactly arranged horizontally flattened cells with cuticle on outer walls. It is continuous all round the oval peripheral layer.

**Mesophyll:** Below epidermis there lie many circular chlorophyllous zones. In between the zones parenchyma cells are present. They are compactly set, colourless and filled up with mucilage. Parenchyma is also present below the chlorophyllous zones and extends up to the central vascular bundle. The chlorophyllous zones consist of 2-3 central sclerenchyma cells encircled by two layers of large palisade cells. The peripheral or outer layer of mesophyll is elongated and contains numerous scattered chloroplasts. The inner layer of mesophyll is funnel shaped and also contains chloroplastids. But in this case the chloroplastids are aggregated at the base of the cells.

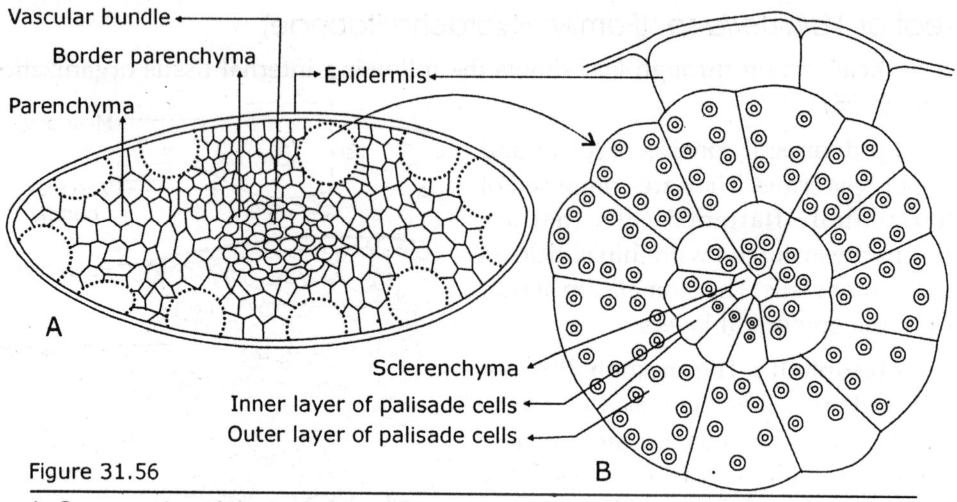


Figure 31.56

A. Cross section of the leaf of *Portulaca*. B. Cells present in the circular region (shown as empty spaces) in A and epidermis. Adapted after Majumder, 1963.

**Vascular bundle:** A central vascular bundle is present. It remains encircled by a layer of border parenchyma. It is colourless. The bundle consists of xylem and phloem.

**Comment:** The leaf shows succulent xerophytic adaptation due to the presence of cuticle on epidermis and parenchyma filled with mucilage. The circular chlorophyllous zone is the photosynthetic tissue. Xylem is the only mechanical cell.

## Petiole of *Cucurbita* sp. (Family: Cucurbitaceae)

Transverse section of the petiole is wavy in outline and shows the following internal tissue organization from periphery towards the centre (Fig. 31.57).

**Epidermis:** It is single layered and composed of horizontally flattened cells. The cells are compactly arranged and have well developed cuticle on their outer walls. Numerous multicellular epidermal outgrowths are present here and there. The epidermis is thrown into ridges and furrows.

**Ground tissue:** The peripheral layer just below the ridges is composed of a few layers of collenchyma. The cells are thickened at corners. Collenchyma occurs as patches and not in a continuous band. The rest of ground tissue is parenchymatous. The cells are thin walled with conspicuous intercellular spaces. Chloroplast may be present in the parenchyma. There is no tissue at the centre and in this region there lies a large air cavity. Vascular bundles remain embedded in the ground tissue. Endodermis and pericycle are absent.

**Vascular bundle:** The bundles are arranged in a ring in the ground tissue. The bundles are large and small, and many in number. The large vascular bundle lies below the ridges whereas the smaller ones occur below the furrows. Each vascular bundle is bicollateral and similar to those present in the stem of *Cucurbita*.

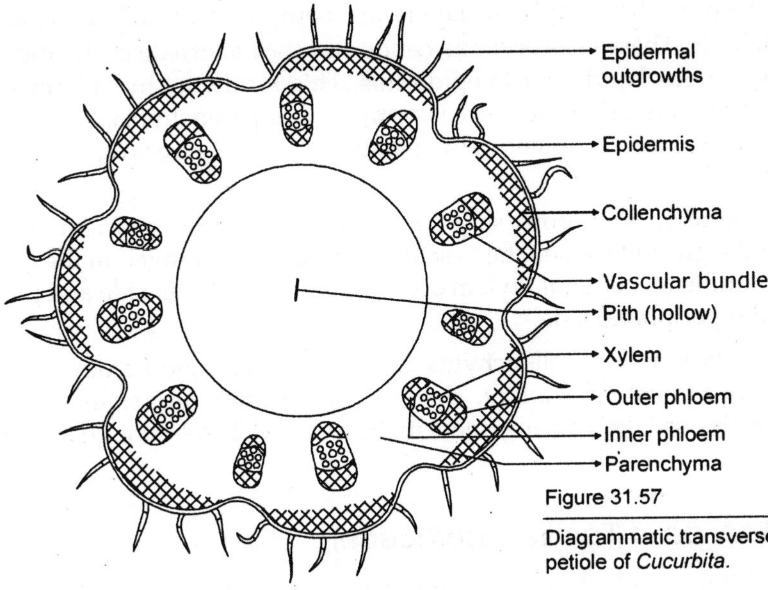


Figure 31.57

Diagrammatic transverse section of petiole of *Cucurbita*.

**Comment:** Collenchyma patches and xylem are the mechanical cells. They form flanges of the composite I-girders. They provide strength against inflexibility. The central large air cavity is meant for aeration. The ring arrangement of bicollateral vascular bundle is the characteristic of *Cucurbita* petiole.

## Petiole of *Helianthus annuus* (Sunflower. Family: Compositae)

Transverse section of petiole shows a prominent groove on one side whereas the other side is round. It shows the following internal tissue organization from periphery towards the centre (Fig. 31.58).

**Epidermis:** It is single layered and composed of horizontally flattened cells. Numerous epidermal outgrowths occur here and there.

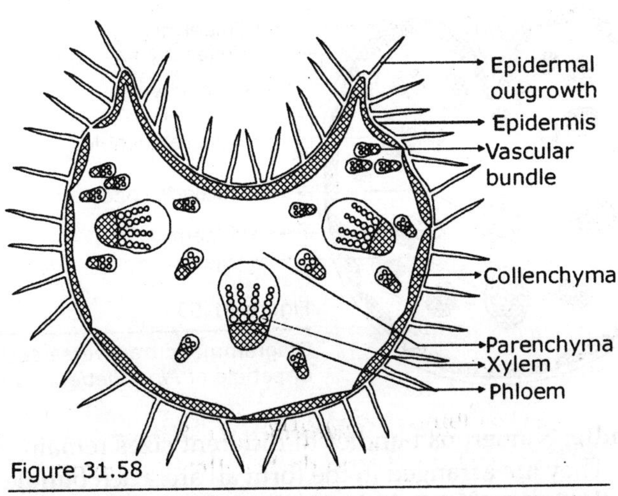


Figure 31.58

Diagrammatic transverse section of petiole of *Helianthus*.

**Ground tissue:** The peripheral layers are composed of collenchyma and chlorenchyma cells. These cells occur as bands and they alternate each other thus forming a continuous band next to epidermis. This forms the hypodermis. The remaining portion of the ground tissue is composed of parenchyma. The cells are isodiametric and enclose conspicuous intercellular spaces. There is no endodermis and pericycle.

**Vascular bundle:** Numerous small and some quite large vascular bundles are present in the ground tissue. The vascular bundles are arranged in the form of a band. Each bundle is collateral. Xylem occurs towards the inner side and phloem lies towards the peripheral side.

**Comment:** Hypodermal collenchyma and xylem are the mechanical cells that provide strength against inflexibility. The presence of numerous small and some quite large bundles and their arrangement in the form of a band is the characteristic of the petiole.

## Petiole of *Piper betle* (Family: Piperaceae)

Transverse section of petiole is cordate in shape, i.e. there occurs a prominent groove on one side whereas the other side is round. The following internal tissue organization is observed from periphery towards the centre (Fig. 31.59).

**Epidermis:** It is single layered and composed of horizontally flattened parenchyma cells. The cells are compactly arranged and cutin deposits on the outer walls.

**Ground tissue:** Peripheral layers are composed of collenchyma cells. It forms hypodermis and is not in a continuous band. The continuity of it is interrupted by the presence of chlorenchyma cells. These chlorophyll-containing cells occur as patches here and there. The remaining portion of the ground tissue is composed of parenchyma cells. At the centre there occurs a large mucilage canal. A few smaller canals are present here and there in the ground tissue. There is no endodermis and pericycle.

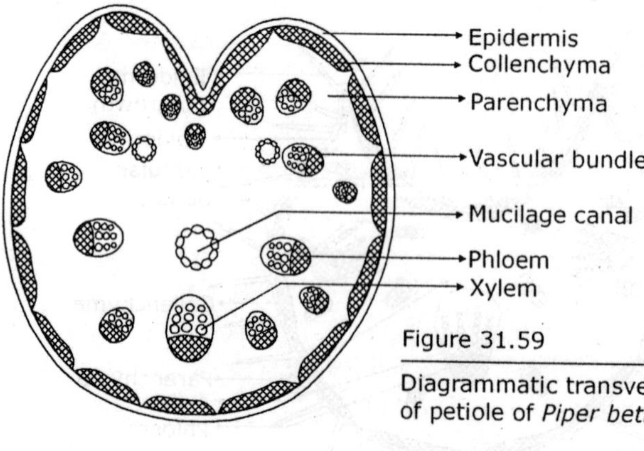


Figure 31.59

Diagrammatic transverse section of petiole of *Piper betle*.

**Vascular bundle:** Numerous bundles of different sizes remain embedded in the ground tissue. They are arranged in the form an arc. Each bundle is collateral. Xylem is towards the inner side and phloem lies on the peripheral side.

**Comment:** Hypodermal collenchyma and xylem are mechanical cells that provide strength against inflexibility. The presence of a large central mucilage canal and a few small canals here and there, and the arc arrangement of the vascular bundle in the ground tissue are the characteristics of the petiole.

## Petiole of *Nymphaea stellata* (Family Nymphaeaceae)

Transverse section of petiole is more or less round and reveals the following tissue organization from periphery towards the centre (Figs. 31.60 and 29.12).

**Epidermis:** It is single layered. The cells are more or less round, thin walled, compactly arranged with thin cuticle on their outer walls. Chloroplastids occur in the epidermal cells. Unbranched multicellular hairs are present on epidermis here and there.

**Ground tissue:** The peripheral layer of it forms hypodermis. It is two or more layered. It is composed of collenchyma cells that are thickened at the corners. The inner layers are composed of parenchyma cells. The cells are thin walled. Many large air spaces occur in this region. Numerous trichoblasts remain attached to the cells that border the air spaces. It is of different shapes and deposited with crystals of calcium oxalate.

**Vascular bundle:** Large number of vascular bundles are present and they are scattered over the ground tissue. They remain on the partitions of air chambers. Each vascular bundle is collateral and closed. Xylem is poorly developed. The elements of xylem break down to form lacunae below the phloem. The lacunae resemble air chambers. Phloem is well developed. The vascular bundles may be large or small. In the large bundles two patches of phloem occur on the peripheral side of xylem in contrast to small bundles that have a single patch of phloem.

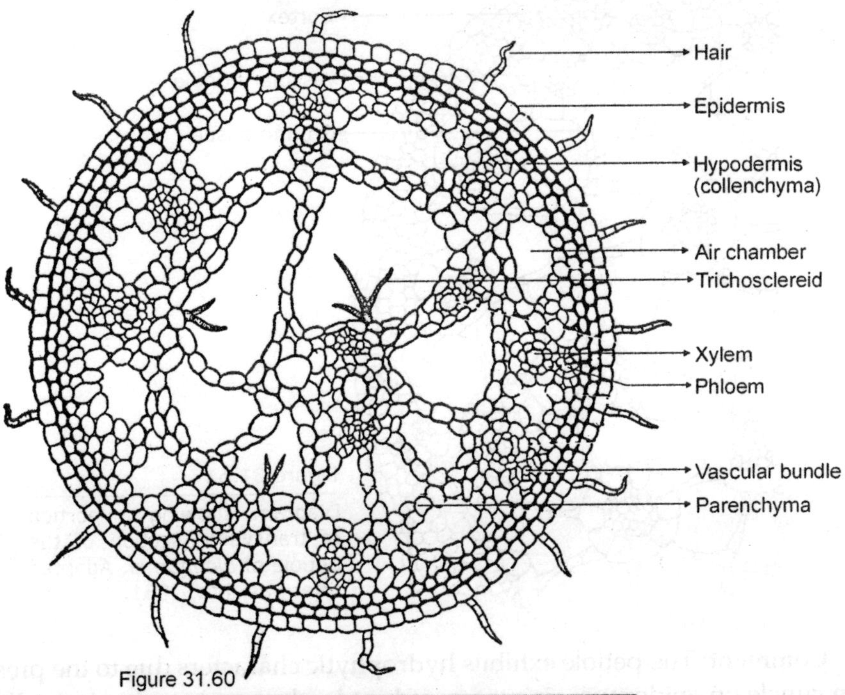


Figure 31.60
Cellular drawing of cross-section of petiole of *Nymphaea*.

**Comment:** Petiole shows hydrophytic nature due to the presence of large air spaces in the ground tissue, poorly developed xylem and thin cuticle on the outer wall of epidermis. Collenchyma at hypodermis is the only mechanical cell that provides strength to the petiole to hold the leaf in position. Anomaly in the petiole is due to the presence of scattered vascular bundles though it is a dicot. Presence of trichosclereids with calcium oxalate deposition is the characteristic of the petiole.

## Petiole of *Eichhornia* sp. (Family: Pontederiaceae)

Transverse section of petiole is more or less oval in outline and shows the following internal tissue organization from periphery towards the centre (Fig. 31.61).

**Epidermis:** It is uniseriate and composed of thin walled compactly arranged cells with thin cuticle on their outer walls.

**Cortex:** The peripheral few layers are composed of thin walled parenchyma cells. It encloses little intercellular spaces. There is no endodermis and pericycle. The rest of the cortex has numerous air spaces. Parenchyma cells of one layer thick form the wall of the spaces. It also forms the partition walls between spaces. This is lacunate cortex. There are many scattered vascular bundles. The bundles are present on the partition walls of lacunae. Each vascular bundle is composed of xylem and phloem and collateral in arrangement. Xylem is towards the centre and phloem occurs towards the peripheral side. The vascular elements are poorly developed.

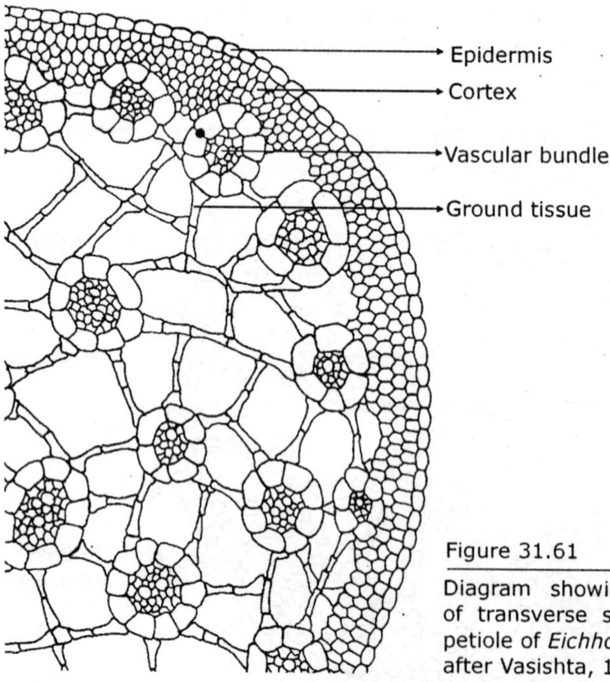


Figure 31.61

Diagram showing a portion of transverse section of the petiole of *Eichhornia*. Adapted after Vasishta, 1981.

**Comment:** The petiole exhibits hydrophytic characters due to the presence of thin cuticle on epidermis, air spaces, reduced xylem and mechanical cells.

CHAPTER **32**

# Applied Aspects of Plant Anatomy

The subject plant anatomy is not studied just for academic purposes. The knowledge of plant anatomy can be put into diverse range of applications, a brief account of which is discussed below.

• *Enables to identify fragmentary plant materials*: Since the time of Linnaeus flowers and fruits provided the characters of identification. Sometimes situation arises where these characters are not available. Example: Fragments of herbarium specimens, leaf, dried and powdered medicinal plants etc. The prerequisite of any botanical research is the proper identification of the specimen. Anyone dealing with plants for food, furniture, building materials, medicine etc. the plant breeders, geneticists and cytologists must have proper identifying characters of their source materials. These characters will identify parallel specimens, if required. They will be in a position to verify whether the parallel specimen is from the same species of the source material. The characters that differentiate a species from other species are considered as of taxonomic significance. Apart from vegetative and reproductive organs plant anatomy provides characters that are of taxonomic significance. Trichome anatomy, wood and leaf anatomy, leaf epidermis and cuticle etc. provide valuable characters in differentiation between species. As for example the different species of *Rhododendron* and *Ficus* can be differentiated by means of trichome characters.

• *Enables to detect adulterants in crude drugs*: The medicinal plants provide the crude drug. Drug can be obtained from all parts of a plant (ex. *Swertia chirata*), leaves (ex. *Adhatoda vasica, Andrographis paniculata* etc.), roots (*Cephaelis ipecacuanha*), rhizome (ex. *Zingiber officinale, Rauwolfia serpentina* etc.), or bark (*Alstonia scholaris*). The crude drugs are imported in dry form and in some cases in dry powdered form. In this condition it becomes difficult to identify the materials by macroscopic appearance only. For this reason the microscopical along with morphological characters of drug materials are studied. They are described and published in pharmacopoeia. The pharmacopoeias may be of official publications. Example: *European Pharmacopoeia, British Pharmacopoeia, British Pharmaceutical Codex, United States Dispensatory, Indian Pharmacopoeia, The Indian Pharmaceutical Codex* etc. A very brief and to the point description of drug material are given in the pharmacopoeias. The characters that will identify the drug specifically are mentioned only. Proper authentication of crude drug material is a prerequisite for importers. They must be sure about the quality, purity and if adulterated, the nature of the adulterants of materials. Samples to be imported are studied morphologically and anatomically. Their authenticity is established by comparing with the descriptions published in pharmacopoeias. A crude drug may also be identified from its chemistry. But the identification with the study of microscopical examination is much easier and quicker than that of chemical analysis. Mention may be made of a few drug plants with their uses and adulterants that can be detected microscopically.

*Swertia chirata* (Family: Gentianaceae), commonly known as chirata is an indigenous drug of India. It is used as stomachic bitter tonic, anthelminthic and in skin diseases. The root is used as a substitute of *Gentiana lutea*, which is used as gastrointestinal tonic, because the root of chirata does not constipate the bowels. The most common adulterant is *Swertia angustifolia* commonly known as pahari chirata. The distinguishing characters between the two species are given below according to Prasad *et al.*, (1960)

| *Swertia chirata* | *Swertia angustifolia* |
|---|---|
| 1. The cortical parenchyma of stem is 3-5 layered. Most of the cells contain calcium oxalate crystals and resin. | 1. The cortical parenchyma of stem is 3-7 layered. Calcium oxalate crystals and resin are present in few cells. |
| 2. The internal phloem is well developed. | 2. The internal phloem is less developed. |
| 3. The palisade cells are poorly developed. | 3. The palisade cells are well developed. |
| 4. Stomata are present on the lower surface of a leaf. | 4. The leaves are amphistomatic. |
| 5. Mature root shows small cavity at the centre and cavities at the central region. | 5. The central region is compact and lignified. |

Apart from *Swertia angustifolia, Enicostema littorale,* roots of *Rubia cordifolia* and *Andrographis paniculata* are found to be mixed with *Swertia chirata. Andrographis paniculata* differs from *Swertia chirata* in having characteristic cystolith on leaves, diacytic type of stoma and phloem on the dorsal side of xylem only.

*Zingiber officinale* (Family: Zingiberaceae), commonly known as ginger is rhizome drug. The rhizome is used as carminative medicine. It is used in digestive disorders. It expels gas from stomach and intestine. It dilates the blood vessels causing a warm feeling. It increases the rate of perspiration and thus lowers the body temperature. It is mainly used as condiment.

The rhizome of *Zingiber officinale* contains abundant starch grains. They remain singly or in groups. Each grain is simple and the shape may be round, oval, oblong and flattened. The hilum is small and terminal. The striations are very faint. The common adulterant is *Zingiber mioga*. It has compound starch grains and thus can be differentiated. Starch grain from wheat flower, *Curcuma* etc. are the other adulterants. The study of starch grains detects them. Adulteration may also occur with 'spent ginger' that is exhausted in the preparation of essence. This can be detected by chemical tests only.

*Cephaelis ipecacuanha* (Family: Rubiaceae) is the root drug and is used in cough mixture. The drug contains abundant starch grains that are mostly compound with 2-4 or five or up to 8 parts. The individual granule is fairly small, not more than 15 μm in diameter (Cutler, 1978). The shape of the granules may be round or oval. The vessels are moderately thick walled with narrow lumen and numerous bordered pits on walls. *Ionidium* (Family: Violaceae) and other roots are the adulterant of *Cephaelis ipecacuanha.* These adulterants have wide vessels and lack the characteristic starch granules. The other adulterant is *Cephaelis acuminata* that have starch granules up to 22 μm in diameter (Cutler *et al.*, 2008).

*Adhatoda vasica* (Family: Acanthaceae) is the leaf drug and is used as an expectorant medicine. It gives relief in bronchitis. The powdered leaf contains fragments of epidermis with diacytic stoma, non-glandular and glandular tichomes. The non-glandular trichomes are elongated, multicellular usually 3-4 celled, conical in shapes with wide bases and the apical cell is pointed. The glandular trichome is sessile, more or less circular in shape from top view; quadricellular and the partition walls have cross like appearance. Cystoliths are present within the palisade tissues that are double layered. The cystoliths are cylindrical and have warty projections on the surface. *Adhatoda vasica* plants grow profusely as a weed and the leaves are collected from natural sources. So they are rarely adulterated. In Kerala *Adhatoda beddomei* is used as substitute. *Adhatoda vasica* leaves are used as adulterant of tea (*Camellia sinensis*).

*Digitalis purpurea* (Family: Scrophulariaceae) is a leaf drug. It is used as stimulant of heart and as diuretics. The drug makes the heart powerful and causes the complete contraction of heart, and thus the circulatory system is toned up. *Digitalis lanata* is also used medicinally. *D. purpurea* can be distinguished from *D. lanata*. In the former the anticlinal walls of abaxial epidermal cells are more beaded than the latter (Cutler *et al.*).

- *Enables to identify wood*: The anatomy of a wood sample reveals many characters that help in the identification of plant from which the wood comes. In each country there are several excellent books on wood anatomy. These books provide not only anatomical description of wood of the plants occurring at the particular part of the world, but also give the salient anatomical features that help in the identification of plants. In India there are several volumes of book written on wood anatomy. Mention may be made of *Indian woods,* published by the Manager of Publications, Government of India, Delhi. In these volumes the anatomical description of woods, mostly from Indian origin, are given and the identifying characters are also mentioned. In India the number of timbers available in large quantities is not more than sixty, among which the following three are most important—*Tectona grandis* (teak), *Shorea robusta* (sal) and *Cedrus deodara* (deodar), (Chowdhury *et al.*, 1958). A brief description of the wood of teak, sal and deodar is given below.

❑ *Tectona grandis* Linn. f. (Family: Verbenaceae): The sapwood and heartwood are sharply demarcated. The colour of sapwood may be white or pale yellow, whereas the heartwood may be of light golden brown or dark brown in colour. The growth rings are conspicuous with ring-porous wood. The vessels of early wood are large in diameter, oval in outline and mostly solitary. The vessels of latewood are large to small in diameter, oval to round in outline and may occur as single, in radial pairs or radial multiples. The vessels have alternate pittings and simple perforation. The vessels are partly filled with tyloses and sometimes with white powdery deposits. The parenchyma cells are vasicentric and form a thin sheath around the vessel. The rays are moderately broad, 1–3 cells wide, heterocellular and uniformly distributed. The fibres are septate.

❑ *Shorea robusta* Gaertn. f. (Family: Dipterocarpaceae): The growth rings are usually absent with diffuse-porous wood. The vessels are moderately large in diameter, moderately few (5-10/$mm^2$) and occasionally numerous (10-20/$mm^2$). The distribution of pores is more or less uniform. The pores are solitary and oval to round in shape and have simple perforation plate. Tyloses are present and they almost completely fill up the pore cavities. Parenchyma vasicentric forming a narrow sheath round the pores or pore groups. The rays are fine to moderately broad, heterocellular. The gum ducts are present and they are vertical canals. The diameter of gum duct is usually smaller than pore. Each duct remains surrounded by thin-walled epithelial cell. The ducts are very irregularly spaced and occur in one or more rows. The ducts are often filled up with white gummy deposits.

❑ *Cedrus deodara* D. Don (Family: Pinaceae): The sapwood and heartwood are sharply demarcated. The sapwood is white to creamy white in colour. The colour of heartwood is yellowish brown turning to purplish-brown on exposure. The wood is non-porous. Growth rings are distinguishable. The transition from early wood to late wood is either gradual or abrupt. Resin canals are present and they are arranged in long tangential rows. Ray cells are numerous and they are arranged in closely-spaced line.

In the wood of *Eucalyptus* (Family: Myrtaceae) the vessels are solitary and occur in radial and oblique multiples. The vessels have simple perforation plates. Tyloses are present in lumen of vessels. Parenchyma cells are mostly amphivasal. The rays are wide and heterocellular.

It was previously mentioned that each country has its own publication of name of species from which the wood comes along with their wood anatomy. It is an authoritative work and provides the identifying characters of woods. Any wood sample in question is studied anatomically and compared with the publications. Thus the identity of wood sample can be established. For correct identification microscopic slides are prepared from wood samples under study and compared with those from reference microscopic slides.

For domestic purposes woods are best used in making furniture and building materials. Craftsman needs the identification of the antique furniture when repairs are necessary. The wood of doors and windows can be checked as declared.

Apart from making furniture and building materials, woods are used for many more specialized purposes. Sometimes normally used species become unavailable. Then search is made for substitute woods. By studying the anatomy of wood that is to be replaced, sometimes it is possible to suggest other species, from which similar properties can be expected. Mention may be made of the followings from where specialized properties are obtained. The cricket bat is produced from *Salix alba* var. *caerulea*. The wood is light and the cell walls are moderately thick. The walls are resilient and after denting recover their shape better than dense wood. The dense wood of *Guaiacum officinale* is heavy and the fibres are thick walled. It is used for making bowls and pulleys. The wood of *Ochroma lagopus* and *O. pyramidale* is very light and the wood of former is lighter than cork. The wood consists of large-thin walled parenchyma cells and thin walled fibres. It is used in life boat industries and making insulating materials in aircraft. The wood of Leguminosae in tangential longitudinal section reveals that the rays and fibres are oriented in regular horizontal rows. This type of figure has decorative value. The wood of *Fraxinus* and *Carya* has straight grain, i.e. the elements of wood are oriented parallel to the longitudinal axis of the plant. The wood is resilient and so it is used in making handles of axe and similar tools. Many woods have the property of resistance against decay, insect and fungus attack. Mention may be made of *Tectona,* where the heartwood is one of the most durable timbers of the world. The wood contains anthraquinones, which is effective in termites and fungi. The heartwood is used in boat building and it is one of the most important timbers in shipbuilding.

- *Enables to identify archaeological plant remains*: The wood-anatomy of present-day-plant provides characters to identify the fragmentary wood. These characters also enable to identify the wood and charcoal preserved in sites from antiquity.

A burnt wood or charcoal sample is collected from the site of excavation. Microscopic slides are prepared and examined thoroughly. The observation shows that the very delicate features like perforation plate and lateral wall pitting are still retained. The wood anatomy of archaeological sample is compared with that of present-day-wood and thus their identity can be detected. After authentication it can be decided whether it was selected for burning purpose only. It may happen that the plants composed the vegetation of that area at that time. The plants, as they grew locally, were obtainable at ease and so were selected for burning purposes.

A wood is best preserved in those localities where continuous wet or dry conditions prevail. A fluctuating dry and wet atmosphere encourages the growth of pathogenic microorganisms that may attack the wood thus causing the wood to decay. It is not yet definitely known when the first use of wood began. It is assumed that the practice of using wood started in Old Stone Age for digging purposes in search of food. For the first time, actual evidences of using wood have been recorded from the sites of New Stone Age. But no sites at this age in India produce any evidence of the use of wood. The Indus Valley civilization revealed the uses of wood for many purposes. The different use of wood was also recorded from the archaeological excavation of the proto-historic period in India namely, the Bronze Age civilization of Harappa and the Copper Age civilization of Hastinapura. It has been reported (Chowdhury *et al.*, p. xix) that the Harappans used the wood of *Cedrus deodara* (Deodar) and *Dalbergia latifolia* (Rosewood) for making coffins. These durable and scented woods are still in use after thousands of years for the same purpose. *Zizyphus* was used as wooden mortar for pounding grains. This wood has the property of shock absorbing and the Harappans were quite aware of the fact. *Dalbergia sissoo* (Sissoo) and *Holarrhena antidysenterica* (Kurchi) – these two timber-yielding plants were found in Hastinapura. These plants provide good fuel woods. It is not known whether these were used as firewood or charcoal. It is assumed that the Copper Age civilization was aware about the woods that have high calorific value.

In Iron Age the species of *Quercus* were used in making buildings and boats. At the site of exacavation at Brigg at South Humberside a boat made up of *Quercus* wood was preserved. It was interesting to find that the main logs of the boat were sewn together with twigs of *Salix*. The twigs were twisted and passed through the regularly made holes on the timbers. No nails were used to hold the timbers (Cutler *et al.* p. 167). *Corylus* were well preserved in waterlogged condition in Somerset at Bronze Age. These were used in building track ways across swampy grounds (Cutler *et al.*).

Apart from wood and charcoal other archaeological plant remains are also preserved. As for example a sandal was preserved in ancient Egypt. Anatomical studies reveal that it is composed of *Cyperus papyrus* and *Borassus* sp. (Cutler *et al.*).

• *Applied aspects of meristem culture*: Meristems may be apical, intercalary and lateral. Each of the meristems is exploited in the improvement of plants.

❑ Apical meristem: Apical meristems occur at the tips of root, leaf and shoot. The shoot apical meristems are particularly used in culture. In culture method the shoot apical meristem is excised out and placed in a glass container, containing nutrient. In a strict botanical sense the cells in the apical dome of shoot apex compose the meristem. In apical meristem culture the sub-millimetre shoot tip with 0.1 to 0.5 mm high apical dome is dissected out and placed in nutrient media. Different media are used for different plants. The media usually consist of a carbohydrate source, minerals, vitamins, aminoacids, growth regulators and the gelling agent agar. Perfect temperature, humidity, filtered air and controlled light are necessary during the culture. Aseptic condition is the prime importance in all the steps of culture to check the introduction of pathogens.

Sometimes circumstances arise when meristem culture becomes necessary for vegetative propagation. As for example the plant under study is infertile as in the case of triploid *Musa* sp., certain varieties of apple, tulips, iris, hyacinths etc. In breeding experiments the hybrid of first filial generation ($F_1$), when heterozygous, never breeds true, i.e. segregation of characters occur in the second filial generation ($F_2$). The meristem culture of $F_1$ plant keeps the progeny alike. In many experiments on plant breeding the hybrid plants fail to produce normal seeds. The seeds are either abortive or nonviable. These hybrids are propagated through meristem culture. The haploid plants produced as a result of anther or pollen culture are always sterile. They become fertile when they are converted to homozygous diploid. The haploid plants are propagated through apical meristem culture. In apical meristems viruses are either absent or present in a very low concentration because the cells of this region have fast mitotic activity. By apical meristem culture a clone of virus-free plant can be obtained. The clones can be multiplied vegetatively by meristem culture. The problems of plant tumors, especially the crown gall can be eradicated by meristem culture. It is to note that a virus-free plant is not virus resistant. So aseptic measures are the dominant prerequisite for obtaining virus-free plant that serves as a source for propagation of other virus-free plants. Sometimes plants are imported or exported to other countries. They are also exchanged in crop improvement programmed. In these cases quarantine laws are applicable. But the quarantine authorities relax the procedure of checking when the plants in question are derived from apical meristem culture.

The apical meristem culture is also useful in micropropagation. Micropropagation is the practice where a stock plant material is multiplied vegetatively into a large number of progeny plants. It is the true-to-type propagation of the selected genotype. Micropropagation gives immediate results in contrast to traditional means of propagation and so it is used to overcome the limitations imposed by long breeding cycle. The practice of micropropagation is extensively used in horticulture, agriculture, forestry and conservation of endangered plants. It was estimated in 1991 that the world production of micropropagated plants are about 600 million (Stefaan *et al.*, 1994 p. 127). In horticulture, floriculture especially in cut-flower industry the registered line of stalk plants is maintained by micropropagation. In cut-flower industry India has bright future. India has more than 35000 hectares of land under flower cultivation (Varshney *et al.*, 1998) and is gradually entering into export market. Traditional methods of plant multiplication are very slow and it takes long time of being commercially available. So micropropagation through apical meristem culture is necessary to meet the high demand of flower, for pathogen-free plants and for hybrid plants etc.

In forestry micropropagation through apical meristem culture is also practiced as it gives immediate result. By traditional means to obtain a homogeneous true seed might take over a hundred years.

The teak wood is obtained from *Tectona grandis.* This tree grows naturally in India. The other naturally growing species of *Tectona* occur in Burma, Thailand and the adjoining areas of Laos. *Tectona grandis* is one of the most important timbers of India. The wood is moderately hard, moderately heavy and strong timber. The heartwood has a reputation of being one of the most durable timbers of the world and is practically immune to fungus and termites. The essential oil citronella is obtained from the leaves of *Eucalyptus citriodora* tree. These plants are micropropagated from where about 500 plants of teak and 1000 plants of *Eucalyptus citriodora* are raised from a single bud (Fig. 32.1) in a year (De, 2004). Apical meristem culture is widely used to raise virus-free plants. Example: *Manihot esculentus,* which is usually infected by Mosaic virus or Streak virus. A single meristem tip of *Manihot esculentus* provides a number of virus-free plants.

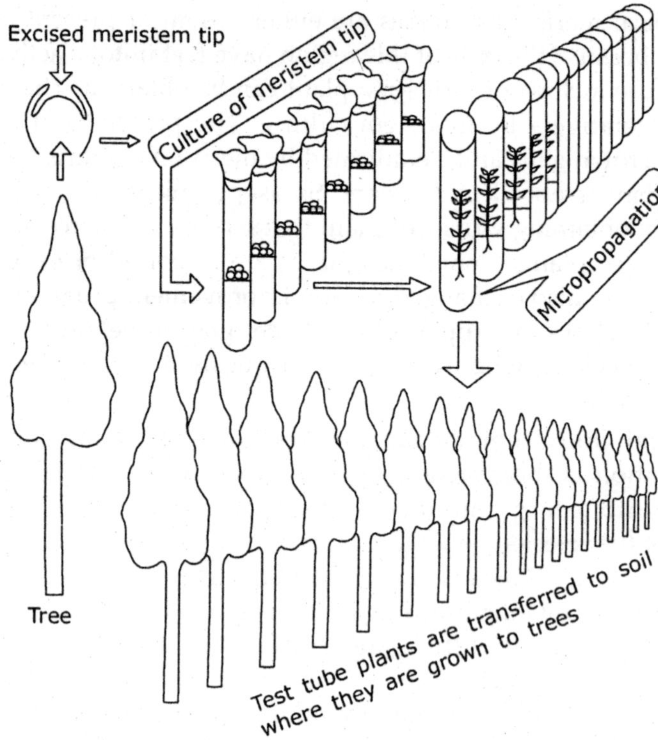


Figure 32.1

Diagram illustrating the meristem tip culture for rapid propagation of plants. Adapted after De, 2004.

In the conservation of endangered species micropropagation technique plays an important role. Through this technique endangered plants are propagated and thus conserved. Mention may be made of the orchids *Liparis loeselii, Cypripedium calceolus* etc., carnivorous plants *Nepenthes, Drosera, Dionaea* etc. and the woody plant *Ramosmania rodriguesii, Hyophorbe lagenicaulis* etc. (Kew information sheet K1).

❑ Intercalary meristem: A large number of plants have meristem adjacent to and just above most nodes. This meristem is derived from apical meristem. During the course of development of apical meristem of a plant, a portion of meristem becomes separated from apical meristem by more or less mature tissues. This is intercalary meristem. In many plants adventitious roots are formed at nodes from this meristem. In horticulture this property is used in the propagation of plants by stem cuttings. Adventitious roots are formed in many plants which failed to grow upright and fell to the ground. Ex. *Triticum, Dianthus* etc. Adventitious roots grew at the nodes and made the plants upright again. The intercalary meristem is capable of producing adventitious roots. This can be demonstrated by the following experiment (Cutler *et al.*). The experimental material is *Dianthus*. The plant is cut just below a node. The node is split longitudinally up to the intercalary zone. The split was kept apart by means of stick. Proper nutrition and aseptic condition was maintained as usual. It was observed that from the split sides adventitious roots develop (Fig. 32.2).

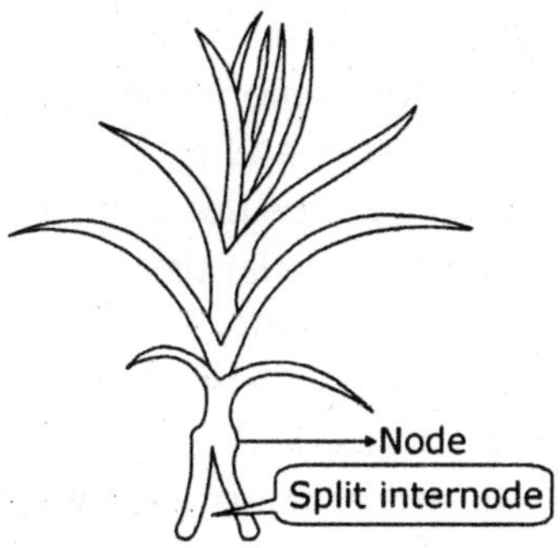


Figure 32.2

Diagram showing a portion of *Dianthus* with split internode. The split side of the internode is the site of adventitious root formation. Adapted after Cutler *et al.* (2008), P. 23.

❑ Lateral meristem: Lateral meristem occurs on the lateral sides of a plant. Example: Phellogen or cork cambium, fascicular cambium that is present between xylem and phloem of a vascular bundle of dicot stem and the interfascicular cambium that develops at the time of secondary growth from the tissues present between the two vascular bundles. The fascicular—and interfascicular cambium unite on lateral sides to form a complete cambium ring. The function of cambium ring and phellogen was discussed in the chapter 20 & 23. These cambia, if wounded, normally can regenerate and form callus cells adjacent to them. The cambial continuity is regained and thus the wound is healed. This wound healing property is employed for commercial purposes.

The phellogen donates on the peripheral side phellem cells that are also known as cork. The bottle cork, which is used as stoppers, is obtained from the phellem. *Quercus suber* provides most of the cork. The bottle corks that are used as stoppers are made sizes by tangential cut to stop leakage from vertically oriented lenticels (Fig. 32.3). The phellogen of *Q. suber* persists indefinitely and forms mass of cork tissue externally. After about twenty years of age of the plant the cork tissues are removed by stripping. This cork, also known as virgin cork (Eames and MacDaniels, 1947), consists of phellem cells only and sometimes phellogen and phelloderm may be present. The virgin corks are nearly useless. After stripping, the underlying phelloderm and cortical cells are exposed and they gradually die. This wound is healed by a cork cambium or phellogen, which originates in the deeper layer of cortex. Phellogen forms new layers of cork, which is harvested after about ten years. This cork is not of good quality but definitely better than the virgin cork. The better quality of cork is obtained from the third and subsequent stripping. The cork layers are harvested at intervals of nine to ten years until the tree becomes 150 or more years old (Eames and MacDaniels). Portugal is the centre of cork industry. The recent use of plastic corks is reducing the demand of oak-cork.

The other lateral meristems, fascicular—and interfascicular cambium are also employed in plant propagation. This is usually done by cuttings and grafting. Cutting is a method where a shoot is cut away from a desired plant and planted to soil for rooting. Thus a new individual is produced. In horticulture this method is employed to propagate plants thus maintaining similar genotype. In stem grafting the cut away shoot of a plant, also called scion, is inserted to the stem of another plants, also called stock. In grafting both scion and stock are wounded and the two are joined in such a way that the exposed meristematic cambial cells of them are brought into close contact. The inherent ability of wounds to heal causes the formation of callus tissue. Through the callus the cambia of stock and scion become continuous and form the secondary vascular tissues as usual. Thus the scion becomes a permanent part of stock. The plants like *Mangifera indica*, apples, *Nephelium litchi* etc. are propagated by grafting. This device can regulate the size of the plant and earlier fruiting can be induced. The gourd root, which is *Verticillicum* wilt resistant, is used as stock where the watermelon with wilt-prone

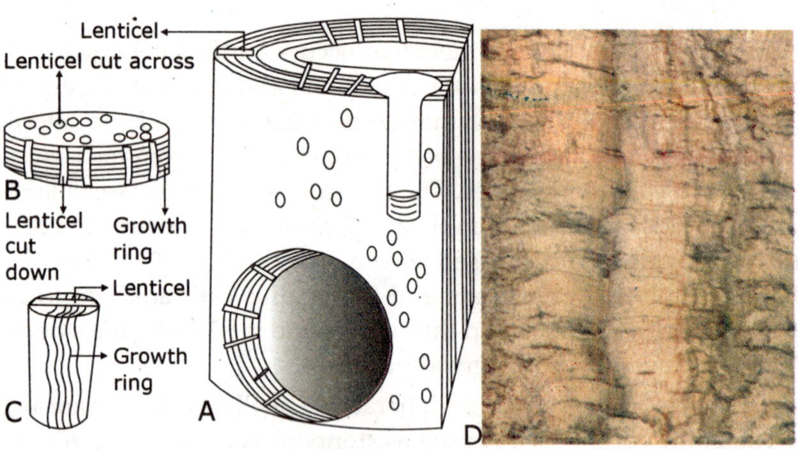


Figure 32.3

A. Diagram showing cavities in a cork sheet from which have been cut. B. A piece of cork indicating the position of lenticel. C. A piece of cork for a bottle. Adapted after Priestley and Scott, 1938, p. 359. D. Photograph of the bark of *Quercus suber*. The tree is present at Hopetoun Gardens, Elsternwick, VIC, Australia.

root is grafted. Thus the wilt disease can be avoided in watermelon. *Juniperus glauca* has vigorous roots whereas the root of *J. virginiana* is weak. The latter is the desirable species and so it is grafted to the stock of *J. glauca* (Cutler *et al.*).

In conifers the growth ring are continuous at an early stage below lower branches. At the branch region the continuity of growth ring is interrupted where gap is present. In forestry there is a practice to remove the lower branches of conifers at an early stage. This enables the growth ring to become continuous over the discontinuous growth ring. Thus a sound new wood can be formed (Cutler *et al.*).

• *Provides evidences in Forensic investigation*: The application of forensic science is indispensable in investigating a crime. Forensic science has many disciplines and forensic botany is one of them. Forensic botany encompasses many sub-disciplines, which include plant anatomy, plant systematic (taxonomy and species identification), palynology (the study of spores and pollen) etc. They refer the use of plant materials in solving crimes or resolving other legal problems.

Plant remains are present everywhere. In a crime scene they may occur in the form of macroscopic pieces (ex. wood, twigs, leaves, flowers, fruits, seeds etc.) or microscopic forms (ex. pollen, spores, trichomes, cell walls in stomach contents etc.). The morphological and anatomical diversity expressed by plant species provide characters to identify plant parts. Species identification is a prerequisite in analyzing botanical evidences for casework. Plant anatomy provides characters such as trichomes, stomata, cuticular pattern, leaf venation, wood anatomy, growth rings etc. to aid in species identification and in performing physical matches of evidence. The identified plant materials help the investigators of criminal cases to determine whether a suspect was present in a crime scene, in which season the crime occurred, how long the body has been buried, whether the body has been moved etc.

The following is an example of one of the earliest and famous case in which botanical evidence was used and accepted in court (Coyle *et al.*, 2001). In 1932, in the evening of March 1st the infant son of famous American aviation hero Charles Lindbergh was kidnapped from his home in Hopewell, New Jersey, US. For the release of their son, the family paid $50,000. But the kidnapper did not return their son. Two months later the dead body of the son was discovered a few mile away from the family home.

The son was kidnapped from a second-story nursery. The kidnapper used wooden ladders to gain the access there. The ladders were the only evidence left at the scene. They were homemade and crude. Xylotomist Arthur Koehler of United state Forest Services in Wisconsin examined the wood of the ladder both morphologically and anatomically. Four years later, when the case finally came to trial, Koehler offered the evidences from plant anatomy, which was ever to be heard and accepted in American court.

The ladders had been constructed in three sections. It was presumed that they were made such for ease of transport. Koehler numbered each piece of rungs and side rails. Koehler identified each piece to species. The following four species were used to construct the ladder – namely Yellow pine, *Pinus ponderosa*, Douglas fir and *Betula* sp. The basis of identification was the microscopic analysis of grain patterns of the woods. Next Koehler analyzed the tool marks left on the wood. Koehler was able to differentiate the woods with mill plane marks and hand plane marks. The hand plane marks were distinct by their dull and nicked appearance. The nick was distinct when the wood was placed in oblique light in a dark room. The wood of rail #16 was very distinct from others. It was from *Pinus* wood with hand planed marks and had four very prominent marks of square nail holes. Moreover the rail #16 was not weathered suggesting that the wood was removed from some interior construction like shed. Finally Koehler analyzed the growth ring patterns and knots present on the wood and especially on the rail #16.

The progress of the case was very slow as no suspect was identified. Sometimes in September of 1934 a gas station received some currency notes and it was detected that those were once used to pay the ransom for the release the infant son. Bruno Richard Hauptman, a carpenter from Bronx, New York city paid the notes to the gas station. Then search was made at Hauptman's house and $14,600 of the ransom was found in the garage. So Hauptman was arrested. Further search was continued for the recovery of more money of the ransom. During investigation it was noticed that one of the joists of floorboards of Hauptman's attic was shorter of about eight feet than the others. This joist had four square nail hole marks and the growth ring patterns and knots were studied. Amazingly it was found that the four square nail hole marks, growth ring patterns and knots matched with rail #16. The hand plane marks of rail #16 matched with the wood of a homemade shelf present in the Hauptman's garage. A hand plane was recovered from the garage and it made identical nick that was present in rail #16 and wood that constructed the shelf.

Hauptman was convicted and executed on 3rd April 1936.

In the above case the features of plant anatomy provided evidences in solving the kidnapping of infant son of Lindbergh and involvement of Hauptman in the crime.

The anatomical features like silica bodies, starch grains, raphides, sclereids, druse etc. found in edible plants are used to identify the stomach contents and last meal of a victim. This is necessary to find the victims' whereabouts and actions prior to death. In the analysis of stomach contents the cell wall provides important identifying characters of many food materials because the cell walls are not easily digested and persist longer time period than other anatomical features. Moreover the characteristic thickening of cell walls is sometimes taxon-specific.

Many dicotyledonous roots show growth rings or annual rings. The number of rings reveals the age of the tree. This property is utilized in forensic anthropology to estimate the time since skeletal remains had been in their present location. In one case living roots were found in the cavity of a skull. The anatomy and growth ring pattern revealed that the root belongs to *Ranunculus ficaria* and the plant was approximately one year old. These findings enabled to determine that the skeleton had been there for at least one year. When a grave is dug roots can be damaged but still continue to grow leaving a permanent lesion. In dicotyledonous roots growth rings are formed as usual and the number of rings formed after the lesion indicates the number of years passed since the damage. So growth rings can indicate the number of years passed since burial. From the number of growth rings or annual rings of root that are in contact with the bones can also indicate minimum time passed since death.

Dendrochronology or tree-ring dating is a method of scientific dating of a tree based on the analysis of annual growth ring patterns. The number of growth ring gives the age of the tree. The growth ring analysts can pinpoint the exact age of a tree and the year when the tree was cut. This enables to detect any art fraud, the provenance of wood arts objects, wood of musical instruments etc. Many European painters used to paint directly on wood. Dendrochronology techniques enable to date the wood and so the year of painting. These techniques are also used to detect when the wood was used to make art objects or musical instruments etc.

Marijuana is a controlled drug and obtained from *Cannabis sativa*. In powdered form it is identified microscopically by the presence of characteristic cystolith and hairs. Drug enforcement confirms the drug by chemical tests in combination with microscopic observations. Apart from *Cannabis sativa* there are large number of plant species, which are used as drugs, substitutes and adulterants. Most of them are used in a very finely powdered form. Quite a lot of time and effort is needed to authenticate them. There exists enormous genetic diversity among plant populations. Forensic scientists take the advantage of the property of genetic diversity to identify plant species by molecular biology techniques. The development of DNA typing methods for plant species enables to identify plant species and thus helps in solving criminal and civil cases. Though the traditional microscopic anatomical identification is not always conclusive, it is still in use for preliminary identification. Moreover the anatomical techniques are simple and inexpensive.

• *Provides characters of taxonomic significance*: It is the prime importance to know exactly to which species a plant specimen belongs. This is necessary for a natural and reliable classification. Most of the plants are classified according to their macro-morphological features. But an accurate classification results when the information from diverse sources are utilized. The sources may be from anatomical features, palynology, biochemistry, embryology, cytogenetics, phytogeography, physiology etc. It is now realized that alpha taxonomy can form a natural, accurate and reliable classification.

Once morphology and anatomy formed the backbone of taxonomy. Anatomical features provide characters to supplement the macro-morphological characters of plant species. The following important anatomical features those are often good indicators of the family, genera and sometimes species are discussed below.

❑ Trichomes: Trichomes are the collective term of hairs and papillae. They occur on all organs of a plant. There exists much morphological diversity among them. This property and the simple means of preparation of slides for study enabled the taxonomists to employ the trichome characters for systematic comparisons and individual identification. Metcalfe and Chalk (1950) provided the diverse types of trichome with their structure, nomenclature and distribution. Trichomes are of taxonomic significance especially at generic and specific level.

The families like Restionaceae and Centrolepidaceae can be recognized by the presence of distinctive type(s) of trichomes that are simple and unbranched. The hair of *Aphelia cyperoides* (Centrolepidaceae) has a boat-hook-shaped end (Fig. 32.4A) that characterizes the species (Cutler *et al.*). *Leptocarpus* (Restionaceae) has flattened, multi-cellular stem hairs with short stalk.

There are two major categories of hairs—the glandular and non-glandular. Each of the categories are sub-divided according to their gross structure, cellular constitution, nature of branching etc. There exist much diversities and varied forms in non-glandular hairs than the glandular hair. A few types of hair are illustrated in Fig 32.4.

A particular type of hair is constant in a species. This property is used as an aid to identify the different species of Oleaceae, *Ficus* and *Rhododendron* to some extent. Presence of T-shaped hair diagnoses the family Malpighiaceae. Trichomes provide the distinguishing characters within the family Icacinaceae where the genus *Ottoschulzia* is segregated from *Poraqueiba* on the basis of trichome characters. Trichome types with their distribution pattern can be correlated with the sub-generic and specific distinction in *Nicotiana.* In Sparganiaceae and Typhaceae, the presence of sessile glandular trichomes provides a link between these families. Trichome anatomy is of immense significance in the separation of species and even varieties in the tropical family Combretaceae (Stace, 1965).

❑ Stomata: Stoma has been shown to have great value in the taxonomy of several taxa. The distribution, morphology and ontogeny of a stoma are of taxonomic significance. Stoma is absent in roots. In exceptional cases it is reported from *Ceratonia siliqua* and *Pisum arvense* seedling roots. Stomata are also absent from the chlorophyll-less parasitic angiosperm like *Monotropa* and *Neottia.* The submerged hydrophytes lack stoma. The floating hydrophytes i.e. *Nymphaea, Victoria* etc. are epistomatic (= stoma is present on upper surface of a leaf). Hypostomatic (= stoma is present on lower surface of a leaf) leaf occurs on those plants that have xeromorphic habit. Amphistomatic (= stoma occurs on both upper and lower surfaces of a leaf) leaf is observed in mesophytes. The stomatal frequency and stomatal index have taxonomic importance. The taxonomic groups and the different species of a genus can be differentiated on the basis of stomatal index.

The different morphologic and ontogenetic types of angiosperm stoma are discussed in the chapter 12. They are of immense taxonomic significance. Metcalfe and Chalk provide a list of angiosperm families where different morphological types of stoma occur. It is shown that paracytic stomata characterize woody Ranales. So it is regarded that such stoma is primitive among dicotyledons (Payne, 1979). Sen and De (1992a) recognized 24 types of stoma in ferns among which polocytic stoma is regarded as the basic form from which the other types have ontogenetically been derived. The polocytic stoma is found in *Diplazium*

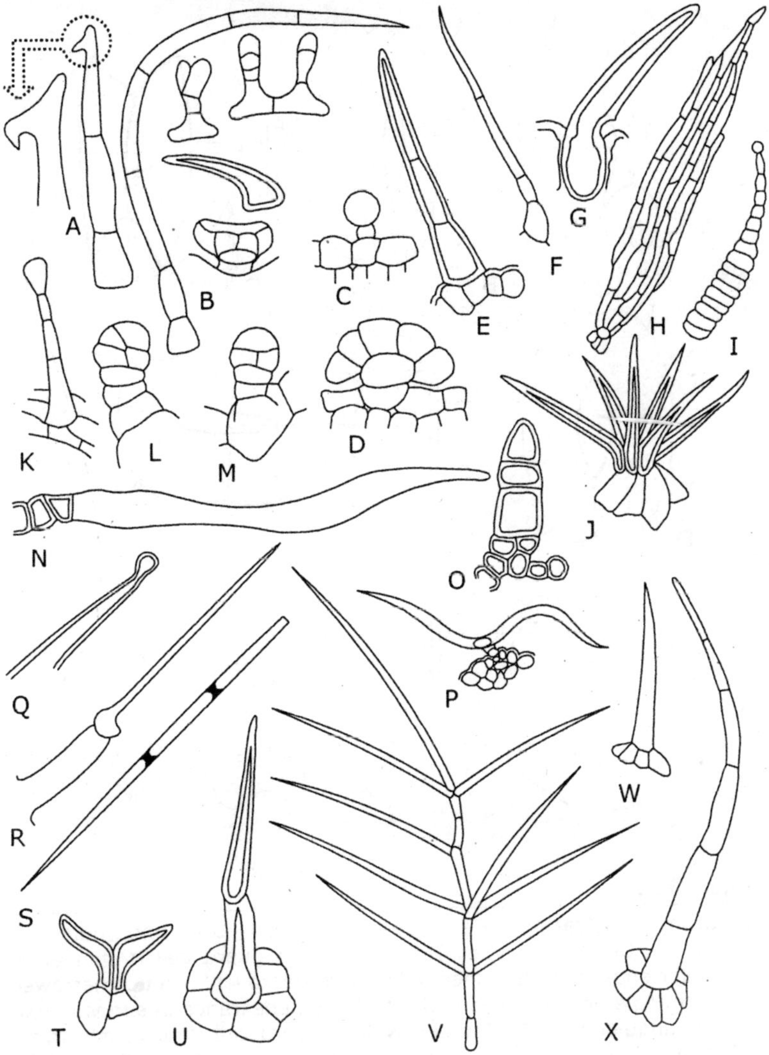


Figure 32.4

A. Hair of *Aphelia cyperoides*; B. Different hair types in *Mentha spicata*; C. Bicellular hair of *Salvia officinalis*; D. Multicellular hair of *Salvia officinalis*; E. Non-glandular hair of *Salvia officinalis*. F. Hair of *Centrolepis*; G. Hair of *Thamnochortus;* H. Surface view of hair of *Leptocarpus*; I. Glandular hair of *Cistus*; J. Dendritic hair of *Cistus*; K. Glandular trichome of *Pelargonium*; L. & M. Glandular trichome of *Cucurbita*; N. Hair of *Loxocarya*; O. & P. Hair of *Artemisia*; Q. & R. Glandular hair of *Urtica dioca*; S. Glandular hair of *Mucuna* containing irritant oil droplets; T. Hair of *Trigonobalanus*; U. Non-glandular hair of *Justicia*; V. Non-glandular hair of *Verbascum*; W. Non-glandular hair of *Coldenia*; X. Hair of *Origanum*. Redrawn after Cutler *et al.*, 2008 and Cutter, 1978.

*polypodioides* (Fig. 32.5G), *Cyathea contaminans* etc. This type of stoma has single subsidiary cell. The stoma is attached to the distal side of the subsidiary cell. Thus the subsidiary cell appears U-shaped or horseshoe shaped. Pant (1965) and Payne provide the different ontogenetic types of stoma. The diameristic and

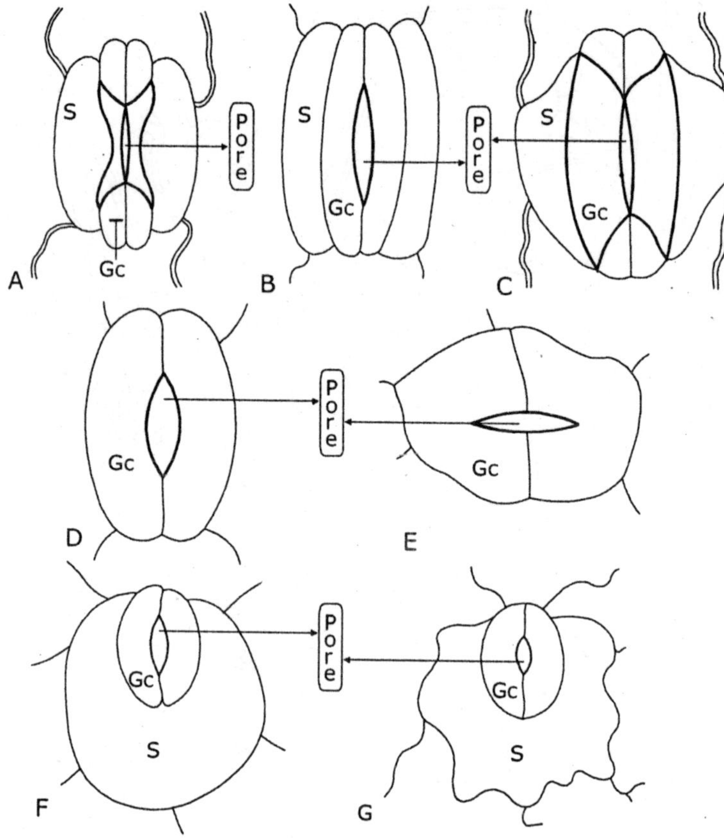


Figure 32.5

Diagram illustrating the types of guard cell and polocytic stoma. A. Dumb-bell type: The cell wall of guard cell is thinner at the polar ends than the other walls. Each guard cell is much narrowed in the middle and so it has dumb-bell like appearance in surface view. B. Rectangular type-A: The cell wall of guard cell is uniformly thick. Each guard cell is of uniform breadth in surface view. C. Rectangular type-B: The cell wall of guard cell is thinner at the polar ends than the rest of the walls. Each guard cell is of uniform breadth. D. Elliptic type: Each guard cell is as broad as long or longer than broad. The cell is reniform or more or less uniformly wide. Pore of stoma occurs parallel to guard cells. E. Right-angle type: Each guard cell is shaped like U or V. The contiguous wall of guard cells determines the stomatal axis, also referred to as polar axis of stoma. The pore of stoma occurs approximately at right angles to stomatal axis. The polar axis is frequently shorter than the equatorial diametre. F. Polocytic stoma (diagrammatic). G. Polocytic stoma of *Diplazium polypodioides*. A-E redrawn after Rajagopal *et al.* (1977). F & G respectively adapted after Van Cotthem (1970) & Sen *et al.* (1992a). Gc = guard cell & S = subsidiary cell.

mesoperigenous stoma are thought to be primitive in the Embryophyta. It is to note that monocotyledons are characterized by the above patterns of stoma. In monocots stoma with two or more subsidiaries appears to be more primitive than those with none (Stebbins and Khush, 1961).

The surface view of guard cells of a stoma can be of taxonomic significance. Rajagopal and Ramayya (1977, p. 59) classified five types of guard cells on the basis of their appearance in surface view under light microscope (Fig. 32.5). The types are (a) dumb-bell type (ex. *Cyanodon*), (b) rectangular type-A (ex. *Eriocaulon*), (c) rectangular type-B (ex. *Cyperus*), (d) elliptic type (ex. *Scilla, Mollugo*) and (e) right-angle type (ex. *Azolla*). The dumb-bell type guard cells characterize the family Gramineae. The monocotyledonous families like Cyperaceae, Restionaceae, Juncaceae, Marantaceae etc. show rectangular type-B guard cells. In dicotyledon rectangular type-B guard cell is reported from *Haloxylon articulatum* (Chenopodiaceae) only (Fahn, 1987). The rectangular type-A guard cells occur in the monocotyledonous families like Xyridaceae, Eriocaulaceae and Palmae (Tomlinson, 1961). The elliptic type of stoma is the characteristic of gymnosperms, dicotyledons, most of pteridophytes and some monocotyledonous families like Dioscoriaceae and Liliaceae. Rajagopal and Ramayya regarded that within the monocotyledons the above types of guard cell might have value at higher taxonomic levels. The leaf of *Pinus pinea* shows circular raised rim above the stoma. The epidermal cells form the rim. It is revealed by the scanning electron micrograph. The raised rim is termed as Florin ring.

❑ Veins: The veins and their innumerable variations in leaf venation pattern provide various characters of taxonomic importance. The anatomical division of Angiosperm into dicotyledon and monocotyledon is based on venation pattern. With a few exceptions dicots have reticulate venation and monocots show parallel venation. The veins are the vascular strands or traces that diverse from the vascular cylinder of stem at the nodes. In dicotyledons they may consist of one, three or many traces. These traces and other accessory strands collectively form diverse patterns or types of venation.

In a dicotyledonous leaf the veins may be primary, secondary and tertiary. The primary vein is the widest vein in the leaf. It originates at or just above the petiole. It is usually symbolized as $1^{o}$ veins. The secondary veins are narrower than the primary. It originates from the primary vein. It is usually symbolized as $2^{o}$ veins. The tertiary veins are narrower than the secondary. It may originate from secondary and primaries. It is usually symbolized as $3^{o}$ veins. (Fig. 32.6). The primary and secondary veins are the structural veins of a leaf. The tertiary veins fill the field of the leaf. The primary and secondary veins gradually taper towards the margin. The tertiary veins connect primary and secondary veins thus forming a more or less regular polygonal field over the leaf area. A polygonal field is usually designated as vein-islet. Apart from above types most dicotyledonous leaves have higher orders of veins, i.e. after tertiary there are $4^{o}$, $5^{o}$ veins category and they may be up to $7^{o}$. In the lamina of many leaves the vein terminates blindly as veinlets in the mesophyll. These veinlets are termed as vein endings. The sieve tube elements are absent from vein endings.

The different categories of vein provide many characters that are very useful in leaf identification. The $1^{o}$ veins may be single, three or more. The $2^{o}$ veins form an angle with $1^{o}$. The angles are constant in a species. The angles may be uniform,

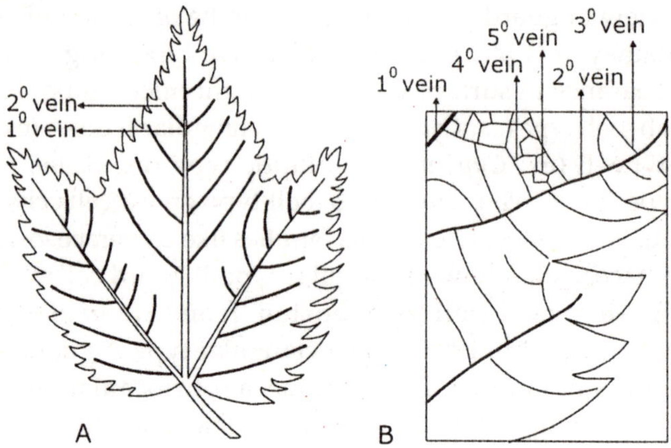


Figure 32.6

A. Diagram of a leaf of *Acer argutum* showing 1° and 2° veins. B. A portion of A enlarged showing 1°, 2°, 3°, 4° and 5° vein. Redrawn after Hickey, 1973.

abruptly increasing towards the base, smoothly decreasing towards base etc. The spacing between 2° veins is also of taxonomic significance. It may be uniform, irregular and increasing or decreasing towards the base. The 3° veins show different angles to 1° (Fig. 32.7A, B, C, & D). The course of 3° veins may be straight, convex and sinuous (Fig. 32.7E, F, G). The number of vein-islets in a unit area is species-specific. Usually four square millimetre area of a leaf is considered as a unit in counting the vein-islet numbers. The ultimate free endings of vein-lets have diagnostic value. They may be unbranched, linear or curved, 1-branched, 2 or more branched etc. The following features of veins provide taxonomic information. The veins may or may not be raised above the two epidermises of a leaf. Bundle sheath may be absent or present in vascular bundle, which is the cross-section of veins. When present it surrounds the vascular bundle and may be one or two layered, parenchymatous or sclerenchymatous. In two-layered bundle sheath both the layers may be parenchymatous or sclerenchymatous, or the inner layer is parenchymatous and the outer layer is composed of sclerenchyma. Chloroplastids may be present or absent from the cells of bundle sheath. Bundle sheath extension may be absent or present and when present it may be composed of parenchyma, collenchyma or sclerenchyma. It is taxon specific. Hickey (1973, 1974) provides terminologies for the description of leaf form and venation. This enables to identify and classify dispersed leaves. The foliar characters may or may not offer conclusive evidences of affinities, but generally they do allow distinguishing the closely related species (Merrill, 1978). Moreover the diverse patterns of venation can be utilized in the morphological interpretation of such organs as bracts, sepals and petals (Foster, 1967). The leaf epidermal characters such as pattern of epidermal cells, cuticular ornamentation, the patterns of leaf wax structure as revealed by means of electron microscopy have, of late, been demonstrated to be of taxonomic importance of many taxa.

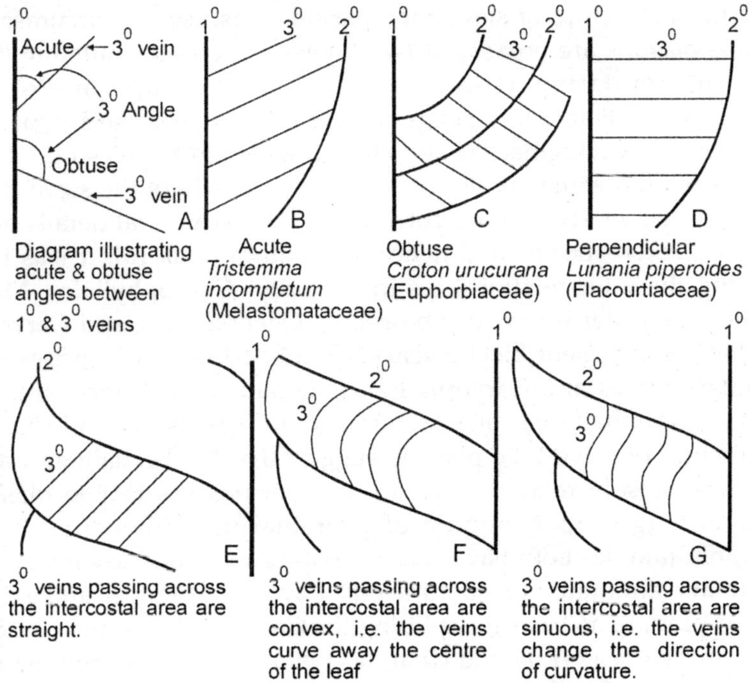


Figure 32.7

Diagram illustrating the angles formed between $1^0$ and $3^0$ veins (A, B, C, D) with terminology and the course of $3^0$ veins in the intercostal areas (E, F, G). Redrawn after Hickey, 1973.

Crystals, especially calcium oxalate crystals, silica bodies on the epidermal cells, foliar sclereids with their types and distribution are taxon specific and hence taxonomically useful.

The measurement called palisade ratio is taxon-specific and hence it is often used to identify a leaf or fragments of leaf. Pharmacognosists frequently use this measurement to authenticate the leaf drug that is mostly obtained in powdered forms. The palisade ratio indicates the number of palisade cells present beneath each epidermal cell. The cells of epidermis and palisade are observed and counted from surface views. A large number of counts are recorded and an average number is determined from the counts. The average number obtained from a statistically sound count provides a fairly reliable identifying datum of materials. Example: The palisade ratio of *Swertia angustifolia*, *S. chirata*, *S. paniculata* and *S. dilatata* is 2.94, 2.06, 3.27 and 6.5 respectively (Gayen & Datta, 1983, p. 17).

The other taxonomically useful anatomical features are stem, leaf and root anatomy, nodal anatomy, wood anatomy, primary and secondary anomalous structure, ultra-structure of sieve tube plastids etc. These were discussed in previous chapters.

Examples where anatomical features solved taxonomic problems: There are numerous examples where the anatomical features were used in solving taxonomic disputes, a few of which are mentioned below.

The ultra-structure of sieve tube plastids was used in circumscribing taxa. Sieve tube plastids are broadly of two types—starch accumulating (S-type) and protein accumulating (P-type). The S-type occurs in Polygonaceae, Plumbaginaceae, Bataceae, Theligoniaceae etc. whereas the P-type is found in Phytolaccaceae, Molluginaceae etc. There are certain families like Rafflesiaceae, Crassulaceae, and some species of Moraceae etc. where the sieve tube plastids accumulate neither starch nor protein. The ultra-structural details of sieve tube plastids are taxon-specific and hence taxonomically useful. It was best used in elucidating the inter-relationships in the order Caryophyllales. Mabry (1976) segregated the order into two sub-orders—Chepodiineae and Caryophyllineae, on the basis of pigment biochemistry. Chenopodiineae comprises the betalain containing families and Caryophyllineae includes the anthocyanin containing families—Caryophyllaceae and Molluginaceae. The nature and character of the ultra-structure of sieve tube plastids suggest that the betalain and anthocyanin families are closely related since all of them possess P-type plastids with a peripheral ring-shaped bundle of proteinaceous filaments. Betalain and anthocyanin families both have characteristic anomalous secondary thickenings and primary anomalous structure. Moreover anatomical and palynological evidences provide no basis for splitting the Caryophyllales into Chenopodiineae and Caryophyllineae (Roy and Chanda, 1996). *Gisekia* has been included under the anthocyanin family Molluginaceae. Unlike other Molluginaceae *Gisekia* possesses betalain instead of anthocyanin. So Takhtajan (1980) preferred to shift the genus to the Phytolaccaceae that has several characters common with *Gisekia*. *Gisekia* shows globular crystalloid P-type plastids in sieve elements like Molluginaceae and Phytolaccaceae. From the available data from plastid structure, pigmentation and palynology its alignment with Phytolaccaceae is indicated. The monotypic species *Halophytum ameghinoi* has often been associated with Chenopodiaceae. Chenopodiaceae and *Halophytum* both possess P-type plastids with ring shaped bundle of filaments. But Chenopodiaceae lacks the crystalloid in the ring-shaped bundle of filaments. In contrast *Halophytum* possesses globular crystalloid in the ring-shaped bundle of filaments. So the structure of sieve element plastids does not favour its association with the Chenopodiaceae. Theligoniaceae had been placed often close to or even into Caryophyllales. The presence of S-type plastid and anthocyanin argues against its incorporation into the Caryophyllales. On the basis of the plastid data *Dysphania* can be included in the Chenopodiaceae. It possesses P-type of plastid with ring-shaped bundle of filament; but it lacks the central crystalloid.

*Anarthria* and *Ecdeiocolea* were included under the family Restionaceae. Later a survey on anatomical and macro-morphological features of these two genera resulted in the recognition of two families—Anarthriaceae and Ecdeiocoleaceae (Cutler *et al.*). The stem anatomy of most Restionaceae shows a continuous parenchymatous cylinder that is absent in the other two families. Below the epidermis there exists a zone of chlorenchyma in all the three families. The stem anatomy of Ecdeiocoleaceae shows the presence of hypodermal sclerenchyma, which is absent from Restionaceae. In Anarthriaceae there exists the sub-epidermal fibre strands associated with vascular bundles.

# Further Reading


BECK, C. B., 2005. *Introduction to Plant Structure and Development*. Cambridge University Press. Cambridge.

BHATTACHARYA, K., MAJUMDAR, M. R. & BHATTACHARYA, S. G., 2006. *A Textbook of Palynology*. New Central Book Agency (P) Ltd. Kolkata 700 009, India.

BRETT, C. and WALDRON, K., 1996. *Physiology and Biochemistry of plant cell walls.* Unwin Hyman, London.

CARLQUIST, S., 1961. *Comparative Plant Anatomy*. Holt, Rinehart and Winston, New. York.

CUTLER, D. F., TED BOTHA and DENNIS Wm. STEVENSON, 2008. *Plant Anatomy: An applied approach*. Blackwell Publishing, U. K.

CUTTER, E. G., 1971. *Plant Anatomy: Experiment and Interpretation*. Part 2: *Organs*. Edward Arnold, London.

CUTTER, E. G., 1978. *Plant Anatomy:* Part 1: *Cells and tissues*. Edward Arnold, London.

EAMES, A. J. and MAcDANIELS, L. H., 1947. *An Introduction to Plant Anatomy*. McGraw-Hill, N. Y. and London.

ERDTMAN, G., 1969. Handbook of Palynology—An Introduction to the Study of Pollen Grains and Spores. Munksgaard, Copenhagen, 486 pp.

EVERT, R. F., 2007. *Esau's Plant Anatomy*. J. Willey, N. Y.

FAHN, A., 1997. *Plant Anatomy*. Pergamon Press, Oxford.

FOSTER, A. S., 1967. *Practical Plant Anatomy*. Affiliated East-West Press Private Ltd. New Delhi - 3.

HABERLANDT, G., 1914. *Physiological Plant Anatomy*. Macmillan, London.

MAUSETH, J. D., 1988. *Plant Anatomy*. The Benjamin/Cummings Publishing Company, Inc.

METCALFE, C. R. and CHALK, L., 1950. *Anatomy of the Dicotyledons*. Vols. I and II. Clarendon Press, Oxford.

RUDALL, P. J., 2007. *Anatomy of Flowering Plants: An Introduction to Structure and Development*. Cambridge University Press, UK.

SPORNE, K. R., 1976. The morphology of Pteridophytes. B. I. Publ., Calcutta.

# References


ABBE, E. C., & PHINNEY, B. O., 1951. The growth of the shoot apex in maize: external features. *Amer. Jour. Bot.* **38:** 737-744.

ABROL, D. P., 1997. *Bees and Beekeeping in India.* Kalyani Publishers, Calcutta-9, India.

AGASHE, S. N., 1995. *Paleobotany: Plants of the Past, Their Evolution, Paleoenvironment and Application in Exploration of Fossil Fuels.* Science Publishers Inc. Lebanon, U.S.A.

AGRAWAL, M. L. & SHASHI JHAMB, 1998. Aeroallergen Surveillance : A Critical Review of Sampling Techniques. In *Current Concepts in Pollen-Spore and Biopollution Research.* Prof. Sunirmal Chanda 60th Birth Anniversary Felicitation Volume (Eds. N. M. Dutta, S. Gupta-Bhattacharya, S. Mandal & K. Bhattacharya). Research Periodicals & Book Publishing House, India.

AHLERS, F., BUBERT, H., STEUERNAGEL, S. & WIERMANN, R., 2000. The nature of oxygen in sporopollenin from the pollen of *Typha angustifolia* L. *Z. Naturforsch.* **55c**, 129-136.

AJELLO, L., 1941. Cytology and cellular interrelations of cystolith formation in *Ficus elastica. Am. J. Bot.*, **28**, 589-594.

AL - TALIB, K. H. & TORREY, J. G., 1959. The aseptic culture of isolated buds of *Pseudotsuga taxifolia. Pl. Physiol., Lancaster*, **34:** 630–637.

AL - TALIB, K. H. & TORREY, J. G., 1961. Sclereid distribution in the leaves of *Pseudotsuga* under natural and experimental conditions. *Am. J. Bot.*, **48**, 71-79.

ANDREWS, H. N., 1961. *Studies in Paleobotany.* John Wiley & Sons, London.

ARBER, A., 1946. Goethe's botany. *Chron. Bot.*, **10**, 63-126.

ARZEE, T., 1953. Morphology and ontogeny of foliar sclereids in *Olea europaea.* I. Distribution and structure. II. Ontogeny. *Am. J. Bot.*, **40**, 680-687 and 745–752.

BAILEY, I. W., 1956. Nodal anatomy in retrospect. *J. Arnold Arb.* **37**, 269-287.

BALL, E., 1941. The development of the shoot apex and of the primary thickening meristem in *Phoenix canariensis* Chaub., with comparisons to *Washingtonia filifera* Wats. and *Trachycarpus excelsa* Wendl. *Am. J. Bot.*, **28**, 820-832.

BANIK, S., BHATTACHARYA, K. & CHANDA, S., 1986. Pollen flora of Dum Dum and The Adjacent Areas of West Bengal with Reference to The Identification of dispersed Palynomorphs. *Transactions of The Bose Research Institute*, Vol. **49**: No. 1-2.

BARBAR, J. T. & STEWARD, F. C., 1968. The proteins of *Tulipa* and their relation to morphogenesis. *Devl Biol.*, **17**, 326-349.

BECK, C. B., 2005. *Introduction to Plant Structure and Development*. Cambridge University Press. Cambridge.

BELL, L. G. E., 1962. Polysaccharides and Cell Membranes. *J. Theoret. Biol.*, **3**(1), 132.

BEHNKE, H. D., 1976. Ultrastructure of sieve-element plastids in Caryophyllales (Centrospermae), Evidence for the delimitation and classification of the order. *Pl. Syst. Evol.*, **126:** 31-54.

BEHNKE, H. D. & TURNER, B. L., 1971. On specific sieve-tube plastids in Caryophyllales: Further investigations with special reference to Bataceae. *Taxon*, **20:** 731-737.

BENNETT, K. D., 1986. The Rate of Spread and Population in Increase of Forest Trees During the Postglacial. *Phil. Trans. R. Soc. Lond.* **B 314**, 523-531.

BERNIER, G., 1966. The morphogenetic role of the apical meristem in higher plants. *Les congrés et Colloques de l'Université de Liège,* **38**, Les phytohormones et l'Organogenèse, 151-211.

BHANDARI, N. N., 1969. Ontogeny and marginal growth in the leaf of *Eubrachion ambiguum*. *Ann. Bot.* **33:** 537-540.

BHAR, D. S. & RADFORTH, N. W., 1969. Vegetative and reproductive development of shoot apices of *Pharbitis nil* as influenced by photoperiodism. *Can. J. Bot.*, **47**, 1403-1406.

BHATTACHARYYA, P. K., 1982. Anatomy of *Caldesia Parnassifolia* (Bassi ex L.) Parl, (=*Alisma reniforme* D. Don) and *Sagittaria guayanensis* H. B. K. ssp. *lappula* (D. Don) Bogin. *Botan. Soc. Bengal*: **36**: 67-76.

BHATTACHARYA, K., MAJUMDAR, M. R. & BHATTACHARYA, S. G., 2006. *A Textbook of Palynology*. New Central Book Agency (P) Ltd. Kolkata 700 009, India.

BLACKMAN, V. H., 1919. The compound interest law and plant growth. *Ann. Bot.* **33:** 353-360.

BLECMAN, C. A., H. M. HULL. & R. W. HOSHAW., 1980. Cuticular ultrastructure of *Prosopis velutina* and *Acacia greggii* leaflets. Bot. Gaz. **141** : 1- 8.

BLOCH, R., 1944. Developmental potency differentiation and pattern in meristems of *Monstera deliciosa*. *Am. J. Bot.*, **31**, 71-77.

BLOCH, R., 1946. Differentiation and pattern in *Monstera deliciosa*. The idioblastic development of the trichosclereids in the air root. *Am. J. Bot.*, **33**, 544-551.

BOKE, N. H., 1948. Development of the perianth in *Vinca rosea* L. *Am. J. Bot.*, **35**, 413-423.

BOWES, B. G. & MAUSETH, J. D., 2008. *Plant Structure: A Colour Guide*. Collinwood Publishers, London.

BRETT, C. & WALDRON, K., 1990. *Physiology and Biochemistry of plant cell walls.* Unwin Hyman, London.

BRETT, C. & WALDRON, K., 1996. *Physiology and Biochemistry of plant cell walls.* Unwin Hyman, London. Second Edition.

BROWN, W. V., HEIMSCH, & W. H. P. EMERY., 1957. The organization of the grass shoot apex and systematics. *Amer. Jour. Bot.* 44: 590 – 595.

BRYANT, V. M., 2003. Invisible Clues to New World Plant Domestication. *Science* 229: 1029-1030.

BRYANT, V. M., JR., 2001. Pollen Contents of Honey. *CAP Newsletter* 24(1): 10-24, 2001. http://www.scirpus.ca/cap/articles/paper17.htm

BRYANT. V. & R. HOLLOWAY, 1985. A Late-Quaternary Paleoenvironmental Record of Texas :-An Overview of the Pollen Evidence. In: *Pollen Records of Late-Quaternary North American Sediments* (eds. V. Bryant and R. Holloway). Pp. 39-71. American Association of Stratigraphic Palynologists Foundation Book Series, Dallas, Texas.

BRYANT, V & HOLLOWAY, R., 1996. Archaeological Palynology. In: *Palynology: Principles and Applications,* (J. Jansonius and D. C. McGregor, eds.), pp 913-918. American Association of Stratigraphic Palynologists Foundation Book Series, Dallas, Texas.

BRYANT Jr., V.M. & D.C. MILDENHALL., 1998. Forensic palynology: a new way to catch crooks. *In* Bryant Jr., V.M. and J.H. Wrenn, editors. (eds.). New Developments in Palynomorph Sampling, Extraction and Analysis. American Association of Stratigraphic Palynologists Contributions Series/ No. 33: 145-155.

BURGESS, J., 1976. Scanning electron microscopy of cell wall formation around isolated plant protoplasts. *Planta*, **131**, 173-178

BURGESS, J. 1985. *Pant Cell development.* Cambridge University Press.

BURGESS, J. 1989. *Pant Cell development.* Cambridge University Press.

BURSTRÖM, H. G. & ODHNOFF, C., 1964. *Vegetative Anatomy of Plants*. Svenska Bokförlaget/Bonniers. Scandinavian-University-Books.

BUSSE, J. S. & R. F. EVERT, 1999. Vascular differentiation and transition in the seedling of *Arabidopsis thaliana* (Brassicaceae). *Int. J. of Plant Sci.* 160: 241-251.

BUVAT, R., 1952. Structure, évolution et fonctionnement du méristème apical de quelques dicotylédones. *Annls Sci. nat. (bot.)*, Sér. **11**, **13**, 199-300.

BUVAT, R., 1955. Le *méristème* apical de la tige. *Année biol.*, **31**, 596-656.

CAJLACHJAN, M. C., 1936. On the Hormonal theory of Plant Development. *Comp. Rend. (Doklady) Acad. Sci.* U.S.S.R. **3**: 443.

CARLQUIST, S., 1961. *Compbrative Plant Anatomy.* Holt, Rinehart and Winston, New. York.

CHAILAKHYAN, M. K., 1966. *Biol. Zh. Armenii.* **19**, 3.

CHAILAKHYAN, M. H., 1968. Mechanism of Flowering. *Ann. Rev. Plant Physiol.*, **19** : 1.

CHAILAKHYAN, M. K., 1970. Flowering and Photoperiodism of Plants. *Plant. Sci. Bull.* **3**(3): 1-7.

CHAMPAGNAT, M., 1961. Recherches de morphologie descriptive et experimentale sur le genre *Linaria. Ann. des. Sci. Nat., Bot. Ser.* **22**, 1-170.

CHANDA, S., 1963. On the pollen morphology of some Scandinavian Caryophyllaceae. *Grana Palynologica.* Vol. 3: 3; pp. 67- 89.

CHANDA, S., 1966. On the pollen morphology of the Centrolepidaceae, Restionaceae and Flagellariaceae, with special reference to taxonomy. *Grana Palynol.*, **6**: 355-415.

CHANDA, S., 1970-71. On Application of Pollen morphology in the studies of Taxonomy and Palynology. *Trans. Bose. Res. Inst.*, **33** & **34** (2-4): 33-39.

CHANDA, S., 1992. Profile Carolus Linnaeus. *Science courier*, Vol. **3**: 2.

CHANDA, S. & SARKAR, P.K., 1972. Pollen grains as a causative agent of respiratory allergy with reference to aeropalynology of Greater Calcutta. *Trans. Bose Res. Inst.* **35**, 61-67.

CHANDA, S., 1994. Pollen grains as aeroallergens. In : *Recent Trends in Aerobiology, Allergy and Immunology.* S. N. Agashe (ed.). Oxford & IBH Publishing Co. Pvt. Ltd., New Delhi, pp 85-92.

CHANDURKAR, P. J., 1983. *Plant Anatomy.* Oxford & IBH Publishing Co. Calcutta.

CHEADLE, V. I., 1937. Secondary growth by means of a thickening ring in certain monocotyledons. *Bot. Gaz.*, **98**, 535-555.

CHOWDHURY, K. A., 1939. The formation of growth rings in Indian trees. Part II. *Ind. For. Rec. (New Series), Util.* **2**: 2, 40-57.

CHOWDHURY, K. A. & GHOSH, S. S., 1958. *Indian Woods, Their Identification, Properties and uses.* Vol. 1. Published by The Manager of Publication, Delhi.

CLOWS, F. A. L., 1961. *Apical Meristems.* Blackwell Scientific Publ., Oxford.

COEN, H.S. & MEYEROWITZ, E.M., 1991. The War of Whorls: Genetic Interactions Controlling Flower Development. *Nature*, **353**: 31-37.

COHN, N. S., 1964. *Elements of Cytology.* Harcourt, Brace & World, Inc. New York.

COMMITTEE ON NOMENCLATURE, INTERNATIONL ASSOCIATION OF WOOD ANATOMISTS, 1957. International glossary of terms used in wood anatomy. *Trop. Woods.* 107: 1-36.

CORSON, G. E., Jr., 1969. Cell division studies of the shoot apex of *Datura stramonium* during transition to flowering. *Am. J. Bot.*, **56**, 1127-1134.

COYLE, H. M., LADD, C., PALMBACH, T. & LEE, H. C., 2001. The Green Revolution: Botanical Contributions to Forensics and Drug Enforcement. *Corat. Med. J.* **42:** 340-345.

CUNNINGHAME, M. E. & LYNDON, R. F., 1986. The relationship between the distribution of periclinal cell divisions in the shoot apex and leaf initiation. *Ann. Bot.* **57:** 737-746.

CUSICK, F., 1956. Studies of floral morphogenesis. 1. Median bisections of flower primordia in *Primula bulleyana* Forrest. *Trans. R. Soc. Edin.*,**63**, 153-166.

CUTLER, D. F., 1978. *Applied Plant Anatomy.* Longman. London.

CUTLER, D. F., TED BOTHA & DENNIS Wm. STEVENSON, 2008. *Plant Anatomy An applied approach.* Blackwell Publishing, U. K.

CUTTER, E. G., 1971. *Plant Anatomy: Experiment and Interpretation*. Part **2**: *Organs.* Edward Arnold, London.

CUTTER, E. G., 1978. *Plant Anatomy:* Part **1**: *Cells and tissues.* Edward Arnold, London.

DATTA, S. C., 1980. *Plant Physiology.* Central Book Depot, Allahabad. 781p.

DELEVORYAS THEODORE, 1966. *Plant diversification.* Amerind Publishing Co. Pvt. Ltd., New Delhi-1.

DERMEN, H., 1947. Periclinal cytochimeras and histogenesis in cranberry. *Am. J. Bot.* **34:** 32-43.

DERMEN, H., 1951. Ontogeny of tissues in stems and leaf of cytochimeral apples. *Am. J. Bot.* **38:** 753-760.

DE. K. K., 2004. *An Introduction to Plant Tissue Culture*. New Central Book Agency (P) Ltd. Kolkata-700009, India.

DENNE, M. P., 1960. Leaf development in *Narcissus pseudonarcissus* L. II. The comparative development of scale and foliage leaves. *Ann. Bot., N. S.*, **24:** 32-47.

DER OHE WERNER VON, LIVIA PERSANO ODDO, MARIA LUCIA PIANA, MONIQUE MORLOT & PETER MARTIN., 2004. Harmonized methods of melissopalynology. *Apidologie*: **35,** S18–S25.

DEV KAPIL, 2005. Significance of Palynology in Forensic Problems. *Indmedica,* vol. 5, No. 1. (2005-01-2005-03).

DEVLIN, R.M., 1969. *Plant Physiology.* Affiliated East-West Press Pvt. Ltd. New Delhi.

DILCHER, D. L., 1979. Early angiosperm reproduction: an introductory report. *Review of Palaeobotany and Palynology.* **27**: 291-328.

DORMER, K. J., 1962. The taxonomic significance of crystal forms in *Centaurea. New Phytol.,* 61: 32-35. 120.

DUCHAIGNE, A., 1955. Les divers types de collenchymes chezles Dicotyledones; leur ontogenie et leur lignification. Annls. *Sci. nat. Bot.* Ser. 11, **16**, 455-479.

DYCUS, A. M. & KNUDSON, L., 1957. The role of velamen of the aerial roots of Orchids. *Bot. Gaz.* **119**: 78 – 87.

EAMES, A. J. & MACDANIELS, L. H., 1947. *An Introduction to Plant Anatomy.* McGraw-Hill, N. Y. and London.

El-GHAZALY, G. & ROWLEY, J. R., 1999. Microspore and tapetum development in *Echinodorus cordifolius*,. Alismataceae. *Nordic Journal of Botany.* **19(1):** 101-120.

ERDTMAN, G., 1943. *An Introduction to Pollen Analysis.* Waltham Mass., 239 pp.

ERDTMAN, G. 1948. Did dicotyledonous plants exist in early Jurassic times? *Geol. Fören. Stockholm Förh.*, 70: 265-271. (*Grana Palynol.* 1).

ERDTMAN, G., 1952. *Pollen Morphology and Plant Taxonomy. Angiosperms.* Almquist and Wiksell, Stockholm, 539 pp.

ERDTMAN, G., 1969. Handbook of Palynology - An Introduction to the Study of Polllen Grains and Spores. Munksgaard, Copenhagen, 486 pp.

ERDTMAN, G. & STRAKA, H., 1961. Cormophyte spore classification. *Geol. Fören. Förenhandl.*, 83, H.1: 65-78.

ERICKSON, R. O. & FRANCIS, J. M., 1957. The Plastochron Index. *American Journal of Botany*, **44:** 297-305.

ERVIN, E. L. & EVERT, T. F., 1970. Observations on sieve elements in three perennial monocotyledons. *Am. J. Bot.*, **57**, 218-224.

ESAU, K., 1940. Developmental anatomy of the fleshy storage organ *Daucus carota. Hilgardia*, **13**, 175-226.

ESAU, K., 1961. *Anatomy of Seed Plants.* John Wiley & Sons, Inc. London.

ESAU, K., 1965. *Plant Anatomy.* Wiley, New York.

ESAU, K., 1977. *Anatomy of Seed Plants.* John Wiley & Sons, Inc. London.

FÆGRI, K., 1975. *Textbook of Modern Pollen Analysis.* Munksgaard. Copenhagen, 3rd ed.

FÆGRI, K. & IVERSEN, J., 1950. *Textbook of Modern Pollen Analysis.* Munksgaard. Copenhagen, 169 pp.

FÆGRI, K. & IVERSEN, J., 1989. IV Edition by Kunt Faegri, Peter Emil Kaland & Kunt Krzywinski. John Willey & Sons, New York.

FAHN, A., 1958. Xylem structure and rhythm of development in trees and shrubs of the desert: I. *Tamarix aphylla, T. jordanis var. negevensis, T. gallica var. maris-mortui. Trop. Woods.* **109**, 81-94.

FAHN, A., 1979. *Secretory Tissues in Plants.* Academic Press, London.

FAHN, A., 1987. *Plant Anatomy.* Pergamon Press, Oxford.

FAHN, A., 1997. *Plant Anatomy.* Pergamon Press, Oxford.

FAHN, A. & BAILEY, I. W., 1957. The nodal anatomy and the primary vascular cylinder of the Calycanthaceae. *J. Arnold Arb.*, **38**, 107-117.

FAHN, A. & LESHEM, B., 1963. Wood fibres with living protoplasts. *New Phytol.* **62**, 91-98.

FLORIN, R., 1933. Studien über die Cycadales des Mesozoikums nebest Erörterungen über die Spattöffnung sapparate der Bennettitales. *K. svenska Vetensk Akad Handt.*; Ser. 3, **12**: 1-134.

FORD, BRIAN J., 1986. Even Plants Excrete. *Nature*, **323:** 763.

FOARD, D. E., 1958. An experimental study of sclereid development in the leaf of *Camellia japonica. Pl. Physiol., Lancaster*, **33** (suppl.), xli.

FOSTER, A. S., 1936. Leaf differentiation in angiosperms. *Bot. Rev.* **2:** 349-372.

FOSTER, A. S., 1937. Structure and behavior of the marginal meristem in the bud scales of *Rhododendron. Am. J. Bot.***24:** 304-316.

FOSTER, A. S., 1947. Structure and ontogeny of the terminal sclereids in the leaf of *Mouriria huberi. Cogn. Amer. Jour. Bot.* 34: 501-514.

FOSTER, A. S., 1949/1967 (Reprint). *Practical Plant Anatomy.* Affiliated East-West Press Private Ltd. New Delhi-3.

FREY-WYSSLING, A., 1976. *The Plant Cell Wall.* Borntraeger, Berlin and Stuttgart.

FREY-WYSSLING, A. & STECHER, H., 1951. *Das Flachenwachstum der pflanzlichen Zellwande. Experientia,* **7** : 420-421.

GALSTON, A. W., 1963. *The Life of the green Plant.* Prentice-Hall of India (Private) Limited. New Delhi.

GARNER, W. W. & ALLARD, H. A., 1920. Effect of Length of Day on Plant growth. *J. Agr. Res.***18**: 553.

GARRISON, R., 1955. Studies in the development of axillary buds. *Amer. Jour. Bot.*, **42**: 257-266.

GAYEN, S. & DATTA, P. C., 1983. *Herba Chirattae* and adulterants. *Bull. Botan. Soc. Bengal*: **37**: 7-21.

GEMMELL, A. R., 1969. *Developmental Plant Anatomy.* Edward Arnold (Publishing) Ltd. London.

GEWIRTZ, M. & FAHN, A., 1960. The anatomy of sporophyte and gametophyte of *Ophioglossum lusitanicum* L. *Phytomorphology,* **10**: 342-351.

GHOUSE, A. K. M. & KHAN, M. I. H., 1977. Seasonal variation in the nuclear number of fusiform cambial initials in *Psidium guajava* L. *Caryologia* **30**: 441-444.

GIFFORD, E. M., Jr. & TEPPER, H. B. 1962. Ontogenetic and histo-chemical changes in vegetative shoot tip of *Chenopodium album. Am. J. Bot.,* **49**, 902-911.

GILLILAND, M. G., BORNMAN, C. H., ADDICOTT, F. T., 1976. Ultrastructure and acid phosphatase in pedicel abscission of *Hibiscus. Amer. J. Bot.* **63**: 925-935.

GRAYUM, M. H., 1986. Correlations between Pollination biology and pollen morphology in the Araceae, with some implications for angiosperm evolution. In *Pollen and Spores: Form and Function*, ed. S. Blackmore and I. K. Ferguson. Academic press. London, p. 313-327.

GUENNEL, G. K., 1952. Fossil spores of the Alleghenian coals of Indiana. Indiana Dept. Conserv., Geol. Surv., Rept. of Prog., 4: 1-40.

GUTTENBERG, H. VON, 1960. *Grundzuge der Histogenese hoherer Pflanzen.* I. *Die Angiospermen.* Gebruder Borntraeger, Berlin.

HABERLANDT, G., 1914. *Physiological Plant Anatomy.* Macmillan, London.

HARA, N., 1957. On the types of marginal growth in dicotyledonous foliage leaves. *Bot. Mag. Tokyo.* **70:** 108-114.

HATCH, M. D. & SLACK, C. R., 1966. Photosynthesis by sugar-cane leaves. A new carboxylation reaction and the pathway of sugar formation. *Biochem. J.*, **101**, 103-111.

HAYWARD, P., 1974. Waxy structures in the lenticels of potato tubers and their possible effects on gas exchange. *Planta,* **120**: 273-277.

HENNING BUBERTB, JÖRG LAMBERTB, STEFAN STEUERNAGELC, FRIEDHELM AHLERSA & ROLF WIERMANNA, 2002. Continuous Decomposition of Sporopollenin from Pollen of *Typha angustifolia* L. by Acidic Methanolysis. *Z. Naturforsch*: **57c**, 1035-1041 (2002).

HESLOP-HARRISON, J., 1964. Sex expression in flowering plants. *Brookhaven Symp. Biol.,* **16**, 109-125.

HESLOP-HARRISON, J. & HESLOP-HARRISON, Y., 1969. *Cannabis sativa* L. In *The Induction of Flowering,* Evans, L. T., 205-226. Cornell Univ. Press, Ithaca, New York.

HESSE, M., HALBRITTER, H., ZETTER, R., WEBER, M., BUCHNER, R., FROSCH-RADIVO, A., & ULRICH, S., 2009. *Pollen Terminology: An Illustrated Handbook.* Springer.

HESSE M., MEIER H. & ZEEH B., 1995. *Spektroskopische Methoden in der organischen Chemie,* Thieme Verlag, Stuttgart.

HICKEY, L. J., 1973. Classification of the architecture of dicotyledonous leaves. *American Journal of Botany,* **60:** 17-33.

HICKEY, L. J., 1974. A revised classification of the architecture of dicotyledonous leaves, pp. 25-39 *in* C. R. Metcalfe and L. Chalk, eds. *Anatomy of the Dicotyledons,* Volume 1, Second Edition. Clarendon Press, Oxford.

HOLMAN, R. M. & ROBBINS, W. W., 1939. *A Text Book of General Botany.* Chapman & Hall, Limited, London.

HOUWINK. A. L. & ROELOFSEN. P. A., (1954). Fibrillar architecture of growing plant cell walls. *Acta bot. neerl.* **3:** 385-95.

HOWELL, S. H., 1998. *Molecular Genetics of Plant Development.* Cambridge University Press.

HYDE, H.A. & WILLIAMS, D.W., 1944. The Right Word. Pollen Anal. Circ., 8: 6.

JEFFREY, E. C., 1917. *The anatomy of woody plants.* Chicago, University of Chicago Press.

KAPP, O. R., 1969. *Pollen and Spores.* Wm. C. Brown Company Publishers. Dubuaue, Lowa.

KAUFMAN, P. B., 1959. Development of the shoot of *Oryza sativa* L. II. Leaf histogenesis. *Phytomorphology.* **9:** 277-311.

KAWASE, M. & TAKAHASHI, M., 1995. Chemical composition of sporopollenin in *Magnolia grandiflora* (Magnoliaceae) and *Hibiscus syriacus* (Malvaceae). *Grana* **34**, 242-245.

KETTUNEN, P. O., 2006. *WOOD: Structure and Properties.* Trans Tech Publications Ltd., Switzerland.

KEW INFORMATION SHEET K1. http://www.rbgkew.org.uk/ksheets/pdfs/k1microprop.pdf

KREMP, G. O. W., 1965. *Morphologic Encyclopedia of Palynology.* Univ. Arizona Press, Tucson.

KRIZEK, B. A. & MEYEROWITZ, E. M., 1996. The *Arabidopsis* homeotic gene *APETALA3* and *PISTILLATA* are sufficient to provide the B class organ identity function. *Development*, **122**: 11-22.

KAWASE, M. & TAKAHASHI, M., 1995. Chemical composition of sporopollenin in *Magnolia grandiflora* (Magnoliaceae) and *Hibiscus syriacus* (Malvaceae). *Grana*, **34**: 242-245.

KIDWAI, P. & ROBARDS, A. W. 1969. On the ultrastructure of resting cambium of *Fagus sylvatica* L. *Planta.* **89**, 361-368.

KRAMER, P. J., 1956. Physical and physiological aspects of water absorption. *In: "Encyclopedia of Plant Physiology"* (W. Ruhland, ed.), **3**, 124-159. Springer-verlag, Berlin.

KRAMER, P. J., 1969. *Plant & soil Water Relationships: A Modern Synthesis.* Tata McGraw-Hill Publishing Company Ltd. New Delhi-110 002.

KRUTZSCH, W., 2004. New studies of Pre-Quaternary malvacean pollen from the Tilioideae, Helicteroideae and Bombacoideae subfamilies. *Palaeontographica. Abteilung B: Palaeophytologie*, 264 (4 – 6).

LANG, A., 1952. Physiology of Flowering. *Ann. Rev. Plant Physiol.* **3**: 265.

LANG, A., 1957. The Effect of Gibberellin upon Flower Formation. *Proc. Natl. Acad. Sci. U. S. A.* **43**: 709-717.

LANG, A., 1965. Physiology of Flower Initiation. In *Encyclopedia of Plant Physiology (Ed. W. Ruhland).* Springer – Verlag, New York. pp 1380-1536.

LARSON, P. R., 1962. The indirect effect of photoperiod on tracheid diameter in *Pinus resinosa. Am. J. Bot.*, **49:** 132-137.

LARSON, P. R., 1964. Some indirect effects of environment on wood formation. In: *The formation of Wood in Forest Trees,* Zimmermann, M. H., 345-365. Academic press, New york.

LAWERENCE, G. H. M., 1967. *Taxonomy of Vascular Plants.* Oxford & IBH Publishing Co. Calcutta.

LEVETIN ESTELLE., 1998. Yearly Fluctuations in Atmospheric Ragweed Pollen concentrations in Tulsa, Oklahoma. In *Current Concepts in Pollen-Spore and Biopollution Research.* Prof. Sunirmal Chanda 60th Birth Anniversary Felicitation Volume (Eds. N. M. Dutta, S. Gupta-Bhattacharya, S. Mandal & K. Bhattacharya). Research Periodicals & Book Publishing House, India.

LEVIN, D. A., 1973. The role of trichomes in plant defence. Q. Rev. Biol. **48**: 3-15.

LIESE, W., 1965. The fine structure of bordered pits in softwoods. In *Cellular Ultrastructure of Woody Plants*, COTE, W. A., Jr(ed.), 271-290. Syracuse Univ. Press. N. Y.

LINNAEUS, C., 1751. Philosophy of Botany. Stockholm, 362 pp.

LOUVEAUX, J., MAURIZIO, A., & VORWOHL, G., 1978. Methods of Melissopalynology. *Bee world* **59**, 139-157. http://www.airborne.co.nz/images/technical/methmels.pdf

LYNDON, R. F., 1968. Changes in volume and cell number in the different regions of the shoot apex of *Pisum* during a single plastochron. *Ann. Bot., N. S.* **32,** 371-390.

LYNDON, R. F., 1990. *Plant Development, The Cellular Basis.* Unwin Hyman, London.

LYNNE A. MILNE, VAUGHN M. BRYANT JR., & DALLAS C. MILDENHALL, 2005. Forensic Palynology 14 in *Forensic Botany: Principles and Applications to Criminal case work.* Edited by Heather Miller Coyle, 2005. CRC Press, p. 217-252, London.

MABRY, T. J., 1976. Pigment dichotomy and DNA-RNA hybridization data for centrospermous families. *Pl. Syst. Evol.*, **126**: 79-94.

MAHESWARI, P., 1930. Contribution to the morphology of *Boerhaavia diffusa. J. Ind. Bot. Soc.* **9**: 41-61.

MAJUMDER, R. B., 1963. *An Introduction to Spermatophyta.* Merit Publishers, Calcutta-6.

MAKSYMOWYCH, R., 1956. Development of lamina in *Xanthium italicum* Moretti represented by plastochron index. (*M. S. thesis*). Univ. Pennsylvania.

MAKSYMOWYCH, R., 1959. Quantitative analysis of leaf development in *Xanthium pensylvanicum. Am. J. Bot.*, **46;** 635- 644.

MAKSYMOWYCH, R., 1973. *Analysis of Leaf Development.* Cambridge University Press.

MARTENS, P., 1937. L'origine des espaces intercellulaires. *Cellule* **46**: 357-388.

MATZKE, E. B., 1956. Progressive configurational changes, during cell division, of cells within the apical ceristem. *Proc. Natn. Acad. Sci.*, **42**: 26-33.

MAUSETH, J. D., 1988. *Plant Anatomy.* The Benjamin/Cummings Publishing Company, Inc.

MAUSETH, J. D., 2009. *Botany: An Introduction to Plant Biology*. Sudbury, Mass: Jones & Bartlett Publishers.

McCAY, S. & MAHLBERG, P. G., 1973. Study of antibacterial activity and bacteriology of latex from *Asclepias syriaca* L. *Antimicrob. Ag. Chemother*. **3**: 247-257.

McHUGHEN, A. 1980. The regulation of floral organ initiation. *Bot. Gaz.* **141:** 389-395.

MEIER-MELIKYAN, N. R., GABARAEVA, N. I., POLEVOVA, S. V., GRIGOR'EVA, V. V., KOSENKO, YA. V. & TEKLEVA, M. V., 2003. Development of Pollen Grain Walls and Accumulation of Sporopollenin. *Russian Journal of Plant Physiology*. **50**(3): 330-338.

MELCHERS, G., 1936. Versuche zur Genetik und Entwicklungsphysiologie der Blühreife. *Biol. Zbl.* **56**: 567.

MERRILL, E. K., 1978. Comparison of mature leaf architecture of three types in *Sorbus* L (Rosaceae). *Botanical Gazette*. **139**: 447-453.

METCALFE, C.R., 1961. The anatomical approach to systematics. General introduction with special reference to recent work on monocotyledons. In *Recent Advances in Botany*, 146-150. Toronto Univ. Press.

METCALFE, C. R. & CHALK, L., 1950. *Anatomy of the Dicotyledons.* Vols. I and II Clarendon Press, Oxford.

METCALFE, C. R. & CHALK, L., (1979). *Anatomy of the Dicotyledons.* Second Edition. Vol. I. Oxford University Press, Oxford.

METCALFE, C. R. & CHALK, L., 1983. *Anatomy of the Dicotyledons.* 2nd ed. Vol. II Clarendon Press, Oxford.

MEYEN, S. V., 1987. *Fundamentals of Palaeobotany*. Chapman and Hall. London.

MEYER, D. E. & DRYDEN, V. M., 1965. *Biological Science: An Inquiry into Life*. National council of Educational Research and Training, New Delhi.

MIA, A. J. & PATHAK, S. M., 1965. Histochemical studies of sclereid induction in the shoot of *Rauwolfia* species. *F. exp. Bot.*, **16:** 177-181.

MICHELINI, F. J., 1955. The use of the plastochron index in studies of morphological and physiological development in *Xanthium italicum* Moretti. *(Ph. D. Dissertation)*. Univ. Pennsylvania.

MITCHELL, J. W. & WORLEY, J. F., 1964. Intracellular transport apparatus of phloem fibers. *Science*, N. Y. **145**: 409-410.

MOORE, P. D., WEBB, J. A. & COLLINSON, M. E., 1991. *Pollen Analysis*. Blackwell Scientific Publications. Oxford.

MOORE, T. C., 1994. *Biochemistry and Physiology of Plant Hormones*. 2nd. ed. Narosa Publishing House, Calcutta.

MOSS, E. H., 1940. Interxylary cork in *Artemisia* with a reference to its taxonomic significance. *Amer. Jour. Bot.* **27:** 762-768.

MUELLER, R. J., 1982. Shoot morphology of the climbing fern *Lygodium* (Schizaeaceae): General organography, leaf initiation, and branching. *Bot. Gaz.* 143: 319-330.

NAIR, P. K. K., 1970. *Pollen Morphology of Angiosperms: A Historical and Phylogenetic Study.* Barnes & Noble, Inc, New York.

NAIR, P. K. K., 1985. *Essentials of Palynology.* Today and Tomorrow Printers, New Delhi.

NAIR, P. K. K., 1998. The New Dimensions of Palynological Sciences. In *Current Concepts in Pollen-Spore and Biopollution Research.* Prof. Sunirmal Chanda 60th Birth Anniversary Felicitation Volume (Eds. N. M. Dutta, S. Gupta-Bhattacharya, S. Mandal & K. Bhattacharya). Research Periodicals & Book Publishing House, India.

NAMBOODIRI, K. K. & BECK, C. B., 1968. A Comparative Study of Primary Vascular System in Conifers. III. Stelar Evolution in Gymnosperm. *Amer. J. Bot.* **55**: 461-472.

NEWMAN, I.V., 1965. Pattern in the meristems of vascular plants. III. Pursuing the patterns in the apical meristem where no cell is a permanent cell. *J. Linn. Soc.* (Bot.), **59**: 185-214.

NOUGARÈDE, A. & BERNIER, G., 1964. The intermediate phase in photoperiodic and cold-requiring plants; its signification. In *Differentiation of Apical Meristems and Some Problems of Plants,* 25-28. Academia, Praha.

OLIEN, W.C., & BUKOVAC, M.J., 1982. Ethephon-induced gummosis in our cherry (*Prunus cerasus* L.). *Plant Physiol.* 70: 547-559.

PALIWAL, G. S., 1969. Stomatal ontogeny and phylogeny. I. Monocotyledons. *Acta bot. neerl.,* **18:** 654-668.

PALIWAL, G. S., PRASAD, N. V. K., SAJWAN, V. S. & AGGARWAI, S. K., 1975. Seasonal activity of cambium in some tropical trees. II. *Polyalthia longifolia. Phytomorphology* **25:** 478-484.

PALMER, P. G., 1976. Grass cuticles: a new palaeoecological tool for East African lake sediments. *Can. J. Bot.,* **54**: 1725-1734.

PANDEY, B. P., 1979. *Plant Anatomy.* S. Chand & Company Ltd. New Delhi-110055.

PANT, D. D., 1965. On the ontogeny of stomata and other homologous structures. *Plant Sci. Ser.* (Allahabad), **I**: 1-24.

PAYNE, W. W., 1979. Stomatal patterns in embryophytes: their evolution, ontogeny and interpretation. *Taxon* **28**: 117-132.

PHILIPSON, W. R., WARD, J. M. & BUTTERFIELD, B. G., 1971. *The vascular cambium.* Chapman &Hall, London.

POPHAM, R. A., 1952. *Developmental plant anatomy.* Long, Columbus, Ohio.

POPHAM, R. A. & CHAN, A. P., 1950. Zonation in the vegetative stem of *Chrysanthemum morifolium* Bailey. *Am. J. Bot.,* **37**: 476-484.

POPHAM, R. A. & CHAN, A. P. 1952. Origin and development of the receptacle of *Chrysanthemum morifolium. Am. J. Bot.* **39**: 329-339.

PRASAD, S., NAYAR, K. C. & BHATTACHARYA, I. C., 1960. Pharmacognistical studies on chirata and its substitutes and adulterants. Part II-*Swertia angustifolia* Buch.-Ham. *J. Sci. Industr. Res.* Vol. 19C. p. 139-145.

PRESTON, R.D., 1974. Plant cell walls. In *Dynamic Aspects of Plant Ultrastructure,* ROBARDS, A. W. (ed), 256-309. McGraw-Hill, London.

PRIESTLEY, D. S. O. & LORNA L. SCOTT., 1938. *An Introduction to Botany.* Longmans, Green and Co. London.

PUNT, W., HOEN, P. P., BLACKMORE, S., NILSSON, S. & LE THOMAS, A., 2007. Glossary of pollen and spore terminology. *Review of Palaeobotany and Palynology,* 143 (1-2): 1-81. http://www.bio.uu.nl/~palaeo/glossary/glos-int.htm

RAJAGOPAL, T. & RAMAYYA, N., 1977. The taxonomic value of guard cells seen in surface view. *Botanical Journal of The Linnean Society.* **74:** 57-61.

RAO, K. R. & JUNEJA, K. B. S., 1971. *A Hand Book for Field Identification of Fifty Important Timbers of India.* The Manager of Publications, Govt. of India, Delhi.

RAO, A. N. & SINGARAYAR, M., 1968. Controlled differentiation of foliar sclereids in *Fagraea fragrans. Experientia,* **24:** 298-299.

RAY, P. M., 1963. *The Living Plant.* Holt, Rinehart & Winston, Inc., U. S. A.

RENNER, O. & VOSS, M., 1942. Zur Entwicklungsgeschichte randpanaschierter Formen von *Prunus, Pelargonium, Veronica, Dracaena. Flora* **35:** 356-376.

ROBERTSON, J. D., 1959. The ultrastructure of cell membranes and their derivatives. *Biochem. Soc. Symp.* **16:** 3-43.

ROELOFSEN, P. A., 1965. Ultrastructure of the wall in growing cells and its relation to the direction of the growth. *Adv. Bot. Res.,* **2:** 69-149.

ROSENE, H. F., 1954. A comparative study of the rates of water influx into the hairless epidermal surface and the root hairs of onion roots. *Physiologia Pl.,* **7:** 676-686.

ROWLEY, J. R., 1959. The fine structure of the pollen wall in the Commelinaceae. *Grana Palynol.,* 2(1): 3-31.

ROWLEY, J. R., FLYNN, J. J. & TAKAHASHI, M., 1995. Atomic force microscope information on pollen exine substructure in *Nuphar. Bot. Acta,* **108**: 300.

ROY, P & CHANDA, S., 1992. Phylogenetic trend in the pollen grains of Centrospermae based on number of apertures. *Journal of Palynology,* **28**: 143-208.

ROY, P. & CHANDA, S., 1996. An evaluation of the segregation of Centrospermae into Betalain and Anthocyanin containing suborders. In *Contemporary Thoughts in Plant Sciences,* pp. 83-87. Ed. P. K. Pal, Academic Staff College, Burdwan University.

RUDALL, P. J., 2007. *Anatomy of Flowering Plants: An Introduction to Structure and Development*. Cambridge University Press, UK.

SAAD, S. I., 1963. Sporoderm stratification: The "medine", a distinct third layer in the pollen wall. *Pollen et Spores*, 5: 17-38.

SAHU, R., 1967. *Botany*. Merit Publishers, Kolkata- 6.

SAKAI, W. S. & HANSON, M., 1974. Mature raphid and raphid idioblast structure in plants of the edible aroid genera *Colocasia, Alocasia* and *Xanthosoma*. *Ann. Bot.*, **38**: 739-748.

SANTRA, S. C., CHATTERJEE, T. P. & DAS, A. P., 2005 *College Botany Practical (Volume-1)*. New Central Book Agency (P) Ltd. Kolkata-9, India.

SCHWABE, W. W., 1959. Some effects of enviromment and hormone treatment on reproductive morphogenesis in the Chrysanthemum. *J. Linn. Soc. (Bot)*. **56**, 254-261.

SEELEY, T. D., 1995. *The Wisdom of the Hive: The Social Physiology of Honey Bee Colonies*. Harvard University Press.

SEN, D. N. & CHAWAN, D. D., 1972. Leafless Euphorbias on Rajasthan rocks. IV. Water relations of seedlings and adult plants. *Vegetatio, Haag* **24**: 193-214.

SEN, U., & DE, B. 1992a. Structure and ontogeny of stomata in Ferns. *Blumea* **37:** 239-261.

SEN, U. & DE, B., 1992b. Stomata in primitive land plants: their ontogeny and evolution. *Indian Fern 1.* **9**: 51-59.

SHAH, J. J., & JAMES, M. R., 1968. Sieve tube elements in the stem of *Neptunia oleracea* lour. *Aust. J. Bot.* **16**: 433-444.

SHARMA, A. 1985. *Chromosomes*. Oxford and IBH Publishing Co. Calcutta.

SHRMA, G. K. & BUTLER, J. 1975. Environmental pollution: leaf cuticular patterns in *Trifolium pratense* L, *Ann. Bot.*, **39**, 1087-1090.

SHARMAN, B. C., 1942. Developmental anatomy of the shoot of *Zea mays* L. *Ann. Bot.*, **6:** 245-282.

SIEBERS, A. M., 1971. Initiation of radial polarity in the interfascicular cambium of *Ricinus communis* L. *Acta bot. neerl.*, **20**, 211-220.

SINGER, S. J., & NICOLSON, G. L., 1972. The fluid mosaic model of the structure of cell membranes . *Science (New York)*. **175:** 720-730.

SINGH, A. B. & KUMAR PAWAN, 2004. Aerial Pollen diversity in India and their Clinical Significance in Allergic Diseases. *Indian Journal of Clinical Biochemistry*, **19 (2)**, 190-201.

SINGH, V., PANDEY, P. C., & JAIN, D. K., (1994). *Anatomy of Seed Plants.* Rastogi Publications, Merrut, India 250 002.

SINGH, A. B. & RAWAT, A., 2000. Aeroallergens in India. In: Advances in Allergy and Asthma. R. Prasad (ed.), Shivam Publications, Lucknow, UP, India, pp 23-46.

SINGH, B. P., SINGH, A. B. & GANGAL, S. V., 1992. Pollen Calendars of Different States of India. CSIR Centre for Biochemicals. Pub., Delhi, India.

SING, V. & JAIN, D. K., 1996. *Taxonomy of Angiosperms.* Rastoghi Publications, Merrut-2, India.

SINNOTT, E. W., 1914. Investigations on the phylogeny of angiosperms: I, The anatomy of the node as an aid in the classification of angiosperms. *Am. J. Bot.*, **1**: 302-322.

SINNOTT, E. W. & BLOCH, R., 1941. Division in vacuolate plant cells. *Amer. Jour. Bot.*, **28**: 225-232.

SIRCAR, S. M. & CHAKRASVERTY, R., 1960. The effect of gibberellic acid on jute (*Corchorus capsularis* Linn). *Sci. Cult., 26, 141-143.*

SIVANNA, K. R., 2003. *Pollen Biology and Biotechnology*. Oxford & IBH Publishing Co. Pvt. Ltd., New Delhi.

SIVANNA, K. R., B. M. JOHRI & D. C. SASTRI., 1975. *Development and physiology of angiosperm pollen.* Today & Tomorrow Printers & Publishers, New Delhi-110005.

SMITH, G. M., 1955. *Cryptogamic Botany* Vol. II. McGraw-Hill, N. Y.

SMITH, D. L. 1966. Development of the inflorescence in *Carex. Ann. Bot., N. S.,* **30**: 475-486.

SOLEREDER, H., 1908. *Systematic Anatomy of the Dicotyledons.* Clarendon Press. Oxford.

SPORNE, K. R., 1976. The morphology of Pteridophytes. B. I. Publ., Calcutta.

SRIVASTAVA, L. M. & SINGH, A. P. 1972. Stomatal structure in corn leaves. *J. Ultrastruct. Res.*, **39**, 345-363.

STACE, C.A., 1963. *A Guide to Subcellular Botany.* Longmans, London.

STACE, C.A., 1965. Cuticular studies as an aid to Plant taxonomy. *Bull. Br. Mus. nat. Hist. (Bot.),* **4:** 1-78.

STEBBINS, G. L. & KHUSH, G. S., 1961. Variation in the organization of the stomatal complex in the leaf epidermis of monocotyledons and its bearing on their phylogeny. *Am. J. Bot.*, **48:** 51-59.

STEFAAN P., WERBROUCK, O. & PIERRE C. DEBERGH., (1994). Micro-propagation. In *plant cell culture: A practcal approch.* Editors: R. A. Dixon & R. A. Gonzales. Oxford University Press. Second editon.

STEINGRAEBER, D.A. & FISHER, J. B., 1986. Indeterminate growth of leaves in *Gaurea* (Meliaceae): a twig analogue. *Am. J. Bot.* **73:** 852-862.

STEVENSON, D. W., 1980. Radial growth in *Beaucarnea recurvata, Am. J. Bot.* **67:** 476-489.

STEWART, W.N., 1983. *Paleobotany and the Evolution of Plants.* Cambridge University Press. Cambridge.

STEWART, W.N., 1993. *Paleobotany and the Evolution of Plants.* Cambridge University Press. Cambridge. Second Edition.

TAKHTAJAN, A., 1980. Outline of the classification of Flowering Plants (Magnoliophyta). *Bot. Rev.* **46:** 225-359.

TAYAL, M. S., 1994. *Plant anatomy.* Rastogi Publications, Meerut.

THANIKAIMONI, G., 1980. Complexities of aperture, columella and tectum. *Proc. IVth Intern. Palynol. Conf.*, **I**: 228-239.

THANIKAIMONI, G., 1984. Omniaperturate Euphorbiaceae pollen with striate spines. *Bull. Jard. Bot. Nat. Belge*, **54**: 305-325.

THOMAS, R. G., 1963. Floral induction and the stimulation of cell division in *Xanthium. Scienc, N. Y.*, **140**, 54-56.

TOMLINSON, P. B., 1961. Palmae. In: *Anatomy of the Monocotyledons* (ed. C. R. Metcalfe). Vol II. Clarendon Press, Oxford.

TRAVERSE, A., 1988. *Paleopalynology.* Unwin Hyman, Boston, 600 pp.

TRAVERSE, A., 2007. *Paleopalynology.* Second Edition, 813 pp. Springer.

TSCHIRCH, A., 1889. *Angewandte Pflanzenanatomie.* Urban & Schwarzenberg, Wien.

TSCHUDY, R. H. & SCOTT, R. A., 1969. *Aspects of Palynology.* Willey – Interscience.

VAN COTTHEM, W. R. J., 1970. A classification of stomatal types. *Bot. J. Linn. Soc.*, **63**: 235-246.

VARSHNEY, A. & DHAWAN, V. 1998. Micropropagation of Ornamental Plants. In *Plant Tissue Culture and Molecular Biology: Applications and Prospects*, pp. 402-528, edited by P. S. Srivastava New Delhi; Narosa Publishing House.

VASISHTA, P. C., 1968. *Plant Anatomy.* First edition. Pradeep Publications. India.

VASISHTA, P. C., 1981. *Plant Anatomy.* Pradeep Publications. India.

VESQUE, M. J., 1889. De l'emploi des caracteres anatomeques dans la classification des végélaux. *Bull. Soc. Bot.* France, **36:** 41-77.

VON DER OHE, W., LIVIA PERSANO ODDO, MARIA LUCIA PIANA, MONIQUE MORLOT & PETER MARTIN, 2004. Harmonized methods of melissopalynology. *Apidologie* 35, S18-S25.

VORONIN, N. S., VORONIN, S. N. & VORONINA, R. A. 1976. Distribution of hydathodes on leaves of some species of the genus *Crassula. Bot. Zb. SSSR* **61**: 621-628.

WARDLAW, C.W. 1961. Growth and development of the inflorescence and flower. In *Growth in Living Systems,* ZARROW, M. X., 491-523. Basic Books, Inc., New York.

WARDROP, A. B. & DAVIES. G. W., 1962. Wart structure of gymnosperm tracheids. *Nature*, **194**: 497-498.

WAREING, P. F., & ROBERTS, D. L., 1956. Photoperiodic control of cambial activity in *Robinia pseudacacia* L. *New Phytol.*, **55:** 356-366.

WAREING, P. F., HANNEY, C. E. A. & DIGBY, J., 1964. The role of endogenous hormones in cambial activity and xylem differentiation. In: *The formation of Wood in Forest Trees* (ed. M. H. Zimmermann). Academic press, New York, pp. 323-344.

WETMORE, R. H., GIFFORD, E. M., Jr. & GREEN, M. C., 1959. Development of vegetative and floral buds. In *Photoperiodism and Related Phenomena in Plants and Animals*, WITHROW, R. B., 255-273.

WIKIPEDIA, 2007. *The free Encyclopedia. http://en.wikipedia. Org/wiki/ The_ABC_Model_of_Flower_Development:*

WILMESMEIER, S. & WIERMANN, R., 1995. Influence of EPTC (S-Ethyl-dipropyl-thiocarbamate) on the composition of surface waxes and sporopollenin structure in *Zea mays. J. Plant Physiol.* **146**, 22-28.

WILLIS, J. C. Late, 1973. *A Dictionary of The Flowering Plants And Ferns* Revised by H. K. AIRY SHAW. Cambridge University Press.

WILLMER, C. M., 1983. *Stomata.* Longman.

WILSON, C. L., 1982. Vestigial structures of the flowers. *Amer. J. Bot.* **69**: 1356-1365.

WINSTON, M. L., 1987. *The Biology of the Honey Bee.* Harvard University Press.

WULLSTEIN, L. H., BRUENING, M. L. & BOLLEN, W. B., 1979. Nitrogen fixation associated with sand grains root sheaths (rhizosheaths) of certain xeric grasses. *Physiol. Plant.* **46**: 1-4.

XING SHUPING, LI BINGSHI, WANG CHEN, HU YUXI & LIN JINXING., 2000. Atomic force microscopic observation on substructure of pollen exine in *Cedrus deodara* and *Metasequoia glyptostroboides. Chinese Science Bulletin.* Vol. **45**, No. 16.

ZETSCHE, F. & VICARI, H., 1931. Untersuchungen über die Membran der Sporen und Pollen III. 2. *Picea orientalis* (L.) Link, *Pinus silvestris* L., *Corylus avellana* L. Helv. Chim. Acta, 14: 62-67.

ZIMMERMANN, W.A., (1928). Histologische Studien am Vegetationspunkt von *Hypericum uralum. Jahrb. Wiss. Bot.,* **68:** 289-344.

ZIMMERMANN, M. H. & TOMLINSON, P. B., 1967. Anatomy of the palm *Rhapis excelsa*: **IV.** Vascular development in apex of vegetative aerial axis and rhizome. *J. Arnold Arb.* **48**: 122-142.

# Index